3/97 LR/LEND/002

Harrison Learning Centre
Wolverhampton Campus
University of Wolverhampton
St Peter's Square
Wolverhampton WV1 1RH
Wolverhampton (01902) 322305

ONE WEEK LOAN

-5 OCT 2009

Telephone Renewals: 01902 321333
Please RETURN this item on or before the last date shown above.
Fines will be charged if items are returned late.
See tariff of fines displayed at the Counter. (L2)

WITHDRAWN

D1356993

BACTERIOPHAGES

Biology and Applications

BACTERIOPHAGES

Biology and Applications

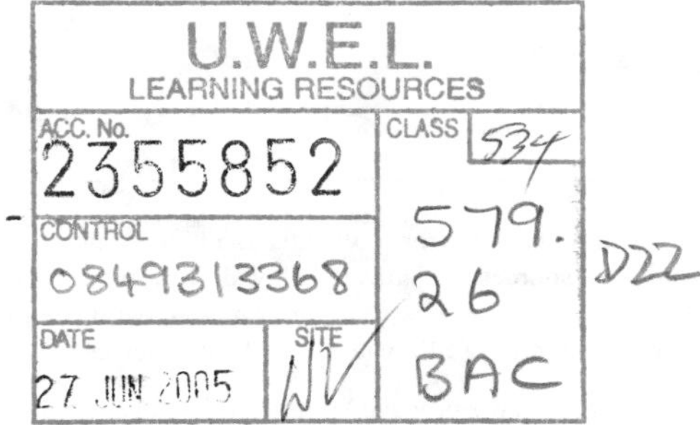

EDITED BY

Elizabeth Kutter
Alexander Sulakvelidze

CRC PRESS

Boca Raton London New York Washington, D.C.

Library of Congress Cataloging-in-Publication Data

Bacteriophages : biology and applications / edited by Elizabeth Kutter, Alexander Sulakvelidze.
p. cm.
Includes bibliographical references and index.
ISBN 0-8493-1336-8
1. Bacteriophages. 2. Genetic vectors. I. Kutter, Elizabeth. II. Sulakvelidze, Alexander.

QR342.B338 2004
579.2'6—dc22 2004058578

This book contains information obtained from authentic and highly regarded sources. Reprinted material is quoted with permission, and sources are indicated. A wide variety of references are listed. Reasonable efforts have been made to publish reliable data and information, but the author and the publisher cannot assume responsibility for the validity of all materials or for the consequences of their use.

Neither this book nor any part may be reproduced or transmitted in any form or by any means, electronic or mechanical, including photocopying, microfilming, and recording, or by any information storage or retrieval system, without prior permission in writing from the publisher.

All rights reserved. Authorization to photocopy items for internal or personal use, or the personal or internal use of specific clients, may be granted by CRC Press, provided that $1.50 per page photocopied is paid directly to Copyright Clearance Center, 222 Rosewood Drive, Danvers, MA 01923 USA. The fee code for users of the Transactional Reporting Service is ISBN 0-8493-1336-8/04/$0.00+$1.50. The fee is subject to change without notice. For organizations that have been granted a photocopy license by the CCC, a separate system of payment has been arranged.

The consent of CRC Press does not extend to copying for general distribution, for promotion, for creating new works, or for resale. Specific permission must be obtained in writing from CRC Press for such copying.

Direct all inquiries to CRC Press, 2000 N.W. Corporate Blvd., Boca Raton, Florida 33431.

Trademark Notice: Product or corporate names may be trademarks or registered trademarks, and are used only for identification and explanation, without intent to infringe.

Visit the CRC Press Web site at www.crcpress.com

© 2005 by CRC Press

No claim to original U.S. Government works
International Standard Book Number 0-8493-1336-8
Library of Congress Card Number 2004058578

Printed in the United States of America 1 2 3 4 5 6 7 8 9 0
Printed on acid-free paper

Foreword

It is a privilege for me to have this opportunity to provide a brief foreword to *Bacteriophages: Biology and Applications* by Elizabeth Kutter and Alexander Sulakvelidze. I was one of many who first became fascinated with the romance of science by reading the book *Arrowsmith* as a teenager. In that novel written by Sinclair Lewis in 1925, an attempt to develop phage therapies against bacterial diseases played a central role. But by the early 1950s, when I read the book, the widespread success of newly introduced antibiotics had seemed to make this alternative approach to the selective killing of bacteria unnecessary.

Instead, a small set of bacteriophages had begun to attract attention as "model organisms"—prime systems for probing the basic chemistry of life. These phages were attractive to scientists, because they were much easier to study with the then-available tools than were more complex life forms such as bacterial or human cells. They had relatively small genomes and multiplied rapidly, making them unusually amenable to genetic analyses that aimed at obtaining multiple mutants in each bacteriophage gene. To enable the essential genes for viral multiplication to be genetically identified, screening techniques were developed that focused on *conditional lethal* mutations—for example, through the identification of "temperature-sensitive" phage mutants that would grow at low—but not high—temperatures. Moreover, because large amounts of infected cells were easy and inexpensive to obtain, biochemical approaches could be readily employed, so that the products of the genes identified by genetic screens could be isolated and characterized in cell-free systems.

The model organism approach worked better than anyone had had a right to expect, in part because the mechanisms that are used to control gene expression and to recombine and replicate DNA genomes turned out to be much more highly conserved across life forms than anyone had suspected. Much of the work was concentrated on several viruses that infect the bacterium *E. coli*—most notably the bacteriophages lambda, T4 and T7. The findings made in multiple laboratories could thereby be combined, yielding results that were immensely important in developing the field of molecular biology, as reviewed in the early chapters of this book.

To give a personal example, for 30 years beginning in 1965, my own laboratory would exploit the combined genetic and biochemical advantages of the T4 virus for study of fundamental DNA replication mechanisms. In the end, the "protein machine" mechanisms revealed at the replication fork through bacteriophage studies turned out to be highly similar to those used to move the replication forks of higher organisms, including those of humans (Alberts, 2003).

In the 1960s and 1970s, many advances were made in a wide range of laboratories studying both bacteriophages and the bacterial cells themselves. The new knowledge of biological mechanisms that resulted soon allowed the development of more powerful research tools (such as DNA cloning). With these new tools, researchers could begin to unravel the molecular mechanisms in more complex cells and organisms.

As a result, by the 1980s most of the action and excitement in molecular biology had moved away from simpler organisms to investigations of mammalian cells.

For several unrelated reasons, we may have come full circle over the course of the last 80 years. First of all, there is an urgent need for new types of antibacterial therapies. We now live in an ever-more crowded, more interconnected world in which resistant strains of microorganisms spread with amazing rapidity. Modern science has increased our ability to design countermeasures to these diseases of humans and animals. The standard countermeasures have been new drug and vaccine developments. But producing a new drug is an enormously expensive endeavor. In addition, market failures have discouraged the development of new vaccines in the private sector. As a result, the world now faces a serious challenge in dealing with a host of microbial threats that were once thought to be defeated rather easily by antibiotics (Institute of Medicine, 2003). As described in Chapters 12 to 14, there is therefore every reason to reintroduce bacteriophage therapies as an additional tool in the war against bacterial diseases.

A second feature of modern biology that is reawakening interest in bacteriophages is our new ability to obtain the DNA sequences of large numbers of organisms inexpensively. From this DNA sequence information, we can determine the relatedness of organisms and attempt to retrace the past history of life on the Earth. The sequencing of bacteriophages is only just beginning. Not only are there immense numbers of novel proteins yet to be discovered among what could be 100 million different bacteriophages in the environment—the vast majority not yet known (the genomes of only about 400 have thus far been completely sequenced)—but it is now suspected that some of the lytic phages carry genes that trace back in evolutionary history to the common ancestor of eukaryotic and prokaryotic cells (see Chapter 5). In summary, bacteriophages represent a huge untapped genetic reservoir that can be productively mined—both by those interested in proteomics and by those who are trying to decipher the mysterious nature of the early cells that predated the split among the three families of cells that are alive today: the archaea, the bacteria, and eukaryotes.

Now that we have access to the complete molecular anatomy of a cell, a third reason for a new focus on bacteriophages stems from the realization—sobering to scientists like myself—that biological systems are so complex that they cannot be understood without new methods of analyzing and conceptualizing them. Thus, for example, the nearly 500 different protein molecules that are encoded by the genome of the simplest known living cell, the small bacterium *Mycoplasma genitalium,* interact with each other and with substrates in an enormous number of ways. Even if we had a complete catalog of all of these interactions and their rate constants, information we are far from achieving today, we could not claim to understand this cell in any deep sense—that is, in the sense of being able to explain how the cell is able to grow and reproduce itself as a chemical system. Living systems are made possible by a huge web of networked chemical reactions, and we presently lack the tools to decipher what is most significant within such complexity. This realization, new to most molecular biologists, raises the question of whether it might be productive to focus once again on one or a few bacterial viruses that could serve as model organisms—far simpler than any free-living cell—for developing new types of complexity analyses. If so, which viruses should be targeted and through what types of experimental strategies?

Finally, the increasingly large role that science and technology will play in driving societal changes in the twenty-first century argues strongly for a new type of science education in our schools. Beginning with 5-year-olds, what is needed is an education that allows students to explore the world around them using evidence and logic, so that they leave school learning to solve problems the way that scientists do. They also need to understand what science is and why it represents a special way of knowing about the natural world, if they are to respect its judgments concerning the many important issues that they will need to decide in their lifetimes—such as whether they should avoid exposures to substances that could adversely affect their health in the future, or whether their nation should make sacrifices to reduce the release of greenhouse gases into the atmosphere.

The National Science Education Standards call for a revolutionary change in science teaching, with an emphasis on teaching science as inquiry (National Research Council, 1996). As the ultimate step in such an education effort, it should be possible for a select group of students to participate in a real scientific investigation in their upper years of high school. It is thus encouraging to find high school students appearing as coauthors of a major publication from the University of Pittsburgh, in which a diverse set of novel bacteriophages that infect mycobacteria have been identified and sequenced (Pedulla et al, 2003).

The National Academy of Sciences has just published the results of an unusual workshop in which 25 leading scientists outside the field were exposed to the biology of the smallpox virus and challenged with the task of suggesting new approaches to antiviral therapies (Harrison et al, 2004). As this exercise made clear, we badly need a new infusion of talent and energy into the field of virology, where there is an enormous opportunity for scientific breakthroughs whose results will be of great practical benefit to human health (Alberts and Fineberg, 2004). What better way to recruit outstanding young people into such fields than to expose them as teenagers to a scientific exploration of the wonderfully rich and diverse world of bacteriophages?

I would like to end by congratulating both the editors and the many contributors to this volume for their dogged persistence in sticking to bacteriophage research over many decades. They have survived their years in the shadows, and now we can all appreciate the strong platform that their work has established for the many exciting years of research ahead.

Bruce Alberts
President, National Academy of Sciences, Washington, D.C.

Professor of Biochemistry and Biophysics, University of California
San Francisco, CA

REFERENCES

Alberts, Bruce, 2003. DNA replication and recombination. *Nature* 421: 431-435.

Alberts, B. and Fineberg, H.V., 2004. Harnessing new science is vital for biodefense and global health. *Proc. Natl. Acad. Sci. USA* 101 (31): 11177.

National Research Council, 1996. *National Science Education Standards.* National Academy Press, Washington, D.C.

Institute of Medicine, 2003. *Microbial Threats to Health: Emergence, Detection, and Response.* Mark S. Smolinski, Margaret A. Hamburg, and Joshua Lederberg, Eds. National Academy Press, Washington, D.C.

Pedulla, M. L., Ford, M. E., Houtz, J. M., Karthikeyan, T., Wadsworth, C., Lewis, J. A., Jacobs-Sera, D., et al., 2003. Origins of highly mosaic mycobacteriophage genomes. *Cell* 113: 171-182.

Harrison, S.C., Alberts, B., Ehrenfeld, E., Enquist, L., Fineberg, H., McKnight, S.L., Moss, B., et. al., 2004. Discovery of antivirals against smallpox. *Proc. Natl. Acad. Sci. USA* 101:11178-11192.

Editors

Elizabeth Kutter, Ph.D. has been a member of the faculty in biophysics and head of the laboratory of phage biology at The Evergreen State College, Olympia, WA, since 1972. She received her B.S. in mathematics at the University of Washington, Seattle, in 1962 and her Ph.D. in biophysics at the University of Rochester, New York, in 1968. Her thesis, under John Wiberg, dealt with bacteriophage T4's substitution of hydroxymethylcytosine for cytosine in its DNA and the transition from bacterial to phage metabolism, and her research has focused on phage biology ever since. From 1969 to 1972, she worked with Rolf Benzinger at the University of Virginia. She took the NIH grant she won there to Evergreen, where she has maintained an active undergraduate research program, funded largely by the National Science Foundation (NSF) and National Institutes of Health (NIH). From 1975–1980, she was a member of the NIH Recombinant DNA Advisory Committee, drafting the initial national guidelines in the field. She has also served on many grant and oversight panels at NSF, NIH, and the Howard Hughes Foundation and as a reviewer for various journals.

From the time of her first talk in 1964 at a Cold Spring Harbor phage meeting, Dr. Kutter was drawn by the highly collaborative nature of the phage field. In 1975 she started holding West Coast T4 Meetings which developed into the biennial Evergreen International Phage Biology meetings; the 15th, in 2003, drew 100 scientists from 14 countries. She spent 1978–1979 as a sabbatical fellow with Bruce Alberts at University of California at San Francisco, where she began work on the T4 genome project—the start of a worldwide collaboration which included 4 months in 1990 in Moscow, Pushchino, Vilnius, and Tbilisi through an exchange program between the U.S. and Soviet Academies of Science. During this visit, she also learned about the extensive Soviet history of phage use as antibiotics. Her initial skepticism about phage therapy gave way over the next few years as she talked and worked with researchers, physicians, and patients in Tbilisi, Republic of Georgia, where it is a standard part of care for purulent and gastrointestinal infections. In 1997, she and Evergreen colleagues established the nonprofit PhageBiotics Foundation to help stimulate worldwide interest in exploring the possibilities and challenges of phage therapy, including the broad studies of phage biology crucial to this endeavor.

Dr. Kutter was an editor and author of *Bacteriophage T4* (ASM, 1983) and *The Molecular Biology of Bacteriophage T4* (ASM, 1994), which grew out of the Evergreen T4 meetings, and of *The Encyclopedia of Genetics* (Academic Press, 2002). She has also co-authored dozens of other papers, book chapters, and popular articles, as well as over 100 meeting presentations related to phage, and participated in programs on CBS's *48 Hours, Dateline Australia, Discover-Canada, NPR Science Friday,* and *Voice of America*. This current book is a natural outgrowth of these various aspects of her work, which has included studies of the molecular mechanisms of host DNA degradation, nucleotide biosynthesis and transcription regulation after phage infection and of phage ecology, genomics and evolution, and has particularly focused on phage as an effective teaching tool.

Alexander Sulakvelidze, Ph.D. is an associate professor of epidemiology and preventive medicine at the University of Maryland School of Medicine, and vice president of research and development and chief scientist of Intralytix, Inc. Dr. Sulakvelidze received his formal training in microbiology in the former Soviet Union, including a B.A. from Tbilisi State University in 1986 and a Ph.D. in microbiology and epidemiology from Tbilisi State Medical University in 1993. He continued his training at the Engelhard Institute of Molecular Biology in Moscow, and, under the auspices of the U.S. National Academy of Sciences, at the University of Maryland School of Medicine in Baltimore.

Dr. Sulakvelidze's research interests are in the broad areas of emerging infectious diseases, molecular epidemiology, pathogenesis of bacterial enteric diseases, and phage therapy. A major focus of his research involves studies of the potential usefulness of bacteriophages in preventing and treating infectious diseases caused by multi-drug-resistant bacteria. In 1998, Dr. Sulakvelidze co-founded Intralytix, Inc., in Baltimore—a pioneering U.S. company involved in the development of therapeutic phage preparations for a variety of agricultural, medical, and environmental applications. One particularly strong research area has focused on developing phage preparations and their application strategies for improving the safety of foods contaminated with various foodborne pathogens. Dr. Sulakvelidze has published extensively on the subject and he is the author of one issued and several pending patents related to this field. His phage therapy research has been featured in several magazines and newspapers (including the *Los Angeles Times*, *Newsweek*, *Science*, *Smithsonian,* and *Wired*), and in various radio programs and television documentaries (including National Public Radio's *Science Friday, BBC Radio,* and *Voice of America* radio programs, and a BBC Horizon documentary about phage therapy).

In addition to his work with bacteriophages, Dr. Sulakvelidze is actively involved with the molecular characterization of various bacterial pathogens, particularly *Listeria, Yersinia,* and *Salmonella.* He maintains an active, extramurally funded laboratory at the University of Maryland School of Medicine, where the epidemiology, virulence traits, and genetic composition of those pathogens are being investigated using multilocus sequence typing and other state-of-the-art approaches. Dr. Sulakvelidze serves as an ad hoc reviewer on such journals as *Antimicrobial Agents and Chemotherapy, Applied and Environmental Microbiology, FEMS Immunology and Medical Microbiology,* and the *Journal of Clinical Microbiology,* and for several funding agencies, including the Civilian Research and Development Foundation, International Science and Technology Center, and National Institutes of Health.

Acknowledgments

We would like to express our appreciation for their hard work and patience to all the authors who contributed to this book. Particular thanks are due to Hans Ackermann, Karin Carlson, Burton Guttman, and Raul Raya for their thoughtful editorial contributions to other chapters in addition to their own. Also, we gratefully acknowledge the invaluable assistance of our many phage colleagues who provided constructive criticism of various chapters and shared much material prior to publication—particularly to Thomas Häusler, author of *Gesund durch Viren,* and to Richard Calendar and Steve Abedon, coeditors of the 2005 *Bacteriophages,* and the many contributors to their book. We thank Arnold Kreger for his editorial assistance with many chapters. We also are very grateful for the patience and support of our families, colleagues, students, CRC editors, and friends throughout the writing process and to the board of the PhageBiotics Foundation for their support. Special thanks also go to editorial assistant Gautam Dutta at Evergreen for his yeoman's work with the figures, formatting, bibliographies, and details of submission, and for maintaining our Web page (www.evergreen.edu/phage), which will include material to complement the book, such as color versions of many figures and a section for sharing and discussion of methods.

Acknowledgments

Contributors

Hans-W. Ackermann
Department of Medical Biology
Faculty of Medicine
Laval University
Québec, Canada

Paul Barrow
Institute for Animal Health
Compton Laboratory
Berkshire, United Kingdom

E. Fidelma Boyd
Department of Microbiology
National University of Ireland
Cork, Ireland

Harald Brüssow
Nestle Research Centre Nutrition and
Health Department, Functional
Microbiology Group
Lausanne, Switzerland

Karin Carlson
Department of Cell and Molecular
Biology
University of Uppsala
Uppsala, Sweden

Vincent A. Fischetti
Laboratory of Bacterial Pathogenesis
Rockefeller University
New York, NY

Ketevan Gachechiladze
Eliava Institute of Bacteriophage,
Microbiology and Virology
Georgian Academy of Sciences
Tbilisi, Georgia

Burton Guttman
Laboratory of Phage Biology
The Evergreen State College
Olympia, WA

Elizabeth Kutter
Laboratory of Phage Biology
The Evergreen State College
Olympia, WA

Martin J. Loessner
Institute of Food Science and Nutrition
Swiss Federal Institute of Technology
Zürich, Switzerland

Céline Lévesque
Department of Biochemistry and
Microbiology
Université Laval
Québec, Canada

Sylvain Moineau
Department of Biochemistry and
Microbiology
Université Laval
Québec, Canada

Raul Raya
Lab of Phage Biology
The Evergreen State College
Olympia, WA

Catherine E.D. Rees
School of Biosciences
University of Nottingham
Loughborough, United Kingdom

David A. Schofield
Medical University of South Carolina
Department of Microbiology and
Immunology
Charleston, SC

Alexander "Sandro" Sulakvelidze
Intralytix, Inc.
Baltimore, MD

William C. Summers
Yale University
New Haven, CT

Caroline Westwater
Medical University of South Carolina
Divison of Oral and Community
Health Sciences
Charleston, SC

Contents

Foreword v
Editors ix
Acknowledgments xi
Contributors xiii

Chapter 1. Introduction 1
Elizabeth Kutter and Alexander Sulakvelidze

Chapter 2. Bacteriophage Research: Early History 5
William C. Summers

Chapter 3. Basic Phage Biology 29
Burton Guttman, Raul Raya, and Elizabeth Kutter

Box 1: Antigenicity of Phages 34
Ketevan Gachechiladze

Chapter 4. Bacteriophage Classification 67
Hans-W. Ackermann

Chapter 5. Genomics and Evolution of Tailed Phages 91
Harald Brüssow and Elizabeth Kutter

Chapter 6. Phage Ecology 129
Harald Brüssow and Elizabeth Kutter

Chapter 7. Molecular Mechanisms of Phage Infection 165
Elizabeth Kutter, Raul Raya, and Karin Carlson

Chapter 8. Bacteriophages and Bacterial Virulence 223
E. Fidelma Boyd

Chapter 9. Phage for the Detection of Pathogenic Bacteria 267

Catherine E.D. Rees and Martin J. Loessner

Chapter 10. Control of Bacteriophages in Industrial Fermentations 285

Sylvain Moineau and Céline Lévesque

Chapter 11. Phage As Vectors and Targeted Delivery Vehicles 297

Caroline Westwater and David A. Schofield

Chapter 12. The Use of Phage Lytic Enzymes to Control Bacterial Infections 321

Vincent A. Fischetti

Chapter 13. Phage Therapy in Animals and Agribusiness 335

Alexander Sulakvelidze and Paul Barrow

Chapter 14. Bacteriophage Therapy in Humans 381

Alexander Sulakvelidze and Elizabeth Kutter

Appendix Working with Bacteriophages: Common Techniques and Methodological Approaches 437

Karin Carlson

Box 2: Electron Microscopy 488

Hans-W. Ackermann

Index 495

1 Introduction

Elizabeth Kutter and Alexander Sulakvelidze

Bacteriophages are ubiquitous in our world—in the oceans, soil, deep sea vents, the water we drink, and the food we eat. They are the most abundant living entities on earth—the estimates range from 10^{30} to 10^{32} in total—and play key roles in regulating the microbial balance in every ecosystem where this has been explored. The more information that is generated about the biology, ecology, and diverse nature of phages, the more exciting the field becomes, and the more obvious it is that we still know surprisingly little about them beyond the intricate details of a few model systems. Since their independent discovery by Frederick Twort and Felix d'Herelle in 1915 and 1917, respectively, bacteriophages have been studied in numerous laboratories worldwide, and they have been used in a variety of practical applications. As interest grows in the possibility of using them as antimicrobial agents in a variety of clinical and agricultural settings, they are increasingly capturing the public's as well as scientists' imagination—and we fully expect that a growing number of students and seasoned investigators from various fields will become involved with this exciting field. Thus, our primary purpose in writing this book was to make phage research more accessible for people with a range of backgrounds by providing an overview of the broad depth of phage knowledge and literature, from research techniques and basic molecular biology, to genomics, to applications in agriculture, human therapy, and biotechnology.

In 1959, Interscience Publishers, Inc., published a book entitled *Bacteriophages* that quickly became the most widely used reference source for novice and experienced phage researchers around the world. The book was largely written by Mark Adams; however, when he died suddenly in 1956, many of the best-recognized names of that era in the biological sciences (including Max Delbrück, Alfred Hershey, Gunther Stent, Jim Watson, Tom Anderson, Seymour Benzer, Francois Jacob, Eduard Kellenberger, and George Streisinger) took on its completion—in an effort that strongly emphasizes both the breadth of interest in phage research and the strength and cohesiveness of the phage community at that time. In his preface to *Bacteriophages*, Max Delbrück wrote: "Phage research, after a fitful history during its first twenty years, had all but died out in the middle 1930's. In the textbooks of bacteriology, the bacteriophages, if they were mentioned at all, figured as a curiosity item, unconnected with the rest and disposed of in a couple of pages at most. Today, phage research is vigorously pursued in many outstanding laboratories" (Adams, 1959). Indeed, during the years surrounding the release of *Bacteriophages*, the results of phage research were very

0-8493-1336-X/05/$0.00+$1.50
© 2005 by CRC Press LLC

important in making some of the most significant discoveries in the history of the biological sciences, including the identification of DNA as the genetic material, the discovery of the transduction phenomenon, the deciphering of the genetic code, and the discovery of messenger RNA, as documented in "Phage and the Origins of Molecular Biology" (Cairns et al., 1966).

The intense collaborative application of physical, genetic, biochemical, physiological, and morphological techniques resulted in excellent books about a few model phages (mainly those infecting *E. coli* and *Salmonella*) which dealt with topics ranging from genetic mechanisms and enzymology to structure-function relationships and the details of complex morphogenesis—including whole books about bacteriophages Lambda (Hershey, 1971; Arber et al., 1983), T4 (Mathews et al., 1983; Karam et al., 1994), and Mu (Symonds et al., 1987). In addition, much of the phage community collaborated to produce *The Bacteriophages* (Calendar, 1988), a 1200-page compendium of detailed information about a variety of phages, a new edition of which is in press (Calendar, 2005). Other books explored such subjects as phage-induced enzymes (Cohen, 1968), phage genetics (Birge, 1981; 2000) and phage taxonomy (Ackermann, 1987), or collected key papers on bacterial viruses (Stent, 1965) from the burgeoning literature in the field. During the 1980s, much of the focus of phage research shifted to the development of new technologies exploiting phages and their enzymes; for example, the phage enzymes and vectors that made genetic engineering possible, phage display technologies, and using phages to detect bacterial pathogens—as discussed in Chapters 9 and 11. However, the distribution and functioning of phages in the natural world, their role in general microbial ecology, and their potential therapeutic applications were largely ignored until relatively recently. In the field of phage ecology, as in molecular biology, the publication of a key monograph played a substantial role in stimulating general interest and new research. In that regard, Goyal et al. (1987) provided the first integrated view of the distribution of phages in a variety of environments (e.g., seawater, soil, food, sewage, and wastewater resulting from various industrial processes), and they outlined common methodological approaches used to detect and analyze phages in those environments. Chapters 5, 6, 10, and 13 explore the current rapid developments in this field, which have been greatly facilitated by new genomic and metagenomic approaches as well as by advances in instrumentation and other methods of detection and analysis.

Today, phage research is experiencing a second renaissance because of the new appreciation of phages' ubiquity and prevalence in nature and because of a rekindled public and scientific interest in potential phage applications against antibiotic-resistant bacteria, as well as in various basic and applied aspects of phage biology. Thus, Max Delbrück's introductory statement is quite possibly as applicable today as it was in 1959, when *Bacteriophages* was first published. Indeed, phage research is currently being pursued in many laboratories worldwide, the number of phage therapy- and basic phage biology-related publications is on the rise, and several state-of-the-art technologies are being developed on the "bacteriophage platform." In addition, recent popular books such as *The Killers Within* (Shnayerson and Plotkin, 2002), *Gesund durch Viren: ein Ausweg aus der Antibiotika-krise* (Häusler, 2003), and *Félix d'Herelle and the Origins of Molecular Biology* (Summers, 1999) have explored the history of phage discovery and early phage therapy applications, as well as the current issues and concerns of antibiotic resistance and potential applications of phages. Public interest has been further increased by documentaries from the BBC and CBC, Dateline Australia, and the Canadian Discover Channel, along with many phage therapy-related articles in the popular press. The American Society for Microbiology has hosted a specialized phage meeting for the first time (entitled "The New Phage Biology") in the summer of 2004—while the Cold Spring Harbor Phage and Microbial Genetics meetings, the Phage and Virus Assembly Meetings, the Evergreen International Phage Biology meetings, and other more specialized gatherings continue to bring together new and old researchers in the field. Furthermore, special sessions dealing with phage applications are being held at many meetings and symposia worldwide, including the 2002 International Virology Conference in Paris, the 2003 International Food Technology Association Meeting, and the Dairy Council's 2003 meeting on agricultural problems of antibiotic resistance and a major German physicians' conference on new approaches to severe wound infections in Dec. 2004. Steve Abedon's Bacteriophage Ecology Group website, www.phage.org, and the Evergreen site, www.evergreen.edu/phage or http://phage.evergreen.edu, provide wide ranges of phage information and links to other sites.

In contrast to the various more specialized phage books, this book integrates the historical phage story with the technicalities of phage research and little-known details about their widespread therapeutic applications in various times and places. Within this general framework, the book also presents basic information about phage biology, phage ecology and genomics, the roles of phages in bacterial evolution, the molecular mechanisms of phage interactions with bacteria and what such studies have taught us about general biological principles, and phage applications in a variety of fields. Thus, our goal has been to focus on integrating the various threads of basic and applied phage research, with the help of colleagues from a range of fields, and to help students and new investigators discover their primary areas of interest and exploration. With this latter objective in mind, we have included an Appendix that details the classical and modern techniques for studying phages—probably the first such compendium since the Adams's book, complementing the more detailed methods described by Carlson and Miller (1994) in *The Molecular Biology of Bacteriophage T4*. Phage research is a field where researchers at all levels can potentially make significant contributions to the body of knowledge. It is relatively easy and inexpensive to isolate and initially characterize phages from a variety of environmental sources; thus, phages are powerful teaching tools for guiding both beginning and advanced students into understanding complex biological systems and their interactions, and they provide many fertile areas for collaborations that may result in exciting and potentially very important practical applications. For example, a joint project between high schools and a leading phage lab that added greatly to our understanding of mycobacterial phages was recently featured on the cover of *Cell* (Pedulla et al., 2003). We hope that this book will help scientists from a variety of backgrounds to better understand the history and excitement of phage research, access the prodigious literature in the field, and explore new applications of phages—the "good viruses" (Radetsky, 1996) that are the most abundant living entities on Earth.

REFERENCES

Ackermann, H.W., and Dubow, M., 1987 *Viruses of Prokaryotes*. CRC Press, Boca Raton, FL.

Adams, M., H., 1959 *Bacteriophages*. Interscience Publishers, New York.

Arber, W., Enquist, L., Hohn, B., Murray, N.E. and Murray, K., 1983 Experimental methods for use with lambda, in *Lambda II*, edited by R.W. Hendrix, J.W. Roberts, F.W. Stahl and R.A. Weisberg. Cold Spring Harbor Laboratory Press, Cold Spring Harbor, NY, pp. 433–466

Birge, E.A. 1981 *Bacterial and Bacteriophage Genetics: An Introduction*. Springer-Verlag, New York.

Birge, E.A. 2000 *Bacterial and Bacteriophage Genetics*, 4th ed. Springer-Verlag, New York.

Cairns, J., Stent, G.S. and Watson, J.D., 1966 *Phage and the Origins of Molecular Biology*. Cold Spring Harbor Laboratory, Cold Spring Harbor, NY.

Calendar, R. (Ed.), 1988 *The Bacteriophages*. Plenum Press, New York.

Calendar, R. (Ed.), 2005 *The Bacteriophages*. Oxford University Press, New York.

Carlson, K. and Miller, E.S., 1994 Working with T4, in *Molecular Biology of Bacteriophage T4*, edited by J. Karam, J.W. Drake, K.N. Kreuzer, G. Mosig, D.H. Hall, F.A. Eiserling, L.W. Black, E.K. Spicer, E. Kutter, K. Carlson and E.S. Miller. American Society for Microbiology, Washington, D.C., pp. 421–426

Goyal, S.M., Gerba, G.P., and Bitton, G., 1987 *Phage Ecology*. John Wiley & Sons, New York.

Häusler, T., 2003 *Gesund Durch Viren: ein ausweg aus der antibiotika-krise*. Piper Verlag, Munchen, GmbH.

Hershey, A.D. (Ed.), 1971 *The Bacteriophage Lambda*. Cold Spring Harbor Laboratory, New York.

Karam, J.D., Drake, J.W., Kreuzer, K.N., Mosig, G., Hall, D.H., Eiserling, F.A., Black, L.W., Spicer, E.K., Carlson, C., and Miller, E.S., (Eds.), 1994 *Molecular Biology of Bacteriophage T4*. American Society for Microbiology, Washington, D.C.

Mathews, C.K., Kutter, E., Mosig, G. and Berget, P.B. (Eds.), 1983 *Bacteriophage T4*. American Society for Microbiology, Washington, D.C.

Pedulla, M.L., Ford, M.E., Houtz, J.M., Karthikeyan, T., Wadsworth, C., Lewis, J.A., Jacobs-Sera, D., et al., 2003 Origins of highly mosaic mycobacteriophage genomes. *Cell* 113: 171–182.

Radetsky, P, 1996 Return of the good virus. *Discover* 17: 50–58.

Shnayerson, M. and Plotkin, M.J., 2002 *The Killers Within: The Deadly Rise of Drug-Resistant Bacteria*. Little, Brown & Company, Boston.

Stent, G. 1965 *Papers on Bacterial Viruses*. Little, Brown and Co., Boston

Summers, W. C., 1999 *Félix d'Herelle and the Origins of Molecular Biology*. Yale University Press, New Haven, CT.

Symonds, N., Toussaint, A., van de Putte, P. and Howe, W.V. (Eds.), 1987 *Phage Mu*. Cold Spring Harbor Press, Cold Spring Harbor, NY.

2 Bacteriophage Research: Early History

William C. Summers
Yale University, New Haven, CT

CONTENTS

2.1. Introduction 5
2.2. Discovery of Bacteriophages 6
2.3. Early Debate Regarding the Nature of Bacteriophages 7
2.4. First Practical Applications: Veterinary and Medicine 9
2.5. Basic Phage Research and Major Scientific Discoveries Associated with Bacteriophages 12
2.5.1. The Chemical and Physical Nature of Phages 12
2.5.2. Max Delbrück and the Development of Molecular Genetics 12
2.5.3. Role of the Phage Community 15
2.5.4. The Nature of the Gene and of Mutations 16
References 23

2.1. INTRODUCTION

The distribution, prevalence, and dramatic manifestations of bacteriophages make it surprising that they were not recognized for almost 40 years after the beginning of serious bacteriological work in the laboratories of Europe and America in about 1880. In retrospect, there are a few reports in the literature that hint at the presence of bacteriophages, but the interpretations in these papers did not suggest useful pathways for further research. Hankin (1896) reported that the waters of the Jumna and Ganges Rivers in India had antiseptic activity against many kinds of bacteria, and against the *cholera vibrio* in particular. This activity was filterable and destroyed by boiling. He concluded that the antiseptic principle was some volatile chemical substance. Emmerich and Löw (1901) reported in that some substance in autolyzed cultures was capable of causing the lysis of diverse cultures, of curing experimental infections, and of providing prophylactic immunity to subsequent inoculations. In addition, there is substantial literature on bacterial autolysis by Gamalieya, Malfitano, Kruse, and Pansini, which was reviewed by Otto and Munter (1923). At this point, however, it is difficult to provide unambiguous interpretations of these early studies. Although some of these observations are compatible with the action of bacteriophages,

0-8493-1336-X/05/$0.00+$1.50
© 2005 by CRC Press LLC

others suggest bacteriocin effects and still others may be attributable to lytic enzyme production. These reports all describe experiments on liquid cultures, and in this period of bacteriology, a culture was conceptualized not in terms of the population dynamics of individual cells, but as an organism in itself. It was not until the 1920s that a significant shift in thinking took place that allowed a reconceptualization of the bacterial cell as the organism rather than the entire culture (Summers, 1991). The first dramatic and clear experiments on bacteriophages employed cultures on solid medium, and they were based on the observation of localized bacteriolysis (i.e., plaques).

2.2. DISCOVERY OF BACTERIOPHAGES

A rather odd observation made by Frederick W. Twort (1915) is usually understood to be the beginning of modern phage research. It was only later, however, after the pioneering work by Félix d'Herelle, that Twort's report was recognized as dealing with bacteriophages. Twort was a student of William Bulloch, a famous British bacteriologist who worked at the Brown Institution (a veterinary hospital) in London (Twort, 1993). While attempting to grow *Vaccinia virus* on agar media in the absence of living cells, Twort noted that many colonies of micrococci grew up, and he interpreted these colonies as bacterial contamination in the vaccinia pulp he was using. His interesting observation was that some of the colonies appeared mucoid, watery or "glassy." This "glassy transformation" could be induced in other colonies by inoculation of the fresh colony with a bit of material from the watery colony. This transformation could be propagated indefinitely. When Twort examined the glassy colonies under the microscope, he noted that the bacteria had degenerated into small "granules" that were colored red with Giemsa stain. Twort's discussion of this phenomenon of glassy transformation was tentative:

> From these results it is difficult to draw definite conclusions. . . . It is quite possible that an ultra-microscopic virus belongs somewhere in this vast field of life more lowly organized than the bacterium or the amoeba. It may be living protoplasm that forms no definite individuals, or an enzyme with power of growth. . . . [I]t [the agent of glassy transformation] might almost be considered as an acute infectious disease of micrococci. (Twort, 1915, p. 1242)

Quite independently, and for quite different reasons, Felix d'Herelle (d'Herelle, 1917) discovered a microbe that was "antagonisitic" to bacteria and that resulted in their lysis in liquid culture and death in discrete patches (he called them plaques) on the surface of agar seeded with the bacteria. D'Herelle conceived of these invisible microbes as "ultraviruses" that invaded bacteria and multiplied at their expense, and so he termed them *bacteriophage.* D'Herelle was working at the Pasteur Institute in Paris under wartime conditions where he was called to investigate an outbreak of bacillary dysentery in a group of French soldiers (Summers, 1999). In addition to his clinical duties, he pursued his research interest in the question of why enteric bacteria were sometimes pathogenic and sometimes not. Basing his approach on previously reported studies on the etiology of hog cholera, which was believed to be caused by a filterable virus but exacerbated by the presence of a bacterium,

Salmonella cholerasuis (Summers, 1998), d'Herelle examined the filtered dysentery samples for invisible viruses that might alter the growth and pathogenicity of the bacteria from the dysentery patients. To his surprise, he noted lysis in liquid culture and the formation of clear spots (later he called them *taches vierges,* "virgin spots") in the confluent bacterial culture that covered the agar slants. Because he was looking for an invisible microbe, his conception of bacteriophage as a parasite of bacteria was quite logical. His investigations proceeded on this assumption, and he noted that the invisible agent multiplied indefinitely, that it needed living cells to multiply, and that cell lysis seemed to be required for the multiplication process. D'Herelle astutely realized that the plaque count provided a way to enumerate these invisible agents, which he conceived as particulate. He was able to show that phage multiplied in waves, or steps, which he interpreted as representing cycles of infection, multiplication, release, and reinfection.

D'Herelle pioneered two important areas of phage research. One was based on his finding that phage titers rose in patients with infectious diseases just as recovery was taking place. From these observations he reasoned that phages represented natural agents of resistance to infectious diseases, and he went on to advocate phages as therapeutic agents in the preantibiotic era. However, his concept of phages as viruses of bacteria was not widely accepted. (The leading authorities up until the early 1940s thought that the lytic phenomena associated with bacteriophages resulted from an autocatalytic activation of an induced endogenous lytic enzyme.)

To counter his critics and establish his priority for the discovery of phages in a long-running dispute with Twort's supporters, d'Herelle's second research program examined the biological nature of the bacteriophage. All his evidence pointed to the conception of phages as organized infectious agents that are obligate intracellular parasites. He found that the antigenic properties and host-range specificity of phages appeared to be characteristic of given "races" of phages. Thus, from the beginning, there were hints that phage might be fruitful organisms for genetic study.

D'Herelle believed that phages were responsible for much of the recovery from infectious diseases. Because he observed increasing titers of phage during the course of recovery from dysentery and typhoid, he concluded that the gradual adaptation of lytic phage to specific pathogens, their subsequent multiplication, and lysis of the pathogen was the mechanism of recovery. He called phages "exogenous agents of immunity (d'Herelle, 1917, p. 375)." This ecological concept of phage and disease supported the effort to employ phages as therapeutic and prophylactic agents in a wide variety of infectious diseases.

2.3. EARLY DEBATE REGARDING THE NATURE OF BACTERIOPHAGES

D'Herelle's view of phages as related to immunity to disease was, in several ways, a direct challenge to the views of many bacteriologists, including Jules Bordet, who had just received a Nobel Prize in 1919 for his work demonstrating the bacteriolytic effects of serum components (*antibody* and *complement* in current terminology). Bordet and his colleagues in Belgium immediately started to work on phage and its possible role in Bordet's bacteriolytic phenomenon. Bordet searched for phage in peritoneal extracts of immunized mice and tested these extracts on bacteria. Bordet's experimental system usually involved injecting bacteria (and sometimes phage) into the host animal, and he emphasized the role of the host-bacteria interaction in producing phage. The colonies of bacteria that he isolated from the infected animals often gave rise to phage upon further in vitro culture or upon treatment with peritoneal extracts. He termed such bacteria *lysogenic*. Bordet's interpretation of this phenomenon was that the phage were produced from the bacterium itself upon induction by some alteration in the metabolic state of the bacterium (such as antibody treatment) and that the phage was a lytic enzyme rather than a particulate ultramicrobe.

Bordet's prestige as a Nobel prizewinner and as the director of the Pasteur Institute in Brussels strongly influenced the general perception of the nature of phage. D'Herelle, on the other hand, was an obscure, unpaid volunteer researcher at the Pasteur Institute in Paris. Nevertheless, Bordet launched an attack on d'Herelle's views about phage. Bordet's attack did not rest on scientific evidence alone. Instead it took the form of a priority dispute (Duckworth, 1976; van Helvoort, 1985; Summers, 1999). In 1923, Twort's paper on glassy transformation came to Bordet's attention, and he immediately challenged d'Herelle's priority to the discovery of bacteriophage. As a result of Twort's wartime responsibilities and his postwar research, he did not follow up on his work on glassy transformation, and, had it not been for Bordet's resurrection of his 1915 paper as a challenge to d'Herelle, the Twort paper might have become only a historical footnote. D'Herelle's response to this challenge was to assert that the phenomenon described by Twort was fundamentally different from the bacteriophage phenomenon. Thus, he conducted a series of experiments to characterize the biological nature of phage, experiments that he might not have undertaken without this challenge by Bordet. For about ten years this controversy raged, and Twort occasionally participated as the reluctant surrogate for Bordet in what became known as the "Twort-d'Herelle Controversy." The arguments were often petty and turned on minor differences in the heat and pH resistance between d'Herelle's phages and Twort's agent of glassy transformation. The controversy finally ended in 1932 with a full-fledged scientific duel: d'Herelle and André Gratia (a protegé of Bordet) agreed to an independent, side-by-side comparison of Twort's material and d'Herelle's material, to be conducted in an independent laboratory by highly respected independent scientists representing the two opposing camps. Paul-Christian Flu, director of the Institute of Tropical Medicine at Leiden, and E. Renaux, professor of microbiology at Liege, conducted the tests and concluded that the Twort and d'Herelle phenomena were the same (Flu and Renaux, 1932).

Whereas the particulate nature of phage seemed clear to d'Herelle and a few others, significantly Frank Macfarlane Burnet in Australia, the majority of microbiologists agreed with Bordet that the "Twort-d'Herelle Phenomenon," or the "bacteriophage phenomenon," was a case of induced autolytic enzyme expression. Part of the controversy was fueled by the occurrence of lysogenic cultures that seemed devoid of phage and then would suddenly lyse and produce bacteriophage. This symbiotic relationship between phage and cell was not understood, especially in the absence of any clear understanding of the nature of the genetic apparatus of bacteria. Since several enzymes such as lysozyme and pyocyanase were known to lyse bacteria, it was logical to try to interpret bacteriophage lysis into this existing paradigm as transmissible inducible autolysis. D'Herelle strongly argued, however, that the only way to explain plaques and the dilution effects on the plaque count was to assume that phages were discrete particles. Burnet and Lush (1936) and Wollman and Wollman (1937) suggested that lysogeny resulted from a transient and reversible interaction between the phage and the hereditary apparatus of the host cell. The direct visualization of phages by electron microscopy, beginning in 1940, led to the widespread adoption of d'Herelle's view of bacteriophage as viruses that infect bacteria.

2.4. FIRST PRACTICAL APPLICATIONS: VETERINARY AND MEDICINE

These investigations on the biological nature of phage were, however, eclipsed by the therapeutic potential of phage. Interest in the understanding of phage as biological objects succumbed to the commercial and medical possibilities of phage. From d'Herelle's first observations that phage titers increased in stool samples from dysentery patients, the role of bacteriophage in the course of infectious disease was of crucial interest. The first report of the therapeutic use of phage was a note by Bruynoghe and Maisin (1921) from Louvain in which they injected a preparation of staphylococcal phage in the local region of cutaneous boils (furuncles). They noted both reduction in swelling and in pain as well as some reduction in fever.

In the summer of 1919, d'Herelle carried out extensive tests of phage as prophylaxis against the natural infection of chickens by *Bacillus gallinarum,* the bacterium that causes avian typhosis. He reported these results in 1921 in his first monograph on phage, *Le bactériophage: Son rôle dans l'immunité* (d'Herelle, 1921). These early studies on the use of phages to control epidemics of avian typhosis appear reasonable, even by current standards. Chickens in certain pens were treated with phage prior to inoculation with *B. gallinarum;* groups of chickens, some phage-treated, were exposed to infected animals in the chicken pen so that the infection would spread under natural conditions. Phage treatment was administered orally, which minimized the possibility that other material in the phage lysate (e.g., bacterial debris), acted as an active immunogen (as it might, had it been injected parenterally). In these experiments, phage offered a high degree of protection. Extending this laboratory approach to actual field trials in rural areas of France where the epidemic was severe, d'Herelle inoculated (either orally or by injection) numerous flocks on farms in several widely separated regions. The overall results suggested that phage-treated flocks had many fewer deaths, the duration of the epidemic was shorter, and second rounds of the infection were prevented. D'Herelle's results were confirmed for the same disease in Holland by a Dutch investigator. The main shortcoming in all of these early phage experiments was the absence of a double-blind design. It should be noted, however, that this level of rigor was very uncommon at the time, and d'Herelle's studies appear to have been conducted according to the best scientific standards of his day.

Phage therapy was also evaluated in field trials against bovine hemorrhagic septicemia (*barbone*) in Indochina. In this disease, too, it appeared that parenteral inoculation of phages specific for this causative bacterium could protect water buffaloes against experimental inoculation with what is now called *Pasteurella multocida*, usually a highly fatal infection (d'Herelle, 1926).

With evidence of therapeutic effectiveness of phage in both gastrointestinal disease (avian typhosis) and septicemic disease (*barbone*), d'Herelle extended his trials to human beings. The procedures for conducting human trials in the 1920s, both scientific and ethical, seem crude and inadequate by current standards, but d'Herelle's approach was typical. He first determined the safety of his phage preparations by self-administration:

> Before undertaking experiments on man I had to assure myself that the administration of suspensions of the Shiga-bacteriophage caused no reaction. First, I ingested increasing quantities of such suspensions, aged from six days to a month, from one to thirty cubic centimeters, without detecting the slightest malaise. Three persons in my family next ingested variable quantities several times without showing the least disturbance. I then injected myself subcutaneously with one cubic centimeter of a forty-day old suspension. There was neither a local nor a general reaction. (d'Herelle, 1926, p. 540)

He also injected his coworkers as well as his family. This procedure was considered sufficient to evaluate the safety of this material: "After being assured that no harmful effects attended the ingestion of the Shiga-bacteriophage, this treatment was applied for therapeutic purposes to patients afflicted with [culture-confirmed] bacillary dysentery" (d'Herelle, 1926, p. 541).

The work that attracted the most attention for phage therapy was probably d'Herelle's report of four cases of bubonic plague which he treated with antiplague phage. While he was stationed at the League of Nations Quarantine Station in Alexandria Egypt, d'Herelle observed four patients on a ship passing through the Suez Canal, all of whom had laboratory-diagnosed bubonic plague. D'Herelle treated all four with antiplague phage preparations by direct injection of phage into the buboes (the infected inguinal and axillary lymph nodes). All four patients recovered in what was considered a remarkable fashion, and this result was reported in the widely read French medical periodical, *La presse médical* (d'Herelle, 1925). On the basis of this work, d'Herelle was invited by the British government to go to India to work on phage therapy of the plague at the Haffkine Institute in Bombay. This short visit led to the later establishment of The Bacteriophage Inquiry in India, under the patronage of the Indian Research Fund Association. This project studied the application of phage therapy in India, especially for cholera epidemics that regularly occurred in association with religious festivals and pilgrimages (d'Herelle, Malone, and Lahiri, 1930; Summers, 1993b).

In some ways, cholera is an ideal test case for therapy with phages: the bacteria are initially confined to the gastrointestinal tract, killing of large numbers of the bacteria reduces the burden of the pathogenic toxin, the mode of transmission and epidemiological characteristics of the disease are well-known, and, at least until recently, good vaccines have not been available. Phage therapy for cholera seems to be established as helpful in the treatment of patients with the disease; its use as a preventive measure, as d'Herelle had hoped when he went around India pouring phage stocks into the drinking water supplies, is less clearly established. These cholera studies are discussed in more detail in Chapter 14.

From the initial reports from India in the 1920s and 1930s (d'Herelle, Malone, and Lahiri, 1930; Summers, 1993b), it seems consistently observed that the severity and duration of cholera symptoms and the overall mortality from the disease were reduced in patients given cholera-specific phage by mouth. In several WHO-sponsored studies in Pakistan in the 1970s (Marcuk, 1971; Monsur, Rahman, Huq, Islam, Northrup, and Hirschhorn, 1970), in which phage was compared with antibiotics (tetracycline), high-dose phage therapy seemed about equivalent to tetracycline in certain aspects of the clinical control of cholera.

Interest in phage therapy has waxed and waned since its inception: first early enthusiasm, followed by critical skepticism and abandonment, and currently renewed interest and reappraisal. The changing attitudes toward phage therapy reflect both scientific and cultural influences. While many early phage therapy trials were reported to be successful, and many of the major pharmaceutical firms sold phage preparations (e.g., Parke-Davis and Lilly in the United States), there were also failures. The Council on Pharmacy and Chemistry, established in 1905 by the American Medical Association to set standards for drugs and lead the battle against quack remedies, undertook the evaluation of phage therapy in the mid-1930s. The voluminous report of the Council (Eaton and Bayne-Jones, 1934) concluded with an ambiguous assessment of the literature on phage therapy, acknowledging that there were both positive and negative results in the literature. The Council was concerned that the biological nature of bacteriophage was poorly understood and that the lack of standards for purity and potency of phage made it impossible to compare most of the published studies. Such a report might have generated more research and new and better understanding, but World War II and the discovery of antibiotics effectively diverted efforts away from further study of phage therapy in the United States. As discussed in Chapter 14, some very good research was conducted during the early 1940s by scientists like Rene Dubos at Harvard (Dubos, Straus, and Pierce, 1943) and physicians fighting typhoid in Los Angeles (Knouf, Ward, Reichle, Bower, and Hamilton, 1946), but it largely disappeared from view until very recently. D'Herelle had returned to France and was held under virtual house arrest in Vichy during the war, and thus the most vigorous advocate for phage therapy was silenced. In Europe, however, there were two major efforts in phage therapy continuing in a decidedly military context: the Soviet Union waged war against the Finns, and there were many battle casualties; phage therapy was extensively used to treat these war-wounded, as well as the many soldiers suffering from dysentery on the southern fronts. Phage therapy continued to be researched extensively with military support after the war and became part of the general standard of care there, particularly in the Republic of Georgia. The German military was also engaged in use of phage therapy; medical kits captured from Rommel's North African forces contained vials of phages that were standard supplies of the German war medic (W. Summers, unpublished).

2.5. BASIC PHAGE RESEARCH AND MAJOR SCIENTIFIC DISCOVERIES ASSOCIATED WITH BACTERIOPHAGES

2.5.1. THE CHEMICAL AND PHYSICAL NATURE OF PHAGES

In the 1930s, some investigators started to examine bacteriophage from a chemical point of view. This research program was intimately linked to the study of other filterable viruses, such as TMV and polio (Creager, 2002). The new techniques of ultracentrifugation, filtration through collodion membranes, and chemical analysis were brought to bear on phages. It became clear from work done by Elford and Andrews (1932) in England and by d'Herelle's colleagues in France (Sertic and Boulgakov, 1935) that the sizes of different isolates of phages differed markedly. These methods for determining the sizes of viruses, however, were indirect and without any accepted standardization. Criteria for purity were lacking, and it was not until Stanley was able to obtain poliovirus and TMV in crystalline form that reliable chemical and physical studies of viruses began to move forward (Creager, 2002). Stanley, of course, was greatly aided by the use of the newly invented ultracentrifuge, not only for analytical purposes, but also to purify and concentrate viruses by differential centrifugations.

Chemical analyses of these purified viruses showed that they were proteinaceous. Soon, however, the presence of phosphorus in these virus preparations suggested a second component, subsequently recognized as nucleic acid. Max Schlesinger (1936), who worked first in Germany and then in England, was nearly the only biochemist who studied the chemistry of phage. He prepared partially purified phage, estimated their sizes, and demonstrated the presence of DNA by means of the Feulgen reaction.

Interestingly, in 1940 one of the first scientific applications of the newly invented electron microscope was to visualize preparations of bacteriophages and to examine their interaction with bacterial cells. This research finally settled the smoldering issue of the nature of bacteriophages by clearly showing their particulate character and by showing that specific phages have characteristic morphologies (Leviditi and Bonet-Maury, 1942; Luria and Anderson, 1942; Pfankuch and Kauche, 1940; Ruska, 1940). The application of electron microscopy to viruses, and phages in particular, was one direction that was advocated as central to the research program of postwar "biophysics" (Rasmussen, 1997). Later research on the morphogenesis and supramolecular assembly of phage, as well as on the conformation of phage DNA molecules, relied heavily on improved electron microscopic methods (Anderson, 1966; Kellenberger, 1966; Wood and Edgar, 1967).

2.5.2. MAX DELBRÜCK AND THE DEVELOPMENT OF MOLECULAR GENETICS

The canonical accounts of the origins of molecular biology cite the entrance of the German physicist Max Delbrück as the major transition event in modern phage research (Cairns, Stent, and Watson, 1966; Fischer and Lipson, 1988). A closer examination of the historical record, however, shows that this view is oversimplified (Summers, 1993a). Delbrück's introduction to phage came about through his meeting with Emory L. Ellis, a postdoctoral researcher at Caltech who had started to study bacteriophages around 1936. Ellis was working with phages because he thought their study would contribute to the understanding of the role of viruses in cancer. Rather than plunge ahead with empirical research, he decided that an understanding of the fundamental biology of viruses was an essential prerequisite to the study of viral carcinogenesis. With this in mind he set out to investigate the process of infection of bacteria by bacteriophage and to see if he could repeat the basic experiments that supported the viral nature of bacteriophage (Summers, 1993a). By 1938 he had developed a research program on the fundamental biology of bacteriophages.

While Delbrück is often credited with "inventing" the so-called one-step growth experiment, Ellis noted that this result formed a cornerstone of d'Herelle's concept of phage multiplication: in d'Herelle's words, "the increase in the number of corpuscles does not take place in a continuous progressive fashion, but by successive liberations" (d'Herelle, 1926, p. 117). Ellis (1980) stated, "My first work was to develop the plaque count technique and show the step-wise growth curves showing that the phage multiplied in the bacterium, not in the solution. These step-growth curves really intrigued Delbrück, and I think were responsible for his wanting to join in the work."

Ellis followed d'Herelle's methods, and from Pasadena sewage he isolated a phage active on *E. coli*. This bacterium was selected because it grew well and was available from Carl Lindegren, one of Morgan's students who had turned to work on microorganisms. Ellis's work on the biology of bacteriophages attracted sufficient attention to be featured in a front-page story in the *Los Angeles Times* of April 30, 1938. At this time Ellis focused on the basic biology of the phage and its role as a model virus. Because Delbrück was not mentioned in this article, it is possible that he had not yet joined Ellis, or that his contributions to the project had not become noteworthy. Indeed, Delbrück later recalled that it was not until "early 1938" that he met Ellis and learned of his phage work (Fischer and Lipson, 1988, p. 113).

Delbrück had been educated as a theoretical physicist in Germany; he had worked on atomic physics with Niels Bohr in Copenhagen and adopted the philosophical outlook of the Copenhagen physicists. In 1932 he entered into a notable collaboration with the Russian geneticist Nicolai V. Timofeev-Ressovsky and the German biophysicist K.G. Zimmer. Another member of this group was the geneticist Herman J. Muller, who spent 1932 in Berlin working with Timofeev-Ressovsky. Muller contributed significantly to the effort to define the understanding of the nature of the gene as *the* critical problem in biology (Keller, 1990). This group discussed ways in which physics might be applied to biology to understand the nature of the gene, and in 1935 this threesome published a famous paper on the nature of gene structure and mutation (Timofeev-Ressovsky, Zimmer, and Delbrück, 1935). Their goal was, in the words of Gunther Stent (1963, p. 18), "to develop a quantum mechanical model of the gene." It was on the basis of this work that T.H. Morgan, the head of the Biology Division at Caltech, invited Delbrück to visit Caltech as a Rockefeller Foundation Fellow. Morgan hoped that the input of a theoretician such as Delbrück would contribute to his program in *Drosophila* genetics. Delbrück himself expected to continue to think about *Drosophila* as he had done in Germany, and he described his first few months at Caltech as a nightmare of trying to comprehend the details and intricacies of the actual experimental genetics of *Drosophila* (Delbrück, 1978).

Delbrück, however, was considering viruses as the right organism for the study of gene duplication even before he left Germany for the United States in 1937. While still in Europe he wrote some preliminary notes for a talk entitled "The Riddle of Life." In these 1937 notes, later translated from German and appended to his Nobel lecture (Delbrück, 1970), he made the point that "we want to look upon the replication of viruses as a particular form of a primitive replication of genes." His collaborator Muller, in particular, had been interested in bacteriophage for a long time, and as early as 1922 he had suggested that bacteriophages ("d'Herelle bodies") were naked genes:

> If these d'Herelle bodies were really genes, fundamentally like our chromosome genes, they would give us an utterly new angle from which to attack the gene problem . . . . It would be very rash to call these bodies genes, and yet at present we must confess that there is no distinction known between the genes and them. Hence we can not categorically deny that perhaps we may be able to grind genes in a mortar and cook them in a beaker after all. Must we geneticists become bacteriologists, physiological chemists and physicists, simultaneously with being zoologists and botanists? Let us hope so. (Muller, 1922, p. 49)

In Emory Ellis, Delbrück found a kindred spirit. Ellis was trained in physical chemistry and thermodynamics, not genetics or microbiology. He appreciated the formal similarities relating phage, Rous sarcoma virus, and TMV and was content to study them as "black boxes." Like Delbrück, Ellis had considered other viruses, especially TMV, and had rejected them for some of the same reasons given by Delbrück.

After a year of collaborative work, Ellis and Delbrück (1939) published their only joint paper. This paper included the step-curves on which Ellis had been working prior to his collaboration with Delbrück. Delbrück's contribution to the paper was a statistical analysis of the plating efficiency of the virus, involving a comparison of the infectivity in liquid culture (d'Herelle's terminal dilution experiment) with the plaque formation on solid media. This analysis was considered crucial because some workers used the ability to lyse a culture (or even the time it took to do so) as a measure of phage concentration, whereas others used the plaque assay. Ellis's name was first on the paper because he had started the project and was considered the senior member of the team. In his report to the Rockefeller Foundation at the end of his fellowship, Delbrück (1939) noted, "The leading idea was the belief that the growth of phage was essentially the same process as the growth of viruses and the reproduction of the gene. Phage was chosen because it seemed to offer the best promise for a deeper understanding of this process through a quantitative experimental approach."

It was Delbrück's focus on the gene that distinguished his interests from those of Ellis, who wanted to understand the biology of the viral life cycle as a whole. Both Ellis and Delbrück clearly understood they were working toward Delbrück's goal of understanding the nature and replication of the gene. Delbrück (1939) noted in his fellowship report that "during the second year Dr. Ellis returned to work on the tumor problem [required by the conditions of his fellowship support] and I carried on alone with the phage-work."

The experimental approach employed by Timofeev-Ressovsky, Zimmer, and Delbrück was almost exclusively based on an experimental methodology derived from atomic physics and known as "target theory" (Lea, 1946). The target theory provided a way to estimate the size of molecules and other "targets," such as genes within cells, from the parameters of the dose-response curves for radiation inactivation experiments. Radiobiology was a widely used tool during this period prior to the introduction of radioisotopes and modern biochemical methods of analysis. The target theory approach to the study of viruses and genes was especially appealing to physicists who saw interesting problems in biology. One such scientist was Salvador Luria (1984), an Italian physician who was trained in radiology and who learned about phage from Geo Rita, an early phage researcher in Rome. Luria moved to Paris in 1938, where he worked on the radiobiology of phage with a well-known French physicist, Fernand Holweck, at the Radium Institute. They used target theory approaches to estimate the size of phages (Wollman, Holweck, and Luria, 1940). When Luria came to the United States in 1940, he worked with radiation biologists at Columbia University but soon made contact with Delbrück, and the two of them initiated a long-term relationship that spawned a generation of phage research in America.

2.5.3. Role of the Phage Community

Concurrently with the work of Ellis and Delbrück and that of Luria, Alfred Hershey was studying phages from the point of view of their physiology (Stahl, 2000). At Washington University in St. Louis, Missouri, he was collaborating with Jacques Bronfenbrenner, who had a long interest in the possible metabolic and structural organization of bacteriophage. Delbrück was a natural organizer and, together with Luria and Hershey, he began to recruit people to work on phage biology. He developed a group of protégés, followers, devotees, and students who were indoctrinated with his viewpoint on the important problems of phage research and the legitimate way to approach these problems (Cairns et al., 1966). Delbrück saw the value in focusing research on a small group of phages so that results from different laboratories could be compared, and so he selected a group of "authorized phages," which became designated the T-phages, T1–T7 (T for type) (Demerec and Fano, 1945).

One of the most influential efforts of Delbrück was the organization of an annual Phage Course, which, starting in 1945, took place at the Cold Spring Harbor Laboratories each summer. This course was designed to provide laboratory instruction in the techniques of phage research, but most importantly it served to recruit a cadre of students to Delbrück's research program on the problem of the gene (Watson, 1966). Students in the Phage Course included both graduate students and senior scholars from major American universities, and, after appropriate education into the mysteries of phage, they returned to their universities to spread the word, give courses, and advocate for the research program of the Delbrück School of phage research. The group of disciples of the Cold Spring Harbor Phage Course formed a loose group of scientists who kept in close communication by a periodic newsletter, the "Phage Information Service," published by Delbrück. On one hand, it provided an alternative to formal publication of results, and, on the other hand, this newsletter enforced a sort of orthodoxy of approach that fit Delbrück's notions of what was acceptable phage research. The annual Phage Course, and the associated Phage Meetings, became a powerful social force in the beginning of molecular biology (Cairns, et al., 1966).

Although Delbrück was suspicious of the phenomenon known as lysogeny, the problem it presented was directly attacked by André Lwoff and his colleagues in Paris in the 1950s. Lysogenic phenomena had been observed from almost the first days of phage research, but the nature of the relationship between phage and host was unclear. Was lysogeny a sort of smoldering, persistent infection with the phage multiplying in a steady state with the growth of the host, or did the phage become truly latent? In the 1930s Eugène and Élizabeth Wollman (1937) suggested that phage in the lysogenic state behaved as part of the cellular hereditary apparatus. Lwoff and Antoinette Gutmann (1950) finally clarified the nature of lysogeny and christened the latent form of the phage *prophage* in their work in which they followed phage induction and release from single cells using direct microscopic observation and sampling with a micromanipulator. Lysogeny, induction, and its regulation became a major focus of phage research in Paris in the 1950s. The thesis of François Jacob was on lysogeny in *Pseudomonas pyocyanea,* and in the 1950s the study of lysogeny provided the groundwork for the operon concept of gene regulation that was developed by Jacob and Monod in the Service de Physiologie Microbienne at the Institut Pasteur (Morange, 2002).

Just as the T-phages were the model organisms for lytic phage research, bacteriophage lambda, discovered in 1951 by Esther Lederberg, became the prototypic lysogenic phage (Lederberg and Lederberg, 1953). Study of lambda has provided a deep understanding of the regulation of gene expression on one hand, and the mechanisms of lysogeny on the other.

While the French phage workers pursued research of a more physiological sort, based as it was on Lwoff's lifelong interest in growth, nutrition, and physiological adaptations, Delbrück's followers, who came to be called the American Phage Group, favored more direct physical approaches.

2.5.4. The Nature of the Gene and of Mutations

The approach of the Phage Group led to a very clear answer to one of the major questions of biology: what is the nature of the genetic material? The initial approaches were based on radiobiological methods for study of phage reproduction. In the immediate postwar period, a French biologist, Raymond Latarjet, came to the United States as part of a government program to reinvigorate and modernize French science. Latarjet had been working on the processes of inactivation of phage by ultraviolet radiation and had developed a method that he believed could be used to measure the number of intracellular infective phage: by measuring the radiobiological target number (i.e., the number of "hits" needed to completely inactivate an infected bacterium so that it could not produce a plaque). Latarjet was introduced to Luria soon after his arrival in America, and because Luria was also involved in phage radiobiology, they began a famous collaboration that resulted in a paradigmatic experiment that came to be called the Luria-Latarjet

Experiment (Luria and Laterjet, 1947). For the first time, there existed a way of measuring the intracellular increase in phage between infection and lysis. Delbrück and his followers saw this approach as providing a way to study the intracellular steps in the reproduction of the gene, and they adopted radiobiological methods as a mainstay in their experimental program.

One phenomenon that attracted much attention was multiplicity reactivation (MR), the phenomenon in which it appeared that phage that had been inactivated by radiation as determined by single infection (i.e., multiplicity of infection of one phage per bacterium) could be reactivated if several "dead" phage could infect a single bacterium. MR was discovered by Luria (1947), and it became a central problem that was investigated by the American Phage Group. As Watson (1966) recalled: "Delbrück remained confident, however, that MR was the key breakthrough which soon should tell us what was what ['what' referring to the self-reproduction of the gene]." It is interesting and important to note that the term used was always *reproduction,* never *replication,* thus situating this research in a very traditional biological context.

While it is often assumed that the Phage Group had a clear vision of the importance of DNA as the genetic material, that was not the case. Luria (1947) developed a detailed model of the gene to explain the results of MR. He proposed that "a phage particle contains a certain number of different self-reproducing 'units' (loci) [he calculated that the T-even phages have 30–50 such units], each capable of undergoing a lethal mutation under the action of radiation." Further, "it indicated that there is no appreciable amount of linkage in the transfer process [the parent to progeny transfer of the 'units']." Luria viewed the reproduction of the gene in a way that was analogous to the ideas about protein synthesis: "We can imagine that each active unit impresses its specificity on a number of elements produced in excess inside the host cell, elements which represent the raw material then utilized to build more phage particles. The units carrying lethal mutations may be unable to mold the substrate in their own image and likeness." He explicitly related this mechanism to the case of type-specific transformation in pneumococci, but as Watson (1966) later recalled, "Then there was also the fact that despite Avery, McCarty and McLeod, we were not at all sure that only the phage DNA carried genetic specificity." Watson's thesis research, carried out under the supervision of Luria, was on the MR phenomenon in x-rayed T7 phage. When he published portions of his thesis he remarked:

> More interesting is the possibility that these agents [other than UV] will cause different types of damage, which will block virus reproduction at different stages. We might therefore be able to reconstruct the successive steps in host-virus interaction by studying the stages in the synthesis at which multiplication of the inactive phage is blocked. (Watson, 1950, p. 697)

He suggested an extension of this general approach "through the use of chemical inactivating agents acting in more specific ways to subdivide reproduction into finer steps."

From the work on MR, Luria's units and the interest in understanding the steps in gene reproduction, the research program of the American Phage Group focused on the intracellular steps in phage growth as the key problem to be solved. Biochemical approaches to this problem, made possible by the newly available radioisotopes of phosphorus, carbon, and sulfur, allowed Frank Putnam and Lloyd Kozloff (1950), working it the Chicago lab of Earl Evans, a specialist in the use of isotopic tracers in metabolic research, to begin the study of the metabolic origins of the components of phage as well as the fate of parental phage material in parent-to-progeny label transfer experiments. These experiments followed earlier rather unsuccessful work of the same sort done in TMV by Stanley. A major problem encountered in these experiments was the failure to find more than a minority of the parental material transferred to the progeny virus. The possibility of two types of DNA, genetic and nongenetic, was discussed. For his postdoctoral research, Watson joined the Danish biochemist Ole Maaløe to follow up on these parental label transfer experiments. Their main question was, "Are there specific macromolecular structures (genes?) that are preserved and passed on intact to the progeny?" Unable to find more than 50 percent of the parental phosphorus appearing in the progeny phage, they wondered: "is it transmitted in the form of large blocks of DNA or after complete breakdown and assimilation?" and "is it in one particle of the progeny, or distributed over all of them?" (Maaløe and Watson, 1951). The precise fate of all parental DNA remained unresolved for several decades (Summers, 1968).

By early 1952, most of the experimental evidence from both the biological experiments on MR and the biochemical experiments on parental-to-progeny label transfer suggested that genes were not conserved in the process of phage reproduction: "Neither phosphorus nor sulfur is transferred from parent to progeny in the form of special hereditary parts of the phage particle" (Hershey, Roesel, Chase, and Forman, 1951), "The transfer experiments, and especially the observation that chemical transfer may occur in the absence of genetic transfer, might, therefore, be viewed as evidence for an obligate and extensive breakdown of the infecting particle soon after adsorption" (Watson and Maaløe, 1953), and "[T]he transfer of parent virus N and P to the progeny is largely independent of genetic units" (Kozloff, 1952). At this time, the nature of the genetic material and its mode of reproduction were anything but clear. It was clear, however, that some new experimental approaches would be needed to make progress. These were to come very soon in the form of the structural studies of Watson and Crick (1953); Franklin and Gosling (1953); Wilkins, Stokes, and Wilson (1953), and the clever extension of the parental-to-progeny label transfer experiment devised by Hershey and Chase (1952).

Hershey noted that the DNA of phage T2 could be separated from the protein "ghosts" of the phage by *in vitro* osmotic shock, and he reasoned that this dissociation might be related to some of the long-sought intracellular steps in reproduction of the phage. Most of the famous paper by Hershey and Chase (1952) is devoted to characterizing this dissociation *in vitro*. They used radioactive-labeled proteins and nucleic acids in the standard parent-to-progeny transfer experiment, but by careful attention to the association of the labeled components with the infectious material, they concluded that DNA in the phage becomes liberated (as determined by slow sedimentation and sensitivity to DNase) when phage adsorb to bacterial debris. Further, this process is accompanied by loss of infectivity. These observations suggested that there is a dissociation of the "membranous ghosts" of the phage from the DNA component as one of the steps in phage reproduction. Turning their attention to adsorption of phage to intact bacteria rather than bacterial debris, they wrote:

> Anderson (1951) has obtained electron micrographs indicating that phage T2 attaches to bacteria by its tail. If this precarious attachment is preserved during the progress of the infection, and if the conclusions reached above are correct, it ought to be a simple matter to break the empty phage membranes off the infected bacteria, leaving the phage DNA inside the cells. p. 46

Hershey and Chase employed hydrodynamic shear (i.e., a Waring food blender) to disrupt "this precarious attachment" of empty phage coats from infected bacteria. Using radioactive labels to follow the fate of protein and nucleic acid, they were able to shear 75% to 80% of the protein label off the infected cells while removing only 21 to 35 percent of the DNA label. The shearing process did not cause appreciable killing of the bacteria. They then combined this sort of analysis with the standard label transfer experiment and found that essentially none of the retained protein label (i.e., the 20%–25%) appeared in the progeny phage, in contrast to the well-known transfer of about half of the nucleic acid label from parent-to-progeny phage. These results, taken together, have been idealized as a textbook case, taken to show "that only the DNA of phage entered the host bacteria. Their surrounding protein coats remained outside and thus could be ruled out as potential genetic material" (Watson, Gilman, Witkowski, and Zoller, 1992, p. 16–17). This result, of course, does not apply to all viruses (most animal cell viruses enter the cell by endocytosis in an intact form) and not even to all phages (e.g., ϕ29 which has protein attached to its DNA and which enters the cell upon infection). While the Hershey-Chase experiment (sometimes known as "the blender experiment") has been taken as the final "proof" that DNA is the genetic material, Alfred Hershey apparently expected the protein of the phage to play a role in the genetic process, right up to the time of these experiments (Szybalski, 2000).

Not all research on phage reproduction was focused on the "problem of the gene." Other biochemical approaches were aimed at a more metabolic description of phage reproduction. From the very early days of phage research, when it was thought that phages were highly organized but very small ultramicrobes, there was an interest in the potential of phage for independent metabolic activity. The nature of their dependence on the host cell as obligate intracellular parasites was of interest. Attempts to detect respiration in phages by means of the standard gasometrical analyses with a Warburg apparatus were fruitless (Bronfenbrenner and Reichert, 1926). The study of the metabolic activities in phage-infected cells was central to the problem of protein and nucleic acid synthesis. Several groups, including that of Earl Evans (1952) and Seymour Cohen (1947), undertook the study of enzyme activities and metabolic changes in phage-infected cells as a way to approach problems of macromolecular biosynthesis. In the immediate post–World War II period, radioactive tracers for biological research were just becoming available. Labeled intermediates and precursor compounds made it possible to investigate metabolism on a new and more refined scale than previously possible, when chemical methods were the only tool for such study. In addition to the parental-to-progeny label transfer experiments previously described, soon a novel DNA base, 5-hydroxymethylcytosine, was discovered in the DNA of the T-even phages (Wyatt and Cohen, 1953). Inhibitors of protein synthesis (amino acid analogues and antibiotics) were observed to block the intracellular development of phage, which was strong evidence that new protein synthesis was needed as part of the process of phage reproduction. The timing of this inhibition suggested that there were proteins needed early in the infectious cycle that were not part of the virus structure (e.g., the coat that could be detected antigenically late in the infectious cycle). The nature of these "early proteins" became clear when Flaks and Cohen (1959) and Kornberg, Zimmerman, Kornberg, and Josse (1959) discovered that these early proteins included enzymes needed for phage DNA biosynthesis (e.g., deoxycytidylate hydroxymethylase, DNA polymerase, glucosyl transferases, DNases, polynucleotide kinase, and thymidylate synthetase, to mention a few of the many now-known phage-encoded enzymes).

However, much of the basic research about fundamental aspects of genetics occurred quite independently of any understanding of the physical nature of the gene. The nature and cause of gene mutations were investigated in the 1930s, and a key question emerged from this work: were mutations caused by "needs" imposed by the selective growth conditions (e.g., the addition of lactose to the medium) or did the mutations occur randomly all the time and their existence become known by imposition of the selective growth conditions? The outcome of this research would have profound implications for the science of genetics, but also for a deeper understanding of evolutionary biology in general.

The major problem in this work was one of experimental design: how to observe a rare event that happened in a huge population prior to the selection for the outcome of that event. In a particularly clear and convincing work, Isaac M. Lewis (1934) examined the mutational change from the inability to ferment lactose to the ability to use this sugar source in the *Escherichia coli* strain *mutabile*. He concluded that the mutations to lactose utilization (lac^- to lac^+) occurred *prior* to the selection, not *as a consequence* of exposure to the selective conditions. His approach did not change many minds, but in the 1940s two related experimental approaches gave results believed to settle this question. Both of these employed bacteriophage as experimental tools. Luria and Delbrück knew from the work of d'Herelle that bacteria often developed heritable resistance to phage. Their routine use of statistical models in their target theory studies, as well as their backgrounds in atomic physics, helped them to devise a statistical approach (fluctuation test) to show that phage-resistant mutants existed in the bacterial population prior to exposure to the lethal effects of phage. The method of Luria and Delbrück (1943) and a related procedure devised in 1949 by Howard B. Newcome were indirect and mathematical. However, Joshua Lederberg (1952) and Esther Lederberg (Lederberg and Lederberg, 1952) devised a simple and direct way to demonstrate that mutations were occurring randomly, independent of the selection procedures. They transferred very large numbers of colonies from one plate to another by the use of velvet cloth as a transfer tool; these replica plates could be used to test colonies in great numbers for mutant properties. They applied this technique to study of phage resistance as well as streptomycin resistance. Again it was clear that the mutants had appeared before the application of the selective agent.

The random nature of mutation and its very low frequency of occurrence suggested that it might be similar to, or governed by, a quantized, two-state process. This model appealed to physicists such as Delbrück, Erwin Schroedinger, and Bohr who thought deeper understanding of this paradoxical behavior might reveal new physical laws of nature (Fischer and Lipson, 1988). Although no new laws of nature seemed to emerge, deeper insight was indeed provided by strictly formal genetic analysis of phage mutations. Hershey and Rotman (1949) had employed plaque morphology mutants (large plaques, interpreted as rapid-lysis mutants, or "r-mutants") to show that phages could be "mated," or crossed, by simultaneous mixed infections, and thus it was possible to carry out formal genetic analysis on phage just as with sexually mating organisms. Seymour Benzer found one class of Hershey's r-mutants of phages T2 and T4 to be particularly interesting: they did not give any plaques at all (less than one in a hundred million) on bacterial hosts carrying the lambda prophage (K12 lambda) but gave the large (r-mutant) plaques on the usual bacterial host (strain B). Benzer exploited this case of conditional expression of a phage mutation to develop the fine structure genetic analysis of the T4 *rII* gene. Because of the strong specificity of the system (the discrimination against the *rII* mutants in host K12 lambda is greater than 10^8), very low frequency wild-type r^+ recombinants or revertants can be detected, and hence, recombination between very close mutations can be observed. Benzer calculated that he could detect recombination between adjacent base pairs in the *rII* gene—still the only known system where that is possible.

While numerological considerations pointed to the need for at least three nucleotides to specify the full set of 20 amino acids, which seemed to show no restrictions on nearest neighbor sequence (Brenner, 1957), it was unclear just what the coding rules were. In the 1950s there was the strong conviction that understanding the process of gene mutation would illuminate many puzzling aspects of gene reproduction and function. DeMars (1953) made the observation that acridine compounds were potent mutagens of phage, but weak mutagens of bacteria. One hypothesis to explain this difference was that acridines acted on actively recombining DNA, prevalent in phage infection, but rare in bacterial growth. Benzer's T4 *rII* phage system allowed some of these ideas to be tested, and for both chemical and genetic reasons, Brenner, Jacob, and Meselson (1961) advanced the notion that acridine mutagenesis occurred by the insertion or deletion of a single base pair. The detailed genetic analysis of these hypothetical deletions and insertion mutations in the T4 *rII* gene soon provided strong evidence for the triplet nature of the genetic code (Crick, Barnett, Brenner, and Watts-Tobin, 1961). The basis for this important work was the ability of a triplet, nonoverlapping, nonpunctuated (commaless) coding scheme to correctly predict the outcome of a large series of genetic crosses between T4 *rII* phage mutants classified as single or double insertions or deletions. Thus, a functional, wild-type *rII* gene was possible only when the parental phage carried T4 *rII* mutations (+1, +2, –1, or –2), which, when recombined, could result in a recombinant phage with a hypothetical gene structure with multiples of 3 (0, +3, or –3). This work, mainly because of its strong internal consistency and coherence, was taken as the first strong experimental evidence for the triplet nature of the genetic code. Biochemical approaches using synthetic messenger RNAs of defined composition or sequence were subsequently employed to determine the "dictionary" that relates specific triplets to specific amino acids.

Another noteworthy example of the research on phage mutations was the discovery by several investigators of the phenomenon that was named *host-induced-modification.* This puzzling phenomenon was discovered when examining various host-range phenomena related to phage mutations and adaptations. It was observed by Luria and Human (1952) for T2 phage, and by Bertani and Weigle (1953) for lambda phage, that the host range, as measured by plating efficiency, was sometimes determined by the particular host strain of bacteria in which the phage had most recently replicated. It appeared as though the host, in some way, modified the phage so as to affect its ability to infect other host strains. This phenomenon was clearly not an example of "classical" mutation, but it appeared to be some sort of short-term heritable change. Phages with the incorrect modification were restricted in their growth in certain bacterial strains. This odd phenomenon was regarded as a minor peculiarity of phage biology for about a decade. Both genetic and biochemical investigation of this aspect of phage biology revealed, of course, that it was the result of sequence-specific nucleases that cleaved DNA lacking certain host-specified modifications, usually methylations or glucosylations on the phage DNA. The exploitation of this arcane bit of phage biology has provided a major tool for both fundamental and technological progress in biology in the last three decades with the development of recombinant DNA technology.

By the 1960s, it became clear that phages represented only one sort of extrachromosomal genetic structure that could exist in bacteria. The fertility factor F, transmissible drug-resistant determinants, as well as prophages, all represented examples of what Joshua Lederberg (1952) termed *plasmids.* It remained to be determined in what form these plasmids existed in the cell. Some experiments suggested that the plasmids (the F-factor in Hfr strains and some prophages) were somehow attached to the cell chromosome. A very fruitful model for how this attachment could occur was proposed by Alan Campbell (1962). He reasoned that because the genetic map of phage T4 was circular, whereas the phage DNA appeared to be linear, the phage might assume a circular intracellular form. This circular form could, with a single reciprocal recombination event, become linearly integrated into the chromosomal DNA. Although it turned out that T4 has a circular genetic map for reasons other than forming a physically circular molecule, Campbell's model was also supported by the genetic findings of Calef and Licciardello (1960) for phage lambda (which does exist in a circular form), and eventually many other plasmids. It has provided the conceptual basis for retrovirus integration and excision in animal cells as well.

Although much work on phage focused on the problem of the reproduction of the gene (i.e., transmission genetics), some phage biologists studied the processes by which phages developed intracellularly, through phage embryology. Biochemical analysis of metabolic changes after phage infection gave some clues to how phage might grow and multiply inside infected cells. A particularly important theme in this research was the role of RNA in phage development. Quite early on, Volkin and Astrachan (1956) found that the RNA base composition in phage-infected cells was much more DNA-like than in uninfected cells. Soon thereafter, to explain the phenomena involved in induced enzyme synthesis in bacteria, Jacob and Monod (1961) hypothesized that an unstable RNA intermediate might provide the material link for information flow from the gene to the site of protein synthesis, the ribosome. Because the great bulk of cell RNA was made up of the stable RNA in the ribosomes, this hypothetical messenger RNA of fleeting existence was difficult to observe experimentally. The earlier results of Volkin and Astrachan (1956) supported the notion that RNA could reflect DNA sequence information, an idea that was more directly tested by Brenner and colleagues (1961). They used the density gradient method pioneered by Meselson, Stahl, and Vinograd (1957) to distinguish between RNA made prior to infection and that made after phage infection. The results supported the messenger RNA hypothesis by showing that the RNA made after infection is not ribosomal RNA, but it does associate with ribosomes made prior to infection. Conclusive evidence that this newly made RNA has the properties expected of a true carrier of information was provided by Hall and Spiegelman (1961) who used the then novel technique of nucleic acid hybridization (Doty, Marmur, Eigner, and Schildkraut, 1960) to show that the RNA made after T2 phage infection was complementary to the phage DNA itself. Thus, phage biology provided the definitive experimental verification of the link in the pathway of genetic information from the gene to the ribosome.

As the study of phage replication and gene expression developed, it became clear that the full understanding of the workings of genes could be seen as a unified process rather than separate domains of transmission genetics on one hand and developmental genetics on the other. This unity was captured by Crick's metaphor of the so-called central dogma of molecular biology. Describing the function of genes in information-theoretic terms, he stated that information flows from DNA to DNA and from DNA to RNA and thence to proteins.

The fruitful collaboration between biochemists, geneticists, and microbiologists has provided detailed descriptions of the mechanisms of phage replication and transcription, of phage morphogenesis and assembly, and of phage adsorption and entry phenomena. Many of these advances are still ongoing and are part of the recent history of phage molecular biology described in the various sections of this book, particularly in Chapter 7.

REFERENCES

Anderson, T.F., Techniques for the preservation of three dimensional structure in preparing specimens for the electron microscope, *Trans New York Acad Sci,* 13, 130, 1951.

Anderson, T.F., Electron microscopy of phages, in *Phage and the Origins of Molecular Biology,* J. Cairns, G.S. Stent, and J.D. Watson, (Eds.), Cold Spring Harbor Laboratory Press, Cold Spring Harbor, NY, 1966.

Benzer, S., Fine structure of a genetic region in bacteriophage, *Proc Natl Acad Sci U S A,* 41, 344, 1955.

Bertani, G. and Weigle, J.J., Host controlled variation in bacterial viruses, *J Bacteriol,* 65, 113–121, 1953.

Brenner, S., On the impossibility of all overlapping triplet codes in information transfer from nucleic acids to proteins, *Proc Natl Acad Sci U S A*, 43, 687, 1957.

Brenner, S., Jacob, F., and Meselson, M., An unstable intermediate carrying information from genes to ribosomes for protein synthesis, *Nature,* 190, 576–581, 1961.

Bronfenbrenner, J. and Reichert, P., Respiration of so-called filterable viruses, *Proc Soc Exptl Biol Med*, 24, 176–177, 1926.

Bruynoghe, R. and Maisin, J., Essais de thérapeutique au moyen du bactériophage du Staphylocoque, *J Compt Rend Soc Biol,* 85, 1120–1121, 1921.

Burnet, F.M. and Lush, D., Induced lysogenicity and mutation of bacteriophage within lysogenic bacteria, *Aust J Exptl Biol Med Sci,* 14, 27–38, 1936.

Cairns, J., Stent, G.S., and Watson, J.D., *Phage and the Origins of Molecular Biology,* Cold Spring Harbor Laboratory, Cold Spring Harbor, NY, 1966.

Calef, E. and Licciardello, G., Recombination experiments on prophage host relationships, *Virology,* 12, 81, 1960.

Campbell, A.M., Episomes. *Adv Genetics,* 11, 101, 1962.

Cohen, S.S., Synthesis of bacterial viruses in infected cells, *Cold Spring Harbor Symp Quant Biol,* 12, 25, 1947.

Creager, A.N.H., *The Life of a Virus: Tobacco Mosaic Virus as an Experimental Model, 1930–1965*, University of Chicago Press, Chicago, IL, 2002.

Crick, F.H.C., Barnett, L., Brenner, S., and Watts-Tobin, R.J., General nature of the genetic code for proteins, *Nature,* 192, 1227, 1961.

Delbrück, M., Report of research done during the tenure of a Rockefeller Fellowship in 1938 and 1939, pp. in *Caltech Archives,* 1939.

Delbrück, M., A physicist's renewed look at biology: Twenty years later, *Science,* 168, 1315, 1970.

Delbrück, M., Oral history, in *Caltech Archives,* 1978, p. 63.

DeMars, R.I., Chemical mutagenesis in bacteriophage T2, *Nature,* 172, 964, 1953.

Demerec, M. and Fano, U., Bacteriophage-resistant mutants in *Escherichia coli, Genetics,* 30, 119, 1945.

d'Herelle, F., Sur un microbe invisible antagoniste des bacilles dysentériques. *Compt Rend Acad Sci,* 165, 373–375, 1917.

d'Herelle, F., *Le bactériophage: Son rôle dans limmunité*, Masson et Cie, Paris, 1921.

d'Herelle, F., Essai de traîtement de la peste bubonique par le bactériophage, *La Presse Méd,* 33, 1393–1394, 1925.

d'Herelle, F., *The Bacteriophage and its Behavior,* Williams and Wilkins, Baltimore, MD, 1926.

d'Herelle, F., Malone, R.H., and Lahiri, M., Studies on Asiatic cholera, *Indian Med Res Memoirs,* 14, 1, 1930.

Doty, P., Marmur, J., Eigner, J., and Schildkraut, C., Strand separation and specific recombination in deoxyribonucleic acids: Physical chemical studies, *Proc Natl Acad Sci U S A,* 46, 461–476, 1960.

Dubos, R., Straus, J.H., and Pierce, C., The multiplication of bacteriophage in vivo and its protective effect against an experimental infection with *Shigella dysenteriae*, *J Exp Med,* 20, 161–168, 1943.

Duckworth, D.H., Who discovered bacteriophage? *Bacteriol Rev,* 40, 793–802, 1976.

Eaton, M.D. and Bayne-Jones, S., Bacteriophage therapy. Review of the principles and results of the use of bacteriophage in the treatment of infections, *J Am Med Assoc,* 23, 1769–1939, 1934.

Elford, W.J. and Andrews, C.H., The sizes of different bacteriophages, *Brit J Exp Path,* 13, 446, 1932.

Ellis, E.L., Letter to Horace Davenport, in *Caltech Archives: Delbrück Papers*, 1980.

Ellis, E.L. and Delbrück, M., The growth of bacteriophage, *J Gen Physiol,* 22, 365–384, 1939.

Emmerich, R. and Löw, O., Die künstliche Darstellung der immunisierenden Substanzen (Nucleasen-Immunproteïdine) und ihre Verwendung zur Therapie der Infektionskrankheiten und zur Schutzimpfung and Stelle de Heilserums, *Zeitsch f Hyg u Infektionskrankh,* 36, 9, 1901.

Evans, E.A., Jr., *Biochemical Studies of Bacterial Viruses,* University of Chicago Press, Chicago, IL, 1952.

Fischer, E.P. and Lipson, C., *Thinking about Science: Max Delbrück and the Origins of Molecular Biology,* Norton, New York, 1988.

Flaks, J.G. and Cohen, S.S., Virus-induced acquisition of metabolic function. I. Enzymatic formation of 5-hydroxymethydeoxycytidine, *J Biol Chem,* 234, 1501, 1959.

Flu, P.C. and Renaux, E., Le phénomène de Twort et la bactériophage, *Ann Inst Pasteur,* 48, 15, 1932.

Franklin, R.E. and Gosling, R.G., Molecular configuration in sodium thymonucleate, *Nature,* 171, 740–741, 1953.

Hall, B.D. and Spiegelman, S., Sequence complementarity of T2-DNA and T2-specific RNA, *Proc Natl Acad Sci U S A,* 47, 137–146, 1961.

Hankin, E.H., L'action bactericide des eaux de la Jumna et du Gange sur le vibrion du cholera, *Ann de l'Inst Pasteur,* 10, 511, 1896.

Hershey, A.D. and Chase, M., Independent functions of viral protein and nucleic acid in growth of bacteriophage, *J Gen Physiol*, 36, 39–56, 1952.

Hershey, A.D., Roesel, C., Chase, M., and Forman, S., Growth and inheritance in bacteriophage, *Carnegie Inst Wash Yearbook,* 50, 195, 1951.

Hershey, A.D. and Rotman, R., Genetic recombination between host-range and plaque-type mutants of bacteriophage in single bacterial cells, *Genetics,* 34, 44, 1949.

Jacob, F. and Monod, J., Genetic regulatory mechanisms in the synthesis of proteins, *J Mol Biol,* 3, 318–356, 1961.

Kellenberger, E., Electron microscopy of developing bacteriophage, in *Phage and the Origins of Molecular Biology,* J. Cairns, G.S. Stent, and J.D. Watson, (Eds.), Cold Spring Harbor Laboratory, Cold Spring Harbor, NY, 1966, pp. 116–129.

Keller, E.F., Physics and the emergence of molecular biology: A history of cognitive and political synergy, *J Hist Biol,* 23, 389, 1990.

Knouf, E.G., Ward, W.E., Reichle, P.A., Bower, A.G., and Hamilton, P.M., Treatment of typhoid fever with type-specific bacteriophage, *J Am Med Assn*, 132, 134–136, 1946.

Kornberg, A., Zimmerman, S.B., Kornberg, S.R., and Josse, J., Enzymatic synthesis of deoxyribonucleic acid VI. Influence of bacteriophage T2 on the synthetic pathway in infected cells, *Proc Natl Acad Sci U S A,* 45, 722, 1959.

Kozloff, L.M., Biochemical studies of virus reproduction. VII. The appearance of nitrogen and phosphorus in the progeny, *J Biol Chem,* 195, 95, 1952.

Lea, D.E., *Actions of Radiations on Living Cells,* Cambridge University Press, Cambridge, 1946.

Lederberg, E.M. and Lederberg, J., Genetic studies of lysogenicity in *Escherichia coli, Genetics,* 38, 51, 1953.

Lederberg, J., Cell genetics and hereditary symbiosis, *Physiol Rev,* 32, 403, 1952.

Lederberg, J. and Lederberg, E.M., Replica plating and indirect selection of bacterial mutants, *J Bacteriol*, 63, 399, 1952.

Leviditi, C. and Bonet-Maury, P., Les ultravirus: Considérés à travers le microscope électronique, *La Presse Méd,* 17, 203, 1942.

Lewis, I.M., Bacterial variation with special reference to behavior of some mutabile strains of colon bacteria in synthetic media, *J Bacteriol,* 26, 619, 1934.

Luria, S.E., Reactivation of irradiated bacteriophage by transfer of self-reproducing units, *Proc Natl Acad Sci U S A,* 33, 253, 1947.

Luria, S.E., *A Slot Machine, a Broken Test Tube,* Harper and Row, New York, 1984.

Luria, S.E. and Anderson, T.F., Identification and characterization of bacteriophages with the electron microscope. *Proc Natl Acad Sci U S A* 28, 127, 1942.

Luria, S.E. and Delbruck, M., Mutations of bacteria from virus sensitivity to virus resistance, *Genetics,* 28, 491–511, 1943.

Luria, S.E. and Human, M.L., A non-hereditary host-induced variation of bacterial viruses, *J Bacteriol,* 64, 557, 1952.

Luria, S. E. and Latarjet, R., 1947 Ultraviolet irradiation during intracellular growth. *J Bacteriol* 53: 149.

Lwoff, A. and Gutmann, A., Recherches sur un *Bacillus megathérium* lysogène, *Ann Inst Pasteur,* 78, 711, 1950.

Maaløe, O. and Watson, J.D., The transfer of radioactive phosphorus from parental to progeny phage. *Proc Natl Acad Sci U S A* 37: 507, 1951.

Marcuk, L.M., Clinical studies of the use of bacteriophage in the treatment of cholera, *Bull World Health Org.* 45, 77–83, 1971.

Meselson, M., Stahl, F.W., and Vinograd, J., Equilibrium sedimentation of macromolecules in density gradients, *Proc Natl Acad Sci U S A,* 43, 581, 1957.

Monsur, K.A., Rahman, M.A., Huq, F., Islam, M.N., Northrup, R.S., and Hirschhorn, N., Effect of massive dosis of bacteriophage on excretion of *Vibrio,* duration of diarrhoea and output of stools in acute cases of cholera, *Bull World Health Org,* 42, 723–732, 1970.

Morange, M., Les mousquetaires de la nouvelle biologie: Monod, Jacob, Lwoff, *Pour la Science,* 10, 3, 2002.

Muller, H.J., Variations due to change in the individual gene, *Am Naturalist,* 56, 48, 1922.

Otto, R. and Munter, H., Bacteriophagie, *Ergebn d Hyg Bakteriol Immunitätsforsch u exp Therapie,* 6, 1, 1923.

Pfankuch, E. and Kauche, G.A., Isolierung und übermikroskopische Abbildung eines Bakteriophagen, *Naturwissenshaften,* 28, 46, 1940.

Putnam, F. and Kozloff, L.M., Biochemical studies of virus reproduction. IV. The fate of the infecting virus particle, *J Biol Chem,* 182, 243, 1950.

Rasmussen, N., *Picture Control: The Electron Microscope and the Transformation of Biology in America, 1940–1960,* Stanford University Press, Stanford, CA, 1997.

Ruska, H., Die sichtbarmachung der bakteriophagen lyse im ubermikroskop, *Naturwissenschaften,* 28, 45, 1940.

Schlesinger, M., The Feulgen reaction of the bacteriophage substance, *Nature,* 138, 508, 1936.

Sertic, V. and Boulgakov, N., Classification et identification des typhiphages, *Comptes rendus Soc Biol Paris,* 119, 1270, 1935.

Stahl, F., *We Can Sleep Later: Alfred D. Hershey and the Origins of Molecular Biology,* Cold Spring Harbor Laboratory Press, Cold Spring Harbor, NY, 2000.

Stent, G.S., *Molecular Biology of Bacterial Viruses,* Freeman, San Francisco, CA, 1963.

Summers, W.C., Equal transfer of both parental T7 DNA strands to progeny bacteriophage, *Nature,* 219, 159–160, 1968.

Summers, W.C., From culture as organism to organism as cell: Historical origins of bacterial genetics, *J Hist Biol,* 24, 171, 1991.

Summers, W.C., How bacteriophage came to be used by the Phage Group, *J Hist Biol,* 26, 255, 1993a.

Summers, W.C., Plague and Cholera in India: The bacteriophage inquiry of 1928–1936, *J Hist Med All Sci,* 48, 275, 1993b.

Summers, W.C., Theorien der Verursachung, ihre Rechtfertigung und de experimentelle Wissenschaft: Daniel E. Salmon und die Schweinepest, in *Strategien der Kausalitaet. Konzeptionen der Krankheitsversursachung im 19 und 20,* Jahrhundert, Gradmann, Christoph, Schlich, and Thomas, (Eds.), Centaurus, Pfaffenweiler, 1998, pp. 79–94.

Summers, W.C., Bacteriophage discovered, in *Felix d'Herelle and the Origins of Molecular Biology,* Yale University Press, New Haven, CT, 1999, pp. 47–59.

Szybalski, W., In memoriam: Alfred D. Hershey (1908–1997), in *We Can Sleep Later: Alfred D. Hershey and the Origins of Molecular Biology,* F. Stahl, (Ed.), Cold Spring Harbor Laboratory Press, Cold Spring Harbor, NY, 2000, pp. 19–22.

Timofeev-Ressovsky, N.W., Zimmer, K.G., and Delbrück, M., Über die natur der genmutation und der genstruktur, *Nachr Ges Wiss Göttingen Math-Phys Kl Fachgruppe,* 6, 190, 1935.

Twort, A., *In Focus, Out of Step: A Biography of Frederick William Twort, F.R.S., 1877–1950,* Alan Sutton, Dover, NH, 1993.

Twort, F.W., An investigation on the nature of ultra-microscopic viruses, *Lancet,* II, 1241–1243, 1915.

van Helvoort, T., Felix d'Herelle en de controverse rond het Twort-d'Herelle Fenomeen in de jaren 1920: Ultrafiltreerbarr virus of lytisch ferment, *Tsch Gesch Gnk Natuurw Wisk Techn,* 9, 118, 1985.

Volkin, E. and Astrachan, L., Phosphorus incorporation in *Escherichia coli* ribonucleic acid after infection with bacteriophage T2, *Virology,* 2, 149, 1956.

Watson, J.D., The properties of x-ray-inactivated bacteriophage. I. Inactivation by direct effect, *J Bacteriol,* 60, 697, 1950.

Watson, J.D., Growing up in the phage group, in *Phage and the Origins of Molecular Biology,* J. Cairns, G.S. Stent, and J.D. Watson, (Eds.), Cold Spring Harbor Laboratory, Cold Spring Harbor, NY, 1966, pp. 239–245.

Watson, J.D. and Crick, F.H.C., A structure for desoxyribose nucleic acids, *Nature,* 171, 737–738, 1953.

Watson, J.D., Gilman, M., Witkowski, J., and Zoller, M., *Recombinant DNA,* Scientific American Books, New York, 1992.

Watson, J.D. and Maaløe, O., Nucleic acid transfer from parent to progeny bacteriophage, *Biochim Biophys Acta,* 10, 432, 1953.

Wilkins, M.H.F., Stokes, A.R., and Wilson, H.R., Molecular structure of deoxypentose nucleic acid, *Nature,* 171, 738–740, 1953.

Wollman, E., Holweck, F., and Luria, S., Effect of radiations on bacteriophage C16, *Nature,* 145, 935, 1940.

Wollman, E. and Wollman, E., Les phases de bactériophages (facteurs lysogènes). *Comptes rendus Soc Biol Paris,* 124, 931, 1937.

Wood, W.B. and Edgar, R.S., Building a bacterial virus, *Sci Am,* 217, 61–66, 1967.

Wyatt, G.R. and Cohen, S.S., The bases of the nucleic acid of some bacterial and animal viruses: The occurrence of 5-hydroxymethylcytosine, *Biochem J,* 55, 774–782, 1953.

3 Basic Phage Biology

Burton Guttman,[1] Raul Raya,[1,2] and Elizabeth Kutter[1]
[1]The Lab of Phage Biology, The Evergreen State College, Olympia, WA
[2]Cerela, Tucuman, Argentina

CONTENTS

3.1. Introduction 30
  3.1.1. The Nature of Bacteriophages 30
  3.1.2. Antigenic Properties of Phages 32
  3.1.3. Susceptibility of Phages to Chemical and Physical Agents 33
3.2. The Host Cell: Gram-Negative and Gram-Positive Bacteria 38
  3.2.1. The Cytoplasmic Membrane 38
  3.2.2. Peptidoglycan 38
  3.2.3. The Outer Membrane and Periplasmic Space of Gram-Negative Bacteria 40
  3.2.4. Cellular Energetics 40
3.3. General Overview of the Infection Process 41
  3.3.1. Adsorption 42
  3.3.2. Penetration 45
  3.3.3. Transition from Host to Phage-Directed Metabolism 45
  3.3.4. Morphogenesis 46
  3.3.5. Cell Lysis 46
3.4. Lysogeny and Its Consequences 46
3.5. Effects of Host Physiology and Growth Conditions 49
3.6. Survey of Tailed Phages 49
  3.6.1. Phages of Gram-Negative Bacteria 49
  3.6.2. Phages of Gram-Positive Bacteria 53
    3.6.2.1. Phages of Streptococci, Staphylococci, Listeria, and the Bacilli 54
    3.6.2.2. Phages of Lactic Acid Bacteria: The Dairy Phages 55
    3.6.2.3. Phages of Mycobacteria 57
  3.6.3. Cyanophages 57
3.7. The Small and Tailless Phages 58
  3.7.1. RNA Phages 58
  3.7.2. Small DNA Phages 59
  3.7.3. Lipid-Containing Phages 60

0-8493-1336-X/05/$0.00+$1.50
© 2005 by CRC Press LLC

3.7.4. Phages of Mycoplasmas 61
3.8. Archaephages 62
Acknowledgements 63
References 63

3.1. INTRODUCTION

3.1.1. THE NATURE OF BACTERIOPHAGES

As discussed throughout this book, bacteriophages are viruses that only infect bacteria. They are like complex spaceships (Fig. 1), each carrying its genome from one susceptible bacterial cell to another in which it can direct the production of more phages. Each phage particle (virion) contains its nucleic acid genome (DNA or RNA) enclosed in a protein or lipoprotein coat, or capsid; the combined nucleic acid and capsid form the nucleocapsid. The target host for each phage is a specific *group* of bacteria. This group is often some subset of one species,[1] but several related species can sometimes be infected by the same phage.

Phages, like all viruses, are absolute parasites. Although they carry all the information to direct their own reproduction in an appropriate host, they have no machinery for generating energy and no ribosomes for making proteins. They are the most abundant living entities on earth, found in very large numbers wherever their hosts live—in sewage and feces, in the soil, in deep thermal vents, and in natural bodies of water, as discussed in Chapter 5. Their high level of specificity, long-term survivability, and ability to reproduce rapidly in appropriate hosts contribute to their maintainance of a dynamic balance among the wide variety of bacterial species in any natural ecosystem. When no appropriate hosts are present, many phages can maintain their ability to infect for decades, unless damaged by external agents.

Some phages have only a few thousand bases in their genome, whereas phage G, the largest sequenced to date, has 480,000 base pairs—as much as an average bacterium, though still lacking the genes for such essential bacterial machinery as ribosomes. Over 95% of the phages described in the literature to date belong to the Caudovirales (tailed phages; see Chapter 4). Their virions are approximately half double-stranded DNA and half protein by mass, with icosahedral heads assembled from many copies of a specific protein or two. The corners are generally made up of pentamers of a protein, and the rest of each side is made up of hexamers of the same or a similar protein. The three main families are defined by their very distinct tail morphologies: 60% of the characterized phages are *Siphoviridae,* with long, flexible tails; 25% are *Myoviridae,* with double-layered, contractile tails; and 15% are *Podoviridae,* with short, stubby, tails. The latter may have some key infection proteins enclosed inside the head that can form a sort of extensible tail upon contact

[1]It is well to keep in mind that the concept of *species* for asexual organisms such as bacteria is quite different from the concept of a sexual species, which can generally be defined as the group of individuals that share a common gene pool. Asexual species must simply be collections of organisms with similar features, probably maintained by continuous selection for adaptation to a particular niche.

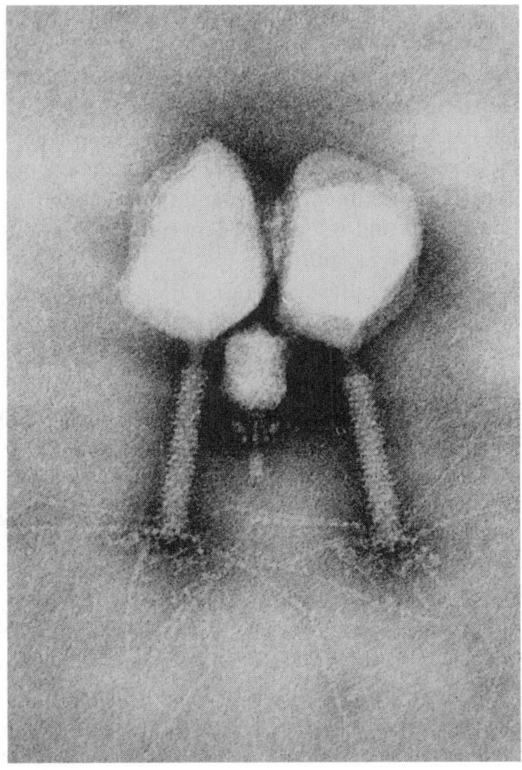

FIGURE 3.1 A "family portrait": bacteriophages ϕ29 and T2. Electron micrograph provided by Dwight Anderson.

with the host, as shown most clearly for coliphage T7 (Molineux, 2001). Archaea have their own set of infecting viruses, often called *archaephages.* Many of these have unusual, often pleiomorphic shapes that are unique to the Archaea, as discussed in Chapter 4. However, many viruses identified to date for the Crenarchaeota kingdom of Archaea look like typical tailed bacteriophages (Prangishvili, 2003); some of these are discussed in section 3.8.

The ten families of tailless phages described to date each have very few members. They are differentiated by shape (rods, spherical, lemon-shaped, or pleiomorphic); by whether they are enveloped in a lipid coat; by having double- or single-stranded DNA or RNA genomes, segmented or not; and by whether they are released by lysis of their host cell or are continually extruded from the cell surface. Their general structures, sizes, nucleic acids, adsorption sites, and modes of release are all described in detail in Chapter 4. The basic infection processes of the *Inoviridae, Leviviridae, Microviridae,* and *Tectiviridae* are further described in section 3.7. Relatively little is known about most of the others, which generally have been isolated under extremes of pH, temperature, or salinity and have only been observed in Archaea.

Phages can also be divided into two classes based on lifestyle: *virulent* or *temperate*. Virulent phages can only multiply by means of a lytic cycle; the phage virion adsorbs to the surface of a host cell and injects its genome, which takes over much of host metabolism and sets up molecular machinery for making more phages. The host cell then lyses minutes or hours later, liberating many new phages. Temperate phages, in contrast, have a choice of reproductive modes when they infect a new host cell. Sometimes the infecting phage initiates a lytic cycle, resulting in lysis of the cell and release of new phage, as previously shown. The infecting phage may alternatively initiate a lysogenic cycle; instead of replicating, the phage genome assumes a quiescent state called a prophage, often integrated into the host genome but sometimes maintained as a plasmid. It remains in this condition indefinitely, being replicated as its host cell reproduces to make a clone of cells all containing prophages; these cells are said to be lysogenized or lysogenic (i.e., capable of producing lysis) because one of these prophages occasionally comes out of its quiescent condition and enters the lytic cycle. The factors affecting the choice to lysogenize or to reenter into a lytic cycle are described in the following material. As discussed by Levin and Lenski (1985), the lysogenic state is highly evolved, requiring coevolution of virus and host that presumably reflects various advantages to both. Temperate phages can help protect their hosts from infection by other phages and can lead to significant changes in the properties of their hosts, including restriction systems and resistance to antibiotics and other environmental insults. As discussed in Chapter 9, they may even convert the host to a pathogenic phenotype, as in diphtheria or enterohemorrhagic *E. coli* (EHEC) strains. Bacteriophages lambda (λ), P1, Mu, and various dairy phages are among the best-studied temperate phages. (Note that mutation of certain genes can create virulent *derivatives* of temperate phages; these are still considered members of their temperate phage families.)

The larger virulent phages generally encode many host-lethal proteins. Some of them disrupt host replication, transcription, or translation; they may also degrade the host genome, destroy or redirect certain host enzymes, or alter the bacterial membrane. The temperate phages, in contrast, generally do much less restructuring of the host, and they carry few if any host-lethal proteins that would need to be kept under tight control during long-term lysogeny. They always encode a repressor protein, which acts at a few operator sites to block transcription of other phage genes. This repressor may be the only phage-encoded protein produced during the lysogenic state, but often a few other genes that may be beneficial to host survival are also expressed from the prophages. The repressor also blocks lytic infection by other phages of the same immunity group—that is, other phages whose genes can be regulated by the same repressor. In this way, a temperate phage generally protects its host bacterium from infection by several kinds of phages.

3.1.2. Antigenic Properties of Phages

Phages, like other protein structures, can elicit an antibody response, and antiphage antibodies have been used in a variety of ways since early phage researchers first showed that rabbits injected with phage lysates produced phage-neutralizing antibody;

d'Herelle (1926) and Adams (1959) reviewed early studies in the field. For most phages, the neutralization follows first-order kinetics to an inactivation of about 99%, with the survivors then generally being more resistant and often having smaller plaque size. The activity of a given preparation at a standard temperature and salt concentration is denoted by a velocity constant, *K*, which is related to the log of the percent of phage still infectious after a given exposure time *t*. As emphasized by Adams, such factors as salt concentration are very important in titering antibody activity, with monovalent cations above 10^{-2} M and divalent cations above 10^{-3} M being inhibitory. He cites an example of anti-T4 serum with an inactivation constant (*K* value) of over 10^5 min^{-1} under optimal conditions but only 600 min^{-1} in broth. Also, somewhat surprisingly, different assay hosts may give different estimates of the fraction of unneutralized phage. Adams suggests this may be due to minor differences in the receptors recognized on various hosts, leading to differences in bonding strength. Once determined for a given set of conditions, the *K* value can be used to calculate the appropriate dilution for any given experiment.

With the aid of noted Russian electron microscopist Tamara Tikhonenko (Tikhonenko, Gachechiladze, Bespalova, Kretova, and Chanishvili, 1976), such techniques as immunoelectronmicroscopy were used to carry out particularly detailed studies of the process of inactivation by antibodies, as presented in Box 1. Implications of phage antigenicity for human therapy are discussed in Chapter 13.

3.1.3. Susceptibility of Phages to Chemical and Physical Agents

Phages vary greatly in their sensitivity to various chemical and physical agents, in ways that are generally unpredictable and need to be determined experimentally in each case. For example, no one has yet explained the stability of phage T1 to drying, a stability that has led to much grief in the biotech industry as well as in phage labs from contamination by ubiquitous T1-like phages. There are, however, certain general principles. For example, all phages are very susceptible to UV light in the range of 260 nm as well as in the far UV; in addition to the general effects of sunlight, there are multiple reports of phage collections being lost to fluorescent-lighted refrigerators. Other factors potentially affecting phage viability include pH, ascorbic acid, urea, urethane, detergents, chelating agents, mustard gas, alcohols, and heat inactivation. The most extensive summary of the known effects of chemical and physical agents on phages is found in Adams (1959); there is also very good information in Ackermann (1987).

Phage generally are stable at pH 5 to 8, and many are stable down to pH 3 or 4; each phage needs to be specifically characterized. Phage are often quite sensitive to protein-denaturing agents such as urea and urethane, but the level of inactivation depends on both concentration and temperature, and differs for different phages. Not surprisingly, detergents generally have far less effect on phage than they do on bacteria, although the few phages that are enveloped in membranes are quite susceptible, while chelating agents have strong effects on some phages but not on others; this apparently depends mainly on cofactor requirements for adsorption. Chloroform has little or no effect on nonenveloped phages; in fact, one needs to be very careful

BOX 1: ANTIGENICITY OF PHAGES

Ketevan Gachechiladze
Eliava Institute, Tbilisi, Republic of Georgia

Every portion of each phage is antigenic, as seen dramatically in Fig. B1. However, a variety of experiments show that only antibodies against particular tail structures inactivate phage infectivity. Hershey (1943) showed that about 4600 molecules of antibody can combine with each T2K phage, while only 1-3 may be sufficient to cause neutralization. We have demonstrated these specificities particularly clearly in the case of T4, T4-like Tbilisi phages DDVI and ϕ1, using antibodies whose specificity against particular structures has been directly confirmed by immunoelectronmicroscopy. For example, antibodies against phage ϕ1 can be seen to interact only with some protein(s) on the bottom surface of the baseplate and to be capable of cross-linking two phages (Fig. B2), which leads to very slow cross-inactivation; no antigen binding to any other parts of the phage can be observed. Specific phage structures can be examined by using mutants or other

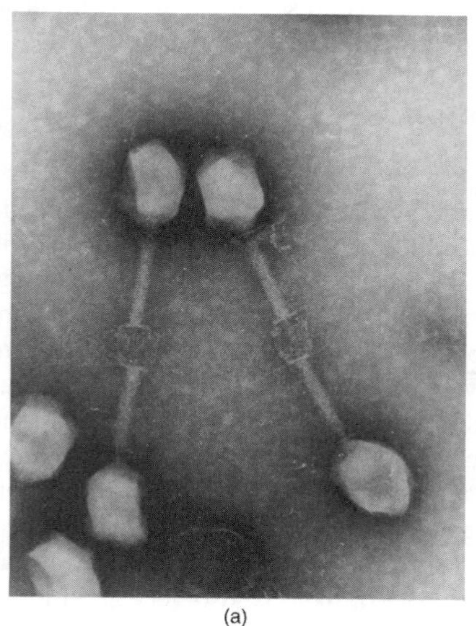

(a)

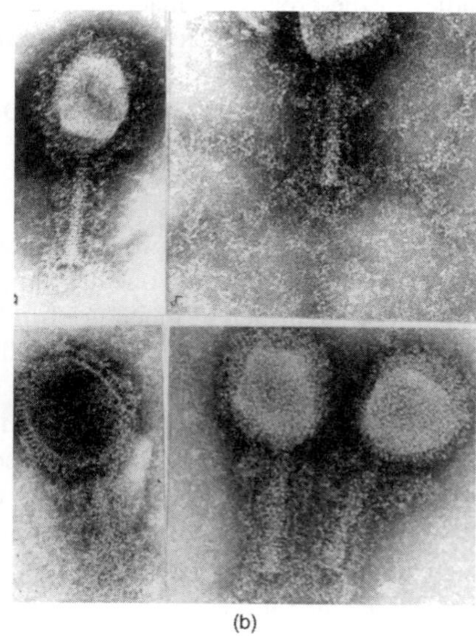

(b)

FIGURE B1 Immunoelectronmicrographs of T4-like phages interacting with heterologous and homologous antiserum. (Gachechiladze, 1981; Tikhonenko et al., 1976)
(a) Shows T4 in the presence of antiserum against T4-like phage ϕ1, which can be seen to interact only with some protein(s) on the bottom surface of the baseplate and is capable of cross-linking two phages, producing very slow cross-inactivation; no nonspecific binding to other parts of the phage can be observed.
(b) In the case of homologous antiserum, every part of the phage is covered with a fringe of antibody molecules.

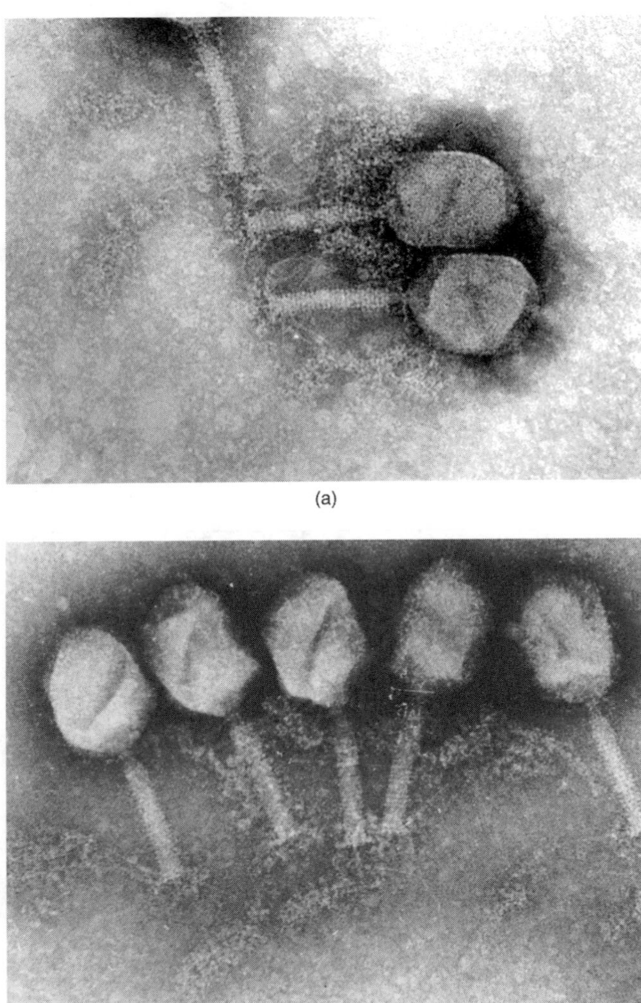

FIGURE B2 Hyperimmune antibody made against phage DDVI was interacted exhaustively with a host-range mutant of phage DDVI (DDVIh), which is not significantly inactivated by that serum, and then allowed to interact with DDVI, which is still inactivated. Only antibodies against the distal part of the DDVI tail fibers remain, as seen in these immunoelectronmicrographs. (A, B)

members of the same family; for example, if antibodies made against phage DDVI are pre-adsorbed on host range mutant DDVIh, the only remaining antibodies are directed against the distal part of the tail fibers (Fig. B2). These residual antibodies can inactivate phage DDVI, whereas DDVIh exhibits very little cross inactivation.

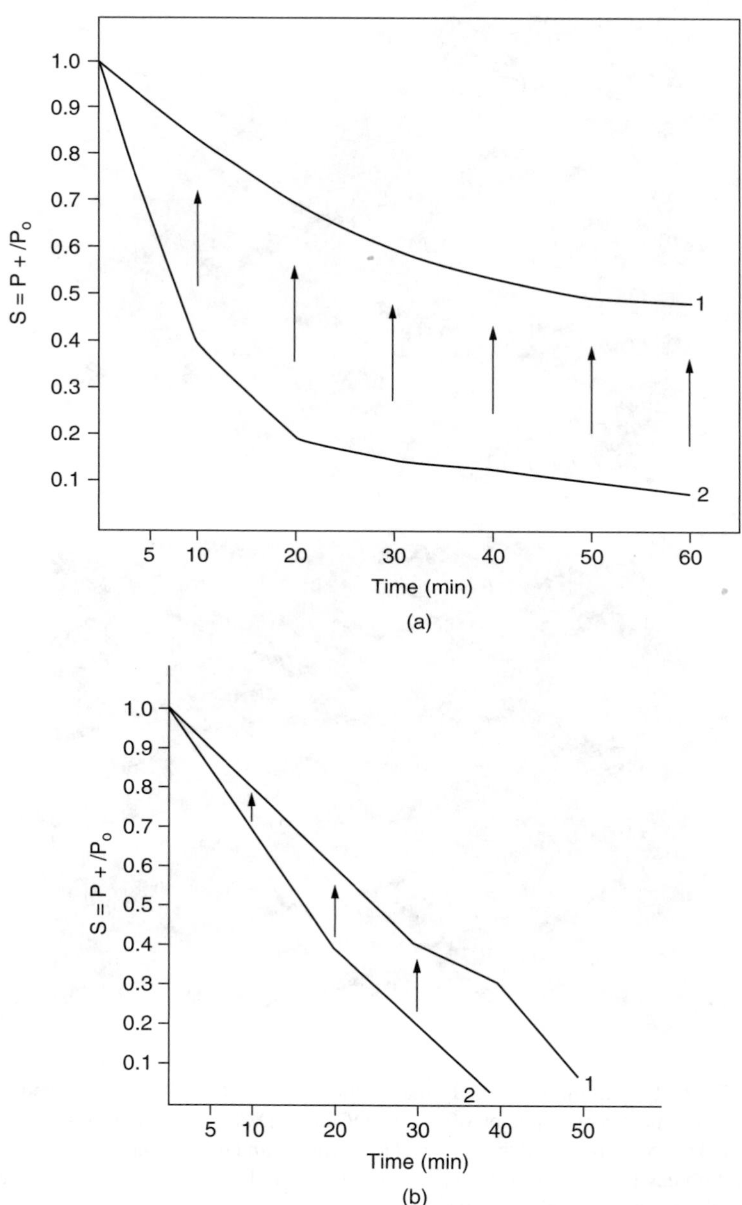

FIGURE B3 Comparison of the reversibility of neutralization between (a) the antiserum against phage DDVI produced by a single injection and (b) the so-called hyperimmune system produced later in the same rabbits after two additional injections spaced 3 weeks apart.

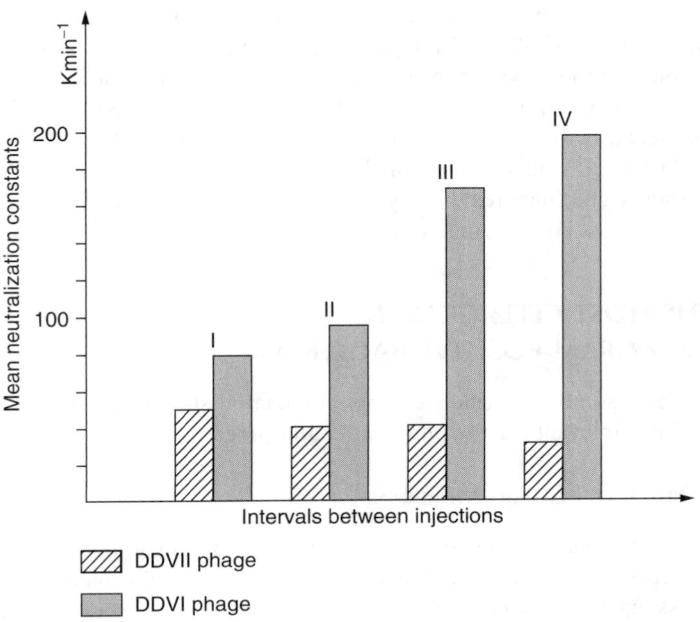

FIGURE B4 The influence of intervals between phage injections on the levels of production of antibodies against siphovirus DDVI and myovirus DDVII.

In general, most phage inactivation is due to interaction of the tail structures with individual antibodies; only 10–15% of the inactivation was found to be related to any sort of phage clumping except at very high antibody levels.

A much higher antibody titer is produced if there is a delay of several weeks between first and subsequent injections than if daily injections are used—10-fold higher for T4-like phages, about 4-fold for T1-like siphovirus DDVII, with its much simpler tail structure. The antibodies produced in the first few days, "earlysera," are largely of the IgM type, and a large fraction of them bind reversibly (Fig. B4a). In contrast, only 10–20% of the binding is reversible for the *hyperimmune* serum (consisting primarily of IgG) that is produced when a secondary immunization is carried out several weeks after the first (Fig. B4b). The nature of the phage also affects the level of inactivating antibodies that are seen. For example, as reported also by Adams (1959), T3- and T4-type phages produce several-fold higher levels of neutralizing antibodies than do T5- and T1-type phages, which have a less complex adhesin structure, and the production of neutralizing antibodies also responds differently to the spacing between successive administrations of phage, as in this rabbit study (Fig. B5). No cross reactions are seen between phages of different morphotypes.

not to get phage contaminants into the chloroform being used to produce phage lysis. However, contaminants toxic to phage in high concentrations are found in some commercial chloroform. Mutagenic agents such as mustard gas, nitric oxide, and ultraviolet light inactivate phage and can induce the lytic cycle in many lysogens. Lytic phages can generally still infect cells recently inactivated with UV or other mutagens, because they do not require ongoing synthesis of host proteins after infection. In fact, UV inactivation of the host just prior to infection is sometimes used to virtually eliminate residual synthesis of host proteins for experiments using pulse labeling to examine patterns of phage protein synthesis.

3.2. THE HOST CELL: GRAM-NEGATIVE AND GRAM-POSITIVE BACTERIA

We review here some key basic elements of bacterial structure and physiology that are particularly relevant for the phage infection process.

3.2.1. The Cytoplasmic Membrane

The bacterial cytoplasmic membrane is quite similar to eukaryotic cell membranes, except that it has no cholesterol. It contains many imbedded proteins that carry out functions assigned to *separate organelles* in eukaryotes—the generation of energy, the export of protein products, the transduction of information about the outside world. It is also responsible for the anchoring of pili and flagella, in addition to the selective transport of nutrients and metabolites. It is also the site of synthesis of the components of the peptidoglycan layer and outer membrane, and of capsid assembly for some phages, as discussed under morphogenesis in Chapter 7. The membranes of *E. coli* and *B. subtilis* have been most thoroughly studied. They are about half protein and half phospholipid, containing 6%–9% of the cell's protein and, for *E. coli,* 65%–75% of the phospholipids—mainly phosphatidylethanolamine (70%–80%) and phosphatidylglycerol (15%–25%). The specific pattern of membrane proteins differs widely under different growth conditions, as different mechanisms of energy production, transmembrane signaling systems, specific transport systems, and other factors change with the cell's available resources and current needs. Many phages encode new integral membrane proteins, and phage infection may also lead to changes in membrane lipid patterns, as reported for T4 (Harper et al., 1994).

3.2.2. Peptidoglycan

The shape-determining rigid wall that protects virtually all bacteria from rupture in media of low osmolarity is actually one enormous molecule, forming a peptidoglycan or murein sack that completely encloses the cell—a sack that grows and divides without losing its structural integrity, and is the site of attack of antibiotics like penicillin. The glycan warp of this sack consists of chains of alternating sugars—*N*-acetylglucosamine (NAG) and *N*-acetyl muramic acid (NAM)—wrapped around the axis of the cylinder for rod-shaped bacteria. These chains are cross-linked to make a two-dimensional sheet by short peptides. The outer wall of gram-positive

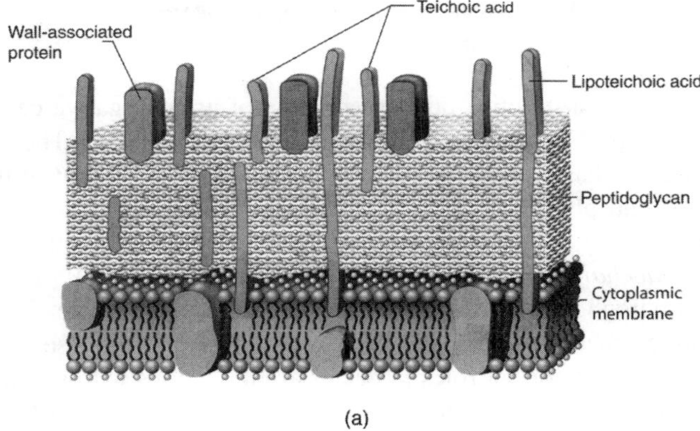

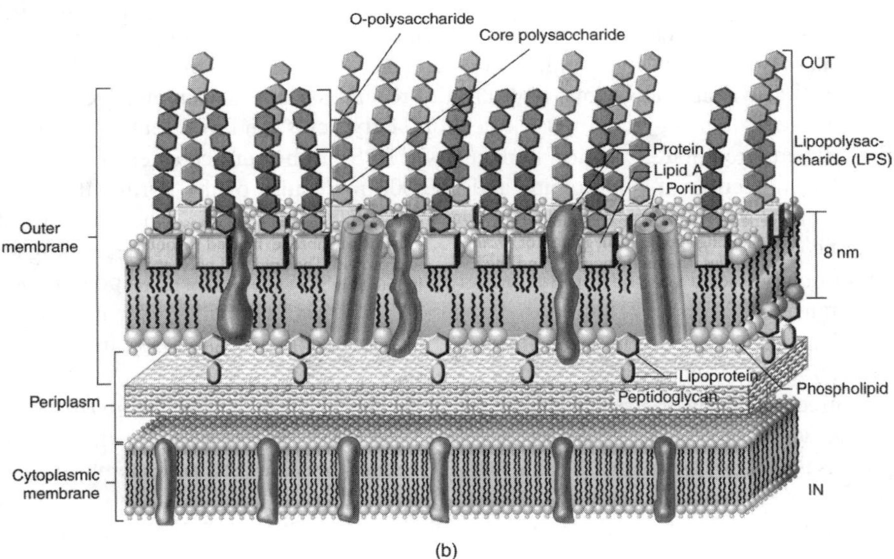

FIGURE 3.2 (a) Gram-positive, and (b) gram-negative bacterial membrane and peptidoglycan structure, adapted with permission from Madigan (2002).

bacteria is a large stack of such sheets with many links between them. Phages of gram-positive bacteria generally recognize one of the molecules embedded in these layers, such as *teichoic acid* (Fig 3.2a). In contrast, gram-negative bacteria have a single murein sheet anchored to an outer membrane (Fig. 3.2b). In both cases, phage release after lytic growth requires significant disruption of the peptidoglycan. Mycoplasmas differ from most bacteria in not having a peptidoglycan layer; their phages (discussed below in section 3.6.4) thus face a unique set of challenges in recognizing and infecting their hosts, and in phage release.

3.2.3. The Outer Membrane and Periplasmic Space of Gram-Negative Bacteria

The properties of the distinctive outer membrane of gram-negative cells have been reviewed in detail by Nikaido (1996; 2003) and Ghuysen (1994). The *inner face* of the outer membrane has a phospholipid composition similar to that of the cytoplasmic membrane, and phospholipids can exchange rapidly between it and the inner membrane. In contrast, the lipid part of the *outer face* is mainly a unique substance called *lipopolysaccharide* (*LPS*), seen nowhere else in nature. LPS is composed of three parts: a hydrophobic *lipid A* membrane anchor and a complex distal polysaccharide O-antigen that are connected by a core polysaccharide (see Chapter 7.1). The O-antigens play important roles in bacterial interactions with mammalian hosts and in virulence; they are indicated in strain names, such as *E. coli* O157. Many common lab strains, including *E. coli* K-12, lack any O-antigen; *E. coli* B, used extensively for phage work, even lacks the more distal part of the LPS core.

The outer membrane contains several families of general *porins*—proteins that form large β-barrel channels with charged central restrictions that support *nonspecific* rapid passage of small hydrophilic molecules but exclude large and lipophilic molecules. It also contains various high-affinity receptor proteins that catalyze *specific* transport of solutes—vitamin B_{12}, catechols, fatty acids, and different iron derivatives. There are about 3 million molecules of LPS on the surface of each cell, plus 700,000 molecules of lipoproteins and 200,000 molecules of the porins; the latter two connect the outer membrane to the peptidoglycan layer. The concentrations of specific other membrane proteins used as receptors vary considerably, depending on environmental conditions and the general need for the particular compound each one transports. The enormous variety seen in the O-antigens and outer membrane proteins of the enteric bacteria presumably is an adaptation to living in animals, protecting them against host immune systems.

Infecting phage must somehow get their DNA safely across the *periplasmic space*, which appears to be a sort of viscous gel that contains many nucleases and proteases. Proteins TonB and TolA, which are anchored to the inner membrane but span the periplasmic space, are particularly important to various uptake systems as well as to infection by many phages, as discussed in Chapter 7.2. The Tol system is also important in retaining periplasmic proteins and in resistance to antibiotics, dyes, and detergents. While the bacterial cytoplasm is a reducing environment, the periplasmic compartment is oxidizing, so disulfide bonds can be formed once proteins move there; this process is assisted by several periplasmic proteins and is important in synthesis of filamentous phages. There are no nucleoside triphosphates (NTPs) in the periplasm, so the periplasmic nucleases and proteases must operate very differently than their cytoplasmic counterparts (Miller, 1996).

3.2.4. Cellular Energetics

Most bacteria used in phage studies use two basic kinds of metabolic pathways to get the energy they need: fermentation and respiration. In both cases, electrons are withdrawn from an oxidizable molecule, freeing energy that is stored as ATP, NADH, or similar coenzymes. These electrons, now at a lower potential, must in turn be removed by reducing some final acceptor. In *fermentation*, electrons removed from substrates such as glucose are transferred to breakdown products of the glucose to produce wastes such as ethanol or lactic acid, and only a small fraction of the energy in the glucose can be used. In *respiration*, the electrons are eventually passed through a membrane-bound *electron transport system* (ETS) and are carried away by reducing oxygen to water (in ordinary aerobic respiration), or, in *anaerobic respiration*, by reducing ferric to ferrous iron, sulfate to sulfide, fumarate to succinate, or nitrate to nitrite or even to N_2.

The electron transport system is oriented in the cell membrane so as to transport protons, H^+, to the *outside* of the cell membrane. This builds up an electrochemical potential, known as the proton motive force (PMF), across the cell membrane; the energy stored in this potential can be trapped chemically by forcing protons to do useful work as they pass back through the membrane, down their potential gradient. They may produce ATP, drive the co-transport of lactose or inorganic ions through specific membrane transport systems, or pass through the basal protein complexes of flagella, powering cell motility. The PMF is also critical for infection by most phages. As discussed by Goldberg (1994), the PMF is separable into two parts: a pH gradient and a membrane potential that is due to the general separation of charges. For many (but not all) phages, DNA transfer into the cell requires the membrane potential, as discussed in Chapter 7 section 7.2.

3.3. GENERAL OVERVIEW OF THE INFECTION PROCESS

The general lytic infection process for tailed phages is presented here, with some examples from model phages; key molecular mechanisms are explored in Chapter 7. More detail on these subjects can be found in Calendar's *Bacteriophages,* a new edition of which is in press, or in Webster's *Encyclopedia of Virology*, as well as in books and articles on the individual phages. The tailless phages are discussed later in the chapter, in section 3.7.

Since the early studies of d'Herelle, the details of phage-host interactions have been studied by use of the *single-step growth curve* (Fig. 3.3a), as systematized by Ellis and Delbrück (1939) and detailed in the Methods appendix in this volume. Phage are mixed with appropriate host bacteria at a low multiplicity of infection. After a few minutes for adsorption, the infected cells are diluted (to avoid attachment of released phage to uninfected cells or bacterial debris) and samples are plated at various times to determine *infective centers*. An infective center is either a single phage particle or an infected cell that bursts on the plate to produce a single plaque. The number of plaques generally remains constant at the number of infected cells for a characteristic time, the *latent period,* and then rises sharply, leveling off at many times its initial value as each cell lyses and liberates the completed phage. The ratio between the numbers of plaques obtained before and after lysis is called the *burst size*. Both the burst size and the latent period are characteristic of each phage strain under particular conditions, but are affected by the host used, medium, and temperature. As first shown by Doermann (1953), if the infected cells are broken open at various times after infection, the phage seem to have all disappeared for a certain period. This *eclipse period* was a mystery until the nature of the phage particle

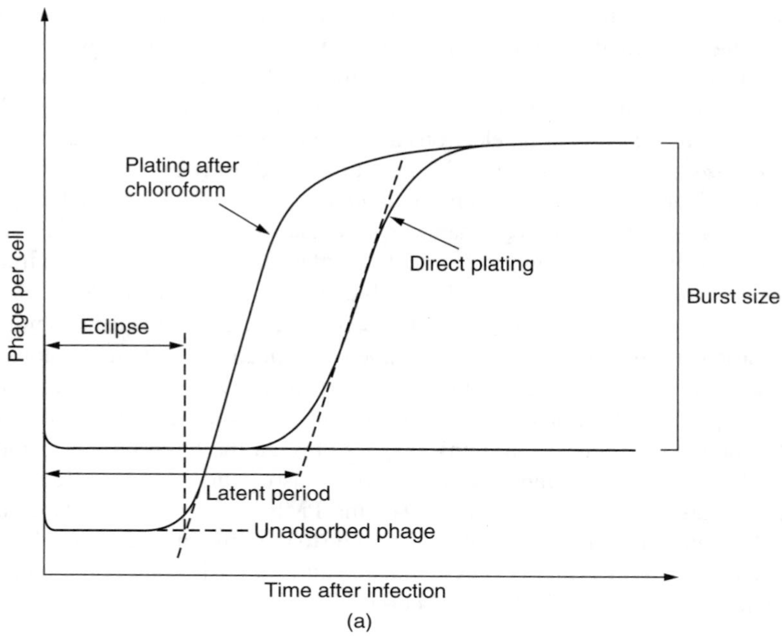

FIGURE 3.3 (a) Classical single-step growth analysis of phage infection (see Appendix).

and of the infection process was determined and it was realized that during this period only naked phage DNA is present in the cell. Using chloroform-induced lysis followed by plating, both the eclipse period and the subsequent rate of intracellular synthesis of viable phage particles are now routinely measured. Infecting at a high multiplicity (5–10 phage/cell) allows one to also measure the effectiveness of killing (i.e., the number of bacterial survivors) and the impact of the phage infection in terms of continued expansion of cell mass and eventual cell lysis, as reflected in the absorbance or optical density (Fig. 3.3b).

The infection process involves a number of tightly programmed steps, as diagramed in Fig. 3.4. The efficiency, timing, and other aspects of the process may be very much affected by the host metabolic state, and in many cases the host can no longer adapt to major metabolic changes once the phage has taken over the cell. Thus, for example, studies on anaerobic infection can only be carried out in host cells that were themselves grown anaerobically.

3.3.1. Adsorption

Infection by tailed phages starts when specialized adsorption structures, such as fibers or spikes, bind to specific surface molecules or capsules on their target bacteria. In gram-negative bacteria, virtually any of the proteins, oligosaccharides, and lipopolysaccharides described above can act as receptors for some phage. The more complex murein of gram-positive bacteria offers a very different set of potential binding sites.

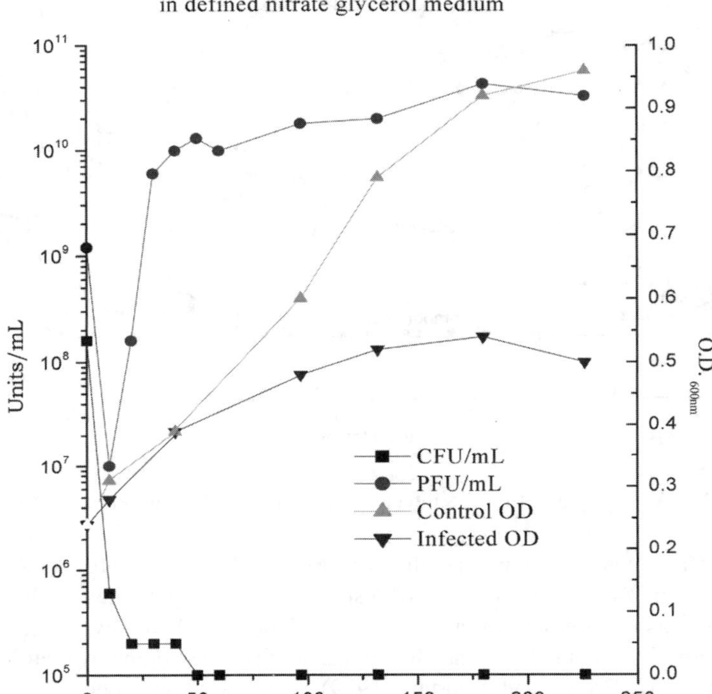

FIGURE 3.3 (*Continued*) (b) Infection of *E. coli* B by T4D at an MOI of 9 in a minimal NG medium (Appendix: section A.6.2.1. and Table 2.5.3.). Note the very rapid initial drop in both surviving bacteria and unadsorbed phage, the doubling of the cell mass even after infection, as measured by absorbance (OD_{600}) and the fact that there is a long delay before lysis actually occurs under these conditions; this phenomenon of "lysis inhibition" is discussed in detail in Chapter 7.

Many phages require clusters of one specific kind of molecule that is present in high concentration to properly position the phage tail for surface penetration. However, coliphage N4 manages to use a receptor, called NfrA, which is only present in a few copies per cell. The attachment of some phages involves two separate stages and two different receptors. T4-like phages, for instance, must bind at least three of their six long tail fibers to primary receptor molecules to trigger rearrangements of baseplate components, which then bind irreversibly to a second receptor. Different members of the family use very different primary receptors, but they seem to all use the heptose residue of the LPS inner core for the secondary receptor.

Adsorption velocity and efficiency are important parameters that may vary for a given phage-host system depending on external factors and host physiological state. For example, the lambda receptor is only expressed in the presence of the sugar

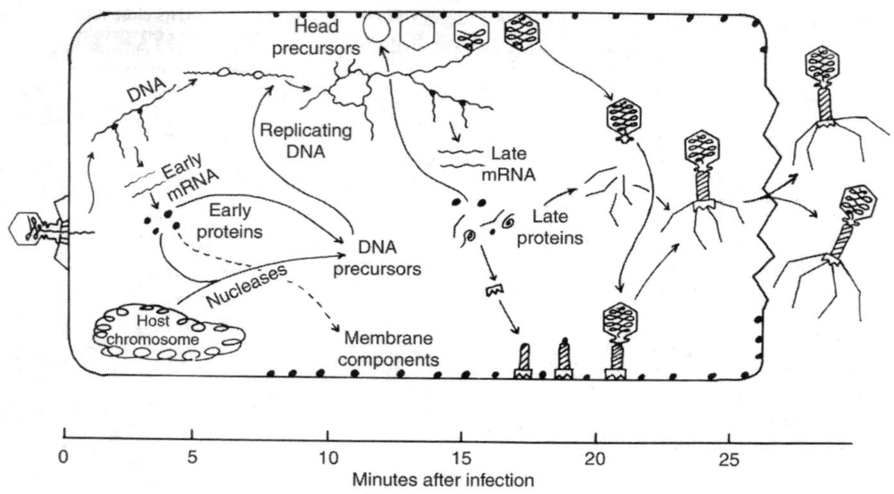

FIGURE 3.4 An overview of the T4 infection cycle. From Mathews et al. (1983)

maltose. Many phages require specific cofactors, such as Ca^{2+}, Mg^{2+} or simply any divalent cation. T4B strains (but not T4D strains) require tryptophan for binding, but can bypass this requirement for a short "nascent" period when they are first released from the prior host and still attached to the inside of the membrane by their baseplates (Wollman, 1952; Brown, 1969; Simon, 1969). A nascent-phage period with broader adsorption properties was also described for a *Staphylococcus* phage by Evans (1940), but the mechanism of nascent-phage interaction to give the extended host range is not known in either system. Bacteria commonly develop resistance to a particular phage through mutational loss or alteration of receptors used by that phage. However, losing some receptors offers no protection against the many other kinds of phage that use different cell-surface molecules as receptors. Furthermore, in most cases, the phage can acquire a compensating adaptation through appropriate *host range mutations*, which alter the tail fibers so they can recognize the altered cell-surface protein or bind to a different receptor. This is likely less efficient than bacteria acquiring resistance, since phage adaptation to new receptors requires establishment of a new functional interaction, while development of host resistance is a "negative" event, requiring loss of function (Lenski, 1984). Still, some such mutants seem to be present in every substantial phage population. Furthermore, some phages, such as P1 and Mu, encode multiple versions of the tail fibers. Others can recognize multiple receptors; for example, T4 tail fibers bind efficiently to an *E. coli* B-specific lipopolysaccharide, to the outer membrane protein OmpC found on K strains, and to OmpF (Tetart et al., 1998). The adhesion regions of the tail fibers are highly variable between related phages, with high rates of recombination that facilitate the formation of new, chimeric adhesins. Not surprisingly, there is much interest in engineering new receptor-recognition elements into the tail fibers of well-characterized phages so they can infect taxonomically distant hosts.

The physiological state of a cell can substantially change the concentration of particular cell-surface molecules and thus the efficiency of infection by certain phages. Many of the surface molecules that particular phages use as receptors are crucial to the bacterial cell, at least under some environmental conditions, so resistance may lead to loss of important functions and reduction in competitiveness.

3.3.2. Penetration

After irreversible attachment, the phage genome passes through the tail into the host cell. This is not actually an "injection" process, as has often been depicted, but involves mechanisms of DNA transfer specific for each phage (see Chapter 7, section 7.2). In general, the tail tip has an enzymatic mechanism for penetrating the peptidoglycan layer and then touching or penetrating the inner membrane to release the DNA directly into the cell; the binding of the tail also releases a mechanism that has been blocking exit of the DNA from the capsid until properly positioned on a potential host. The DNA is then drawn into the cell by processes that generally depend on cellular energetics, as mentioned above, but are poorly understood except for a few phages; for example, in T7 the entry of the DNA is mediated by the process of its transcription. Once inside the cell, the phage DNA is potentially susceptible to host exonucleases and restriction enzymes. Therefore, many phage circularize their DNA rapidly by means of sticky ends or terminal redundancies, or have the linear ends protected. Many also have methods to inhibit host nucleases (T7, T4) or use an odd nucleotide in their DNA such as hydroxymethyldeoxyuridine (hmdU: SPO1) or hydroxymethyldeoxycytidine (hmdC: T4) for protection. In other cases, their genomes have been selected over evolutionary time to eliminate sites that would be recognized by the restriction enzymes present in their common hosts (Staphylococcal phage Sb-1, coliphage N4).

3.3.3. Transition from Host to Phage-Directed Metabolism

The initial step generally involves recognition by the host RNA polymerase of very strong phage promoters, leading to the transcription of *immediate early* genes. The products of these genes may protect the phage genome and restructure the host appropriately for the needs of the phage; they may inactivate host proteases and block restriction enzymes, directly terminate various host macromolecular biosyntheses, or destroy some host proteins. The large virulent phages such as T4, SPO1, and Sb-1 encode many proteins which are lethal to the host even when cloned individually that appear to participate in this process of host takeover. A set of *middle genes* is often then transcribed, producing products that synthesize the new phage DNA, followed by a set of *late genes* that encode the components of the phage particle. For some phages, these transitions involve the synthesis of new sigma factors or DNA-binding proteins to reprogram the host RNA polymerase; other phages encode their own RNA polymerase. Degradation of the host DNA and inhibition of the translation of host mRNAs are other mechanisms that can contribute to reprogramming the cell for the synthesis of new phage.

3.3.4. Morphogenesis

The DNA is packaged into preassembled icosahedral protein shells called *procapsids*. In most phages, their assembly involves complex interactions between specific scaffolding proteins and the major head structural proteins, followed by proteolytic cleavage of both the scaffolding and the N-terminus of the main head proteins. Before or during packaging, the head expands and becomes more stable, with increased internal volume for the DNA. Located at one vertex of the head is a portal complex that serves as the starting point for head assembly, the docking site for the DNA packaging enzymes, a conduit for the passage of DNA, and, for myoviruses and siphoviruses, a binding site for the phage tail, which is assembled separately. The assembly of key model phages provides the best-understood examples from all nature of morphogenesis at the molecular level, as extensively discussed in Chapter 7. It also has provided the models and components for several key developments in nanotechnology.

3.3.5. Cell Lysis

The final step—lysis of the host cell—is a precipitous event whose timing is tightly controlled, as discussed in Chapter 7, section 7.8, and by Wang et al. (2003). If lysis happens too quickly, too few new phages will have been made to effectively carry on the cycle; if lysis is delayed too long, opportunities for infection and a new explosive cycle of reproduction will have been lost (Abedon, 1990; 2005). The tailed phages all use two components for lysis: a *lysin*—an enzyme capable of cleaving one of the key bonds in the peptidoglycan matrix—and a *holin*—a protein that assembles pores in the inner membrane at the appropriate time to allow the lysin to reach the peptidoglycan layer and precipitate lysis. The timing is affected by growth conditions and genetics; mutants with altered lysis times can be selected. The general mechanisms and wide variety of holins and lysins are discussed in detail in Chapter 7. The tailless phages encode a variety of single-protein lysis-precipitating proteins that subvert host peptidoglycan-processing enzymes in various ways.

3.4. LYSOGENY AND ITS CONSEQUENCES

The concept of lysogeny has had a checkered history. Early phage investigators in the 1920s and 1930s claimed to find phages irregularly associated with their bacterial stocks and believed that bacteria were able to spontaneously generate phage, which were thus long considered by many to be some sort of "ferment" or enzyme rather than a living virus. When Max Delbrück and his companions began their work, they confined themselves to the classical set of coliphages designated T1-T7, none of which showed this property, and Delbrück attributed the earlier reports to methodological sloppiness. However, the phenomenon could no longer be denied after the careful work of Lwoff and Gutmann (1950); through microscopic observation of individual cells of *Bacillus megatherium* in microdrops, they demonstrated that cells could continue to divide in phage-free medium with no sign of phage production, but that an occasional cell would lyse spontaneously and liberate phage. Lwoff named the hypothetical intracellular state of the phage genome a *prophage* and showed later that by treating lysogenic cells with agents such as ultraviolet light, the prophage could be uniformly *induced* to come out of its quiescent state and initiate lytic growth. The prophage carried by a bacterial strain is given in parentheses; thus, K(P1) means bacterial strain K carrying prophage P1.

Esther Lederberg (1951) showed that strains of *E. coli* K-12 carried such a phage, which she named lambda (λ). Meanwhile, Jacob and Wollman (1961) had been investigating the phenomenon of conjugation between donor (Hfr) and recipient (F^-) strains of *E. coli*. They found that matings of F^- strains carrying λ and non-lysogenic Hfrs proceeded normally, but that reciprocal matings yielded no recombinants and, in fact, produced a burst of lambda phage. A mating of Hfr(λ) by non-lysogenic F^- would proceed normally if it were stopped before transfer of the *gal* (galactose metabolism) genes, but if conjugation proceeded long enough for the *gal* genes to enter the recipient cell, the prophage would be induced (*zygotic induction*). These experiments indicated that the λ prophage occupies a specific location near the *gal* genes; that a lysogenic cell maintains the prophage state by expression of one λ gene, encoding a specific *repressor protein*, which represses expression of all other λ genes; and that if the *gal* genes—and thus the λ prophage—are transferred into a non-lysogenic cell during mating, the prophage finds itself in a cytoplasm lacking repressor and therefore expresses its other genes and enters the lytic cycle. Typical plaques made by phage λ are turbid, due to lysogenization of some bacteria within the plaque. Mutants of λ producing clear plaques are unable to lysogenize. Analysis of these mutants revealed three genes, designated *cI, cII,* and *cIII,* whose products are required for lysogeny. The *cI* gene encodes the repressor protein. Allen Campbell demonstrated that the sequence of genes in the prophage is a circular permutation of their sequence in the phage genome. He therefore postulated that the prophage is physically inserted into the host genome by circularization of the infecting genome followed by crossing-over between this genome and the (circular) bacterial genome. We now know that the genome in a λ virion has short, complementary single-stranded ends; as one step in lysogenization, it circularizes through internal binding of these ends and is then integrated (by means of a specific integrase) at a point between the *gal* and *bio* (biotin biosynthesis) genes.

Temperate phages can carry host genes from one bacterial cell to another (*transduction*). Many temperate phages have their own specific sites of integration, and when their prophages are induced during the transition to lytic growth, the prophage DNA is sometimes excised mistakenly along with a piece of host DNA. This produces a *transducing phage*, which can effect a *specialized transduction* by carrying this particular piece of host DNA into other cells, thus changing their genomes through recombination with the transduced segment. Other phages, like Mu, integrate randomly into the genome and then always carry some host DNA with them, so they are able to effect a *generalized transduction*. Both temperate and virulent phages may also produce generalized transduction by mistakenly producing particles filled with host rather than phage DNA; those are generally rare and cannot reproduce, but may still contribute substantially to bacterial genetic exchange in nature. Transduction plays important roles in bacterial genetics, both in the lab and in the wild. Briani et al. (2001) have emphasized the importance of temperate phages as "replicons endowed with horizontal transfer capabilities," and they have probably been major factors in bacterial evolution by moving segments of genomes into new organisms. Broudy and Fischetti (2003) have recently reported in vivo lysogenic conversion of Tox$^-$ *S. pyogenes* to Tox$^+$ with lysogenic streptococci or free phage in a mammalian host.

The mechanism of the molecular decision between lytic and lysogenic growth has been worked out in most detail for phage λ (Little, 2005), but Dodd and Egan (2002) found great similarities for the unrelated temperate phage 186. In each case, lysogeny is governed by a repressor protein, CI, which binds to a set of operators in the lysogenic state and represses the expression of all genes except its own (Fig. 3.5). Both phages have critical promoter regions, one promoting lysogeny and the other lysis, which are closely associated, and lysogeny is promoted by binding of the CI protein to these sites in such a way as to inhibit lytic growth. The CI proteins of both phages are strongly cooperative, forming tetramers or octomers. In the λ case, CI is involved in a molecular competition with another protein, Cro, which promotes the lytic cycle. The transition toward lysogeny in λ is also promoted by two other proteins, CII and CIII, which bind to critical promoters and stimulate transcription of the *cI* gene and others. The stability of CII is determined by factors that measure the cell's energy level. A cell with sufficient energy has little cyclic AMP (cAMP); when the cell is energy-starved (for instance, intracellular glucose is low), the cAMP concentration is high. A high level of cAMP promotes CII stabilization and thus lysogeny. It is clearly adaptive for a phage genome entering a new cell to sense whether there is sufficient energy to make a large burst of phage or whether the energy level is low, so its best strategy for survival is to go into a prophage state.

Notice that the critical event in establishing lysogeny is regulation of the phage genes. Integration of the phage genome into the host is secondary, and so it is understandable that other temperate phages, such as P1, can establish lysogeny with their prophages functioning as plasmids in the cytoplasm. Many interesting variations on lysogeny and temperate phages have been discovered. Some are discussed below in the phage survey section.

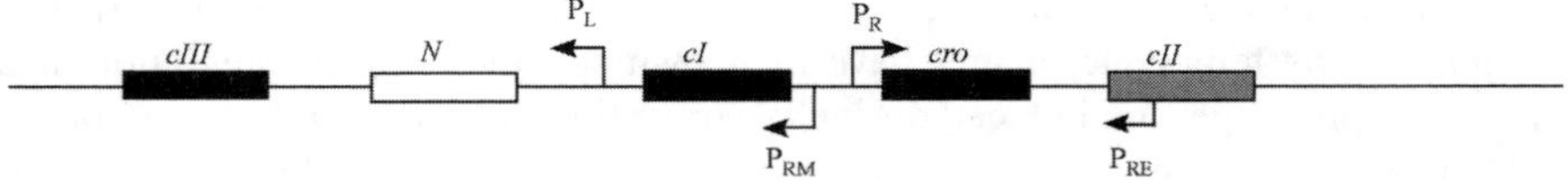

FIGURE 3.5 The regulatory region of the phage λ genome where the decision is made between lysis and lysogeny. The heart of the matter is competition between two proteins, CI and Cro. Initially, lysogeny is promoted by the combined action of the cII and cIII proteins, which act at P_{RE} (promoter right for establishment) and promote transcription of the *cI* gene; the cI protein then binds to various operator/promoter sites and represses them. In particular, CI prevents transcription of the *N* gene whose product acts at other sites in the genome to promote expression of genes required for lytic growth. CI also binds to P_{RM} (promoter right for maintenance) and promotes further transcription of its own gene. These lysogeny-directed processes are in competition with transcription from P_R of the *cro* gene and other genes to the right needed for lytic growth. Cro and CI compete for binding sites in the complex promoter/operator region that includes both P_{RM} and P_R where the lysis/lysogeny decision is ultimately made.

3.5. EFFECTS OF HOST PHYSIOLOGY AND GROWTH CONDITIONS

The lytic phage infection process is very much dependent on the host metabolic machinery, so in most cases it is highly affected by what the host was experiencing shortly prior to infection, as well as by the energetic state and what nutrients and other conditions are present during the infection process itself (Kutter et al., 1994). Note that phages that efficiently turn off host gene function, such as coliphage T4 and *B. subtilis* phage SPO1, generally render their hosts incapable of responding to substantial environmental changes after infection. The energetic state of the cell also generally has significant effects on the probability of establishing lysogeny and, in some cases, on the re-activation of phages from lysogens. The ability of various phages to infect or survive anaerobically, in sporulating or stationary-phase cells, and in biofilms is discussed in some detail in Chapter 6.

3.6. SURVEY OF TAILED PHAGES

This section is designed to give a better idea of the variety and versatility of phages and the general features of those commonly referred to in this book and elsewhere. We include here brief overviews of some that have played key roles in our current understanding of phage and of molecular biology, have interesting features or applications, or are the "type phages" for phage taxa. We have chosen alphabetical listing under broad host groups as the most accessible approach for readers without detailed knowledge of phage taxonomy and nomenclature. We have not tried to give citations for general properties of the more widely studied phages, since so many labs have often been involved and they are broadly accessible. More in-depth general resources include Webster's *Encyclopedia of Virology* (1999), Calendar's *The Bacteriophages* (1988; 2005), Ackermann and DuBow (1987) and Adams (1959). The complete genomic sequences of a number of the phages have now been determined and are available in the bacteriophage section of the genome sites at the National Center for Biotechnology Information http://www.ncbi.nlm.nih.gov and the Pittsburgh Phage Institute http://pbi.bio.pitt.edu, as are the sequences of some of the prophages determined in the course of microbial genome projects. Phages of *E. coli* and *B. subtilis* have generally been the best studied with regard to genetics and the physiology of the infection process, taking advantage of—and contributing to—the broad knowledge about these two key model organisms. Note that in general the designation of phage names has been quite arbitrary, giving no information about phage properties or relationships; in Chapter 4, Ackermann discusses ideas for making future phage names more informative.

3.6.1. Phages of Gram-Negative Bacteria

Lambda (λ), a 48,502-bp siphovirus, is the best-studied of the temperate coliphages; its key early role in understanding gene regulation and integration in lysogeny has been explored earlier in this chapter. From the beginning, it was a primary vector in genetic engineering work. Its efficient system for generalized recombination is the basis for the *recombineering* system that has revolutionized the construction of transgenic systems, from bacteria to mice (Court et al., 2002). It was one of the first phages sequenced (Sanger et al., 1982), and the intensely explored lambdoid phage group has strongly influenced our view of phage evolution (Chapter 5). Many λ-like phages carry toxin genes and other genes associated with pathogenicity, as discussed in Chapter 8.

Mu is a temperate generalized transducing myovirus with a broad host range that includes *E. coli, Salmonella, Citrobacter,* and *Erwinia*. It integrates into the genome at quite random sites, which led to its being called *Mu*, for mutator, and takes with it stretches on both sides when it is packaged, as discussed in Chapter 7. Mu is unusual in that its DNA replication does not occur in a free state, but rather is associated with transposition to random sites within the host chromosome. The first round (lysogenization) occurs by conservative transposition. Further rounds of replication are replicative, with transposed copies of Mu DNA interspersed within host DNA sequences. The mature Mu particle has 37 kbp of conserved Mu-specific DNA flanked on the left by 50–150 bp of host DNA and on the right by 0.5 to 3.0 kbp of host DNA. These random samples of host DNA sequence are incorporated during packaging of the phage DNA, which occurs directly from the host-integrated form. This property is also seen in coliphage D108, a number of *Pseudomonas* phages including phage D3112, and the levivirus *Vibrio cholera* phage VcA1. The transposable *Pseudomonas* prophages are all inducible by DNA-damaging agents, whereas Mu and D108 are not. Mu and D108 expand their host ranges by making two forms of the tail fibers responsible for host recognition; the orientation of an invertible stretch of DNA called the G-loop determines which is expressed. Genes expressed in one orientation are responsible for adsorption to *E. coli* K-12; genes expressed in the other orientation permit adsorption to other gram-negative bacteria such as *Erwinia* and *Citrobacter freundii.*

N4 is a 72-kbp podovirus that infects *E. coli* K-12. Its sequential use of three different DNA-dependent RNA polymerases, the first of them carried in the virion, has led to its extensive use in the study of transcriptional regulation. While it inhibits host DNA replication, host transcription and translation are largely unimpaired, and the host polymerase is actually used to transcribe late genes. Its DNA has been selected over the eons to have no enzyme-recognition sequences for a wide variety of common restriction enzymes, including *Eco*R1, *Eco*RV, *Hind*III, *Nco*I, *Pst*I, *Sal*II, and *Xho*I. The receptor to which N4 adsorbs is present in at most five copies on the host surface. It is one of the few tailed phages that exits the host by simply overfilling and bursting the cell rather than by means of a timed lysis mechanism involving a holin-lysin pair.

N15 is a lambdoid coliphage with an unusual form of lysogeny. Its prophage is a linear plasmid—a rarity in bacteria—complete with genes controlling the plasmid mode of replication and a system controlling partitioning between daughter cells. The virion DNA has the usual sticky ends that circularize it after entering the host cell, but instead of having an integration protein and attachment site, it has protelomerase enzyme, TelN, that opens the dsDNA at a specific site, *tos*, and seals the resultant ends to make covalently closed hairpins. In effect, the plasmid is thus a single-stranded DNA circle that anneals to form a double strand throughout its length but has solved the end-replication problem that all linear DNAs must confront. Upon induction, these ends are opened and resealed to the opposite ends to recreate the *tos* site.

P1 is a well-studied temperate myovirus of enteric bacteria that infects a broad range of gram-negative hosts. P1 and its sister, P7, are unusual in that the prophage most commonly exists free as a plasmid rather than integrated into the host chromosome. It is particularly useful because it can carry out generalized transduction of genes between strains of *E. coli* and *Shigella* and can also transduce *Myxococcus,* although it cannot replicate there; it can also transduce other plasmid DNAs. It has an 85-nm icosahedral head, but some particles with 65-nm and 47-nm heads are also produced, depending on phage and host strain and on conditions. The small phage acquire partial P1 genomes during their "head-full" packaging from a long concatameric DNA; these cannot complete infections on their own but can recombine intracellularly with other P1 DNA. Two different origins of replication are used: *oriR* in the prophage state, *oriL* in the lytic phase. Though present in only 1–2 copies per cell, the P1 prophage is only lost once per 100,000 divisions because it encodes its own effective partition proteins that ensure proper division of the daughter phage chromosomes during cell division. The P1-encoded restriction-modification system, studied by Nobelist Werner Arber, played a major role in determining the biology of such systems and thus in laying the foundation for recombinant DNA work. Another unusual feature is that P1 encodes two different versions of its tail-fiber genes on an invertible "C-segment" that is largely homologous to the smaller G-segment of phage Mu. The two versions seem to be specific for different particular lipopolysaccharides, and induced prophages seem to be randomly distributed between the two.

P2 and P4 are generally considered as a pair because P4 has no genes for structural proteins of its own. Rather, it has the ability to instruct the main head protein of P2 and related temperate phages to assemble into a particle one-third the normal size—the right size for the P4 genome (11,624 bp) instead of for the P2 genome (33,593 bp). P4 is a true parasite of a parasite, absolutely requiring its helper phage despite having virtually no sequence homology and no organizational similarity to P2. Phages of the P2 type, widespread in nature, are myoviruses that have a pair of discs, one inside the head and one outside, to attach the head to the inner tail tube, while an outer contractile sheath attaches to a baseplate that has six tail fibers and a single tail probe. P2 and P4 can each infect cells lysogenic for the other, while lytic development of either induces the other. Induction of P4 requires the *cox* gene of P2, which activates transcription from the P4 lytic promoter. Replication of P2 occurs via a rolling-circle mechanism that has its own site-specific initiation functions but relies otherwise on host genes. In contrast, P4 replicates bidirectionally from a unique origin that requires a second site nearby and several phage proteins, but only two host proteins—PolIII and Ssb, the single-strand-DNA binding protein. P2 DNA is packaged from monomeric circles rather than from linear DNA and it requires a specific 125-bp region, including a site that is cleaved to give 19-bp cohesive ends. The P2-related phage 186 also encodes a protein that depresses host replication and thus increases the phage burst size, but is not essential.

P22, a generalized transducing phage of *Salmonella typhimurium,* was the phage involved in the initial discovery of transduction by Zinder and Lederberg (1952). P22 is a member of the *Podoviridae*. The 41,724-bp genome includes 64 genes and unidentified ORFs. Its genes are clustered by function in the same general order as in lambda, with which it can exchange certain blocks of nonstructural genes. In P22, as in many other phages, the DNA is packaged by headfulls starting from a specific *pac* site and proceeding unidirectionally along a long replicative DNA concatamer, with packaging of the next prohead then starting wherever the previous one finishes. Generalized transduction is thought to be a consequence of the occasional packaging of host DNA starting at some *pac*-like site and continuing through headfuls of bacterial DNA. The P22 DNA packaging apparatus and *pac* site have been used very effectively in building cloning vectors, as discussed in Chapter 11. Upon entry into the host, the DNA circularizes and then either integrates into a specific chromosomal site to form a prophage or replicates via a rolling-circle process to form a new concatemer. P22 can be very advantageous to its host. In addition to exclusion of phages in the same immunity group through the repressor involved in maintaining lysogeny, P22 prophages express genes that interfere with DNA injection by related phages, that alter the O-antigen structure to interfere with further P22 adsorption, and that abort the lytic cycle of some other *Salmonella* phages.

ϕKZ is the best-studied member of a broad and distinct worldwide family of giant *Pseudomonas* phages with 120-nm isometric capsids and contractile 200 nm tails. Its 280-kbp genome, recently sequenced by Mesyanzhinov et al. (2002), has over 300 open reading frames, only 59 of which encode proteins that resemble proteins of known function in the databases, emphasizing how much we still have to learn about the genomic space of the large DNA phages. It encodes its own RNA polymerase subunits and enzymes of nucleotide metabolism, but no DNA replication-associated proteins have yet been identified. It is stable, easy to grow, and highly virulent. It can carry out generalized transduction (at very low frequency) and contains proteins with weak similarities to proteins from a variety of bacteria, including some pathogens, but no pathogenicity islands or genes related to establishing lysogeny have been found.

T1 is a virulent siphovirus belonging to the original set of seven *T-phages* infecting *E. coli* B chosen by Delbrück for use in the early development of molecular biology, as described in Chapter 2. It is best known for remaining viable even when it dries out. It can therefore wreak havoc on a microbiology laboratory if it is once brought in, and relatively few studies have been done with it. Its 50.7-kbp circularly permuted, terminally redundant genome has recently been sequenced and 77 ORFs identified, including many small proteins that show no homology with known phage or other proteins (Andrew Kropinski, personal communication). The tail genes show homology to those of N15, but the head genes are unique.

T4, another classic coliphage, is a 169-kbp lytic myovirus that substitutes 5-hydroxymethylcytidine (hmdC) for cytidine in its DNA. It played a central role in the development of molecular biology and is one of the most thoroughly studied biological entities on earth (along with its close relatives T2 and T6—the *T-even phages*). T4-like phages are prime components of a number of therapeutic cocktails and several other members of its broad family, infecting a range of gram-negative bacteria, have recently been sequenced. Many details of its enzymes, regulatory mechanisms, structure, and infection process are discussed in Chapter 7.

T5 is a siphovirus from the original set of *T-phages*. Its 121-kbp DNA has 10-kbp terminal repeats, unique ends, and, strangely, four conserved nicks at precise sites in one strand. The DNA enters the cell in a two-step process. The left terminal repeat enters first, and products from its pre-early genes completely shut off host replication, transcription and translation, block host restriction systems, and degrade the host DNA to free bases and deoxyribonucleosides that are ejected from the cell. This first-step-transfer DNA segment also encodes genes needed for the rapid entry of the rest of the genome once this process is complete; for shutoff of the pre-early genes; and for the orderly expression of early and then late genes from the rest of the genome. T5 encodes a variety of genes for enzymes of nucleotide metabolism and DNA synthesis and for modifiers of RNA polymerase. Precut genomes containing both terminal repeats are inserted into the preformed heads. About 25 kbp of the genome, in three large blocks, is in principle deletable; this includes genes for tRNAs for all 20 amino acids as well as a number of ORFs. However, deletion of more than 13.3 kbp without compensating insertion(s) prevents DNA packaging.

T7 and its close relative T3 are the final two members of the original T-phage set of coliphages; similar 40-kbp podoviruses are found infecting virtually all gram-negative bacteria, and individual broad host range variants have been found that can, for example, infect both *E. coli* and *Yersinia pestis*. New phage genomes are cut from a DNA concatamer by nicking it at the terminal redundancy site; because of this mechanism, a phage deletion mutant carries less DNA than a normal phage. T7 has an unusual replicative cycle, determined in part by the slow injection of its genome, which takes about 10 minutes; the RNA polymerase is involved in pulling the DNA into the cell. Transcription of the genome occurs in three stages. Early transcription uses the host RNA polymerase, but later transcription uses a T7-encoded RNA polymerase that only recognizes specific T7 promoters and is the most-studied member of the single-subunit family of RNA polymerases, found also in other viruses and in mitochondria and related to retroviral reverse transcriptases. It is 10 times as fast as the multisubunit cellular RNA polymerases and requires no auxiliary proteins, leading to its very wide use in biotechnology; it has also been a powerful tool in studying the mechanism of transcription in intricate detail, as discussed in Chapter 7.

3.6.2. Phages of Gram-Positive Bacteria

The phages of gram-positive bacteria have less variety of potential binding sites than do those of gram-negative bacteria, which can use many different structures in the elaborate outer membranes of their hosts. Most characterized phages of gram-positive hosts seem to bind to the generally species-specific glucosylated teichoic acids that are imbedded in the peptidoglycan layer. The genomes of all the bacteria that have been characterized contain multiple prophages, which generally seem to belong to only a few families of closely related phages. For example, the temperate phages of *Bacillus subtilis*, all viruses with long tails, have been classified into four groups, I to IV, with genomes of about 40, 40, 126, and 60 kbp, respectively. A number of the group III phages, in particular, encode their own versions of such enzymes as DNA polymerase and thymidylate synthase.

3.6.2.1. Phages of Streptococci, Staphylococci, Listeria, and the Bacilli

C_1, one of the first phages isolated (Clark, 1926), is a lytic 16,687-bp podovirus infecting group C streptococci that was involved in the early development of phage typing (Evans, 1936). Functions have only been assigned to 9 of its 20 ORFs (Nelson, 2003). Like ϕ29, it uses a protein-primed mechanism of initiating replication, along with its own, but there are no homologies between the terminal proteins used by the two phages. There are also no database matches for its major head protein, suggesting that it is a novel podovirus. Its polyhedral head sits directly on a base plate made from a single 36-kDa protein, to which are attached three short tail fibers about 30 nm long. Its host-specific lysozyme belongs to a family being developed as therapeutic agents against gram-positive bacteria, as discussed in Chapter 12.

G is an enormous lytic myovirus that infects *B. megaterium*—the largest characterized virus, with more genetic information than many bacteria. The 667 ORFs identified in its 498-kbp genome encode a variety of different interesting homologues of known metabolic pathways, but like the other large phage species sequenced to date, the bulk of its genes are unique and of unknown function; 74% of the predicted proteins had no database match considered significant. Phage G also encodes 17 of its own tRNAs. It appears that G encodes its own DNA polymerase alpha subunit, but in the form of multiple ORFs from which the final protein is assembled by means of a combination of mRNA splicing and removal of *inteins*—segments that get spliced out at the protein level after translation. Its DNA was initially reported to be highly glucosylated—nearly two residues per cytosine—but no obvious enzymes for substituting a modified base that can be glucosylated, or for the glucosylation itself, have been found in the isolate just sequenced by the Pittsburgh phage group.

MM1, a siphovirus with a 40,248-bp genome, is the first temperate *Streptococcus pneumoniae* phage to be sequenced (Obregon et al., 2003). It was isolated from the multiple-antibiotic-resistant strain Spain^{23F}-1. This strain accounts for 40% of 328 penicillin-resistant clinical isolates analyzed in 38 U.S. states and is a leading cause of middle ear infections in children and of pneumonia in the elderly. MM1 has a number of structural and replicative genes in common with various different *S. pyogenes* phages; other structural proteins relate to phages of *L. monocytogenes, L. lactis,* and *S. thermophilus.* No genes related to known pathogens have been identified, but only 26 of the 53 ORFs could be assigned probable functions on the basis of homologies with known genes in the databases. (Confusion can be caused by a coli myovirus also called MM1, underscoring the potential confusion engendered by the naming problems discussed in Chapter 4.)

ϕ11, the best studied of the temperate phages of *Staphylococcus aureus,* is a transducing siphovirus. Its sequenced 45-kbp genome is terminally redundant and circularly permuted; host protein synthesis is shut down 30–40 minutes after prophage induction. Its *lysozyme* has D-alanyl-glycyl endopeptidase as well as *N*-acetylmuramyl-L-alanyl amidase activity.

ϕ29 is a rather small virulent phage (19,285 bp) infecting *B. subtilis*. It is characterized by a terminal protein (TP) covalently linked at the 5′ ends of its genome via a phosphoester bond, leading to a very interesting mechanism of replication that has been extensively studied, as have its morphogenesis and DNA packaging (see Chapter 7). The phage encodes the same B-type DNA polymerase characteristic of T4, PRD1, and adenoviruses. A special (and so far unique) 174-base phage-encoded packaging RNA (pRNA) is essential for in vitro DNA packaging; six copies are found attached to the head-tail connector. The tail connector protein seems to have an important role in giving the head its prolate shape.

ϕ105 and SPβ are the most-used transducing phages for *B. subtilis*. They have genomes of 39.2 and 120 kbp, respectively; the latter has a 27-kbp deletable region, allowing it to carry large inserts.

Sb-1 is a virulent phage infecting *S. aureus*. Staphylococcal infections are among the major infections susceptible to phage treatment, and Sb-1 and its relatives are major components of the Eliava Institute pyophage therapeutic phage preparations against purulent infections, as well as being effectively used in monophage formulations. This myovirus is the only phage that has been used intravenously for therapeutic applications in the Republic of Georgia, after very extensive purification and testing of each batch in animals. It infects virtually all clinical staph isolates, and host resistance is almost never seen. It is currently being sequenced by the lab of Rezo Adamia in Tbilisi; no signs of genes involved in host pathogenesis or prophage formation have been detected. Its 120-kbp genome contains no recognition sites for various staph restriction enzymes, presumably contributing to its virulence and broad host range.

SPO1 and its relatives are large, virulent *B. subtilis* myoviruses (145 kbp, including a 12.4-kbp terminal redundancy) in which hydroxymethyldeoxyuridine (hmdU) replaces thymidine. Despite the potentially identifying presence of the unusual base, host DNA is not degraded during infection (as it is during T4 infection) and there is no indication that the substitution of hmdU for T is involved in, or required for, the shutoff of host DNA synthesis and transcription. The hmdU does seem to enhance middle-mode SPO1 transcription and the binding of TF1, an SPO1-specific DNA binding protein made in large quantities after infection, which enhances—but is not required for—replication. SPO1 is replicated as a long concatamer with a single copy of the terminal redundancy between monomers; this concatamer is then cleaved in staggered fashion, leaving overhanging 5′ends, which are then replicated. A large gene cluster is involved in shutting off host replication and gene expression, and inhibiting cell division. However, SPO1 does not shut off host ribosomal RNA synthesis nor degrade the host DNA. Like T4, SPO1 has a very complex capsid, involving at least 53 different polypeptides and almost half its genome. However, there is no indication of any relationship between SPO1 and T4. The SPO1 sequence has just been completed at the Pittsburgh Phage Institute, but little is known about the morphogenesis or infection process of SPO1, or the functions of individual genes except for some of those involved in its complex regulatory processes and DNA replication.

3.6.2.2. Phages of Lactic Acid Bacteria: The Dairy Phages

Various lactic acid bacteria have been used for centuries in the preservation and production of fermented foods and feeds of plant and animal origins. One of the most critical problems in these processes is the contamination of the starters by bacteriophages that cause bacterial lysis, spoilage, and significant economic losses, as explored further in Chapters 5 and 10. Numerous reports have described the isolation and characterization of both virulent and temperate *Lactobacillus* and *Lactococcus* phages; only a few of them have been studied in great detail. *Lactobacillus delbrueckii* subsp. *bulgaricus* temperate and virulent siphovirid phages are closely related based in morphology, host range, serology and DNA-DNA hybridization analysis (Mata and Ritzenthaler 1988) as are the temperate and virulent *L. helveticus* phages (Sechaud et al., 1988). These data suggest that many of the "virulent" *Lactobacillus* phages may actually have originated from temperate phages rather than being "professional" lytic phages (see Chapter 7). *L. delbrueckii* subsp. *bulgaricus* and subsp. *lactis* are widely employed as starter cultures in manufacturing yogurt (*bulgaricus*) and cheeses (both *bulgaricus* and *lactis*); the apparent temperate origin of most of the lytic phages seen here may represent particular selective pressures in the dairy industry.

A2, a siphovirus isolated from a homemade cheese whey sample, is today the best characterized *Lactobacillus casei* and *Lactobacillus paracasei* phage. It has a 60-nm isometric head and a 280-nm tail that ends in a baseplate with a protruding spike. The functions of half of the 61 ORFs detected in its 43,411-bp DNA have been determined; they are involved in DNA replication and packaging, morphogenesis, cell lysis, and the lysis-lysogeny decision, as well as encoding the integrase. Integration of A2 occurs at a highly conserved nucleotide sequence of the $tRNA_{Leu}$ gene. A replication-thermosensitive suicide vector was constructed with the integrase function and phage attachment (*AttP*) sequences of A2.

J-1 and the distantly related siphovirus **PL-1** were isolated from abnormal fermentation processes of a lactic acid beverage named Yakult. Phage J-1 has a 55-nm head and a 290-nm tail with no tail fibers. Phage PL-1 has a 63-nm head and a 275-nm tail with a 60-nm fiber attached to a small baseplate One-step growth curves with PL-1 showed a lag period of about 75 min, a rise period of 100 min, and a burst size of 180–200 particles per cell. Although isolated as a lytic phage for ATCC 27092, PL-1can lysogenize strain ATCC 334 and confers superinfection immunity to the homologous phages, 393 and ϕ4lk. PL-1, which requires calcium and is inhibited by L-rhamnose, has been used as a model system in the study of the mechanism of infection by *Siphoviridae* (Watanabe, 1985).

LL-H, isolated in Finland from faulty cheese fermentation processes (Alatossava et al., 1987), is the best characterized of the "virulent" siphoviruses. LL-H has a 50-nm isometric head and a 180-nm tail with a small base plate and tail fiber. Calcium ions are required to stabilize the phage particles from inactivation in Tris buffer and to improve adsorption and the efficiency of DNA penetration. Phage LL-H's 34,659-bp DNA is circularly permuted and terminally redundant; the integrase, structural proteins, lysin and early genes, as well as DNA packaging sequences, have been located. A functional group I intron was detected in the large subunit terminase gene, *terL*. Phage LL-H shows 50% identity to phage mv4 and limited DNA similarity with three small regions of phage JCL1032, a *L. delbrueckii lactis* phage with a prolate head and a long, cross-barred tail. JCL1032 protects its genome by having few recognition sites for a variety of restriction enzymes. *L. salivarius* phage 223

(Tohyama, et al., 1972) and *L. delbrueckii lactis* phage 0235 (Sechaud et al., 1988) are also siphoviruses with prolate heads.

***ϕ*adh**, a temperate *Lactobacillus gasseri* siphovirus, has been sequenced and used very extensively for comparative genomic analysis; its lytic and lysogenic cycles of phage replication has been studies, as they have also for phages PLS-1 and *ϕ*FSW. In these cases, prophage-cured derivative strains were isolated and found sensitive to the phage lytic growth, not inducible with mitomycin C, and able to serve as host where lysogeny could be reestablished.

Sfi11 and Sfi21 define the two quite distinct groups of temperate siphoviruses of *Streptococcus thermophilus*, as discussed in Chapter 5.

***L. brevis* phage 6107 and *L. helveticus* phage 0241** are among the rather rare myoviruses of lactic acid bacteria.

3.6.2.3. Phages of Mycobacteria

Mycobacterial phages have been studied since 1954 and have been used extensively for such purposes as rapid detection of tuberculosis. However, until very recently only four of them had been characterized genetically. All four are siphoviruses with 40-kbp genomes and are closely related. Now, nine more siphoviruses have been sequenced, two of them with a different kind of unusual elongated heads (Pedulla et al., 2003); it will be very interesting to learn more about their general biology. Their G+C content ranges from 57% to 69 %, and they form a range of sizes. Only one has significant sequence similarity with the original set, and most of them vary substantially from each other, though the structural gene region can be recognized in most cases. In a PsiBlast comparative analysis of all 1659 ORFs against the public databases, about 50% are unrelated to those of other characterized mycophages or ORFs from any other previously sequenced organisms, and three-quarters of the rest match only other mycophage genes. There is clearly extensive exchange among the various phages and all of the phages share at least some genes. The relationships are reticulate and they cannot be ordered into any single hierarchical phylogeny, but two primary groups stand out—one closely knit, incorporating L5, D29, Bxb1, and Bxz2, and the other more diverse, including Che8, Che9d, Cjw1, and Omega as well as the two elongated-head phages, Corndog and Che9c. Most appear to be temperate, but TM4, Barnyard, and Corndog have no genes related to lysogeny or integration.

Only one lytic myovirus, Bxz1, was found and studied. Its 156-kbp genome encodes 26 tRNA genes, with anticodons for 15 amino acids and a putative suppressor, and a number of proteins related to DNA replication and recombination, but most of its genes match nothing else in the databases. No genes related to pathogenicity or to lysogeny were identified.

3.6.4. Cyanophages

The organisms once called "blue-green algae" are actually phototrophic gram-negative bacteria. Studies have been carried out with a number of phages, belonging to all three major phage morphological categories, which infect either the unicellular or the filamentous cyanobacteria (Suttle, 2000); they are discussed further in

Chapter 6. Cyanophages are being used extensively to explore the complex physiology of these interesting, important oxygen-producing organisms. Phage infecting filamentous forms generally cause rapid invagination and destruction of the host's photosynthetic membranes, whereas such destruction is only seen very late in the infection cycle with those that infect unicellular cyanobacteria, for which successful infection seems to depend on ongoing photosynthesis. The *Cyanomyoviridae* are very diverse, with G+C content from 37% to 55%, 14 to 30 reported capsid proteins, tail-structure variations, and genome sizes ranging from 37 to 100 kbp. The studies of their genomics and relationships to each other are still in the early stages, but four distinct morphotypes have been identified in a single water sample. Similar phages are found in very different locations; they are the most common form that has been found in marine waters. **Cyanophage AS-1,** for example**,** is the type species for a group of myoviruses infecting unicellular fresh-water *Synecococcus* and *Anacystis* species. **Cyanopodovirus LPP-1,** the first cyanophage isolate, has a 59-nm head, a 15–20-nm tail, and a 42-kbp genome, with G+C content in the low 50% range; related phages have come from sites such as waste stabilization ponds in the United States and the Ukraine, infecting several bacteria assigned to the *Lyngbya, Plectonema, and Phormidium* subgroup of cyanobacteria.

3.7. THE SMALL AND TAILLESS PHAGES

There are ten very small classes of phages with quite different structures and ways of multiplying, as mentioned above and discussed in detail in Chapter 4. Here we provide an overview of the properties of the best-studied of these, with specific examples. Since they are little discussed in other chapters of this book, they are presented in somewhat more detail here than are the various tailed phages. The tailless phages are also extensively represented in the soon-to-be released edition of Calendar's *The Bacteriophages,* with chapters on the *Microviridae* (Fane et al., 2005), filamentous phage (Russell and Model, 2005), lipid-containing phages PRD1 and PM2 (Grahn et al., 2005; Bamford and Bamford, 2005), single-stranded RNA phages (van Duin and Tsareva, 2005), and phages with segmented double-stranded RNA genomes (Mindich, 2005).

3.7.1. RNA Phages

Zinder (1961) suspected that "male-specific" phages might exist—that is, phages specific to F^+ or Hfr strains of *E. coli,* and indeed identified such phages. Members of one class have uniformly small, spherical virions with RNA genomes. The best-known are the *Leviviridae,* including R17, f2, MS2, and Qβ. While subgroups can be distinguished by differences in size and other features, their virions generally are about 25 nm in diameter, with single-stranded RNA genomes only a few thousand bases long. Each virion has an icosahedral capsid made of 180 copies of a single coat protein, arranged as 20 hexamers plus 12 pentamers that form the vertices.

Some of these small RNA phages are specific for "male" bacteria because they adsorb to the pili specifically synthesized by such strains; others use other types of pili as their receptors. The pili are long filamentous extensions from the cell surface, made of multiple copies of the protein pilin. Electron micrographs show that the small RNA phages adsorb along the sides of these pili, and bacteria to which hundreds or thousands of phage have adsorbed look as if their pili were tripled in diameter, until the individual phage particles are resolved. However, the phage genomes do not enter the cell through the pili; of the thousands that might adsorb to a cell, infection is restricted to the very few at the bases of the pili, which are also in contact with the cell surface. The mechanism of uncoating and transfer of the RNA into the cell remains obscure. Mutants of these phages have been easily collected. Genetic analysis has been complicated because they don't undergo recombination, but complementation studies indicate that they have three or four genes. Gene 1 (also designated C) encodes an RNA replicase; gene 2, or A, encodes a so-called maturation protein; and gene 3, or B, encodes the coat (capsid) protein. Some phages have a fourth gene encoding an enzyme that precipitates lysis. The infecting RNA is a *plus strand*, the strand that actually encodes proteins, and thus it serves as its own messenger and is translated by the host protein-synthesis apparatus. The RNA replicase has two functions: first, to convert infecting RNA genomes into double-stranded replicative forms (RFs) and, second, to synthesize new plus strands from these RFs. The new plus strands are then variously used as messengers for protein synthesis, converted into new RFs, or combined with coat proteins to form new virions. Mutants lacking the maturation protein produce defective virions that are unable to adsorb to a new host and are susceptible to RNase.

While these phages were discovered in *E. coli*, similar phages of other bacteria are known. Some of them infect the stalked bacterium *Caulobacter*. When a stalked cell divides, one daughter cell becomes a motile "swarmer" cell, which swims off by means of its terminal flagellum; the flagellum is soon replaced by a new stalk. Since phage of this type only adsorb to filamentous cell extensions like the flagellum, *Caulobacter* is only susceptible during the swarmer stage, which means that most cells in an exponentially growing culture are immune.

The family *Cystoviridae* is represented by ϕ6, a small (60-nm) enveloped virus whose genome consists of three polycistronic pieces of double-stranded RNA of 6374, 4057, and 2948 base pairs. Its RNA is encased in a dodecahedral polymerase complex, an icosahedral capsid, and then a membrane, with a lytic enzyme between the membrane and capsid. The membrane is half lipid and half phage-encoded proteins, including an adsorption-fusion complex. This virus is also pilus-specific and infects the plant pathogen *Pseudomonas syringae pv. Phaseolicola*. After infection, a viral transcriptase transcribes all three segments. The largest segment encodes a polymerase-procapsid complex, which takes in one of each of the three mRNAs and replicates them to double-stranded form. Genes for the capsid and membrane proteins are then transcribed within the procapsid until it has been encased within the capsid.

3.7.2. Small DNA Phages

Small phages with single-stranded DNA genomes and spherical (icosahedral) nucleocapsids constitute the *Microviridae*. They have been found infecting bacteria from *E. coli* to *Chlamydia* and *Spiroplasma* and are probably more widespread in nature than we realize. Only two distinct subfamilies have been found to date

(Brentlinger et al., 2002); one of them is represented by ϕX174, infecting various γ-proteobacteria, while the other is represented by Chp2 and SpV4, infecting *Chlamydia* and *Spiroplasma,* and ϕMH2K, which infects the intracellular parasitic ϕ-proteobacterium *Bdellovibrio bacteriovorus.*

The microviruses all multiply by means of a double-stranded replicative form (RF), as has been described in detail for ϕX174. The single-stranded, covalently closed infecting genome is a plus strand, which is converted into a double-stranded form (RFI) by host enzymes. The newly-synthesized minus strand is the template strand, and it is transcribed by host RNA polymerase to form messenger RNAs for synthesis of viral proteins. One of these, the A protein, binds to the RFI and nicks its plus strand at the origin of replication, thus forming a complex called RFII. Host enzymes then cause displacement of the plus strand and replication (through a rolling-circle mechanism) of a new plus strand. The displaced plus strand can then be converted into a new RFI. Thus, many new RFs are created. Eventually, when enough new capsid proteins have accumulated, newly formed plus strands combine with these proteins to make new virions. The infected cell lyses after about 40 minutes, producing on the order of 500 new phages.

The small DNA phages of the other class constitute the family *Inoviridae* (*ino-* = thin). They have the form of long filaments, their capsids having helical symmetry. (Thus, they are similar to one of the most classical viruses, tobacco mosaic (TMV), and many other plant viruses.) Their genomes are all single-stranded, covalently closed DNA molecules; the nucleocapsid is then made of a helical array of many copies of the major coat protein, with a few copies of minor proteins located at the ends. Inoviruses are another group of "male-specific phages," like the RNA phages discussed above, but they bind to the F-pilus *tip,* in contrast to the small RNA phages. Adsorption of a phage stimulates retraction of the pilus, which brings one end of the phage particle into the periplasm. Then the nucleocapsid disassembles and the phage DNA is translocated into the cytoplasm. DNA replication occurs through formation of a double-stranded RF. The new single-stranded genomes that are eventually formed combine with capsid proteins, but assembly occurs in the bacterial membrane with extrusion of the completed phage from the cell surface. Thus, phage production continues without cell lysis and it has been observed to continue for over 300 generations. Inoviruses do produce plaques on a bacterial lawn (otherwise, they might never have been discovered). But while classical phages produce plaques because their hosts lyse, inovirus plaques are formed because their hosts grow more slowly than do uninfected cells, thus forming a less turbid region in the lawn. However, these plaques must be observed at the right time; while bacteria in the rest of the lawn stop growing after they have exhausted the local nutrients, infected bacteria can continue to grow for about another generation, wiping out the plaque.

3.7.3. Lipid-Containing Phages

Coliphage PRD1 is the type virus of the *Tectiviridae*; similar phages have been found in bacilli and in *Thermus*. It attaches to the sex pili that plasmids such as N, P, and W create for their conjugative transfer. The 14,925-bp genome of PRD1 encodes 22 genes and is enclosed in a membrane layer that is in turn surrounded by a protein shell. The membrane includes phage-encoded proteins involved in adsorption, DNA injection, and DNA packaging. The phage also encodes its own DNA polymerase as well as an initiator protein that is bound covalently to the start of each DNA strand. The shell is formed and then lined with membrane (taken from the host plasma membrane) before it is filled with DNA; several hundred viral particles are liberated upon cell lysis. Corticovirus PM2 has a similar structure

3.7.4. Phages of Mycoplasmas

As described above, mycoplasmas are small bacteria that lack the typical cell walls of other bacteria. This makes them resistant to antibiotics such as penicillin, which attack cell-wall formation, but the lack of a wall makes them susceptible to osmotic lysis. The lack of cell walls also makes them extremely pleomorphic, and they vary from spheres to elongated filaments. They are found in diverse habitats, including soils, compost piles, and many kinds of plants and animals—where they are known to cause various diseases, especially pneumonias. It has been estimated that about 10% of cell cultures are infected with mycoplasmas. Genomic studies indicate that they arose by a kind of degenerate evolution from streptococci. Some species have genomes of only 600-800 kbp.

Three mycoplasma genera (*Acholeplasma, Spiroplasma,* and *Mycoplasma*) are known to be infected by phages, but their lack of cell walls means that their viruses cannot adsorb to the components attacked by phages of other bacteria. Therefore, mycoplasma phages, like animal viruses, must adsorb directly to cell membranes. All known mycoplasma phages use DNA. Some have circular, single-stranded DNA, with filamentous, icosahedral, and quasi-spherical morphology. One of the best known is *Acholeplasma* phage L51, an inovirus with rather unusual filamentous morphology; its virions are non-enveloped, bullet-shaped particles, 14 nm × 71 nm, with one end rounded and the other irregularly shaped or flat. The 4.5-kbp genome encodes four proteins. The phage adsorbs to unknown cell-membrane receptors but infects very inefficiently; although 300 virions can adsorb to a cell, only 10–20 attach to receptors that allow them to infect. Replication and production of new phage genomes follows the typical pattern discussed above for filamentous phages, with new phages extruded through the cell surface, so the infected cell does not lyse and continues to produce phage indefinitely. Each infected cell produces about 150–200 new phages by 2–3 hours after infection.

One enveloped quasi-spherical mycoplasma phage containing circular double-stranded DNA has been isolated and characterized. However, most of the few known mycoplasma phages with double-stranded DNA appear to be podoviruses with linear genomes having unusual features. *Acholeplasma* phage L3 is a well-known mycoplasma podovirus with a 39.4-kbp circularly permuted genome and 8% terminal repeat. L3 virions have a 60-nm icosahedral head, an 8nm × 16 nm collar to which fibers are attached, and, a short tail 10 nm wide and 20 nm long. As adsorbed L3 virions diffuse on the cell membrane, the fibers behave like polyvalent ligands, cross-linking mobile receptors and producing capping or clustering of adsorbed virions. This capping process is dependent on host-cell energy metabolism. Adsorbed L3 virions can also mediate cell fusion by cross-linking receptors on different cells, thus producing the giant cells seen in L3-infected cultures. Although L3 kills its host and produces clear plaques, it does not do so through the usual cell lysis. Instead, mature phages accumulate in infected cells and phages are then released continuously over many hours. By 8 hr after infection, the inside of the cell membrane is covered by L3 virions, oriented radially with tails facing the membrane. They apparently escape from the cell through the budding of extracellular membrane-bound vesicles enclosing progeny phage; these eventually break down to release non-enveloped L3 virions. Other phages, with these morphologies or long-tailed phage morphology, have been reported but have not been propagated or characterized.

3.8. ARCHAEPHAGES

As discussed in Chapter 4, a large fraction of archaephages have highly interesting, often pleiomorphic morphologies very different from those seen for bacteriophages. However, many of the viruses isolated to date against members of the Archaeal kingdom Euryarchaeota look like typical members of the *Myoviridae* and *Siphoviridae* (Dyall-Smith et al., 2003). Both temperate and lytic archaephages were found; in some cases, terminal redundancy and circular permutation of the genomes indicate a headful mechanism of packaging from a concatameric replication form. Rachel et al. (2002) have recently reported the first members of the *Siphoviridae* and *Podoviridae* to be seen in hydrothermal environments in a culture inoculated from the Yellowstone Obsidian Pool into an anaerobic chemostat maintained at 85°C. The specific hosts have not yet been identified; several hyperthermic bacterial species were identified there, in addition to a wide range of Archaea, by means of 16S rDNA sequencing data.

HF1 and HF2 are related virulent myoviruses with different host specificities isolated from an Australian salt lake. They infect a range of haloarchaea—*Haloferax, Halobacterium, Haloarcula, Natrialba,* and *Halorubrum.* Their 77 kbp encodes 121 ORFs, 90% of which did not have any matches in the sequence databases, but there was some homology to coliphage Mu as well as to bacteria like *Listeria* and *Bacillus.*

ΦH is a temperate *Halobacterium* myovirus with a 59-kbp genome whose prophage form persists as a circular episome, as for P1, rather than integrating into the genome.

ψM2 infects the methanogenic archaeon *Methanobacterium thermoautotrophicum.* The 31 ORFs encoded in its 26,111-bp genome includes two recognized as probable phage structural proteins and three as DNA packaging proteins, an integrase and a pseudomurein endoisopeptidase, a novel enzyme involved in cell lysis (Pfister et al., 1998).

A much wider variety of interesting morphologies have been seen among the archaephages invading the hyperthermophiles of the Crenarchaeota (Rice et al., 2001; Rachel et al., 2002). Bettstetter et al. (2003) have recently analyzed the genomics, structure, and infection process of a very interesting enveloped filamentous virus, AFV1 (with claws on both ends that apparently latch onto the pili of the some strains of the hyperthermophile *Acidianus*, have a latent period of about 4 hours, are extruded rather than lysing the cell, and can establish a persistent state).

Unfortunately, too little is generally known about the genomics, infection processes, or even hosts of most archaephages to warrant their detailed inclusion in this book; they clearly offer a range of new frontiers to explore, as discussed by Prangishvili (2003). Viruses of haloarchaea are found at 10^7 per ml in natural hypersaline environments and tend to be primarily virulent; a majority of these are lemon-shaped in appearance like Prangishvili's well-studied *Sulfolobus* virus SSVI, though without sequence similarities.

Rachel, Bettstetter, et al. (2002) also found a surprisingly wide range of new pleiomorphic phages in both the Obsidian Pool sample and one from a very acidic spring (pH 2.0, 85°C) from Yellowstone's Crater Hills. To date, all of the *specifically identified* hosts of viruses from hot habitats are members of the hyperthermophilic genera *Sulfolobus, Thermoproteus,* and *Acidianus,* belonging to the Crenarchaeota. Prangishvili and Garrett (2004) describe the unprecedented morphological diversity of viruses from hydrothermal environments >80°C.

ACKNOWLEDGEMENTS

We are particularly grateful to the many members of the phage community who have shared their insights and suggestions here. In particular, we appreciate the willingness of Richard Calendar and a number of the authors from the second edition of his *Bacteriophages* to share chapters with us prior to publication; this is an extremely valuable resource for information on many of the phages described here. The aid and support of Hans Ackermann and Karin Carlson have also been especially important, as has the input from the current students in the Evergreen phage biology lab. Special thanks go to Gautam Dutta for his technical assistance with the figures and references.

REFERENCES

Abedon, S.T., "Phage Ecology," in *The Bacteriophages,* R. Calendar (Ed.). Oxford University Press, New York, 2005.

Abedon, S.T. Selection for lysis inhibition in bacteriophage. *J Theor Biol* 146: 501–511, 1990.

Ackermann, H.W. and Dubow, M., *Viruses of Prokaryotes*. CRC Press, Boca Raton, FL.

Adams, M.H. (1959) *Bacteriophages*. Interscience Publishers, New York, 1987.

Alatossava, T., Jutte, H. and Seiler, H., Transmembrane cation movements during infection of *Lactobacillus lactis* by bacteriophage LL-H. *J Gen Virol* 68 (Pt 6): 1525–1532, 1987.

Bamford, D.H. and Bamford, J.K.H., Lipid-Containing Bacteriophage PM2, the Type-Organism of *Corticoviridae*, in *The Bacteriophages*, R. Calendar (Ed.). Oxford University Press, New York, 2005.

Bettstetter, M., Peng, X., Garrett, R.A. and Prangishvili, D., AFV1, a novel virus infecting hyperthermophilic archaea of the genus acidianus. *Virology* 315: 68–79, 2003.

Brentlinger, K.L., Hafenstein, S., Novak, C.R., Fane, B.A., Borgon, R., McKenna, R. and Agbandje-McKenna, M., Microviridae, a family divided: Isolation, characterization, and genome sequence of ϕMH2K, a bacteriophage of the obligate intracellular parasitic bacterium *Bdellovibrio bacteriovorus*. *J Bacteriol* 184: 1089–1094, 2002.

Briani, F., Deho, G., Forti, F. and Ghisotti, D., The plasmid status of satellite bacteriophage P4. *Plasmid* 45: 1–17, 2001.

Broudy, T.B. and Fischetti, V.A., In vivo lysogenic conversion of Tox(-) *Streptococcus pyogenes* to Tox(+) with Lysogenic Streptococci or free phage. *Infect Immun* 71: 3782–3786, 2003.

Brown, D.T. and Anderson, T.F., Effect of host cell wall material on the adsorbability of cofactor-requiring T4. *J Virol* 4: 94–108, 1969.

Calendar, R., *The Bacteriophages.* 2nd Ed. Oxford University Press, New York, 2005.

Calendar, R. (Ed.), *The Bacteriophages.* 1st Ed. Plenum Press, New York, 1988.

Clark, P.F. and Clark, A.S., A "Bacteriophage" active against a hemolytic *Streptococcus. J Bacteriol* 11: 89, (1926).

Court, D.L., Sawitzke, J.A. and Thomason, L.C., Genetic engineering using homologous recombination. *Annu Rev Genet* 36: 361–388, 2002.

d'Herelle, F., *The Bacteriophage and Its Behavior.* The Williams and Wilkins Company, Baltimore, Maryland, 1926.

Dodd, I.B. and Egan, J.B., Action at a distance in CI repressor regulation of the bacteriophage 186 genetic switch. *Mol Microbiol* 45: 697–710, 2002.

Doermann, A.H., The vegetative state in the life cycle of bacteriophage: Evidence for its occurence and its genetic characterization. *Cold Spring Harbor Symp Quant Biol* XVIII: 3–11, 1953.

Dyall-Smith, M., Tang, S.L. and Bath, C., Haloarchaeal viruses: How diverse are they? *Res Microbiol* 154: 309–313, 2003.

Ellis, E.L. and Delbrück, M., The growth of bacteriophage. *J Gen Physiol* 22: 365, 1939.

Evans, A.C., The potency of nascent *Streptococcus* bacteriophage B. *J Bacteriol* 39: 597–604, 1940.

Evans, A.C., Studies on hemolytic streptococci. I. Methods of classification. *J Bacteriol* 31: 423–437, 1936.

Fane, B.A., Brentlinger, K.L., Burch, A.D., Chen, M., Hafenstein, S., Moore, E. and Novak, C.R., ϕX174 *et al.*, the *Microviridae*, In press in *The Bacteriophages,* R. Calendar (Ed.). Oxford University Press, New York, 2005.

Gachechiladze, K., *Antigenic Characteristics of Structure Elements of Bacterial Viruses of the Coli-Dysenterial Group,* p. 331. Bacteriophage Institute, Tbilisi, Republic of Georgia, 1981.

Ghuysen, J.M. and Hackenbeck, R., *Bacterial Cell Wall.* Elsevier, Amsterdam, New York, 1994.

Goldberg, E., Grinius, L. and Letellier, L., "Recognition, attachment, and injection," in *Molecular Biology of Bacteriophage T4,* J.D. Karam, J.W. Drake, K.N. Kreuzer, G. Mosig, D.H. Hall, F.A. Eiserling, L.W. Black, et al. (Eds.). American Society for Microbiology, Washington, D.C., 1994.

Grahn, A.M., Butcher, S.J., Bamford, J.K.H. and Bamford, D.H., "PRD1 Dissecting the Genome, Structure and Entry," In press in *The Bacteriophages,* R. Calendar (Ed.). Oxford University Press, New York, 2005.

Harper, D., Eryomin, V., White, T. and Kutter, E., "Effects of T4 infection on membrane lipid synthesis," pp. 385–390 in *Molecular Biology of Bacteriophage T4,* J. Karam, J.W. Drake, K.N. Kreuzer, G. Mosig, D.H. Hall, F.A. Eiserling, L.W. Black, et al. (Eds.). American Society for Microbiology, Washington, D.C., 1994.

Hershey, A.D., Kalmanson, G. and Bronfenbrenner, J., Quantitative relationships in the phage-antiphage reaction: Unity and homogeneity of the reactants. *Journ of Immunol* 46: 281–299, 1943.

Jacob, F. and E.L. Wollman, *Sexuality and the Genetics of Bacteria*. Academic Press, New York, 1961.

Kutter, E., Stidham, T., Guttman, B., Kutter, E., Batts, D., Peterson, S., Djavakhishvili, T., et al., "Genomic map of bacteriophage T4," pp. 491–519 in *Molecular Biology of Bacteriophage T4,* J. Karam, J.W. Drake, K.N. Kreuzer, G. Mosig, D.H. Hall, F.A. Eiserling, L.W. Black, et al. (Eds.). American Society for Microbiology, Washington, D.C., 1994.

Lederberg, E.M., Lysogenicity in *E. coli* K12. *Genetics* 36: 560, 1951.

Lenski, R.E., Coevolution of bacteria and phage: Are there endless cycles of bacterial defenses and phage counterdefenses? *J Theor Biol* 108: 319–325, 1984.

Levin, B.R. and Lenski, R.E., "Bacteria and phage: A model system for the study of the ecology and co-evolution of hosts and parasites," pp. 227–241 in *Ecology and Genetics of Host-Parasite Interactions*. The Linnean Society of London, 1985.

Little, J.W., "Gene Regulatory Circuitry of Phage l, "In press in *The Bacteriophages*, R. Calendar (Ed.). Oxford University Press, New York, 2005.

Loeb T. and Zinder, N.D., A bacteriophage containing RNA. *Proc Natl Acad Sci U S A* 47: 282–289, 1961.

Lwoff, A. and Gutmann, A., (1950) Recherches sur un *Bacillus megatherium* lysogene. *Ann InstPasteur* 78: 711–739, 1961.

Madigan, M.T., Martinko, J.M. and Parker, J. (Eds.) *Brock Biology of Microorganisms*. Prentice Hall, Upper Saddle River, 2002.

Mata, M. and Ritzenthaler, P., Present state of lactic acid bacteria phage taxonomy. *Biochimie* 70: 395–400, 1988.

Mathews, C.K., Kutter, E., Mosig, G. and Berget, P.B. (Eds.) *Bacteriophage T4*. American Society for Microbiology, Washington, D.C., 1983.

Mesyanzhinov, V.V., Robben, J., Grymonprez, B., Kostyuchenko, V.A., Bourkaltseva, M.V., Sykilinda, N.N., Krylov, V.N., et al., The genome of bacteriophage ϕKZ of *Pseudomonas aeruginosa*. *J Mol Biol* 317: 1–19, 2002.

Miller, C.G., "Protein Degradation and Proteolytic Modification," pp. 936–954 in *Escherichia coli and Salmonella*, F.C. Neidhardt (Ed.). American Society for Microbiology, Washington, D.C., 1996.

Mindich, L., "Phages with Segmented Double-Stranded RNA Genomes," In press in *The Bacteriophages,* R. Calendar (Ed.). Oxford University Press, New York, 2005.

Molineux, I.J., No syringes please, ejection of phage T7 DNA from the virion is enzyme driven. *Mol Microbiol* 40: 1–8, 2001.

Nelson, D., Schuch, R., Zhu, S., Tscherne, M., Fischetti, V.A., Genomic sequence of C1, the streptococcal phage. *J Bacteriol* 185: 3325–3332, 2003.

Nikaido, H., Molecular basis of bacterial outer membrane permeability revisited. *Microbiol Mol Biol Rev* 67: 593–656, 2003.

Nikaido, H., "Outer Membrane," pp. 29–47 in *Escherichia coli and Salmonella,* F.C. Neidhardt (Ed.). American Society for Microbiology, Washington, D.C., 1996.

Obregon, V., Garcia, J.L., Garcia, E., Lopez, R. and Garcia, P., Genome organization and molecular analysis of the temperate bacteriophage MM1 of *Streptococcus pneumoniae*. *J Bacteriol* 185: 2362–2368, 2003.

Pedulla, M.L., Ford, M.E., Houtz, J.M., Karthikeyan, T., Wadsworth, C., Lewis, J.A., Jacobs-Sera, D., et al., Origins of highly mosaic mycobacteriophage genomes. *Cell* 113: 171–182, 2003.

Pfister, P., Wasserfallen, A., Stettler, R. and Leisinger, T., Molecular analysis of Methanobacterium phage psiM2. *Mol Microbiol* 30: 233–244, 1998.

Prangishvili, D., Evolutionary insights from studies on viruses of hyperthermophilic archaea. *Res Microbiol* 154: 289–294, 2003.

Prangishvili, D. and Garrett, R.A., Exceptionally diverse morphotypes and genomes of crenarchaeal hyperthermophilic viruses. *Biochem Soc Trans* 32: 204–208, 2004.

Rachel, R., Bettstetter, M., Hedlund, B.P., Haring, M., Kessler, A., Stetter, K.O. and Prangishvili, D., Remarkable morphological diversity of viruses and virus-like particles in hot terrestrial environments. *Arch Virol* 147: 2419–2429, 2002.

Rice, G., Stedman, K., Snyder, J., Wiedenheft, B., Willits, D., Brumfield, S., McDermott, T., et al., Viruses from extreme thermal environments. *Proc Natl Acad Sci U S A* 98: 13341–13345, 2001.

Russell, M. and Model, P., "Filamentous Phage," In press in *The Bacteriophages,* R. Calendar (Ed.). Oxford University Press, New York, 2005.

Sanger, F., Coulson, A.R., Hong, G.F., Hill, D. F. and Petersen, G.B., Nucleotide sequence of bacteriophage lambda DNA. *J Mol Biol* 162: 729–773, 1982.

Sechaud, L., Cluzel, P.J., Rousseau, M., Baumgartner, A. and Accolas, J.P., Bacteriophages of lactobacilli. *Biochimie* 70: 401–410, 1988.

Simon, L.D., The infection of *Escherichia coli* by T2 and T4 bacteriophages as seen in the electron microscope. III. Membrane-associated intracellular bacteriophages. *Virology* 38: 285–296, 1969.

Suttle, C.A., "Cyanophages and Their Role in the Ecology of Cyanobacteria," pp. 563–589 in *The Ecology of Cyanobacteria*: *Their Diversity in Time and Space,* B.A. Whitton and Potts, M. (Eds.), Kluwer Academic Publishers, Boston, 2000.

Tetart, F., Desplats, C. and Krisch, H.M., Genome plasticity in the distal tail fiber locus of the T-even bacteriophage: Recombination between conserved motifs swaps adhesin specificity. *J Mol Biol* 282: 543–556, 1998.

Tikhonenko, A.S., Gachechiladze, K.K., Bespalova, I.A., Kretova, A.F. and Chanishvili, T.G., Electron-microscopic study of the serological affinity between the antigenic components of phages T4 and DDVI. *Mol Biol* (Mosk) 10: 667–673, 1976.

Tohyama K.S.T., Arai H., Oda A., Studies on temperate phages of *Lactobacillus salivarius.* I. Morphological, biological, and serological properties of newly isolated temperate phages of *Lactobacillus salivarius. Jpn J Microbiol* 16: 385–395, 1972.

van Duin, J. and Tsareva, N.A., "Single-Stranded RNA Phages," in *The Bacteriophages*, R. Calendar (Ed.). Oxford University Press, New York, 2005.

Wang, I.N., Deaton, J. and Young, R., Sizing the holin lesion with an endolysin-beta-galactosidase fusion. *J Bacteriol* 185: 779–787, 2003.

Watanabe, K.K., N.; Sasaki, T, Mechanism of in vitro inactivation of Lactobacillus phage PL-1 by D-glucosamine. *Agriculture and Biol Chem* 49: 63–70, 1985.

Wollman, E.L. and Stent, G.S., Studies on activation of T4 bacteriophage by cofactor. IV. Nascent activity. *Biochemica Et Biophysica Acta* 9: 538–550, 1952.

Zinder, N.D. and Lederberg, J., Genetic exchange in *Salmonella. J Bacteriol* 64: 679–699, 1952.

4 Bacteriophage Classification

Hans-W. Ackermann
Dept. of Medical Biology, Laval University, Quebec, Canada

CONTENTS

4.1. Development of Bacteriophage Taxonomy 68
4.1.1. The Early Stage 68
4.1.2. The ICTV 69
4.1.3. Purpose of Phage Classification 69
4.2. Current Phage Classification 70
4.2.1. Tailed Phages (Order *Caudovirales*) 70
4.2.2. Polyhedral, Filamentous, and Pleomorphic Phages 73
4.2.2.1. Polyhedral DNA Phages 73
Microviridae (ssDNA) 73
Corticoviridae (dsDNA) 74
Tectiviridae (dsDNA) 74
4.2.2.2. Polyhedral RNA Phages 74
Leviviridae (ssRNA) 74
Cystoviridae (dsRNA) 74
4.2.2.3. Filamentous Phages 74
Inoviridae (ssDNA) 74
Lipothrixviridae (dsDNA) 78
Rudiviridae (dsDNA) 78
4.2.2.4. Pleomorphic Phages 79
Plasmaviridae (dsDNA) 79
Fuselloviridae (dsDNA) 79
Sulfolobus SNDV-like viruses (dsDNA) 81
4.3. Taxonomic Problems 81
4.3.1. New Taxa and New Techniques 81
4.3.2. The λ-P22 Controversy 82
4.3.3. Sequence-Based Taxonomy 83
4.3.3.1. The Phage Proteomic Tree 83
4.3.3.2. Modular Classification 83
4.4. Links Between Phages and Other Viruses 85
References 86

0-8493-1336-X/05/$0.00+$1.50
© 2005 by CRC Press LLC

4.1. DEVELOPMENT OF BACTERIOPHAGE TAXONOMY

4.1.1. The Early Stage

Felix d'Hérelle, one of the two discoverers of bacteriophages (for more details on the discovery of bacteriophages please refer to Chapter 1), believed that there was only one bacteriophage with many races, the "*Bacteriophagum intestinale*" (d'Hérelle, 1918). Subsequently, bacteriophages active against many bacterial species were found in numerous habitats, and the concept of a single bacteriophage became indefensible. Burnet (1933) showed that enterobacterial phages were heterogeneous and could be subdivided by serology, particle size as determined by filtration, host range, and stability tests. Ruska (1943) pioneered the use of electron microscopy for virus classification in 1943, and he distinguished three morphological types of bacteriophages (an English translation of Ruska's paper is included in Appendix II). Holmes (1948) proposed that bacteriophages should be grouped in the order *Virales,* and that they should constitute the suborder *Phagineae* containing a single family (*Phagaceae*) and one genus, *Phagus.* The classification called for 46 species (with Latin or Greek names) defined by host range, plaque and particle size, and resistance to urea and heat. However, the proposed scheme was not accepted by the scientific community, and the notion that all viruses should be classified by their physicochemical properties gained increased recognition. In 1962, Lwoff, Horne, and Tournier proposed a classification scheme based on the nature of the virus' nucleic acid (DNA or RNA), capsid shape, presence or absence of an envelope, and number of capsomers. The scheme, later known as the LHT system, included phage ϕX174 and two tailed phages. In 1965, it was adopted by the Provisional Committee on Nomenclature of Viruses (PCNV) and was expanded to include ssRNA phages and filamentous phages. In addition, tailed phages were given order rank and the name *Urovirales* (PCNV, 1965). The PCNV system was attacked by virologists who opposed its hierarchical structure and Latin names (Gibbs et al., 1966); they proposed that viruses should have two names, an English vernacular and a "*cryptogram*" containing four pairs of characters, namely:

Nucleic acid (NA) type	:	NA molecular weight	:	Particle outline	:	Host
Strandedness		NA percent		Nucleocapsid shape		Vector

Accordingly, the "cryptogram" for phage T4 would read D/2: 169/40 : X/X : B/0, where D signifies DNA, X indicates a complex shape, and 0 means that no vector is known. This type of name was clearly unpronounceable, and the idea of a cryptogram was soon abandoned. Almost simultaneously, Bradley (1967) proposed a phage classification scheme based on nucleic acid type and gross morphology. His approach grouped phages into six basic types (A to F) consisting of (i) phages with contractile, long noncontractile, and short tails, (ii) cubic phages with DNA or RNA, and (iii) filamentous phages. Tikhonenko (1968) subsequently proposed a similar scheme, and Bradley's scheme was later expanded to include various cubic, filamentous, and pleomorphic phages of recent discovery (Ackermann and Eisenstark, 1974).

4.1.2. The ICTV

In 1966, the PCNV became the International Committee for Nomenclature of Viruses (ICNV) and a Bacterial Virus Subcommittee was created. The ICNV became the International Committee on Taxonomy of Viruses (ICTV) in 1973 (Matthews, 1983). Although the ICTV did not adopt the old LHT scheme, it conserved most of its taxa. At a meeting in Mexico City in 1970, six phage genera were approved, corresponding to T-even phages and groups exemplified by phages λ, PM2, ϕX174, MS2, and fd (Wildy, 1971). Subsequently, the ICTV met at every International Congress of Virology and issued six reports updating virus classification.

New families or genera were created when the need arose. The genera approved during the Mexico meeting were upgraded to families, and seven new phage families were added over the years. A notable development was the adoption of the *polythetic* species concept, meaning that a species is defined by set of properties, some of which may be absent in a given member (Van Regenmortel, 2000). Finally, during the 1998 Virology Congress in Sydney, tailed phages received order rank and the name *Caudovirales* (Ackermann, 1999), and 15 genera of tailed phages were recognized (Ackermann, 1999; Maniloff and Ackermann, 1998, 2000). The present edifice of viral taxonomy includes 3 orders, 61 families, 214 genera, and more than 3600 species (Van Regenmortel et al., 2000). In addition, it has 20 "floating" genera, mostly of plant viruses, that cannot be related to other taxa because of lack of data. Although emphasis is on the nature of nucleic acids, particle structure, and nucleotide or amino acid sequences, other kinds of information may be used for classification. The ICTV system is neither hierarchical nor phylogenetic because taxonomical structures (orders and phyla) above the family level have not yet been developed for most viruses (Van Regenmortel et al., 2000). The ICTV is essentially concerned with classifying viruses into high taxa (i.e., orders, families, and genera). The names of the orders, families, and genera generally reflect characteristic properties, are typically derived from Latin or Greek roots, and end in *-virales, -viridae,* and *-virus,* respectively. Most genera of polyhedral, filamentous, and pleomorphic phages have latinized names.

4.1.3. Purpose of Phage Classification

The classification of bacteriophages has eminently practical purposes. For example, it is impossible to memorize the properties of some 5000 individual phages, and it is pointless to list them. The purpose of classification is to condense data, to summarize and categorize information, and to simplify complicated situations by establishing groups. An intended consequence of phage classification is their precise and simple identification. Also, classification should, ideally, explain evolutionary relationships among phages. On a more specific level, classification is needed for (i) teaching, (ii) identification of novel phages, (iii) detecting relationships among phages, (iv) maintaining phage databases and collections, (v) identification of phages with therapeutic and industrial applications, and (vi) identification of harmful phages in biotechnology and fermentation industry for control and eradication purposes. Furthermore, classification is a prerequisite for the newest branch of phage research, comparative genomics, because it facilitates conclusions above the level of individual viruses. As an added benefit, classification helps generate useful nomenclature. For example, "myovirus T4" is much shorter—and more practical to use—for identification than "phage T4 with a long, contractile tail."

4.2. CURRENT PHAGE CLASSIFICATION

Bacteriophages infect Eubacteria and Archaea; thus, they may also be defined as "viruses of prokaryotes." They infect more than 140 bacterial genera (Ackermann, 2001), including those consisting of (i) aerobes and anaerobes, (ii) exospore- and endospore-formers, (iii) cyanobacteria, spirochetes, mycoplasmas, and chlamydias, (iv) budding, gliding, ramified, stalked, and sheathed bacteria, (v) extreme halophiles and methanogens, and (vi) hyperthermophilic Archaea growing at 100°C. Coliphage T7-like particles have also been found in bacterial endosymbionts of paramecia and insects. Reports of tailed phages in eukaryotes, such as in cultures of the green alga *Chlorella* and the mold *Penicillium,* have not been confirmed, and they probably describe laboratory contaminants (Ackermann and DuBow, 1987).

Phages are extremely heterogeneous in their structural, physicochemical, and biological properties, which suggests that they are polyphyletic in origin. Virions are tailed, polyhedral, filamentous, and pleomorphic (Fig. 4.1), and phage families may be listed in this order for convenience. The vast majority contain dsDNA; however, ssDNA, ssRNA, or dsRNA are found in small groups of phages. All DNA phages contain a single molecule of DNA. Several types of phages have lipid-containing envelopes or internal vesicles. More than 5000 bacteriophages have been examined by electron microscopy (Ackermann, 2001). They are, by far, the largest category of viruses studied by that method, which indicates the strong efforts invested in characterizing phages. Bacteriophages are classified into 1 order, 13 families, and 31 genera. Families are chiefly defined by the nature of phage nucleic acid and overall virion morphology. Although approximately 40 criteria are used for classification, there are no universal criteria for genera and species.

4.2.1. Tailed Phages (Order *Caudovirales*)

Tailed phages (Ackermann, 1999) constitute the largest and most widespread group of bacterial viruses; e.g., at least 4950 tailed phages have been observed in the electron microscope (Ackermann, 2001). They may also be the oldest viruses, dating back perhaps 3.5 billion years—which is the approximate age of the oldest known microbial fossils (Schopf, 1992; Ackermann, 1999). Typical tailed phages infect both Eubacteria and Archaea (Table 4.5), which suggests that their appearance antedates the separation of those bacterial kingdoms (Ackermann and DuBow, 1987; Hendrix, 1999). Virions consist of a protein shell and linear dsDNA only. Phage particles are not enveloped, and they are said to have a "binary" symmetry (Lwoff et al., 1962), since their heads (capsids) have cubic symmetry and their tails are helical. Their heads are icosahedra, either regular (85% of tailed phages) or prolate. Capsid proteins are organized into capsomers (electron microscopically visible assemblies of 5 or 6 protein subunits), but capsomers are difficult to observe and capsids usually appear smooth. Phage tails are true helices or consist of stacked disks, and they usually

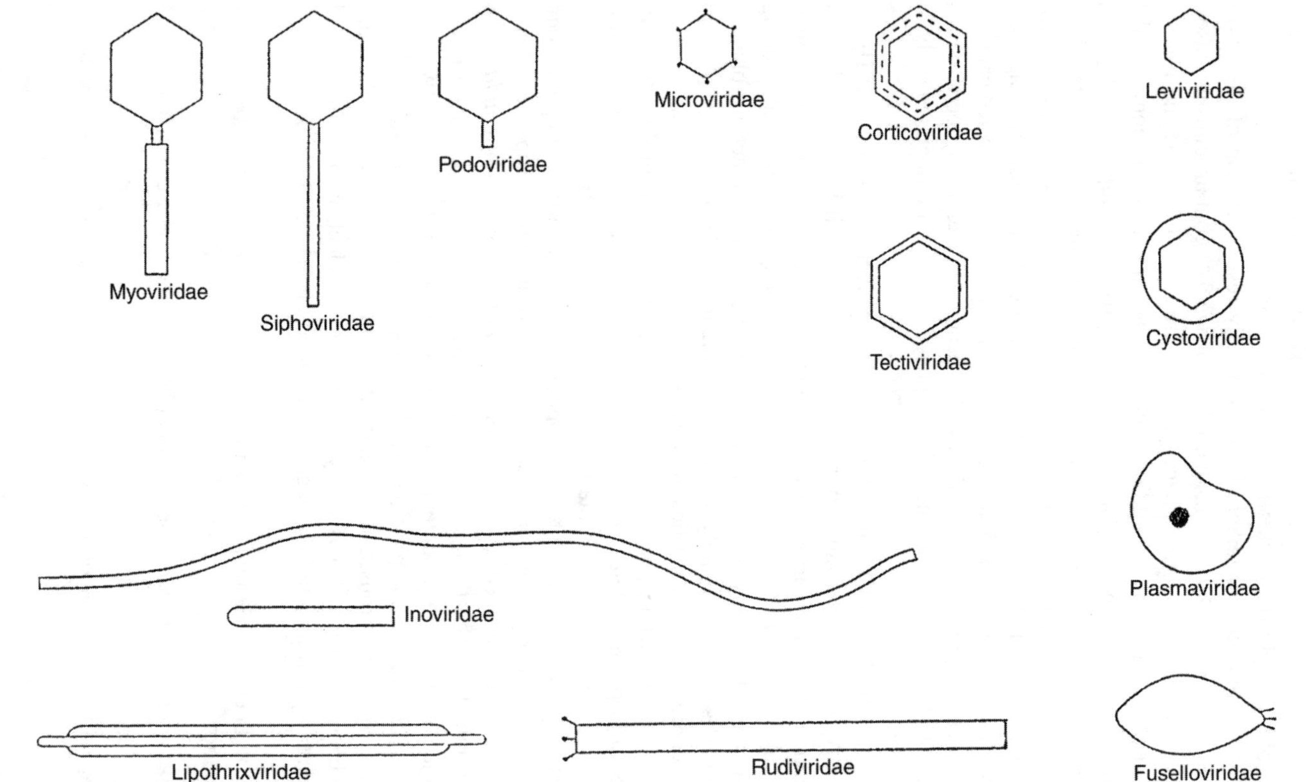

FIGURE 4.1. Basic bacteriophage morphotypes (Modified from Ackermann, H.-W. and DuBow, M.S., in *Viruses of Prokaryotes*, Vol. 1, CRC Press, Boca Raton, 1987, 16; with permission).

possess terminal adsorption structures such as base plates, spikes, or fibers. Tail-like structures occur in a few other viruses, particularly in tectiviruses (*infra*) and some algal viruses, but they are inconstant and are clearly different from the permanent, regular tails of tailed phages (Ackermann, 1999).

The DNA composition of tailed phages generally resembles that of their host bacteria. However, some DNAs (e.g., that of coliphage T4) contain unusual bases, such as 5-hydroxymethylcytosine. Phage genomes are large, complex, and usually organized into interchangeable building blocks or *modules*. The genome of the largest bacteriophage known, *Bacillus* phage G, comprises 498 kbp and 684 genes, and it is larger than the genome of some mycoplasmas (Pedulla et al., 2003). As a rule, genes for related functions cluster together. Replicating DNA tends to form large, branched intermediates or *concatemers,* which are much longer than a single genome. The DNA is then cut into unit lengths and inserted into preformed capsids. Virions are assembled via separate pathways for heads, tails, and tail fibers, which are added as a last step of maturation, and new phage particles are liberated by lysis of the host cells.

As explained further in Chapters 3 and 7, tailed phages can be virulent (lytic) or temperate. Virulent phages multiply rapidly and destroy their hosts, which burst during liberation of progeny virions. However, temperate phages are able to establish a condition called *lysogeny* by going into a latent or *prophage* state and reproducing as their hosts reproduce, only occasionally generating a burst of infectious phages. Temperate phages (which are likely to represent more than 50% of all tailed phages) are able to confer novel properties to infected bacteria, a phenomenon called *lysogenic conversion*. Latent phage genomes or prophages persist in their hosts in an integrated or a plasmid state. Both forms of latency exist elsewhere in the viral world, but are infrequent in eukaryote viruses (Ackermann, 1999).

Tailed phages seem to constitute a monophyletic evolutionary group possessing clearly related morphologic, physicochemical, and physiological properties; therefore, they have been classified as a single order, *Caudovirales* (Latin *cauda,* tail). At the same time, however, their properties are extremely varied; e.g., their dimensions and fine structure, DNA content and composition, nature of constitutive proteins, serology, host range, and physiology. Thus, tailed phages are the most diverse, as well as the most numerous and widespread, of all viruses. Based on their tail structure, a handy criterion that is easily determined via electron microscopy and reflects assembly pathways, tailed phages are divided into three families:

Myoviridae, with contractile tails consisting of a sheath and a central tube (at least 1250 observations, ca. 25% of tailed phages) (Ackermann, 2001).

Siphoviridae, with long, noncontractile tails (3000 observations, ca. 61% of tailed phages).

Podoviridae, with short, noncontractile tails (700 observations, ca. 14% of tailed phages).

The three families show considerable overlap in their physicochemical properties and cannot be differentiated on that basis. However, the capsids of myoviruses tend to be larger and to contain more DNA than those of the other two families (Ackermann, 1999). *Siphoviridae* and *Podoviridae* seem to be closely related, and the validity of the family *Podoviridae* may be questioned (*infra*) because the main difference between the two families, tail length, may be due to the presence or absence of a tail length ruler molecule. Arguments in favor of its validity are that podoviruses apparently have no separate pathway for tail assembly, and that there are no intermediate lengths between long and short tails. Also, as a practical consideration, siphoviruses and podoviruses are readily differentiated by electron microscopy, and maintenance of the *Podoviridae* family eliminates ca. 700 phages out of a collection of nearly 4000 viruses.

Fifteen genera have been defined among tailed phages (Table 4.5), based on criteria related to genomic structure and replication. The criteria include the presence or absence of *cos* or *pac* sites, terminal redundancy and circular permutation, DNA or RNA polymerases, unusual bases, nucleotide sequence, and concatemer formation (Maniloff and Ackermann, 1998, 2000). The 15 genera correspond to phage groups with vernacular names derived from those of the type species (e.g., T4 or λ), which have not been internationally approved. Tailed phage classification into genera is still beginning, and more genera are likely to be defined in the future. The present genera may be considered as islands in an ocean, around which unclassified phages are expected to aggregate. In addition, approximately 250 species are presently recognizable, mostly on the basis of morphology, DNA-DNA hybridization and nucleotide sequencing, and serology.

4.2.2. Polyhedral, Filamentous, and Pleomorphic Phages

The tailless phages only include about 190 known viruses, corresponding to less than 4% of currently recognized bacterial viruses (Ackermann, 2001). They are classified into 10 small families which sometimes include only a single genus and species. Families differ in fundamental properties and seem to constitute many independent phylogenetic groups. The phages, which are enveloped and unenveloped, are of three types: polyhedral phages that are icosahedra or related bodies with cubic symmetry, filamentous phages with helical symmetry, and a few types of variable shape without obvious symmetry axes. Some types contain lipids, which results in a low buoyant density and ether- and chloroform-sensitivity of virions. Most families have narrow host ranges. Many or all phages of the *Fuselloviridae, Inoviridae, Lipothrixviridae,* and *Plasmaviridae* families are *lysogenic*. The mode of lysogeny used by the lipothrixviruses is unknown. As with many tailed phages, lysogeny in the *Fuselloviridae* and *Plasmaviridae* is integrase-mediated (Ackermann, 1999). Phages in the genus *Inovirus* integrate into the host genome by means of host recombinases (Huber and Waldor, 2000). Lysogeny is therefore not limited to tailed phages.

4.2.2.1. Polyhedral DNA Phages

Microviridae (ssDNA)

Virions are unenveloped icosahedra (ca. 27 nm in diameter) with 12 capsomers and contain circular ssDNA. They infect evolutionary very diverse hosts (enterobacteria,

Bdellovibrio, Chlamydia, and *Spiroplasma*) and are classified into four genera (Table 4.1). Their DNA replicates, via the "rolling-circle" model, as a double-stranded replicative form (RF). Genome sequencing data have been used (Brentlinger et al., 2002) to divide the family into two subfamilies, one for phages propagating in proteobacteria and the other for *Chlamydia* and *Spiroplasma* phages.

Corticoviridae (dsDNA)

The only member of this family, phage PM2, was isolated from seawater and was the first bacteriophage in which the presence of lipids was demonstrated. A recent study (Kivelä et al., 2002) indicates that it has, like tectiviruses, a protein capsid with an internal phospholipoprotein vesicle. Two similar, poorly characterized phages were isolated from seawater (Ackermann, 2001).

Tectiviridae (dsDNA)

The viral particle has a rigid protein capsid containing a thick, flexible, lipoprotein vesicle. In addition, *Bacillus* tectiviruses have apical spikes. After the phages adsorb to their host bacteria, or after they are shaken with chloroform, the vesicle transforms itself into a tail-like tube (ca. 60 nm long) which serves as a nucleic acid ejection device. This is an interesting example of convergent evolution because the tube has the same function as the tails of tailed phages. Despite the family's small size, tectiviruses infect apparently unrelated bacteria (Table 4.5).

4.2.2.2. Polyhedral RNA Phages

Leviviridae (ssRNA)

The leviviruses are unenveloped and morphologically similar to polioviruses. Their genome consists of four partially overlapping genes, and their RNA acts as mRNA and is therefore positive-stranded. Most known leviviruses are plasmid-specific coliphages that adsorb to F pili (i.e., sex pili). They have been divided, by serology and genome structure, into two genera. Several unclassified leviviruses infect hosts other than enterobacteria (Table 4.4).

Cystoviridae (dsRNA)

This family has a single "official" member; however, eight related viruses have recently been found (Mindich et al., 1999). They are specific for the phytopathogenic bacterium *Pseudomonas syringae.* Cystoviruses have icosahedral capsids surrounded by lipid-containing envelopes, and they are unique among bacteriophages because they contain a dodecahedral RNA polymerase complex (Bamford et al., 1993) and three molecules of dsRNA. The capsids of infecting cystoviruses enter the space between the bacterial cell wall and the cytoplasmic membrane.

4.2.2.3. Filamentous Phages

Inoviridae (ssDNA)

The family consists of two genera with different particle morphologies and host ranges. The phage DNA replicates as a double-stranded form (RF) via a rolling-circle mechanism. This mode of replication seems to reflect the single-stranded

TABLE 4.1
Overview of Prokaryote Viruses

Shape	Nucleic acid	Family	Genera	Particulars	Derivation of Name	Example	Members
Tailed	dsDNA (L)	*Myoviridae*	6, see Table 5	Tail contractile	Greek *mys, myos,* "muscle"	T4	1,243
		Siphoviridae	6	Tail long, noncontractile	Greek *siphon,* "hollow tube"	λ	3,011
		Podoviridae	3	Tail short	Greek *pous,* podos, "foot"	T7	696
Polyhedral	ssDNA (C)	*Microviridae*	*Microvirus* *Bdellomicrovirus* *Chlamydiomicrovirus* *Spiromicrovirus*	Conspicuous capsomers	Greek *mikros,* "small" *Bdellovibrio* *Chlamydia* *Spiroplasma*	ϕX174	40
	dsDNA (C, S)	*Corticoviridae*	*Corticovirus*	Double capsid, lipids	Latin *cortex,* "bark, crust"	PM2	3?
	dsDNA (L)	*Tectiviridae*	*Tectivirus*	Double capsid, pseudo-tail, lipids	Latin *tectus,* "covered"	PRD1	18
	ssRNA (L)	*Leviviridae*	*Levivirus* *Allolevirirus*		Latin *levis,* "light" Greek allos, "different"	MS2	39
	dsRNA (L, M)	*Cystoviridae*	*Cystovirus*	Envelope, lipids	Greek *kystis,* "bladder, sack"	ϕ6	1
Filamentous	ssDNA (C)	*Inoviridae*	*Inovirus*	Long filaments	Greek *is, inos,* "fiber"	fd	42
			Plectrovirus	Short rods	Latin *plectrum,* "small stick"	MV-L51	15
	dsDNA (L)	*Lipothrixviridae*	*Lipothrixvirus*	Envelope, lipids	Greek *lipos,* "fat"; *thrix,* "hair"	TTV1	6
	dsDNA (L)	*Rudiviridae*	*Rudivirus*	Stiff rods, no envelope, no lipids	Latin *rudis,* "small rod"	SIRV-1	2
Pleomorphic	dsDNA (C, S)	*Plasmaviridae*	*Plasmavirus*	Envelope, no capsid, lipids	Greek *plasma,* "shaped product"	L2	7?
	dsDNA (C, S)	*Fuselloviridae*	*Fusellovirus*	Lemon-shaped, envelope, lipids	Latin *fusellum,* "little spindle"	SSV1	7?
	dsDNA (C, S)	*Guttaviridae*	*Guttavirus*	Droplet-shaped	Latin *gutta,* "drop"	SNDV	1

Modified from Ackermann, H.-W. and DuBow, M.S., in *Viruses of Prokaryotes,* Vol. 1, CRC Press, Boca Raton, 1987, 16. With permission). C, circular; L, linear, M, multipartite; S, supercoiled. Numbers indicate phages examined by electron microscopy, excluding phage-like bacteriocins and known defective phages. Computed in October 2000; from Ackermann (2001).

TABLE 4.2
Dimensions and Physicochemical Properties of Prokaryote Viruses

Phage Group	Virion Capsid Size, nm	Tail Length, nm	Weight, Mr × 10^6	Buoyant Density, g	Lipids, %	%	Nucleic Acid MW, kb	GC, %
Tailed phages	67	153	100	1.49	—	46	79	48
Range	30–160	10–800	29–470	1.4–1.54		30–62	17–498[a]	27–72
Microviridae	27		6–7	1.36–1.41	—	26	4.4–6.1	44
Corticoviridae	60		49	1.28	13	14.3	9.0	43–44
Tectiviridae	63		70	1.29	15	14	15–16	51
Leviviridae	23		3.6–4.2	1.46	—	30	3.5–4.3	51
Cystoviridae	75–80		99	1.27	20	10	13.4	56
Inoviridae (*Inovirus*)	760–1950 × 7		12–34	1.30	—	6–21	5.8–7.3	40–60
(*Plectrovirus*)	85–250 × 7			1.37	—		4.5–8.3	
Lipothrixviridae	400–2400 × 20–40		33	1.25	22	3	16–42	33
Rudiviridae	780–900 × 23			1.26	—		33–36	25
Plasmaviridae	80				11		12	32
Fuselloviridae	85 × 55			1.24	10		15	40
Guttaviridae	110–180 × 70–95						20	

Modified from Ackermann, H.-W., Bacteriophage taxonomy in 1987, *Microbiol. Sci.*, 4, 214–218, 1987. With permission of Blackwell Scientific Publications Ltd., Oxford, England. Buoyant density is g/ml in CsCl. GC, guanine-cytosine content; kb, kilobases; Mr, relative molecular mass; MW, molecular weight; nm, nanometers; –, absent.

[a]Pedulla et al, 2003.

TABLE 4.3
Comparative Biological Properties of Prokaryote Viruses

Phage Group	Adsorption Site	Fixation Structure	Infection by	Assembly Site	Release	Host-Virus Relationship
Tailed phages	Cell wall, capsule, pili, flagella	Tail tip	DNA	Nucleoplasm and periphery	Lysis	Lytic or temperate
Microviridae	Cell wall	Spikes	DNA	Nucleoplasm	Lysis	Lytic
Corticoviridae	Cell wall	Spikes		Plasma membrane	Lysis	Lytic
Tectiviridae	Pili, cell wall	Spikes, "tail"	DNA	Nucleoplasm	Lysis	Lytic
Leviviridae	Pili	Apical protein	RNA	Cytoplasm	Lysis	Lytic
Cystoviridae	Pili, cell wall	Envelope	Capsid	Nucleoplasm	Lysis	Lytic
Inovirus genus	Pili	Virus tip	Virion	Plasma membrane	Extrusion	Carrier state or temperate
Plectrovirus genus	Plasma membrane	Virus tip	Virion?	Plasma membrane	Extrusion	Carrier state
Lipothrixviridae	Pili	Virus tip			Lytic	Temperate
Rudiviridae	Cell wall	Virus tip				Carrier state
Plasmaviridae	Plasma membrane	Envelope	DNA?	Plasma membrane	Budding	Temperate
Fuselloviridae		Apical spikes		Plasma membrane	Extrusion	Temperate or carrier state
Guttaviridae		Apical fibers				Carrier state

Modified from Ackermann, H.-W. and DuBow, M.S., in *Viruses of Prokaryotes,* Vol. 1, CRC Press, Boca Raton, 1987, 75. For data on archaeal viruses, see Prangishvili et al. (2001).

TABLE 4.4
Host Range of Prokaryote Viruses

Phage Group	Bacterial Group or Genus
Tailed phages	a. Eubacteria
	b. Euryarchaeota (extreme halophiles and methanogens)
Microviridae	Enterobacteria, *Bdellovibrio, Chlamydia, Spiroplasma*
Corticoviridae	*Alteromonas*
Tectiviridae	a. Enterobacteria, *Acinetobacter, Pseudomonas, Thermus, Vibrio*
	b. *Bacillus, Alicyclobacillus*
Leviviridae	Enterobacteria, *Acinetobacter, Caulobacter, Pseudomonas*
Cystoviridae	*Pseudomonas*
Inoviridae: Inovirus	Enterobacteria, *Pseudomonas, Thermus, Vibrio, Xanthomonas*
Plectrovirus	*Acholeplasma, Spiroplasma*
Plasmaviridae	*Acholeplasma*
Lipothrixviridae	Crenarchaeota: *Acidianus, Sulfolobus, Thermoproteus*
Rudiviridae	Crenarchaeota: *Sulfolobus*
Fuselloviridae	a. Crenarchaeota: *Acidianus, Sulfolobus*
	b. Euryarchaeota: *Methanococcus, Pyrococcus?*
Guttaviridae	Crenarchaeota: *Sulfolobus*

Modified from Ackermann, H.-W., Bacteriophage observations and evolution, *Res. Microbiol.*, 154, 245–251, 2003. With permission of Editions Elsevier, Paris.

nature of the phage DNA rather than a common origin of the two genera. The 42 phages of the genus *Inovirus,* which have been classified into 29 species (Day and Maniloff, 2000), are long, rigid or flexible filaments whose length reflects the size of their genomes. They infect enterobacteria and their relatives, the genus *Thermus,* clostridia, and propionibacteria (Table 4.5). Virions are sensitive to chloroform and sonication, and they are very resistant to heat. The genus *Plectrovirus,* whose virions are short, straight rods, includes 15 members that infect only mycoplasmas. Progeny inoviruses are extruded from host cells, which survive and may produce phages indefinitely.

Lipothrixviridae (dsDNA)

This family includes six viruses of extremely thermophilic archaebacteria. Viral particles are characterized by the combination of a rod-like shape, a lipoprotein envelope, and a nucleosome-like core. In contrast to inoviruses, the progeny virions are released by lysis. Lipothrixviruses may be taxonomically heterogeneous and form two subfamilies (Arnold et al., 2000b).

Rudiviridae (dsDNA)

This family includes two viruses of different lengths, which were isolated from the thermophilic archaeon *Sulfolobus*. Particles are straight, rigid rods without envelopes, have conspicuous fixation structures at one end, and resemble the tobacco

TABLE 4.5
Tailed Phage Genera

Family	Genus	Type Species	Species	Members	Principal Hosts
Myoviridae	T4 like viruses	T4	7	47 (+100)	Enterobacteria
	P1-like viruses	P1	3	12	Enterobacteria
	P2-like viruses	P2	2	16	Enterobacteria
	Mu-like viruses	Mu	1	2	Enterobacteria
	SPO1-like viruses	SPO1	1	13	*Bacillus*
	ΦH-like viruses	ΦH	1	2	*Halobacterium*[a]
Siphoviridae	λ-like viruses	λ	1	7	Enterobacteria
	T1-like viruses	T1	1	11 (+50)	Enterobacteria
	T5-likeviruses	T5	1	5 (+20)	Enterobacteria
	L5-like viruses	L5	1	4 (+15)	*Mycobacterium*
	c2-like viruses	c2	1	5 (+200)	*Lactococcus*
	ψM1-like viruses	ψM1	1	3	*Methanobacterium*[a]
Podoviridae	T7-like viruses	T7	3	26	Enterobacteria
	P22-like viruses	P22	1	11	Enterobacteria
	ϕ29-like viruses	ϕ29	4	12	*Bacillus*

Parentheses indicate approximative numbers of poorly characterized isolates that may or may not represent independent species.

[a]Archaea.

mosaic virus. Rudiviruses and lipothrixviruses have genome homologies, which suggests that they form a superfamily (Peng et al., 2001).

4.2.2.4. Pleomorphic Phages

Plasmaviridae (dsDNA)

The family has only one definite member, *Acholeplasma* virus MVL2 or L2. Other plasmavirus-like isolates have been found, but they cannot be classified with certainty at the present time. Virions do not have capsids, but they possess an envelope and a dense nucleoprotein granule. Like enveloped vertebrate viruses, plasmaviruses infect their hosts by fusion of the viral envelope with the mycoplasmal cell membrane, and progeny viruses are released by budding.

Fuselloviridae (dsDNA)

Fuselloviruses are spindle-shaped with short spikes at one end. The best-studied member of the family, SSV1, is harbored in the archaean *Sulfolobus shibatae* as a plasmid and as an integrated prophage. It has not been propagated because of the absence of a suitable host, but it is inducible by UV light and mitomycin C. The coat consists of two hydrophobic proteins and host lipids, and is disrupted by chloroform. Fuselloviruses are liberated by extrusion from the host cell.

TABLE 4.6
Phages and Eukaryote Viruses

Phages	Eukaryote Viruses	Similarities	References
Tailed phages	*Herpesviridae*	Large capsid, high DNA molecular weight	Ackermann, 1999
		Replication: infecting DNA circularizes, three waves of transcription, rolling circle replication, formation of concatemers	
		Capsid maturation: prohead maturation by proteolytic cleavage, scaffolding protein, DNA enters capsid through unique opening, headful packaging	
		Latency state as plasmid or integrated DNA	
		Order of capsid and capsid assembly genes; sequence of DNA packaging terminase proteins?	Hendrix, 1999
		Portal protein	Newcomb et al., 2001
Microviridae	*Parvoviridae, Picornaviridae*	Particular type of β-barrel in capsid proteins	McKenna et al., 1993
Tectiviridae	*Adeno-, Asfar-, Como-, Irido-, Phycodna-, Picornaviridae*	Capsomer structure and lattice	Nandhagopal et al., 2002
	Adenoviridae	DNA has inverted terminal repeats with proteins at 5' termini; hexons are arranged on a pseudo-T=25 lattice	Bamford et al., 2002
		Protein-primed type B DNA polymerase	Bamford et al., 1995
Leviviridae	*Flaviviridae, Tombusviridae*	Type of RNA polymerase	Koonin, 1990
Cystoviridae	*Reoviridae, Birnaviridae*	Capsid structure, lifestyle	Bamford et al., 2002
Rudiviridae	*Asfar-, Pox-, Phycodnaviridae*	Nucleotide sequence relationships, genome organization, mechanism of DNA replication	Peng et al., 2001

Sulfolobus SNDV-like viruses (dsDNA)

This awkward name designates a virus-like particle, SDNV, found in a *Sulfolobus* isolate from a solfatara in New Zealand. The particle is droplet-shaped and has a unique beehive-like structure with a "beard" of fibers at its pointed end (Arnold et al., 2000a). The terms *Guttaviridae* and *Guttavirus* have been proposed as family and genus names, respectively. SNDV may be reclassified in the future as a member of the fuselloviruses.

4.3. TAXONOMIC PROBLEMS

4.3.1. New Taxa and New Techniques

Since 1959, when negative staining was introduced as a method in electron microscopy, new bacteriophages have been described at a rate of about 125 per year (Ackermann, 2001). This trend is likely to continue; therefore, any taxonomic system for phages must be sufficiently flexible to accommodate such an influx of viruses. Several taxonomic changes, already hinted at above, lie ahead. For example, a rearrangement of the *Microviridae* family is likely, four families of cubic and filamentous phages (*Corticoviridae* and *Tectiviridae; Lipothrixviridae* and *Rudiviridae*) may be fused together or rearranged in supergroups, and several types of archaeal viruses are waiting for classification. The virions of one of these types have roughly the shape of an arrow with heads of 130–150 × 56–70 nm and tails 100–620 nm in length. Though repeatedly found in hypersaline habitats and volcanic springs (Rachel et al., 2002), they have not been propagated. In addition, archaeal viruses of novel morphology, which may represent three novel families, have recently been isolated from volcanic hot springs (Prangishvili, personal communication).

Genome sequencing, greatly stimulated by the introduction of rapid sequencers, has ushered in a revolution called the "Age of Genomics." Prior to that age, the genome of phage λ was published in 1983, and remained for years the only fully-characterized genome of a tailed phage. However, genome sequencing has recently generated a series of papers of fundamental importance to the classification of phages, some of which are even aimed at viruses in general (Lawrence et al., 2002). A recent, rather optimistic forecast predicts thousands of sequenced phage genomes in a few years (Brüssow and Hendrix, 2002). Genome sequencing is confirming or disconfirming existing taxa and revolutionizing our ideas about virus evolution, but it has also introduced new taxonomic problems caused by (i) using a single phage property, the genome sequence, for classification, and (ii) dependency on incomplete databases.

Classification by a single criterion is evident in the recent proposal to disregard phenetic properties as unreliable and to rely entirely on genome sequences (Lawrence et al., 2002). This approach overlooks the fact that all phenetic properties are ultimately an expression of a phage's genome, and that many characteristics of phage DNA (e.g., structure, composition, replication strategy), virion (e.g., size, fine structure, assembly, mass, composition), and constitutive proteins (e.g., antigenicity) are not evident in gene or nucleotide sequences. The danger of relying on a single criterion became apparent about 15 years ago, when plant virologists tried to classify all viruses, including tailed phages, by their DNA or RNA polymerases (Rybicki, 1990; Ward, 1993). When applied to tailed phages, the classification attempt yielded two phyla complete with subphyla, classes, orders, and six families for only seven phages (Ward, 1993). However, this classification is problematic because many tailed phages do not have DNA polymerase and very rarely have RNA polymerase.

The dependency of comparative genomics on databases is a potential problem because a database is only as good as the data therein. Unfortunately, the currently available genomic databases include (i) numerous phages of which nothing is known except a name and a DNA base or protein sequence, and (ii) a series of prophages, defective or not, which are not identified as such and are just labeled "phages." Database administrators are not phage taxonomists and they cannot be expected to attribute novel phages to specific phage families or genera. Thus, the current situation is rather detrimental to the comparison of phage genomes.

4.3.2. The λ-P22 Controversy

In the present ICTV classification of viruses, coliphage λ and *Salmonella* phage P22 are assigned to different families, the *Siphoviridae* and *Podoviridae,* respectively. However, both phages have been proposed to be closely related because they form hybrids, may exchange genes in the process, and have genomic maps with a similar gene order. This relationship is considered a major argument against the ICTV phage classification; indeed, phage P22 has often been called a "lambdoid" phage (Casjens et al., 1992).

Botstein and Herskowitz (1974) created viable hybrids between λ and P22, which resulted in λ-like phages containing the P22 *immC* region. Susskind and Botstein (1978) also observed that λ and P22, though different in many respects (morphology, morphogenesis, DNA metabolism and packaging, antirepressor, adsorption, injection), had similar genome organization and lysogeny regulation genes. They concluded that P22 is a member of the λ family and proposed the modular theory of phage evolution. In essence, the theory proposes that certain phages are constructed of exchangeable modules (genes or gene blocks) and are able to evolve by exchange of the modules.

There are several arguments against the above-described idea of a close relationship between phages λ and P22:

1. The λ-P22 hybrids are not natural. They were derived from a λ-infected *E. coli* strain mated with *S. typhimurium,* into which a P22-carrying episome had been inserted (Botstein and Herskowitz, 1974).
2. Phage hybridization is not unusual. Both λ and P22 form viable and nonviable hybrids with many other, morphologically unrelated phages of enterobacteria and *Bacillus* (λ with Mu, P1, P4, T4 and SPP1; P22 with P1, Fels1, Fels2 and P221). In addition, there exist hybrids between myovirus phages Mu and P1 (reviewed in Ackermann and DuBow, 1987).
3. The exchanged gene region is small and limited to the *immC* region (Susskind and Botstein, 1978).

4. As shown by whole-genome DNA-DNA hybridization, λ and P22 have only 13.5% total genome homology (Skalka and Hanson, 1972).
5. The P22 genome is a mosaic of genes occurring in at least 17 different phages, even including a *Lactococcus* phage and an inovirus. Also, their genomic similarities are limited to a small number of specific genes (*c1-3, ead, nin, ral, 13, 23*), most of which have very high (up to 96%) identity with λ genes (Vander Byl and Kropinski, 2000). This indicates recent acquisition of λ-type genes by an ancestor of P22.
6. Genome maps with a λ-like gene order are found in many phages of gram-positive and gram-negative bacteria, even in an archaeal virus, *Methanobacterium* phage ΨM2 (Brüssow and Desiere, 2001; Desiere et al., 2002). These observations indicate that gene order is an ancestral property (Hendrix et al., 1999), conserved over eons, which may be compared to the backbone of vertebrates.

In conclusion, gene acquisition alone is no proof that the ancestors of phages λ and P22 were related (it only indicates that their descendants are or *became* related), phage P22 appears to be a chimaera, and the existence of limited genomic relationships between phages λ and P22 is an unsatisfactory argument against the ICTV phage classification.

4.3.3 Sequence-Based Taxonomy

4.3.3.1. The Phage Proteomic Tree

A total of 105 completely sequenced phage genomes from GenBank (including genomes of ssRNA phages, microviruses, and inoviruses) were recently analyzed by distance- and tree-generating computer programs (PROTDIST, FITCH) (Rohwer and Edwards, 2002). In addition, it was attempted to find a gene common to all phages, analogous to 16S mRNA in bacteria, but no such gene was identified. However, a neighbor-joining tree was produced that featured several ICTV phage families and phage groups already established in the literature, but also comprised some improbable results. For example, tectivirus PRD1 was clustered with the ϕ29 phage group ("PZA-like podophage") of tailed phages. The stated reason was that both types of viruses shared the same protein-primed DNA replication machinery (i.e., they both utilized a type B DNA polymerase). However, the same type B DNA polymerase also occurs in adenoviruses (Bamford et al., 1995), and the other genes of ϕ29 and PRD1 are apparently not considered. This is a case of reliance on a single criterion, and duplicates earlier mistakes in the classification of phages by DNA polymerases (Ward, 1993).

4.3.3.2. Modular Classification

The concept of "modular classification" was derived from the theory of modular phage evolution (Susskind and Botstein, 1978), and it was developed in a series of recent publications (Hendrix et al., 1999; Brüssow and Hendrix, 2002; Hendrix, 2002; Lawrence et al., 2002). The concept postulates that tailed phages have access to a common gene pool and acquire genes or groups of genes by horizontal exchange. This results in mosaic genomes and a web of relationships among phages (Hendrix, 2002). Thus, according to the modular classification concept, it is impossible to represent the history of tailed phages by a simple branching phylogeny (Brüssow and Hendrix, 2002). To solve this problem, it was proposed (Lawrence et al., 2002) to abandon the ICTV edifice of orders, families, genera, and species, and to classify all viruses as reticulate groups defined by *modi,* a *modus* being a characteristic module or phenetic property. Modi may overlap, meaning that they occur repeatedly in different viruses; therefore, the same virus may belong to several modi.

The concept is intriguing; however, it also has some important limitations. Perhaps the most important problem with this proposed classification scheme is its limited scope. The system was derived from a study of the genomes of phage λ and other small temperate phages of *E. coli* and *Shigella*. Lytic tailed phages, large temperate phages, and eukaryotic DNA viruses were neither discussed nor classified. For example, the system does not include coliphages T4, T7, and P1, *Bacillus* phage SPβ, and *Pseudomonas* phage ϕKZ. Moreover, it apparently does not account for the enormous amount of horizontal gene transfer that is evident in virus genomes and links phages to the rest of the living world. Indeed, large phages and eukaryotic dsDNA viruses have mosaic genomes too and some of their genes have extraordinarily wide host ranges. A few examples will include:

1. T4 genes occur in bacteria, yeasts, protozoa, and vertebrates. For example, T4-like lysozyme and thymidylate synthase occur in humans, making T4 a "relative" of *Homo sapiens* (Ackermann, 1999; Bernstein and Bernstein, 1989).
2. *P. aeruginosa* phage ϕKZ has a genome of 280 kbp with 305 ORFs. Relatives of ϕKZ genes occur in 22 bacterial species, 6 phages (including T4), protozoa, nematode worms, eukaryotic viruses, and insects (*Drosophila*) (Mesyanzhinov et al., 2002).
3. EsV-1, a phycodnavirus of the brown alga *Ectocarpus,* has a genome of 336 kbp with 231 major genes. It shares genes with 85 other entities including vertebrates (humans, rats, cows), plants, worms, insects (*Drosophila*), protozoa, fungi, 36 bacterial species, 7 phages, and 9 eukaryote viruses (Delaroque et al., 2001).

The complexity of large viruses would ensure a prodigious number of modules and cross-links, which would increase exponentially with the number of viruses under study, and may create a system of astronomic complexity. An example of such cross-links has been described by Hendrix et al. (1999). Such a situation is not new in virus classification. It previously arose when the genomes of positive-strand ssRNA viruses of plants and vertebrates were found to consist of similar genes (e.g., RNA polymerase, protease, and capsid proteins) that occurred repeatedly in apparently distinct virus groups (Goldbach and Wellink, 1988). Although these genomes have only three to six genes, the ICTV realized that a genome-based classification of RNA viruses would lead to a classification of proteins rather than viruses, and it abandoned the approach.

4.4. LINKS BETWEEN PHAGES AND OTHER VIRUSES

The present ICTV classification edifice of 61 families includes only three orders (*Caudovirales, Mononegavirales, Nidovirales*); i.e., the vast majority of virus families and about 20 "floating genera" exist in apparent isolation. Thus, there is much incentive to look for interviral relationships. Recent advances indicate that several bacteriophage groups have structural, genomic, or biological similarities to viruses of eukaryotes; e.g., viruses of vertebrates, insects (iridoviruses), plants (comoviruses and tombusviruses), fungi, algae, and protozoa (phycodnaviruses, totiviruses). For example, many properties of herpesviruses resemble those of the tailed phages. None of them is found in *all* tailed phages or is, considered alone, direct proof of a relationship. It is only when tailed phages and herpesviruses are compared with the rest of the viral world that their similarities are striking. For example, herpesviruses resemble tailed phages in their modes of replication and morphogenesis, and both are able to become latent by integration and as plasmids. Also, herpesviruses have portal proteins corresponding to the head-tail connector of tailed phages (Newcomb et al., 2001).

In several cases, similarities between phages and other viruses are limited to morphological principles; i.e., although the virions may look different, they are made of similar building blocks. For example, the similarity of the major capsid F protein of phage ϕX174 to capsid proteins of a human rhinovirus and a canine parvovirus suggests that all these proteins share a common ancestral structure (McKenna et al., 1993). Similarly, it appears that the capsid proteins of adenoviruses and tectivirus phages evolved from a common ancestor that had already acquired a particular sixfold organization (Bamford et al., 2002).

A comparison of RNA polymerases suggests that leviviruses may be distantly related to plant tombusviruses and to togaviruses of vertebrates and insects (Koonin, 1991), but this idea has been contested (de A. Zanotto et al., 1996). Curiously, although the capsids of leviviruses and many vertebrate and plant ssRNA viruses have an identical structure, the coat protein of levivirus MS2 differs from the coat proteins of other known RNA viruses (Valegård et al., 1990). Thus, the relationship between leviviruses and other ssRNA viruses is obscure. On the other hand, there are definite structural and biological similarities between cystoviruses and the *Reoviridae* and *Birnaviridae*. Viruses of all three families have segmented genomes with 2 to 12 dsRNA molecules, and they contain RNA polymerase. In addition, their virions have one or more concentric, structurally similar capsids. In addition, based on our current state of knowledge, all dsRNA viruses have the same peculiar lifestyle: (i) Viral capsids enter the cell, (ii) ssRNA copies are synthesized from each gene, leave the capsid and act as mRNA for synthesis of procapsids, RNA polymerase, and ssRNA genomic precursors, and (iii) the precursors enter the procapsids and are used to synthesize new genomic dsRNA within the capsid. These particulars strongly suggest that dsRNA viruses have a common ancestry (Bamford, 2000; Bamford et al., 2002).

Finally, the genomes of the archaeal rudiviruses share multiple features with those of large eukaryal viruses; i.e., poxviruses, the African Swine Fever virus (ASFV, *Asfarviridae*), and the *Chlorella* virus PBCV-1 (*Phycodnaviridae*). Similarities include covalently closed DNA ends, long inverted terminal repeats, terminal location of hotspots for recombination, and partial nucleotide sequence identity in the 20%–57% range. In addition, 14 of 54 rudivirus ORFs show sequence similarity to poxvirus genes; thus, the replication machineries of rudiviruses and poxviruses, ASFV and PBCV-1, are believed to have a common origin (Peng et al., 2001).

REFERENCES

Ackermann H.-W., 1999, Tailed bacteriophages. The order *Caudovirales. Adv. Virus Res.,* 51, 135–201.

Ackermann H.-W., 2001, Frequency of morphological phage descriptions in the year 2000, *Arch. Virol.,* 146, 843–857.

Ackermann, H.-W. and DuBow, M.S., 1987, *Viruses of Prokaryotes,* Vol. 1, *General Properties of Bacteriophages,* CRC Press, Boca Raton, pp. 2, 33, 183, 173.

Ackermann, H.-W. and Eisenstark, A., 1974, The present state of phage taxonomy, *Intervirology,* 3, 201–219.

Arnold, H.P., Ziese, U., and Zillig, W., 2000a, SNDV, a novel virus of the extremely thermophilic and acidophilic archaeon *Sulfolobus, Virology,* 272, 409–416

Arnold, H.P., Zillig, W., Ziese, U., Holz, I., Crosby, M., Utterback, T., Weidmann, J.F., Kristjanson, J.K., Klenk, H.P., Nelson, K.E., and Fraser, C.M., 2000b, A novel lipothrixvirus, SIFV, of the extremely thermophilic Crenarchaeon *Sulfolobus, Virology,* 267, 252–266.

Bamford, D.H., 2000. Virus structures: Those magnificent molecular machines, *Curr. Biol.,* 10, R558–R561.

Bamford, J.K.H., Bamford, D.H., Li, T., and Thomas, G.J., 1993, Structural studies of the enveloped dsRNA bacteriophage ϕ6 of *Pseudomonas syringae* by Raman spectroscopy. II. Nucleocapsid structure and polymerase complex, *J. Mol. Biol.,* 230, 473–482.

Bamford, D.H., Burnett, R.M., and Stuart, D.I., 2002, Evolution of viral structure, *Theoret. Popul. Biol.,* 61, 461–470.

Bamford, D.H., Caldentey, J., and Bamford, J.K.H., 1995, Bacteriophage PRD1: A broad host range dsDNA tectivirus with an internal membrane, *Adv. Virus Res.,* 45, 281–319.

Bernstein, H. and Bernstein, C., 1989, Bacteriophage T4 genetic homologies with bacteria and eukaryotes, *J. Bacteriol.,* 171, 2265–2270.

Botstein, D. and Herskowitz, I., 1974, Properties of hybrids between *Salmonella* phage P22 and coliphage λ, *Nature (London),* 251, 584–589.

Bradley, D.E., 1967, Ultrastructure of bacteriophages and bacteriocins, *Bacteriol. Rev.,* 31, 230–314.

Brentlinger, K.L., Hafenstein, S., Novak, C.R., Fane, B.A., Borgon, R., McKenna, R., and Agbandje-McKenna, M., 2002, *Microviridae,* a family divided: isolation, characterization, and genome sequence of ϕMH2K, a bacteriophage of the obligate intracellular parasitic bacterium *Bdellovibrio bacteriovorus, J. Bacteriol.,* 184, 1089–1094.

Brüssow, H. and Desiere, F., 2001, Comparative phage genomics and the evolution of *Siphoviridae*: Insights from dairy phages, *Mol. Microbiol.,* 39, 213–222.

Brüssow, H. and Hendrix, R.W., 2002, Phage genomics: Small is beautiful, *Cell,* 108, 13–16.

Burnet, F. M., 1933, The classification of coli-dysentery bacteriophages. III. A correlation of the serological classification with certain biochemical tests. *J. Pathol. Bacteriol.,* 37, 179–184.

Casjens, S., Hatfull, G., and Hendrix, R., 1992, Evolution of dsDNA tailed bacteriophage genomes, *Sem. Virol.,* 3, 383–397.

Day, L. and Maniloff, J., 2000, Inoviruses, in Van Regenmortel and coll., pp. 267–498.

Delaroque, N., Müller, D.G., Bothe, G., Pohl, T., Knippers, R. and Boland, W., 2001, The complete DNA sequence of the *Ectocarpus siliculosus* virus EsV-1 genome, *Virology,* 287, 112–132.

Desiere, F., Lucchini, S., Canchaya, C., Ventura, M., and Brüssow, H., 2002, Comparative genomics of phages and prophages in lactic acid bacteria, Antonie Leeuwenhoek, 82, 73–91.

de A. Zanotto, P.M., Gibbs, M.J., Gould., E.A., and Holmes, E.C., 1996, A reevaluation of the higher taxonomy of viruses based on RNA polymerases, *J. Virol.,* 70, 6083–6096.

d'Hérelle, F., 1917, Sur un microbe antagoniste des bacilles dysentériques. *C.R. Acad. Sci. Ser. D,* 165, 373–375.

d'Hérelle, F., 1918, Technique de la recherche du microbe filtrant bactériophage (*Bacteriophagum intestinale*), *C.R. Soc. Biol.,* 81, 1160–1162.

Gibbs, A.J., Harrison, B.D., Watson, D.H., and Wildy, P., 1996, What's in a virus name? *Nature (London),* 209, 450–454.

Goldbach, R. and Wellink, R., 1988, Evolution of plus-strand RNA viruses, *Intervirology,* 29, 260–267.

Hendrix, R.W., 1999, The long evolutionary reach of viruses, *Curr. Biol.,* 9, R914–R917.

Hendrix, R.W., 2002, Bacteriophages: Evolution of the majority, *Theoret. Popul. Biol.,* 61, 471–480.

Hendrix, R.W., Smith, M.C.M., Burns, R.N., Ford, M.E., and Hatfull, G.F., 1999, Evolutionary relationships among diverse bacteriophages and prophages: All the world's a phage, *Proc. Natl. Acad. Sci. U.S.A.,* 96, 2192–2197.

Holmes, F.O., 1948, Order Virales; the filterable viruses, in *Bergey's Manual of Determinative Biology,* 6th ed., Breed, R.S. Murray, E.G.D., and Hitchens, A.P., Eds., Williams & Wilkins, Baltimore, pp. 1126–1144.

Huber, K.E. and Waldor, M.K., 2000, Filamentous phage integration requires the host recombinases XerC and XerD, *Nature (London),* 417, 656–659.

Kivelä, H.M., Kalkkinen, N., and Bamford, D.H., 2002, Bacteriophage PM2 has a protein capsid surrounding a spherical proteinaceous core, *J. Virol.,* 76, 8169–8178.

Koonin, E.V., 1991, The phylogeny of RNA-dependent RNA polymerases of positive-strand viruses, *J. Gen. Virol.,* 72, 2197–2206.

Lawrence, J.G., Hatfull, G.F., and Hendrix, R.W., 2002, Imbroglios of viral taxonomy: Genetic exchange and failings of phenetic approaches. *J. Bacteriol.,* 184, 4891–4905.

Lwoff, A., Horne, R.W., and Tournier, P., 1962, A system of viruses, *Cold Spring Harbor Symp. Quant. Biol.,* 27, 51–62.

Maniloff, J. and Ackermann, H.-W., 1998, Taxonomy of bacterial viruses: Establishment of tailed virus genera and the order *Caudovirales. Arch. Virol.,* 143, 2051–2063.

Maniloff, J. and Ackermann, H.-W., 2000, Order *Caudovirales,* in Van Regenmortel and coll., pp. 63–68.

Matthews, R.E.F., 1983, The history of viral taxonomy, *In A Critical Appraisal of Viral Taxonomy,* Matthews, R.E.F., Ed., CRC Press, Boca Raton, 1–35.

McKenna, R., Xia, D., Willingmann, P., Ilag, L.L., Krishnaswamy, S., Rossmann, M.G., Olson, N.H., Baker, T.S., and Incardona, N.L., 1993, Atomic structure of single-stranded DNA bacteriophage ΦX174 and its functional implications, *Nature (London),* 355, 137–143.

Mesyanzhinov, V.V., Robben, J., Grymonprez, B., Kostyuchenko, V.A., Bourkaltseva, M.V., Sykilinda, N.N., Krylov, V.N., and Volckaert, G., 2002, The genome of bacteriophage ϕKZ of *Pseudomonas aeruginosa, J. Mol. Biol.,* 317, 1–19.

Mindich, L., Qiao, X., Qiao, J., Onodera, S., Romantchuk, M., and Hoogstraaten, D., 1999, Isolation of additional bacteriophages with genomes of segmented double-stranded RNA, *J. Bacteriol.,* 181, 4505–4508.

Nandhagopal, N., Simpson, A.A., Gurnon, J.R., Yan, X., Baker, T.S., Graves, M.V., van Etten, J.L., and Rossmann, M.G., 2002, The structure and evolution of the major capid protein of a large, lipid-containing DNA virus, *Proc. Natl. Acad. Sci. U.S.A.,* 99, 14758–14763.

Newcomb, W.W., Juhas, R.M., Thomsen, D.R., Homa, F.L., Burch, A.D., Weller, S.K., and Brown, J.C., 2001, The UL6 gene product forms the portal for entry of DNA into the herpes simplex virus capsid, *J. Virol.,* 75, 10923–10932.

P.C.N.V., 1965, Proposals and recommendations of the Provisional Committee on Taxonomy of Viruses (P.C.N.V.), *Ann. Inst. Pasteur (Paris),* 109, 625–637.

Pedulla, M.L., Lewis, J.A., Hendrickson, H.L., Ford, M.E., Houtz, J.M., Peebles, C.L., Lawrence, J.G., Hatfull, G.F., and Hendrix, R.W. 2003. Bacteriophage G: Analysis of a bacterium-sized genome. 103rd General ASM Meeting, Washington, DC, May 18–22.

Peng, X., Blum, H., She, Q., Mallok, S., Brügger, K., Garrett, R.A., Zillig, W., and Prangishvili, D., 2001, Sequences and replication of genomes of the archaeal rudiviruses SIRV1 and SIRV2: Relationships to the archaeal lipothrixvirus SIFV and some eukaryal viruses, *Virology,* 291, 226–234.

Prangishvili, D., Stedman, K., and Zillig, W., 2001, Viruses of the extremely thermophilic archaeon *Sulfolobus, Trends Microbiol.,* 9, 39–43.

Rachel, R., Bettstetter, M., Hedlund, B.P., Häring, M., Kessler, A., Stetter, K.O., and Prangishvili, D., 2002, Remarkable morphological diversity of viruses and virus-like particles in hot terrestrial environments, *Arch. Virol.,* 147, 2419–2429.

Rohwer, F. and Edwards, R., 2002, The Phage Proteomic Tree: A genome-based taxonomy for phage, *J. Bacteriol.,* 184, 4529–4535.

Ruska, H., 1943, Versuch zu einer Ordnung der Virusarten, *Arch. Ges. Virusforsch.,* 2, 480–498.

Rybicki, E., 1990, The classification of organisms at the edge of life or problems with virus systematics, *South African J. Sci.,* 86, 182–186.

Schopf, J.W., 1992, The oldest fossils and what they mean, in *Major Events in the History of Life,* Schopf, J.W., Ed., Jones and Bartlett, Boston, pp. 29–63.

Skalka, A. and Hanson, P., 1972, Comparisons of the distribution of nucleotides and common sequences in deoxyribonucleic acid from selected bacteriophages, *J. Virol.,* 9, 583–593.

Susskind, M. and Botstein, D., 1978, Molecular genetics of bacteriophage P22, *Microbiol. Rev.,* 42, 385–413.

Tikhonenko, A.S., 1966, *Ultrastructure of Bacterial Viruses,* Izdadelstvo "Nauka", Moscow; Plenum Press, New York, 1970, p. 30.

Twort, F.W., 1915. An investigation on the nature of ultra-microscopic viruses. *Lancet,* 2, 1241–1243.

Valegård, K., Liljas, L., Fridborg, K., and Unge, T., 1990, The three-dimensional structure of the bacterial virus MS2, *Nature (London),* 345, 36–41.

Van Regenmortel, M.H.V., 2000, Introduction to the species concept in virus taxonomy, *In* Van Regenmortel and coll., pp. 3–16.

Van Regenmortel M.H.V., Fauquet, C.M., Bishop, D.H.L., Carstens, E.B., Estes, M.K., Lemon, S.M., Maniloff, J., Mayo, M.A., McGeoch, D.J., Pringle, C.R., Wickner, R.B., Eds., 2000, *Virus taxonomy: Classification and Nomenclature of Viruses. Seventh Report of the International Committee on Taxonomy of Viruses*, Academic Press, San Diego, p. 53.

Vander Byl, C. and Kropinski, A.M., 2000, Sequence of the genome of *Salmonella* bacteriophage P22, *J. Bacteriol.,* 182, 6472–6481.

Ward, C.W., 1993, Progress towards a higher taxonomy of viruses, *Res. Virol.,* 144, 419–453.

Wildy, P., 1971, *Classification and Nomenclature of Viruses. First Report of the International Committee on Nomenclature of Viruses,* Monographs in Virology, Vol. 5, S. Karger, Basel.

5 Genomics and Evolution of Tailed Phages

Harald Brüssow[1] *and Elizabeth Kutter*[2]
[1]Nestle Research Center, Lausanne, Switzerland
[2]Lab of Phage Biology, The Evergreen State College, Olympia, Washington

CONTENTS

5.1. Introduction 92
5.2. Temperate Phages 95
5.2.1. Overview 95
5.2.2. Lambdoid Coliphages 95
5.2.3. Dairy Phages 96
5.2.3.1. Overview 96
5.2.3.2. Comparative Genomics of *S. thermophilus* Phages 97
5.2.3.3. Genetic Relatedness among Sfi21-Like *Siphoviridae* 100
5.2.3.4. Sfi11-Like *Siphoviridae* 101
5.2.4 λ Supergroup of *Siphoviridae:* Synthesis of Data from Dairy and Lambdoid Phages 101
5.2.5. Other Temperate Phages 103
5.2.5.1. Overview 103
5.2.5.2. P2 Phage Group 103
5.2.5.3. Mu Phage Group 104
5.2.5.4. Inoviridae 104
5.2.5.5. Large Temperate Phages 104
5.2.6. Prophage Genomics and Phage Evolution 106
5.3. Professional Lytic Phages 110
5.3.1. Overview 110
5.3.2. Medium-Sized Virulent Phages 110
5.3.2.1. The T7 Group 110
5.3.2.2. The Φ29 Family 111
5.3.3. Large Virulent Phages 111
5.3.3.1. Overview 111
5.3.3.2. *P. aeruginosa* Bacteriophage ΦKZ 111
5.3.3.3. *Lactobacillus* Phage LP65 112
5.3.3.4. *Staphylococcus* Phage K 113
5.3.3.5. *B. subtilis* Phage SP01 113
5.3.3.6. The T4-Like Phages 114

0-8493-1336-X/05/$0.00+$1.50
© 2005 by CRC Press LLC

5.4. Early Evolution of Phage Metabolic Enzymes..117
5.5. Summary and Outlook..120
Acknowledgments..122
References...122

5.1. INTRODUCTION

Large numbers of bacteriophages have been found wherever potential host bacteria are found; e.g., in soils, lakes, hot springs and deep sea vents, or associated with commensal bacteria inhabiting plants and animals; phage ecology is discussed in some depth in Chapter 6. About 10 times as many phages as bacteria have been detected in seawater samples, leading to estimates of about 10^{32} phages on earth (Bergh et al., 1989; Wommack and Colwell, 2000). Phages may be incredibly varied in their properties; e.g., their hosts, genetic content, regulatory mechanisms, physiological effects, and most of the genes in the larger phages correspond to nothing yet described in the currently available databases. There is wide interest in how viruses arose; i.e., whether they are representatives of early pre-cellular forms of life or are sophisticated forms of selfish genes derived from modern genomes, how they acquired their special properties and genes, and how they relate to each other and to cellular genomes. The evidence so far indicates that the tailed phages, at least, are of very ancient origin. As discussed below, some phage-encoded enzymes like T4 thymidylate synthase appear to have diverged from the precursor of their bacterial and eukaryotic relatives before those two diverged from each other. So far, we have only begun to examine an infinitesimal sample of the phages infecting common, culturable bacteria. We can only look with fascination at the variety of tailed phages in aqueous and soil habitats, where most of the bacteria cannot yet be cultured to use as hosts for isolating phages. The various groups of tailless phages have been less widely studied; they seem to be less numerous, but they also are harder to distinguish from nonviral elements and algal viruses in the environment.

At the present time, only about 200 complete phage genome sequences are available, most of them from a few families such as lambdoid and dairy-phage siphoviridae and/or prophages found in bacterial genomes. However, complete sequences are available for members of all 13 phage families, and the number of T7- and T4-like phages and of other phages >100-kb in size is growing significantly. Also, bits of sequences have been derived from a range of other phages, as well as from random clones derived from phage pools prepared using feces, seawater and marine sediments, many of which may be infecting non-culturable hosts (Breitbart et al., 2002, 2003, 2004; Rohwer et al., 2000). The growing data confirmed several ideas resulting from extensive genetic, biochemical, and physical studies of a few model phages; they have also brought many surprises. About half of the sequenced bacterial genomes are known to have prophages, sometimes several of them (Canchaya et al., 2003; Casjens, 2003)—and, as discussed below, the evolutionary pressures are quite different for phages residing in bacterial chromosomes than they are during lytic infection cycles. Although the presently available genome sequence data still reflect only a frustratingly minute fraction of the phage biosphere, a framework is being built (Lawrence et al., 2002; Proux et al., 2002; Rohwer and Edwards, 2002) for incorporating data fragments and new whole phage genomes into patterns of relationships. We can expect a marked evolution of our understanding as the number of sequenced phages increases, but information from phage genomics is already impacting our views about early evolutionary patterns and about the general importance of ongoing lateral gene exchange processes.

Recent genomics-based ideas concerning phage evolution are dominated by two perspectives: in one view—summarized by the phrase "All the World's a Phage" (Hendrix, 2002)—all dsDNA tailed phage genomes are mosaics with access, to varying extents, to an enormous common genetic pool (Hendrix et al., 1999). In this model, horizontal gene transfer dominates over vertical evolution, totally masking it, and all current efforts to determine viral phylogenies are meaningless (Lawrence et al., 2002); thus, only some sort of reticular approach can work. This is in many ways an updated version of the classical modular theory of phage evolution developed 30 years ago on the basis of heteroduplex mapping and genetic analysis with lambdoid coliphages (Botstein, 1980; Casjens et al., 1992), and it does seem to be true, to a significant degree, for temperate phages from a wide range of hosts. Thus, some relationships observed between phages of Gram-positive and Gram-negative hosts might be explained by long-distance horizontal gene transfer in the distant past.

The second idea is based on the observation that, despite significant horizontal transfer, vertical lines of phage evolution play key roles in many phage families and can be effectively used to explore ancient origins and relationships as well as current properties. For example, although the few large virulent phages that have been sequenced show relatively few homologies beyond their immediate families, the members of the very large and widespread family of T4-like phages are clearly related, as are the T7-like phages, as judged by sequence comparisons of their genes encoding capsid and DNA replication proteins. The clear vertical transmission of the genes for these two very different kinds of complex molecular machines is key to the usefulness of these phages for many kinds of evolutionary studies.

The two previously described viewpoints mainly reflect the different balance for horizontal and vertical elements in phage evolution between the temperate phages and the "professionally lytic" phages. However, even among the large virulent phages of each family, the nucleotide sequence homologies of their capsid and DNA replication genes decreases from 99% to 30%, and fewer other genes are similar as one looks at increasingly distant relatives. For example, comparing coliphage T4 with vibriophage KVP40, or with coliphage RB49, there are no observable relationships for a large fraction of their early and middle mode genes. It is probable that most of these "nonessential" genes, which are largely deletable under laboratory conditions yet shared between close relatives, play important roles in enabling efficient adaptation to particular ecological niches and to rapid fluctuations in physiological conditions; they may provide insights into ancient contacts as we come to understand them better. However, one must keep in mind that vertical evolutionary relationships generally do not apply to the whole phage genome (which is indeed a patchwork of DNA segments with distinct evolutionary histories), although it may have meaning for a definable core for each phage family.

The balance between horizontal and vertical inheritance appears, not surprisingly, to be quite different for temperate vs. lytic phage families. The temperate phages generally have many opportunities for exchange with various prophages and other infecting phages, as well as with host genes, while residing in somewhat different variants of their host cells. However, virulent phages can only recombine during simultaneous infection with another phage, with homologous sequences in resident prophages on the host chromosome, or with complementary sequences on plasmids. In view of the apparent antiquity of phages (discussed in more detail below), the masking accumulation of point mutations may eliminate obvious sequence similarities between more distantly related phages (except for those caused by recently ongoing horizontal transfer). Comparative phage genomics can reveal protein sequence or gene map similarities even in the absence of DNA sequence similarities between widely diverged phages. (Note that since the genetic code is degenerate, especially in the third codon position, and functional constraints operate on the protein, not the DNA, evolutionary relationships can often be detected in the protein sequence far longer than they are clear in the DNA.). For even more distant relationships, data from structural studies can be informative. However, even using such methods, only about 10% of T4 genes show significant homologies beyond the T4-like phages (Miller et al., 2003). This pattern of largely unique ORFS has also been noted for each family of large eukaryotic viruses, such as the Baculo-, Irido-, Phycodna-, Pox-, and Herpesviruses, which also have a few proteins with obvious relationships to the large phages. As discussed in some detail in section 5.4, Patrick Forterre has suggested that some of these viral proteins may actually reflect a large gene pool predating the last common cellular ancestor. He and others have also proposed that the various families of DNA polymerases, and even the use of DNA rather than RNA as the genetic material, may actually have evolved in such early viruses and only later been adopted by their cellular hosts. Evidence actually supports the idea that the enzymes responsible for DNA replication evolved independently more than once (Leipe et al., 1999).

In this chapter, we discuss the assorted sequence information available to date. We treat the phage evolution issue separately for "temperate" and "virulent" phages because these two types of phages are under distinct selection constraints, as is discussed in the section on prophages. However, there are some limitations to this scheme. For example, it is not clear whether there is always a clear-cut division between "temperate" and "virulent" phages per se; rather, some phages' life cycles may be predominantly "virulent," and vice versa. Infection by temperate phages generally leads to a productive infection cycle ending with cell lysis, and only a small minority of the infected cells actually goes into the lysogenic state. However, as Hershey and Dove (1983) state, "The frequency with which an infecting lambda particle initiates prophage replication rather than productive phage growth depends on the temperature, the state of the bacteria, and the genotypes of the phage and the host. For wild-type lambda, the frequency of lysogenization can vary from nearly zero to near unity, depending only on the conditions of infection." Furthermore, in some ecological settings (e.g., streptococcal dairy phages), the population is dominated by phages that are phenotypically virulent even though they are direct evolutionary derivatives of temperate phages. In addition, bioinformatic analysis of sequenced bacterial genomes has revealed a few prophages that share close sequence relationships with phages previously known as obligate virulent phages (e.g., a T7-like prophage in *Pseudomonas putida* and a bIL170-like prophage in *Streptococcus agalactiae*). The very large temperate phages such as the P1 family have many properties usually ascribed to lytic phages, such as their own replication genes. Furthermore, for many phage groups just one or two phages have been sequenced thus far, so little meaningful comparative analysis is possible. Here, we will largely focus on phage groups with enough sequenced representatives to allow some generalizations.

5.2. TEMPERATE PHAGES

5.2.1. Overview

The bulk of the phage genomes sequenced to date have been temperate, due largely to (i) the importance of such phages in the dairy industry and for generalized transduction, (ii) extensive studies of mycobacterial and lambdoid phages at the Pittsburgh Phage Institute, and (iii) prophage data collected from various bacterial genome projects. Canchaya et al. (2003) and Casjens (2003) have extensively reviewed what can be deduced to date from prophage genomics.

5.2.2. Lambdoid Coliphages

The basic ideas about phage evolution were developed for lambdoid coliphages in the pre-genomics era, as mentioned above, primarily on the basis of heteroduplex mapping (Botstein, 1980; Casjens et al., 1992). According to the modular theory of phage evolution, the product and unit of phage evolution is not a given virus, but a family of interchangeable genetic elements (modules), most of which are multigenic and can be considered as independent functional units, each with its own regulatory sites. For each module, homologous functions can be fulfilled by any one of a number of distinct DNA segments that lack any sequence similarity. One module can be exchanged for another through recombination among phages that can infect the same host, and most recombination events appear to have occurred at the junction between these modules. This method of analysis allowed the division of the lambda genome into 11 modules, each represented by a number of alleles. Some of these (e.g., the head gene cluster) covered a large DNA segment, were quite homogeneous, and were represented by only a few distinct alleles. Others occupied only a small genomic region and had many allelic forms (e.g., late gene control regulators) or showed further subdivision into smaller subunits (e.g., early gene control). Comparative genomics of four quite differently sequenced lambdoid phages (λ, HK022, HK097 and N15) identified a colinear gene map for their structural genes, and confirmed the presence of four distinct alleles of the head module (Clark et al., 2001; Juhala et al., 2000; Ravin et al., 2000). Occasional small inserts of genes under independent transcriptional control were observed, and they were designated "morons" (Juhala et al., 2000). Phages HK097 and HK022 share extensive DNA sequence identity in this mosaic-like modular fashion. The structural genes of phages λ and N15 also show DNA sequence relationships (Ravin et al., 2000). However, the two groups of lambdoid coliphages represented by HK097 and λ were not even linked by protein sequence similarity, suggesting that they represent distinct evolutionary lineages of phage structural modules (the "Sfi21-" and "Sfi11-like" lineages, respectively, described later), both derived from a very ancient and shared ancestral module that has diversified beyond sequence recognition. Interestingly, *Pseudomonas* phage D3 shares a similar structural gene map and much protein sequence similarity with HK97 and HK22, although there is no obvious DNA relationship (Kropinski, 2000), which suggests that D3 is a more distant relative of HK97 within the Sfi21-like phage lineage. D3 is an interesting evolutionary linker, since several of its head proteins are sequence-related to phages from Gram-positive bacteria such as the dairy streptococcal phage Sfi21 (Desiere et al., 2001b). In addition, the location of the D3 lysin gene between the tail genes and integrase is typical of low GC-content temperate phages infecting Gram-positive bacteria. Similar observations were made with a Burkholderia phage (Woods et al., 2002).

The right arm of the vegetative lambdoid coliphage genomes encodes nonstructural genes. Over this region, phages λ, HK022, and HK097 show similar gene organization and patchwise DNA sequence identity in modular fashion. Phage N15 deviates substantially from this gene map, which presumably reflects its peculiar lifestyle (Ravin et al., 2000); i.e., instead of integrating into the genome, it persists in the lysogenic cell as a low-copy linear plasmid with closed hairpin telomeres. Comparative genomic data constrain hypotheses for the mechanisms of modular exchange. Transition zones from homology to heterology in lambdoid phages were mostly observed in intergenic regions, and the few observed intragenic transition zones frequently separated protein domains. A recent publication (Clark et al., 2001) reported some short regions of sequence homology between distinct nonstructural gene modules in lambdoid coliphages, and those linker sequences could promote modular exchanges through homologous or site-specific recombination. In an alternative model, nonhomologous recombination occurs indiscriminately and pervasively across the genome of lambdoid phages, followed by stringent selection for functional phages (Juhala et al., 2000). Thus, most products of nonhomologous recombination within coding regions are eliminated, thereby giving the overall process an undeserved appearance of order and purpose.

5.2.3. Dairy Phages

5.2.3.1. Overview

Due to the economic impact of bacteriophages for the dairy industry, *Streptococcus, Lactococcus,* and *Lactobacillus* phages have become among the best documented phage groups with respect to comparative genomics data (see Fig. 5.1) (Brüssow, 2001). Those phages tend to be more closely interrelated than are the lambdoid phages discussed earlier, perhaps because of their rather specific ecological niche. Brüssow and Desiere (2001) have reported strong elements of vertical evolution in the structural gene clusters of phages that infect dairy bacteria that have not been erased by horizontal gene transfer events.

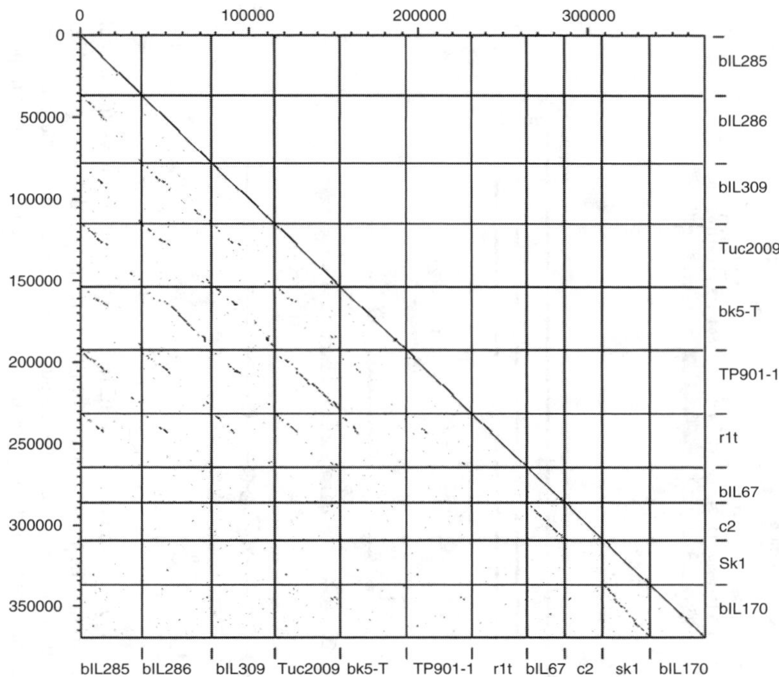

FIGURE 5.1 DNA sequence comparison between phages from *Lactococcus lactis,* the major bacterial starter used in the cheese industry. The DNA sequences of the indicated phages were compared in a dotplot matrix. A dot indicates high level of DNA sequence similarity at the indicated genome region. Closely related phages belonging to the same phage lineage yield a diagonal line only interrupted by regions of sequence diversification. The temperate phages (isolates bIL285 to r1t) present as a swarm of more or less related phage genomes. Sequence relatedness ranged from individual modules to essentially the entire phage genome (e.g. Tuc2009-TP901-1 comparison). In contrast, two groups of virulent phages (c2/ bIL67 and sk1/ bIL170) belong to two distinct phage lineages.

5.2.3.2. Comparative Genomics of *S. thermophilus* Phages

The genomes of six *S. thermophilus* phages have been completely sequenced, and they appear to be quite similar to one another (Brüssow and Desiere, 2001). The genomes can be subdivided into four large segments, each with its own mechanism for creating diversity (Fig. 5.2). They are:

1. *A regulatory segment* covering the rightmost 5-kb of the genome, which gives rise to early transcripts encoding a protein necessary for middle and late transcription (Ventura et al., 2002). Diversity is created by insertion/ deletion processes, while the core DNA sequence is highly conserved (Lucchini et al., 1999a).
2. *A DNA replication segment* found in one of two distinct gene constellations; i.e., the Sfi21-like and the 7201-like DNA replication modules. The

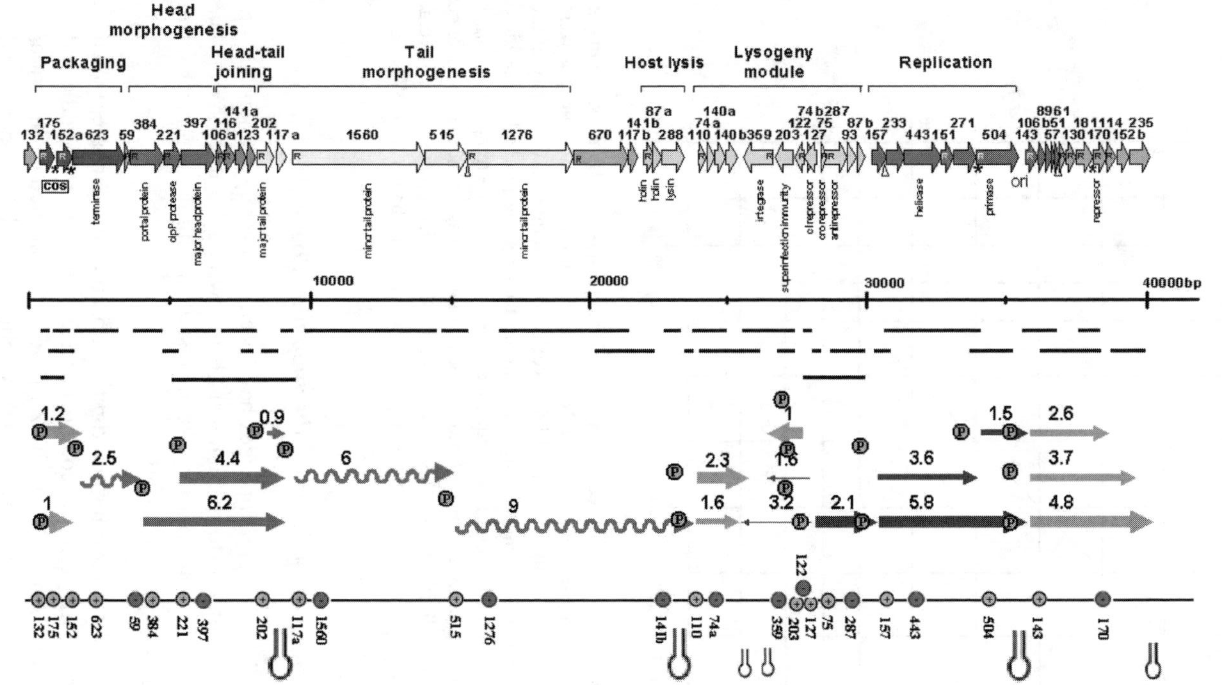

FIGURE 5.2 Genome organization and transcription pattern of the *Streptococcus thermophilus* phage Sfi21. The upper panel shows the modular organization of the 41-kb-long dairy Siphovirus genome, given in the vegetative form. The likely functions of the different modules are given above the brackets. The lower panel gives important additional functional information (from top to bottom): the length scale, the location of the probes used in Northern-blot analyses, the locations and approximate concentrations of the transcripts (including positive and negative circles indicating primer extension results), and the location of putative terminators (hairpins).

majority of the isolated phages have an Sfi21-like replication module, which is unusually conserved (Desiere et al., 1997). Even at the third codon position <1% sequence diversity was observed in independent phage isolates, which suggests that this module recently spread horizontally between the *S. thermophilus* phages.

3. *A late gene segment* extending from the DNA packaging genes to the tail genes, which is represented by two unrelated configurations related to the signal and mechanism involved in initiating DNA packaging (Lucchini et al., 1999a). One configuration is characteristic of the *cos*-site phages (the prototype is phage Sfi21), in which the DNA is cut at specific sites in order to generate single-stranded cohesive ends. The other configuration is typical for *pac*-site phages (the prototype is phage Sfi11), in which headfulls of DNA are packaged from existing ends. Both mechanisms are discussed in Chapter 7. The two structural gene clusters are not related to each other even at the protein sequence level, and each cluster diversifies by accumulating point mutations (Lucchini et al., 1999b). Pair-wise comparisons within each cluster have revealed, on average, 10% to 20% DNA sequence differences.
4. *A tail fiber, lysis, and lysogeny segment* whose diversity is created by insertion, deletion, and replacement of DNA segments and, to a lesser degree, by point mutations (Desiere et al., 1998). When the lysogeny modules from two temperate *S. thermophilus* phages were aligned (Neve et al., 1998), an alternation of conserved and variable DNA segments suggested that recombination processes underlie the acquisition of different types of superinfection immunity and repressor binding specificity in the genetic switch region. Some transition zones were exactly at gene borders, whereas others were in the middle of genes and separated protein domains. Also, deletions in spontaneous or repressor-selected phage mutants were located close to the transition zones. Related phages with a lytic phenotype, which dominate the *S. thermophilus* phage population, apparently are derived from temperate phages by a combination of rearrangements and deletions in the lysogeny module (Lucchini et al., 1999c), rather than belonging to truly virulent phage families. Recombination clearly plays a role in creating diversity for the putative tail fiber genes. Variable and conserved DNA segments alternate in the gene encoding the phage protein that probably interacts with the phage receptor on the host cell (Lucchini et al., 1999a). In addition, Desiere (1998) observed spontaneous deletions that started and ended in DNA repeats encoding collagen-like protein motifs. Phages differing in host range showed completely unrelated variable regions, while phages with overlapping host ranges shared highly related variable regions. Swapping of variable domains between *S. thermophilus* phages resulted in corresponding host range changes in the recombinant phage (Duplessis and Moineau, 2001), which supports the possibility of engineering phages with particularly desired host ranges.

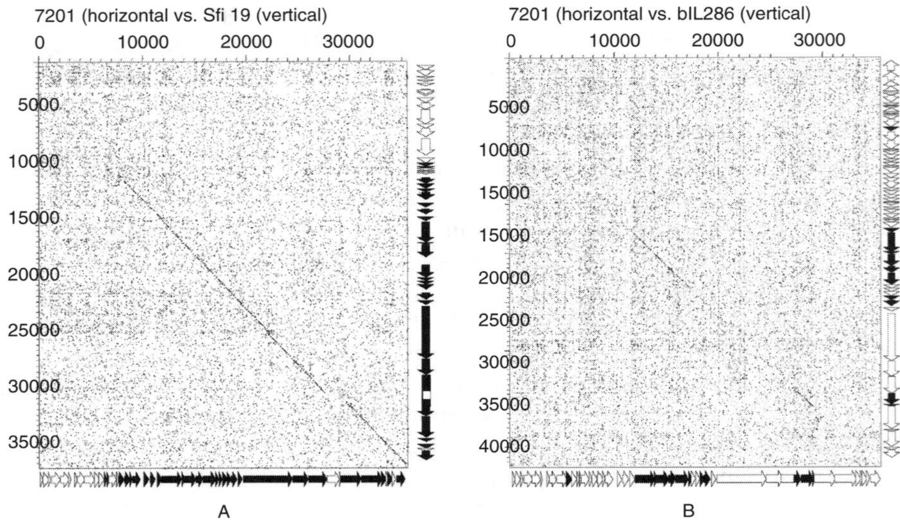

FIGURE 5.3 Graded relatedness between Sfi21-like Siphoviridae from dairy bacteria (DNA level). Dot plot alignment of the two Sfi21-like *S. thermophilus* phages Sfi19 and 7201 (left panel) and *S. thermophilus* phage 7201 and *Lactococcus lactis* phage bIL286 (right panel). The genes that share significant DNA sequence identity are marked in black on the phage gene maps along the horizontal and vertical axis of the plots.

5.2.3.3. Genetic Relatedness among Sfi21-Like *Siphoviridae*

In an effort to differentiate between vertical and horizontal aspects of phage evolution, comparisons have been made between temperate *Siphoviridae* infecting distinct but evolutionary related genera of host bacteria. Complete genome sequences are now available for nine Sfi21-like, *cos*-site *Siphoviridae* that infect six distinct genera of low GC content, Gram-positive bacteria and share an identical modular genome organization: DNA packaging-head-tail-tail fiber-lysis-lysogeny-DNA replication-transcriptional regulation (Desiere et al., 2001a). The patterns of relatedness of the phages suggest some coevolution with their hosts. Taking *S. thermophilus cos*-site phage Sfi21 as a reference point, its closest relatives were other *cos*-site *S. thermophilus* phages, which share more than 80% DNA sequence identity with phage Sfi21. At the next level of relatedness is *Lactococcus* phage BK5-T, whose DNA packaging and head morphogenesis genes share 60% DNA identity with those of Sfi21. At the protein level, the similarity between the phages extends essentially over the entire morphogenesis module (Desiere et al., 2001a). The DNA packaging and head and tail morphogenesis genes of the next most closely related phage, *Lactobacillus* phage adh, share about 40% protein identity with those of Sfi21, although no DNA similarity was detected. Lower levels of protein sequence identity were seen between individual proteins of phage Sfi21, *Bacillus* phage phi-105 and *Staphylococcus* phage PVL.

This "gradient" of relatedness seems to support a simple model of phage coevolution with their hosts. However, the data suffer from a projection bias caused by examining a single group of phages—in this case, *S. thermophilus* phages. A different pattern emerges when using *Lactococcus lactis* phages as the reference point. *L. lactis* is the major industrial starter in cheese production, and lactococcal phages have been intensely investigated due to their economic impact on the cheese industry. Although "gradients of relatedness" are observed among those phages, there are also differences seen that are unrelated to evolutionary differences between current hosts. For example, Sfi21-like phages have been found that only infect *L. lactis*, which are closely related at the DNA-sequence level, or are only related at the protein-sequence level, or share only a common genome map with no obvious nucleotide sequence similarity (Proux et al., 2002). This observation suggests substantial cross-species infection, in an evolutionay time scale, among lactic acid bacteria. All known dairy phages are species and are, in many cases, strain-specific. However, ecological surveys of *Lactobacillus* phages in spontaneous vegetable fermentation and in commensals of the vagina have identified phages that cross the species barriers within the genus *Lactobacillus*.

5.2.3.4. Sfi11-Like *Siphoviridae*

Sfi11-like, *pac*-site phages also are not limited to *S. thermophilus*. For example, structural genes whose nucleotide sequences are very similar to those of Sfi11 have been detected in several *pac*-site phages that infect Gram-positive bacteria with low GC content. They include, in decreasing order of relatedness, *S. pyogenes* prophages, *Lactococcus* phages (TP901-1 and Tuc2009), phages infecting three *Lactobacillus* species (*L. plantarum*:φg1e, *L. delbrueckii*: LL-H, and prophages from *L. johnsonii*), a *Bacillus* phage (SPP1), and a *Listeria* phage (A118) (Desiere et al., 2000). Nucleotide sequence similarity between *pac*-site *S. thermophilus* phages infecting the same host species is high and extends over large regions of the genome. DNA sequence similarity has also been detected between *S. thermophilus* and *S. pyogenes* phages; however, the degree of similarity was lower than between those infecting the same species and was restricted to part of the structural genes. At the protein level, a complex but extensive network links the genomes of *pac*-site phages infecting various bacterial species (Desiere et al., 2000). The structural gene clusters of Sfi11-like phages show an almost identical gene map; i.e., only phage SPP1 differs from the other phages, in that it has supplementary genes inserted at two positions. Also, Sfi11-like phages differ from Sfi21-like phages by having two distinct major head proteins instead of only one, by using a scaffold protein to aid assembly, and by not proteolytically processing their major head proteins.

5.2.4. λ Supergroup of *Siphoviridae*: Synthesis of Data from Dairy and Lambdoid Phages

Broad genomic similarities link the lambdoid coliphages infecting Gram-negative bacteria with phages infecting the Gram-positive lactic acid bacteria and *Bacillus* species. Analysis of genome organization has led to the definition of a lambda supergroup of *Siphoviridae*. The DNA packaging, head and tail modules of Sfi21-like phages show clear similarities with those of *E. coli* phage HK97 (Brüssow and Desiere, 2001) and *Pseudomonas* phage D3 (Desiere et al., 2001a). Also, in addition to their aforementioned modules having nearly identical gene maps, several proteins of phage D3 and dairy phages have similar amino acid sequences. The amino acid sequences of the major head proteins of Sfi21 and HK97 phages are not similar, but their predicted secondary structures as well as proteolytic processing at specific amino acid positions are identical. Also, the major head proteins of phage λ and Sfi11-like phages exhibit still weak sequence similarities (Desiere et al., 2000). Interestingly, the gene map of Sfi11-like phages resembles that of phage λ as closely as Sfi21-like phages resemble coliphage HK97. This observation suggests that the head modules of Sfi21/HK97-like and Sfi11/λ-like phages represent different, very ancient lineages of *Siphoviridae* head modules. In that regard, Casjens and Hendrix (1974) have proposed that the intensive protein-protein interactions during virion morphogenesis impeded very extensive gene exchanges between those phages. However, it is less clear what selective forces maintained the conserved gene order after the postulated split of the ancestral module into the Sfi21/HK97 and Sfi11/λ module lineages. These phages also have other, more distant relatives. For example, the structural genes of *Streptomyces* phage ϕC31 resemble those of the Sfi21 supergroup, even though their nonstructural genes markedly differ.

Although this hypothesis that there are many forbidden exchanges within a cluster of genes encoding interacting proteins is logically appealing, not all of the comparative genomics data support it. For example, several pairs of phages have many related DNA packaging and head morphogenesis genes, but a single gene (mostly the major head gene) differs between them. The discordant head gene sometimes has a nucleotide sequence which is similar to that of a head gene from a third phage with otherwise unrelated sequence. One might argue that the head protein mainly interacts with itself when building the phage capsid. However, it must also interact with the other capsid, DNA-packaging, and tail-attachment proteins. Therefore, the rules underlying the conservation of structural gene order in the λ supergroup of phages are still unclear. The nonstructural genes seem to have evolved according to other constraints; i.e., modular exchanges are much more intensive, which results in significant reshuffling of genes. It is also clear that distinct structural and nonstructural phage gene clusters are relatively free to combine. For example, nonstructural gene clusters that share DNA sequence identity have been found (Brüssow, 2001) in dairy phages with three distinct structural gene clusters (*Lactococcus lactis* phages BK5-T, Tuc2009, and r1t).

Interestingly, lambdoid supergroup phages infect a wide range of hosts, including low GC-content Gram-positive bacteria (streptococci, lactococci, listeriae, staphylococci, bacilli, and clostridia), high GC-content Gram-positive bacteria (*Streptomyces,* corynebacteria and mycobacterial species), and Gram-negative bacteria (γ-proteobacteria). Unfortunately, the true phylogenetic spread of this phage group is currently unknown because too few sequences are documented outside of the aforementioned bacteria. Remarkably, one subgroup of the Archaea is infected by phage-like viruses which resemble the Siphoviruses and have the same structural gene cluster map as the λ supergroup (Pfister et al., 1998).

The chimeric origin of phage lambda and many other phages is highly apparent when analyzing their GC contents, which often show a clear division between structural and nonstructural gene clusters. This analysis reemphasizes the fact that when we speak of phage evolution, we must keep in mind that this term can not be applied to the entire phage genome (which is a patchwork of DNA segments with distinct evolutionary histories), and that it only has meaning for individual DNA modules. However, the expanding phage database at least allows an outline of evolutionary analysis to be made for the structural gene clusters in the temperate phage families (expressly excluding the tail fiber genes). The comparative genomics of the structural gene cluster provides one rational basis for a phage taxonomy that reflects natural relationships (Proux et al., 2002). Most large virulent phages encode many of their own genes for nucleic acid metabolism (i.e., genes encoding DNA and RNA polymerases and their many auxiliary proteins), which can also be examined for evolutionary relationships (see below), in contrast to the temperate phages, which are largely or totally dependent on the host for their nucleic acid metabolism.

5.2.5. Other Temperate Phages

5.2.5.1. Overview

Siphoviridae of the supergroup are not the only phages that can establish lysogeny. At least two different groups of *Myoviridae* (P2- and Mu-like phages) and *Inoviridae* have developed this capacity.

5.2.5.2. P2 Phage Group

A fairly detailed picture can be painted for the evolution of the P2-like *Myoviridae*. The structural gene clusters of coliphages P2 and 186 have similar nucleotide sequences, except for two extra genes in P2 (one of which is a lysogenic conversion gene) (Dodd, 1999). In contrast, the right arms of the genomes, which consist of nonstructural genes, are not obviously related by DNA sequence, although four of the proteins they encode have relationships that can be detected at the level of amino acid sequence. The structural gene clusters of P2 and *Pseudomonas* phage ϕCTX have a weak but clear DNA sequence similarity; however, their nonstructural genes are not similar (Nakayama et al., 1999). Vibriophage K139 lacks any apparent sequence similarity with P2, in spite of its conserved genome map. In contrast, the amino acid sequences of the head proteins of the coliphages P2 and 186 and *Haemophilus* phage HP1 are very similar (Dodd, 1999). These data demonstrate the conservation of a characteristic genome map, with the structural gene clusters showing a "gradient" of relatedness. Interestingly, the genetic organization of the lysogeny modules of the P2-like *Myoviridae* and the temperate *Siphoviridae* which infect low GC-content Gram-positive bacteria are similar, and this similarity extends, in some cases, to the protein sequence level. As also suggested by the conservation of structural gene order between the P22-like *Podoviridae* and the λ-like *Siphoviridae*, an ancient lateral gene transfer of structural and non-structural phage modules between distinct phage families (*Myoviridae, Siphoviridae, Podoviridae*) is a distinct possibility.

5.2.5.3. Mu Phage Group

Phage Mu is the only sequenced freestanding phage representative of its group. However, Mu-like phages are increasingly being identified in bacterial genome sequences. For example, the γ-proteobacterium *Shewanella* contains two Mu-like prophages, in addition to one that is lambda-like (Heidelberg et al., 2002). *Shewanella* prophage MuSo2 shares structural gene organization and sequence relatedness with coliphage Mu; however, neither duplication of tail fiber genes nor an invertase-encoding gene were found near the right end of the former's genome. Also, although the early genes were organized as in Mu-like phages, half of the genes lacked database matches. A potential secretion activator and a TraR-like protein are encoded upstream of the DNA packaging genes. Also, although *Shewanella* prophage MuSo1 is a Mu-like phage, the synteny is punctuated by deletions of tail genes that are replaced by additional transposase and methylase genes and by a four-orf insertion in the head gene cluster. In addition, transposases are inserted near the lysis cassette. Mu-like prophages have also been detected in the γ-proteobacterium *Haemophilus*, and even in *Neisseria* and *Deinococcus*, belonging to increasingly distant bacterial groups: β-proteobacteria and the *Deinococcus-Thermus* group, respectively. Interestingly, the similarity between the Mu-like phages co-varied with the phylogenetic distance separating their bacterial host species. Similarities ranged from varying degrees of protein sequence identity (γ-proteobacteria > β-proteobacteria) to gene map similarity in the near absence of sequence similarity (for *Deinococcus*) (Morgan et al., 2002).

5.2.5.4. Inoviridae

The well-investigated, filamentous, single-stranded DNA phages (e.g., M13 and fd) establish a persistent infection, which results in constant production of progeny phage. Other filamentous phages integrate as double-stranded DNA into the bacterial chromosome, using bacterial recombination enzymes to establish a true lysogenic state. The paradigm for this lysogeny is the cholera prophage ϕCTX, whose role in pathogenicity is discussed in Chapter 8. Bacterial genomics revealed that filamentous prophages are relatively common (Canchaya et al., 2003). They share a basic genome map organization with M13 and fd phages. However, no sequence similarity with M13 was observed, and similarity with other filamentous phages or prophages—beyond those infecting the same host species—were quite limited, thus precluding formulation of any hypotheses concerning the evolution of this phage group.

5.2.5.5. Large Temperate Phages

Nucleotide sequence data have been obtained from members of three groups of large temperate phages. Their analysis has revealed many genes lacking database matches, demonstrating large uncharted regions in the phage DNA sequence space.

P1 and P7: Lobocka et al. (2004) have recently sequenced the 93,601-bp genome of bacteriophage P1, a widely studied generalized transducing phage frequently used in molecular biological applications (see Chapters 3 and 7), but one that most generally propagates as a low copy number plasmid rather than being incorporated into the host genome (Lehnherr, 2005). Lobocka has also sequenced the related phage P7, thus facilitating the comparative analysis of both phages. As has been observed with the other large phages that have been sequenced, relatively few genes from either have significant homologues in the databases. Interestingly, while the head genes seem unrelated to those of other phages, the tail tube and portal proteins and some baseplate proteins show a relationship to those of T4-like phages, while the tail assembly chaperonin and the adhesion region are Mu-like. P1 has homologues of the *E. coli* DNA helicases and DNA repair enzyme polV, the *Salmonella* single-stranded-DNA-binding protein, and various methylases and restriction enzymes. However, in contrast to the large virulent phages, it does not have a replicative polymerase or nucleotide synthesizing enzymes of its own.

SPβc2: Lazarevic et al. (1999) have sequenced the 134,416-bp genome of *Bacillus* temperate siphovirus SPβc2, which belongs to a family of prophages that are rare examples of large temperate phages encoding several enzymes required for nucleotide metabolism. Its ribonucleotide reductase-encoding genes also harbor group IA2 introns. Like most of the other large phages, the majority of SPβ's predicted ORFs do not have significant homology with genes characterized in the current databases. However, it has six extended regions, containing ca. 7000-bp, that have over 90% identity to its *Bacillus subtilis* host. In contrast, none of the large virulent phages has any apparent close host homology.

Mycobacteriophages: Four *Siphoviridae* phages (L5, D29, Bxb1, and TM4) infecting the high GC-content Gram-positive mycobacteria had been sequenced and characterized in some molecular detail. L5 is a temperate phage, whereas the virulent phage D29 is an apparent deletion mutant from a temperate L5-like phage. Although these phages were isolated at different times and at different locations, they shared many characteristics, including morphology and genome size (49–53 kb). In a recent seminal paper, Pedulla et al. (2003) reported the genome sequences of ten new morphologically diverse phages. With one exception, all of the new mycobacteriophage genomes are larger than the four previously described ones; i.e., their genome sizes go up to 156-kb. Strikingly, half of the ORFs in the new phage genomes lack matches to genes in the current databases; the other half show matches mainly restricted to other mycobacteriophages, thus suggesting a large unexplored sequence space for phage genomes. Three of the originally characterized mycobacteriophages (L5, D29 and Bxb1) have similar structural gene maps, and two of them (L5 and D29) share DNA sequence identity in a patchwise pattern over major parts of their genomes. Only one of the newly sequenced phages, phage Bxz2, shows sequence identity with the previously characterized phages. Among the 10 newly characterized phages, only two phages (Che8 and Che9d) demonstrate DNA sequence identity over about half of their genomes' lengths, while the rest of the phages only share identity over small genome segments. The nucleotide sequence comparisons in this large dataset identified numerous potential modular exchanges. Their analysis suggested illegitimate recombination as the major motor of phage evolution for this class.

5.2.6. Prophage Genomics and Phage Evolution

The current phage databases are still too small to allow definitive conclusions concerning the mechanisms of diversification and evolution of temperate phages. Bacterial genome sequencing projects provide an important additional source of data for analyses of phage genomics and evolution. Data deposited in public databases indicate that many of the genomes contain prophages, which may constitute a sizable part of the total bacterial DNA (Fig. 5.4). At the present time, the most extreme example is the food-borne pathogen *Escherichia coli* O157:H7 strain Sakai, which contains 18 prophage genome elements responsible for ca. 16% of its genome's total length. Less extreme, but still impressive examples are represented by various strains

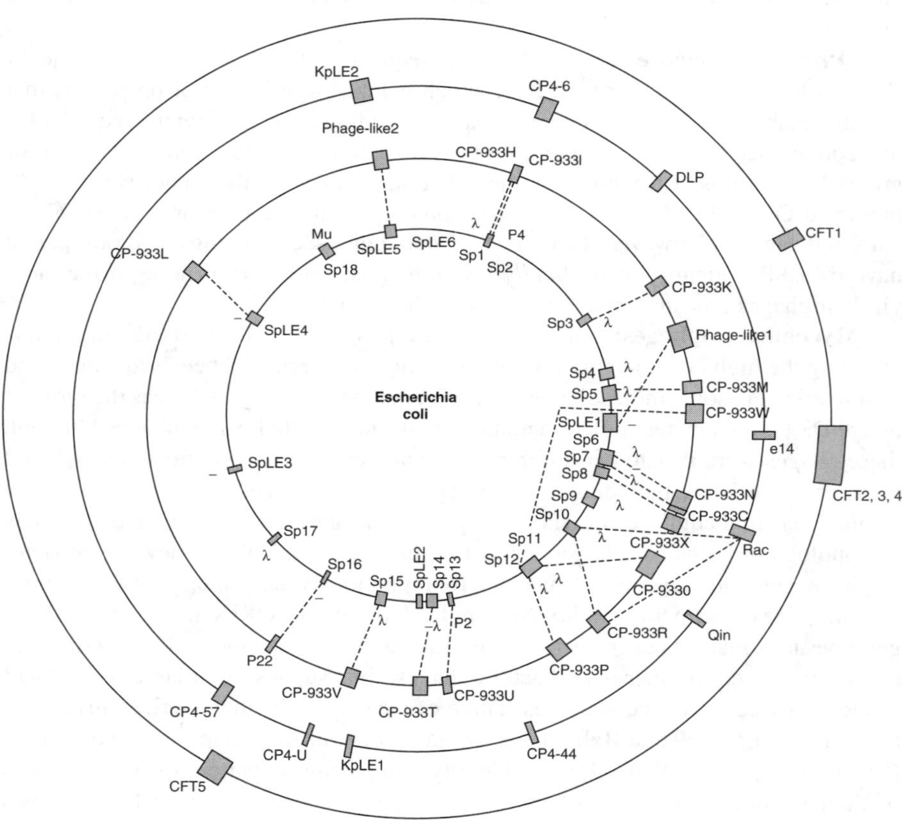

FIGURE 5.4 Prophage content of the sequenced *E. coli* strains. The rectangles on the circular genome maps denote prophages and prophage remnants. From center outwards: the enteropathogenic O157:H7 strains Sakai and EDL933, the laboratory K-12 strain, and the uropathogenic strain CFT073. Prophages sharing DNA sequence identity between the different genomes are linked with dotted lines. Not indicated are the substantial DNA sequence identities between the prophages residing in the same O157 strain.

of *S. pyogenes,* with four to six prophages responsible for ca. 12% of the bacterial DNA content. These prophages do not represent exotic phage types; the O157 prophages resemble the well-known temperate coliphages λ, P2, P4, and Mu (Ohnishi et al., 2001), and the *S. pyogenes* prophages belong to the Sfi11-, Sfi21-, and r1t-like *Siphoviridae*.

More than 200 prophage sequences have been retrieved from data obtained during bacterial genome projects, which doubled the phage sequence database. However, the importance of these sequences goes well beyond this substantial increase in sheer data. Studying the evolution of temperate phages only from the viewpoint of the extracellular genomes of replication-competent phages neglects the other side of their life history (i.e., their becoming part of the bacterial genome). In fact, the peculiar lifestyle of temperate phages makes them important model systems to address a number of fundamental questions in evolutionary biology. Phage DNA undergoes different selective pressures when replicated during a lytic infection cycle compared to prophage DNA maintained in the bacterial genome during the lysogenic state. Darwinian considerations and the "selfish gene" concept have led to interesting conjectures (Boyd and Brüssow, 2002; Brüssow et al., 2004; Canchaya et al., 2003). Theory predicts that temperate phages encode functions that increase the fitness of the lysogen. Therefore, depending on the selective value of the phage genes, the lysogenic cell will be maintained or even overrepresented in the bacterial population. The immunity (phage repressor) and super-infection exclusion genes of prophages that protect against extra phage infection confer obvious selective advantages for the lysogenic host as well as for the prophage. However, prophage genomic studies have revealed many temperate phage-encoded lysogenic conversion genes that have no obvious link to phage functions even though they increase host fitness. Classic examples of such phage-encoded genes include diphtheria, cholera, botulinum, and tetanus toxins, as well as the nonessential λ genes *bor* and *lom*, which confer serum resistance and increased survival in macrophages, respectively, of the *E. coli* lysogen (Barondess and Beckwith, 1990).

However, even when phages confer selective advantage traits to the cell, phage-host relationships retain the character of an arms race. Benefits from prophages carrying genes that increase host fitness are short-lived from a bacterial standpoint if the resident prophage ultimately destroys the bacterial lineage. In this way, prophages may be considered dangerous "molecular time bombs" that can kill the lysogenic cell upon their eventual induction (Lawrence et al., 2001). Therefore, one would expect evolution to select for lysogenic bacteria with prophage mutations that inactivate the induction process. Selection also should lead to large-scale deletion of prophage DNA, in order to decrease the metabolic burden of extraneous-DNA replication, thus leading to gradual decay of prophage sequences in bacterial genomes. A sequential series of steps has been postulated to lead from fully inducible prophages to (i) prophages with crucial point mutations, (ii) prophage remnants that have lost a major part of their genomes, and (iii) elimination of prophage DNA, with the exception of those prophage genes that confer a selective advantage on the bacterial cell (Lawrence and Ochman, 1998). Lawrence et al. (2001) have proposed that a high genomic deletion rate is instrumental in removing "genetic parasites" from the bacterial genome. These deletion processes could explain why bacterial genomes do not generally increase in size despite constant bombardment with parasitic DNA over evolutionary periods. Streamlined bacterial chromosomes containing few pseudogenes have been proposed to be a consequence of this deletion process for parasitic DNA, and prophage genomics data are very compatible with that scenario. Thus, it would be shortsighted to restrict the analysis of phage evolution only to the phages isolated as replication-competent viruses.

Phage DNA is one of the main vectors for lateral gene transfer between bacteria (Bushman, 2002), and an increasing number of observations demonstrate that phages may be major stimulants of bacterial evolution. Not surprisingly, numerous bacterial virulence factors are phage-encoded (Boyd and Brüssow, 2002). Furthermore, prophages account for a major part of interstrain genetic variability in *S. aureus* (Baba et al., 2002) and *S. pyogenes* (Smoot et al., 2002). When genomes from closely related bacteria (e.g., *L. monocytogenes/innocua, S. typhi/typhimurium,* and *E. coli* O157/K12) were compared by dot plot analyses, prophage sequences accounted for a substantial, if not major part of the differences in the genomes (Glaser et al., 2001). Microarray analysis (Porwollik et al., 2002) and PCR scanning (Ohnishi et al., 2002) also demonstrated that prophages contributed a large part of the strain-specific DNA. Finally, when microarrays were used to characterize mRNA expression patterns in lysogenic bacteria undergoing physiologically relevant changes in growth conditions, prophage genes figured prominently in the mRNA species changing expression patterns (Smoot et al., 2001; Whiteley et al. 2001). Subtractive mRNA hybridization analysis has also demonstrated that prophage genes are prominent components of the *E. coli* genes upregulated when the bacteria invade the lungs of birds (Dozois et al., 2003). Such data demonstrate that prophages are not a passive genetic component of the bacterial chromosome, but are active players in bacterial physiology. There are also good indications that prophage acquisition actually shapes the epidemiology of some important bacterial pathogens, as discussed in Chapter 8. As a striking example, nucleotide sequencing, microarray, and epidemiological data indicate that recently emerged, highly virulent *S. pyogenes* strains arose through sequential acquisition of three prophages possessing distinct, phage-encoded virulence factors. The possibility of polylysogeny (i.e. the acquisition of multiple prophage genomes), and the multitude of molecularly distinct virulence genes found in temperate phages infecting some pathogenic bacteria, suggests that the evolution of bacterial pathogenicity may be a function of what could be called *combinatorial biology with phages.*

The evolution of pathogenicity is not the only consequence of prophage DNA in bacterial genomes. Prophages also can become weapons of the lysogenic cell in the fight for its ecological niche. For example, *P. aeruginosa* phages experienced genome reduction until only phage tail genes (bacteriocins) remained (Nakayama et al., 2000). In still other cases, prophages became gene transfer particles for their host bacteria. Examples are the *B. subtilis* prophage PBSX, which has lost head genes without losing the capacity to build phage particles (Wood et al., 1990). The smaller-than-normal phage heads are incapable of packaging the prophage genome, but they pick up random pieces of bacterial DNA. Similarly, size-reduced phages of *Rhodobacter capsulatus* have become generalized gene transfer agents in the service of the bacterial cell (Lang and Beatty, 2000). In other cases, the mini-phage particles transfer specific cellular DNA segments, such as the *S. aureus* pathogenicity island (Ruzin et al., 2001). However, it would be erroneous to view prophages only in the context of their service to the bacterial cell. Being, at times, part of the bacterial genome also has advantages for the prophage, which is subject to a selection-induced decay process. For example, its passive replication by the bacterial cell frees the prophage from the selective pressure to produce a replication-competent phage. Also, prophage sequences can freely recombine when they are in multiple copies within a bacterial cell (Nakagawa et al., 2003). A number of bacterial genomes exhibit substantial sequence identity between prophages residing in the same genome, e.g. over large prophage genome segments in *E. coli* O157, *Xylella fastidiosa* and *L. plantarum* and over small prophage genome segments, as in *S. pyogenes* (Canchaya et al., 2003). It is unclear whether these repetitions are the basis for, or the consequence of, recombination processes. The repeats allow prophage recombination, thus creating potentially new prophages and host strains with various genome arrangements (Nakagawa et al., 2003). Several sequencing projects have demonstrated that bacterial strains differ little in overall sequence diversity, but they do show rearrangements of the bacterial genome, sometimes using repeat sequences between prophages (Van Sluys et al., 2003; Canchaya et al., 2003).

Prophage genomics revealed various new combinations of phage DNA elements resulting in hybrid phages. For example, prophages with λ-like head genes and P2-like tail genes were observed in *Shigella* and *E. coli* (Allison et al., 2002). These chimeric phages appear morphologically as *Myoviridae,* and they are fully replication competent. This observation demonstrates the great power of modular exchanges, since in these phages elements from two distinct phage families are combined. Even more complex combinations of prophages have been observed in nature; e.g., a *Salmonella* prophage combined Mu-like head genes with P2-like tail genes and λ-like tail fiber genes (Canchaya et al., 2003). Other prophages have demonstrated smaller, but perhaps not less consequential, gene acquisitions. For example, an Inovirus prophage from *Pseudomonas* acquired an integrase and a reverse transcriptase-like gene next to its prophage genome (Canchaya et al., 2003). Even more astonishing was a *P. putida* prophage that had the genome organization of the obligate virulent phage T7 with sequence similarities to a phage from a γ-proteobacterium (*Yersinia*) (Pajunen et al., 2001) and a marine phage from cyanobacteria, phylogenetically very distant relatives of Gram-negative bacteria (Chen and Lu, 2002). Here, a gene with a link to integrases was identified downstream of the morphogenesis genes.

It is likely that the major steps of phage evolution by new gene acquisition (e.g., integrases and virulence factors), or by reshuffling of phage genomes via modular exchanges, largely occur within lysogenic cells during superinfection with extracellular phages having appropriate homologies. Thus, lysogenic bacteria may be the "melting pot" of phage evolution. This is nicely demonstrated by the two sequenced *E. coli* O157 strains, both of whose genomes contain a number of closely related lambdoid coliphages. However, despite the recent sharing of a common ancestor, none of the compared prophages turned out to be molecularly identical. Many small modular exchange reactions were demonstrated (Yokoyama et al., 2000).

5.3. PROFESSIONAL LYTIC PHAGES

5.3.1. Overview

Compared to the temperate phages considered above, surprisingly little information is available for professional lytic phages. The reason is not obvious. Historically, phages T1 to T7 belonged to the set of phages chosen by Max Delbrück for detailed investigation. Despite the ease of DNA sequencing, not even all of these type phages have been sequenced. One technical reason might be the presence of many host-lethal genes that complicate the sequencing process in professional lytic phages. However, these problems can be circumvented by the sequencing of PCR products. The T4 and T7 phage families were sources of key enzymes for biotechnology, including T7 RNA polymerase and T4 DNA and RNA ligases, phosphatase, and polynucleotide kinase—and T7 was one of the first fully sequenced phages (Dunn and Studier, 1983). The T4 sequence, in contrast, took 20 years to complete and depended on the development of special vectors and genomic sequencing techniques to deal with regions containing many small host-lethal genes (Kutter et al., 1994; Kutter, 1996). In view of the limited numbers of sequences from related phages, comparative genomics of these lytic phages is still an underdeveloped area. However, genomics studies are finally gaining in momentum for the T4- and T7-like phages from a variety of bacteria and regions and for several families of virulent phages of low-GC-content Gram-positive bacteria (e.g., phages related to c2-like and sk1-like *Lactococcus* phages c2 and sk1, *Bacillus* phage φ29, and a group of 138-kb related *Myoviridae* from *Listeria, Staphylococcus, Bacillus,* and *Lactobacillus*). Serious comparative genomic studies of a range of major virulent phage families should soon be possible.

5.3.2. Medium-Sized Virulent Phages

5.3.2.1. The T7 Group

The widely studied bacteriophage T7 is the type phage for a genus of *Podoviridae,* and is one of the classic coliphages used to develop many molecular biological approaches (see Chapter 2). Its 39,937-bp genome was the first virulent phage genome to be sequenced (Dunn and Studier, 1983). Most of the proteins encoded in its genome have known functions, as is true for most of the small phages. Bartel, Roecklein et al. (1996) used a yeast two-hybrid system to identify protein-protein interactions and produce the first genome-wide protein linkage map of T7 phage. Other recently sequenced members of the T7 family include *P. putida* phage gh-1 (Kovalyova and Kropinski, 2003) and *Yersinia* phages φYeO3-12 (Pajunen et al., 2001) and φA1122 (Garcia et al., 2003). T7 can be aligned with *Yersinia* phage φYeO3-12 over essentially the entire genome if one allows for numerous isolated gene replacements. Phage φA1122, which infects virtually all isolates of *Yersinia pestis* and has long been used by the Centers for Disease Control as a diagnostic reagent, is also closely related to T7.

More distant, sequenced relatives of T7 include roseophage SIO1 (Rohwer, 2000), cyanophage P60 (Chen and Lu, 2002) and *P. aeruginosa* phage PaP3 (NCBI #

NC_004466). Interestingly, cyanophage P60 is more closely related to coliphages T3 and T7 than it is to the marine roseophage SIO1, which infects an evolutionary relative of *Rhodobacter* (an α-proteobacterium; whereas *E. coli* and *Yersinia* both belong to the γ-subclass). SIO1 shows a similar genome organization and protein sequence similarities with T7, but no detectable DNA sequence relatedness (Rohwer, 2000). PaP3, isolated in China, has only limited relationships to any of these other phages, but shows over 90% identity at the protein level for such genes as its DNA polymerase, exonuclease, and portal protein to a phage called PEV1 that was isolated at Evergreen against the *P. aeruginosa* from local cystic fibrosis patients (Kutter lab, unpublished). These results emphasize the lack of simple correlations of phage taxonomy with either geography or host phylogeny.

5.3.2.2. The Φ29 Family

The *B. subtilis* Φ29 family of podoviruses with prolate icosahedral heads includes *Bacillus* phages B103, M2, PZA, and GA-1—and, more distantly, streptococcal phage Cp-1 and *Mycoplasma* phage P1. Their genomes are about half the size of the T7 family; e.g., the Φ29 genome contains only 19,285 bp. They still encode their own DNA polymerases, the sequences of which have been used to construct a general phylogenetic map of the Φ29 and T7 groups of podoviruses (Chen and Lu, 2002). Most of the genomes of *B. subtilis* phages GA-1 and PZA can be aligned in a mosaic fashion, except for one DNA inversion, and PZA's inversion shares DNA sequence similarity with *Streptococcus* phage CP-1.

5.3.3. Large Virulent Phages

5.3.3.1. Overview

A number of virulent phages have genomes containing >100-kb. Scientists at the Pittsburgh Phage Institute recently reported sequencing the 498-kbp genome of *B. megatarium* phage G; the sequence is available on their Web site but no annotation has yet been published. At the present time, full genomic sequence is available for only a few of the large phages, but the T4-like phages and other phages that contain >100-kbp genomes show a very different pattern than those discussed to date. They have virtually no genes in common with the families of the phages discussed earlier (but see LP65 that follows). Also, they generally encode their own complexes of enzymes involved in such processes as DNA synthesis, but they have few other similarities with anything in the database outside their own families.

5.3.3.2. *P. aeruginosa* Bacteriophage ΦKZ

The largest phage whose genome sequence has been published to date is the isometric-headed myovirus ΦKZ, which contains ca. 280-kbp (Mesyanzhinov et al., 2002). Widespread in nature, it belongs to a family of *Pseudomonas* phages now assigned to two new genera (represented by phages ΦKZ and EL) and three species (Burkal'tseva et al., 2002). The results of studies of those phages further emphasize how disparate phages can be. Only 11 of ΦKZ's predicted 306 encoded proteins showed any similarity to proteins in other phages (*Bacillus, Lactococcus,* and *Streptomyces* phages); four are simply members of the homing-endonuclease families commonly associated with phage genomes. Only one ΦKZ protein is related (distantly) to the T4 family, to an uncharacterized ORF, *arn.3*. All together, 59 matched proteins in the current databases, most of them of unknown function. Thus, approximately 80% of the proteins encoded by ΦKZ appear not to have been characterized in any other life form. Surprisingly, ΦKZ seems to generally lack homologues of known families of replication-related proteins. In that regard, Burkal'tseva et al. (2002) reported that phages ΦKZ and EL do not have detectable homology, at the DNA level, in a set of clones from those phages that they examined, although some protein homology was observed.

5.3.3.3. *Lactobacillus* Phage LP65

An interesting case is presented by the myovirus LP65 isolated from *Lactobacillus plantarum,* a lactic acid bacterium used in a variety of food fermentation processes and a member of the commensal flora of the human oral cavity. Pulse field electrophoresis revealed a genome of about 130 kbp, which fits the 131,573-bp of nonredundant DNA determined by sequence analysis (Chibani-Chennoufi et al., 2004). Despite some similarity in the tail contraction process, its head and baseplate morphology were clearly distinct from *Myoviridae* of Gram-negative bacteria. Rather, they closely resembled *Myoviridae* from a wide range of low GC-content Gram-positive bacteria (*Bacillus subtilis* phage SPO1, *Staphylococcus aureus* phages Twort and Sp-1, *Listeria* phage A511, *Enterococcus* phage 1). The LP65 genome contains about 160 ORFs. N-terminal sequencing and mass spectrometric analysis of its major structural proteins identified its structural module, which is related to those of A511 (Loessner lab Web site: http://www.food-microbiology.net/) and to SPO1 (Pittsburgh Phage Institute, Web site: http://pbi.bio.pitt.edu/). This observation raises the possibility that we deal here with different members of a new family of *Myoviridae* restricted to this group of phylogenetically related bacteria. The phages share sequence identity at the protein level (25% to 56%), but not at the DNA level. Notably, the protein sequence identity between LP65 and SPO1 was not limited to the structural module, but was also observed over DNA replication genes. Another observation was notable: The overall genetic organization of the structural gene cluster from phage LP65 (the putative terminase-, portal-, minor head-, scaffold-, major head-, neck-, major tail-, tail tape measure-, side tail fiber-, and lysin-encoding genes) resembled that of *Siphoviridae* isolated from the same group of host bacteria. In fact, some of the genes still shared weak but significant protein sequence identity. Scattered over the LP65 genome are also genes that share protein sequence similarity with genes from the host bacterium. However, the lack of DNA sequence identity excludes recent horizontal gene transfer between phage and host genomes. Downstream of the LP65 lysin gene lies a cluster of 13 tRNA genes, followed by a DNA replication module. The matches included a DNA polymerase-encoding gene related to that of *B. subtilis* phage SPO1 and endonuclease-, exonuclease-, helicase-, and primase-encoding genes. For the latter, the best matches were with T4-like *Myoviridae* from *E. coli;* no other links to T4-like phages were detected. More than 50% of the ORFs lacked database matches. As with many other phages of its family, LP65 contains several H-N-H endonuclease genes.

5.3.3.4. *Staphylococcus* Phage K

Phage K has a wide host range on *Staphylococcus aureus* strains, including many clinical isolates, which makes it an ideal candidate for phage therapy approaches. It was therefore important to establish its genome sequence for the assessment of its safe biological use. Phage K showed a linear DNA genome of 127,395 bp organized into 118 orfs. The genome falls neatly into two unequal parts. One segment covers the first 30 kbp of the genome with exclusively leftward transcribed genes. Except for the presence of a lysis cassette, mainly hypothetical genes without database matches were found here. The remainder of the genome contains exclusively rightward oriented orfs. The two arms of the phage K chromosome are separated by a 4-kbp-long DNA segment containing two divergently oriented promoters. This 4-kbp region contains tRNA, but no protein-encoding genes. The right arm shows a structural gene cluster closely related to *Listeria* phage A511, a long DNA replication module assuring the phage a relatively great autonomy from the cellular DNA replication machinery, and a 20-kbp segment containing nearly exclusively hypothetical genes. From the similarity of the structural gene clusters of phage K and A511, the authors suggested an ancient intergenus horizontal gene transfer. However, further phage genome sequences make it more likely that we deal here with a lineage of *Myoviridae* that is linked by vertical evolution (see 3.3.6.). The phage K genome shows two further particularities, which make it interesting from an evolutionary viewpoint. One is the complete absence of GATC sites, making it insensitive to a number of restriction enzymes from *S. aureus,* which possibly represents an adaptive feature of phage K evolved in parallel with its broad host range. The other aspect is the invasion of its genome by introns harboring HNH endonucleases. As in dairy phages and phage T4 the introns target critical enzyme-encoding genes, namely the lysin and DNA polymerase genes.

5.3.3.5. *B. subtilis* Phage SP01

SPO1 is a 140-kbp member of the *Myoviridae* with a 12-kbp terminal redundancy. Its genome contains hydroxymethyluracil (HMU) instead of thymine. Isolated from soil in Osaka, Japan in the 1960s, it is the best studied of the large phages of Gram-positive bacteria, in terms of both genetics and physiology. It has many similarities to T4, in terms of size, structure, life cycle, very active recombination, the presence of introns, and the substitution of an unusual pyrimidine, and a number of closely related phages have been described, such as SP82 and φE. SPO1 also has several DNA metabolic enzymes, some of them unrelated to the comparable T4 enzymes. (As discussed below, its DNA polymerase is in a completely different family). The determination of its sequence is just being completed by the Pittsburgh Phage Institute group. This sequence revealed that it is much more closely related to lactobacillus phage LP65 and listeria phage A511 than to T4. This should allow advantage to be taken of the fairly extensive genetic work done in this system, including the analysis of host-lethal proteins and the extensive transcription studies, for the analysis of the other related phages from low GC-content Gram-positive bacteria.

Comparative genomics in the SPO1 phage group already allows some tentative inferences. As in the case of *Siphoviridae* from low GC Gram-positive bacteria, *Myoviridae* from the same group of bacteria showed graded relatedness. Myophages LP65 and fri infecting the same bacterial species (*L. plantarum*) share extensive DNA sequence identity, as demonstrated by cross-hybridization in Southern analysis. *Lactobacillus* phage LP65 shares extensive protein sequence identity with *Bacillus* phage SPO1 and *Staphylococcus* phage K, mainly over the structural genes, but this relationship is no longer obvious at the DNA sequence level. Finally, LP65's genome map is similar to that for the *Siphoviridae* from the lambda superfamily over the structural genes, but only weak protein sequence identity can be observed.

As in the case of other phage groups (e.g. the Sfi21-like *Siphoviridae,* the T4-like *Myoviridae*), the structural genes form the most conserved genome cluster in SPO1-like *Myoviridae,* followed by the DNA replication genes. Some gene insertions, deletions, and replacements are observed in this central core of the SPO1-like genomes, as well as gene relocations into other genome regions. In contrast, large nonconserved gene clusters are found at the right and left arms of the linear SPO1-like phage genomes. In fact, the LP65 gene clusters downstream of the DNA replication module have virtually no links to phages K and SPO1, while a number of links were observed with the bacterial host and its prophages. This region might thus code for genus- or even species-specific gene functions. The gene cluster upstream of the LP65 structural genes shows more matches with entries from the database than the other LP65 genome arm, but no function could be proposed for this segment. The corresponding genome arm in phage K largely lacks database matches.

5.3.3.6. The T4-Like Phages

Although bacteriophage T4 played a central role in the development of molecular biology, its nucleotide sequence was only completed in 1994 (Kutter et al. 1994; Kutter, 1996). A detailed analysis of its genome was recently published (Miller et al., 2003), tying the genetic and physiological work of the last 50 years together with new insights into biochemical pathways, control regions, protein structures, predicted membrane proteins, and the infection process (e.g., phage assembly and release). Heteroduplex analysis of the original T-even phages T2, T4, and T6 showed that their genomes are colinear but about 10% of the DNA is heterologous (i.e., insertions, deletions, and substitutions are present as blocks ranging in size from 200-bp to 3-kbp) (Kim and Davidson, 1974). Phages with T4 morphology infect many different Gram-negative bacteria (Ackermann and Krisch, 1997), in many places around the world; e.g., sewage plants, zoos, coastal and offshore seawater, and stools of diarrhea patients in the former Soviet Union and the United States (Russell, 1974; Matsuzaki et al., 1992; Zachary, 1978; Eddy and Gold, 1991; Kutter, 1996; Ackermann and Krisch, 1997; Hambly et al., 2001). The genomic regions encoding capsid and DNA replication proteins of many of the phages have been amplified by PCR and sequenced. Additional data for other regions of various phages have also been obtained by PCR analysis or localized cloning and sequencing (Jozwik and Miller, 1995; Repoila et al., 1994; Monod et al., 1997; Yeh et al., 1998).

The essential components of T4-like phages, such as capsid and tail proteins and enzymes involved in DNA replication, generally show conserved amino acid sequences. Sequencing analysis of selected capsid genes has demonstrated "graded" classes of relatedness, leading to the definition of the T-even phages (sharing DNA sequence identity with phage T4 over major parts of the genomes) and of more distantly related pseudo-, schizo-, and exo-T-even phages (sharing a decreasing degree of sequence identity with T4). For example, the amino acid sequence similarities of head protein gp23 of T-even and pseudo-, schizo-, and exo-T-even phages are greater than 90%, 60%, 50%, and 30%, respectively (Hambly et al., 2001; Tetart et al., 2001). Also, different core genes (the main head protein gp23 and single-strand-DNA-binding protein gp32) lead to similar phylogenetic trees, which suggests that vertical evolution dominates over horizontal evolution for the morphogenetic and replication genes of T4-like phages, with the notable exception of tail fiber proteins (Haggard-Ljungquist et al., 1992).

Complete sequence data are now also available (http://www.phage.bioc.tulane.edu) for phage RB69. The gene maps of the T4 and RB69 genomes are largely colinear, and most of their genes are clearly related. In fact, in a dot plot alignment, large parts of the two genomes could still be aligned at the DNA level (the structural genes were the most prominently aligned genes). As for the T2, T4, and T6 phages, they differ from each other by insertion/deletion of, or by replacement of, small groups of genes. The sequence of the schizo-T-even phage KVP40, which infects several *Vibrio* species and *Photobacterium,* has been determined by Miller et al. (2003). Its DNA replication- and accessory protein-encoding genes align with those of phage T4, the gene products' amino acid sequences show 47%–70% identity, and the structural protein block is very similar, as is the mechanism of the control of late transcription. However, KVP40 does not have anything similar to the unique T4 early promoters or middle-mode transcription factors. Also, there are some major differences in its genome organization; its DNA does not have any modified bases, and only about 20% of its ORFs are similar to those of T4. In addition, the functions of ca. 60% of its ORFs are not yet known.

T-even phages have been isolated only from Enterobacteriaceae; however, pseudo-T-even, schizo-T-even, and exo-T-even phages have been isolated from a larger range of γ-proteobacteria, from α-proteobacteria, and from cyanobacteria, respectively (Hambly et al., 2001; Tetart et al., 2001). The nucleotide sequences have recently been completed and are being annotated for the genomes of *E. coli* pseudo-T-even phage RB49 and *Aeromonas* phages Aeh1 and 44RR2, and studies are well under way for the genomes of other T4-like phages that infect γ-proteobacteria (*Aeromonas* phage 65), α-proteobacteria (*Acinetobacter johnsonii* phage 133 and *Burkholderia cepacia* phage 42), and cyanobacteria (*Synechococcus* phage S-PM2) (Desplats et al., 2002; Hambly et al., 2001; Matsuzaki et al., 1998; Miller et al., 2003; Tetart et al., 2001; Petrov and Karam, 2003). Most have dsDNA genomes similar in size to that of T4, but some have longer heads filled by substantially larger genomes; e.g., 245-kbp for KVP40 (Miller et al., 2003). ClustalW and neighbor-joining analyses of capsid and tail tube proteins of the T4-like phages studied to date have yielded similar phylogenetic trees, with relationships largely paralleling those of the host bacteria (Tetart et al., 2001; Hambly et al., 2001). The results of current sequencing projects have confirmed that the genomes of the T-even phages infecting *E. coli* are arranged much like those of T4. However, group I introns, the *seg* and *mob* homing endonuclease genes, and some other characterized nucleotide metabolism genes are absent, and there are occasional changes in other genes. The results of PCR and limited nucleotide sequencing studies have shown that many of the "nonessential" genes, deletable in T4 under standard laboratory conditions, are largely conserved in the T-even phages, but fewer of them are found in phage RB69, the most distant relative to date that is included in this group. The T4-like coliphages in the pseudo-T-even class, such as RB49 and Aeh1, contain even fewer of the "nonessential" genes found in T4; instead, they have more uncharacterized ORFs (Desplats et al., 2002). Conservation of the DNA replication and virion structural proteins is evident in RB49 and RB43, but those phages' DNA does not contain the replacement of cytosine by hydroxymethylcytosine that is seen for most T4-like coliphages and, thus, they have no genes for making HMdCTP or for blocking transcription of cytosine-containing DNA. Interestingly, though, the use of hydroxymethylcytosine is not limited to coliphages; these genes are also found in *Aeromonas* phages 31 and 44RR2 (see http://phage.bioc.tulane.edu/).

The conservation of protein sequence identity in the genomes of T4-like phages infecting host bacteria with such large phylogenetic distances separating them is fascinating. Nearly all sequence similarity has been lost between lambda-like phages infecting α-proteobacteria (Gram-negative bacteria) and low GC-content-containing Gram-positive bacteria (the large phylogenetic distance separating them is similar to that separating α-proteobacteria and cyanobacteria); only their structural gene maps are related.

Relatively distant members of the T4-like phage family can infect *E. coli*. For example, a set of phages recently isolated from diarrheal patients in Bangladesh consisted almost exclusively of T4-related phages, some of which had broad host ranges against pathogenic strains (Chibani-Chennoufi et al., 2005). The standard primers developed by Krisch's lab for the diagnosis of T-even phages (Tetart et al., 2001) yielded a PCR product for only half of the 50 new T4-like isolates. Several failed to hybridize either with T4 or with each other, suggesting substantial sequence diversity in T4-like coliphages from the stools of diarrhea patients. Phage JS98, which still hybridizes with T4 DNA in Southern blots but failed to amplify the diagnostic g32 and g23 PCR products, was sequenced (Chibani-Chennoufi et al., in press). Its major head protein shared only 65% protein sequence identity with T4. Despite that difference, the gene maps of the structural gene clusters were identical except for T4 mobile endonuclease genes *segC, D,* and *E* and duplication in JS98 of *g24,* a head vertex protein. Within the T-even phages, diversity seems to be achieved by the accumulation of point mutations and occasional gene duplications. The modular exchanges so frequently observed in *Siphoviridae* seem to be the exception in T4-like phages. Within the T-even group, modular exchanges were up to now only observed across a few regions, e.g. the 4-kbp segment at the 25-kbp genome map position of T4, encoding superinfection exclusion and cytosine modifying enzymes. When more distant relatives like the schizo-T-even vibriophage KVP40 were compared to T4, a conservation of several core genome regions (structural genes, DNA replication genes) was still observed at the protein though not the DNA level. Gene duplications were also a prominent feature in the KVP40/T4 comparison (Miller et al., 2003). KVP40 has a single baseplate gene cluster in contrast to T4's separate groups of baseplate wedge and hub genes preceding and following, respectively, the head and other tail genes. In both phages, the tail fiber genes are separated from the other structural genes; in KVP40 the tail fiber genes are even split between two different genome regions. KVP40 displays long segments of intercalated DNA containing many genes that lack not only homologies with T4, but also other entries of the database. Insertion even of long new DNA segments is thus an important mode of evolution in this phage group, favored by the tolerance of the capsid to adapt to a much larger genome size (245 vs. 168 kbp in KVP40 and T4, respectively).

In summary, sequence analyses of T4-like phages suggest that T4 and its relatives are largely in a group by themselves, undergoing few exchanges with other phage families; homing endonucleases occasionally even restrict exchanges between closely related members of the T4 family (Liu et al., 2003). Amino acid sequence homology has been observed between the receptor binding regions of the tail fibers of the T-even group and other phages, apparently reflecting selection for new receptor-binding capabilities and the special recombinogenic properties of that region (Tetart et al., 1998). A few DNA replication genes are shared between T4-like phages and *Lactobacillus* phage LP65 and some amino acid homologies in the tail tube and baseplate proteins were observed between T4 and P1, but overall, there is not much evidence for ongoing exchanges with other unrelated phages. The high frequency of recombination observed for T4, its lytic developmental program, and the presence of multiple promoters throughout the genome apparently allow for many independent exchanges of small genome segments within this group of phages, but seem to preclude exchanges of large modular units with different phage groups.

5.4. EARLY EVOLUTION OF PHAGE METABOLIC ENZYMES

For explorations of the overall paths of early evolution, T4 and some of the other larger phages have a major advantage over most of the temperate phages, in that they encode many of their own enzymes for nucleic acid metabolism. Determining the evolutionary relationships among those enzymes is a second major way to characterize the taxonomic status of phages, in that it complements approaches based primarily on structural proteins and morphological features. The amino acid sequences of phage DNA polymerases, aerobic and anaerobic ribonucleotide reductases, thymidylate synthase, topoisomerase II, and other such enzymes typically are similar to those of functionally related enzymes of other organisms across the phylogenetic tree, including large eukaryotic viruses and cellular organisms. Sequence data and analysis are still sparse, but a variety of quite different patterns are already apparent and broader studies are highly warranted. T7 RNA polymerase and T4 DNA and RNA ligase and polynucleotide kinase have been widely used in molecular biology since its inception, and it is likely that many more useful phage enzymes will be found as more phages are sequenced and their enzymes are studied.

Surprisingly, the five DNA polymerase families so far discovered seem to have little or no relationship with each other. T4 DNA polymerase aligns with the B family, which initially was thought to be limited to eukaryotes and their viruses, but is now known to also include archaeal and a few bacterial enzymes (Filee et al., 2002). Included in this group are the pol II enzymes of γ-proteobacteria like *E. coli* (involved in some types of DNA repair), and the DNA polymerases of *Saccharomyces* and of herpes and chlorella viruses. This relationship is also seen in the crystal structure of the DNA polymerase of the T4-related phage RB69 (Wang et al., 1997; Karam and Konigsberg, 2000), and a mutation introduced into yeast DNA polymerase (Pol3) on the basis of the mutator properties of an altered T4 DNA polymerase has led to a yeast mutator phenotype (Hadjimarcou et al., 2001). Interestingly, the polymerases of T4-like phages are most closely related to polymerases of the Archaeal halophile *Halobacterium* and those encoded by two of its viruses, HF1 and HF2 (Filee et al., 2002), but the relationship is clearly very ancient. The DNA polymerases of SPO1, T5, T7, and a number of other phages are members of the A family. In addition, although temperate phages usually rely on their hosts' replicative machinery, the temperate mycobacteriophage L5 also has its own DNA polymerase, which is of the A type. No homologues of the known DNA polymerase families have been found in the 288-kb phage ϕKZ, implying either that yet another polymerase family exists or that this is the first of the large phages not to have its own DNA polymerase-encoding gene.

Thymidylate synthase (called td or TS; see Fig. 5.5) is a particularly interesting example of the ability to explore evolutionary history by looking at individual proteins. A number of amino acid stretches are highly conserved between the T4 TS and all characterized viral, bacterial, and eukaryotic TSs, facilitating precise alignment and analysis. However, several regions are totally different between the T4 family of enzymes and all other known thymidylate synthases; these regions are largely hydrophobic in T4, but hydrophilic in other members of the family. They lie on the surface of the enzyme, where they are presumably involved in the interaction between TS and other enzymes of T4's nucleotide synthesis complex (see Chapter 7, Fig. 7.11). When these segments are excluded and the core regions are used for alignment, the phylogenetic relationship shown in Fig. 5.6 is obtained. The tree suggests that the T4 enzyme branched off before the split between bacteria and eukaryotes. The apparently ancient branch point is not just due to accelerated evolution of viral proteins, since, for example, Herpes TSs branch just before the separation between the human and rat-mouse lines. Also, T4 TS has several regions that are characteristic of the eukaryotic enzymes, intermixed with others that seem to be unique to the bacterial sequences. T4 TS also has one sequence near the N-terminus that is otherwise unique to archaeal thymidylate synthases, which are sufficiently different from those of bacteria and eukaryotes that they are more difficult to align unequivocally. The ϕKZ TS gene also has two extended inserts that are very different than the general families of thymidylate synthases (Mesyanzhinov et al., 2002);

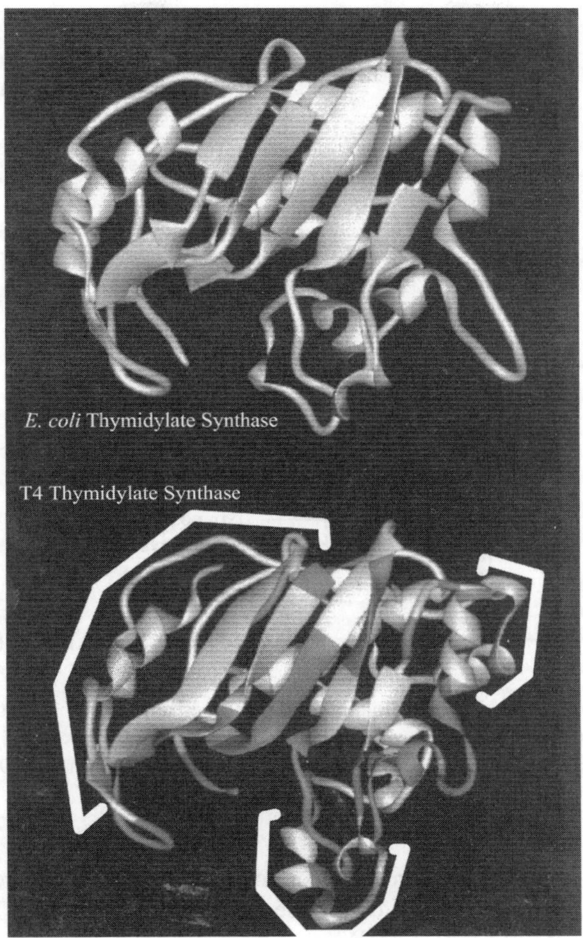

FIGURE 5.5 Structure of thymidylate synthase encoded by bacteriophage T4 in comparison with that of *E. coli*. The enzyme coordinates come from public databases; the analysis was done in the Kutter lab. The dark regions on the T4 enzyme indicate regions of identity between virtually all thymidylate synthases. The brackets indicate regions on the surface of the molecule that are generally hydrophilic in known thymidylate synthases but are hydrophobic in the T4 enzyme, reflecting its participation in a nucleotide synthesizing complex.

it is not clear whether these also reflect special properties of enzyme interactions or some other factor. Figure 5.6 also shows the (very distant) relationship between thymidylate synthases and T4 dCMP HMase—a not-surprising relationship, since both of these enzymes catalyze the transfer of the methyl (or for HMase hydroxymethyl) group to the same position on a pyrimidine nucleotide.

The T4 dihydrofolate reductase and the three topoisomerase II components (gp 39, 52, and 60) also appear to have diverged before the separation of prokaryotes and eukaryotes, as seen in both the clade patterns and the clear interspersion of

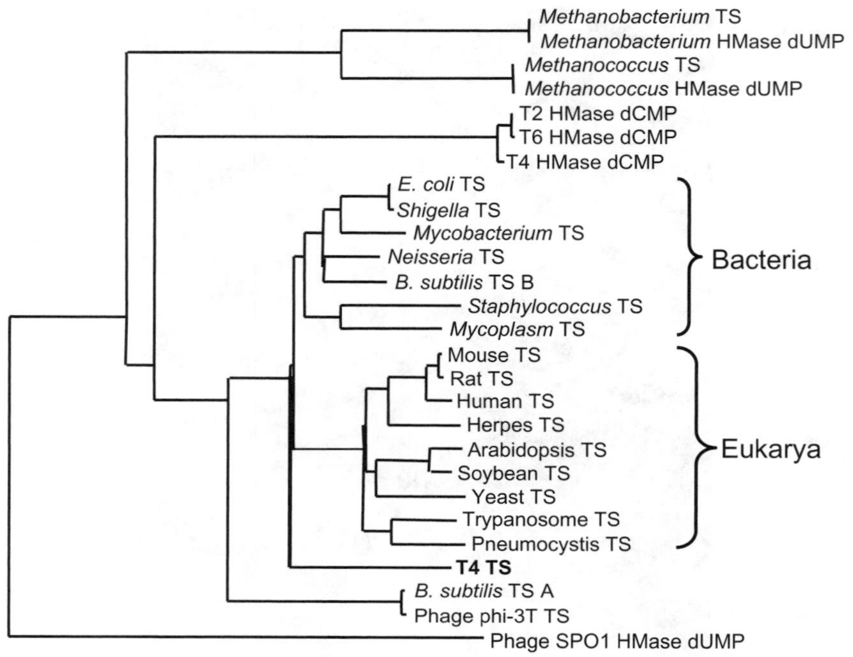

FIGURE 5.6 Phylogenetic tree of thymidylate synthases and deoxynucleotide hydroxymethylases. All protein sequences were obtained from the public databases. Alignment and tree construction were carried out using the methods of Feng and Doolittle (1996). From Miller et al., 2003b.

sequences uniquely conserved among eukaryotes between ones that are characteristic of prokaryotes. Relationships involving these and other proteins of T4-like phages are discussed by Miller et al. (2003).

5.5. SUMMARY AND OUTLOOK

The accelerating number of phage sequencing studies has resulted in phage genomics entering a critical phase. Either we will discover (i) recurrent themes leading to unifying principles and some sort of inclusive taxonomic principles, or (ii) an extremely large number of conceptually new phage genome types and a breakdown of our current phage taxonomy. Many marine microbiologists believe we can expect an explosion of the number and variety of different phages and kinds of phage genes as researchers expand the range of hosts being studied. Some of their arguments are theoretical. About 50 phage types have been found that infect *E. coli;* this number might not be typical, but let us assume at least 10 phage types for most hosts. Conservative estimates suggest that there are about 6 million free-living microbial species, to which, perhaps, the same number of animal- and plant-associated bacteria could be added. Multiplying 10 phage types per species by roughly 10 million microbial species yields the stunning number of 100 million potentially different phages. It is clear that current taxonomic tools cannot deal with anything approaching this number, even if only a small fraction represent new phage genome types and many phages infecting distinct host species are closely connected.

At the present time, the data available do not provide an unequivocal indication of future trends; however, some recurrent themes have been noted. For example, independent sequencing projects of *Myoviridae* in low GC-content Gram-positive bacteria, motivated by phage therapy, food fermentation, and basic research goals, revealed closely related phages that all fell within the genus of a previously established phage group: the SPO1-like genus defined in *B. subtilis* phages (see 3.3.5.). The close similarity of the structural and DNA replication modules of phages SPO1, K, and A511, and the more distant links of the structural gene cluster between SPO1-like phages and lambda-like *Siphoviridae* from the same group of hosts, suggest the existence of widely conserved lineages of phages. We therefore do not expect to find a nearly endless variation of new phage types, at least within this relatively intensively investigated group of bacteria. There are further arguments that speak in favor of a limited variability of phages within a given bacterial group. For example, temperate phages from lactic acid bacteria are closely related over their structural genes irrespective of the ecological specialization of their bacterial hosts, which include important pathogens, commensals, and free-living bacteria. The phylogenetic relationship of the host correlates with a substantial similarity of their phages, which still fall into a limited number of lineages despite intensive phage surveys conducted, for example, in the dairy field. However, these observations represent just a small window into the bacterial diversity in nature and we may still expect many fundamentally new viral genomes when further exploring prokaryotic diversity. A lively illustration is the analysis of viruses from Archaea: not only are exotic forms of viruses found in this group of prokaryotes (Rachel et al., 2002), but the sequencing of several of these phages revealed that their genes literally matched nothing in the current database (Bettstetter et al., 2003, Prangishvili and Garrett, 2004, 2003). However, as these viruses differ fundamentally from tailed phages, let's come back to the diversity of phages from Eubacteria.

Mycobacterial phage sequencing data provide another test case for the diversity of phage genomes (Pedulla et al., 2003). While the initial analysis of four mycobacterial phage genomes had suggested that there was limited diversity, with some relationships to *L. lactis* phage r1t, only one of the ten new genomes analyzed showed close sequence relatedness to those of previously characterized mycobacteriophages. Two of the other nine belonged to a new, closely knit group.

Marine ecologists also have experimentally addressed the issue of phage diversity, by conducting random sequencing projects of viral genomes contained in seawater. About 100-liter samples yielded estimates of up to 7000 different viral sequences (Rohwer, 2003), and more than 65% of the sequences were not significantly similar to previously reported sequences. The most common similarities were observed with the major families of double-stranded DNA phages and some algal viruses. Statistical models based on the aforementioned observations predict that perhaps 2 billion different phage-encoded ORFs remain to be detected. However, the sequencing of a handful of marine phages revealed only one new genome type;

the majority turned out to be (sometimes distant) relatives of T7-, T4-, and Sfi21-like phages. This same group has carried out metagenomic analysis of uncultured viral communities from a variety of sources, such as one from human feces (Breitbart et al., 2003) that led to a prediction of at least 1200 genotypes of phages even in the feces from one individual; these studies are discussed in Chapter 6. The large number of ORFs lacking any database matches in newly sequenced phages with genome sizes over 100 kbp seem to support such statistical approximations. In contrast, the prophage sequences retrieved from sequenced bacterial genomes do not give high estimates of fundamentally new phage types, suggesting that the widespread special group of temperate phages have experienced selection pressures leading to streamlined genomes and substantial convergence, in contrast to virulent phages. As discussed in the chapter on Phage Ecology, the jury is thus still out with respect to the expected diversity of phage genomes in the ecosphere. We are clearly entering a very interesting and productive era in the field of phage genomics, with many lessons and applications to be expected.

ACKNOWLEDGMENTS

We would like to particularly thank Andrew Kropinski and Jim Karam for access to unpublished data and Hans Ackermann, Fumio Arisaka, Burton Guttman, Raul Raya, and Gautam Dutta for their comments and insights.

REFERENCES

Ackermann, H.W. and Krisch, H.M., A catalogue of T4-type bacteriophages. *Arch Virol* 142: 2329–2345, 1997.

Allison, G.E., Angeles, D., Tran-Dinh, N. and Verma, N.K., Complete genomic sequence of Sf V, a serotype-converting temperate bacteriophage of *Shigella flexneri*. *J Bacteriol* 184: 1974–1987, 2002.

Baba, T., Takeuchi, F., Kuroda, M., Yuzawa, H., Aoki, K., Oguchi, A., Nagai, Y., et al., Genome and virulence determinants of high virulence community-acquired MRSA. *Lancet* 359: 1819–1827, 2002.

Barondess, J.J. and Beckwith, J., A bacterial virulence determinant encoded by lysogenic coliphage lambda. *Nature* 346: 871–874, 1990.

Bartel, P.L., Roecklein, J.A., SenGupta, D. and Fields, S., A protein linkage map of *Escherichia coli* bacteriophage T7. *Nat Genet* 12: 72–77, 1996.

Bergh, O., Borsheim, K.Y., Bratbak, G. and Heldal, M., High abundance of viruses found in aquatic environments. *Nature* 340: 467–468, 1989.

Bettstetter, M., Peng, X., Garrett, R.A. and Prangishvili, D., AFV1, a novel virus infecting hyperthermophilic archaea of the genus acidianus. *Virology* 315: 68–79, 2003.

Botstein, D., A theory of modular evolution for bacteriophages. *Ann NY Acad Sci* 354: 484–491, 1980.

Boyd, E.F. and Brüssow, H., Common themes among bacteriophage-encoded virulence factors and diversity among the bacteriophages involved. *Trends Microbiol* 10: 521–529, 2002.

Breitbart, M., Hewson, I., Felts, B., Mahaffy, J.M., Nulton, J., Salamon, P. and Rohwer, F., Metagenomic analyses of an uncultured viral community from human feces. *J Bacteriol* 185: 6220–6223, 2003.

Breitbart, M., Salamon, P., Andresen, B., Mahaffy, J.M., Segall, A.M., Mead, D., Azam, F., et al., Genomic analysis of uncultured marine viral communities. *Proc Natl Acad Sci U S A* 99: 14250–14255, 2002.

Breitbart, M., Wegley, L., Leeds, S., Schoenfeld, T. and Rohwer, F., Phage community dynamics in hot springs. *Appl Environ Microbiol* 70: 1633–1640, 2004.

Brüssow, H., Phages of dairy bacteria. *Annu Rev Microbiol* 55: 283–303, 2001.

Brüssow, H. and Desiere, F., Comparative phage genomics and the evolution of Siphoviridae: Insights from dairy phages. *Mol Microbiol* 39: 213–222, 2001.

Burkal'tseva, M.V., Krylov, V.N., Pleteneva, E.A., Shaburova, O.V., Krylov, S.V., Volkart, G., Sykilinda, N.N., et al., [Phenogenetic characterization of a group of giant ϕKZ-like bacteriophages of *Pseudomonas aeruginosa*]. *Genetika* 38: 1470–1479, 2002.

Bushman, F., *Lateral DNA Transfer: Mechanisms and Consequneces*. Cold Spring Harbor Laboratory Press, Cold Spring Harbor, New York, 2002.

Canchaya, C., Fournous, G. and Brüssow, H., The impact of prophages on bacterial chromosomes. *Mol Microbiol* 53: 9–18, 2004.

Canchaya, C., Proux, C., Fournous, G., Bruttin, A. and Brüssow, H., Prophage genomics. *Microbiol Mol Biol Rev* 67: 238–276, 2003.

Casjens, S., Prophages and bacterial genomics: What have we learned so far? *Mol Microbiol* 49: 277–300, 2003.

Casjens, S., Hatfull, G. and Hendrix, R., Evolution of dsDNA tailed-bacteriophage genomes. *Semin in Virol* 3: 383–397, 1992.

Casjens, S. and Hendrix, R., Comments on the arrangement of the morphogenetic genes of bacteriophage lambda. *J Mol Biol* 90: 20–25, 1974.

Chen, F. and Lu, J., Genomic sequence and evolution of marine cyanophage P60: A new insight on lytic and lysogenic phages. *Appl Environ Microbiol* 68: 2589–2594, 2002.

Chibani-Chennoufi, S., Canchaya, C., Bruttin, A., Brüssow, H., (2005a) Comparative genomics of the T4-like *Escherichia coli* phage JS98: Implications for the evolution of T4 phages. *J Bacteriol.*

Chibani-Chennoufi, S., Dillmann, M.L., Marvin-Guy, L., Rami-Shojaei, S., Brüssow, H., (In press) *Lactobacillus plantarum* bacteriophage LP65: a new number of the SPO1-like genus of the family Myoviridae. *J Bacteriol.*

Clark, A.J., Inwood, W., Cloutier, T. and Dhillon, T.S., Nucleotide sequence of coliphage HK620 and the evolution of lambdoid phages. *J Mol Biol* 311: 657–679, 2001.

Desiere, F., Lucchini, S. and Brüssow, H., Evolution of *Streptococcus thermophilus* bacteriophage genomes by modular exchanges followed by point mutations and small deletions and insertions. *Virology* 241: 345–356, 1998.

Desiere, F., Lucchini, S., Bruttin, A., Zwahlen, M.-C. and Brüssow, H., A highly conserved DNA replication module from *Streptococcus thermophilus* phages is similar in sequence and topology to a module from *Lactococcus lactis* phages. *Virology* 234: 372–382, 1997.

Desiere, F., Mahanivong, C., Hillier, A. J., Chandry, P.S., Davidson, B.E. and Brüssow, H., Comparative genomics of lactococcal phages: insight from the complete genome sequence of *Lactococcus lactis* phage BK5-T. *Virology* 283: 240–252, 2001a.

Desiere, F., McShan, W.M., van Sinderen, D., Ferretti, J.J. and Brüssow, H., Comparative genomics reveals close genetic relationships between phages from dairy bacteria and pathogenic Streptococci: Evolutionary implications for prophage-host interactions. *Virology* 288: 325–341, 2001b.

Desiere, F., Pridmore, R.D. and Brüssow, H., Comparative genomics of the late gene cluster from *Lactobacillus* phages. *Virology* 275: 294–305, 2000.

Desplats, C., Dez, C., Tetart, F., Eleaume, H. and Krisch, H.M., Snapshot of the genome of the pseudo-T-even bacteriophage RB49. *J Bacteriol* 184: 2789–2804, 2002.

Dodd, I.B. and Egan, J.B., P2, 186 and Related Phages (Myoviridae), pp. 1087–1094 in *Encyclopedia of Virology,* R.A.G. Webster (Ed.), A. Academic Press, London, 1999.

Dozois, C.M., Daigle, F. and Curtiss, R., 3rd, Identification of pathogen-specific and conserved genes expressed in vivo by an avian pathogenic *Escherichia coli* strain. *Proc Natl Acad Sci U S A* 100: 247–252, 2003.

Dunn, J.J. and Studier, F.W., Complete nucleotide sequence of bacteriophage T7 DNA and the locations of T7 genetic elements. *J Mol Biol* 166: 477–535, 1983.

Duplessis, M. and Moineau, S., Identification of a genetic determinant responsible for host specificity in *Streptococcus thermophilus* bacteriophages. *Mol Microbiol* 41: 325–336, 2001.

Eddy, S.R. and Gold, L., The phage T4 nrdB intron: A deletion mutant of a version found in the wild. *Genes Dev.* 5: 1032–1041, 2001.

Feng, D.F. and Doolittle, R.F., Progressive alignment of amino acid sequences and construction of phylogenetic trees from them. *Methods Enzymol.* 266: 368–382, 1996.

Filee, J., Forterre, P., Sen-Lin, T. and Laurent, J., Evolution of DNA polymerase families: Evidences for multiple gene exchange between cellular and viral proteins. *J Mol Evol* 54: 763–773, 2002.

Garcia, E., Elliott, J.M., Ramanculov, E., Chain, P.S., Chu, M.C. and Molineux, I.J., The genome sequence of Yersinia pestis bacteriophage A1122 reveals an intimate history with the coliphage T3 and T7 genomes. *J Bacteriol* 185: 5248–5262, 2003.

Glaser, P., Frangeul, L., Buchrieser, C., Rusniok, C., Amend, A., Baquero, F., Berche, P., et al., Comparative genomics of Listeria species. *Science* 294: 849–852, 2001.

Hadjimarcou, M.I., Kokoska, R.J., Petes, T.D. and Reha-Krantz, L.J., Identification of a mutant DNA polymerase delta in *Saccharomyces cerevisiae* with an antimutator phenotype for frameshift mutations. *Genetics* 158: 177–186, 2001.

Haggard-Ljungquist, E.C., Halling, C. and Calendar, R., DNA sequence of the tail fiber genes of bacteriophage P2: Evidence for horizontal gene transfer of tail fiber genes among unrelated bacteriophages. *J Bacteriol* 174: 1462–1477, 1992.

Hambly, E., Tétart, F., Desplats, C., Wilson, W.H., Krisch, H.M. and Mann, N.H., A conserved genetic module that encodes the major virion components in both the coliphage T4 and the marine cyanophage S-PM2. *Proc Natl Acad Sci U S A* 98: 11411–11416, 2001.

Heidelberg, J.F., Paulsen, I.T., Nelson, K.E., Gaidos, E.J., Nelson, W.C., Read, T.D., Eisen, J.A.,et al., Genome sequence of the dissimilatory metal ion-reducing bacterium *Shewanella oneidensis*. *Nat Biotechnol* 20: 1118–1123, 2002.

Hendrix, R.W., Bacteriophages: evolution of the majority. *Theor Popul Biol* 61: 471–480, 2002.

Hendrix, R.W., Smith, M.C., Burns, R.N., Ford, M.E. and Hatfull, G.F., Evolutionary relationships among diverse bacteriophages and prophages: All the world's a phage. *Proc Natl Acad Sci USA* 96: 2192–2197, 1999.

Hershey, A.D. and Dove, W., Introduction to Lambda, pp. 3–11 in *Lambda II*, R. W. Hendrix, J. W. Roberts, F. W. Stahl, and R. A. Weisberg (Eds.). Cold Spring Harbor Laboratory Press, Cold Spring Harbor, New York, 1983.

Jozwik, C.E. and Miller, E.S., RNA-protein interactions of the bacteriophage RB69 RegA translational repressor protein. *Nucleic Acids Symp Ser* 33: 256–257, 1995.

Juhala, R.J., Ford, M.E., Duda, R.L., Youlton, A., Hatfull, G.F. and Hendrix, R.W., Genomic sequences of bacteriophages HK97 and HK022: pervasive genetic mosaicism in the lambdoid bacteriophages. *J Mol Biol* 299: 27–51, 2000.

Karam, J.D. and Konigsberg, W.H., DNA polymerase of the T4-related bacteriophages. *Prog Nucleic Acid Res Mol Biol* 64: 65–96, 2000.

Kim, J.-S. and Davidson, N., Electron microscope heteroduplex study of sequence relations of T2, T4, and T6 bacteriophage DNAs. *Virology* 57: 93–111, 1974.

Kovalyova, I.V. and Kropinski, A.M., The complete genomic sequence of lytic bacteriophage gh-1 infecting *Pseudomonas putida*—evidence for close relationship to the T7 group. *Virology* 311: 305–315, 2003.

Kropinski, A.M., Sequence of the genome of the temperate, serotype-converting, *Pseudomonas aeruginosa* bacteriophage D3. *J Bacteriol* 182: 6066–6074, 2000.

Kutter, E., Analysis of bacteriophage T4 based on the completed sequence data., pp. 13–28 in *Integrative Approaches to Molecular Biology*, J. Collado-Vides, B. Magasanik, and T. Smith (Eds.). MIT Press, Cambridge, Massachusetts, 1996.

Kutter, E., Stidham, T., Guttman, B., Kutter, E., Batts, D., Peterson, S., Djavakhishvili, T., et al., Genomic map of bacteriophage T4, pp. 491–519 in *Molecular Biology of Bacteriophage T4*, J. Karam, J.W. Drake, K.N. Kreuzer, G. Mosig, D.H. Hall, F.A. Eiserling, L.W. Black, et al. (Eds.). American Society for Microbiology, Washington, D.C., 1994.

Lang, A.S. and Beatty, J.T., Genetic analysis of a bacterial genetic exchange element: The gene transfer agent of rhodobacter capsulatus. *Proc Natl Acad Sci U S A* 97: 859–864, 2000.

Lawrence, J.G., Hatfull, G.F. and Hendrix, R.W., Imbroglios of viral taxonomy: Genetic exchange and failings of phenetic approaches. *J Bacteriol* 184: 4891–4905, 2002.

Lawrence, J.G., Hendrix, R.W. and Casjens, S., Where are the pseudogenes in bacterial genomes? *Trends Microbiol* 9: 535–540, 2001.

Lawrence, J.G. and Ochman, H., Molecular archaeology of the *Escherichia coli* genome. *Proc Natl Acad Sci U S A* 95: 9413–9417, 1998.

Lazarevic, V., Dusterhoft, A., Soldo, B., Hilbert, H., Mauel, C. and Karamata, D., Nucleotide sequence of the *Bacillus subtilis* temperate bacteriophage SPbetac2. *Microbiology* 145: 1055–1067, 1999.

Lehnherr, H., Bacteriophage P1, in *The Bacteriophages,* R. Calendar (Ed.). Oxford University Press, New York, 2005.

Leipe, D.D., Aravind, L. and Koonin, E.V., Did DNA replication evolve twice independently? *Nucleic Acids Res* 27: 3389–3401, 1999.

Liu, Q., Belle, A., Shub, D.A., Belfort, M. and Edgell, D.R., SegG endonuclease promotes marker exclusion and mediates co-conversion from a distant cleavage site. *J Mol Biol* 334: 13–23, 2003.

Lobocka, M., Rose, M., Samojedny, A., Lehnherr, H., Yarmolinski, M.B. and Blattner, F.R., The Genome of Bacteriophage P1. *J. Bacteriol,* Submitted, 2004.

Lucchini, S., Desiere, F. and Brüssow, H., Comparative genomics of *Streptococcus thermophilus* phage species supports a modular evolution theory. *J Virol* 73: 8647–8656, 1999a.

Lucchini, S., Desiere, F. and Brüssow, H., The genetic relationship between virulent and temperate *Streptococcus thermophilus* bacteriophages: Whole genome comparison of cos-site phages Sfi19 and Sfi21. *Virology* 260: 232–243, 1999b.

Lucchini, S., Desiere, F. and Brüssow, H., Similarly organized lysogeny modules in temperate siphoviridae from low GC content gram-positive bacteria. *Virology* 263: 427–435, 1999c.

Matsuzaki, S., Inoue, T., Kuroda, M., Kimura, S. and Tanaka, S., Cloning and sequencing of major capsid protein (mcp) gene of a vibriophage, KVP20, possibly related to T-even coliphages. *Gene* 222: 25–30, 1998.

Matsuzaki, S., Tanaka, S., Koga, T. and Kawata, T., A broad-host-range vibriophage, KVP40, isolated from sea water. *Microbiol Immunol* 36: 93–97, 1992.

Mesyanzhinov, V.V., Robben, J., Grymonprez, B., Kostyuchenko, V.A., Bourkaltseva, M.V., Sykilinda, N.N., Krylov, V.N., et al., The genome of bacteriophage φKZ of Pseudomonas aeruginosa. *J Mol Biol* 317: 1–19, 2002.

Miller, E.S., Heidelberg, J.F., Eisen, J.A., Nelson, W.C., Durkin, A.S., Ciecko, A., Feldblyum, T. V., et al., Complete genome sequence of the broad-host-range vibriophage KVP40: Comparative genomics of a T4-related bacteriophage. *J Bacteriol* 185: 5220–5233, 2003a.

Miller, E.S., Kutter, E., Mosig, G., Arisaka, F., Kunisawa, T. and Ruger, W., Bacteriophage T4 genome. *Microbiol Mol Biol Rev* 67: 86–156, 2003b.

Monod, C., Repoila, F., Kutateladze, M., Tétart, F. and Krisch, H.M., The genome of the pseudo T-even bacteriophages, a diverse group that resembles T4. *J Mol Biol* 267: 237–249, 1997.

Morgan, G.J., Hatfull, G.F., Casjens, S. and Hendrix, R.W., Bacteriophage Mu genome sequence: Analysis and comparison with Mu-like prophages in Haemophilus, Neisseria and Deinococcus. *J Mol Biol* 317: 337–359, 2002.

Nakagawa, I., Kurokawa, K., Yamashita, A., Nakata, M., Tomiyasu, Y., Okahashi, N., Kawabata, S., et al., Genome sequence of an M3 strain of *Streptococcus pyogenes* reveals a large-scale genomic rearrangement in invasive strains and new insights into phage evolution. *Genome Res* 13: 1042–1055, 2003.

Nakayama, J.-i., Klar, A.J.S. and Grewal, S.I.S., A chromodomain protein, Swi6, performs imprinting functions in fission yeast during mitosis and meiosis. *Cell* 101: 307–317, 2000.

Nakayama, K., Kanaya, S., Ohnishi, M., Terawaki, Y. and Hayashi, T., The complete nucleotide sequence of phi CTX, a cytotoxin-converting phage of Pseudomonas aeruginosa: Implications for phage evolution and horizontal gene transfer via bacteriophages. *Mol Microbiol* 31: 399–419, 1999.

Neve, H., Zenz, K.I., Desiere, F., Koch, A., Heller, K.J. and Brüssow, H., Comparison of the lysogeny modules from the temperate *Streptococcus thermophilus* bacteriophages TP-J34 and Sfi21: Implications for the modular theory of phage evolution. *Virology* 241: 61–72, 1998.

Ohnishi, M., Kurokawa, K. and Hayashi, T., Diversification of *Escherichia coli* genomes: Are bacteriophages the major contributors? *Trends Microbiol* 9: 481–485, 2001.

Ohnishi, M., Terajima, J., Kurokawa, K., Nakayama, K., Murata, T., Tamura, K., Ogura, Y., et al., Genomic diversity of enterohemorrhagic *Escherichia coli* O157 revealed by whole genome PCR scanning. *Proc Natl Acad Sci U S A* 99: 17043–17048, 2002.

Pajunen, M.I., Kiljunen, S.J., Soderholm, M.E. and Skurnik, M., Complete genomic sequence of the lytic bacteriophage φYeO3-12 of *Yersinia enterocolitica* serotype O:3. *J Bacteriol* 183: 1928–1937, 2001.

Pedulla, M.L., Ford, M.E., Houtz, J.M., Karthikeyan, T., Wadsworth, C., Lewis, J.A., Jacobs-Sera, D., et al., Origins of highly mosaic mycobacteriophage genomes. *Cell* 113: 171–182, 2003.

Petrov, V. and Karam, J.D., Functional Genomics Of The T4-Like Myoviridae, in *15th Evergreen International Phage Biology Meeting*, Olympia, Washington, 2003.

Pfister, P., Wasserfallen, A., Stettler, R. and Leisinger, T., Molecular analysis of Methanobacterium phage psiM2. *Mol Microbiol* 30: 233–244, 1998.

Porwollik, S., Wong, R.M. and McClelland, M., Evolutionary genomics of Salmonella: Gene acquisitions revealed by microarray analysis. *Proc Natl Acad Sci U S A* 99: 8956–8961, 2002.

Prangishvili, D., Evolutionary insights from studies on viruses of hyperthermophilic archaea. *Res Microbiol* 154: 289–294, 2003.

Prangishvili, D. and Garrett, R.A., Exceptionally diverse morphotypes and genomes of crenarchaeal hyperthermophilic viruses. *Biochem Soc Trans* 32: 204–208, 2004.

Proux, C., van Sinderen, D., Suarez, J., Garcia, P., Ladero, V., Fitzgerald, G.F., Desiere, F., et al., The dilemma of phage taxonomy illustrated by comparative genomics of Sfi21-like Siphoviridae in lactic acid bacteria. *J Bacteriol* 184: 6026–6036, 2002.

Rachel, R., Bettstetter, M., Hedlund, B.P., Haring, M., Kessler, A., Stetter, K.O. and Prangishvili, D., Remarkable morphological diversity of viruses and virus-like particles in hot terrestrial environments. *Arch Virol* 147: 2419–2429, 2002.

Ravin, V., Ravin, N., Casjens, S., Ford, M.E., Hatfull, G.F. and Hendrix, R.W., Genomic sequence and analysis of the atypical temperate bacteriophage N15. *J Mol Biol* 299: 53–73, 2000.

Repoila, F., Tétart, F., Bouet, J.-Y. and Krisch, H.M., Genomic polymorphism in the T-even bacteriophages. *EMBO J* 13: 4181–4192, 1994.

Rohwer, F., Global phage diversity. *Cell* 113: 141, 2003.

Rohwer, F. and Edwards, R., The phage proteomic tree: A genome-based taxonomy for phage. *J Bacteriol* 184: 4529–4535, 2002.

Rohwer, F., Segall, A., Steward, G., Seguritan, V., Breitbart, M., Wolven, F. and Azam, F., The complete genomic sequence of the marine phage Roseophage SIO1 shares homology with non-marine phages. *Limnology and Oceanography* 45: 408–418, 2000.

Russell, R.L., Comparative genetics of the T-even bacteriophages. *Genetics* 78: 967–988, 1974.

Ruzin, A., Lindsay, J. and Novick, R.P., Molecular genetics of SaPI1—a mobile pathogenicity island in *Staphylococcus aureus*. *Mol Microbiol* 41: 365–377, 2001.

Smoot, J.C., Barbian, K.D., Van Gompel, J.J., Smoot, L.M., Chaussee, M.S., Sylva, G.L., Sturdevant, D.E., et al., Genome sequence and comparative microarray analysis of serotype M18 group A Streptococcus strains associated with acute rheumatic fever outbreaks. *Proc Natl Acad Sci U S A* 99: 4668–4673, 2002.

Smoot, L.M., Smoot, J.C., Graham, M.R., Somerville, G.A., Sturdevant, D.E., Migliaccio, C.A., Sylva, G.L., et al., Global differential gene expression in response to growth temperature alteration in group *A Streptococcus*. *Proc Natl Acad Sci U S A* 98: 10416–10421, 2001.

Tetart, F., Desplats, C. and Krisch, H.M., Genome plasticity in the distal tail fiber locus of the T-even bacteriophage: recombination between conserved motifs swaps adhesion specificity. *J Mol Biol* 282: 543–556, 1998.

Tetart, F., Desplats, C., Kutateladze, M., Monod, C., Ackermann, H.W. and Krisch, H.M., Phylogeny of the major head and tail genes of the wide-ranging T4-type bacteriophages. *J Bacteriol* 183: 358–366, 2001.

Van Sluys, M.A., de Oliveira, M.C., Monteiro-Vitorello, C.B., Miyaki, C.Y., Furlan, L.R., Camargo, L.E., da Silva, A.C., et al., Comparative analyses of the complete genome sequences of Pierce's disease and citrus variegated chlorosis strains of *Xylella fastidiosa*. *J Bacteriol* 185: 1018–1026, 2003.

Ventura, M., Foley, S., Bruttin, A., Chennoufi, S.C., Canchaya, C. and Brüssow, H., Transcription mapping as a tool in phage genomics: The case of the temperate *Streptococcus thermophilus* phage Sfi21. *Virology* 296: 62–76, 2002.

Wang, J., Sattar, A.K.M.A., Wang, C.C., Karam, J.D., Konigsberg, W.H. and Steitz, T.A., Crystal structure of a pol α family replication DNA polymerase from bacteriophage RB69. *Cell* 89: 1087–1099, 1997.

Whiteley, M., Bangera, M.G., Bumgarner, R.E., Parsek, M.R., Teitzel, G.M., Lory, S. and Greenberg, E.P., Gene expression in *Pseudomonas aeruginosa* biofilms. *Nature* 413: 860–864, 2001.

Wommack, K.E. and Colwell, R.R., Virioplankton: viruses in aquatic ecosystems. *Microbiol Mol Biol Rev* 64: 69–114, 2000.

Wood, H.E., Dawson, M.T., Devine, K.M. and McConnell, D.J., Characterization of PBSX, a defective prophage of *Bacillus subtilis*. *J Bacteriol* 172: 2667–2674, 1990.

Woods, D.E., Jeddeloh, J.A., Fritz, D.L. and DeShazer, D., *Burkholderia thailandensis* E125 harbors a temperate bacteriophage specific for *Burkholderia mallei*. *J Bacteriol* 184: 4003–4017, 2002.

Yeh, L.-S., Hsu, T. and Karam, J.D., Divergence of a DNA replication gene cluster in the T4-related bacteriophage RB69. *J Bacteriol* 180: 2005–2013, 1998.

Yokoyama, K., Makino, K., Kubota, Y., Watanabe, M., Kimura, S., Yutsudo, C.H., Kurokawa, K., et al., Complete nucleotide sequence of the prophage VT1-Sakai carrying the Shiga toxin 1 genes of the enterohemorrhagic *Escherichia coli* O157:H7 strain derived from the Sakai outbreak. *Gene* 258: 127–139, 2000.

Zachary, A., An ecological study of bacteriophages of *Vibrio natriegens*. *Can J Microbiol* 24: 321–324, 1978.

6 Phage Ecology

Harald Brüssow[1] *and Elizabeth Kutter*[2]

[1]Nestlé Research Center, Lausanne, Switzerland
[2]Lab of Phage Biology, The Evergreen State College, Olympia, WA

CONTENTS

6.1. Introduction ... 130
6.2. General Principles ... 131
  6.2.1. Phage Numbers in the Natural World ... 131
  6.2.2. Dynamic Phage-Host Relationships ... 132
  6.2.3. Co-Evolution of Phages and Their Host Bacteria ... 135
  6.2.4. Effects of Host Physiology and Nutritional Status ... 136
  6.2.5. Biofilms ... 137
6.3. Marine Phage Ecology ... 139
  6.3.1. Marine Phage Impacts on the Food Web ... 139
  6.3.2. Marine Phage Characterization ... 141
    6.3.2.1. Cyanophages ... 141
    6.3.2.2. Genomic Analysis of Marine Phages ... 142
  6.3.3. Marine Phage Host Specificity and Horizontal Gene Transfer ... 142
6.4. Soil and Plant-Associated Phages ... 144
6.5. Animal-Associated Phages ... 145
  6.5.1. Introduction ... 145
  6.5.2. Enteric Phages ... 146
  6.5.3. Other Phage Sites of Interest: Oral Cavity and Vagina ... 149
6.6. Industrial Phage Ecology ... 150
  6.6.1. Dairy Industry ... 150
  6.6.2. Non-Dairy Food Fermentation: Sauerkraut and Sausage ... 155
6.7. Outlook ... 155
Acknowledgements ... 156
References ... 156

0-8493-1336-X/05/$0.00+$1.50
© 2005 by CRC Press LLC

6.1. INTRODUCTION

Felix d'Herelle, co-discoverer of phage, had a strikingly modern approach to biology. Nearly 100 years ago, he used living organisms to control pests (diarrhea-causing bacteria to halt locust epidemics) and disease (phage therapy of diarrheal diseases). His approaches reflected ecological insights before this branch of biology became an established scientific discipline. In fact, one might have predicted that phage research would become a springboard for studies of microbial ecology. Instead, studies of phage ecology were largely ignored and phage research became the cradle of molecular biology. This turn in the history of biological research is not explained by any critical technical breakthrough, but rather by a number of biographical reasons in the lives of a handful of scientists. The second generation of Western phage researchers concentrated on a few phages from *E. coli*, the workhorse of bacterial genetics, in order to better understand the basic nature of phages and of the phage infection process and use this knowledge to explore fundamental aspects of biology at the molecular level. Phage ecology was not within their conceptual framework. The diversity of phages was better appreciated by medical microbiologists, who used phages for the typing of clinical isolates of bacterial pathogens. However, in that field phages were exclusively used as tools without intrinsic interest in their ecology or molecular characteristics. Consequently, the first monograph on the distribution and behavior of bacterial viruses in the environment appeared only in 1987 (Goyal).

Since the appearance of Goyal's book, the study of phage ecology has fundamentally changed. One drastic reminder of the importance of phage in the ecosystem was the surprising discovery of very large numbers of phage-like particles in the ocean (Bergh et al., 1989). Phage ecology quickly became an intensively investigated branch of marine microbiology, as documented by a recent review listing hundreds of publications, mostly from the last decade (Wommack and Colwell, 2000). Recently, general reviews have appeared on various major aspects of phage ecology (Abedon, 2005; Ashelford et al., 2003; Azam and Worden, 2004; Breitbart et al., 2002; 2003; Chibani-Chennoufi et al. 2004; Paul and Kellogg, 2000; Suttle, 2000b).

Scientists in two fields have developed a particularly keen interest in phage ecology. One is the food industry, where fermentation techniques are used to transform milk, vegetables, or meat into processed foods like cheese, sauerkraut, or salami. This food production relies either on spontaneous fermentation or, in the case of milk, on fermentation initiated by the addition of industrial bacterial starters. Phages that infect these starters are the major cause of fermentation failures in the dairy industry (Chapter 10). The high economic losses associated with phage infection there motivated intense research into phages from lactic acid bacteria, which are the major dairy starter organisms. As the dairy factory is a man-made environment, it did not so much attract the interest of ecologists, but rather that of more technologically oriented microbiologists, focusing on the design of efficient starter rotation systems and the construction of phage-resistant starter cells. In doing that, dairy microbiologists had to investigate the factory ecology of phage infections, leading to large systematic collections of dairy phages. However, these phages were characterized more by sequence analysis and molecular biology than by classical ecological approaches. The other emerging field is linked to the rekindling of interest in phage therapy, as discussed in Chapters 13 and 14. The successful application of this approach to the growing problem of antibiotic resistance depends on the availability of large collections of phages and a detailed knowledge of phage-host interactions in different physiological compartments that necessitates sound ecological knowledge of the interactions between the phage, bacteria, and plant or animal host. It is also increasingly extending to interest in potential applications against plant pathogens, bacteria in biofilms, and other complex real-world situations.

6.2. GENERAL PRINCIPLES

6.2.1. Phage Numbers in the Natural World

A short 1989 *Nature* paper by a Norwegian group reported 10^7 phage-like particles per ml of coastal and open ocean water, and even 20-fold higher concentrations in a pre-alpine lake (Bergh et al., 1989). Two independent *Nature* papers in the following year confirmed the findings and provided experimental evidence that these viruses limit the primary productivity of cyanobacteria, the major oceanic photosynthetic bacteria (Suttle, 1990; Proctor, 1990). Goyal's 1987 *Phage Ecology* monograph still started the marine chapter with a quote suggesting that phages isolated from the oceans are not indigenous to the marine environment but are transported to the sea by rivers or sewage. Now we know that phage are universally observed in the open and coastal ocean all over the world; in surface water and in great depth (Cochlan, 1993); in ocean ice (Maranger, 1994) and in ocean sediment. Even higher phage concentrations have been detected in marine sediments than in the water columns above them. Counts of up to 10^9 phage particles per ml of sediment were reported by Danovaro and Serresi (2000). Meaningful studies of phage distribution are much more difficult in terrestrial ecosystems, but such techniques as electron microscopy suggest concentrations of the order of 10^7 viruses/gram in soil (Ashelford et al., 2003) and in the feces of ruminants (Furuse, 1987). There seems now to be broad agreement that phage are the most abundant life form on earth; the total number is generally estimated at 10^{30} to 10^{32}. Most of our detailed quantitative data still comes from marine environments, where extensive ongoing mixing makes meaningful sampling possible, but many of the principles and insights also seem applicable to other environments.

In all but the most extreme environments, large numbers of different bacteria and phages are found; there is growing suspicion that phages may represent the largest unexplored reservoir of sequence information in the biosphere. For example, in the case of Chesapeake Bay, Wommack et al. (1999) have estimated that there are about 100–300 phage strains (with many variants of each), infecting 10–50 different bacterial species. Random sequencing of viral DNA from two uncultured 100-liter marine water samples suggested that they contained between 400 and 7000 different phage types (Breitbart et al., 2002); a similar analysis shows that the human gut also contains hundreds of different phage genotypes (Breitbart et al., 2003).

Marine phage biologists have developed precise data about *in situ* burst sizes (Borsheim, 1993), since this figure is essential for calculations of virus production in the given environment and the level of virus-mediated mortality of bacterioplankton;

such determinations have not generally been possible in other natural habitats. The burst size in the environment is generally smaller than that determined in the laboratory by one-step growth experiments, reflecting the smaller size of bacteria in most natural settings (Robertson and Button, 1989; Weinbauer and Peduzzi, 1994). The nutrient level and temperature of the water sample and the morphology of the host were the most important determinants of the burst size. In situ burst-size determination is mostly done by microscopic observation of virus particles within bacterial cells (Hennes, 1995; Weinbauer et al., 1993). Alternatively, the *in situ* burst size is calculated by balancing viral production with viral decay (Suttle and Chan, 1994), assuming that if the phage concentrations are relatively stable over a given (short) time period, then the rates of phage production and phage decay must be equal. This lets one look at specific viable phage infecting a particular host. Quantitative data on virus decay are important for other fields of research as they provide a measure of the tenacity of a virus in the environment. Such information is crucial for such diverse questions as public health evaluations of viral contamination of water samples, source tracking, the persistence of phages in industrial environments, and the half-life of therapeutic phages.

6.2.2. Dynamic Phage-Host Relationships

Major fluctuations in phage and host numbers are observed over various periods. For example, there is substantial seasonality in marine phage titers (Bratbak, 1990; Hennes, 1995; Mathias, 1995; Cochran and Paul, 1998). The levels are often 10-fold lower in the winter than in the summer months, even though the bacterial densities do not appear to fluctuate by a factor of more than 2–3, with no seasonal trend; the reasons behind this apparent uncoupling of bacterial and phage populations are not yet clear, but one explanation could be increased induction of prophages in the summer months due to higher sunlight exposure. Another possibility is that phage production itself is much more temperature sensitive, in general, than is bacterial growth. At the same time, the UV in sunlight causes up to 5% loss in viable phage per hour for surface water due to production of thymine dimers (Wommack, et al., 1996); this causes far more loss in summer than in winter at more polar latitudes. A strong shift in balance was seen over 6 months between the two major phage types infecting the rhizosphere of sugar beets: from *Siphoviridae* with long latent period and big burst sizes to *Podoviridae* with short latent periods and small burst sizes, apparently reflecting changes with season in the availability and physiological state of the host bacteria and plants (Ashelford et al., 1999). Variability was also shown over shorter time periods, in one marine study even over half-hour time intervals, demonstrating a highly dynamic relationship between phages and their hosts (Bratbak, 1996).

As discussed in Chapters 3 and 7 and by Abedon (2005), the phage lytic life cycle has at least 4 key steps that are very relevant to phage ecology:

1. An extracellular search that is limited by diffusion rates and thus dependent on host concentration.
2. A phage adsorption step that combines reversible phage binding, irreversible attachment, and genome transfer into the host, which typically occurs

rapidly following productive collision between a phage particle and a phage-susceptible bacterium.
3. An infection step, during which host physiology is appropriately restructured, the phage genome is replicated, and phage particles are assembled.
4. For *temperate* phages, an indefinite period of *lysogeny* may be inserted, during which the phage is inserted in the host genome and replicates with it or replicates synchronously as a plasmid and the genes responsible for lytic growth are repressed, followed by a step in which phage progeny are released from the infected bacterium; except for filamentous phages, this process involves cell lysis, which usually is carefully timed.

The attachment and lysis steps are generally rapid. Thus, most of the time the phage is either:

1. Free during the extracellular search,
2. Trapped in some compartment where no bacteria are readily available, such as bound to relatively inert particulate matter,
3. Actively infecting a bacterium, leading eventually to cell lysis or phage secretion, or
4. Existing as a prophage

For the first two, the phage may become inactivated through capsid or genome damage, but it also may remain infectious for many years, depending on environmental conditions. The lytic infection phase seldom lasts more than a few minutes to a day, though there are conditions like the one we call "hibernation" in which the phage genome remains benignly inside until nutrients become available and the host resumes more active growth. For virulent phages like T4, this differs markedly from lysogeny, though, in that the phage genome then takes over and the only possible outcome is host-cell death, accompanied by eventual phage production if nutrients become available, as discussed in 2.4; no colonies are formed.

Our understanding of the phage infection process comes primarily from experiments in which the laboratory researcher mixes a single phage strain with a single bacterial strain at about 10^8 cells/ml. In contrast, in near coastal water, for example, bacterioplankton concentrations are typically 10^6 cells/ml and the population normally consists of 100 different bacterial host species, yielding a mere 10^4 cells/ml for the average host species (Murray and Jackson, 1992). Is this enough to maintain an infection cycle? The answer is apparently yes, since no marine water samples devoid of phage were ever reported, and enrichment procedures generally permit the isolation of phage against any given host from the particular ecosystem. Yet phage replication is clearly sensitive to effective cell concentration. Laboratory experiments with T4 and *Bacillus* and *Staphylococcus* phages showed no phage production until the host cell concentration reached 10^4 cells/ml (Wiggins and Alexander, 1985). However, studies using *Pseudomonas* phages (Kokjohn et al., 1991) showed evidence of lytic infection at cell concentrations as low as 10^2 cells/ml. In some studies in natural marine environments, no intracellular phage were observed when the number of rod-shaped bacteria fell below 10^5 cells/ ml (Steward et al., 1992;

Weinbauer and Peduzzi, 1994). However, some marine viruses replicated efficiently down to 10^3 specific host cells/ml (Suttle and Chan, 1994; 1993). The ups and downs of host cell concentration over a time series allowed the approximation that cyanophage replication still occurred when the host cell concentration fell to 10^2 cells/ml (Waterbury, 1993).

Such theoretical concerns related to the reproduction of virulent phages at low host densities led to the hypothesis that temperate phages should outnumber virulent phages in the ocean, since the production of temperate phages is independent of host cell density. It has been proposed that lysogeny becomes the preferred strategy when the cell density falls below the lower limit necessary for maintenance of the phage density by repeated cycles of lytic infections (Stewart and Levin, 1984). Lysogens might out-compete the non-lysogenic congeners by the selective advantage conferred by lysogenic conversion genes contributed by many temperate phages. Some of these are relatively universal, such as immunity functions and superinfection-exclusion genes. Other prophages contribute genes that make the lysogen competitive under special ecological situations, such as the serum resistance conferred to the lysogen by the phage lambda *bor* gene under blood growth of *E. coli*. This phenomenon is very marked in lysogenic bacterial pathogens, where many virulence factors are encoded by prophages. However, even laboratory phages like P1, P2, lambda, and Mu led to higher metabolic activity and faster and longer growth than seen in non-lysogens (Edlin et al., 1975; Lin et al., 1977). With this selective advantage even under laboratory conditions and the intrinsic difficulties with the lytic life style, the prediction is a high concentration of lysogens in the oceans. Indeed, two marine surveys revealed 40% mitomycin-C-inducible cells; similar proportions of lysogens were identified in *Pseudomonas* colonies from lakes (Ogunseitan et al., 1992). The surveys showed the trend for lysogeny to be more prevalent in oligotrophic environments (Jiang, 1994; 1997). This observation fits with theory since this setting is dominated by low densities of slow-growing bacteria. However, other data contradict this explanation. Surveys in estuarine waters showed a seasonal development of lysogeny with highs in the summer months when eutrophic conditions were prevalent and lows in the winter months when cells were at their minimum (Cochran and Paul, 1998). There are further contradictions of expectations. First, spontaneous induction of prophages is generally low (10^{-2} to 10^{-5} phage per bacterium per generation) (Stewart and Levin, 1984). This release can only account for <1% of the phage concentrations in the ocean (Jiang, 1997). Second, large phage surveys in the North Sea revealed that only 10% of the phage isolates are temperate (Moebus, 1983).

Paul and Kellogg (2000) extensively explored the available research related to the frequency of lysogeny in natural environments, the factors that can induce such lysogens, and the roles of phages in bacterial genetic exchange in various ecosystems. A variety of approaches have indicated that lysogeny and even polylysogeny are common; the various microbial genome projects to date have confirmed this, with at least half of the sequenced bacteria carrying prophages, and prophages or defective prophages are responsible for many differences between isolates of the same species. The extent of lysogeny varies between different kinds of bacteria; for example, nearly 100% of naturally occurring *Pseudomonas* are lysogenic. Prophages could be induced from most bacteria in eutrophic lakes and estuaries, while induction was far less common in offshore and northern-lake environments. The degree to which this reflects prophage presence vs. metabolic state is not clear. Inducibility of lysis and phage release by means of mitomycin C or UV are the most common criteria for the presence of lysogeny; however, the two may not induce the same prophages, and many prophages are not induced by either of them. Other inducing agents that have been explored on natural isolates include sunlight, temperature, pressure, polynuclear aromatic hydrocarbons (PAHs), fuel oil, and trichloroethylene. PAHs, a PCB mixture, and Arochlor 1248 were the most efficient agents, giving effective induction of prophages in 75% of the tested samples, vs. 50% for 254-nm UV radiation and for mitomycin C. Raising the temperature to as little as 30 degrees for 30 min could induce lysis; Paul and Kellogg suggest that this may help explain the 10-fold summer increase in free phage. Induction also generally works less well when the host cells are at a lower metabolic state, since most inducing agents act on replicating DNA.

In summary, it appears that a key aspect of the high prevalence of virulent phages in the oceans is that the numbers of any given phage-host pair are constantly fluctuating in any natural setting, probably coupled with the continual mixing and astronomical total numbers involved. Phage replicate most rapidly on the most abundant, fastest-growing host population in a given setting at a given time, where new hosts are found most rapidly, thus, for example, terminating microbial blooms (cf. Hennes and Simon, 1995). Various approaches in different aquatic environments suggest that about 15% of the bacterioplankton are lysed by phages daily, leading to the release of nutrients important to marine ecosystems. The fact that most phage will not replicate to a significant degree at host concentrations below 10^3–10^5 per ml assures the maintenance of microbial diversity despite the presence of phages that can infect each potential host. A high fraction of the bacteria in the oceans, as in many other habitats, also harbour one or more prophages, but free temperate phages do not generally contribute substantially to the high concentrations of phages observed in both marine and terrestrial environments.

6.2.3. Co-Evolution of Phages and Their Host Bacteria

Classical experiments with *E. coli* and its phages generally showed a rapid outgrowth of phage-resistant bacterial strains. The coexistence of susceptible bacteria and their corresponding phages in the environment was thus a somewhat surprising observation. Further classical experiments showed a co-evolution, with bacteria and their phages involved in a continuous cycle of resistance and counter-resistance mutations. Theoretical ecologists pointed to an asymmetry in this relationship, since receptor mutations arise more easily in bacteria than anti-receptor mutations in phages (Lenski 1984; 1985). Since there are far more phages than bacteria in many ecological settings, phages may balance these differences out. The development of phage-resistant cyanobacteria was demonstrated in the field (Waterbury, 1993). Other researchers argued that virulent phages might survive due to the maintenance of sensitive parental strains in the population if the parental strains have a slight growth advantage. Most discussions of these issues use the terms *resistant* and *sensitive* as if they are absolutes. In nature, one actually sees many cases in which a particular host can be infected by a given phage, but with low efficiency; this can also lead to co-survival of phage and host. For example, in spot testing nearly 100 phages against a battery of hosts, 10 were identified as able to infect *E. coli* O157; however, the EOP turned out to be only about 10^{-4} for most of them (Kutter lab, unpublished). Only RB69 plated as efficiently on O157 as on B or K12. Parameters affecting the efficiency of plating are discussed in the Appendix, section A.4.1.2.

6.2.4. Effects of Host Physiology and Nutritional Status

The normal state of a marine bacterium was predicted to correspond to the nutritional state of a laboratory bacterium under stationary-phase conditions (Kolter et al., 1993). While gram-negative bacteria do not sporulate, as do some gram-positive bacteria, they do undergo a variety of metabolic and structural changes in stationary-phase conditions that contribute to long-term survival in hostile environments. These are mediated by a new sigma factor, σ^s, which controls at least 30 genes that are expressed during starvation and at the transition into stationary phase. *RpoS* mutants survive much less well under laboratory conditions of carbon or nitrogen starvation and fail to develop starvation-mediated cross protection to oxidative, osmotic, and heat stresses (McCann et al., 1991). From laboratory studies, it has been widely accepted that most phages cannot productively infect stationary-phase bacteria, leading to great surprise at the observed phage concentration levels in the ocean and illustrating gaps in our knowledge when we try to transfer our experiences from laboratory phage-host interactions to the ecological situation.

Clearly part of the problem was the limited number of phage-host systems on which the accepted model had been based. Woods (1976) had actually found that *Pseudomonas* phages could infect starved host cells maintained for 40 days in natural riverine conditions. Under these conditions, the latent period was lengthened and the burst size greatly reduced when compared to logarithmic-phase infection. Schrader et al. (1997) showed that coliphage T7 and the three *Pseudomonas* phages he tested could replicate well in cells that were starved or had entered stationary phase; the variety of patterns seen emphasizes the importance of avoiding generalizations. Phage ACQ could even infect *P. aeruginosa* maintained in starvation conditions for 5 years; the burst size was reduced whether the starvation was short or long, and the latent period was extended severalfold. For T7, the latent period was also lengthened severalfold, but, interestingly, the burst size *increased* from about 50 to 450 when the host had been starved for 24 hours. T4 cannot produce a burst in stationary-phase cells. However, T4 can enter and persist in long-term stationary-phase cells, but as long as the stationary-phase sigma factor is present, the infection process is suspended at an early stage. Whereas T4 normally blocks all host-gene transcription and translation within minutes, it enters what we call "hibernation mode" and produces a stable infective center in these starved cells (Kutter lab, unpublished results; Fig. 6.1); when nutrients become available, the usual host outgrowth proteins are produced, but then the phage program takes over, all further host protein synthesis is blocked, and progeny phage rather than colonies are produced after resumption of cell growth. This phenomenon could help explain the persistence of many virulent phages within populations of non-growing cells.

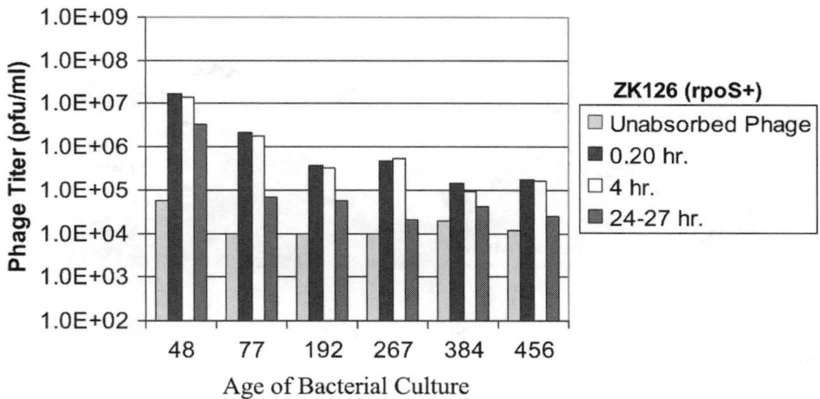

FIGURE 6.1 Ability of bacteriophage T4 to survive in stationary-phase *E. coli* for extended periods and still respond to form an infective-center plaque when transferred to nutrient-rich plates. After growth in TSB for varying times (expressed in hours), 10^7 phage/ml were added to samples of the culture. This ability to form stable infective centers in starved cells depended on the presence of the stationary-phase sigma factor, rpoS; in otherwise-isogenic rpoS cells, few infective centers were seen even at 4 hours (Kutter, 2001).

Complementary observations have been made in *sporulating* bacteria, illustrating some of the complexities of phage-host interactions under growth-limited conditions in the wild. For example, phages φe and φ29 are lytic when infecting growing *B. subtilis*. However, each phage shows a narrow window during sporulation when it can become entrapped in the spore and persist in a quiescent state which allows later germination of the spore followed by formation of an infective center rather than a bacterial colony (Sonenshein and Roscoe, 1969). Also, the complex life cycle of the gram-negative bacterium *Myxococcus xanthus* includes the formation of fruiting bodies containing relatively stable myxospores. Bacteriophage MX-1 is virulent on exponential-phase cells but will not bind to myxospores (Burchard and Dworkin, 1966). For an hour after the transition has been induced, MX-1 still binds and injects its DNA but the phage genome becomes trapped in a persistent state; again in this case, germination begins normally upon nutrient addition, but within an hour the phage program takes over (Burchard and Voelz, 1972).

6.2.5. Biofilms

Most studies of the phage infection process have been carried out with bacteria suspended free in liquid culture. However, at interfaces between solid surfaces and aqueous environments, bacteria frequently aggregate to form complex attached communities called *biofilms* (Fig. 6.2). Such biofilms are widespread and pervasive—on river rocks, in pipes, industrial equipment and medical implants, lining the colon, as dental plaque, in the lungs of cystic fibrosis patients (cf. Costerton, 1999). Microorganisms undergo profound phenotypic developmental changes during the

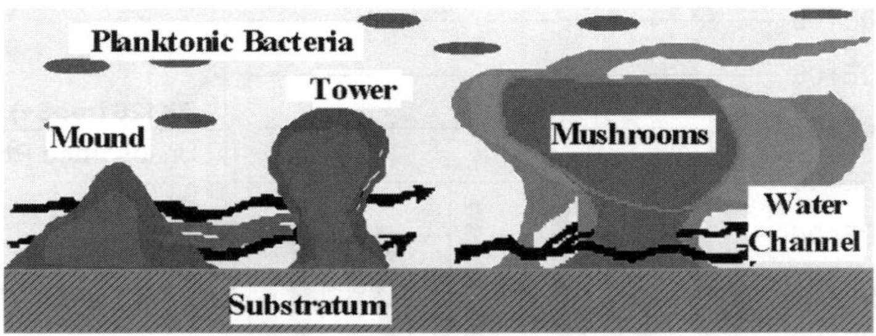

FIG 6.2 Complex structures and water channels proposed for biofilm architecture. Drawn by Kalai Mathee.

transition from free-floating planktonic to biofilm growth (O'Toole et al., 2000). Due to these inherent physiological changes, their slow growth rates and the fact that the cells are embedded in an extensive exopolysaccharide matrix with heterogeneous spatial distribution, cells living in biofilms tend to be at least 10-fold more resistant than planktonic cells to most antibiotics and other antimicrobial treatments (Anwar et al., 1992; Bagge et al., 2004).

Interest is growing in the natural roles of bacteriophages in modulating biofilms and, particularly, in the potential for their use in controlling biofilms in a variety of settings. Adams and Park (1956) first explored the properties of a phage polysaccharide depolymerase, while Lindberg (1977) reported EM observation of these enzymes as spikes attached to the phage baseplate. Here, the capsular material acts as a secondary receptor to which the phage can bind, degrading the polymer until it can reach its outer-membrane receptor and infect the cell (see Chapter 7, Fig. 7.4). In exploring the role of cell death in *Pseudomonas* biofilm development, Webb (1996) found evidence that an induced *prophage* actually played a significant role in the process. They found that after a few days up to 50% of the microcolony structures within biofilms formed by various bacteria showed areas of killing and lysis in their centers, sculpting internal structures and releasing free-swimming bacteria; this was paralleled by the release into the medium of a temperate phage that seemed to be responsible for this process.

Studies have been carried out of phage interactions with biofilms involving such bacteria as *Pseudomonas aeruginosa, E. coli, Listeria monocytogenes, Enterobacter agglomerans,* and *Staphylococcus aureus.* Doolittle et al. (1995; 1996) demonstrated lytic infection of *E. coli* biofilms by bacteriophage T4 and used fluorescent probes to track the interactions of the phage with the biofilms. Hughes (1998), exploring biofilm bacteria from a food processing factory, showed that *E. agglomerans* phage SF153b has a polysaccharide depolymerase that can disrupt biofilms through exopolysaccharide (EPS) degradation even when a phage mutation blocks cell infection and lysis. They partially purified the enzyme, an endoglycanohydrolase, and confirmed that it could work in isolation and was specific for the *Enterobacter* EPS; it could not attack similar biofilms formed by *Serratia.* As demonstrated by Hanlon et al. (2001), appropriate phages could diffuse through alginate gels and produce a 2-log reduction in bacterial numbers in 20-day-old *P. aeruginosa* biofilms, reducing the viscosity by as much as 40% despite the presence of EPS. Corbin et al. (2001) used a chemostat coupled to a modified Robbins chamber and scanning confocal microscopy and saw clear T4 effects on glucose-limited biofilms, at least at very high multiplicities of infection. Here and in several of the previous cases, there was no evidence for involvement of a degradative enzyme and it is not clear how the phages got through the EPS layer to reach their receptors. McLean et al. (2001) have explored useful techniques for studying the parameters affecting biofilm growth and phage-biofilm interaction.

6.3. MARINE PHAGE ECOLOGY

6.3.1. Marine Phage Impacts on the Food Web

A number of parameters of the phage-bacterium interaction have been determined in quantitative detail for marine phages, making them currently the best-characterized phage ecology system. The production and distribution of marine viruses is, not surprisingly, determined by the productivity and density of the host populations (Boehme, 1993; Jiang, 1994; Weinbauer, 1995). The usual virus-to-bacterium ratio falls between 3 and 10 and depends clearly on the nutrient level: bacterioplankton produce more phages under environmental conditions favouring fast bacterial growth and productivity (Hara, 1996; Maranger, 1994; Steward, 1996). The quantitative relationships led to numerical calculations of the degree of bacterial mortality caused by marine phages (Binder, 1999) and mathematical modeling of the energy flow in marine food webs. Phage predation of marine bacteria now enters into models of global biogeochemical cycling of carbon (Proctor, 1990). Bacterial and algal viruses are established members of the microbial loop in the oceans with profound effects on the cycling of carbon and nitrogen and on the marine food web (Fuhrman, 1999). The change from discovery of their existence to such a prominent place in ecological modelling could not be more dramatic.

When a bacterium is lysed by phage infection, probably 99% of the cell contents enter the dissolved organic matter pool (Fuhrman, 1992). Phage are thus efficient drivers in the biomass-to-dissolved-organic-matter conversion. They are important sources of bacterial mortality in the sea, along with bacterial grazing by zooplankton. The result of phage lysis of marine bacteria is, paradoxically, a stimulation of bacterial growth. The rationale is the following: The presence of phage stimulates the growth of bacteria in comparison to a situation where phages are lacking. Predation of bacteria by protists results in the transfer of bacterial biomass into the next layer of the food web with no feedback (in the literal sense) to bacteria. Phage lysis, in contrast, releases nutrients from the lysed cells that become available to the bacterial community (Middelboe et al., 1996). In one model, viral lysis causes a net loss of 25% in nanozooplankton production (Fuhrman and Suttle, 1993). To complicate the matter further, some protists like dinoflagellates also prey on viral particles.

Substantial theoretical and experimental research efforts were undertaken to determine the quantitative degree of virus-mediated bacterial mortality and to assess the ratio of phage lysis versus grazing by higher organisms. Not surprisingly, the greatest impact of phage lysis was in oligotrophic environments. Furthermore, the ratio of lysis versus grazing changed with water depth both in the ocean and in lakes (Steward, 1996; Weinbauer and Hoefle, 1998). To summarize a large body of literature, different approaches in different environments yielded a remarkably stable rate of virus-mediated bacterioplankton mortality of about 15% per day (Suttle, 1994). The rate seems to be higher for heterotrophic bacteria than for the autotrophic cyanobacteria (Suttle, 1994). Even if these figures seem to suggest only a modest effect of phages on bacteria, a 15% bacterial mortality can still have a profound effect on the relative proportions of different species or strains in a community. Models of virioplankton that control host community diversity have been developed. For a eutrophic estuarine environment, the basic numbers were approximated as follows: bacteria at a density of 10^6/ml with about 50 different species and viruses with a concentration of 10^7/ml and 200 different strains (Wommack et al., 1999). One concept is that of "killing the winner populations," i.e. phage expand on the fastest-growing host population in the given ecological setting (Thingstad and Lignell, 1997). The epidemic ceases when the diminished host population no longer supports efficient phage replication. Blooms have been observed where up to 80% of the total bacterial population is represented by a single bacterioplankton strain, although lower peak levels are more frequent. There is strong indirect evidence that some bloom collapse is mediated by viral lysis. The most convincing data are from the Lake of Constance, where transient increases in bacterial abundance were closely followed by peaks in the frequency of infected bacteria and free phage (Hennes, 1995).

Phage infection in the ocean leads to better retention of nutrients in the euphotic zone because more organic material remains in non-sinkable bacterial form (Murray and Eldridge, 1994; Thingstad et al., 1993). In contrast, lesser phage infection allows a transfer of organic material upwards in the food chain, into organisms that eventually either sink themselves or are compacted in the fecal pellets of organisms with guts (Fuhrman, 1992). These processes transport organic masses from the euphotic zone to the deep sea. They have substantial impact on global climate models via CO_2 fixation from the atmosphere into marine biomass and eventual transfer to the ocean sediment, with the net result of a reduction of this important greenhouse gas (Wilhelm and Suttle, 1999); phage infection helps counteract this. Marine viruses may have an additional effect on the shaping of the global climate by inducing the release of dimethyl sulfide (DMS) from lysed phytoplankton (Malin et al., 1998). DMS is a gas that nucleates cloud formation and thus affects the radiative properties of the atmosphere.

As discussed in section 6.2.1, much quantitation has been carried out in the marine environment, but extrapolation of data from the marine field to others is often problematic. Ocean water often contains substances that protect viruses from inactivation, such as adsorption to clay particles (LaBelle and Gerba, 1982; Smith et al., 1978), as well as heat-labile uncharacterized virucidal substances that show geographical variation (Suttle and Chen, 1992). In principle, two processes must be differentiated: the destruction of phage particles and the loss of infectivity. There is a consensus that unattenuated sunlight is the dominant factor controlling the decay of viral infectivity in surface waters (Garza and Suttle, 1998). Inactivation by sunlight was significant down to a depth of 200 m. UV-A (320 to 400 nm) has the greatest impact (Murray and Jackson, 1993). UV-mediated dimer formation of adjacent pyrimidines was the principal photodamage (Wilhelm et al., 1998). However, host- and phage-encoded photorepair systems could still recover some of the lost infectivity (Bernstein, 1981). Phages with smaller capsid size turned out to be more sensitive than phages with capsids >60 nm (Heldal and Bratbak, 1991; Mathias, 1995). One to five percent infectivity loss per hour was the average, and no marked differences were observed between marine phages and reference laboratory phages (Wommack et al., 1996); under surface light conditions, a one-log loss of infectivity was observed over a day of natural sunlight exposure. Differences were observed between distinct phage isolates infecting the same host, demonstrating that environmental persistence is a trait particular to a given phage strain. Non-native phages experienced a greater sunlight inactivation than native phages, suggesting adaptation of phages to local conditions (Noble and Fuhrman, 1997).

6.3.2. Marine Phage Characterization

6.3.2.1. Cyanophages

Cyanobacteria are among the most important primary producers and nitrogen fixers on earth, responsible for much of the primary production in oceans, lakes, and other aqueous environments. They are ubiquitous, found in extreme environments from hot springs to polar lakes as well as in ponds for waste stabilization and for raising fish. They can be differentiated into marine and freshwater forms, into those using phycoerythrin vs. those using phycocyanin as their primary photosynthetic pigment, and into those that are unicellular vs. those that grow as filaments. While they are clearly bacteria, their ecological roles are more closely tied to those of eukaryotic algae than to those of heterotrophic bacteria (Suttle, 2000a; Suttle, 2000b).

Cyanophages infecting filamentous freshwater cyanobacteria were first isolated by Safferman and Morris (1963), and stimulated the hope of using such viruses to control cyanobacterial blooms. Phage infecting filamentous cyanobacteria generally cause rapid invagination and destruction of the host's photosynthetic membranes, whereas such destruction is only seen very late in the infection cycle with those infecting unicellular cyanobacteria, for which successful infection seems to depend on ongoing photosynthesis.

Demuth et al. (1993) reported that nearly all of the phages they saw in Lake Plu*β*see had tails, the majority being *Siphoviridae*; in those studies, the phage levels were so high that they simply floated the TEM grids on the samples and let the phages adsorb. Half of the phages seen in Chesapeake Bay by Wommack et al. (1992) also had tails. *Myoviridae* are the most common characterized from marine waters and are also often isolated from fresh-water species. Five *Synechoccus* cyanophages isolated by Wilson et al. (1993) included two myoviruses and one siphovirus. Head proteins of one group of marine phages even have clear sequence relationships to T4-like coliphages, as discussed below. The similarities are particularly interesting since cyanobacteria diverged from other bacteria billions of years ago.

6.3.2.2. Genomic Analysis of Marine Phages

Various genomics approaches have given new insights into the field of phage ecology. A large-scale random sequencing effort of two uncultured marine water samples demonstrated that 65% of the sequences lacked matches to the database (Breitbart et al., 2002). The database hits were mostly with viruses, covering all major families of tailed phages and some algal viruses. A careful statistical analysis of the sample revealed between 400 and 7000 different viral types in the two 100-litre samples, with the most abundant type representing 3% of the total viral population. Over 200 *Vibrio parahaemolyticus* phages were isolated from various locations and seasonal periods in Florida and Hawaii (Paul and Kellogg, 2000). All observed isolates were *Myoviridae* and shared some genetic determinants, giving 83%–100% identity for one sequenced 500 bp region, but on the basis of restriction patterns they could be divided into at least 7 groups for the Florida isolates, one of which was consistently dominant (71%), plus a pair of Hawaii isolates.

In striking contrast to the careful ecological work performed with marine viruses, only a handful of marine phages have actually been sequenced. However, these few examples delivered surprises. A *Pseudoalteromonas* phage with a 10 kb genome became the type phage of a new family, called *Corticoviridae*, of lipid-containing phages. Cyanophage P60 and phage SIO1, infecting the marine heterotroph *Roseobacter*, resembled coliphage T7 closely in their genome organization. Three *Synechoccus* cyanophages were found to share distant head-gene sequence relationships with coliphage T4 (Fuller et al., 1998). Zhong et al. (2002) then designed primers to amplify capsid assembly protein gp20 from both isolated marine cyanophages and natural virus communities and looked at 114 different gp20 homologues. He found these cyanophages to be a highly diverse family, with 65%–96% sequence similarity among the cyanophage gp20's (and 50%–55% similarity with T4). They fall into 9 different phylogenetic groups (Fig. 6.3), with up to 6 clusters and 29 genotypes found in a single sample.

6.3.3. Marine Phage Host Specificity and Horizontal Gene Transfer

In the large majority of the phages investigated in the laboratory, host species specificity is the rule (Ackermann, 1987); phages generally also display strain specificity within a host species due to using a variety of different receptors. Phages with broad host ranges have been described, but they are the exceptions. There was some expectation that marine phages might show broader host range to facilitate their multiplication at the frequently low host densities seen in ocean environments, as discussed above. However, the results to date indicate that most of them are host-species specific; many also demonstrate strain-specificity (Baross et al., 1978a; Bigby and Kropinski, 1989; Koga et al., 1982; Moebus, 1992). Striking host range differences were seen between phages recovered east or west from the Azores islands in the Atlantic Ocean (Moebus and Nattkemper, 1981). Broad host range was more prevalent in cyanophages, but Hennes (1995) demonstrated that fluorescence-labelled cyanophages attached specifically only to their known host and not to other

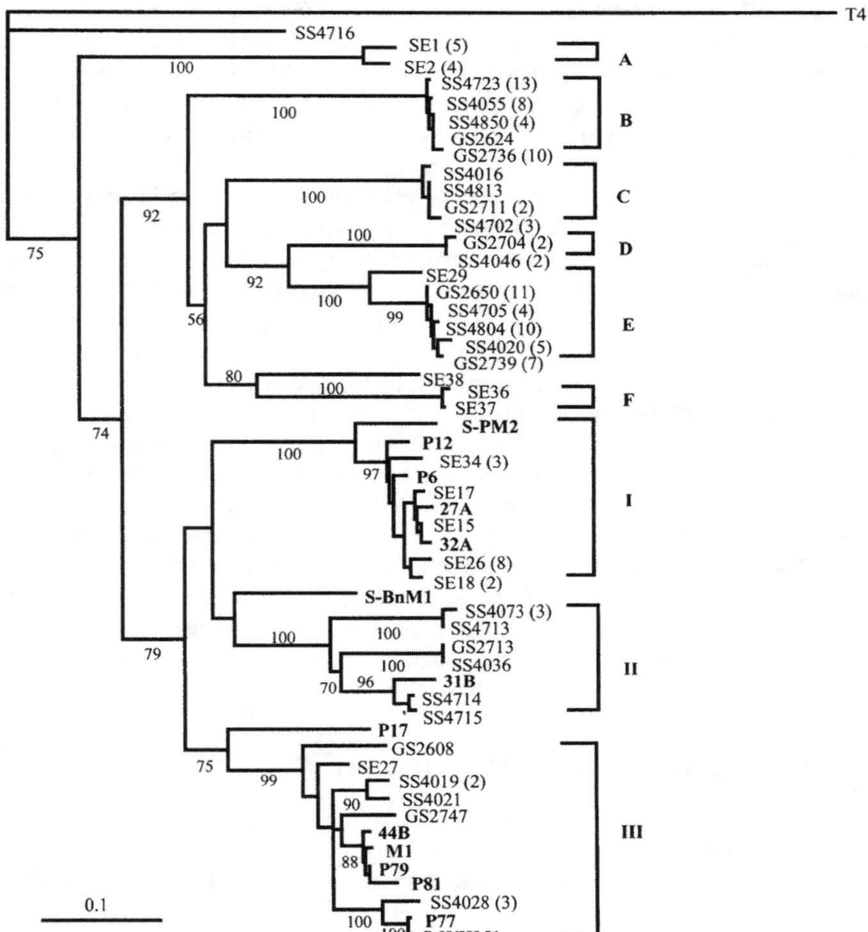

FIGURE 6.3 Neighbor-joining tree showing the phylogenetic affiliation of cyanophage isolates and representative clones from all six of the natural virus communities studied. The tree was constructed on the basis of a 176-amino-acid sequence alignment with T4 as the outgroup. Each value in parentheses is the number of different nucleotide sequences in the same cluster and same community as the representative clone. Clusters A through F and I through III were assigned on the basis of phylogenetic relatedness. Bootstrap values of less than 50 were not shown. The scale bar indicates 0.1 substitution per site.

bacteria of the natural consortium. However, some observations suggested that bacteriophages isolated from very low-nutrient marine habitats showed a trend toward increased breadth of host range. If confirmed, this could represent an adaptation to the low host cell concentrations.

The still-unsettled subject of possible selection for broader host specificity in nutrient-poor marine environments is of substantial scientific relevance. If phages with unusually broad target ranges are indeed more widely distributed in the marine environment than we expect from laboratory infections, transfer of genetic material between marine bacterial species via transduction might occur at even higher frequency than currently anticipated on the basis of the high concentrations of phages and bacterial cells and the tremendous volume of water in the oceans. Transduction, the accidental transfer of host DNA via a phage particle, occurs at varying rates for different phages, but is about once in every 10^8 phage infections in several well-studied systems. A mathematical treatment of experimental data from the estuary of the size of the Tampa Bay led to an estimation of 10^{14} transduction events occurring annually (Jiang and Paul, 1998). If even a minute fraction of this DNA is travelling between different bacterial species, it is clear that marine phages open up enormous possibilities for horizontal DNA transfer. The probability of *transformation* between unrelated species may well also be increased by the liberation of free bacterial DNA during lysis of infected cells. Current bacterial genomic analyses underline the important impact of lateral gene transfer, but despite the theoretical importance of transduction, few transduction studies have been conducted in the marine environment. Experiments with *Pseudomonas* revealed that higher transduction rates were obtained when both the donor and the recipient were lysogenic (Morrison et al., 1978). The reason is that a lysogenic strain can accept foreign DNA from a transducing phage, but is protected from lysis due to the immunity functions of the resident prophage if the two phages are related. In freshwater samples, suspended particles increased the transduction frequency since they allow adsorption of the donor and recipient cells on a solid phase.

6.4. SOIL AND PLANT-ASSOCIATED PHAGES

The constant mixing found in aqueous contexts plays a major role in our ability to make generalizations about marine phages. Phages are also present at high levels in a variety of soils, but are much harder to study in detail and such work is in its infancy. Conditions in terrestrial environments vary far more drastically than they do in marine environments. One is really talking about enormous numbers of microenvironments, affected by position within soil particles, patterns of rain, drought, and temperature, and diurnal and seasonal variations, often with low and variable levels of exchange between them. Bacteria have many physiological adaptations for dealing with these changes, and these in turn affect host-phage interactions in ways that we are scarcely beginning to understand. In addition, phage have the challenge of finding a new bacterial host under conditions where the soil may often be only partially hydrated and the phage may be trapped in biofilms, bound to clay, or inactivated by acidity or other properties of the soil.

High phage concentrations comparable to those in marine environments have also been reported in terrestrial environments. For example, rhizosphere soil from a sugar beet field revealed 10^7 phage per gram by using transmission electron microscopy (Ashelford, 2003) and reconstitution experiments suggested that this figure underestimates the true number by nearly a factor of 10 for technical reasons. Phage from non-pathogenic bacteria like thermophilic *Bacillus* species are easily isolated from soil, compost, silage, and rotting straw, suggesting a tight association of phage with plants (Sharp et al., 1986). A wide variety of phages were observed, most of them strain-specific within a given *Bacillus* species. Hybridization experiments with *Serratia* and *Pseudomonas* colonies from the soil showed that at least 5% of the bacteria are actually phage-infected. There are also reports of plants and plant extracts that can induce bacterial lysogens (Erskine, 1973; Gvozdyak, 1993; Sato, 1983). Soil phages, like their aquatic counterparts, are thus likely to be important in controlling bacterial populations and mediating gene transfer.

Various reports indicate that phage-host interactions can be quite complicated in the soil. For example, phages had only a minimal impact on net growth of Streptomycetes in the soil (Burroughs et al., 2000). In a combination of experimental observations and mathematical modeling, spatial heterogeneity in phage-host interaction, and temporal changes in susceptibility to phages were explored as determinants in bacterial escape from phage lysis in the soil. It turned out that germinating spores were more susceptible to phage infection than hyphae of developed mycelia. Mature resistant mycelia adsorb most of the *Streptomyces*-specific soil bacteria and thus protect younger susceptible hyphae from infection.

A number of labs have explored phages for biocontrol of plant pathogenic bacteria, as discussed in Chapter 13. One popular candidate is *Erwinia amylovora*, the cause of fire blight disease of apple and pear trees. Fire blight has generally been fought with limited success by antibiotics. Biological control by apathogenic *Pseudomonas* or by *Erwinia* phage Ea1 has been explored. *Erwinia*-specific phages like Ea1 were prevalent in orchards affected by fire blight, demonstrating a wide distribution of this phage; other genetically distinct phages, some with very broad host ranges on the fire blight pathogen, were also detected (Schnabel and Jones, 2001). The logistics of studying phage treatment of tree pathogens has been very challenging, but work is also going on in Guelph, Tbilisi, and elsewhere in applying phage against *Erwinia* species that infect important but more experimentally tractable model systems such as potatoes, carrots and ornamental flowers. In all cases, it has been possible to isolate a range of relevant phages from the wild, once again emphasizing their ubiquity in the environment. Phages against *Leuconostoc* and *Lactobacillus* have been isolated from numerous spontaneous fermentation processes, including coffee, pickled cucumbers, sauerkraut, cereals, and wine. These presumably represent phage that are normally associated with the respective plants being used for the fermentation processes.

6.5. ANIMAL-ASSOCIATED PHAGES

6.5.1. Introduction

When the draft sequence of the human genome arrived at the finishing line, it provided only a small part of the genetic material that makes up a human being. In fact, we harbour in our gut more bacterial cells than we have human cells. Not surprisingly, these gut bacteria are associated with their specific phage communities (Breitbart et al., 2003). This situation is not peculiar to humans; phage concentrations up to 10^9 per gram of feces were detected in cattle and sheep (Furuse, 1987). Other vertebrates (birds, fish) also contained appreciable phage concentrations in the gut content. Phages have also been isolated from fecal pellets of many invertebrates belonging to diverse taxonomical groups (earthworms, bees, flies, mussels), as discussed by Ackermann and Dubow (1987). Oysters are filtering water and retain material suspended in the water, so it is no surprise that up to 10^6 pfu of vibriophages were found per g of oyster tissue (Baross et al., 1978b).

A number of studies have reported large phage populations in the rumens of sheep and cattle, affecting the complex balance among the bacteria that convert grass into nutrients to support ruminant growth. This is one area where research can be carried out at a level of sophistication comparable to that seen in the marine and the dairy environments. Animals with permanent rumen cannulae have facilitated non-invasive sampling and allowed studies over time that give new insight into rumen ecology—an important field in science-based livestock husbandry, which aims to optimize the efficiency of converting feed into meat and milk. Klieve and Bauchop (1988) partially purified phages from fluid samples collected through a nylon stocking from the rumens of cattle and sheep and studied them by electron microscopy. They found mainly tailed phages, with a high range of diversity as determined by head shape and size and tail morphologies: at least 14 different kinds of isometric-headed myoviridae, 4 podoviruses, and 4 isometric and 2 prolate-headed siphoviruses, including an astonishing giant phage with a head measuring 85 × 238 nm. The tails ranged from 25 to 1050 nm long. The ruminal phage DNA varied in size from 10 to 850 kbp. Klieve and Swain (1993) saw discrete bands of a wide variety of sizes, each essentially homogeneous, that differed in distribution from sample to sample, against a broad background of DNAs between 30 and 200 kbp; they showed that the latter represented a large, mixed population of intact DNAs, not degraded DNA from larger bands. The total phage population was determined to be about 10^{10} per ml—significantly higher than most earlier estimates. Klieve et al. (1989) also explored the incidence of mitomycin C-inducible temperate phage in the rumen. Of the 38 different ruminal bacteria that they analyzed, only 9 produced phage-like particles, all but one of them siphoviridae.

Phage can be isolated from the feces of most animals. In fact, an extended scientific discussion deals with the question of whether phages can be used as surrogate measures of fecal contamination levels in the environment. Since F^+-specific *E. coli* phages were mainly isolated from animal feces and *Bacteroides fragilis* phages only from human feces, specific phage detection methods can potentially differentiate the origin of environmental fecal contamination and phages were used as tracers to follow the intrusion of polluted surface waters into groundwater. The recent excitement about phage therapy of human and animal diseases has renewed the interest of microbiologists in the ecology of phage-bacterium interaction in the context of their hosts. Here, we will discuss three ecological niches: the gut, the oral cavity, and the skin.

6.5.2. Enteric Phages

The human gut is a complex ecosystem, colonized by about 400–500 microbial species, 30–40 of which account for 99% of the total population. A first understanding of human colonic phages comes from recent metagenomic analyses of an uncultured

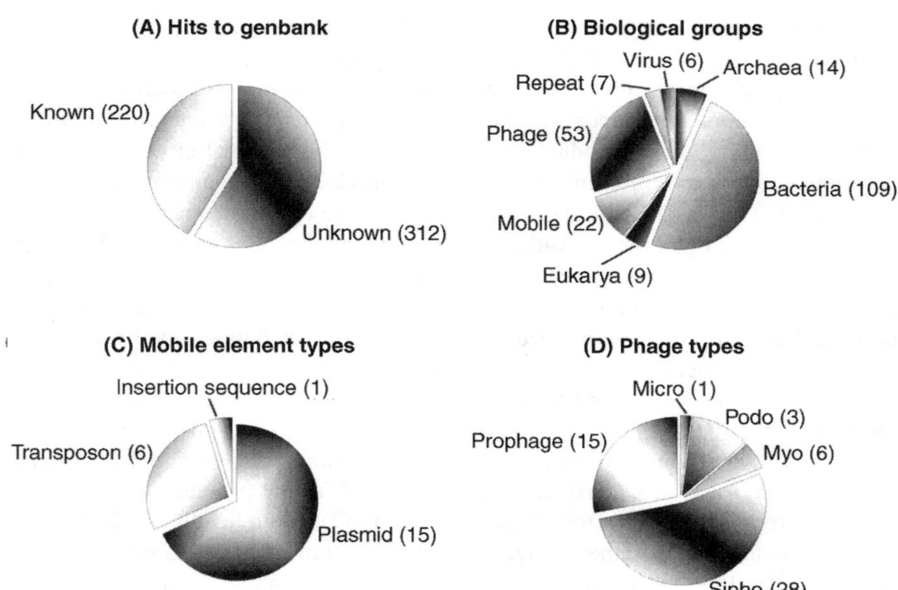

FIGURE 6.4 Genomic overview of uncultured viral community from human feces based on TBLASTX sequence similarities (Breitbart, 2003).

phage community from human feces of a single subject (Fig. 6.4) (Breitbart et al., 2003). DNA analysis using pulse field gel electrophoresis showed major bands at 15 and 90 kbp with minor bands at 30, 40, and 60 kbp and an average size of about 30 kbp—a significantly different distribution than that observed in seawater or the rumen, as described above; the dominant band at only 15 kbp was especially unusual. Extracted DNA was cloned and sequenced. No significant GenBank hits were seen for 59% of the 532 sequences. Half of the positive hits were related to bacterial genes, while a quarter were with phage genes, mainly to structural proteins and terminases; one might expect that a large fraction of the sequences without GenBank hits were from phages, since more information is available from bacterial genome projects and sequenced phages often show a very large fraction of unique sequences. There were few matches to T7-like podoviridae or to λ-like siphoviridae, which were the most abundant species observed in the marine environment. Many were to phages infecting gram-positive bacteria, in agreement with the observation that 62% of the cells detected in human feces with specific rRNA probes were gram-positive bacteria. Since earlier studies exploring gut phages against *E. coli, Salmonella* and *Bacteroides* have shown substantial differences among individuals that did not correlate with age or sex, it will be very important to carry out more studies of this nature and also look for correlations with such factors as dietary patterns and prior antibiotic use.

E. coli and its phages belong for molecular biologists to the most carefully investigated phage systems, yet surprisingly few reports have investigated the gut ecology of coliphages or even the ability of coliphages to replicate under *anaerobic* conditions. Kutter et. al. (1994) reported that phage T4 can successfully infect anaerobically as long as the bacteria are already *pre-adapted* to anaerobic growth. In this connection, it is important to distinguish *anaerobic fermentation*, in which bacteria use an organic byproduct of metabolism as the terminal electron acceptor, from *anaerobic respiration*, in which they use an electron acceptor such as nitrate or fumarate. The former is found, for example, in the rumen of grass-eating mammals, where much glucose is available, while the latter predominates in the colon, where virtually no fermentable substrates remain. For T4-like phages, we find quite different patterns of infection in comparing anaerobic respiration, anaerobic fermentation and aerobic respiration, even when largely similar defined media are used; burst sizes are generally lower anaerobically and lysis is often delayed for many hours, even though cells lysed with chloroform at 50 minutes after infection show good phage production.

In one large study, stool samples from 600 healthy patients and 140 patients suffering from traveller's diarrhea were investigated for the presence of coliphages on 10 different *E. coli* indicator strains (Furuse, 1987). From healthy subjects, 34% of the stool samples contained phages but only 1% showed high amounts. Most of them were classified as temperate phages related to phages phi80, lambda, and phi28. In comparison, 70% of the stools from diarrhea patients (half of whom might have had an enterotoxigenic *E. coli* infection) contained phages, 18% in high concentrations. About half of these isolates were composed of virulent phages related to T4 and T5. This change of phage composition between healthy subjects and diarrhea patients seemed to reflect some disturbance of their intestinal bacterial flora. In another study with stool samples from 160 pediatric diarrhea patients from a clinic in Dhaka Bangladesh, a third of the patients were phage-positive when tested on two *E. coli* indicator cells (Chibani-Chennoufi et al., in preparation). Notably, in this clinic about 30% of cases are caused by *E. coli* (Albert et al., 1995). The phages were nearly exclusively T4-like phages by genome size and morphology. However, restriction analysis, diagnostic PCR and partial sequencing demonstrated substantial genetic diversity between the phage isolates. The stool samples from the Bangladeshi children in the convalescent phase yielded no higher phage counts than the stool samples from the same patients during the acute diarrhea episode, but the phages were not screened on the infecting *E. coli* serotypes. Strain-specific phages might thus suffer from underreporting. In animal experiments (Chibani-Chennoufi et al., 2004) and recent human adult volunteer trials at the Nestle Research Centre (Bruttin et al., manuscript in preparation), T4 and T4-like phages were added to the drinking water. The phages survived the gastric passage and were recovered from the feces in titres comparable to the dose orally fed to the subjects.

This suggests that these phages transit through the entire gut relatively unscathed, but no significant replication on the endogenous *E. coli* gut flora was observed despite the fact that many fecal strains were susceptible to the phage infection *in vitro*; fecal counts of the endogenous *E. coli* flora also remained relatively unchanged both in mouse and in man. These experiments raise doubts about too-simple models of phage-host interaction in the mammalian gut. The metabolically active *E. coli* dwelling as microcolonies in the intestinal mucus layer covering the mucosa might be physically protected against lumenal phage (Poulsen, et al., 1995; Krogfelt, et al. 1993). Mouse experiments suggested that freshly added *E. coli* applied by mouth were susceptible to luminal phage. Actually, we know relatively little about the ecology of *E. coli* in the human gut. *E. coli* might even be a misnomer since the pathogenic strains cause their diarrhea effects by interacting with the small intestine and not the colon. Mucosa-associated *E. coli* in the small intestine might acquire sufficient oxygen from the blood vessels to facilitate host and phage growth. Here again, research areas at the borderline between microbiology, ecology and physiology are key. It is clearly very important to develop a better general understanding of phage infection under such conditions and of the roles of phage in gut and rumen ecology. These ecological considerations are important for phage therapy approaches in the alimentary tract. (The Hungate technique, an inexpensive and fairly simple system for carrying out anaerobic phage infection studies in vitro, is described in the Appendix.)

E. coli and its phages can be easily isolated from the environment. For RNA coliphages, the most common sources were sewage, both from domestic drainage and sewage treatment plants, followed by feces of man, domesticated animals (cows, pigs), and zoo animals. Much less rich sources were environmental water samples (Furuse, 1987). This led to the proposal that coliphages could be a surrogate measure for fecal contamination of recreational waters or other waters of public health interest (el-Abagy et al., 1988). Recently it has become technically possible to screen for human viruses in water samples. However, the majority of the medically important human enteroviruses are RNA viruses, with the single exception of adenovirus. The most sensitive PCR techniques thus cannot be applied. Testing in coastal waters in California impacted by urban run-off water revealed that four out of twelve sites contained adenovirus. However, coliform bacteria and coliphages did not correlate with the adenovirus, calling for a reevaluation of both indicator organisms for the monitoring of recreational waters. In contrast F-specific RNA coliphages showed a good correlation with adenoviruses (Jiang et al., 2001). There is some ecological knowledge on these RNA coliphages in the environment (Furuse, 1987). They are found with strikingly variable prevalence in domestic drainage from different geographical areas. These phages were also found in the feces of humans and domesticated animals. The fact that the feces from cows and pigs contained large amounts of RNA coliphages suggested that these phages were actually propagated in the intestines of the animals. Different groups of RNA coliphages were found with distinct frequencies in humans and animals, suggesting some specificity that probably reflected the distinct composition of the gastrointestinal microbial flora. This fact supports the idea that the intestine of mammals may constitute one of the natural habitats of coliphages despite the fact that *E. coli* represents only a minor constituent of the normal bacterial flora in the human alimentary tract

6.5.3. Other Phage Sites of Interest: Oral Cavity and Vagina

Bacteriophages have also been isolated from other parts of the alimentary tract, for example the oral cavity. This anatomic site is richly colonized with many mainly commensal bacteria, but pathogenic strains like *Streptococcus pneumoniae* and

S. pyogenes were also detected. The actual species composition varies from precise anatomic site to site. Bacteriophages play an important role for these pathogens, since a majority of the clinical isolates of *S. pneumoniae* are lysogenic (Severina et al., 1999; Ramirez et al., 1999) and in the case of *S. pyogenes* the prophages contribute a set of virulence factors to the cell that directly influence the epidemiology of the clinical isolates (Beres et al., 2002). Phage lysins applied to the oral cavity can potentially diminish the degree of colonization of the oral cavity with these pathogenic bacteria and the seeding of these pathogens into the respiratory tract (see Chapter 12). In about 3% of dental patients, dental plaque yielded both *Actinomyces* and the corresponding phages (Tylenda et al., 1985). The phages could be re-isolated from most of the patients up to a month later, suggesting that they belonged to the local microbial community. In another study, about 10% of the oral washings from dental patients allowed the isolation of virulent phages directed against *Veillonella* strains, a resident constituent of the oral cavity. *Enterococcus faecalis* phages were isolated from human saliva (Bachrach et al., 2003), but the ecological role of all these oral phages is still unsettled.

Recent data suggest that phages may play an important role in the ecology of the vagina as well, in a way that is attracting significant medical attention (Kilic et al., 2001). Lactobacilli constitute the dominant vaginal bacterial flora and are beneficial to women's health, since they inhibit the growth of harmful microorganisms by producing lactic acid, hydrogen peroxide and other antimicrobial substances (Redondo-Lopez et al., 1990). Bacterial vaginosis, linked to various medical conditions, is observed when anaerobic bacteria outnumber lactobacilli in the vagina. About 30% of lactobacilli isolated from healthy women from the United States or Turkey were lysogenic. This rate was 50% in women with bacterial vaginosis. Many of these lysogens could be induced by mitomycin C, releasing infectious phage, some at high titer, that could lytically infect lactobacilli belonging to multiple species (*L. crispatus, jensenii, gasseri, fermentum,* and *vaginalis*). The authors note further that smoking is a risk factor for bacterial vaginosis, and the mutagen benzopyrene, which is created by smoking tobacco, could induce phages from lysogenic lactobacilli at the concentrations found in vaginal secretions of smoking women (Pavlova and Tao, 2000). They suggest that smoking may reduce vaginal lactobacilli by promoting phage induction, leading to a replacement of lactobacilli by anaerobic bacteria and precipitating bacterial vaginosis.

6.6. INDUSTRIAL PHAGE ECOLOGY

6.6.1. Dairy Industry

Phage contamination is a constant problem for fermentation-dependent industries such as the dairy industry. In fact, phages are the primary cause for fermentation delays in yogurt and cheese production and can in extreme cases lead to the loss of the product. These cost considerations were a powerful incentive for the dairy industry to design elaborate phage control measures. An overview of this activity is provided in the flow scheme of Fig. 6.5. The first steps in these control measures are ecological surveys of phages in the factory environment. The industry needs to

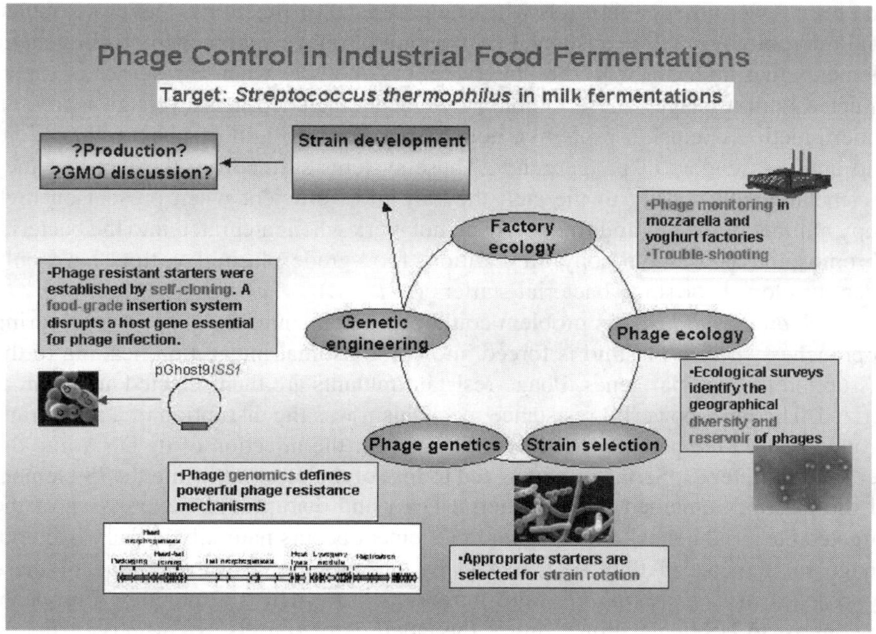

FIGURE 6.5 Phage control in industrial food fermentation. The flow scheme illustrates the approach in the company of one of the authors using *Streptococcus thermophilus* as an example. The different steps depicted in this flow diagram have been the subject of a number of publications. A recent review summarizing this work can be found in Brüssow (2001).

know the extent of the problem, the prevalence and titers of the phages, and their distribution in space and time. The dairy industry has accumulated substantial knowledge in the field of phage factory ecology, but these data are generally not published as the ecology of phages in a man-made environment has not attracted many academic ecology-oriented microbiologists. However, industrial microbiologists soon realized that it is not sufficient to explore the distribution of phages in the factory. The space around the factory and the material delivered into the factory (milk and starters) had to be studied as well. Classically, the major focus of this industrially-oriented work is the selection of appropriate combinations of bacterial starters showing non-overlapping phage susceptibility patterns to allow the design of an efficient starter strain rotation system, which is still the most-used means of coping with the industrial phage problem.

Over the last decade, substantial efforts have been conducted to design phage-resistant starter strains (see also Chapter 10). In the case of cheese production using *Lactococcus lactis* as a starter, this task was facilitated by the availability of many natural phage resistance systems in this bacterium. This allowed the construction of phage-resistant starters by the transfer of plasmids using methods that are not considered genetic engineering. The situation is different for *Streptococcus thermophilus*, the bacterial starter used in yogurt production. Few strains contain plasmids and even fewer possess plasmids with phage-resistance functions. Therefore, industrial microbiologists did substantial sequencing of phage genomes to obtain genetic elements that interfere with the phage replication cycle when the starter is superinfected with a phage. These inhibitory elements included genes (phage repressor, superinfection exclusion proteins), non-coding DNA (origin of phage replication) and anti-messengers of phage genes. These systems suffer from the fact that they are metabolically costly to the cell; they are most efficient when present on high copy number plasmids, and frequently do not work when integrated into the bacterial chromosome as a single copy. In addition, food-grade plasmid vectors have only been developed for some bacterial starters (*L. lactis*).

In *S. thermophilus*, this problem could be circumvented by genetic engineering approaches where a plasmid is forced into chromosomal integration, leading to the disruption of bacterial genes. Phage-resistant mutants are then selected and characterized. The most powerful resistance mechanism was the disruption of a membrane protein that was apparently used by the phage for the injection of its DNA into the cell (see Chapter 7). Serial passaging led to loss of the plasmid, while the IS element of the plasmid remained in the bacterial DNA and disrupted the expression of the targeted bacterial gene. As the plasmid IS element occurs naturally in dairy bacteria, phage resistance can thus be achieved by self-cloning. However, none of these approaches in *S. thermophilus* were introduced into industrial practice. European, in contrast to US, legislation requires the labelling of starters as GMO (genetically modified organisms) that were modified by self-cloning and thus contain only species-specific DNA. The persistent scepticism of the European consumer towards GMO in food production led to a substantial reduction of research activity in the dairy sector, both in industry and in EU-funded public research. Nevertheless, the search for practical solutions to the industrial phage problems made dairy phages one of the best-investigated phage systems with respect to both phage genomics analysis and phage ecology. Pertinent ecological features are analysed in the following paragraphs. As this research was conducted in the private sector, only part of the data has been published.

Two basic kinds of ecological situations can be distinguished in the industrial food-fermentation environment, as characterized by the yogurt factory and the cheese factory. Even if the same bacterial starter is used (yogurt and mozzarella cheese fermented by added *S. thermophilus*), the two dairy factories differ in several basic respects. Milk for yogurt production undergoes treatment at 90°C, which kills all phages (Quiberoni et al., 1999), while raw or pasteurised milk is used in cheese fermentation. Furthermore, yogurt production is a relatively aseptic process where the fermented product has minimal exposure to the factory environment. In contrast, during cheese making the factory experiences a massive daily aerosol contamination during cheese whey separation (Budde-Niekiel et al., 1985). In Europe, industrial yogurt factories are generally smaller than cheese factories (about 50,000 vs. 500,000 liters of milk processed daily per factory). Phages are thus seldom seen in yogurt production, though they may unknowingly be introduced into the factory either by the starter cells or by interventions that compromise the physical barrier separating the product from the environment. Phage problems are still sufficiently frequent to make them the primary source of fermentation failure in yogurt production—mostly in the form of fermentation delays or product alterations, but occasionally also as complete product loss. Cheese factories, in contrast, are characterized by the coexistence of phage in the milk and bacterial starter and thus this is a constant problem.

Yogurt samples in a factory reporting occasional fermentation delays yielded phage-positive samples with titres up to 800 pfu/ml. Aerosols containing phage were the likely vehicle for phage transmission within the factory. In the literature, two potential sources of phage were identified: raw milk (Bruttin et al., 1997) and lysogenic starter strains (Heap et al., 1978). These strains spontaneously release phage that can lead to phage amplification if susceptible starter cells are used in the same factory. Fermentation becomes prolonged when the phage titres mount beyond a critical threshold of 1,000 or 10,000 pfu/ml. When a rotation system is used, regular fluctuations of the phage titres synchronized with the starter strain rotation can be observed. This situation can be maintained over some time. When the titre of the phage rises beyond 10^6 pfu/ml, a fermentation failure is the likely consequence and cannot be buffered by a rotation system. The milk samples inoculated with the starter will no longer coagulate and the product is altered in its technological properties or entirely lost. The production line must then be carefully cleaned to eliminate the phages. If available, starters insensitive to the phage of the fermentation failure are sometimes used to prevent a recontamination of the production line by residual phage in the factory.

Persistent phage infection is frequently observed in cheese factories. In mozzarella fermentation using a complex mixture of *S. thermophilus* starter strains, phage titres in the cheese whey were normally 10^5 pfu/ml or higher (peak titres: 10^7 pfu/ ml) (Bruttin et al., 1997). All cheese whey samples contained between four and eight different phage strains. Some phage types were frequently observed and showed high titres while others were only occasionally seen, at low titres. No apparent regularities could be deduced from the temporal cycling of the specific phage titres in the whey samples. In fermentation simulations, cell counts dropped when phage were added at an moi (multiplicity of infection) of 0.01 and cells were lost and no coagulation occurred when the cells were infected at a moi of 0.1. At low moi, two successive waves of phage replication were observed. High yields of progeny phage were only obtained with cells in active growth. Phage yields dropped by 5 orders of magnitude when cells were infected in the late logarithmic or stationary phase. However, when these cells were resuspended in fresh medium and growth resumed, renewed phage replication was observed, leading to lysis and phage release. In contrast, infected cells maintained in the stationary phase failed to produce sizable amounts of progeny phage.

No phage-resistant cells are selected in the factory ecology, since the fermentation process is restarted each time with a frozen standard starter culture. A repetition of the classical Delbrüeck-Luria phage challenge experiment (Luria and Delbrück, 1943) yielded only very few outgrowing cells that had lost the capacity to adsorb the challenge phage. These mutant cells grew poorly, suggesting a metabolic cost for mutation to phage receptor loss in *S. thermophilus* (unpublished results).

S. thermophilus is found in raw milk after enrichment techniques (heating at 60°C) and has never been isolated from any sources not in contact with milk.

In contrast, its closest relative, *S. salivarius*, is an oral commensal. However, total bacterial counts in uncontaminated raw milk are relatively low: about 1000 cfu/ml. It is thus not surprising that *S. thermophilus* phages can only be isolated from raw milk in low titers, ranging from undetectable (<10/ml) in most samples to a maximum of 130 pfu/ml (Bruttin et al., 1997). This poses a dilemma: How can phages be maintained in the environment when they have only such a small pool of susceptible cells? The most logical explanations might be lysogeny and wide host range. However, the analysis of hundreds of *S. thermophilus* phage isolates from cheese factories showed very narrow host ranges. All but two phages were only able to infect the strain on which they were isolated (Bruttin et al., 1997). This pattern of strain-specificity was also observed in larger industrial strain collections (Le Marrec et al., 1997). Furthermore, lysogenic strains are rare in industrial strain collections; only about 1% of strains can be induced by mitomycin C, for example. Southern hybridization with DNA of the two major classes of *S. thermophilus* phages confirmed a low lysogeny rate. Only a single survey reported a 10% rate of lysogenic cells by hybridization (Fayard et al., 1993). In addition, less than one per cent of *S. thermophilus* phages from major strain collections are temperate phages (Brüssow et al. 1994; Le Marrec et al., 1997; Lucchini et al., 1999b). However, the genome maps of the major virulent *S. thermophilus* phages betray their origin from temperate parental phages (Lucchini et al., 1999a; Bruttin and Brüssow, 1996). The preponderance of virulent phages in our collections might therefore represent an adaptation to the abundance of host cells in the dairy environment. In fact, serial passage of a temperate *S. thermophilus* phage quickly resulted even in the laboratory in its replacement by a virulent derivative deletion mutant. The rare raw milk *S. thermophilus* phages are nevertheless the source of phage contaminations in the cheese factory. In a large intervention trial, one starter combination was replaced by a second that was insensitive to the resident phages of the factory. The intervention resulted in a nearly immediate disappearance of the resident phages: 70% of the milk samples lacked any phages and 30% contained phages detectable only on the old starters (apparently a washout from the previous high phage contamination level since the titres were low). However, by 5–7 days after the intervention the first three phages infecting the new starters were detected. Restriction analysis of the phage DNA traced the origin of the new phages to the rare raw milk samples delivered to the factory during the intervention period.

Cheese factories using *Lactococcus lactis* as a starter frequently yielded high levels of phages; one study reported up to 10^9 pfu/ml whey without great fluctuation. Phages were ubiquitous in the factory, and up to 10^5 pfu/m^3 phages were detected in the factory air (Neve et al., 1994). Like some streptococcal phages, lactococcal phages were reported to survive pasteurisation. *L. lactis* strains were frequently detected in the raw milk while lactococcal phages were only a rare observation. Many lactococcal phages showed very restricted host ranges, limited to one or a few starter strains within the *L. lactis* species. A clear difference from *S. thermophilus* phages is the much greater morphological and genetic variability of lactococcal phages; in some (unpublished) surveys, shifts in the predominant morphological types of phages were observed in the whey samples that correlated with the change of the starter culture.

6.6.2. Non-Dairy Food Fermentation: Sauerkraut and Sausage

Like most vegetable fermentations, sauerkraut fermentation is spontaneous and relies on bacterial epiphytes present on cabbage. Food ecological studies demonstrated a succession of two groups of lactic acid bacteria. In the initial heterofermentative stage, *Leuconostoc* species dominate the fermentation. When the pH decreases, they are followed by the more acid-tolerant *Lactobacillus* species; *L. plantarum* eventually becomes the most abundant species. An ecological survey also demonstrated a succession of two phage populations corresponding to the replacement of *Leuconostoc* by *Lactobacilli* (Yoon et al., 2002; Barrangou et al., 2002; Marchesini et al., 1992; Lu et al., 2003b). The *Leuconostoc* phages represented a number of distinct *Sipho*- and *Myoviridae,* while the *Lactobacillus* phages included in addition *Podoviridae*. The *Myoviridae* showed genome sizes in the 40–50-kbp range. The *Leuconostoc* phages showed very narrow host ranges, infecting only selected strains of *L. fallax* species. In contrast one *Lactobacillus* phage could infect both of the ecologically-related species *L. brevis* and *L. plantarum*. *Leuconostoc* and *Lactobacillus* phages have been isolated from numerous other spontaneous fermentation processes: coffee, pickled cucumbers, cereals, and wine (Lu et al., 2003a; Lu and Dahlquist, 1992). Apparently, phage infections are not limited to food fermentation initiated with bacterial starter cultures.

Lactobacillus plantarum is also used as an industrial starter in meat fermentation. Phage infections were documented in salami production, but they were of no industrial impact. After an initial rise, the phage titres dropped and the initial phage-sensitive starter population was replaced by a phage-insensitive mutant derivative strain. From meat fermentation, a *L. plantarum* myovirus with large genome size was isolated (Chibani-Chennoufi et al., 2004). The 130-kbp genome showed close sequence similarity with *Bacillus* phage SPO1 at the protein level in regions encoding structural and DNA replication proteins. Both have been recently shown to be related to phages infecting other gram-positive bacteria such as *Staphylococcus* and *Listeria*, defining a broad new genus of *Myoviridae,* as discussed in Chapter 5. Phage infections were without consequence in salami fermentation using *Staphylococcus carnosus* starters that do not quickly develop phage resistance. Two reasons were quoted for the limited impact of phages on meat fermentation: The staphylococcal starter showed only a modest growth and the solid food matrix prevented the spread of phage in the food product (Marchesini et al., 1992).

6.7. OUTLOOK

All in all, recent ecological surveys in a variety of environments have underlined the notion that we live in a sea of bacteriophages, as they are common biological parts of the soil we stand on, air we breathe, water we drink, and food we eat. There is also an increased appreciation that phages play a key role in controlling bacterial population levels everywhere in the environment, and in the genetic diversification of bacterial strains and species. These observations clearly have profound basic- and applied-science implications, leading to important theoretical insights and concepts for ecology and medicine. With the addition of such new techniques as metagenomic analyses, it is clear that this field will explode in the next few years, providing many new insights.

ACKNOWLEDGEMENTS

We express special appreciation to Steve Abedon, Jason Gill, Sandra Chibani-Chennoufi, Mya Breitbart, Gautam Dutta, Raul Raya, and Burton Guttman.

REFERENCES

Abedon, S.T., "Phage Ecology," in *The Bacteriophages,* R. Calendar (Ed.). Oxford University Press, New York, 2005.

Ackermann, H.W. and DuBow, M.S., *Viruses of Prokaryotes: General Properties of Bacteriophages*. CRC Press, Boca Raton, FL, 1987.

Adams, M.H. and Park, B.H., An enzyme produced by a phage-host cell system. II. The properties of the polysaccharide depolymerase. *Virology* 2: 719–736, 1956.

Albert, M.J., Faruque, S.M., Faruque, A.S., Neogi, P.K., Ansaruzzaman, M., Bhuiyan, N.A., Alam, K., et al., Controlled study of *Escherichia coli* diarrheal infections in Bangladeshi children. *J Clin Microbiol* 33: 973–977, 1995.

Anwar, H., Strap, J.L. and Costerton, J.W., Establishment of aging biofilms: Possible mechanism of bacterial resistance to antimicrobial therapy. *Antimicrob Agents Chemother* 36: 1347–1351, 1992.

Ashelford, K.E., Day, M.J., Bailey, M.J., Lilley, A.K. and Fry, J.C., In situ population dynamics of bacterial viruses in a terrestrial environment. *Appl Environ Microbiol* 65: 169–174, 1999.

Ashelford, K.E., Day, M.J. and Fry, J.C., Elevated abundance of bacteriophage infecting bacteria in soil. *Appl Environ Microbiol* 69: 285–289, 2003.

Azam, F. and Worden, A.Z., Oceanography. Microbes, molecules, and marine ecosystems. *Science* 303: 1622–1624, 2004.

Bachrach, G., Leizerovici-Zigmond, M., Zlotkin, A., Naor, R. and Steinberg, D., Bacteriophage isolation from human saliva. *Lett Appl Microbiol* 36: 50–53, 2003.

Bagge, N., Hentzer, M., Andersen, J.B., Ciofu, O., Givskov, M. and Hoiby, N., Dynamics and spatial distribution of beta-lactamase expression in *Pseudomonas aeruginosa* biofilms. *Antimicrob Agents Chemother* 48: 1168–1174, 2004.

Baross, J.A., Liston, J. and Morita, R.Y., Ecological relationship between *Vibrio parahaemolyticus* and agar-digesting vibrios as evidenced by bacteriophage susceptibility patterns. *Appl Environ Microbiol* 36: 500–505, 1978a.

Baross, J.A., Liston, J. and Morita, R.Y., Incidence of *Vibrio parahaemolyticus* bacteriophages and other Vibrio bacteriophages in marine samples. *Appl Environ Microbiol* 36: 492–499, 1978b.

Barrangou, R., Yoon, S.S., Breidt Jr., F., Jr., Fleming, H.P. and Klaenhammer, T.R., Characterization of six *Leuconostoc fallax* bacteriophages isolated from an industrial sauerkraut fermentation. *Appl Environ Microbiol* 68: 5452–5458, 2002.

Barrow, P., Lovell, M. and Berchieri, A., Jr., Use of lytic bacteriophage for control of experimental *Escherichia coli* septicemia and meningitis in chickens and calves. *Clin Diagn Lab Immunol* 5: 294–298, 1998.

Beres, S.B., Sylva, G.L., Barbian, K.D., Lei, B., Hoff, J.S., Mammarella, N.D., Liu, M.Y., et al., Genome sequence of a serotype M3 strain of group A Streptococcus: Phage-encoded toxins, the high-virulence phenotype, and clone emergence. *Proc Natl Acad Sci U S A* 99: 10078–10083, 2002.

Bergh, O., Borsheim, K.Y., Bratbak, G. and Heldal, M., High abundance of viruses found in aquatic environments. *Nature* 340: 467–468, 1989.

Bernstein, C., Deoxyribonucleic acid repair in bacteriophage. *Microbiological Reviews* 45: 72–98, 1981.

Bigby, D. and Kropinski, A.M.B., Isolation and characterization of a *Pseudomonas aeruginosa* bacteriophage with a very limited host range. 35: 630–635, 1989.

Binder, B., Reconsidering the relationship between virally induced bacterial mortality and frequency of infected cells. *Aquat Microb Ecol* 18: 207–215, 1999.

Boehme, J., Frischer, M.E., Jiang, S.C., Kellogg, C.A., Prichard, S., Rose, J.B., Steinway, C., Paul, J.H., Viruses, bacterioplankton, and phytoplankton in the southeastern Gulf of Mexico: Distribution and contribution to oceanic DNA pools. *Mar Ecol Prog Ser* 97: 1–10, 1993.

Borsheim, K.Y., Native Marine Bacteriophages. *FEMS Microbiol Ecol* 102: 141–159, 1993.

Bratbak, G., Heldal, M., Norland, S., Thingstad, T.K., Viruses as partners in spring bloom microbial trophodynamics. *Appl Environ Microbiol* 56: 1400–1405, 1990.

Bratbak, G., Heldal, M., Thingstad, T.F., Tuomi, P., Dynamics of virus abundance in coastal seawater. *FEMS Microbiol Ecol* 19: 263–269, 1996.

Breitbart, M., Hewson, I., Felts, B., Mahaffy, J.M., Nulton, J., Salamon, P. and Rohwer, F., Metagenomic analyses of an uncultured viral community from human feces. *J Bacteriol* 185: 6220–6223, 2003.

Breitbart, M., Salamon, P., Andresen, B., Mahaffy, J.M., Segall, A.M., Mead, D., Azam, F., Genomic analysis of uncultured marine viral communities. *Proc Natl Acad Sci U S A* 99: 14250–14255, 2002.

Brüssow, H., Phages of dairy bacteria. *Annu Rev Microbiol* 55: 283–303, 2001.

Brüssow, H., Fremont, M., Bruttin, A., Sidoti, J., Constable, A. and Fryder, V., Detection and classification of *Streptococcus thermophilus* bacteriophages isolated from industrial milk fermentation. *Appl Environ Microbiol* 60: 4537–4543, 1994.

Bruttin, A. and Brüssow, H., Site-specific spontaneous deletions in three genome regions of a temperate *Streptococcus thermophilus* phage. *Virology* 219: 96–104, 1996.

Bruttin, A., Desiere, F., d'Amico, N., Guerin, J.P., Sidoti, J., Huni, B., Lucchini, S., Molecular ecology of *Streptococcus thermophilus* bacteriophage infections in a cheese factory. *Appl Environ Microbiol* 63: 3144–3150, 1997.

Budde-Niekiel, A., Moller, V., Lembke, J. and Teuber, M., Ecology of bacteriophages in a fresh cheese factory. *Milchwissenschaft* 40: 477–481, 1985.

Burchard, R.P. and Dworkin, M., A bacteriophage for *Myxococcus xanthus*: Isolation, characterization and relation of infectivity to host morphogenesis. *J Bacteriol* 91: 1305–1313, 1966.

Burchard, R.P. and Voelz, H., Bacteriophage infection of *Myxococcus xanthus* during cellular differentiation and vegetative growth. *Virology* 48: 555–566, 1972.

Burroughs, N.J., Marsh, P. and Wellington, E.M., Mathematical analysis of growth and interaction dynamics of streptomycetes and a bacteriophage in soil. *Appl Environ Microbiol* 66: 3868–3877, 2000.

Chibani-Chennoufi, S., Sidoti, J., Bruttin, A., Dillmann, M., Kutter, E., Qadri, F., Sarker, S.A., Brüssow, H. (2004) Isolation of *Escherichia coli* bacteriophages from the stool of pediatric diarrhea patients in Bangladesh. *J Bacteriol* (In Press).

Chibani-Chennoufi, S., Dillmann, M., Marvin-Guy, L., Rami-Shojaei, S., Brüssow, H., *Lactobacillus plantarum* bacteriophage LP65: a new member of the SPO1-like genus of the family Myoviridae. *J Bacteriol* (In Press).

Cochlan, W.P., Wikner, J., Stewar, G.F., Smith, D.C., Azam, F., Spatial distribution of viruses, bacteria, and chlorophyll a in neritic, oceanic and estuarine environments. *Mar Ecol Prog Ser* 92: 77–87, 1993.

Cochran, P.K. and Paul, J.H., Seasonal abundance of lysogenic bacteria in a subtropical estuary. *Appl Environ Microbiol* 64: 2308–2312, 1998.

Corbin, B.D., McLean, R.J. and Aron, G.M., Bacteriophage T4 multiplication in a glucose-limited *Escherichia coli* biofilm. *Can J Microbiol* 47: 680–684, 2001.

Costerton, J.W., Stewart, P.S. and Greenberg, E.P., Bacterial biofilms: a common cause of persistent infections. *Science* 284: 1318–1322, 1999.

Danovaro, R. and Serresi, M., Viral density and virus-to-bacterium ratio in deep-sea sediments of the Eastern Mediterranean. *Appl Environ Microbiol* 66: 1857–1861, 2000.

Demuth, J., Neve, H. and Witzel, K.-P., Direct electron microscopy study on the morphological diversity of bacteriophage populations in Lake Pluásee. *Appl Environ Microbiol* 59: 3378–3384, 1993.

Doolittle, R.F., Of Archae and Eo: What's in a name? *Proc Natl Acad Sci U S A* 92: 2421–2423, 1995.

Doolittle, W.F., At the core of the Archaea. *Proc Natl Acad Sci* 93: 8797–8799, 1996.

Edlin, G., Lin, L. and Kudrna, R., Lambda lysogens of *E. coli* reproduce more rapidly than non-lysogens. *Nature* 255: 735–737, 1975.

el-Abagy, M.M., Dutka, B.J., Kamel, M. and el Zanfaly, H.T., Incidence of coliphage in potable water supplies. *Appl Environ Microbiol* 54: 1632–1633, 1988.

Erskine, J.M., Association of virulence characteristics of *Erwinia amylovora* with toxigenicity of its phage lysates to rabbit. *Canadian Journal of Microbiology* 19: 875–877, 1973.

Fayard, B., Haefliger, M. and Accolas, J.-P., Interactions of temperate bacteriophages of *Streptococcus salivarius* subsp. thermophilus with lysogenic indicators affect phage DNA restriction patterns and host ranges. *Journal of Dairy Research* 60: 385–399, 1993.

Fuhrman, J.A., Bacterioplankton roles in cycling of organic matter: The microbial food web, pp. 361–383 in *Primary Productivity and Biogeochemical Cycles in the Sea,* P.G. Falkowski and A.D. Woodhead (Eds.). Plenum, New York, 1992.

Fuhrman, J.A., Marine viruses and their biogeochemical and ecological effects. *Nature* (London) 399: 541–548, 1999.

Fuhrman, J.A. and Suttle, C.A., Viruses in marine planktonic systems. *Oceanography* 6: 50–62, 1993.

Fuller, N.J., Wilson, W.H., Joint, I.R. and Mann, N.H., Occurrence of a sequence in marine cyanophages similar to that of T4 g20 and its application to PCR-based detection and quantification techniques. *Appl Environ Microbiol* 64: 2051–2060, 1998.

Furuse, K., Distribution of coliphages in the environment: General considerations, pp. 87–124 in *Phage Ecology,* S.M. Goyal, G.P. Gerba and G. Bitton (Eds.). John Wiley & Sons, New York, 1987.

Garza, D.R. and Suttle, C.A., The Effect of Cyanophages on the Mortality of Synechococcus spp. and Selection for UV Resistant Viral Communities. *Microb Ecol* 36: 281–292, 1998.

Goyal, S.M., Gerba, G.P., and Bitton, G., *Phage Ecology.* John Wiley & Sons, New York, 1987.

Gvozdyak, R.I., Plant, phage, bacterium: A new hypothesis on their interrelation. *Mikrobiologicheskii Zhurnal* (Kiev) 55: 92–94, 1993.

Hanlon, G.W., Denyer, S.P., Olliff, C.J. and Ibrahim, L.J., Reduction in exopolysaccharide viscosity as an aid to bacteriophage penetration through *Pseudomonas aeruginosa* biofilms. *Appl Environ Microbiol* 67: 2746–2753, 2001.

Hara, S., Koike, I., Terauchi, K., Kamiya, H., Tanoue, E., Abundance of viruses in deep oceanic waters. *Mar Ecol Prog Ser* 145: 269–277, 1996.

Heap, H.A., Limsowtin, G.K.Y. and Lawrence, R.C., Contribution of *Streptococcus lactis* strains in raw milk to phage infection in commercial cheese factories. 13: 16–22, 1978.

Heldal, M. and Bratbak, G., Production and decay of viruses in aquatic environments. M*arine Ecology Progress Series* 72: 205–212, 1991.

Hennes, K.P. and Simon, M., Significance of bacteriophages for controlling bacterioplankton growth in a mesotrophic lake. *Appl Environ Microbiol* 61: 333–340, 1995.

Hennes, K.P., Suttle, C.A. and Chan, A.M., Fluorescently labeled virus probes show that natural virus populations can control the structure of marine microbial communities. *Appl Environ Microbiol* 61: 3623–3627, 1995.

Jiang, S., Noble, R. and Chu, W., Human adenoviruses and coliphages in urban runoff-impacted coastal waters of Southern California. *Appl Environ Microbiol* 67: 179–184, 2001.

Jiang, S.C. and Paul, J.H., Seasonal and diel abundance of viruses and occurence of lysogeny/bacteriocinogeny in the marine environment. *Mar Ecol Prog Ser* 104: 163–172, 1994.

Jiang, S.C. and Paul, J.H., Significance of lysogeny in the marine environment: Studies with isolates and a model of lysogenic phage production. *Microb Ecol* 35: 235–243, 1997.

Jiang, S.C. and Paul, J.H., Gene transfer by transduction in the marine environment. *Appl Environ Microbiol* 64: 2780–2787, 1998.

Kilic, A.O., Pavlova, S.I., Alpay, S., Kilic, S.S. and Tao, L., Comparative study of vaginal Lactobacillus phages isolated from women in the United States and Turkey: Prevalence, morphology, host range, and DNA homology. *Clin Diagn Lab Immunol* 8: 31–39, 2001.

Klieve, A.V. and Bauchop, T., Morphological diversity of ruminal bacteriophages from sheep and cattle. *Appl Environ Microbiol* 54: 1637–1641, 1988.

Klieve, A.V., Hudman, J.F. and Bauchop, T., Inducible bacteriophages from ruminal bacteria. *Appl Environ Microbiol* 55: 1630–1634, 1989.

Klieve, A.V. and Swain, R.A., Estimation of ruminal bacteriophage numbers by pulsed-field gel electrophoresis and laser densitometry. *Appl Environ Microbiol* 59: 2299–2303, 1993.

Koga, T., Toyoshima, S. and Kawata, T., Morphological varieties and host range of Vibrio parahaemolyticus bacteriophages isolated from seawater. *Appl Environ Microbiol* 44: 466–470, 1982.

Kokjohn, T.A., Sayler, G.S. and Miller, R.V., Attachment and replication of *Pseudomonas aeruginosa* bacteriophages under conditions simulating aquatic environments. *Journal of General Microbiology* 137: 661–666, 1991.

Kolter, R., Siegele, D.A. and Tormo, A., The stationary phase of the bacterial life cycle. *Annu Rev Microbiol* 47: 855–874, 1993.

Krogfelt, K.A., Poulsen, L.K. and Molin, S., Identification of coccoid *Escherichia coli* BJ4 cells in the large intestine of streptomycin-treated mice. *Infect Immun* 61: 5029–5034, 1993.

Kutter, E., Kellenberger, E., Carlson, K., Eddy, S., Neitzel, J., Messinger, L., North, J., et al., Effects of bacterial growth conditions and physiology on T4 infection., pp. 406–418 in *Molecular Biology of Bacteriophage T4.*, J. Karam, J.W. Drake, K.N. Kreuzer, G. Mosig, D.H. Hall, F.A. Eiserling, L.W. Black, et al. ASM Press, Washington, D.C., 1994.

LaBelle, R. and Gerba, C.P., Investigation into the protective effect of estuarine sediment on virus survival. *Water Research* 16: 469–478, 1982.

Le Marrec, C., van Sinderen, D., Walsh, L., Stanley, E., Vlegels, E., Moineau, S., Heinze, P., et al., Two groups of bacteriophages infecting *Streptococcus thermophilus* can be distinguished on the basis of mode of packaging and genetic determinants for major structural proteins. *Appl Environ Microbiol* 63: 3246–3253, 1997.

Lenski, R.E., Coevolution of bacteria and phage: Are there endless cycles of bacterial defenses and phage counterdefenses? *J Theor Biol* 108: 319–325, 1984.

Lenski, R.E., Levin, B.R., Constraints on the coevolution of bacteria and virulent phage: A model, some experiments and predictions for natural communities. *Am Nat* 125: 585–602, 1985.

Lin, L., Bitner, R. and Edlin, G., Increased reproductive fitness of *Escherichia coli* lambda lysogens. *J Virol* 21: 554–559, 1977.

Lindberg, A.A., "Bacterial surface carbohydrates and bacteriophage adsorption," in *Surface Carbohydrates of the Prokaryotic Cell,* I. Sutherland (Ed.). Academic Press, 1977.

Lu, J. and Dahlquist, F.W., Detection and characterization of an early folding intermediate of T4 lysozyme using pulsed hydrogen exchange and two-dimensional NMR. *Biochemistry* 31: 4749–4756, 1992.

Lu, Z., Breidt, F., Jr., Fleming, H.P., Altermann, E. and Klaenhammer, T.R., Isolation and characterization of a *Lactobacillus plantarum* bacteriophage, phiJL-1, from a cucumber fermentation. *Int J Food Microbiol* 84: 225–235, 2003a.

Lu, Z., Breidt, F., Plengvidhya, V. and Fleming, H.P., Bacteriophage ecology in commercial sauerkraut fermentations. *Appl Environ Microbiol* 69: 3192–3202, 2003b.

Lucchini, S., Desiere, F. and Brüssow, H., Comparative genomics of *Streptococcus thermophilus* phage species supports a modular evolution theory. *J Virol* 73: 8647–8656, 1999a.

Lucchini, S., Desiere, F. and Brüssow, H., The genetic relationship between virulent and temperate *Streptococcus thermophilus* bacteriophages: Whole genome comparison of cos-site phages Sfi19 and Sfi21. *Virology* 260: 232–243, 1999b.

Luria, S.E. and Delbrück, M., Mutations of bacteria from virus sensitivity to virus resistance. *Genetics* 28: 491–511, 1943.

Malin, G., Wilson, W.H., Bratbak, G., Liss, P.S. and Mann, N.H., Elevated production of dimethylsulfide resulting from viral infection of cultures of *Phaeocystis pouchetii. Limnology and Oceanography* 43: 1389–1393, 1998.

Maranger, R., Bird, D.F., Juniper, S.K., Viral and bacterial dynamics in Arctic sea ice during the spring algal bloom near Resolute, N.W.T., Canada. *Mar Ecol Prog Ser* 111: 121–127, 1994.

Marchesini, B., Bruttin, A., Romailler, N., Moreton, R.S., Stucchi, C. and Sozzi, T., Microbiological events during commercial meat fermentations. *J Appl Bacteriol* 73: 203–209, 1992.

Mathias, C.B., Kirschner, A.K.T., Velimirov, B., Seasonal variations of virus abundance and viral control of the bacterial production in a backwater system of the Danube river. *Appl Environ Microbiol* 61: 3734–3740, 1995.

McCann, M.P., Kidwell, J.P. and Matin, A., The putative sigma factor KatF has a central role in development of starvation-mediated general resistance in *Escherichia coli. J Bacteriol* 173: 4188–4194, 1991.

McLean, R.J., Corbin, B.D., Balzer, G.J. and Aron, G.M., Phenotype characterization of genetically defined microorganisms and growth of bacteriophage in biofilms. *Methods Enzymol* 336: 163–174, 2001.

Middelboe, M., Jorgensen, N.O. G. and Kroer, N., Effects of viruses on nutrient turnover and growth efficiency of noninfected marine bacterioplankton. *Appl Environ Microbiol* 62: 1991–1997, 1996.

Moebus, K., Further investigations on the concentration of marine bacteriophages in the water around helgoland with reference to the phage-host systems encountered. *Helgol Meeresunters* 46: 275–292, 1992.

Moebus, K., Lytic and inhibition responses to bacteriophages among marine bacteria, with special reference to the origin of phage-host systems. *Helgol Meeresunters* 36: 375–391, 1983.

Moebus, K. and Nattkemper, H., Bacteriophage sensitivity patterns among bacteria isolated from marine waters. *Helgol Meeresunters* 34: 375–385, 1981.

Morrison, W.D., Miller, R.V. and Sayler, G.S., Frequency of F116-mediated transdution of *Pseudomonas aeruginosa* in a freshwater environment. *Appl Environ Microbiol* 36: 724–730, 1978.

Murray, A.G. and Eldridge, P.M., Marine viral ecology: Incorporation of bacteriophage into the microbial planktonic food web paradigm. *Journal of Plankton Research* 16: 627–641, 1994.

Murray, A.G. and Jackson, G.A., Viral dynamics II: A model of the interaction of ultraviolet light and mixing processes on virus survival in seawater. *Marine Ecology Progress Series* 102: 105–114, 1993.

Murray, A.G. and Jackson, G.A., Viral dynamics: A model of the effects of size, shape, motion, and abundance of single-celled planktonic organisms and other particles. *Marine Ecology Progress Series* 89: 103–116, 1992.

Neve, H., Kemper, U., Geis, A. and Heller, K.J., Monitoring and characterization of lactococcal bacteriophages in a dairy plant. Kieler Milchwirtschaftliche Forschungsberichte 46: 167–178, 1994.

Noble, R.T. and Fuhrman, J.A., Virus decay and its cause in coastal waters. *Appl Environ Microbiol* 63: 77–83, 1997.

Ogunseitan, O.A., Sayler, G.S. and Miller, R.V., Application of DNA probes to analysis of bacteriophage distribution patterns in the environment. *Appl Environ Microbiol* 58: 2046–2052, 1992.

O'Toole, G., Kaplan, H.B. and Kolter, R., Biofilm formation as microbial development. *Annu Rev Microbiol* 54: 49–79, 2000.

Paul, J.H. and Kellogg, C.A., "The Ecology of Bacteriophages in Nature," p. 538 in *Viral Ecology*, C. Hurst (Ed.). Academic Press, 2000.

Pavlova, S.I. and Tao, L., Induction of vaginal Lactobacillus phages by the cigarette smoke chemical benzo[a]pyrene diol epoxide. *Mutation Research* 466: 57–62, 2000.

Poulsen, L.K., Licht, T.R., Rang, C., Krogfelt, K.A. and Molin, S., Physiological state of *Escherichia coli* BJ4 growing in the large intestines of streptomycin-treated mice. *J Bacteriol* 177: 5840–5845, 1995.

Proctor, L.M., Fuhrman, J.A., Viral mortality of marine bacteria and cyanobacteria. *Nature* 343, 1990.

Quiberoni, A., Suarez, V.B. and Reinheimer, J.A., Inactivation of *Lactobacillus helveticus* bacteriophages by thermal and chemical treatments. *J Food Prot* 62: 894–898, 1999.

Ramirez, M., Severina, E. and Tomasz, A., A high incidence of prophage carriage among natural isolates of *Streptococcus pneumoniae*. *J Bacteriol* 181: 3618–3625, 1999.

Redondo-Lopez, V., Cook, R.L. and Sobel, J.D., Emerging role of lactobacilli in the control and maintenance of the vaginal bacterial microflora. *Rev Infect Dis* 12: 856–872, 1990.

Robertson, B.R. and Button, D.K., Characterizing aquatic bacteria according to population, cell size, and apparent DNA content by flow cytometry. *Cytometry* 10: 70–76, 1989.

Safferman, R.S. and Morris, M.E., Algal virus: Isolation. *Science* 140: 679–680, 1963.

Sato, M., Phage induction from lysogenic strains of *Pseudomonas syringae* pathovar mori by the extract from mulberry leaves. *Annals of the Phytopathological Society of Japan* 49: 259–261, 1983.

Schnabel, E.L. and Jones, A.L., Isolation and characterization of five *Erwinia amylovora* bacteriophages and assessment of phage resistance in strains of *Erwinia amylovora*. *Appl Environ Microbiol* 67: 59–64, 2001.

Schrader, H.S., Schrader, J.O., Walker, J.J., Wolf, T.A., Nickerson, K.W. and Kokjohn, T.A., Bacteriophage infection and multiplication occur in *Pseudomonas aeruginosa* starved for 5 years. *Can J Microbiol* 43: 1157–1163, 1997.

Severina, E., Ramirez, M. and Tomasz, A., Prophage carriage as a molecular epidemiological marker in *Streptococcus pneumoniae*. *J Clin Microbiol* 37: 3308–3315, 1999.

Sharp, R.J., Ahmad, S.I., Munster, A., Dowsett, B. and Atkinson, T., The isolation and characterization of bacteriophages infecting obligately thermophilic strains of Bacillus. *J Gen Microbiol* 132 (Pt 6): 1709–1722, 1986.

Smith, E.M., Gerba, C.P. and Melnick, J.L., Role of sediment in the persistence of enteroviruses in the estuarine environment. *Appl Environ Microbiol* 35: 685–689, 1978.

Sonenshein, A.L. and Roscoe, D.H., The course of phage phi-e infection in sporulating cells of *Bacillus subtilis* strain 3610. *Virology* 39: 265–275, 1969.

Steward, G.F., Smith, D.C., Azam, F., Abundance and production of bacteria and viruses in the Bering and Chukchi seas. *Mar Ecol Prog Ser* 131: 287–300, 1996.

Steward, G.F., Wikner, J., Cochlan, W.P., Smith, D.C. and Azam, F., Estimation of virus production in the sea: II. Field results. 6: 79–90, 1992.

Stewart, F.M. and Levin, B.R., The population biology of bacterial viruses: why be temperate. *Theor Popul Biol* 26: 93–117, 1984.

Suttle, C.A., "Cyanophages and Their Role in the Ecology of Cyanobacteria," pp. 563–589 in *The Ecology of Cyanobacteria*, Whitton, B.A. and Potts, M (Eds.). Kluwer Academic Publishers, Boston, 2000a.

Suttle, C.A., "The ecology, evolutionary and geochemical consequences of viral infection of cyanobacteria and eukaryotic algae," pp. 248–286 in *Viral Ecology*, C.J. Hurst (Ed.). Academic Press, New York, 2000b.

Suttle, C.A., The significance of viruses to mortality in aquatic microbial communities. *Microbiology Ecology* 28: 237–243, 1994.

Suttle, C.A. and Chan, A.M., Dynamics and distribution of cyanophages and their effect on marine Synechococcus spp. *Appl Environ Microbiol* 60: 3167–3174, 1994.

Suttle, C.A. and Chan, A.M., Marine cyanophages infecting oceanic and coastal strains of Synechococcus: Abundance, morphology, cross-infectivity and growth characteristics. *Marine Ecology Progress Series* 92: 99–109, 1993.

Suttle, C.A., Chan, A.M., Cottrell, M.T., Infection of phytoplankton by viruses and reduction of primary productivity. *Nature* 347: 467–469, 1990.

Suttle, C.A. and Chen, F., Mechanisms and rates of decay of marine viruses in seawater. *Appl Environ Microbiol* 58: 3721–3729, 1992.

Thingstad, T.F., Heldal, M., Bratbak, G. and Dundas, I., Are viruses important partners in pelagic food webs? *Trends in Ecology and Evolution* 8: 209–213, 1993.

Thingstad, T.F. and Lignell, R., Theoretical models for control of bacterial growth rate, abundance, diversity and carbon demand. *Aquatic Microbial Ecology* 13: 19–27, 1997.

Tylenda, C.A., Calvert, C., Kolenbrander, P.E. and Tylenda, A., Isolation of Actinomyces bacteriophage from human dental plaque. *Infect Immun* 49: 1–6, 1985.

Waterbury, J.B., and Vaolois, F.W., Resistance to co-occuring phages enables marine *Synechoccocus* communities to coexist with cyanophages abundant in seawater. *Appl Environ Microbiol* 59: 3393–3399, 1993.

Webb, J.L., King, G., Ternent, D., Titheradge, A.J.B. and Murray, N.E., Restriction by *Eco*KI is enhanced by co-operative interactions between target sequences and is dependent on DEAD box motifs. *The EMBO Journal* 15: 2003–2009. 1996

Weinbauer, M.G., Fuks, D. and Peduzzi, P., Distribution of viruses and dissolved DNA along a coastal trophic gradient in the northern Adriatic Sea. *Appl Environ Microbiol* 59: 4074–4082, 1993.

Weinbauer, M.G., Fuks, D., Puskaric, S., Peduzzi, P., Diel, seasonal and depth-related variability of viruses and dissolved DNA in the Northern Adriatic sea. *Microb Ecol* 30: 24–41, 1995.

Weinbauer, M.G. and Hoefle, M.G., Size-specific mortality of lake bacterioplankton by natural virus communities. *Aquatic Microbial Ecology* 15: 103–113, 1998.

Weinbauer, M.G. and Peduzzi, P., Frequency, size and distribution of bacteriophages in different marine bacterial morphotypes. *Marine Ecology Progress Series* 108: 11–20, 1994.

Wiggins, B.A. and Alexander, M., Minimum bacterial density for bacteriophage replication: Implications for significance of bacteriophages in natural ecosystems. *Appl Environ Microbiol* 49: 19–23, 1985.

Wilhelm, S.W. and Suttle, C.A., Viruses and nutrient cycles in the sea. *BioScience* 49: 781–788, 1999.

Wilhelm, S.W., Weinbauer, M.G., Suttle, C.A., Pledger, R.J. and Mitchell, D.L., Measurements of DNA damage and photoreactivation imply that most viruses in marine surface waters are defective. *Aquatic Microbial Ecology* 14: 215–222, 1998.

Wilson, W.H., Joint, I.R., Carr, N.G. and Mann, N.H., Isolation and molecular characterization of five marine cyanophages propagated on Synechoccus sp. strain WH7803. *Appl Environ Microbiol* 59: 3736–3743, 1993.

Wommack, K.E. and Colwell, R.R., Virioplankton: Viruses in aquatic ecosystems. *Microbiol Mol Biol Rev* 64: 69–114, 2000.

Wommack, K.E., Hill, R.T., Kessel, M., Russek-Cohen, E. and Colwell, R.R., Distribution of viruses in the Chesapeake Bay. *Appl Environ Microbiol* 58: 2965–2970, 1992.

Wommack, K.E., Hill, R.T., Muller, T.A. and Colwell, R.R., Effects of sunlight on bacteriophage viability and structure. *Appl Environ Microbiol* 62: 1336–1341, 1996.

Wommack, K.E., Ravel, J., Hill, R.T. and Colwell, R.R., Hybridization analysis of Chesapeake Bay Virioplankton. *Appl Environ Microbiol* 65: 241–250, 1999.

Woods, D.R., Bacteriophage growth on stationary phase *Achromobacter* cells. *J Gen Virol* 32: 45–50, 1976.

Yoon, S.S., Barrangou-Poueys, R., Breidt, F., Jr., Klaenhammer, T.R. and Fleming, H.P., Isolation and characterization of bacteriophages from fermenting sauerkraut. *Appl Environ Microbiol* 68: 973–976, 2002.

Zhong, Y., Chen, F., Wilhelm, S.W., Poorvin, L. and Hodson, R.E., Phylogenetic diversity of marine cyanophage isolates and natural virus communities as revealed by sequences of viral capsid assembly protein gene g20. *Appl Environ Microbiol* 68: 1576–1584, 2002.

7 Molecular Mechanisms of Phage Infection

Elizabeth Kutter,[1] Raul Raya,[1,2] and Karin Carlson[3]
[1]Lab of Phage Biology, The Evergreen State College, Olympia, WA
[2]Cerela, Tucuman, Argentina
[3]Dept. of Cell and Molecular Biology, University of Uppsala, Uppsala, Sweden

CONTENTS

7.1. Introduction 166
7.2. Host Recognition and Adsorption of Tailed Phages 167
7.2.1. Phage Recognition of Gram-Negative Bacteria 168
7.2.2. Phage Recognition of Gram-Positive Bacteria 172
7.2.3. Host-Range Variation 173
7.2.3.1. Mutation and Intraspecies Recombination 174
7.2.3.2. Adhesin Restructuring through Cross-Genus Recombination 174
7.2.3.3. Phage Encoding Multiple Different Adhesins 176
7.2.4. Summary 176
7.3. The Process of DNA Transfer into the Cell 177
7.3.1. Overview 177
7.3.2. Podovirus T7 177
7.3.3. Myovirus T4 179
7.3.4. Three Siphoviridae That All Use the *FhuA* Receptor but Have Very Different Mechanisms 180
7.3.5. Phages PRD1 and PM2 181
7.4. Phage DNA Replication, Repair and Recombination 182
7.4.1. The Process of Replication 182
7.4.2. DNA Repair, Recombination, and Mutagenesis 185
7.4.3. Specific Examples 186
7.4.3.1. Cooption of Host Replicative Machinery: Phage λ 186
7.4.3.2. Complex Phage-Encoded Replication, Recombination, and Repair Machinery: Phage T4 186

0-8493-1336-X/05/$0.00+$1.50
© 2005 by CRC Press LLC

7.4.3.3. Replicative Transposition: Phage Mu 189
7.4.3.4. Protein-Primed Replication Initiation: Phage ϕ 29........ 189
7.4.3.5. Replication of a Linear Chromosome with Closed Ends: Phage N15 190
7.5. Transcription 190
7.5.1. Overview 190
7.5.2. Examples of Transcriptional Regulatory Complexities 193
7.5.2.1. Host Polymerase Used Early, Phage Enzyme Late: *E. coli* Podovirus T7 193
7.5.2.2. A Capsid-Borne RNA Polymerase, and More: *E. coli* Podovirus N4 194
7.5.2.3. Host Polymerase throughout, with Modifications: λ, SPO1, and T4 195
7.6. Phage-Related RNA: Structures and Unexpected Functions...................... 197
7.6.1. Structure and Efficiency of Ribosome Binding Sites 197
7.6.2. Intron Ribozymes 198
7.6.3. Translational Bypassing... 199
7.6.4. RNA-Based Regulatory Systems: Aptamers and SELEX 200
7.7. Morphogenesis and DNA Packaging 201
7.7.1. Head Formation 202
7.7.2. DNA Packaging 203
7.7.3. Tail Morphogenesis 204
7.8. Phage Lysins and Lysis Systems... 206
7.8.1. Overview 206
7.8.2. The Single-Protein Lysis Systems of the Small Phages................ 207
7.8.3. Holins, Endolysins, and Lysis Timing 208
7.8.4. T-Even Phages: Lysis and Lysis Inhibition.................................. 211
Acknowledgements 213
References 213

7.1. INTRODUCTION

Molecular studies of the process of phage infection have laid the basis for much of our understanding of biology, from the mechanisms of DNA replication and transcription to the assembly of complex structures. This chapter provides a general overview and variety of examples of some of the principles involved in phage-host interactions at the molecular level. More detailed information about various specific phages and steps involved in these processes can be found in *The Bacteriophages* (Calendar, 2005; 1988), the *Encyclopedia of Virology* (Granoff and Webster, 1999) and *Molecular Biology of Bacteriophage T4* (Karam et al., 1994) as well as other books about individual phages and the exploding original literature and review articles in the field (Miller et al., 2003b). Some areas, such as host recognition and the infection process, are explored in particular depth because of the paucity of reviews.

7.2. HOST RECOGNITION AND ADSORPTION OF TAILED PHAGES

Tailed phages recognize their host cells through the interaction between attachment sites on their tails and host surface molecules. Some phages use a single, central fiber as their receptor recognition element or *adhesin*; others use a cluster of 3, 6, or 12 fibers associated with the tail structure. Adsorption may involve multiple steps, such as *reversible* binding by fibers responsible for initial host recognition and proper baseplate positioning, followed by *irreversible* binding of a different tail protein to a secondary receptor molecule, as seen for coliphage T1 and for the T4-like phages. The potential phage receptors are quite different on gram-negative bacteria (various lipopolysaccharide (LPS) components and outer membrane proteins such as porins and transport proteins (Fig. 7.1 and Chapter 3, Fig. 3.2b)) and gram-positive bacteria (peptidoglycan elements, embedded teichoic acids and lipoteichoic acids, and

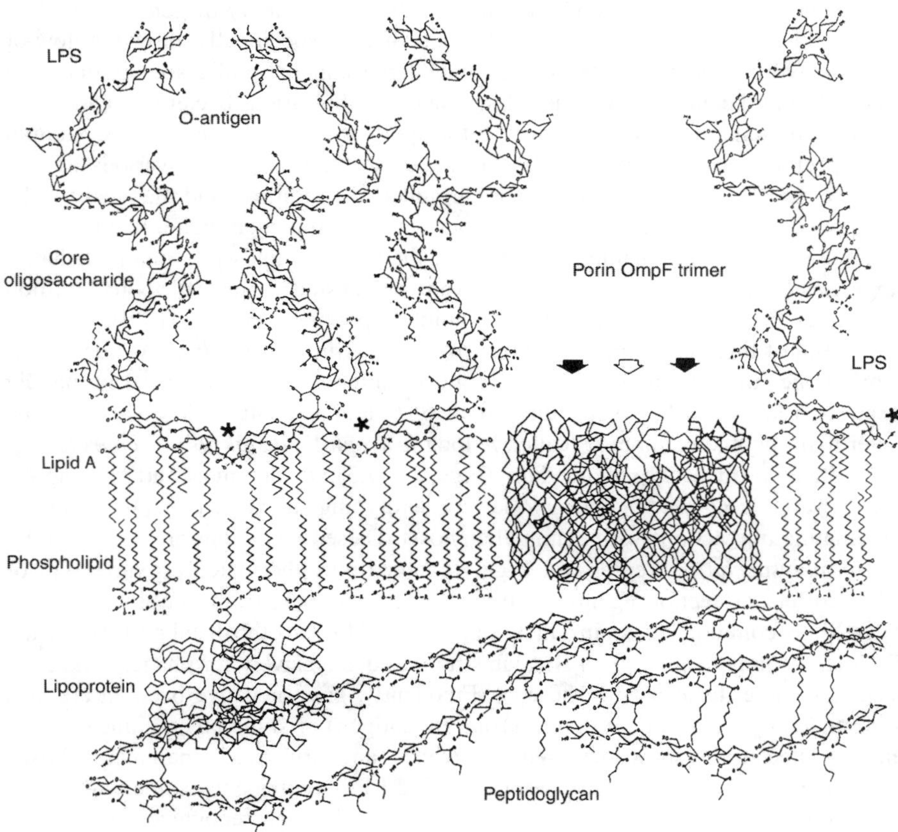

FIGURE 7.1 Details of the outer membrane structures of gram-negative bacteria. Hancock et al., 1994, with permission.

associated proteins (Fig. 7.2)). Components of any secreted external matrix, such as polysaccharide K antigens and proteinaceous S layers, can also be initial recognition targets. The specific phage adhesins often are highly complex and versatile, able to recognize quite different receptors on different bacteria much as a master key can recognize many different locks, as discussed below. In addition, phages can develop very different host specificities through mutation and both intra- and interspecies recombination, including recombination with cryptic prophages. The adhesin regions are thus by far the most variable parts of phage genomes. Understanding the complexities of their interaction with potential host receptor molecules requires a detailed understanding of the nature of the host surface.

7.2.1. Phage Recognition of Gram-Negative Bacteria

The distinctive outer membrane of gram-negative bacteria was reviewed in detail by Nikaido (1996; 2003) and by Ghuysen and Hakenbeck (1994). It is much more permeable than the inner membrane, mainly due to its variety of general *porins*—each cell has up to 200,000 of these channels that nonspecifically allow the flow of smaller hydrophilic molecules but exclude large and lipophilic solutes due to a charged central constriction. In addition, phage use the various bacterial *high-affinity receptor* proteins that catalyze specific transport of catechols, fatty acids, vitamin B_{12}, specific iron derivatives, and so forth. Concentrations of the various ligand-specific receptor proteins depend on environmental conditions and the need for the compounds they transport. While the inner leaflet of this membrane, like the cytoplasmic membrane, is mainly composed of the phospholipids phosphatidylethanolamine and phosphatidylglycerol, the outward-facing side has instead about 3 million molecules (in a well-growing cell) of a unique lipopolysaccharide (LPS), which is composed of three parts (Figs. 7.3). Its *proximal, hydrophobic "lipid A"* region has a backbone structure made of N-acetyl glusosamine (NAG)—also used in making peptidoglycan—which is substituted with 6–7 saturated fatty acids that form the outer hydrophobic membrane leaflet. A long *hydrophilic distal polysaccharide* segment often protrudes into the medium (Figs. 7.1, 7.3), its outermost part forming the highly variable *O-antigen*—a main key in serotyping (as in *E. coli* O157) and in modulating interactions with mammalian hosts. These two regions are connected by a *core polysaccharide* region, the highly charged phosphorylated proximal part of which offers protection against hydrophobic compounds, from antibiotics to bile salts. Many common lab strains, including *E. coli* K12, lack the distal polysaccharide O antigen components; *E. coli* B, a lab strain used extensively for phage work, even lacks the more distal part of the core. Frequently, gram-negative bacteria are also surrounded by a *capsular polysaccharide* (K antigen) or slime layer; some of these encapsulated strains show increased invasiveness and virulence in mammalian hosts.

Some phages interact with various specific LPS components, some with membrane proteins, and some have complex adhesins that can recognize receptors of either type. For example, OmpA is a receptor for T4-like phages K3, Ox2 and M1. LamB, the inducible *E. coli* maltose uptake protein, serves as the receptor for the usual lab strain of phage lambda (λ), which therefore only adsorbs well when maltose is present. T5, T1, and lambdoid phage ϕ80 bind to FhuA (TonA), a 79-kDa minor

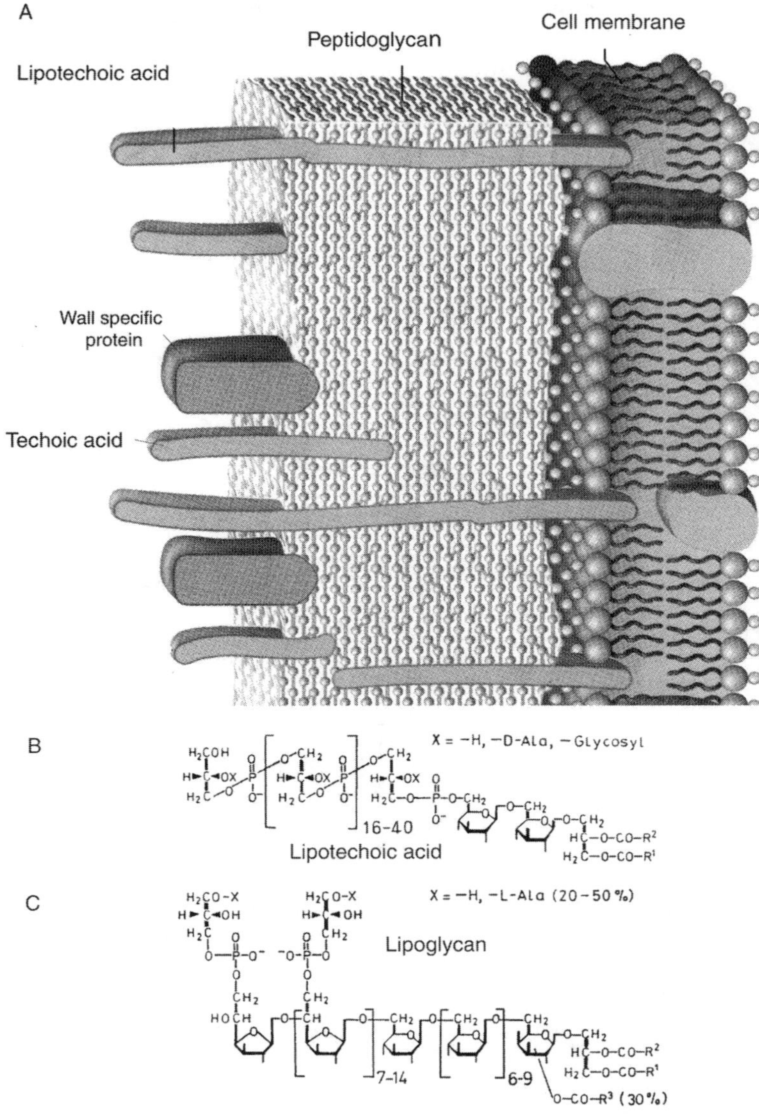

FIGURE 7.2 (A) General structure of the cell wall of gram-positive bacteria. (B) Basic structure of the most widespread lipotechoic acid, the main membrane-linked polyionic component of low-GC gram-positive bacteria. (C) Structure of a lipoglycan, which plays a comparable role in high-GC gram-positive bacteria. Together, the wall-linked teichoic acids and membrane-associated lipoteichoic acids or lipoglycans form a polyionic network between the membrane and surface of the cell which appears to act as a reservoir of divalent cations for membrane-associated enzymes and those involved in peptidoglycan restructuring during growth. They bind extensively with both cellular and humoral components in mammalian hosts, potentially affecting pathogenicity, in addition to affecting phage interactions.

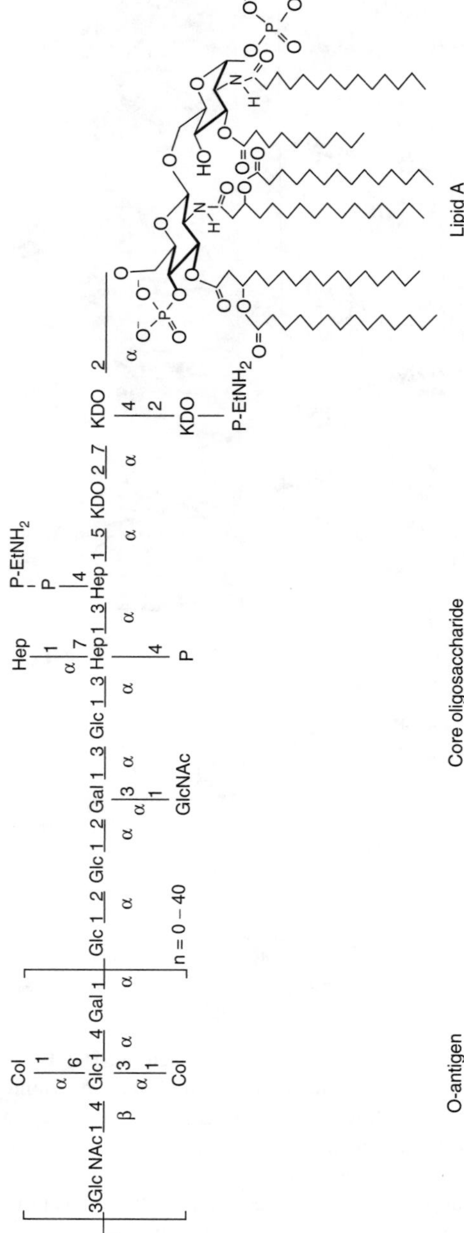

FIGURE 7.3 Detailed general molecular structure of LPS from *E. coli.* Hancock *et al.*, 1994, with permission.

outer membrane protein present in only about 10^3 copies/cell which catalyzes the high-affinity transport of ferric siderophores such as ferrichrome and albomycin across the outer membrane, deriving the energy through a complex protein connection with the inner membrane. Even a protein site present in only about 5 copies per cell functions as a receptor—NfrA, whose function is unknown, is the only receptor for coliphage N4 (Kazmierczak and Rothman-Denes, 2005). More study and awareness is needed of the degree to which various environmental and physiological factors can affect phage binding even when the receptor is present. For instance, Gabig et al. (2002) found that adsorption of λ, P1, and T4 is strongly *inhibited* by physiological concentrations of bile salts and carbohydrates when production of Ag43, the major *phase-variable E. coli* cell surface protein, is turned off; the inhibition disappears when Ag43 is expressed from a plasmid, or in the presence of excess Mg^{++} for λ and T4 and Ca^{++} for P1, emphasizing the important role that divalent cations play in adsorption by many phages. Ag43 is involved in bacterial aggregation and biofilm formation; modulation of the susceptibility of intestinal bacteria to phage infection may be at least one biological function of its ability to undergo phase variation, and should be taken into consideration in efforts to develop phage therapy for enteric bacteria, as discussed by Wegrzyn and Thomas (2002). As they suggest, this inhibition might even be useful in using phage to treat such problems as uropathogenic *E. coli* without disturbing normal gut flora.

There are also phages that specifically recognize the polysaccharide K antigens of the glycocalyx. These commonly are podoviruses with enzymatically-active tail spikes that can depolymerize the polysaccharide, making a path into the host membrane which contains the receptor for irreversible binding, as seen in Fig. 7.4 for phage ϕ infecting encapsulated *E. coli* K29 (Lindberg, 1977). The O antigens can either participate in adhesin recognition or interfere with access to receptor sites that lie deeper, slowing down or blocking phage adsorption and substantially affecting plating efficiency. *Salmonella* phage P22 gets around this problem by having a tail spike that is actually an endoglycosydase against LPS, not only binding to but cleaving the O antigen and facilitating its attachment to the core part of the membrane. A few tailed phages, such as *Salmonella* phage χ, even use flagella and pili as their receptors, as discussed by Ackermann and DuBow (1987); these organelles are very commonly used as receptors by filamentous and other non-tailed phages.

Many phages of Gram-negative bacteria have pleiotropic adhesins that can recognize a variety of different outer membrane molecules. For example, OmpF is a receptor for T4-like phages Tula and T2; the latter can also recognize the fatty acid transporter protein FadL. Receptors recognized by a given adhesin can even be of very different types. For example, T4 binds efficiently to the terminal glucose of *E. coli* B-type LPS. T4 also efficiently binds to the outer membrane protein OmpC, found on K but not B strains of *E. coli*. *E. coli* O157 has a different form of OmpC that is *not* recognized by T4 but *is* recognized by the T4-like phages PPO1 (Oda et al., 2004) and AR1 (Yu et al., 2000). Neither PPO1 nor AR1 binds to standard K12 or B lab strains, and they were long considered to be "O157-specific," but Goodridge et al. (2003) showed that AR1 can infect about half of the 72 members of the *E. coli* collection of reference (ECOR), selected to represent the breadth of environmental *E. coli* isolates (Ochman and Selander, 1984). The Kutter lab found

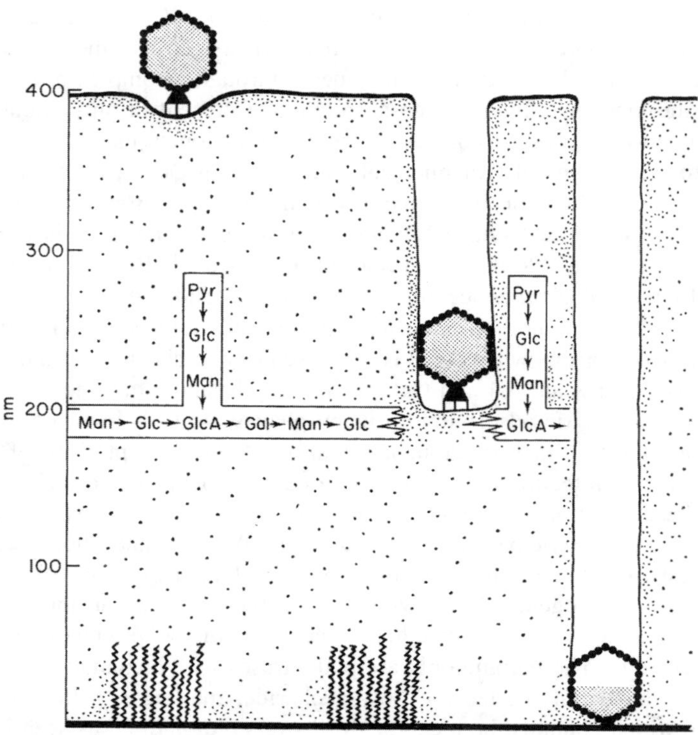

FIGURE 7.4 Process of infection of encapsulated *E. coli* by K-antigen-specific phage. Lindberg, (1977), with permission.

that PPO1 also infects 9/72 ECOR strains (Mueller et al., 2003). Presumably other receptors in addition to OmpC are being recognized, at least for AR1.

7.2.2. Phage Recognition of Gram-Positive Bacteria

Little is yet known about the specific adhesins of phage infecting Gram-positive bacteria or about the receptors to which they bind. There are differences in the exposed peptidoglycan layer due to a large number of possible variations of the amino acids used for the peptidoglycan peptides for different bacteria and the interpeptide bridges that cross-link peptides from different glycan strands in different sheets of the thick stack; there are also variations in the teichoic acids. A variety of cell-wall-associated proteins are bound to the cell either via a specific C-terminal anchor sequence or as N-terminal lipoproteins and interact in various ways with the environment; staph protein A, which binds the constant region of IgG and has been implicated in pathogenesis, is the most famous of these.

Duplessis and Moineau (2001) published the first identification and characterization of a phage gene involved in recognition of gram-positive bacteria, for the fully sequenced phage DT1 and a group of six related virulent phages with different host ranges on *Streptococcus thermophilus*. They confirmed that *orf18* encodes the adhesin, which has a conserved N-terminal domain of nearly 500 amino acids, a collagen-like sequence, and another largely conserved C-terminal domain with an internal 145-AA variable region (VR2). Two of their other phages had an additional 400-bp segment inserted into the collagen-like region. They generated chimeras between *orf18* of phage DT1, which infected 10 of their strains, and that of MD4, which infected one different strain. Using phage DT1 to infect a host in which MD4 genes *17–19* had been cloned, they generated five recombinant phages that now had the MD4 host range; all five contained the MD4 version of the VR2 region, but were otherwise largely like DT1. Ravin et al. (2002) showed that the specificity of adsorption of the isometric-headed phage LL-H and the prolate-headed phage JCL1032 to their host *Lactobacillus delbrueckii* involves a conserved C-terminal region in LL-H Gp71 and JCL 1032 ORF 474.

Stuer-Lauridsen (2003) has used a similar approach to study the host-recognition element of four virulent prolate-headed phages (c2 species) of *Lactococcus lactis*. Again, these phages show overlapping but different host ranges. They first *reversibly* recognize a specific combination of carbohydrates in the outer cell wall. They then bind *irreversibly* with a host surface protein called PIP (phage infection protein), with no other known function, triggering DNA release from the capsid. Using the fully sequenced phages ϕc2 and ϕbIL67 and the sequenced least-conserved late regions of the other phages, the authors showed that the host-range-determining element is in the central 462-bp segment of a gene designated *115, 35,* and *2* in three different phages and showing 65%–71% pairwise similarity between them. While most genes in this region of the genome show no homology with unrelated phages, parts of this host-range-determining element show homologies with genes from phages of *S. thermophilus* and *L. lactis* that have also been implicated in host range determination processes. The product of a neighboring gene, *110/31/5,* with 83%–95% similarity, had earlier been shown to bind to the tail spike and apparently to be involved in the infection process; they could not clone this second gene in their shuttle vector, perhaps because it is the one responsible for binding to PIP and helping transfer the DNA into the cell. This is clearly a promising system for future work. For phage ϕ29, Guo et al. (2003) have shown that gene *12,* whose protein product forms 12 dimers around the neck of the phage, is required for making phage that can bind to the receptor and appears to encode an adhesin.

7.2.3. Host-Range Variation

One often finds phages of any particular species that are virtually identical except that they have quite different ranges of host specificity. Hypervariability and recombinogenicity of the adhesin regions is widespread, explaining the ready development of variants that infect different bacterial strains or otherwise resistant bacteria. This fact suggests that it may be particularly easy to develop efficient phages for therapeutic purposes through natural selection mechanisms and, potentially, through genetic engineering. A variety of experimental observations give insight into the mechanisms that so readily generate new adhesins and alter host specificity.

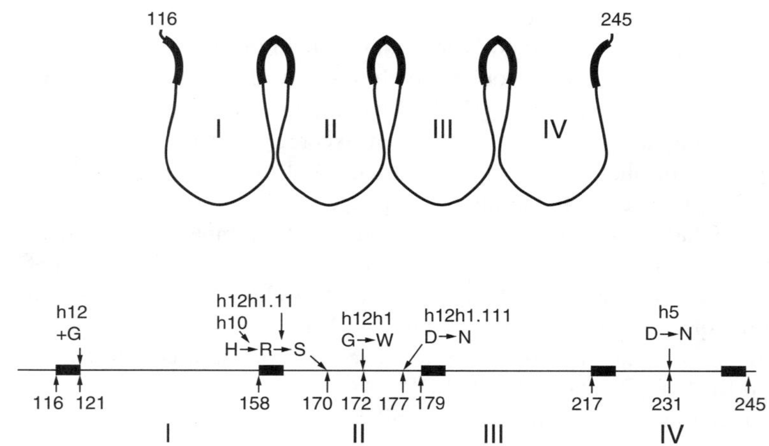

FIGURE 7.5 Detailed structure of the hypervariable adhesin region of T2-like phage Ox2 gp38. All of the mutations conferring changes in host range map in this region, as indicated. The thick lines represent the conserved oligoglycene stretches. From Henning and Hashemolhosseini (1994), with permission.

7.2.3.1. Mutation and Intraspecies Recombination

For many T4-like phages, the receptor-binding region is on a small protein, gp38, which binds to the distal tip of gp37, as reviewed by Henning and Hashemolhosseini (1994). Highly conserved sequences flank a hypervariable region here; through a series of mutations in this hypervariable region, phage Ox2 could switch its receptor recognition from OmpA to OmpC to OmpP to the *E. coli* B LPS (which has two terminal glucoses) to a different LPS with a single glucose. Interestingly, some residual weaker affinities for the other receptors were also retained at various stages along the way. Most of the changes occurred in the second of four potential "omega loops" (Fig. 7.5) formed between conserved stretches of up to nine glycines in the hypervariable portion. These loops are hypothesized to form the actual receptor-binding region; they show interesting similarities to features in the antigen-binding sites of antibodies which appear to be responsible for a plasticity in those sites that can improve complementarity of the interacting surfaces (Mian et al., 1991). A variety of other experiments have generated new receptor specificities through recombination between these regions of T4-like phages and through making combinatorial libraries in these hypervariable regions; the possibilities here seem endless (Tetart et al., 1996; Tetart et al., 1998).

7.2.3.2. Adhesin Restructuring through Cross-Genus Recombination

As discussed in Chapter 5, genomic analysis suggests that many phages have acquired their adhesins through cross-genus exchange. Haggard-Ljungquist (1992) showed that the coliphage P2 tail fiber contains regions very similar to tail-fiber proteins from the highly diverse phages P1, Mu, λ, K3, and T2; a number of others have made similar observations (Schwarz et al., 1983; Sandmeier et al., 1992; 1994). Similarities are seen both in P2 gene *H*, which encodes the tail fiber protein, and gene *G*, whose product is involved in tail-fiber assembly. The relationship can also be seen in functional terms; though P1 and P2 are otherwise unrelated, their tail fibers can be inactivated by the same antibodies, and bacterial strains resistant to P2 are usually also resistant to Mu and P1, but not to most other phages.

Martha Howe (personal communication) pioneered applying this approach for making phages with extended host ranges. She took Mu strains with amber mutations in the tail-fiber genes and plated them on amber-suppressing hosts lysogenic for P1, selecting for rare plaque-forming phage. These recombinant phages were all different, as reflected in efficiencies of plating on a spectrum of hosts. The best was called Mu hP1 (Csonka et al., 1981).

For bacteriophage T4, the adhesin site is located near the tip of the distal half of the long tail fibers, encoded by gene *37* (Fig 7.6). An ancient recombinational event apparently generated this adhesin by replacing the distal end of the original gene *37* and all of gene *38* (as seen in Ox2, T2, and most other related phages) with DNA orthologous to the P2 tail-fiber gene and to a pair of ORFs seen in bacteriophage

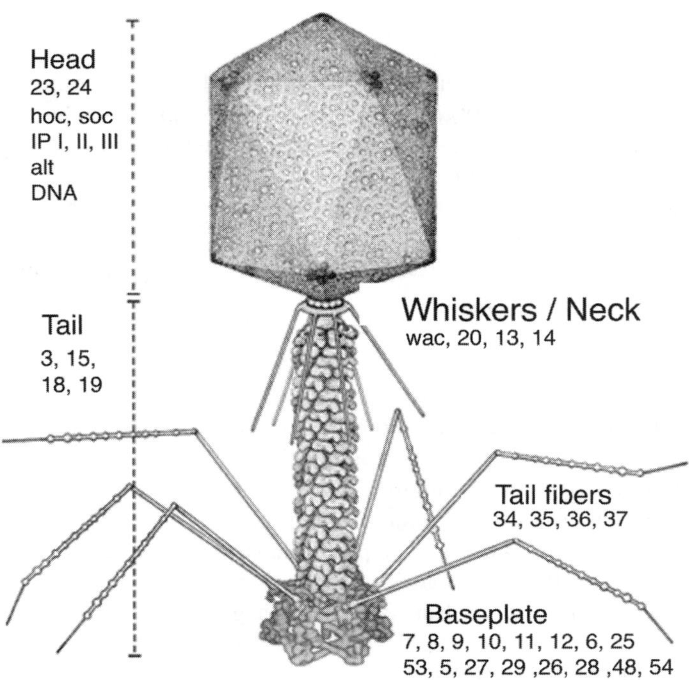

FIGURE 7.6 Bacteriophage T4; the numbers indicate the genes encoding each structure. Jones *et al.* (2001). with permission.

λ—*tfa* (tail fiber assembly) and *stf* (short tail fiber). Hendrix and Duda (1992) provide evidence that the tail fibers of the originally isolated λ were actually encoded by a gene that extended through this region and led to faster adsorption than that of the currently used λ, where a frame shift has led to the appearance of there being two genes and to truncation of the protein product. Today's λ appears to have been selected by early researchers for its larger plaques and uses a totally different tail protein, recognizing the host maltose-binding protein, as its receptor.

7.2.3.3. Phage Encoding Multiple Different Adhesins

Some phages actually encode two alternate forms of the tail-fiber adhesin and can switch between expressing one or the other on the phage particle, substantially broadening their potential flexibility in host range. For example, Scholl et al. (2001) isolated a *coli* podovirus they call K1-5, which recognizes hosts expressing either the K1 or K5 form of the capsular polysaccharide K antigen. K1-5 was found to have tandem genes encoding different activities that can attack the capsid: an endosialidase that can degrade the K1 capsular α-2,8-linked poly-*N*-acetylneuraminic acid carbohydrate and a lyase that can attack the *N*-acetyl heparosin capsid of K5; both were functional, broadening its host range at least within this very narrow family.

Bacteriophage Mu is an isometric-headed myovirus, and has the typical 6 tail fibers that help properly position the baseplate and trigger contraction. It has a very broad host range compared to most known phages, thanks in part to actually switching between a pair of possible versions of its tail-fiber gene by a process of inversion of a 3000-bp segment called G (Sandulache and Prehm, 1985; 1984). As reviewed by Harshey (1988) and by Koch et al. (1987), G(+) phage can infect *E. coli* K12 and *Salmonella arizonae*, whereas G(–) Mu phage instead infect strains of *Citrobacter freundii, Shigella sonei, Erwinia, Enterobacter* and *E. coli* C strains. A specific phage protein, Gin (for G inversion) is responsible for the switching, which occurs over time in the prophage state. Phage P1 also has an invertible segment, and the responsible P1 enzyme can substitute for Gin to invert the G segment of Mu *in vivo*.

7.2.4. Summary

All in all, it is clear that phages of various families have enormous ability to change adhesins by intra- and inter-species recombination and by mutation, and that such changes routinely lead to the ability to recognize new host receptors. This phenomenon plays important roles in the natural ecology of phage-host interactions, can be exploited in the selection of therapeutic phages having desired "target ranges" from nature and in specifically manipulating phage genomes using molecular techniques. In this context, workers at the Eliava Bacteriophage Institute in Tbilisi have for decades taken advantage of the natural variations by using phage isolation techniques that select for increasingly effective binding to the targeted hosts and then by co-infecting hosts with several related phages known to have different specificities, taking advantage of resultant natural variations to produce phage cocktails likely to have a desired, broader range of potential binding specificities.

7.3. THE PROCESS OF DNA TRANSFER INTO THE CELL

7.3.1. Overview

Appropriate positioning of the phage tail on the cell surface triggers irreversible events leading to DNA delivery into the host. There are few phages where much is known about the mechanism or energetics of this process. A detailed review by Letellier, Boulanger, et al. (2004) focuses primarily on three coliphages that have been best studied in this regard—myovirus T4, siphovirus T5, and podovirus T7. For each, a very long polyanionic genome—50 times the cell length for T4—must penetrate unscathed across two hydrophobic barriers—the outer and inner membranes—as well as the peptidoglycan sheath and the nuclease-containing periplasmic space. As they discuss, the rate of transfer can be as high as 3000–4000 base-pairs/second, in contrast to the 100 base-pairs/second seen for conjugation and natural transformation. Furthermore, the very high efficiency of infection for phages like T4 indicates that the process almost always occurs without significant damage to the infecting DNA molecule.

The mechanisms are highly varied; often, ATP, a membrane potential, or enzyme action is involved, but some phage, such as T5, can enter cells in the absence of metabolic energy sources. It is clear that the volume of the rigid phage capsids does not change as the DNA is injected into the cell, and the simple release of energy from metastable packing in the particle does not provide sufficient energy to transport a DNA molecule several micrometers long through the very narrow channel inside the tail tube. For *B. subtilis* phage SP82G, McAllister (1970) showed that the rate of DNA transfer into the cell is constant over the whole genome and is highly temperature dependent, and the second-step transfer of the major part of coliphage T5 DNA, described below, proceeds normally even in experiments where the DNA has already been released from its capsid (Letellier et al., 2004). Though the myoviruses have contractile tails, it does not appear that their tail tubes actually pierce the inner membrane and a potential gradient is still required for DNA entry into the cell. All in all, as discussed by Molineux (2001), it is clear that the widely-invoked "hypodermic syringe" injection metaphor is generally inaccurate and the mechanisms providing energy for the transfer differ from phage to phage.

7.3.2. Podovirus T7

Molineux (2001; 2005) has provided the most detailed analysis to date of the transfer process of DNA from a tailed phage into its host—an analysis which is very instructive, even though the mechanisms are arguably quite different for most other phages. The T7 infection process involves major rearrangement of tail-related proteins that are actually packed inside the head until the time of infection (Fig. 7.7)—a sort of protected "extensible tail," in contrast to the contractile tails of the Myoviridae. Three internal proteins form a prominent 26×21 nm cylindrical structure within the 60-nm T7 head. This core, around which the DNA is wrapped, appears to contain three copies of gp16 (144 kDa each), 12 of gp15 (84 kDa), and 18 of gp14 (21 kDa); two other essential proteins, gp13 and gp7.3, are also somewhere in the head. The bottom of the core is attached to the head-tail connector, gp8, which in turn binds the stubby, 23-nm tail composed of gp11 and 12. A central channel extends through the core, connector, and tail.

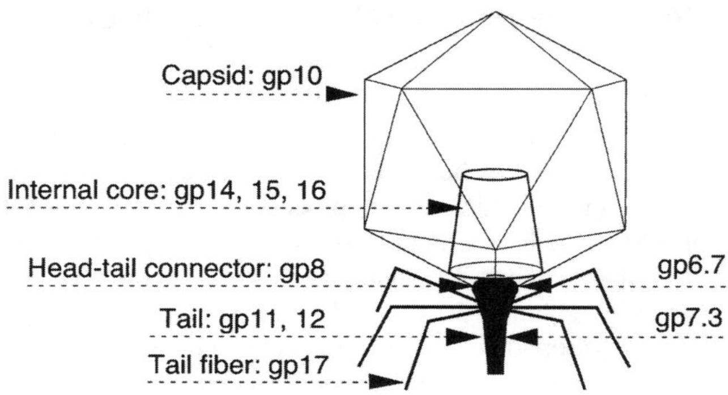

FIGURE 7.7 Detailed structure of bacteriophage T7. Molineux (2001), with permission.

Once interactions between multiple T7 tail fibers and the LPS squarely position the tail against the outer membrane, a signal is transduced to the phage head to initiate the irreversible infection steps; it appears that both the tail and tail fibers are involved in this signaling, in addition to gp13. The evidence for this is that gp13 mutants look normal but will not bind irreversibly to cells; however, second-site revertants of gp13 mutants have been found in the tail fiber, tail, and connector genes. The irreversible process begins with the degradation of gp13 and gp7.3, and the transport of the three core proteins, gp14-16, through the central channel and into the target cell. Presumably, some unfolding is necessary for the core proteins to get through this very narrow channel, particularly for the 1318-amino-acid gp16. Gp14 remains in the outer membrane, while gp15 and gp16 are found in both soluble and membrane-bound forms; Molineux (2001) suggests that the three together form a channel across the entire cell envelope. The N-terminal residues of gp16 show homology to the C-terminal region of the host soluble lytic transglycosylase, sltY, which plays a major role in *E. coli* peptidoglycan turnover. The sltY active-site glutamate is conserved in T7 gp16; if other amino acids are substituted there, the phage cannot efficiently infect cells where the peptidoglycan is believed to be *more highly cross-linked* (growing at 20°C, or in late exponential phase at 30°C) unless sltY is *overexpressed.*

The mechanism by which T7 DNA normally enters the cell is slower than that observed for other phages, allowing the details to be more easily studied. When DNA-labeled phage-host complexes were sonicated at various times after infection to shear DNA that had not yet entered the protection of the cell, DNA from the genetic left end of the genome was sequentially lost from the released virions. Rifampicin treatment showed that loss of all but the leftmost part of the genome was dependent on transcription by *E. coli* polymerase; only the first 850 bp, containing T7's main three *E. coli* promoters and a T7 promoter, normally enter the cell without active transcription. The slow entry of the rest of the DNA allows the potent restriction-enzyme inhibitor encoded by gene 0.3 to be produced before sensitive restriction sites enter the cell. Mutations affecting a 130-amino-acid central region of gp16 define a region that clamps the entering DNA and makes further entry dependent on transcription; a transcribing RNA polymerase is the strongest molecular motor known (Wang et al., 1998) and it presumably overcomes the clamp's resistance to DNA translocation. Under transcription-*independent* conditions (i.e., using the gp16 mutant), the T7 genome enters at a constant rate of about 75 bp/sec. During the gp16 *mutant's* transcription-*independent*, high-speed entry, the incoming DNA is indeed degraded by nucleases unless those have also been mutationally inactivated. In summary, Molineux's evidence suggests that gp15 and gp16 together may form a sort of molecular motor that can rapidly ratchet the DNA into the cell, but the gp16 clamp component slows this process after the first 850 bp unless active transcription is occurring.

7.3.3. Myovirus T4

The genomes of many phages enter the cell rapidly; this is particularly important for those that are circularly permuted, since the whole genome must enter before the developmental program can be counted on starting. T4 DNA transfer has been estimated at 3–10 kbp/s (Grinius and Daugelavicius, 1988; Letellier et al., 1999).

For all the T4-like phages, the reversible binding of the long tail fibers to their specific receptors is only the first step, though the one primarily responsible for host specificity. The initial energy for T4 infection is provided by the baseplate, an exquisite "cocked" mechanical device (Fig 7.6). Once three or more of the long tail fibers position the baseplate parallel to the host surface, the baseplate shifts from a hexagon to a star shape. Then six short tail fibers, encoded by gene *12,* are deployed (Crawford and Goldberg, 1980), and they bind irreversibly to the "second phage receptor"—the heptose residue of the LPS inner core (Riede et al., 1985; Montag et al., 1987). The conformational change in the baseplate simultaneously triggers contraction of the tail sheath, pushing the inner tube through the outer membrane. Contraction of the tail sheath advances as a wave of compression transmitted through the helix-like arrangement of tail sheath annuli.

To cross the peptidoglycan layer, the tail tube uses a lysozyme inserted into baseplate protein gp5, at the tip of the tube. The recently determined three-dimensional structure of this complex clearly shows the lysozyme domain and a needlelike C-terminal domain (Fig. 7.8) (Kanamaru et al., 2002). When the tail sheath contracts and the tail tube protrudes from the bottom of the baseplate, the C-terminal triple-stranded beta-helix seems to act as a needle and position the lysozyme domain so it can cut the murein layer. This lysozyme domain of gp5 is a homologue of the small T4 gene *e* lysozyme, apparently reflecting an ancient duplication. The lysozyme activity is activated only when the *needle* domain is detached from the rest of gp5. The tail tube does not actually penetrate the inner membrane during infection. It appears that the signal for T4 DNA release through the tail involves contact with phosphatidyl glycerol in the inner membrane, and that this release is normally coupled to transport across the membrane by a still-unknown mechanism that requires the electrochemical potential across the inner membrane (reviewed by Goldberg et al., 1994). Without that potential, the DNA remains in the periplasm and is eventually degraded.

As reviewed by Abedon (1994), initial infection by T4-like phages leads to poorly-understood membrane alterations involving the products of the T4 *imm* (immunity) and *sp* (spackle) genes, which set up a phenomenon called *superinfection*

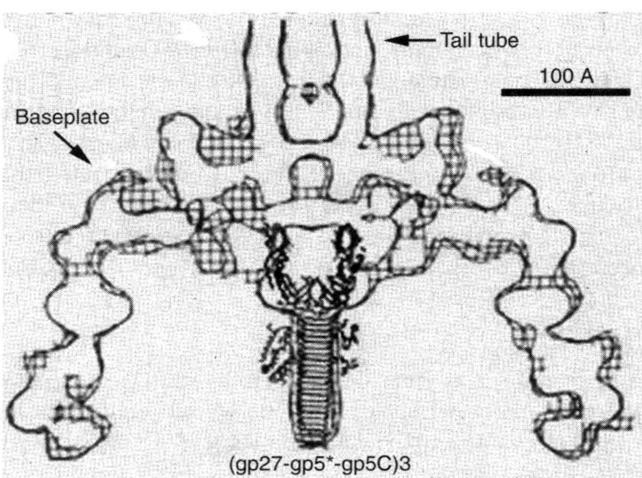

FIGURE 7.8 Structure of T4 tail lysozyme and needle as determined by X-ray crystallography of a $(gp27)_3(gp5)_3$ hetero-hexameric complex. Adapted from Kanamaru, *et al.*, 2002, with permission.

exclusion by about 4 minutes after the initial infection (Lu and Henning, 1989; Lu et al., 1993). Here, T4 or related phages that attempt to infect the already-infected cell release their DNA into the periplasm, where it is degraded. The mechanism whereby these gene products block DNA transfer across the cytoplasmic membrane has not been determined. However, the same genes appear to be involved in establishing post-infection resistance to *lysis from without*—a phenomenon common to many phage-host systems, in which simultaneous adsorption of high numbers of phage particles causes an immediate decrease in turbidity of the culture and liberation of internal proteins (Delbrück, 1940). Cells infected by sp^+ imm^+ T4-like phage rapidly develop a resistance that allows them to safely adsorb hundreds of secondary phage. X-ray-inactivated phage cause lysis from without even more readily than do viable phage, suggesting that gene expression by the first infecting phage rapidly produces products that can immediately lower the susceptibility to lysis from without.

Treatment of T4 with urea produces contraction of the tail and formation of an activated state, but with the DNA remaining inside; successful infection can then occur upon contact with spheroplasts of *E. coli* or any of a number of related bacteria, but the efficiency is very low (Wais and Goldberg, 1969). DNA release can also be triggered by contact with microspheres of phosphatidylglycerol (but not phosphatidylethanolamine); in this case, the DNA is released into the outside medium.

7.3.4. Three Siphoviridae That All Use the *FhuA* Receptor but Have Very Different Mechanisms

Letellier et al. (2004), Christensen (1994), and German (2005) review infection by multiple siphoviruses that use the same receptor yet employ quite different infection mechanisms. The outer membrane protein FhuA, a beta-barrel structure with a central plug, is responsible for iron-ferrichrome transport into the cell. One hydrophilic loop of the barrel, facing the external medium, acts as receptor for phages T1, ϕ80, and T5. Both the transport of iron across the gram-negative outer membrane and the irreversible binding of T1 and ϕ80 are energy-dependent processes, yet no energy sources are available in the outer membrane or periplasm (Hancock and Braun, 1976). The energy derived from the electrochemical gradient of protons generated by the electron transfer chain in the cytoplasmic membrane is transduced to the outer membrane by the cytoplasmic membrane-anchored protein complex TonB-ExbB-ExbD. The TonB protein, which is anchored in the inner membrane by its uncleaved N-terminal signal sequence, is thought to interact with the outer membrane receptor FhuA by its C-terminal domain. The mechanism of the energy transduction and how phage binding depends on TonB has not yet been determined (see above). (The mechanism of actual transfer is not known, but the resultant fall in cellular proton motive force [PMF] and ATP leads to a GTP drop and to inhibition of the initiation of translation of host proteins; other ionic changes may also be involved. Within a few minutes, some sort of membrane "resealing" prevents further PMF drop and phage reproduction proceeds efficiently.)

In contrast, T5 infection is *not* energy-dependent; the phage irreversibly binds to FhuA by means of pb5, a protein located at the distal end of the phage tail. Interaction of phage T5 with purified FhuA leads to virtually instantaneous and rapid release of its DNA into the surrounding medium, in the absence of other factors, whereas this same interaction does *not* release the DNA from T1 or ϕ80 (Letellier et al., 2004). Plançon et al. (1997) have shown that T5 can bind to liposomes into which purified FhuA protein has been reconstituted, and this binding induces the release of the phage DNA and its transfer into the vesicles. This demonstrates that FhuA alone is adequate for T5 DNA release and transfer across a membrane, though it still does not explain the mechanism of its transfer through the *E. coli* peptidoglycan layer and inner membrane. Even for T5, the situation is more complicated than it would seem from this simple experiment. A unique feature of phage T5 is that its DNA transfer process involves two steps; after 8% of the DNA enters the cytoplasm, there is a 4-minute pause while several proteins encoded there are made. They degrade the host DNA and shut off their own transcription before the rest of the phage DNA is pulled in. Some sort of channel must protect the DNA from periplasmic enzymes during this hiatus; this appears to involve the T5 straight fiber protein, pb2, which contains a hydrophobic domain resembling that of viral fusion peptides that may fuse the outer and cytoplasmic membranes, thus facilitating the DNA's transfer across the whole bacterial envelope.

7.3.5. Phages PRD1 and PM2

This chapter largely focuses on tailed phages and thus DNA delivery processes of non-tailed phages are not generally reviewed here. It should be noted, however, that they can be very different from those described for tailed phages. One particularly interesting example is the well-studied DNA delivery process of tectivirus PRD1, a very unusual 15-kbp double-stranded-DNA phage with an internal membrane that

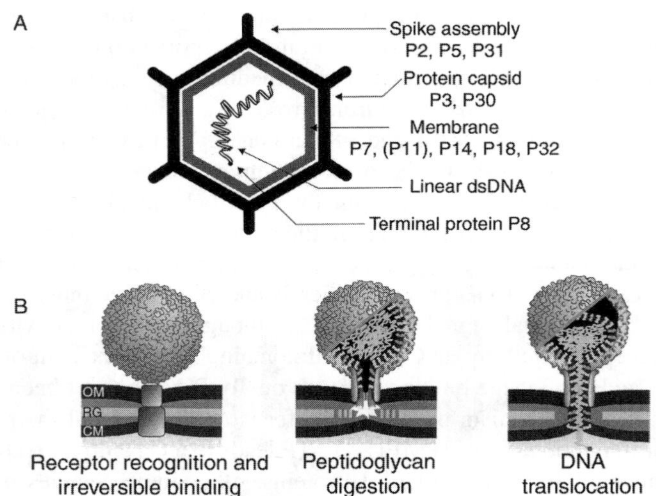

FIGURE 7.9 Bacteriophage PRD1. (A) structure and (B) model of the infection process. Modified from Grahn *et al.*, 2002, with permission.

underlies its icosahedral capsid (Grahn et al., 2002). PRD1 infects *E. coli, P. aeruginosa,* and *S. enterica* carrying conjugative IncP plasmids, making use of the plasmid transfer genes for attachment and entry—specifically the mating pair formation (Mpf) phage receptor complex, which bridges the cytoplasmic and outer membranes and belongs to the type IV secretion-system family. It can bind via the spikes at any of its corners rather than having a single tail (Fig. 7.9a). During infection, the phage's internal membrane becomes part of a tube-like structure that appears to cross the host cell wall and fuse with the inner membrane (Fig. 7.9b). About 10 of its 20 structural proteins appear to be embedded in the membrane; various mutants were used to assign several of them to particular stages in the depicted infection process.

PRD1's mechanism of infection appears to be very different from that of PM2, a phage of the Corticoviridae that infects marine *Pseudoalteromonas* species (see Chapter 4 and Bamford and Bamford, 2005). PM2 also has an internal lipid bilayer and is known to adsorb to the cell wall, but the details of its infection mechanism have not yet been described.

7.4. PHAGE DNA REPLICATION, REPAIR AND RECOMBINATION

7.4.1. The Process of Replication

Accurate replication of its genetic material is one of the most significant challenges facing any organism. The great diversity of molecular mechanisms phage have evolved for replication, recombination, and repair have provided the bulk of our general knowledge about the fundamental processes of DNA synthesis, both where the phage adapts host mechanisms and where the phage has evolved its own distinctive replication apparatus. Most temperate phages depend totally or almost so on the host replication machinery for their lytic as well as prophage replication cycles. They generally use recombination or annealing of staggered ends to circularize after infection—a necessary step for recombination into the host chromosome if they are going to lysogenize—and in their lytic mode, they also use the standard bacterial "theta" mode for circular-DNA replication. Some later generate a long linear concatamer for packaging using a *rolling-circle* mechanism. In contrast, most of the larger lytic phages encode their own DNA polymerases and replicate as linear molecules, solving the problem of replicating the ends by *recombining* to make concatenates many genomes long from which headfuls of DNA are packaged.

DNA replication usually initiates at a defined *origin of replication*. Special features of such *ori* regions include binding sites for specific initiator proteins and adjacent stretches of AT-rich regions that aid in unwinding DNA. Some phages contain a unique *ori* (phage λ), while others contain multiple replication origins, which may use different mechanisms and may be of particular use under different environmental conditions (phages T4, T7, and P4). At the origin, a short RNA primer can be synthesized either by a special primase or by the RNA polymerase, or Rep proteins can generate a DNA primer by introducing a nick. If primer synthesis is mediated by primase, initiator proteins promote the assembly of the pre-primosome at *ori* sites or at preexisting forked templates, such as those created during DNA recombination by homologous strand exchange. If RNA polymerase is used, this mechanism of initiation may be blocked as changes are made to the polymerase in the course of the developmental program; T4, for example, switches to using recombination intermediates, forming a huge, highly branched replication complex. Loading of the DNA polymerase onto the primer requires ATP hydrolysis and is generally mediated by *brace* proteins and a sliding clamp protein; topoisomerases are also involved in unwinding the DNA during the initiation process.

DNA polymerases can only add nucleotides to the free hydroxyl group of an appropriate polynucleotide or protein primer appropriately positioned on a template. This is necessary because the base complementarity responsible for making an accurate duplicate DNA chain depends on the precise spacing and positioning of each nucleotide. Furthermore, DNA replication always proceeds in the 5′ to 3′ direction, using 5′ deoxynucleoside triphosphates (dNTPs) so that the energy is carried by the incoming substrate, not the growing chain; this means that special mechanisms are needed for synthesis of the strand in the other direction—the lagging strand (see Fig. 7.10). This generally involves discontinuous synthesis of *Okazaki fragments* which are then joined together. The process of DNA synthesis involves constant editing to keep the number of spontaneous mutations at an acceptable level despite the inherent challenges of precisely selecting the next base. Thus, all DNA polymerases have a built-in $3' \Rightarrow 5'$ exonuclease activity.

Replication is of necessity very highly *processive*; synthesis of the leading strand needs to go to completion, without ever stopping, letting go, and restarting. In the case of phages whose DNA is concatenated, this may mean processivity over a molecule many times the length of a single genome, while synthesis of the lagging strand, using the same polymerase, terminates every thousand or so base pairs. This and a number

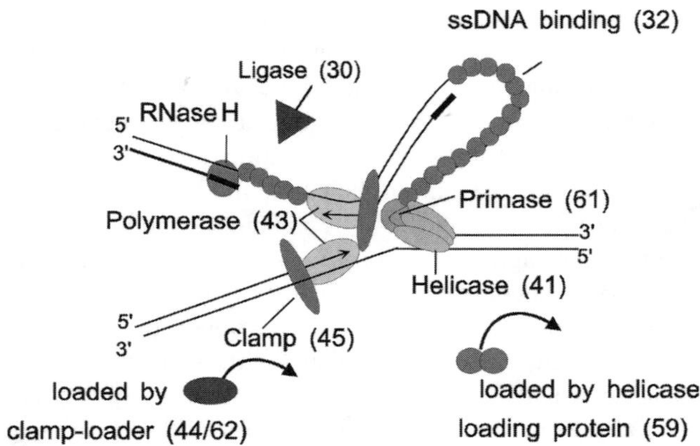

FIGURE 7.10 Bacteriophage T4 replication complex, illustrating leading and lagging strand synthesis. Miller *et al.*, 2003, with permission.

of other questions as to the detailed mechanisms involved have been answered using the combination of genetic, biochemical, and biophysical techniques possible in the bacteriophage T4 system, in which leading and lagging strand syntheses are tightly coupled (Fig. 7.10). Here, it has been shown that only the polymerase and a clamp that encircles the DNA are required for high-processivity leading-strand synthesis; the clamp must, however, be reloaded with the initiation of each new Okazaki fragment (Trakselis et al., 2001; Pietroni et al., 2001). Lagging-strand synthesis requires, in addition, a single-strand-binding protein, a helicase to unwind the DNA, the primases involved in initiating the Okazaki fragments, a $5' \Rightarrow 3'$ exonuclease to specifically remove the primers (either an RNaseH or a host enzyme like *E. coli* DNA polymerase I [pol I]), and a DNA ligase to seal the Okazaki fragments together.

As discussed in Chapter 5, the DNA-dependent DNA polymerases found to date fall into at least six superfamilies or classes, with highly complex phylogenies and patterns of distribution across kingdoms (Filee et al., 2002). The six are so different in size, shape, and sequence as to suggest that DNA polymerases evolved more than once during the transition from the RNA to the DNA world, perhaps initially to provide protective modifications of viral RNA (Koonin, 2003), but much lateral exchange has also occurred among them over evolutionary time. All of the known *phage* DNA polymerases fall into classes A and B, while the Pol III family of bacterial replication enzymes falls into class C. Family A, B, and C polymerases all have distinct domains for the polymerization and 3′–5′ exonuclease (editing) activities that determine replication fidelity and spontaneous mutation rate; A-family proteins have in addition a third domain with 5′–3′ exonuclease activity. *E. coli* Pol I, the most-studied member of the A family, mainly carries out DNA repair and primer removal, but homologues catalyze replication in many phages. The B or alpha family mainly includes the primary replicative polymerases of eukaryotes and their viruses and of archaea, but some unusual repair enzymes in gamma proteobacteria also are included here. Interestingly, *E. coli* Pol II and the T4 DNA polymerase both belong to this class, but the T4 enzyme is much more similar to the B-type polymerases of the archaeal halophile *Halobacterium* and its viruses HF1 and HF2 than it is to the *coli* enzyme (Filee et al., 2002).

7.4.2. DNA Repair, Recombination, and Mutagenesis

The main mechanisms destroying phage in the natural environment involve damage to their DNA from radiation or reactive chemicals. Phages such as T4 have provided the main well-characterized genetic systems used to study fundamental aspects of mutagenesis and repair, including both mechanisms of environmental mutagenesis and the factors determining the spontaneous mutation frequencies that are tied to the process of DNA replication (Drake and Ripley, 1994). Many phages are largely or totally dependent on host mechanisms for recombination and for repair of DNA damage, but a number have their own mechanisms such as photoreactivation, excision repair, and recombination-dependent DNA replication through a damaged site (Cech et al., 1981; Kreuzer and Drake, 1994). For example, the *E. coli* enzyme responsible for photoreactivation converts pyrimidine dimers, which are caused by UV light, to the monomeric state. In contrast, endonuclease V, the product of T4 *denV*, is the prototype of a base excision repair protein. It has both *N*-glycosylase and basic lyase activities, which incise the DNA (forming a covalently linked protein-DNA intermediate). Together these activities remove pyrimidine dimers to allow resynthesis by DNA polymerases like *E. coli* Pol I, with help of the phage single-stranded-DNA-binding protein. The profound difference in UV sensitivities of T4 and T2 is due to the presence of *denV* in T4 but not in T2. The T4 *uvsX* gene encodes a homologue of RecA, the major protein required for invasion of single-stranded DNA into homologous double-stranded DNA; *uvsX* mutants are radiation-sensitive because they cannot engage in DNA replication that uses recombination to repair or bypass DNA lesions. Some of the repair mechanisms are *error prone*; they rescue the phage from lethality at the cost of a high probability of introducing altered information. Both radiation mutagenesis and most chemical mutagenesis occur primarily through error-prone repair in cellular organisms as well as in phage and other viruses (Drake and Ripley, 1994).

The *Red* general recombination system of lambdoid bacteriophages is particularly interesting. The Exo and Beta proteins of the λ Red system carry out recombination that repairs double-strand breaks, requiring as little as 30 bp of sequence homology, and the Gam protein inhibits the cell's RecBCD exonuclease, which otherwise destroys single-stranded fragments of DNA in *E. coli*. These special properties of λ's general Red recombination system have been used to develop a powerful and widely used new technique called *Recombineering* (Court et al., 2002), in which constructs can be engineered on plasmids or directly on the *E. coli* chromosome from PCR products or synthetic oligonucleotides by homologous recombination. This is possible because the Red system efficiently recombines sequences with homologies as short as 35 to 50 base pairs, in contrast to the far longer sequences required by the systems from cellular organisms. Although several methods allow one to put circular plasmid DNA into *E. coli* and other bacteria, linear DNA is rapidly degraded by RecBCD; the inability to use linear DNA in transformation experiments makes it difficult, for example, to create gene knockouts. Court's lab and others have engineered systems that transiently express the Red system. These Red^+ cells efficiently recombine incoming linear DNA pieces with as little as 30–50 bp of added terminal homologous sequences into the bacterial chromosome. Since replacement occurs in about 0.1% of the cells, knockouts and other manipulations can be done even without a selectable marker like chloramphenicol resistance (Cm^R).

7.4.3. Specific Examples

7.4.3.1. Cooption of Host Replicative Machinery: Phage λ

In its lytic cycle, the linear dsDNA from the λ virion or formed by prophage excision circularizes to form a nicked circle, which is closed by host DNA ligase and supercoiled by host DNA gyrase. Replication initiation relies on the host RNA priming system. The unique λ origin of replication is specifically bound by the λ O protein, which opens it and loads a complex of host DnaB (helicase) and λ P protein. DnaB unwinds λ *ori;* the single-stranded regions thus generated are stabilized by host Ssb. Then, primase DnaG binds to the DNA-helicase complex and synthesizes the RNA primer, which is polymerized by host DNA Pol III. Initially, DNA replication is θ; it proceeds bidirectionally to generate two daughter circles. In a late phase, most molecules replicate as rolling circles and generate a concatemeric DNA, which becomes the substrate for terminase in the packaging mechanism. Replication of λ is strongly dependent on host chaperon proteins (dnaK, dnaJ, and GroEL) and is stimulated by transcription of surrounding promoters.

7.4.3.2. Complex Phage-Encoded Replication, Recombination, and Repair Machinery: Phage T4

In T4, dNTP synthesis, replication, recombination, and late transcription are tightly coupled (Fig. 7.11). A nucleotide precursor complex produces dNTPs in the precise ratios needed for T4 DNA synthesis, with 2/3 AT (as compared with *E. coli*'s 50% AT) and with hydroxymethyl dCTP (hmdCTP) in place of dCTP. These complexes funnel the nucleotides directly into the DNA replication forks through interactions with the DNA polymerase and other elements. The basic T4 replisome, which couples leading- and lagging-strand synthesis, includes the phage-encoded DNA polymerase, sliding clamp loader, sliding clamp, DNA helicase, primase, and single-stranded-DNA binding protein. Host DNA polymerase I and DNA ligase contribute to DNA repair; T4 RNaseH (*rnh*) and DNA ligase remove the RNA primers and seal Okazaki fragments and other DNA interruptions. The host RNA polymerase makes the primer for leading-strand initiation while the T4 primase synthesizes the primers for lagging-strand replication.

The first round of T4 DNA replication is initiated from one of four major origins, with different attributes that lead them to be preferentially used under different conditions. As soon as the first replication forks have reached an end, recombination intermediates compete effectively for assembly of replisomes; most T4 replication forks are initiated using recombination intermediates as DNA primers at positions

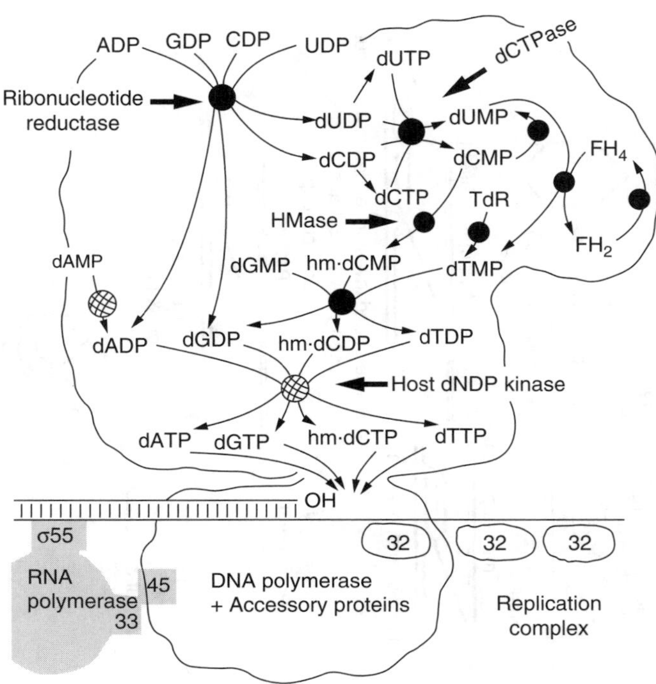

FIGURE 7.11 Bacteriophage T4: coupling of nucleotide synthesis, replication and late transcription. Figure constructed by Gautam Dutta, with nucleotide-synthesizing complex compliments of Chris Mathews.

throughout the genome. Origin-dependent DNA replication is inactivated by the RNA polymerase changes that initiate late transcription. Late T4 transcription is closely coupled to DNA replication, balancing the production of capsids with making the DNA to be encapsidated in a unique way, as described in sec. 7.5.2.3.

T4 has played a major role in exploring the potential complexities of recombination mechanisms and their potential links with DNA replication and repair (Fig. 7.12). Two main recombination-dependent T4 DNA replication modes, join-copy and join-cut-copy, have been described (Mosig et al., 2001). Early T4 join-copy recombination is correlated with *origin-dependent replication* and is initiated from the single-stranded DNA at the unreplicated end of the template for lagging-strand synthesis. In damaged T4 chromosomes, similar mechanisms can be used to repair broken replication forks. The UvsX strand-invasion protein is a structural and functional homologue of the *E. coli* RecA protein and the eukaryotic Rad51 and Rad54 proteins important to repair and recombination. In contrast, endoVII and terminase, the main proteins required for *join-cut-copy recombination-replication*, are expressed late. Together, these two mechanisms generate the complex branched, concatemeric intracellular T4 DNA. This DNA is debranched during DNA packaging by T4 endonuclease VII and the largest (70-kDa) subunit of the heteromeric terminase, as described in section 7.7.

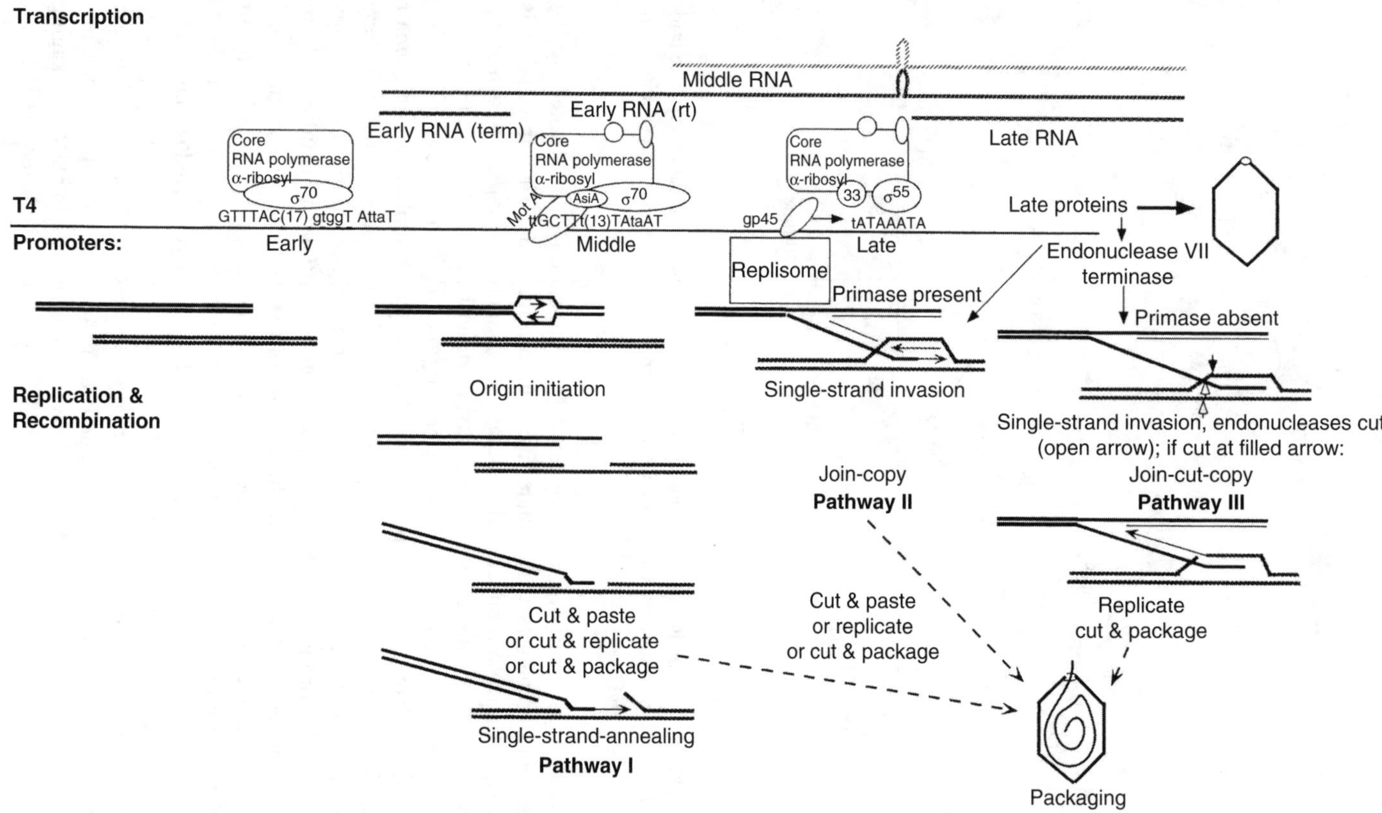

FIGURE 7.12 Major modes of recombination in bacteriophage T4 and their involvement in DNA replication and repair. Mosig *et al.* (2001), with permission.

7.4.3.3. Replicative Transposition: Phage Mu

During its lytic cycle, phage Mu employs a highly unusual replication mechanism known as *replicative transposition*; the circularized Mu genome always enters the host chromosome and specific phage replication occurs without excision of the prophage, as reviewed by Chaconas (1987), Harshey (1988) and Paolozzi and Ghelardini (2005). About 100 copies of the Mu DNA are produced at new sites throughout the host chromosome within 6–8 minutes after infection. Replicative transposition requires the products of the Mu A and B genes, both ends of the Mu genome, and several host-encoded proteins. The ends of each integrated phage genome define both origin and terminus of replication of Mu. Single-strand nicks are introduced at each end of the prophage on opposite strands, and a staggered double-strand cleavage is introduced at the target site. Ligation of DNA from both ends of the prophage to the target DNA produces an intermediate, the strand transfer complex (STC), to which the Mu transposase (MuA) remains tightly bound. This intermediate is identical to a normal replication fork, in which the 3′ end of the host DNA can serve as a primer. DNA synthesis to form the Mu cointegrate from one or both of the prophage termini requires host DNA polymerase III, DnaB, DnaC, DNA gyrase, and replication factors α and β. The last two factors are required to remove MuA from the STC and initiate DNA synthesis. PrtA is also required; it is a 3′–5′ helicase that recognizes DNA forks and promotes loading of the replisome assembly to restart DNA replication.

7.4.3.4. Protein-Primed Replication Initiation: Phage ϕ29

Rather than using a nucleic acid primer to initiate replication, phage ϕ29 uses a phage-encoded terminal protein, TP, that remains covalently attached at each 5′ end of its 19,285-bp linear dsDNA (Salas, 2005). Four phage-encoded proteins are required for replication: DNA polymerase, TP, DBP, and ssb. DBP is a histone-like protein that binds to the TP-containing DNA ends and helps melt the DNA there, leading to binding of a stable TP-DNA polymerase heterodimer. The DNA polymerase then catalyzes the formation of a phosphoester bond between 5′-dAMP and the Ser^{232} hydroxyl group of the new, polymerase-bound TP. This first inserted nucleotide is actually complementary to the *second* nucleotide at the 3′ end of the template, so the TP-dAMP complex slides back one nucleotide before proceeding with forward polymerization. The DNA polymerase dissociates from its complex with the new terminal TP after about 10 nucleotides have been synthesized. Phage ϕ29 DNA polymerase does not require accessory proteins or helicases. It belongs to the B-type superfamily of DNA polymerases, with 3′–5′ exonuclease and pyrophosphorolysis activities, and shows high processivity. Replication is symmetric, from both ends, and is coupled to strand displacement. The displaced single-strand DNA molecules are protected from degradation by the ssb protein, p5. Binding of p5 to ssDNA also prevents nonproductive binding of the DNA polymerase to the ssDNA, as well as the formation of secondary structures in the displaced ssDNA strand, and contributes to increased elongation velocity of the DNA polymerase.

Other phage genomes with a TP covalently linked to their 5′ ends include *S. pneumoniae* phage Cp-1 and phage PRD1, infecting enteric bacteria.

7.4.3.5. Replication of a Linear Chromosome with Closed Ends: Phage N15

N15 is a temperate coliphage, half of whose 46.4-kbp genome is similar to the λ sequence (Ravin et al., 2000; Ravin, 2005). The viral chromosome is a circular permutation of its prophage chromosome. However, instead of integrating into the host genome, the prophage is a linear plasmid with covalently closed hairpin ends. This form is the product of an intramolecular site-specific recombination reaction mediated by protelomerase TelN (or telomerase resolvase) at the *telRL* site (Ravin et al., 2003). Replication of DNA molecules with hairpin telomeres is initiated in the telomeric regions, with strand displacement, resulting in the production of head-to-head and tail-to-tail concatemeric arrays. Telomere resolution is also mediated by TelN, a protein with limited sequence homology to the HK022 integrase, of the tyrosine recombinase family. Linear chromosomes like N15 are also found in *Yersinia* phage PY54, plasmids of *Borrelia,* some mitochondrial DNAs, pox virus, and a *Chlorella* virus.

7.5. TRANSCRIPTION

7.5.1. Overview

Most of the regulation of phage development occurs at the level of transcription. Genes transcribed immediately after infection often serve to restructure the host in ways large or small to facilitate the phage infection process; these range from those involved in the lytic/lysogenic decision pathway for temperate phages to the many genes that help shut off host transcription, translation, and replication in T4 and SPO1. A group of *delayed early/middle-mode genes* are then transcribed that largely direct phage DNA replication, followed by *late genes* that encode the phage capsid and associated proteins. Many phages encode one or more new RNA polymerases of their own to help direct this process; others simply make a series of modifications to the host RNA polymerase (RNAP). Phage regulatory mechanisms are found that act at each of the three well-defined stages of transcription—initiation, elongation, and termination—and the study of these mechanisms has contributed much to our understanding of bacterial transcription.

For all known bacteria, transcription involves a multi-subunit core enzyme (msRNAP) composed of subunits β, β', and two copies of α. In addition, a family of σ subunits provide promoter recognition and initiation specificity. These σ factors may be denoted by their size or by their function. For example, the 70-kd σ^{70} regulates transcription of most general housekeeping genes, while the general-stress or stationary-phase σ, σ^{S}, controls major changes in the cell wall, biosynthetic machinery and enzymes such as catalase that support long-term survival of gram-negative bacteria under conditions of nutrient exhaustion and environmental stress (Hengge-Aronis, 2002a). The relative strengths of the promoters are determined by the precise sequence of the primary recognition elements (regions around positions −10 and −35 relative to the starting nucleotide for transcription for σ^{70} and σ^{S}), by additional sequences such as an extended −10 sequence, or the *upstream element* that contributes to the very high activities of ribosomal RNA and protein σ^{70} promoters, and by nearby binding sites for other regulatory proteins; the C-terminal parts of the α subunits are involved in these interactions (Hengge-Aronis, 2002b).

During initiation, the polymerase-bound factor binds to an appropriate promoter sequence on the double-stranded DNA molecule, forming a *closed promoter complex*. The DNA must then be melted to form an *open promoter complex*, with a bubble extending from about position −11 to +2. The β and β' subunits of the bacterial enzyme form a cleft containing the active site and an 8–9 bp DNA:RNA hybrid, as well as side channels accommodating the non-transcribed DNA strand and the elongating RNA beyond where the transcription "bubble" recloses. Young et al. (2004) showed that a complex of *coli* σ^{70} amino acids 94–507 and β' amino acids 1–314 is sufficient for both binding and promoter melting. After insertion of the first 10–15 nucleotides, the σ factor is released from the complex, allowing the polymerase to enter *elongation phase* and move away from the promoter. At some promoters, a number of 2–6-nucleotide abortive initiation fragments are sometimes formed before the complex succeeds in moving into elongation mode. Elongation is totally processive; pausing may occur, but the transcript remains firmly bound to the polymerase. If the elongation phase gets stalled at some point, the polymerase does not just fall off; it is able to move backwards along the template, cleaving off the most recently added group of nucleotides, and try again; small cellular proteins GreA and GreB can aid in this process. Transcript termination and polymerase release involve either a stem-loop structure followed by an extended T-rich sequence or the action of a specific termination protein called rho. Phage cooption of the host polymerase can include new σ factors, other activators and repressors of transcription initiation, covalent modifications of the host polymerase such as ADP ribosylation, and factors that modify the elongation and termination of specific transcripts. However, little is known about the actual mechanistic details of the reaction or the specific effects of these phage-directed modifications.

A number of phages encode their own RNA polymerases, which have been extremely useful both as reagents of molecular biology and in determining the fundamental mechanisms of transcription. These single-subunit polymerases are much smaller and faster-moving than the multi-subunit polymerases (msRNAPs) of bacteria and eukaryotes, are highly specific for particular promoters, and require no additional protein factors to identify their promoters or carry out transcription. Similar RNA polymerases are found in eukaryotic viruses and in mitochondria (Cermakian et al., 1996; Rousvoal et al., 1998), raising interesting questions as to their origin and evolution; they are also related to reverse transcriptase and to the Pol I (A) family of DNA polymerases. Importantly, the biochemical mechanisms of initiation, elongation, and termination appear similar to those of msRNAPs. The best studied is the 883-aa polymerase encoded by bacteriophage T7, whose transcription rate is over 200 bases/sec at 37°C—5 times that of *E. coli* RNAP. The basic structure has been likened to a right hand, with the active site located in the palm. The promoter it recognizes extends from −17 to −5, where it binds

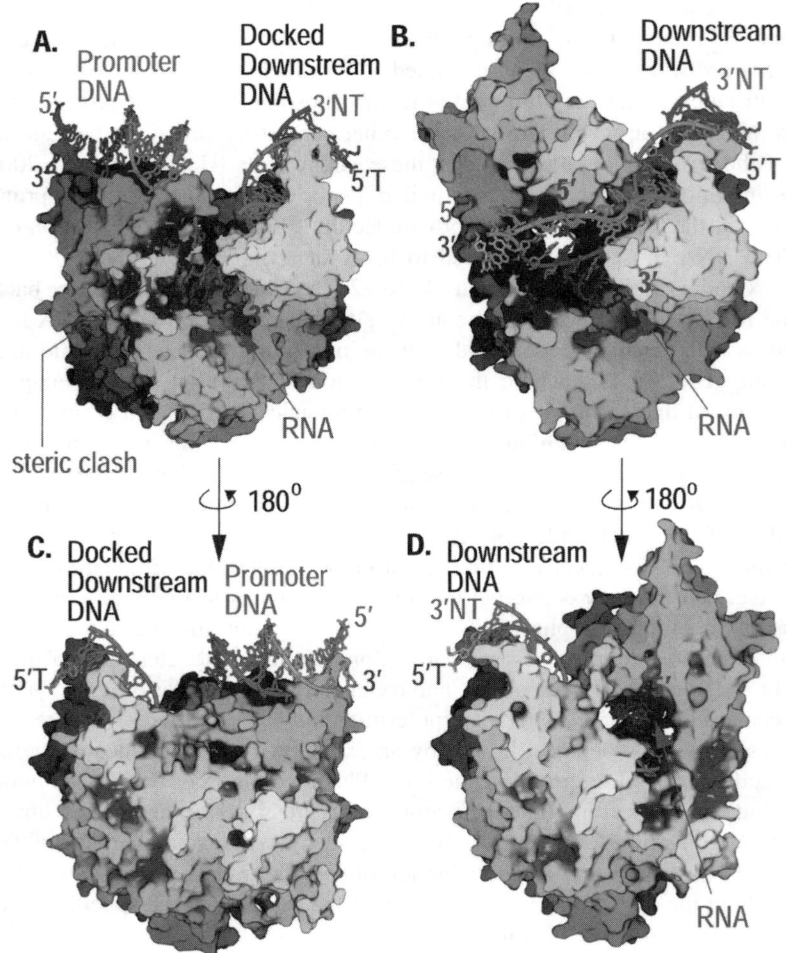

FIGURE 7.13 Crystal Structure of T7 RNA polymerase. A and C show the initiation complex in contest to B and D which shows the elongation complex. (Yin and Steitz, 2002)

to one face of the closed duplex promoter by means of a pair of loops that determine specificity.

Elegant crystallographic studies have explored the detailed mechanism of this sophisticated "molecular motor," using specific combinations of oligonucleotides to mimic the interacting nucleic acid structures expected at each stage. Cheetham and Steitz (1999) and Yin and Steitz (2002) described the structure of the initiation complex (Fig. 13A and C). Upon binding, which involves a *specificity loop* at RNAP residues 740–770, the "fingers" domain partially closes over the DNA. A hairpin at residues 230–240 is able to separate the DNA strands at the upstream end of the transcription bubble and the first few nucleotides are added, but forming a DNA/RNA duplex longer than 3 nucleotides is sterically hindered. As shown (Fig. 13B and D), the transition to a stable elongation complex by the time the nascent transcript is 12–14 nucleotides long involves a major conformational change which releases the sigma factor, creates a tunnel for the elongating RNA chain and provides the structural basis for the very high processivity of the transcription process. The mechanism responsible for the high accuracy of base selection, as well as for the exclusion of dNTPs, was determined by Temiakov (2004), who showed that the pairing with the template DNA base occurs while the enzyme is still in an *open state*, prior to formation of the catalytically active closed conformation; this is very different from what is seen with the related DNA polymerases, but comparison with what is known of the structures of the msRNAPs suggests that the mechanism is similar there. Studies of T7 RNAP are also playing a major role in determining the energetic mechanisms driving this key molecular machine. Comparing the structure in which the incoming NTP is trapped bound to a transcription elongation complex *prior* to catalysis with one trapped after nucleotide addition but before release of the cleaved PPi from the complex, Yin and Steitz (2004) have provided evidence that conformational change of the polymerase associated with dissociation of the PPi is what leads to its translocation along the DNA and to separation of the DNA strands.

One complicating factor during viral infections is that replication and transcription are occurring simultaneously at very different rates on the same template. Replication is generally 10 times as fast as transcription, and the latter is also highly processive; a transcript can only be initiated at a promoter, and large numbers of incomplete transcripts could generate a variety of problems. Liu and Alberts (1995) showed that at least for T4 the replisome is even capable of passing actively transcribing RNA polymerase, going in either the same or the opposite direction, without either one falling off or even slowing down very much. In the case of a head-on collision, they found that the RNA polymerase actually transfers to the newly synthesized strand rather than staying on its original template, now the lagging strand, which would be single stranded until the new Okazaki fragment is made.

7.5.2. Examples of Transcriptional Regulatory Complexities

7.5.2.1. Host Polymerase Used Early, Phage Enzyme Late: *E. coli* Podovirus T7

Like most phages, coliphage T7 uses the host RNA polymerase to transcribe its early genes, initiating at three major A promoters at the left end of its genome, which are among the strongest promoters known. A rho-independent terminator (TE) downstream of gene 1.3 defines the end of the early region of the genome, which encodes proteins such as an inhibitor of host restriction systems and a broad-specificity protein kinase that also inactivates host-catalyzed transcription. TE termination efficiency is dependent on phosphorylation of RNAP by this kinase. Termination here is also *enhanced* by a novel host termination factor, Tau, which is *required* for termination here in the related phage T3. Additional rather inefficient termination sites are found downstream of the terminator for the T7 RNAP and near the promoter it recognizes, at position 3.8.

T7 encodes its own polymerase to transcribe its late genes, as described earlier. Interestingly, initiation by T7 polymerase is modulated by T7 lysozyme, which preferentially inhibits the class II (middle-mode) promoters and thus directs the switch to late gene expression. At least 6 groups of T7-like phages, infecting a wide range of gram-negative bacteria, have been identified on the basis of the specific promoter sequences recognized by their RNA polymerases. A few that seem otherwise to belong to this family actually lack this polymerase (Molineux, 2005).

7.5.2.2. A Capsid-Borne RNA Polymerase, and More: *E. coli* Podovirus N4

Three DNA-dependent RNA polymerases are involved in infection by N4, two of them encoded by N4 (Davydova et al., 2003; Kazmierczak and Rothman-Denes, 2005). N4 is highly unusual in that a virion-encapsulated RNA polymerase (vRNAP), injected with the phage DNA, is responsible for early transcription. The N4 RNA polymerase is a single polypeptide of 3500 amino acids, which is rifampicin- and chloramphenicol resistant. A central, stable domain of 1106 amino acids has the same initiation, elongation, termination, and product-displacement properties as the full-length enzyme and appears to be a very distant relative of the T7 polymerase, despite few sequence identities; the extra C-terminal domain is involved in vRNAP packaging in the virion and the N-terminus in injection with the DNA. The three promoters recognized by vRNAP have a hairpin structure and a conserved GC-rich heptamer at the –12 position; they require the host single-stranded-DNA-binding (SSB) protein for activation. Host replication stops but host DNA is not degraded, and host transcription continues unaffected except for cAMP-dependent promoters, which are somehow silenced after N4 infection.

Middle-mode transcription requires three N4 early proteins, p4, p7, and p17, encoded by genes *16, 15,* and *2*. The 40-kDa p4 and 30-kDa p7 together form the rifampicin-resistant, chloramphenicol-sensitive N4 RNA polymerase II. The p7 subunit is homologous to the N-terminal domain, thumb and part of the palm of the T7 RNA polymerase, while p4 has the rest of the palm and the fingers, including all the key residues in domains A, B, and C. The 128-aa p17 binds RNAPII to the inner membrane/N4DNA complex where N4 replication takes place; p17 can be released from the complex by 0.5M salt and binds nonspecifically to single-stranded DNA. Middle promoters show two conserved sequences, t/aTTTAa/t, at the initiation site, and At/aGACCTGt/a, anywhere from 12–20 bp upstream. The mechanism of promoter recognition is unclear.

Late N4 transcription requires the host RNA polymerase and begins 12 minutes after infection–shortly after replication initiation. It is not dependent on concurrent or previous N4 DNA replication, but both N4 replication and late transcription require the N4-encoded single-stranded-binding protein (N4 SSB). The 265-aa N4 SSB, which shows no similarity to other SSBs, interacts with purified *E. coli* RNAP and can be crosslinked with a terminal 108 aa peptide of the ß′ subunit; its mechanism of activation of promoter recognition is unknown. Late N4 promoters show a weak homology with the –35 and –10 regions of *E. coli* promoters but no other conserved potential recognition elements.

7.5.2.3. Host Polymerase throughout, with Modifications: λ, SPO1, and T4

Lambda

Phage Lambda (λ) uses the *E. coli* RNAP and σ^{70} throughout infection in both lysogenic and lytic modes. In lytic mode, two phage-encoded proteins, N and Q, regulate the temporal gene expression by a positive *antitermination* transcriptional control mechanism. These proteins modify host RNA polymerase so that transcription initiating at the early λ promoters overcomes rho-dependent and intrinsic terminator signals. N is a small, highly basic protein, containing an arginine-rich motif (ARM) at its N-terminus. In its absence, transcription of λ is reduced by 98%. N interacts with N-utilization (*nut*) sites located in the nascent mRNA, stabilizing the RNA into a hairpin and allowing the formation of a terminator-resistant ribonucleoprotein complex formed by the elongating enzyme, N, and host factors including NusA, NusB and NusE.

The choice between the lytic and temperate modes of infection depends primarily on the action of the *CI repressor*, which represses virtually all prophage genes, as well as on *int* (which encodes the integrase responsible for site-specific integration); this mechanism was described in the basic biology chapter. The mechanisms involved have been explored in great detail in a variety of places; just in the new edition of *The Bacteriophages* (Calendar, 2005), Little (2005) explores *Gene Regulatory Circuitry of Phage λ* while Friedman and Court (2005) review *Regulation of λ Gene Expression by Transcription Termination and Antitermination* and Allan Campbell (2005) writes about *General Aspects of Lysogeny.*

SPO1

Six RNA-polymerase-binding proteins are encoded by *B. subtilis* phage SPO1. A cluster of 24 genes constitutes a "host take-over module," which occupies much of the SPO1 terminal redundancy and is responsible for shutting off transcription and translation of host genes as well as subverting other host functions (Stewart et al., 1998). Chelm et al. (1982) developed a system that let them study in detail the factors affecting the transition from early-gene transcription, which uses the major host σ, σ^{55}, to use of a delayed-early SPOI gene product, gp28, which directs transcription of the middle genes. In turn, SPO1 gp33 and gp34 are produced to direct transcription of the late genes. Gp28 is not efficient enough at redirecting the polymerase to shut off early transcription *in vivo*; SPO1 transcription modulators gp 44, 50, and 51 are also required for the normal transition from early to middle transcription patterns, as well as for the shutoff of host DNA, RNA and protein synthesis (Sampath and Stewart, 2004). Their effects are very complex and appear to be regulatory rather than directly causative, affecting the expression of all known early and middle SPO1 genes in one way or another.

T4

Throughout the infection cycle, T4 uses the host RNA polymerase, with a variety of interesting modifications to the normal patterns. No DNA-binding transcriptional repressor proteins have been identified. Non-essential modifications to the host RNAP help modulate early transcription, as reviewed by Kutter et al. (1994). The 76-kDa T4 Alt protein adds an ADP-ribosyl group to residue Arg-265 of one of the two α subunits—the same residue shown to play a key role in catabolite repression. This change enhances relative recognition of most of the 39 characterized phage early promoters—those containing upstream AT-rich UP elements like those responsible for the very high strength of the host ribosomal promoters in addition to well-conserved −10 and −35 regions. This modification gives these T4 promoters greater strength than is seen for any *coli* promoters, but is not required for efficient phage infection. Transcription and replication of host DNA are rapidly blocked, the host DNA is bound to the edges of the cell, and by 1 min virtually no production of host proteins is observed. One of the first proteins made is a 19-kDa protein, Alc, which blocks elongation of transcription of all cytosine-containing DNA, including that of the host and of any other co-infecting phages. It causes termination at particular groups of bases just ahead of the replication fork; that termination is blocked if cytosines in that cluster are methylated on the *non-transcribed* strand or if transcription is progressing very slowly (Kashlev et al., 1993). The latter may help explain the observation that T4 can go into a σ^s-mediated "hibernation" mode during stationary-phase infection, still permitting production of some host proteins in response to nutrient addition, in contrast to the total shutoff of host protein synthesis usually seen after phage infection (Kutter et al., unpublished). However, no further host proteins are produced and the cell converts totally to making new phage, as discussed in Chapter 6.

T4 also directs the synthesis of nucleases to degrade the host DNA, but endonuclease II, which initiates the process, is a middle-mode enzyme and the degradation is gradual, releasing the nucleotides in parallel with phage DNA synthesis. The cell is thus not swamped with excess nucleotides and they are used efficiently, in contrast to the T5 situation, in which DNA degradation happens during a pause after the initial 8% of the DNA enters the cell (see sec. 3.4) and the nucleotides are not retained for use. Several minutes after infection, the T4 ModA protein (also nonessential) catalyzes ADP-ribosylation at Arg265 of *both* α subunits of the host RNAP; this modifies DNA-protein as well as protein-protein interactions, decreases the affinity of RNAP for sigma 70, and inhibits transcription from Pe promoters with the UP element. In the middle stage, two activator proteins (MotA, a DNA-binding protein that specifically recognizes 30 middle-mode promoters by means of a conserved GCTT at −30) and SigA (an anti-sigma-70 factor) direct the modified RNAP to the Pm promoters and activate their transcription (Pande et al., 2002; Pal et al., 2003).

Finally, in the late transcription stage responsible for the synthesis of major and minor structural proteins as well as virion assembly factors, a phage-encoded sigma protein and several other accessory factors direct the host RNAP to transcribe 50 late-mode promoters, in a process concurrent with DNA replication. The promoters show an extensive conserved −10 region, which is recognized by RNAP modified by a phage-encoded sigma factor, gp55. Transcription of late promoters by RNAP-σ^{55} requires the sliding clamp encoded by T4 gene *45* ("mobile enhancer") and the T4 gp33 protein, a co-activator that mediates interactions between σ^{55} and the homotrimeric gp45 protein. Other phage-encoded proteins required are the "clamp-loader" proteins gp44 and gp62 and RpbA which binds to the core of the RNAP.

7.6. PHAGE-RELATED RNA: STRUCTURES AND UNEXPECTED FUNCTIONS

Until fairly recently, RNA has primarily been considered an informational macromolecule, though it was known that ribosomal RNA provides structure to the ribosome and tRNAs form the complex adapters for the highly specific translation process. This view began to change with the discovery of *ribozymes*—RNAs that carry out enzyme-like catalytic functions (Cech et al., 1981)—which led to many changes in the thinking about the early evolution of life. Now, we are seeing an explosion of information about a "modern RNA world."

7.6.1. Structure and Efficiency of Ribosome Binding Sites

Most of the translational-level control of phage protein synthesis involves RNA secondary structures that affect access to the ribosome-binding sites; in many cases, environmental factors can affect these structures. Coliphages have played particularly important roles in studying the factors involved. At least six elements can modulate the formation of the mRNA-30S ribosomal subunit—fMet-tRNA$_{f}$-translational inititiation complex. These include the initiation codon, which is preferentially AUG but can also be GUG, UUG, or AUU; the Shine-Dalgarno (SD) sequence, some subset of UAAGGAGGU, which is complementary to the 3′ end of 16S rRNA; the SD-initiation codon *spacing*, which is 5–13 nucleotides, with an optimum of 6–8; other bases in the vicinity, including *enhancer* sequences that can pair with other 16S rRNA regions; coupling of transcription with that of an upstream gene, with the stop and new start codons overlapping or separated by only a few bases; the identity of the *second* codon, most commonly AAA and GCU; and occlusion of the transcription initiation region (TIR) through specific mRNA secondary structures.

In the case of the three cistrons of RNA phages Qß and f2, the coat protein, replicase, and A protein are produced in a ratio of 20:5:1. The differences reflect secondary structure variations of the mRNA at the initiation sites for each cistron. RNA structure is also involved in the mechanism of temporal control of gene expression. Ribosomal protein S1 represses the expression of coat protein and is also a subunit of the f2 replicase. The association between replicase and S1 releases the negative control and allows full expression of the coat protein.

Both RNA secondary structure and protein repressors can affect translational yields from many T4 RNAs, as reviewed by Miller et al. (1994, 2003). For example, several genes, including that for lysozyme, are transcribed from both early and late promoters but their early transcripts form stem-loop structures that occlude the TIR; the late promoter is in the middle of that stem loop, so the late transcript is translated with high efficiency. The low level of lysozyme produced early is, however, also functionally important; if anything depletes the potential energy gradient across the cell wall before the programmed end of the phage infection cycle, it causes the bacterium to lyse and release any already-completed progeny. This is why chloroform lyses phage-infected cells, but only kills uninfected cells without breaking them open. Three T4 proteins are known to directly control their own translation: the DNA polymerase, the single-strand-DNA binding protein, and RegA, which also

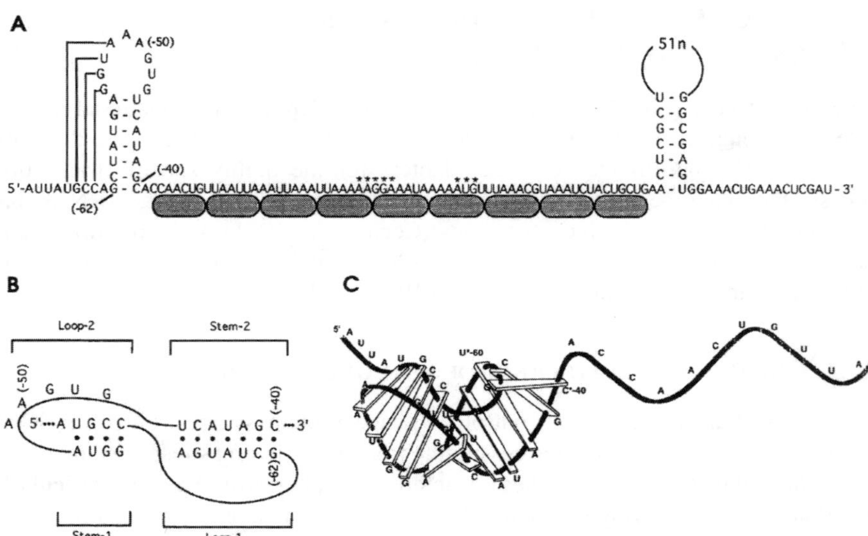

FIGURE 7.14 The operator region for T4 gene 32. (A). Secondary structures flank a stretch of unstructured RNA, which includes the initiating AUG codon and Shine-Dalgarno sequence (marked with asterisks) and can potentially bind 9 monomers of gp32 (stippled ellipsoids). (B) and (C) model the predicted pseudoknot structure; as shown by Konigsberg (personal communication), the sequences of Stem-2 and Loop-2 appear to provide important determinants for binding of the initial gp32. Reprinted with permission from Miller *et al.*, 1994.

regulates synthesis of a number of other proteins at the translational level. Each of these proteins is able to bind to a specific structure in the mRNA and thus inhibit translation. The RNA structures seen are quite different between the three cases. In the case of gene *32*, a complex pseudoknot structure can be formed of the sequence from –67 to –40, which is followed by an unstructured region that can accommodate 9 molecules of cooperatively bound gp32 (Fig. 7.14); this competes with ssDNA, which binds gp32 somewhat more tightly, and thus can control the level of gp32 over a wide range to be able to protect any single-stranded DNA.

7.6.2. Intron Ribozymes

A number of cases are known where RNAs fold into structures that have specific, complex enzymatic activities, previously thought to be properties ascribable only to proteins. These were first seen as RNA endonucleases imbedded in the mRNA of self-splicing group I introns in the mitochondria of filamentous fungi, cutting out long segments of the transcribed RNA to generate the final protein-encoding sequence. As reviewed by Shub et al. (1988), the first intron to be found in the bacterial kingdom was in the thymidylate synthase (*td*) gene of T4, identified in 1984 as the *td* sequence was completed because this particular enzyme was so well characterized biochemically. Other introns were soon identified in T4 and a number of other

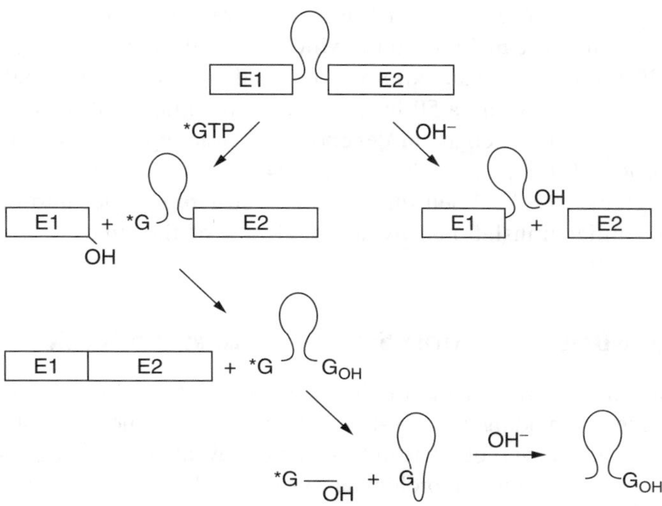

FIGURE 7.15 Self-splicing pathway of group I introns. Splicing is initiated by attack of the 3′ hydroxyl of a free guanosine nucleotide (*GTP) at the 5′ end of the splice site. The G at the 3′ end of the intron and the OH liberated at the end of the first exon (E1) then undergo the reverse reaction, resulting in ligation with exon E2 and release of the linear intron. Radioactive labeling of the uncoded G is a common way to detect new group I introns. Shub *et al.*, 1994, with permission.

bacteriophages, but they are almost nonexistent in bacteria. All of them contain sequences similar to those conserved in the fungal introns and a splicing process in which the 3′-OH of a GTP attacks the 5′ splice site (Fig. 7.15). As described in section 7.7.2, a different sort of ribozyme—an RNA ATPase—has been shown to play a key role in DNA packaging for phage P22.

Despite their ability to self-disseminate, introns are very unevenly distributed among phages (Edgell et al., 2000; Landthaler and Shub, 1999). The adaptation of coliphage T4 intron-encoded endonucleases to the codon usage of their host (Edgell et al., 2000), as well as the exquisite regulation of their expression (Gott et al., 1988), suggest they have been residents of the phage genome for a very long time, yet they are absent from many close relatives of phage T4 (Quirk et al., 1989), suggesting some barriers to their dissemination. The homing endonucleases encoded by many of these introns may contribute partial or complete exclusion of coinfecting phages, facilitating marker exclusion by promoting gene conversion around their homing sites (Belle et al., 2002; Liu et al., 2003); one pair of such endonucleases even favors heterologous DNA (Goodrich-Blair and Shub, 1996).

7.6.3. Translational Bypassing

A number of phages use occasional translational shifting or bypassing of one or two nucleotides at a particular site as a means of generating alternate versions of a particular protein; for example, T7 uses a −1 frameshift to produce gp10B, a variant of the major capsid protein which is itself a minor capsid protein (Molineux 2005). However, an extremely unusual case of translational bypassing is seen in T4 gene 60, where a 50-base stretch in the middle of the coding region is not translated. This mechanism depends on cis-acting signals in the mRNA, ribosomal protein L9, a pair of GGA codons 47 bases apart, and the structure of the cognate glycyl tRNA. When this particular structure is put into other genes, it generates similar translational bypassing 100% of the time, as seen in gp60 (Herr et al., 1999).

7.6.4. RNA-Based Regulatory Systems: Aptamers and SELEX

The concept of *aptamers*—pieces of single-stranded DNA or RNA that fold up to form highly specific binding agents—first emerged from studies of a T4 system; the development of this technology has in turn helped reveal key new regulatory mechanisms that span the range of biological systems, which is why we include this discussion here.

Binding of the DNA polymerase to a specific loop in its mRNA just before the translation initiation region (TIR) inhibits its translation. Tuerk and Gold (1990) randomized 8 nucleotides in the region encoding this loop, transcribed the construct using T7 RNA polymerase, and studied the specificity of T4 DNA polymerase binding to the members of this large RNA pool—calculated to contain 65,536 variants. They found that RNA containing the wild-type T4 sequence bound extremely well, as did one other that differed by 4 nucleotides, but all other sequences bound much less well or not at all. This led to the founding of a company to explore and exploit Systematic Evolution of Ligands by Exponential Enrichment, or SELEX, and to a wide range of successful applications that show affinities higher than those of single-chain antibodies. The random RNA or DNA regions now used for selection are about 40 nucleotides long, between constant fixed-ends primers which are then used to amplify all selected sequences by PCR—about 10^{15} sequence possibilities. Using this approach, it is possible to create short ss-DNA or ss-RNA molecules that will specifically bind to the active site of any enzyme, for example, acting as a powerful inhibitor or activator. Not only can such molecules be selected from an enormous number of possible variants, but once selected, they can be immediately amplified by PCR to very high levels. Similarly, sequences can be identified and amplified that specifically recognize a particular small-molecule ligand and can be cloned into the control region for any given gene, allowing that gene to be turned on or off by that particular ligand. Discovery of this phenomenon has led to the identification of the mechanisms naturally controlling global cellular responses to such ligands as S-adenosyl methionine, guanine, adenine, thiamine pyrophosphate, flavine mononucleotide, and vitamin B_{12} in a wide variety of organisms (Koizumi and Breaker, 2000; Soukup et al., 2000; Mandal et al., 2003; Mandal and Breaker, 2004; Nahvi et al., 2004). Shu and Guo (2003) have shown that an RNA encoded by phage ϕ29 selectively binds ATP and is crucial to the packaging process, which is described in sec. 7. They report that the ATP-binding aptamers selected by SELEX are very similar to the *central part* of this ϕ29 packaging RNA (pRNA).

7.7. MORPHOGENESIS AND DNA PACKAGING

Phages have long provided key model systems for studying the assembly of molecular components into highly complex functional structures, combining genetic, biochemical, electron microscopy, cryo-electron microscopy, and X-ray crystallographic and NMR approaches. The larger Myoviridae build some of the most complex virus particles known, rather resembling lunar landing modules. T4, the best studied, devotes more than 40% of its genetic information to the synthesis and assembly of its prolate icosahedral heads, tails with contractile sheaths, and six tail fibers (Fig. 7.16). Twenty-five proteins are involved in head morphogenesis, 22 in tail morphogenesis, and 7 for tail fibers; 5 of these 54 proteins are assembly catalysts rather than components of the final phage particle. Some other large myoviruses are similarly complex; more than 40 proteins were detected on gels of purified ϕKZ phages, whose 280-kb genome was recently sequenced (Mesyanzhinov et al., 2002), and 53 have been identified by PAGE of the subassemblies of *B. subtilis* phage SP01, whose morphogenetic pathways have not yet been extensively studied. Only a few phages have been studied in detail; T4 and P22 are among the best explored. We describe here general morphogenetic principles that have been determined and some observed variations on the themes. Details of the pathways are reviewed for T4 in a number of chapters of Karam (1994) and in Miller et al. (2003b) and for various other specific phages in Webster (1999) and Calendar (2005).

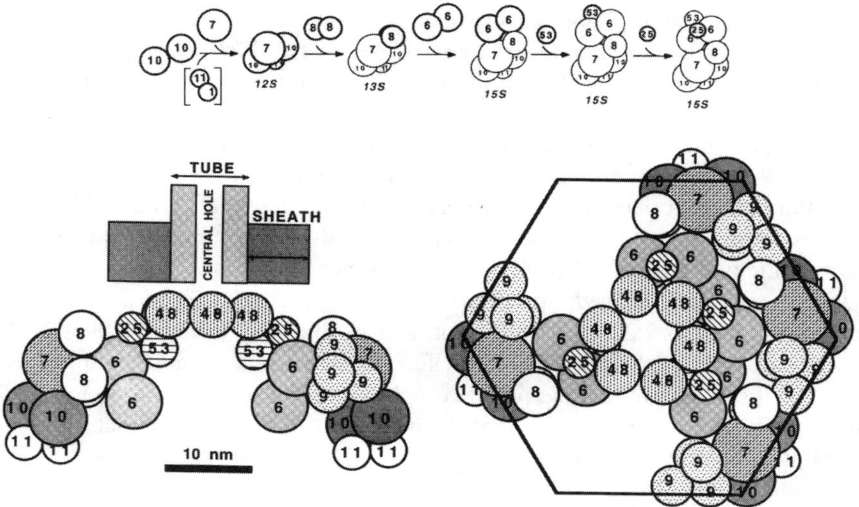

FIGURE 7.16 Bacteriophage T4 baseplate wedge assembly pathway and model structure of the baseplate as determined by chemical cross-linking. Coombs and Arisaka (1994), with permission.

7.7.1. Head Formation

For tailed phages, head morphogenesis is a complex, multistage procedure which has not yet been reproduced *in vitro*. The first head-like structure observed is generally a thick-shelled prohead, formed around a core or protein scaffold. A capsid maturation step usually follows that involves proteolysis of the scaffold as well as cleavage of some virion proteins from larger precursors. The head is filled with DNA before being joined to appropriate preassembled tail structures to form infectious particles. DNA packaging often starts from a long, linear concatemer; it terminates either when specific sequences are recognized or when the head is full. The latter mechanism, in particular, may lead to some level of formation of generalized transducing phage, carrying host DNA. For such phages as SPP1, P22, T3, T7, λ, and T4, researchers have managed to isolate heads in states that are competent for foreign or phage DNA packaging followed by assembly into viable phage or transducing particles, taking advantage of mutants defective in certain stages of the head completion and packaging process. This has helped distinguish between true intermediates in head formation and abortive complexes, and has also been put to use for a variety of exciting technical applications, as described in Chapter 11.

Head assembly generally begins with formation of a *portal protein initiator complex*, upon which a *scaffold* and the *major head protein(s)* are assembled. For some phages, such as T4, this assembly occurs bound to the membrane via phage and/or host proteins. The portal complex comes to form 1 of the 12 vertices of the final icosahedral head. It serves as the docking site for the DNA packaging enzyme complex, the conduit through which the DNA enters and leaves the head, and the attachment site for the phage tail. The scaffold proteins do not generally form icosahedral cores, but rather interact with the major head protein in a metastable complex that lets the head find the appropriate relationships and folding patterns for each position in the icosahedral head as it is completed. P22 heads assemble free in the cytoplasm. The P22 scaffold protein is essential for incorporation of the portal protein complex, as well as the other minor head proteins normally associated with procapsids, and there is other indirect evidence for involvement between the scaffold protein and portal complex in properly initiating assembly, but no direct biochemical interaction has been demonstrated (Moore and Prevelige, 2002). The head protein of P22 is unusual in that it can assemble into a normal-looking structure even without a portal initiation complex. However, in the absence of scaffold protein, P22 often assembles aberrantly shaped structures in addition to empty closed polyhedra. In some cases, such as P22, the scaffold protein is able to exit and function in the assembly of additional heads; in others, like T4, it is cleaved and discarded. For several well-studied phages, the portal complex is a ring of 12 portal protein molecules, yet it takes the place of one of the pentamers of coat protein that generally make up each vertex; Roger Hendrix (1978) suggested that this "symmetry mismatch" might allow easier portal rotation during DNA packaging.

For most phages, various structural proteins are cleaved in conjunction with expansion and maturation of the head into a far more stable final structure; this happens either prior to or in conjunction with DNA packaging. Similar processes are seen for the herpes viruses and adenovirus. Helgstrand et al. (2003) have determined the head shell crystal structure for the λ-like phage HK97—the first for a tailed dsDNA phage. The HK97 shell is remarkable in that neighboring subunits of gp5, its major head protein, become covalently joined in pentameric and hexameric rings (Duda, 1998). The rings also loop through rings formed by neighboring pentamers and hexamers, generating a sort of "protein chain mail." Covalent cross-linking has also been observed in other tailed phages such as *Pseudomonas* phage D3. An interesting issue is whether other phages that do not have covalently joined head proteins form similar interlocking rings.

7.7.2. DNA Packaging

One of the most extraordinary aspects of viral assembly is the packing of the very long DNA molecule into the prohead in a fashion that is highly stable yet allows rapid exit of the DNA, with no tangling, once the infection process is triggered. Jardine and Anderson (2005) have thoroughly reviewed what is known to date about the process, focusing on the structure and mechanism of the motor and the nature of the chemo-mechanical energy conversion.

The substrate for packaging is always a linear molecule, which is packaged processively. Some phages, such as ϕ29 and P2, produce a unit-length chromosome prior to encapsidation, but most known dsDNA phages produce head-to-tail concatameric DNA during replication that is cleaved during the packaging process. In the case of phages like T7 and λ, unique ends are produced by cleaving at specific recognition sites; the cuts are staggered, generating a short terminal redundancy that is sometimes left single-stranded and facilitates circularization or renewed concatamer formation after infection. In other cases, such as P22, P1, and T4, the heads are filled by a "headful" mechanism that pulls in a piece of DNA somewhat longer than the genome size. This generates a significant terminal redundancy and results in a set of circularly permuted DNA molecules distributed among the emerging phage. Some such phages initiate this packaging at a unique "pac" site and produce relatively few phage from each molecule. Others, like T4, can start from any free end and also undergo very extensive recombination, generating a highly branched rather than linear replicative DNA structure that has about 50 phage equivalents of DNA during normal phage infection and many hundreds if phage assembly is mutationally blocked. As the DNA is packaged, a processing enzyme, gp49, clips off any branches, and all nicks and gaps are repaired. In the case of Mu, the phage DNA is packaged while still integrated into the host genome. The host DNA is cleaved a few hundred bases before the start of the Mu genome and packaging continues for a few thousand bases beyond the genome's end, meaning that every Mu phage engages in generalized transduction.

Terminase ATPases convert chemical to mechanical energy. The machinery for packaging follows a general pattern, but with individual variations. A symmetric-ring connector complex with a central channel binds to the portal vertex. DNA-bound molecules of a phage-encoded terminase ATPase bind to this connector. The terminase recognizes and nicks the concatameric dsDNA substrate, binds the DNA to the prohead, and initiates the ATP-driven DNA translocation of packaging. A second nuclease cleavage allows DNA packaging to be completed. Generally, recognition and binding of terminase to the DNA is mediated by a small subunit, while the endonuclease and ATPase activities are located in a large subunit. These are usually both proteins, and are among the most highly conserved of any phage proteins.

In phages of the ϕ29 family, a unique 120-nucleotide RNA, called pRNA, substitutes for one of these two proteins and functions as an ATPase. The pRNA forms a hairpin structure, with a "head loop" region at the top and loops protruding out to the right near the top and to the left partway down. During DNA packaging (only), a hexamer of the folded pRNA binds to the head-tail connector via the head loop region—the region which matches the ATP-binding RNA aptamer selected by the SELEX technique, as described in section 7.6.4. Each of the phages studied to date has its own slightly different pRNA and they cannot cross-function. If even one of the six pRNAs is replaced with an inactive pRNA, DNA packaging is completely blocked, and a single base change can completely inactivate the packaging. In vitro packaging with purified components is very efficient, with no background, and gives substantial insight into the packaging process. The mechanism appears to involve a rotation of the connector/pRNA complex relative to the pentameric head vertex, with the DNA sequentially bound to the outer ends of the pRNA molecules.

7.7.3. Tail Morphogenesis

About 60% of the phages studied to date belong to the Siphoviridae, whose tails are long, flexible tubes made of many copies of a major tail protein, built up on an initiator complex to which one or more fibers may be attached. The tail's precisely defined length is determined by a "tape-measure" protein and the completed tail is stabilized by a terminator protein, which then interacts with the completed head. Lambda tail formation, for example, involves at least 11 genes. GpJ, the attachment fiber, assembles with gpK, gpI, gpH, gpM, and gpL into a cone-shaped initiator, onto which are stacked 32 hexameric disks of the major tail subunit, gpV, forming the hollow 9-mm tube through which the DNA passes during infection. The number is dictated by gpH, which extends up the central hole and acts as a tape measure (Katsura and Hendrix, 1984). Details of assembly are less well understood for the Podoviridae, which may have as few as nine types of proteins in their completed virions. Their baseplates and tail spikes attach directly below the head rather than assembling first into a separate tail, while the other proteins that eventually help the DNA cross the wall into the bacterial cell are internalized within the head (Fig. 7.6).

For Myoviridae, the many components of the contractile tail are assembled in a highly ordered fashion. In the case of T4, the baseplate is formed through associations of six "wedges" around a "hub" to form a dome-shaped baseplate (Figs. 7.16, 7.17). Each wedge consists of seven kinds of proteins that are assembled in a strictly ordered manner, except that gp11 can associate at any time. The hub contains such intriguing components as a tape-measure protein, gp29, and the trimeric tail lysozyme, gp5, discussed in section 7.3.3. During assembly, gp5 is cleaved between Ser351 and Ala352, but the resultant fragments remain associated until infection as a heterohexamer. Gp51 has long been known to function catalytically for the formation of the dome-shaped baseplate. Six trimeric short tail fibers, $(gp12)_3$, are nested beneath the basplate; the head of each short tail fiber associates with the tail

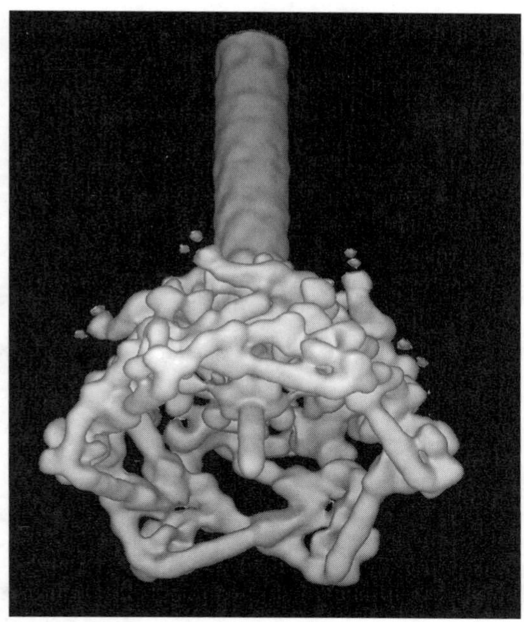

FIGURE 7.17 T4 baseplate structure as determined by X-ray crystallography. Figure supplied by Fumio Arisaka (2004), with permission.

of the next to form a garland (Fig. 7.17; Kostychenko et al., 2003). The short tail fiber interacts with $(gp11)_3$, forming a kink of almost 90°. Gp57A, a phage-encoded molecular chaperone, facilitates the formation of both short and long tail fibers. Polymerization of the tail tube protein, gp19, initiates after gp48 and gp54 join and stops when the gp29 tape-measure protein extends fully, giving a cylinder of 23 stacked hexameric gp19 rings. Gp3 then associates at the tip of the tube. Formation of the contractile sheath follows tube polymerization, with molecules of gp18 arranged in 1:1 correspondence to the tube subunits. Formation of the head-tail connector, a hexameric ring of gp15, completes the tail assembly.

T4 baseplate morphogenesis was shown by Simon (1969) to occur in association with the cell membrane. (Fig. 7.18A, frontispiece). He found that the baseplates remain attached to the membrane by 300-Å fibers from their six corners during the remainder of phage assembly and through the time of cell lysis, as shown by electron microscopy. This attachment is seen even in mutants lacking gene *12,* which encodes the short tail-fiber involved in irreversible phage binding during infection. Interestingly, gp7 has a predicted membrane-spanning domain near its C-terminus, suggesting a possible mechanism for this attachment. Studies by Brown and Anderson (1969) show that the newly released phage carry along with them a fragment of membrane that can actually enhance infectivity for a short time under certain circumstances (Fig. 7.18B)—a little-studied phenomenon called "nascent phage" by Wollman and Stent (1952) and possibly related to the enhanced potency of newly synthesized or "nascent" streptococcal bacteriophage B reported by Evans (1940).

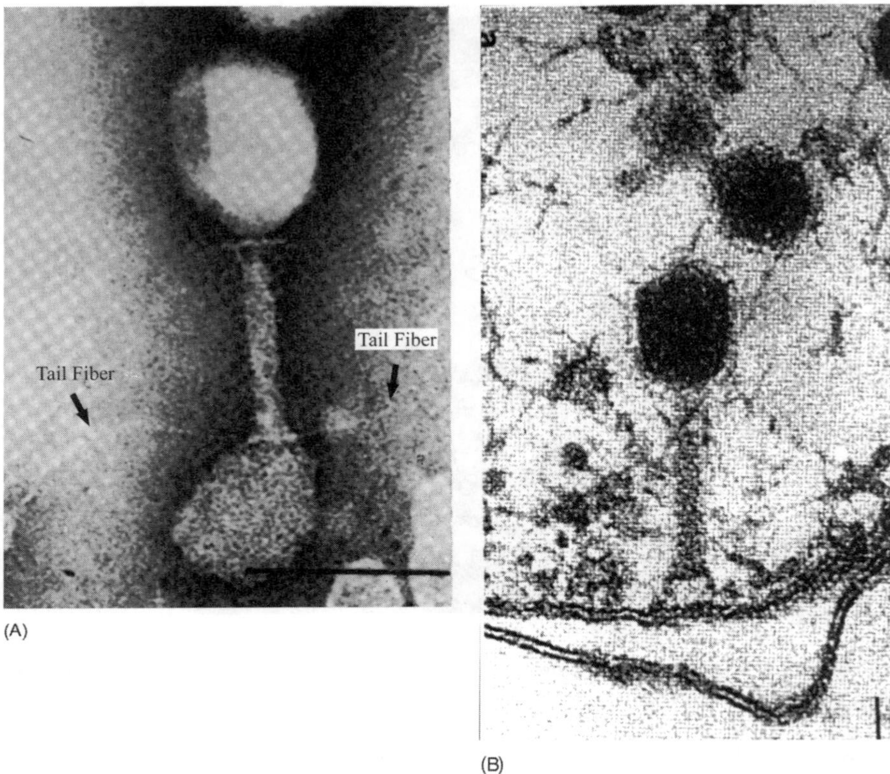

FIGURE 7.18 (A) Nascently adsorbable T4 with a fragment of cell wall bound to the baseplate. (B) Nascent phage shown still attached by tail pins to the inner membrane of a lysed cell. (Brown and Anderson 1969; Simon 1969), with permission.

7.8. PHAGE LYSINS AND LYSIS SYSTEMS

7.8.1. OVERVIEW

Most popular discussions of phage infection imply that progeny phage are made until the cell simply bursts from the pressure, while somewhat more sophisticated versions say that the late appearance of a lysozyme facilitates phage release and the infection of new cells. The whole lysis process is in fact far more intricate, especially for the double-stranded-DNA phages; for them, it is a carefully programmed and timed event, key to the relative ability of each phage to compete in the real world. In fact, Wang and Deaton (2003) argue that "the decision of when to terminate the infection and lyse the host is the *only* major decision made in the vegetative cycle; all other processes proceed at rates which maximize the intracellular accumulation of infective virions." In almost all of the hundreds of cases examined to date for such phages, two main proteins are involved: an *endolysin* to attack the peptidoglycan layer and a *holin* to let that endolysin move from the cytoplasm into the periplasmic space in carefully timed fashion. In most cases, some form of *antiholin* also plays a role in regulating the process. The small, single-stranded-DNA and RNA phages do not encode their own endolysins or holins or control the timing of lysis so carefully, but they have evolved various small "protein antibiotics" that co-opt host enzymes to disrupt the peptidoglycan layer. Phage lysis systems have been discussed in great detail in a variety of reviews (see Young, 1992; Wang et al., 2000); we focus primarily on major principles here.

7.8.2. The Single-Protein Lysis Systems of the Small Phages

Various small lytic single-stranded-DNA and RNA phages use a strategy for lysis induction that depends on subverting host enzymes normally involved in peptidoglycan expansion and using them to facilitate lysis and phage release. These "protein antibiotics" have been dubbed *amurins*; different small phages produce different amurins, affecting a variety of steps in murein synthesis, which potentially provide models for new antibiotics (Wang et al., 2003).

The best-studied of these amurins is the ϕX174 *E*-gene product, a small, 91-amino acid transmembrane protein encoded in a gene embedded in an alternate reading frame within the gene for the capsid scaffolding protein. Both the membrane domain and the lytic function lie in the first 35 aa; in fact, replacement of the entire C-terminal cytoplasmic domain by GFP or β-galactosidase does not inhibit host lysis (Blasi and Lubitz, 1985). Mutations inactivating the host gene *slyD*, encoding a nonessential peptidyl–prolyl cis-trans isomerase (PPIase), abolish E-mediated lysis (Maratea et al., 1985), but this block is overcome by ϕX *Epos* mutations, which increase the synthesis of E by 10-fold. The cloning of an *Epos* allele in an expression vector in turn permitted selection of lysis-resistant host mutants. These mapped to two transmembrane domains of *mraY,* which catalyzes the first lipid-linked step in peptidoglycan biosynthesis—the first demonstration of amurin inhibition of a specific enzymatic step in murein production (Wang et al., 2003). Lysis by E requires cell division, as is also the case for the antibiotic mureidomycin, which also specifically inhibits MraY, apparently by blocking septation. It does not appear that E production or action is regulated in any way except by the degree to which the host is actively growing.

In the case of Leviviridae like Qβ, no separate small amurin is produced. Instead, amurin function seems to be yet another activity of the *main capsid protein* A_2, which is also responsible for receptor binding to the male pilus, protection of the 3′ RNA terminus, and RNA binding during morphogenesis (Winter et al., 1987; Winter and Gold, 1983). Host *rat* mutants, which are resistant to Qβ lysis, all showed a Leu138→Gln mutation in MurA, which catalyzes the first committed step of murein biosynthesis, and high-titer Qβ stocks inhibit purified MurA but not the *rat* mutant protein. Qβ *por* mutants, which plate on rat, map to the N-terminal 120 amino acids of A_2—a region lacking any similarity to the orthologous A protein of other Leviviridae, such as ϕX174. As suggested by Young and Wang (2005), "this provides an obvious control mechanism; when virions accumulate to sufficient levels to block the available MurA, cell wall synthesis stops and lysis ensues if the cells are in growth phase"—and if late log cultures are infected, grossly higher virion yields can be obtained.

RNA phage MS2 gene L, overlapping the end of the *coat* gene and the beginning of the *rep* gene, also encodes an amurin (75 aa). In this case, its lytic ability appears to lie in its 32-aa C-terminal region; the part overlapping the coat gene and intergenic space seems to be there for purposes related to the regulation of its production rather than serving a catalytic function (Berkhout et al., 1985). Regulation of its expression by means of local RNA secondary structure is crucial to the phage, as shown by Klovins (1997).

7.8.3. Holins, Endolysins, and Lysis Timing

Predator-prey considerations have been effectively used to explore the concept that there is an optimal lysis time for any particular phage, host, and set of environmental conditions, balancing the certainties of ongoing linear phage production within the existing host against the possibilities for exponential expansion once the current progeny are released to pursue new prey (Levin and Lenski, 1983; Abedon, 1989; Abedon et al., 2001; Wang et al., 1996; Young and Wang, 2005). It is now clear that the holin-endolysin combinations facilitate the evolution of several key properties that affect the ability of phage to compete:

1. The integrity and productivity of the system must not be impaired until the actual onset of lysis.
2. At the preprogrammed time, the release of infectious progeny should be very rapid and efficient.
3. The timing mechanism for phage release should be responsive to environmental events, such as attacks on the viability of the host, in which case it should efficiently release those progeny that have already been synthesized.
4. Variants acquired by mutation or recombination should readily allow constant selection for more optimal timing for a particular phage and set of conditions.

The T4-like phages have taken property 3 to its ultimate with the development of a property called *lysis inhibition*. Here, attempted superinfection by additional T4-like phages signals a paucity of available hosts in the environment and, as a consequence, lysis is delayed, allowing ongoing production of the phage in that host and providing protection for those phage until outside conditions may be more hospitable (see section 7.8.4).

The holins are proteins that can aggregate to form structures in the inner membrane which allow the endolysin out to degrade the cell wall and thus control the timing of lysis during phage infection. They are highly varied, fairly small membrane proteins; over 250 probable holin sequences, of three different classes and belonging to more than 50 families, had been described at the time of the review by Wang and Deaton (2003). Class I holins (prototype: λ S protein; 95–130 aa) have three potential transmembrane domains, with a short, positively charged intracellular C-terminal tail. Class II (prototype: lambdoid phage ϕ21 S protein; 65–95 aa) have only two transmembrane domains, with both ends cytoplasmic. T4-like phages have been shown to have a very distinctive new type, Class III (218 aa), with a single transmembrane domain at residues 30–49 and a large hydrophilic periplasmic C-terminal tail that is required for lysis inhibition (Ramanculov and Young, 2001b).

Four main classes of phage endolysins attack different bonds in the peptidoglycan layer, as depicted in Fig. 7.19. This rigid wall that protects most bacteria from rupture in environments of low osmolarity is essentially one giant molecule that forms a sack enclosing the cell—a sack which is capable of growth and division without losing its structural integrity, and which is the site of attack for antibiotics like penicillin (Park 1996). The glycan "warp" of this peptidoglycan or murein sack consists of an alternating chain of the sugars *N*-acetylglucosamine (NAG) and

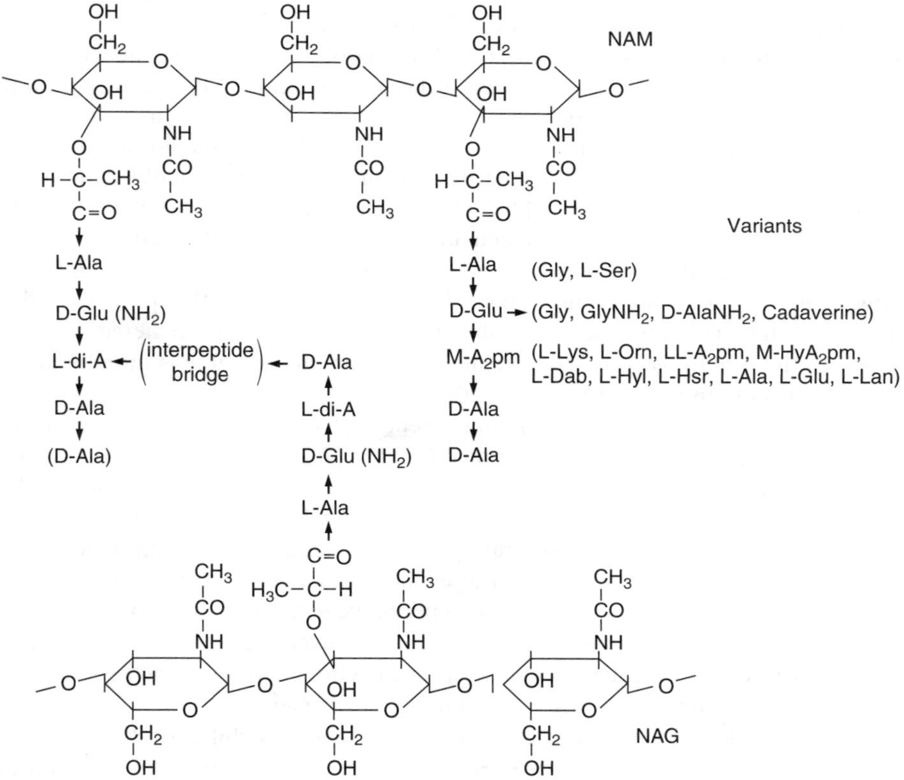

NAM

Variants

L-Ala (Gly, L-Ser)

D-Glu → (Gly, GlyNH$_2$, D-AlaNH$_2$, Cadaverine)

M-A$_2$pm (L-Lys, L-Orn, LL-A$_2$pm, M-HyA$_2$pm, L-Dab, L-Hyl, L-Hsr, L-Ala, L-Glu, L-Lan)

L-di-A ← (interpeptide bridge) ← D-Ala

NAG

FIGURE 7.19 Peptidoglycan basic structure; diA indicates diamino acid. Over 100 possible variations have been identified in different bacteria. Most differences occur in the muropeptide subunit, as indicated below, involving replacement by the amino acids in parentheses; (Hy) A$_2$pm, 2,6-diamino(3-hydroxy)pimelic acid; Dab, diaminobutyric acid; L-Hyl, L-hydroxylysine; Hsr, homoserine; Lan, lanthionine. *E. coli* has no interpeptide bridge, while *S. aureus* has a chain of 5 glycines, for example. The individual chains are generally not very long, averaging about 20 disaccharide units in *E. coli*. Adapted from Labischinski and Maidhof (1994), with permission.

N-acetyl muramic acid (NAM), hooked together with β-1→4 linkages—the same linkage used for NAG polymerization to form chitin in invertebrate exoskeletons. Short muropeptides form the woof cross-linking these glycan chains to make a 2-dimensional sheet; there are significant differences in the details for various groups of bacteria. For gram-positive bacteria, the outer layer consists of a large stack of such sheets that needs to be bridged for phage release.

The *endo-β-N-acetylglucosaminidases* and *N-acetylmuramidases* (lysozymes) act on the two different major bonds of the NAM-NAG sugar backbone. The *endopeptidases* act on the peptide cross-bridges, while *N-acetylmuramyl-L-alanine amidases* (or *amidases*—the most widely found endolysins) hydrolyze the amide bond connecting the sugar and peptide moieties. The endolysin accumulates throughout the period of late-protein synthesis or sometimes even earlier; for T7, it is encoded by an early gene, and for T4, there is a low level of early synthesis but a much higher level later, as discussed in section 7.5. However, the endolysin has no effects on the cell's integrity until the preprogrammed lysis time (or until something disrupts the electrical potential across the membrane, indicating that the cellular phage factory is in serious trouble). At that point, the holin molecules assemble to form transmembrane structures that allow the endolysin to pass through to quickly disrupt the peptidoglycan layer and release the phage, simultaneously terminating respiration and disrupting the associated proton motive force. There is still much debate as to the exact form of these transmembrane structures (cf. Young and Wang, 2005). The abrupt nature of this process at the appointed time has been beautifully shown in films of flagellum-tethered bacteria in which the λ holin gene is induced and the proton motive force is monitored by the rate of rotation (Gründling et al., 2001).

Gram-negative bacteria have a single peptidoglycan layer over most of their surface, anchored to their outer membrane (see Fig. 7.1). In contrast, gram-positive bacteria have a substantial cross-linked stack of such sheets as their outside wall, but no outer membrane, as discussed in sec. 7.2.2 (Fig 7.2). To avoid diffusing away from the cell, most endolysins of gram-positive bacteria have an extra C-terminal component, linked to whichever form of endolysin they employ, that acts to tether them to some component of the specific peptidoglycan layer of their host. The high specificity of this extra tether has been used by several groups to examine the possibility of developing novel treatments for infections involving such bacteria as streptococci and anthrax, as discussed in detail in Chapter 12.

In the simplest cases, lysis is governed only by the availability of the holin and endolysin. Many phages have evolved a more sophisticated mechanism for controlling lysis timing, which employs an additional protein, an *antiholin* that counteracts hole formation and allows the accumulation and then abrupt utilization of significant levels of holin. Antiholins have evolved by at least three independent mechanisms:

1. Expression of an antiholin from the holin gene using a closely spaced alternative start codon (λ, P22, N15, ϕ29)
2. Duplication and alteration of the holin gene
3. Employing an antiholin gene that is not related to a holin (P2 lysA, integral membrane; T4 rI and P1 lydB—soluble).

Bacteriophage λ's lysis system has been the most extensively studied and was the prototype for the dual-component holin/endolysin systems used by all the tailed phages. It was also the system in which the use of holin/antiholin mechanisms was discovered and explored (Raab et al., 1988; Wang et al., 2000). In λ, the *S* (holin) gene has two start codons, separated by a single lysine codon. The longer resulting protein, S_{107}, serves as the antiholin, interacting with the shorter (S_{105}) holin and preventing its assembly into a membrane-perforating structure (hole) until the appropriate lysis time. If the antiholin form is eliminated, the time until lysis is shortened by several minutes and the burst size is decreased. Selection of later- or earlier-lysing lambda mutants gives two kinds of mutations: some affect the translation rate of either the antiholin or holin, reflecting the role of the ratio between the two in determining the timing of lysis, while others affect the structure of the holin/antiholin gene itself. The system is well designed to be rapidly optimized through mutation and selection during adaptation to a new host or environmental conditions. Furthermore, if the membrane potential is disrupted prematurely by any sort of insult, the antiholin no longer inhibits hole formation; thus, any lysozyme present quickly enters the periplasm, leading to rapid release of whatever phage have already been completed. This is the mechanism by which *chloroform* leads to lysis of phage-infected cells as long as lysozyme and holin have been produced, while it kills uninfected cells without causing lysis. Since superinfection by other phages generally causes a brief membrane depolarization, this mechanism also lets the initially infecting phage lyse the cell at that point rather than permitting its cooption by another phage, as long as sufficient holin has already been produced.

Though the lysin and holin genes are often located adjacent to each other, they can equally well be far apart and there is no specificity to the holin-lysin relationship; in lambdoid phages, for instance, both class I and class II holins have been found, coupled with enzymatically different kinds of endolysins, and some phages even encode multiple endolysins with different specificities. The very useful online table that accompanied Wang (2000) summarizes the endolysin and holin properties and genes for a large number of phages. In a variety of cases, including experiments involving the very different λ and T4 systems, cloned heterologous holins function just fine to release the lysin in a specifically timed manner (Rennell and Poteete, 1985; Bonovich and Young, 1991; Ramanculov and Young, 2001a). As will be discussed in detail hereafter, this has been very useful in efforts to understand the mechanisms involved in such phenomena as T4's lysis inhibition.

7.8.4. T-Even Phages: Lysis and Lysis Inhibition

Normally, 100 to 300 phage have accumulated in the cell by the time T-even phage lysis occurs. As in other tailed phages, two proteins are involved in the lysis process. The T4 lysozyme is encoded by gene *e* (for endolysin) (Streisinger et al., 1964); its structure has been extensively studied by Mathews and colleagues (Wozniak et al., 1994; Yang et al., 2000) and it has played key roles in the explorations of protein structure-function relationships, facilitated by the ease of generating site-directed mutations (Rennell et al., 1991). The T4 holin is encoded by gene *t;* in analogy with other well-studied phage lysis systems, it creates a hole in the inner membrane so that lysozyme can reach the peptidoglycan layer from the cytosol. In the absence of either the lysozyme or the holin, lysis does not occur.

The T4 holin looks totally different than any of the holins already discussed, and defined a new class (III) of much larger holin proteins with a very long C-terminal periplasmic tail. This tail makes possible *lysis inhibition*, a phenomenon unique to date to the T-even phages that allows them to sense when there are more phage around than bacteria and respond by delaying lysis, thus maximizing use of their current host and potentially awaiting the accumulation of additional sensitive bacteria (Paddison et al., 1998). For T4, lysis is extensively delayed if more phage attack the infected cell at any time between 5 minutes after infection and the actual start of lysis; the length of the delay depends on the frequency of such additional attacks. This signal is somehow mediated by the rI protein, which in turn regulates assembly of the t holin in the membrane (Paddison et al., 1998). No other proteins besides rI and t are required. Ramanculov and Young (2001a; 2001b) substituted the T4 *t* gene for the *S* gene of bacteriophage λ and showed that these hybrid λ phage could infect normally, lysing at 20 min after infection, and could produce lysis inhibition if the *E. coli* they infected carried a functioning cloned T4 *rI* gene. T4 gprIII further extended lysis inhibition under these circumstances, but no additional T4 proteins were required. We still do not know the specific signal that gprI senses to establish lysis inhibition; possibilities include the (degrading) DNA or the internal proteins from the superinfecting phage, both of which are injected into the periplasm rather than the cytoplasm under these circumstances.

The extent of lysis inhibition appears to be a function of the specific T-even phage, the host strain, and the environmental conditions (Abedon and Thomas, personal communication). While the normal period for T4 infecting B is 4–6 hours in well-aerated broth media, periods as long as 27 hours have been observed under anaerobic respiration conditions, with a somewhat more drawn out final drop in the optical density of the infected cells (Kutter lab, unpublished). The eventual collapse of lysis inhibition and release of phage also appears to be a regulated phenomenon. It also generally happens abruptly, with its timing affected by the frequency of secondary adsorption and by the genetic properties of the primary phage (Abedon, 1999).

Ironically, the classic *rII* genes, which were involved in defining the phenomenon of lysis inhibition when T4 was propagated on *E. coli* B strains (Doermann, 1948), are now seen to have nothing to do with lysis inhibition per se (Paddison et al., 1998; Miller et al., 2003a). Rather, T4 *rII* mutants provide an example of patterns of phage exclusion by genes of resident prophages. It had long been known that T4 *rII* mutant phage are completely excluded by the λ *rexA* and *rexB* genes expressed in λ lysogens of K12 strains. This phenomenon was elegantly used in Benzer's fine-structure analysis of the gene (1957). The molecular mechanism of this exclusion is still poorly characterized. It has been proposed that RexB forms an ion channel that is opened after infection with T4 *rII* mutants or with various other phages, leading to loss of ions and cellular energy (Parma et al., 1992); some enzymes involved in T4 recombination seem to also be involved. But how the rII proteins enable bypassing of this process is still unclear (reviewed in Snyder and Kaufmann, 1994). Paddison et al. (1998) reviewed the evidence that gprIIA and gprIIB are also primarily responsible for protecting T4-infected *E. coli* B cells against damage due to a defective resident prophage related to phage P2 and present in many B strains (Rutberg and Rutberg, 1965). It turns out that *rII* mutants suffer less severe consequences in the P2-like case than when they infect K-12 strains lysogenic for λ; DNA replication is not affected and lysis does not occur until about 25 minutes after infection, so a reasonable burst is produced. This occurs just before lysis inhibition would normally set in, thus preventing the lysis-inhibition response and giving *rII* mutants their large plaque size on the lysogenic *E. coli* B strains (Paddison et al., 1998). If the host B cell has been cured of this prophage, *rII* mutants show normal lysis inhibition and small plaques.

In summary, the primary role of the *rII* genes seems to be related to cellular energetics, not directly to lysis. The apparent lysis inhibition seen on lysogenic B strains appears to be due to the fact that for some unknown reason cell energetics breaks down near the normal lysis time for *rII* mutants infecting lysogenic B cells rather than at the 12 minutes after infection that produces lethality in K-12(λ) hosts.

We have discussed this drastic change in the paradigm related to one of the primary genes used in the development of molecular biology in some detail to emphasize the importance of always looking carefully at the evidence and being willing to change interpretations, no matter how generally accepted a particular concept or principle may be. This is particularly clear when working with phage, where genetic, biochemical, biophysical, physiological, ecological, genomic, and other approaches can be so well integrated in exploring the system due to the relatively manageable size and complexity, but clearly applies to all aspects of doing science.

ACKNOWLEDGEMENTS

We express our appreciation to the many members of the phage community who have helped us to detail the complex molecular mechanisms of the infection cycle, particularly to the many authors of Calendar's *Bacteriophages II* who shared their chapters with us prior to publication and to Burton Guttman, Fumio Arisaka, Pascale Boulanger, Peixuan Guo, and members of our lab Sarah Perigo and Seamus Flynn, for all their scientific and editorial assistance. Special thanks go to Gautam Dutta for his hard work combing the annals and for assembling the figures and bibliography, in addition to his contributions to the content.

REFERENCES

Abedon, S.T., (1999). Bacteriophage T4 resistance to lysis-inhibition collapse. *Genet Res* 74: 1–11.

Abedon, S.T., (1994). Lysis and the interaction between free phages and infected cells, pp. 397–405 in *Molecular Biology of Bacteriophage T4*, J. Karam, J.W. Drake, K.N. Kreuzer, G. Mosig, D.H. Hall, F.A. Eiserling, L.W. Black, et al. (Eds.). *Amer Soc Microbiol*, Washington, D.C.

Abedon, S.T., (1989). Selection for bacteriophage latent period length by bacterial density: A theoretical examination *Microb Ecol* 18: 79–88.

Abedon, S.T., Herschler, T.D. and Stopar, D., (2001). Bacteriophage latent-period evolution as a response to resource availability. *Appl Environ Microbiol* 67: 4233–4241.

Ackermann, H.W., and Dubow, M., (1987). *Viruses of Prokaryotes*. CRC Press, Boca Raton, FL.

Bamford, D.H. and Bamford, J.K.H., (2005). Lipid-Containing Bacteriophage PM2, the Type-Organism of *Corticoviridae*, in *The Bacteriophages*, R. Calendar (Ed.). Oxford University Press, New York.

Belle, A., Landthaler, M. and Shub, D.A., (2002). Intronless homing: site-specific endonuclease SegF of bacteriophage T4 mediates localized marker exclusion analogous to homing endonucleases of group I introns. *Genes Dev* 16: 351–362.

Benzer, S., (1957). The elementary units of heredity, pp. 70–93 in *The Chemical Basis of Heredity*, W.D. McElroy and B. Glass (Eds.). The Johns Hopkins Press, Baltimore.

Berkhout, B., de Smit, M.H., Spanjaard, R.A., Blom, T. and van Duin, J., (1985). The amino terminal half of the MS2-coded lysis protein is dispensable for function: Implications for our understanding of coding region overlaps. *Embo J* 4: 3315–3320.

Blasi, U. and Lubitz, W., (1985). Influence of C-terminal modifications of ϕX174 lysis gene E on its lysis-inducing properties. *J Gen Virol* 66 (Pt 6): 1209–1213.

Bonovich, M.T. and Young, R., (1991). Dual start motif in two lambdoid S genes unrelated to λ S. *J Bacteriol* 173: 2897–2905.

Brown, D.T., and Anderson, T.F., (1969). Effect of host cell wall material on the adsorbability of cofactor-requiring T4. *J Virol* 4: 94–108.

Calendar, R., (2005). *The Bacteriophages*. 2nd Ed, Oxford University Press, New York.

Calendar, R. (Ed.), (1988). *The Bacteriophages.* 1st Ed. Plenum Press, New York.

Campbell, A.M., (2005). "General Aspects of Lysogeny," in *The Bacteriophages*, R. Calendar (Ed.). Oxford University Press, New York.

Cech, T.R., Zaug, A.J. and Grabowski, P.J., (1981). In vitro splicing of the ribosomal RNA precursor of Tetrahymena: Involvement of a guanosine nucleotide in the excision of the intervening sequence. *Cell* 27: 487–496.

Cermakian, N., Ikeda, T.M., Cedergren, R. and Gray, M.W., (1996). Sequences homologous to yeast mitochondrial and bacteriophage T3 and T7 RNA polymerases are widespread throughout the eukaryotic lineage. *Nucleic Acids Res* 24: 648–654.

Chaconas, G., (1987). "Transposons of Mu DNA in Vivo," pp. 137–157 in *Phage Mu*, M. Symonds, A. Toussaint, P. van de Putte and M.M. Howe (Eds.). Cold Spring Harbor Laboratory, New York.

Cheetham, G.M. and Steitz, T.A., (1999). Structure of a transcribing T7 RNA polymerase initiation complex. *Science* 286: 2305–2309.

Chelm, B.K., Duffy, J.J. and Geiduschek, E.P., (1982). Interaction of Bacillus subtilis RNA polymerase core with two specificity-determining subunits. Competition between sigma and the SPO1 gene 28 protein. *J Biol Chem* 257: 6501–6508.

Christensen, J.R., (1994). "T1 bacteriophage," pp. 1371–1376 in *Encyclopedia of Virology*, R.G. Webster and A. Granoff (Eds.). Academic Press, San Diego, CA.

Court, D.L., Sawitzke, J.A. and Thomason, L.C., (2002). Genetic Engineering Using Homologous Recombination. *Annu Rev Genet* 36: 361–388.

Crawford, J.T. and Goldberg, E.B., (1980). The function of tail fibers in triggering baseplate expansion of bacteriophage T4. *J Mol Biol* 139: 679–690.

Csonka, L.N., Howe, M.M., Ingraham, J.L., Pierson, L.S., 3rd and Turnbough, C.L., Jr., (1981). Infection of Salmonella typhimurium with coliphage Mu d1 (Apr lac): Construction of pyr::lac gene fusions. *J Bacteriol* 145: 299–305.

Davydova, E.K., Kazmierczak, K.M. and Rothman-Denes, L.B., (2003). Bacteriophage N4coded, virion-encapsulated DNA-dependent RNA polymerase. *Methods Enzymol* 370: 83–94.

Delbrück, M., (1940). Adsorption of bacteriophage under various physiological conditions of the host. *J Gen Physiol* 23: 631–642.

Doermann, A.H., (1948). Lysis and lysis inhibition with *Escherichia coli* bacteriophage. *J Bacteriol* 55: 257–275.

Drake, J.W. and Ripley, L.S., (1994). "Mutagenesis," pp. 98–124 in *Molecular Biology of Bacteriophage T4*, J.D. Karam, J.W. Drake, K.N. Kreuzer, G. Mosig, D.H. Hall, F.A. Eiserling, L.W. Black, et al. (Eds.). Amer. Soc. Microbiol., Washington, D.C.

Duda, R.L., (1998). Protein chainmail: Catenated protein in viral capsids. *Cell* 94: 55–60.

Duplessis, M. and Moineau, S., (2001). Identification of a genetic determinant responsible for host specificity in *Streptococcus thermophilus* bacteriophages. *Mol Microbiol* 41: 325–336.

Edgell, D.R., Belfort, M. and Shub, D.A., (2000). Barriers to intron promiscuity in bacteria. *J Bacteriol* 182: 5281–5289.

Evans, A.C., (1940). The Potency of Nascent Streptococcus Bacteriophage B. *J Bacteriol* 39: 597–604.

Filee, J., Forterre, P., Sen-Lin, T. and Laurent, J., (2002). Evolution of DNA polymerase families: Evidences for multiple gene exchange between cellular and viral proteins. *J Mol Evol* 54: 763–773.

Friedman, M. and Court, D.L., (2005). "Regulation of λ Gene Expression by Transcription Termination and Antitermination," in *The Bacteriophages*, R. Calendar (Ed.). Oxford University Press, New York.

Gabig, M., Herman-Antosiewicz, A., Kwiatkowska, M., Los, M., Thomas, M.S. and Wegrzyn, G., (2002). The cell surface protein Ag43 facilitates phage infection of *Escherichia coli* in the presence of bile salts and carbohydrates. *Microbiology* 148: 1533–1542.

German, G.J., Misra, R. and Kropinski, A.M., (2005). "The T1-like Bacteriophages," in *The Bacteriophages*, R. Calendar (Ed.). Oxford University Press, New York.

Ghuysen, J.M., and Hakenbeck, R., (Eds.), (1994). *Bacterial Cell Wall*. Elsevier, New York.

Goldberg, E., Grinius, L. and Letellier, L., (1994). "Recognition, attachment, and injection," in *Molecular Biology of Bacteriophage T4*, J.D. Karam, J.W. Drake, K.N. Kreuzer, G. Mosig, D.H. Hall, F.A. Eiserling, L.W. Black, et al. (Eds.). Amer. Soc. Microbiol., Washington, D.C.

Goodrich-Blair, H. and Shub, D.A., (1996). Beyond homing: Competition between intron endonucleases confers a selective advantage on flanking genetic markers. *Cell* 84: 211–221.

Goodridge, L., Gallaccio, A. and Griffiths, M.W., (2003). Morphological, host range, and genetic characterization of two coliphages. *Appl Environ Microbiol* 69: 5364–5371.

Gott, J.M., Zeeh, A., Bell-Pedersen, D., Ehrenman, K., Belfort, M. and Shub, D.A., (1988). Genes within genes: Independent expression of phage T4 intron open reading frames and the genes in which they reside. *Genes Dev* 2: 1791–1799.

Grahn, A.M., Daugelavicius, R. and Bamford, D.H., (2002). The small viral membrane-associated protein P32 is involved in bacteriophage PRD1 DNA entry. *J Virol* 76: 4866–4872.

Granoff, A. and Webster, R. (Eds.), (1999). *Encyclopedia of Virology*. Academic Press, London.

Grinius, L. and Daugelavicius, R., (1988). Depolarization of *Escherichia coli* cytoplasmic membrane by bacteriophages T4 and λ: Evidence for induction of ion-permeable channels. 19: 235–245.

Gründling, A., Manson, M.D. and Young, R., (2001). Holins kill without warning. *Proc Natl Acad Sci U S A* 98: 9348–9352.

Guo, S., Shu, D., Simon, M.N. and Guo, P., (2003). Gene cloning, purification, and stoichiometry quantification of ϕ29 anti-receptor gp12 with potential use as special ligand for gene delivery. *Gene* 315: 145–152.

Haggard-Ljungquist, E.C., Halling, C. and Calendar, R., (1992). DNA sequence of the tail fiber genes of bacteriophage P2: Evidence for horizontal gene transfer of tail fiber genes among unrelated bacteriophages. *J Bacteriol* 174: 1462–1477.

Hancock, R.E.W., Karunaratne, D.N., and Bernegger-Egli, C., (1994). Molecular Organization and Structural role of Outer Membrane Macromolecules, pp. 263–279 in *The Bacterial Cell Wall*, J.M. Ghuysen, Hackenbeck, R. (Eds.). Elsevier, Amsterdam.

Hancock, R.W. and Braun, V., (1976). Nature of the energy requirement for the irreversible adsorption of bacteriophages T1 and ϕ80 to *Escherichia coli*. *J Bacteriol* 125: 409–415.

Harshey, R.M., (1988). "Phage Mu," pp. 193–234 in *The Bacteriophages*, R. Calendar (Ed.). Plenum, New York.

Hashemolhosseini, S., Montag, D., Kramer, L. and Henning, U., (1994). Determinants of receptor specificity of coliphages of the T4 family. A chaperone alters the host range. *J Mol Biol* 241: 524–533.

Helgstrand, C., Wikoff, W.R., Duda, R.L., Hendrix, R.W., Johnson, J.E. and Liljas, L., (2003). The refined structure of a protein catenane: The HK97 bacteriophage capsid at 3.44 Å resolution. *J Mol Biol* 334: 885–899.

Hendrix, R.W., (1978). Symmetry mismatch and DNA packaging in large bacteriophages. *Proc Natl Acad Sci U S A* 75: 4779–4783.

Hendrix, R.W. and Duda, R.L., (1992). Bacteriophage λ PaPa: not the mother of all λ phages. *Science* 258: 1145–1148.

Hengge-Aronis, R., (2002a). Signal transduction and regulatory mechanisms involved in control of the σ^S (RpoS) subunit of RNA polymerase. *Microbiol Mol Biol Rev* 66: 373–395, table of contents.

Hengge-Aronis, R., (2002b). Stationary phase gene regulation: What makes an *Escherichia coli* promoter σ^S-selective? *Curr Opin Microbiol* 5: 591–595.

Herr, A.J., Atkins, J.F. and Gesteland, R.F., (1999). Mutations which alter the elbow region of tRNA2Gly reduce T4 gene 60 translational bypassing efficiency. *Embo J* 18: 2886–2896.

Jardine, J.J. and Anderson, D.W., (2005). "DNA Packaging in DS DNA Phages," In press in *The Bacteriophages*, R. Calendar (Ed.). Oxford University Press, New York.

Jones, C.E., Mueser, T.C., Dudas, K.C., Kreuzer, K.N. and Nossal, N.G., (2001). Bacteriophage T4 gene 41 helicase and gene 59 helicase-loading protein: A versatile couple with roles in replication and recombination. *Proc Natl Acad Sci U S A* 98: 8312–8318.

Kanamaru, S., Leiman, P.G., Kostyuchenko, V.A., Chipman, P.R., Mesyanzhinov, V.V., Arisaka, F. and Rossmann, M.G., (2002). Structure of the cell-puncturing device of bacteriophage T4. *Nature* 415: 553–557.

Karam, J.D., Drake, J.W., Kreuzer, K.N., Mosig, G., Hall, D.H., Eiserling, F.A., Black, L.W., et al. (Eds.), (1994). *Molecular biology of bacteriophage T4*. Amer. Soc. Microbiol., Washington, D.C.

Kashlev, M., Nudler, E., Goldfarb, A., White, T. and Kutter, E., (1993). Bacteriophage T4 Alc protein: a transcription termination factor sensing local modification of DNA. *Cell* 75: 147–154.

Katsura, I. and Hendrix, R.W., (1984). Length determination in bacteriophage λ tails. *Cell* 39: 691–698.

Kazmierczak, K.M. and Rothman-Denes, L.B., (2005). "Bacteriophage N4 in *The Bacteriophages*," In press R. Calendar (Ed.). Oxford University Press, New York.

Klovins, J., Tsareva, N.A., De Smit, M.H., Berzins, V. and Van, D., (1997). Rapid evolution of translational control mechanisms in RNA genomes. *J of Mol Biol* 265: 372–384.

Koch, C., Mertens, G., Rudt, F., Kahmann, R., Kannar, R., Plasterk, H.A., van de Putte, P., et al. (1987). "The Invertible G Segment," pp. 75–91 in *Phage Mu*, M. Symonds, A. Toussaint, P. van de Putte and M.M. Howe (Eds.). Cold Spring Harbor Laboratory, New York.

Koizumi, M. and Breaker, R.R., (2000). Molecular recognition of cAMP by an RNA aptamer. *Biochemistry* 39: 8983–8992.

Koonin, E.V., and Galperin, M.Y., (2003). *Sequence-Evolution-Function: Computational Approaches in Comparative Genomics*. Kluwer Academic Publishers, Boston.

Kreuzer, K.N. and Drake, J.W., (1994). "Repair of lethal DNA damage," pp. 89–97 in *Molecular Biology of Bacteriophage T4*, J. Karam, J.W. Drake, K.N. Kreuzer, G. Mosig, D.H. Hall, F.A. Eiserling, L.W. Black, et al. (Eds.) Amer. Soc. Microbiol., Washington, D.C.

Kutter, E., White, T., Kashlev, M., Uzan, M., McKinney, J. and Guttman, B., (1994). "Effects on host genome structure and expression," pp. 357–368 in *Molecular Biology of Bacteriophage T4*, J. Karam, J.W. Drake, K.N. Kreuzer, G. Mosig, D.H. Hall, F.A. Eiserling, L.W. Black, et al. (Eds.). Amer. Soc. Microbiol., Washington, D.C.

Labischinski, H. and Maidhof, H. (1994). "Bacterial peptidoglycan: overview and evolving concepts," pp. 23–38 in *Bacterial Cell Wall,* Ghuysen, J.-M., and Hakenbeck, R. (Eds.). Elsevier Science, New York.

Landthaler, M. and Shub, D.A., (1999). Unexpected abundance of self-splicing introns in the genome of bacteriophage twort: Introns in multiple genes, a single gene with three introns, and exon skipping by group I ribozymes. *Proc Natl Acad Sci USA* 96: 7005–7010.

Letellier, L., Boulanger, P., Plancon, L., Jacquot, P. and Santamaria, M., (2004). Main features on tailed phage, host recognition and DNA uptake. *Front Biosci* 9: 1228–1339.

Letellier, L., Plancon, L., Bonhivers, M. and Boulanger, P., (1999). Phage DNA transport across membranes. *Res Microbiol* 150: 499–505.

Levin, B. and Lenski, R.E., (1983). "Coevolution in Bacteria and Their Viruses and Plasmids," pp. 99–127 in *Coevolution*, D. Futuyama and M. Slatkin (Eds.). Sinauer Associates, Sunderland, MA.

Lindberg, (1977). "Bacterial surface carbohydrates and bacteriophage adsorption," in *Surface Carbohydrates of the Prokaryotic Cell*, I. Sutherland (Ed.). Academic Press.

Little, J.W., (2005). "Gene Regulatory Circuitry of Phage λ," in *The Bacteriophages*, R. Calendar (Ed.). Oxford University Press, New York.

Liu, B. and Alberts, B.M., (1995). Head-on collision between a DNA replication apparatus and RNA polymerase transcription complex. *Science* 267: 1131–1137.

Liu, Q., Belle, A., Shub, D.A., Belfort, M. and Edgell, D.R., (2003). SegG endonuclease promotes marker exclusion and mediates co-conversion from a distant cleavage site. *J Mol Biol* 334: 13–23.

Lu, M.J. and Henning, U., (1989). The immunity (*imm*) gene of *Escherichia coli* bacteriophage T4. *J Virol* 63: 3472–3478.

Lu, M.J., Stierhof, Y.D. and Henning, U., (1993). Location and unusual membrane topology of the immunity protein of the *Escherichia coli* phage T4. *J Virol* 67: 4905–4913.

Mandal, M., Boese, B., Barrick, J.E., Winkler, W.C. and Breaker, R.R., (2003). Riboswitches control fundamental biochemical pathways in *Bacillus subtilis* and other bacteria. *Cell* 113: 577–586.

Mandal, M. and Breaker, R.R., (2004). Adenine riboswitches and gene activation by disruption of a transcription terminator. *Nat Struct Mol Biol* 11: 29–35.

Maratea, D., Young, K. and Young, R., (1985). Deletion and fusion analysis of the phage ϕX174 lysis gene E. *Gene* 40: 39–46.

McAllister, W.T., (1970). Bacteriophage infection: Which end of the SP82G genome goes in first? *J Virol* 5: 194–198.

Mesyanzhinov, V.V., Robben, J., Grymonprez, B., Kostyuchenko, V.A., Bourkaltseva, M.V., Sykilinda, N.N., Krylov, V.N., et al. (2002). The genome of bacteriophage ϕKZ of Pseudomonas aeruginosa. *J Mol Biol* 317: 1–19.

Mian, I.S., Bradwell, A.R. and Olson, A.J., (1991). Structure, function and properties of antibody binding sites. *J Mol Biol* 217: 133–151.

Miller, E.S., Heidelberg, J.F., Eisen, J.A., Nelson, W.C., Durkin, A.S., Ciecko, A., Feldblyum, T. V., et al. (2003a). Complete genome sequence of the broad-host-range vibriophage KVP40: Comparative genomics of a T4-related bacteriophage. *J Bacteriol* 185: 5220–5233.

Miller, E.S., Karam, J.D. and Spicer, E., (1994). "Control of translational initiation: mRNA structure and protein repressors," pp. 193–208 in *Molecular Biology of Bacteriophage T4*, J. Karam, J.W. Drake, K.N. Kreuzer, G. Mosig, D.H. Hall, F.A. Eiserling, L.W. Black, et al. (Eds.). Amer. Soc. Microbiol., Washington, D.C.

Miller, E.S., Kutter, E., Mosig, G., Arisaka, F., Kunisawa, T. and Ruger, W., (2003b). Bacteriophage T4 genome. *Microbiol Mol Biol Rev* 67: 86–156.

Molineux, I.J., (2001). No syringes please, ejection of phage T7 DNA from the virion is enzyme driven. *Mol Microbiol* 40: 1–8.

Molineux, I.J., (2005). "The T7 Group," in *The Bacteriophages*, R. Calendar (Ed.). Oxford University Press, New York.

Montag, D., Riede, I., Eschbach, M.L., Degen, M. and Henning, U., (1987). Receptor-recognizing proteins of T-even type bacteriophages. Constant and hypervariable regions and an unusual case of evolution. *J Mol Biol* 196: 165–174.

Moore, S.D. and Prevelige, P.E., Jr., (2002). A P22 scaffold protein mutation increases the robustness of head assembly in the presence of excess portal protein. *J Virol* 76: 10245–10255.

Mosig, G., Gewin, J., Luder, A., Colowick, N. and Vo, D., (2001). Two recombination-dependent DNA replication pathways of bacteriophage T4, and their roles in mutagenesis and horizontal gene transfer. *Proc Natl Acad Sci U S A*: 8306–8311.

Mueller, M., Dutta, G., Hoyle, N., Rudinsky, S., Black, A., Bruncke, D., Machowek, M., et al. (2003). "Ranges of Infectivity of Various *E. coli* Phages," pp. P-18 in *15th Evergreen International Phage Biology Meeting*, Olympia, WA.

Nahvi, A., Barrick, J.E. and Breaker, R.R., (2004). Coenzyme B12 riboswitches are widespread genetic control elements in prokaryotes. *Nucleic Acids Res* 32: 143–150.

Nikaido, H., (2003). Molecular basis of bacterial outer membrane permeability revisited. *Microbiol Mol Biol Rev* 67: 593–656.

Nikaido, H., (1996). "Outer Membrane," pp. 29–47 in *Escherichia coli and Salmonella*, F.C. Neidhardt (Ed.). Amer. Soc. Microbiol., Washington, D.C.

Ochman, H. and Selander, R.K., (1984). Standard reference strains of *Escherichia coli* from natural populations. *J Bac*teriol 157: 690–693.

Oda, M., Morita, M., Unno, H. and Tanji, Y., (2004). Rapid detection of *Escherichia coli* O157:H7 by using green fluorescent protein-labeled PP01 bacteriophage. *Appl Environ Microbiol* 70: 527–534.

Paddison, P., Abedon, S.T., Dressman, H.K., Gailbreath, K., Tracy, J., Mosser, E., Neitzel, J., et al. (1998). The roles of the bacteriophage T4 *r* genes in lysis inhibition and fine-structure genetics: A new perspective. *Genetics* 148: 1539–1550.

Pal, D., Vuthoori, M., Pande, S., Wheeler, D. and Hinton, D.M., (2003). Analysis of regions within the bacteriophage T4 AsiA protein involved in its binding to the σ^{70} subunit of *E. coli* RNA polymerase and its role as a transcriptional inhibitor and co-activator. *J Mol Biol* 325: 827–841.

Pande, S., Makela, A., Dove, S.L., Nickels, B.E., Hochschild, A. and Hinton, D.M., (2002). The bacteriophage T4 transcription activator MotA interacts with the far-C-terminal region of the σ^{70} subunit of *Escherichia coli* RNA polymerase. *J Bacteriol* 184: 3957–3964.

Paolozzi, L. and Ghelardini, P., (2005). "The Bacteriophage Mu," In press in *The Bacteriophages*, R. Calendar (Ed.). Oxford University Press, New York.

Park, J.T., (1996). The convergence of murein recycling research with beta-lactamase research. *Microb Drug Resist* 2: 105–112.

Parma, D.H., Snyder, M., Sobolevski, S., Nawroz, M., Brody, E. and Gold, L., (1992). The Rex system of bacteriophage λ: Tolerance and altruistic cell death. *Genes Dev* 6: 497–510.

Pietroni, P., Young, M.C., Latham, G.J. and von Hippel, P.H., (2001). Dissection of the ATP-driven reaction cycle of the bacteriophage T4 DNA replication processivity clamp loading system. *J Mol Biol* 309: 869–891.

Plancon, L., Chami, M. and Letellier, L., (1997). Reconstitution of FhuA, an *Escherichia coli* outer membrane protein, into liposomes. Binding of phage T5 to Fhua triggers the transfer of DNA into the proteoliposomes. *J Biol Chem* 272: 16868–16872.

Quirk, S.M., Bell-Pedersen, D., Tomaschewski, J., Ruger, W. and Belfort, M., (1989). The inconsistent distribution of introns in the T-even phages indicates recent genetic exchanges. *Nucleic Acids Res* 17: 301–315.

Raab, R., Neal, G., Sohaskey, C., Smith, J. and Young, R., (1988). Dominance in λS mutations and evidence for translational control. *J Mol Biol* 199: 95–105.

Ramanculov, E. and Young, R., (2001a). Functional analysis of the phage T4 holin in a λ context. *Mol Genet Genomics* 265: 345–353.

Ramanculov, E. and Young, R., (2001b). Genetic analysis of the T4 holin: Timing and topology. *Gene* 265: 25–36.

Ravin, N.V., (2005). "N15: The Linear Plasmid Prophage," in *The Bacteriophages*, R. Calendar (Ed.). Oxford University Press, New York.

Ravin, N.V., Kuprianov, V.V., Gilcrease, E.B. and Casjens, S.R., (2003). Bidirectional replication from an internal ori site of the linear N15 plasmid prophage. *Nucleic Acids Res* 31: 6552–6560.

Ravin, V., Raisanen, L. and Alatossava, T., (2002). A conserved C-terminal region in Gp71 of the small isometric-head phage LL-H and ORF474 of the prolate-head phage JCL1032 is implicated in specificity of adsorption of phage to its host, *Lactobacillus delbrueckii*. *J Bacteriol* 184: 2455–2459.

Ravin, V., Ravin, N., Casjens, S., Ford, M.E., Hatfull, G.F. and Hendrix, R.W., (2000). Genomic sequence and analysis of the atypical temperate bacteriophage N15. *J Mol Biol* 299: 53–73.

Rennell, D., Bouvier, S.E., Hardy, L.W. and Poteete, A.R., (1991). Systematic mutation of bacteriophage T4 lysozyme. *J Mol Biol* 222: 67–88.

Rennell, D. and Poteete, A.R., (1985). Phage P22 lysis genes: nucleotide sequences and functional relationships with T4 and genes. *Virology* 143: 280–289.

Riede, I., Degen, M. and Henning, U., (1985). The receptor specificity of bacteriophages can be determined by a tail fiber modifying protein. *Embo J* 4: 2343–2346.

Rousvoal, S., Oudot, M., Fontaine, J., Kloareg, B. and Goer, S.L., (1998). Witnessing the evolution of transcription in mitochondria: The mitochondrial genome of the primitive brown alga *Pylaiella littoralis* (L.) Kjellm. Encodes a T7-like RNA polymerase. *J Mol Biol* 277: 1047–1057.

Rutberg, B. and Rutberg, L., (1965). Role of superinfecting phage in lysis inhibition with phage T4 in *Escherichia coli*. *J Bacteriol* 90: 891–894.

Salas, M., (2005). "Phage ϕ29 And Its Relatives," in *The Bacteriophages*, R. Calendar (Ed.). Oxford University Press, New York.

Sampath, A. and Stewart, C.R., (2004). Roles of genes 44, 50, and 51 in regulating gene expression and host takeover during infection of *Bacillus subtilis* by bacteriophage SPO1. *J Bacteriol* 186: 1785–1792.

Sandmeier, H., (1994). Acquisition and rearrangement of sequence motifs in the evolution of bacteriophage tail fibres. *Mol Microbiol* 12: 343–350.

Sandmeier, H., Iida, S. and Arber, W., (1992). DNA Inversion Regions Min of Plasmid p15B and Cin of Bacteriophage P1: Evolution of Bacteriophage Tail Fiber Genes. *J Bacteriol* 174: 3936–3944.

Sandulache, R. and Prehm, P., (1985). Structure of the core oligosaccharide from lipopolysaccharide of Erwinia carotovora. *J Bacteriol* 161: 1226–1227.

Sandulache, R., Prehm, P. and Kamp, D., (1984). Cell wall receptor for bacteriophage Mu G(+). *J Bacteriol* 160: 299–303.

Scholl, D., Rogers, S., Adhya, S. and Merril, C.R., (2001). Bacteriophage K1–5 encodes two different tail fiber proteins, allowing it to infect and replicate on both K1 and K5 strains of *Escherichia coli*. *J of Virol* 75: 2509–2515.

Schwarz, H., Riede, I., Sonntag, I. and Henning, U., (1983). Degrees of relatedness of T-even type *E. coli* phages using different or the same receptors and topology of serologically cross-reacting sites. *Embo J* 2: 375–380.

Shamoo, Y., Tam, A., Konigsberg, W.H. and Williams, K.R., (1993). Translational repression by the bacteriophage T4 gene 32 protein involves specific recognition of an RNA pseudoknot structure. *J Mol Biol* 232: 89–104.

Shub, D.A., Gott, J.M., Xu, M.Q., Lang, B.F., Michel, F., Tomaschewski, J., Pedersen-Lane, J., et al. (1988). Structural conservation among three homologous introns of bacteriophage T4 and the group I introns of eukaryotes. *Proc Natl Acad Sci USA* 85: 1151–1155.

Simon, L.D., (1969). The Infection of *Escherichia coli* by T2 and T4 Bacteriophages as Seen in the Electron Microscope. III. Membrane-Associated Intracellular Bacteriophages. *Virology* 38: 285–296.

Snyder, L. and Kaufmann, G., (1994). "T4 phage exclusion mechanisms," pp. 391–396 in *Molecular Biology of Bacteriophage T4*, J. Karam, J.W. Drake, K.N. Kreuzer, G. Mosig, D.H. Hall, F.A. Eiserling, L.W. Black, et al. (Eds.). Amer. Soc. Microbiol., Washington, D.C.

Soukup, G.A., Emilsson, G.A. and Breaker, R.R., (2000). Altering molecular recognition of RNA aptamers by allosteric selection. *J Mol Biol* 298: 623–632.

Stewart, C.R., Gaslightwala, I., Hinata, K., Krolikowski, K.A., Needleman, D.S., Peng, A.S., Peterman, M. A., et al. (1998). Genes and regulatory sites of the "host-takeover module" in the terminal redundancy of *Bacillus subtilis* bacteriophage SPO1. *Virology* 246: 329–340.

Streisinger, G., Mukai, F., Dreyer, W.J., Miller, B. and Horiuchi, S., (1964). Mutations affecting the lysozyme of phage T4. Cold Spring Harbor Symp. Quant. Biol. 26: 25–30.

Stuer-Lauridsen, B., Janzen, T., Schnabl, J. and Johansen, E., (2003). Identification of the host determinant of two prolate-headed phages infecting *Lactococcus lactis*. *Virology* 309: 10–17.

Temiakov, D., Patlan, V., Anikin, M., McAllister, W.T., Yokoyama, S. and Vassylyev, D.G., (2004). Structural basis for substrate selection by T7 RNA polymerase. *Cell* 116: 381–391.

Tetart, F., Desplats, C. and Krisch, H.M., (1998). Genome plasticity in the distal tail fiber locus of the T-even bacteriophage: Recombination between conserved motifs swaps adhesin specificity. *J Mol Biol* 282: 543–556.

Tetart, F., Monod, C. and Krisch, H.M., (1996). Bacteriophage T4 host range is expanded by duplications of a small domain of the tail fiber adhesin. *J Mol Biol* 258: 726–731.

Trakselis, M.A., Alley, S.C., Abel-Santos, E. and Benkovic, S.J., (2001). Creating a dynamic picture of the sliding clamp during T4 DNA polymerase holoenzyme assembly by using fluorescence resonance energy transfer. *Proc Natl Acad Sci USA* 98: 8368–8375.

Tuerk, C. and Gold, L., (1990). Systematic evolution of ligands by exponential enrichment: RNA ligands to bacteriophage T4 DNA polymerase. *Science* 249: 505–510.

Wais, A.C. and Goldberg, E.B., (1969). Growth and transformation of phage T4 in *Escherichia coli* B-4, *Salmonella, Aerobacter, Proteus*, and *Serratia. Virology* 39: 153–161.

Wang, I.N., Deaton, J. and Young, R., (2003). Sizing the holin lesion with an endolysin-beta-galactosidase fusion. J Bacteriol 185: 779–787.

Wang, I.-N., Dykhuizen, D.E. and Slobodkin, L.B., (1996). The evolution of phage lysis timing. *Evolutionary Ecology* 10: 545–558.

Wang, I.N., Smith, D.L. and Young, R., (2000). Holins: The protein clocks of bacteriophage infections. *Annu Rev Microbiol* 54: 799–825.

Wang, M.D., Schnitzer, M.J., Yin, H., Landick, R., Gelles, J. and Block, S.M., (1998). Force and velocity measured for single molecules of RNA polymerase. *Science* 282: 902–907.

Wegrzyn, G. and Thomas, M.S., (2002). Modulation of the susceptibility of intestinal bacteria to bacteriophages in response to Ag43 phase variation—a hypothesis. *Med Sci Monit* 8: HY15–18.

Winter, R.B. and Gold, L., (1983). Overproduction of bacteriophage Q beta maturation (A2) protein leads to cell lysis. *Cell* 33: 877–885.

Winter, R.B., Morrissey, L., Gauss, P., Gold, L., Hsu, T. and Karam, J., (1987). Bacteriophage T4 regA protein binds to mRNAs and prevents translation initiation. *Proc Natl Acad Sci USA* 84: 7822–7826.

Wollman, E.L., and Stent, G.S., (1952). Studies on Activation of T4 Bacteriophage by Cofactor. IV. Nascent Activity. *Biochemica Et Biophysica Acta* 9: 538–550.

Wozniak, J.A., Zhang, X.-J., Weaver, L.H. and Matthews, B.W., (1994). "Structural and genetic analysis of the stability and function of T4 lysozyme," pp. 332–339 in *Molecular Biology of Bacteriophage T4*, J. Karam, J.W. Drake, K.N. Kreuzer, G. Mosig, D.H. Hall, F.A. Eiserling, L.W. Black, et al. (Eds.) American Society for Microbiology, Washington, D.C.

Yang, G., Cecconi, C., Baase, W.A., Vetter, I.R., Breyer, W.A., Haack, J.A., Matthews, B.W., et al. (2000). Solid-state synthesis and mechanical unfolding of polymers of T4 lysozyme. *Proc Natl Acad Sci U S A* 97: 139–144.

Yin, Y.W. and Steitz, T.A., (2002). Structural basis for the transition from initiation to elongation transcription in T7 RNA polymerase. *Science* 298: 1387–1395.

Yin, Y.W. and Steitz, T.A., (2004). The structural mechanism of translocation and helicase activity in t7 RNA polymerase. *Cell* 116: 393–404.

Young, B.A., Gruber, T.M. and Gross, C.A., (2004). Minimal machinery of RNA polymerase holoenzyme sufficient for promoter melting. *Science* 303: 1382–1384.

Young, R., (1992). Bacteriophage lysis: Mechanism and regulation. *Microbiol Rev* 56: 430–481.

Young, R. and Wang, I., (2005). "Phage Lysis," in *The Bacteriophages*, R. Calendar (Ed.). Oxford University Press, New York.

Yu, S.L., Ko, K.L., Chen, C.S., Chang, Y.C. and Syu, W.J., (2000). Characterization of the distal tail fiber locus and determination of the receptor for phage AR1, which specifically infects *Escherichia coli* O157:H7. *J Bacteriol* 182: 5962–5968.

8 Bacteriophages and Bacterial Virulence

E. Fidelma Boyd
Department of Microbiology, UCC,
National University of Ireland, Cork, Ireland

CONTENTS

8.1. Introduction 224
8.2. Bacteriophage-Encoded Virulence Factors 227
8.2.1. Extracellular Toxins 227
8.2.1.1. *Streptococcus* and *Staphylococcus* 228
8.2.1.2. *E. coli* 229
8.2.1.3. *V. cholerae* 231
8.2.2. Proteins Involved in Bacterial Attachment/Colonization 234
8.2.3. Proteins Required for Host Immune Avoidance 236
8.2.4. Proteins Essential for Bacterial Invasion of Host Cells 238
8.2.5. Proteins Involved in the Survival of Intracellular Bacteria 241
8.2.6. Putative Virulence Proteins 241
8.3. Mechanisms of Virulence Gene Acquisition by Bacteriophages 243
8.3.1. Imprecise Prophage Excision 243
8.3.2. Transferable Virulence Gene Cassettes 244
8.3.3. Integral Components of the Bacteriophage Genome 244
8.4. Role of Bacteriophages in the Pathogenesis of Bacterial Infections 245
8.4.1. Indirect Involvement of Bacteriophages 245
8.4.2. Direct Involvement of Bacteriophages 246
8.5. Roles of Phage-Phage Interactions in Bacterial Virulence 247
8.5.1. Helper Phage Facilitation of Bacteriophage Mobilization 248
8.5.1.1. *V. cholerae* CTXϕ Supplies Morphogenesis Genes for RS1ϕ 248
8.5.1.2. *S. aureus* ϕ80 Generalized Transduction of the tst Element 249
8.5.1.3. *V. cholerae* CP-T1 Generalized Transduction of VPI 250
8.5.1.4. *V. cholerae* CP-T1 Generalized Transduction of CTXϕ 250
8.5.2. Helper Phage Source of Bacteriophage Host Receptor 252
8.5.3. Helper Phage Requirement for Virulence Gene Expression 253

0-8493-1336-X/05/$0.00+$1.50
© 2005 by CRC Press LLC

8.5.3.1. RS1ϕ Encodes an Antirepressor Involved in CT Production Mediated by CTXϕ 253
8.5.3.2. The VPI Encodes Regulatory Genes Required for CT Expression 253
8.5.3.3. Interactions of *S. enterica* Gifsy-1 and Gifsy-2 Phages 253
8.5.3.4. *E. coli* Prophage Enhancement of *eib* Expression 254
8.6. Bacteriophages and Pathogenicity Islands (PAIs) 254
8.7. Conclusions 257
Acknowledgments 258
References 258

8.1. INTRODUCTION

In general, differences in pathogenic potential among bacterial species and among strains within a species are due to the presence and expression of virulence genes (i.e., genes that encode virulence factors) in pathogenic strains and to their absence in related nonpathogenic strains. Many virulence factor-encoding genes of facultative pathogens (e.g., *Escherichia coli, Listeria monocytogenes*, *Pseudomonas aeruginosa, Streptococcus* spp., *Staphylococcus aureus*, and *Vibrio cholerae*) are present in various mobile genetic elements, such as transposons, plasmids, bacteriophages, and pathogenicity islands. In that regard, it has become increasingly evident that one group of mobile genetic elements, the bacteriophages, plays an important role in the evolution and emergence of pathogenic bacteria. In a process called phage conversion, bacteriophage-encoded virulence genes can convert their bacterial host from a nonpathogenic strain to a virulent strain, or to a strain with increased virulence, by providing mechanisms for the invasion of host tissues and the avoidance of host immune defenses (Table 8.1). Indeed, the loss of those bacteriophages may render the bacteria nonpathogenic. However, the majority of bacteriophages possessing virulence genes are temperate bacteriophages that form stable lysogens. Therefore, lysogenic conversion does not result in bacterial lysis but, instead, allows both vertical and horizontal gene transfer—which may confer a selective advantage to the host and result in clonal expansion. This hypothesis is supported by the recent epidemic spread of cholera from Asia to South America after nearly a century of absence. The *ctxAB* genes, which encode cholera toxin (CT), reside in the genome of a filamentous bacteriophage CTXϕ (Waldor and Mekalanos, 1996). Therefore, *V. cholerae*'s lysogenic conversion by the CTXϕ enables it to produce CT, the main cause of the secretory diarrhea so characteristic of cholera, thus facilitating the spread of *V. cholerae* to new hosts and enhancing its pandemic spread.

The location, within the bacteriophage genome, of numerous genes encoding bacterial virulence factors also suggests an evolutionary advantage to the bacteriophage, by enhancing survival of the lysogen. Hence, the ecological success of lysogenic bacteria contributes to the dissemination of bacteriophage genes. Also, the phenomenon may be considered to indicate the co-evolution of bacteriophages and bacteria. In this chapter, the various classes of bacteriophage-encoded virulence factors and the evolutionary distribution and diversity of the bacteriophages among pathogenic bacterial

TABLE 8.1
Bacteriophage-Encoded Virulence Factors Involved in Various Stages of Bacterial Pathogenesis

Bacterial Host	Bacteriophage	Virulence Factor (gene)	Reference
Proteins required for host attachment			
E. coli	λ	OMP (*lom*)	(Pacheco et al., 1997)
M. arthritidis	MAV1	OMP (*vir*)	(Voelker and Dybvig, 1999)
S. mitis	SM1	Coat protein (*pblA, pblB*)	(Bensing et al., 2001b)
V. cholerae	VPIϕ	TCP pilin (*tcp*)	(Karaolis et al., 1999)
Proteins that alter antigenic recognition			
N. meningitidis	Mu-like	Membrane proteins	(Masignani et al., 2001)
S. enterica	ε^{34}	O-antigen (*rfb*)	(Wright, 1971)
S. enterica	P22	O-antigen (*gtr*)	(Vander Byl and Kropinski, 2000)
S. flexneri	Sf6	O-antigen (*oac*)	(Clark et al., 1991)
S. flexneri	SfII, SfV, SfX	O-antigen (*gtrII*)	(Allison and Verma, 2000)
Proteins involved in cellular invasion			
S. enterica	SopEϕ	Type III effector (*sopE*)	(Mirold et al., 1999)
S. enterica	Gifsy-1	Type III effector (*gogB*)	(Figueroa-Bossi et al., 2001)
S. enterica	Gifsy-1	IS-like sequence (*gipA*)	(Figueroa-Bossi et al., 2001)
S. enterica	Gifsy-2	Type III effector (*gtgB*)	(Figueroa-Bossi et al., 2001)
S. enterica	Gifsy-3	Type III effector (*sspH1*)	(Figueroa-Bossi et al., 2001)
S. pyogenes	H4489A	Hyaluronidase (*hylP*)	(Hynes and Ferretti, 1989)
Proteins required for intracellular survival			
E. coli	λ	OMP (*bor*)	(Barondess and Beckwith, 1990)
E. coli	λ-like	OMP (*eib*)	(Sandt et al., 2002)
E. coli O157	Sp4, 10	Superoxide dismutase (*sodC*)	(Ohnishi et al., 2001)
S. enterica	Gifsy-2	Superoxide dismutase (*sodCI*)	(Figueroa-Bossi et al., 2001)
S. enterica	Fels-1	Superoxide dismutase (*sodCIII*)	(Figueroa-Bossi et al., 2001)

(Continued)

TABLE 8.1
(Continued)

Bacterial Host	Bacteriophage	Virulence Factor (gene)	Reference
Extracellular toxins			
C. botulinum	C1	Neurotoxin (c1)	(Barksdale and Arden, 1974)
C. diphtheriae	ß-phage	Diphtheria toxin (*tox*)	(Freeman, 1951)
E. coli	H-19B	Shiga toxins (*stx1,2*)	(O'Brien et al., 1984)
E. coli	ϕFC3208	Enterohemolysin (*hly*)	(Beutin et al., 1993)
V. cholerae	ϕCTX	Cytotoxin (*ctx*)	(Nakayama et al., 1999)
V. cholerae	CTXϕ	Cholera toxin (*ctxAB*)	(Waldor and Mekalanos, 1996)
V. cholerae	CTXϕ	Ace toxin (*ace*)	(Trucksis et al., 1993)
V. cholerae	CTXϕ	ZOT toxin (*zot*)	(Fasano et al., 1991)
S. aureus	NA	Enterotoxin (*see, sel*)	(Betley and Mekalanos, 1985)
S. aureus	ϕN315	Enterotoxin P (*sep*)	(Kuroda et al., 2001)
S. aureus	ϕ13	Enterotoxin A (*entA*)	(Coleman et al., 1989)
S. aureus	ϕMu50A	Enterotoxin A (*sea*)	(Kuroda et al., 2001)
S. aureus	ϕ13	Staphylokinase (*sak*)	(Coleman et al., 1989)
S. aureus	ϕETA	Exfoliative toxin A (*eta*)	(Yamaguchi et al., 2000)
S. pyogenes	T12	Toxin type A (*speA*)	(Weeks and Ferretti, 1984)
S. pyogenes	CS112	Toxin type C (speC)	(Goshorn and Schlievert, 1989)
Putative virulence factors			
S. enterica	Fels-1	Neuraminidase (*nanH*)	(Figueroa-Bossi et al., 2001)
S. enterica	Gifsy-1,2	Hemolysin (*ehly*)	(McClelland, 2001)
S. enterica	Gifsy-1,2	Serum-resistance (*ail*)	(McClelland, 2001)
S. enterica	Gifsy-2, Fels-1	Antivirulence gene (*grvA*)	(Ho and Slauch, 2001)
V. cholerae	K139	G-protein-like (*glo*)	(Reidl and Mekalanos, 1995)

TABLE 8.2
Mechanisms of Virulence Gene Acquisition by Bacteriophages

Mechanism	Virulence Gene	Bacteriophage	Bacterium
Imprecise excision	*see, sea, sek*		*S. pyogenes*
Transferable module	*stx1, stx2*	Sp5, Sp15	*E. coli* O157
Phage component	*pblA, pblB*	SM1	*S. mitis*
	Ace, zot	CTXϕ	*V. cholerae*

isolates are discussed. The virulence genes are categorized on the basis of the gene products' roles in the various stages of pathogenesis: attachment and colonization, host immune avoidance, cellular invasion, intracellular survival, and toxin production (Table 8.1). The prevalence and evolutionary distribution of the bacteriophages and virulence genes are examined in terms of their occurrence within a species. Studies of the diversity of bacteriophages encoding virulence factors indicate that some are morphologically diverse and consist of members of a number of viral families, while others have mosaic genomes possessing characteristics of more than one viral family (reviewed in Canchaya et al., 2003). Three possible mechanisms by which bacteriophages acquire virulence genes are also examined; they (i) are acquired by imprecise prophage excision from the bacterial host genome, (ii) are transferable modules within the bacteriophage genome, or (iii) are integral components of the bacteriophage genome (Table 8.2). The indirect and direct roles that bacteriophages play in bacterial virulence (i.e., as passive vectors for the transfer of virulence genes, and as active components in the regulation of bacterial pathogenesis, respectively) also are explored. In addition, the role of phage-phage interactions in bacterial pathogenesis is examined, including the requirement for a helper phage for the mobilization and functioning of another bacteriophage. Finally, the relationship and common features between bacteriophages and pathogenicity islands are discussed.

8.2. BACTERIOPHAGE-ENCODED VIRULENCE FACTORS

In this section, several categories of bacteriophage-encoded virulence factors are discussed based on their roles in bacterial pathogenesis (Table 8.1). The bacteriophages include members of the viral families *Inoviridae* (filamentous), *Myoviridae* (contractile tails), *Podovirdae* (short tail stubs), and *Siphoviridae* (long flexible tails); (for detailed descriptions of the phage families/phage taxonomy, please refer to Chapter 4).

8.2.1. Extracellular Toxins

The best characterized group of bacteriophage-encoded virulence factors are extracellular toxins expressed in various Gram-negative and Gram-positive bacteria, including *E. coli*, *Shigella* spp., *P. aeruginosa*, *V. cholerae*, *S. aureus*, and *Streptococcus pyogenes*. The bacteriophage-encoded toxins are functionally diverse and include some of the most potent bacterial toxins ever described; e.g., tetanus toxin, botulinum toxin, diphtheria toxin, which are encoded on prophages harbored by *Clostridium tetani, C. botulinum*, and *Corynebacterium diphtheriae*, respectively.

The phage-encoded bacterial toxins may be cytotoxic, enterotoxic, or neurotoxic, and they enable the bacteria to cause an astonishing array of diseases ranging from mild gastrointestinal disease to life-threatening toxemia and sepsis. Another remarkable feature of the toxins is that although all of them were acquired by lateral gene transfer from unknown sources, all are controlled by bacterial host chromosomal regulatory factors, which suggests some common ancestry.

8.2.1.1. *Streptococcus* and *Staphylococcus*

Streptococcal and staphylococcal isolates contain bacteriophage genes encoding superantigen toxins, exfoliative toxin A, enterotoxin P, hyaluronidase, and staphylokinase. *S. pyogenes* causes a wide range of infections (e.g., fasciitis, rheumatic fever, pharyngitis, pyoderma, scarlet fever, and toxic shock syndrome), and the diverse disease syndromes it elicits probably reflects the various mechanisms of action of its phage-encoded toxins. In view of the variety of bacteriophage-encoded superantigen toxins produced by *S. pyogenes*, it appears likely that at least part of the specific pathogenic potential of distinct clinical *S. pyogenes* isolates is determined by a specific combination of lysogenic conversion genes (Desiere et al., 2001; Banks et al., 2002), located between the phage lysin gene and the right attachment site of the prophages closely related to *cos*-site and *pac*-site *Siphoviridae* (Ferretti et al., 2001; Smoot et al., 2002). *S. pyogenes* M1 and M18 serotype strains associated with invasive wound infections and rheumatic fever, respectively, have highly homologous genomes, and the differences between the strains are accounted for by their bacteriophage content (Desiere et al., 2001). This finding is consistent with the recent observation in Japan that the replacement of older strains by newer M3 strains can be traced to the acquisition of a prophage encoding a new combination of superantigens (Inagaki et al., 2000). A similar scenario is responsible for the diversity among *S. aureus* isolates, which cause a range of illnesses with diverse clinical outcomes (Table 8.3). For example, various staphylococcal phage-encoded genes are required for the bacteria to express (i) enterotoxins that cause acute gastroenteritis, (ii) leucocidin causing leukocytolysis, (iii) SPEA toxin that causes scarlet fever and tissue necrosis, (iv) exfoliative

TABLE 8.3
Bacteriophage- and Phage-Like Element-Encoded *S. aureus* Toxins and Their Associated Diseases

Toxin	Disease	Element	Size
TSST-1	Toxic Shock Syndrome (TSS)	PAI	15 kb (ϕ-like)
Enterotoxin A	Food poisoning, TSS	Phage	30 kb
Enterotoxin B,C	Food poisoning, TSS	PAI	15 kb (ϕ-like)
Enterotoxin D	Food poisoning, TSS	Plasmid	30 kb
Exfoliatin A	Blistering skin disease	Phage	45 kb
Exfoliatin B	Scalded-skin syndrome	Plasmid	30 kb
Leukocidin	Necrotizing pneumonia	Phage	45 kb
SPEA, C	Scarlet fever	Phage	45 kb

toxin causing "scalded-skin" syndrome, and (v) TSST toxin that elicits toxic shock syndrome (Table 8.3). Comparative genomic analysis of *S. aureus* has revealed closely related bacterial genomes, with differences between strains confined mostly to their bacteriophage content (Kuroda et al., 2001). Interestingly, the bacteriophages involved are related to *cos*-site and *pac*-site temperate *Siphoviridae* from streptococci and lactobacilli, which indicates that a pool of bacteriophages may circulate among the Gram-positive bacteria with low GC content. All of the lysogenic conversion genes present in the bacteriophages—including those encoding exfoliative toxin A, enterotoxin P, staphylokinase, superantigen toxins, and leukocyte toxins (some related to streptococcal bacteriophage-encoded proteins)—are located near the right bacteriophage attachment site (Kaneko et al., 1998; Yamaguchi et al., 2000).

8.2.1.2. *E. coli*

Diversity in the number and type of toxins among strains, as well as in the bacteriophages involved, is not restricted to streptococci and staphylococci. *E. coli*, whose natural habitat is the gastrointestinal tract of warm-blooded animals, is the most common facultative anaerobe in the human intestine. However, many pathogenic *E. coli* strains cause enteric diseases (e.g., severe watery diarrhea, dysentery, and hemorrhagic colitis) and extraintestinal infections (e.g., cystitis, septicemia, and meningitis). In enterohemorrhagic *E. coli* strains (EHEC, including O157:H7) and Shiga toxin *E. coli* (STEC) strains responsible for hemorrhagic colitis and hemolytic uremic syndrome, the shigatoxins (Stx1, Stx2) and enterohemolysins (Hly) are encoded by a diversity of lambda-like bacteriophages (Wagner et al., 1999; Unkmeir and Schmidt, 2000; Johansen et al., 2001; Recktenwald and Schmidt, 2002). Stx1 and Stx2 are A-B toxins consisting of one active A-subunit and five identical B-subunits. The B subunits bind to glycolipids on the host cells, and the A-subunit is taken up by the cell and causes apoptosis by disrupting protein synthesis. A recent study (Wagner et al., 1999) examining seven different Stx2-encoding bacteriophages isolated from multiple STEC isolates discovered striking differences in bacteriophage titers and in the amount of toxin produced. Also, structural analysis of bacteriophage-borne *stx1, stx2* and flanking sequences in *E. coli* O157, STEC and *Shigella dysenteriae* type 1 strains demonstrated significant amounts of bacteriophage genomic variation, which indicated that the shiga toxin genes were encoded on unrelated bacteriophages (Unkmeir and Schmidt, 2000; Recktenwald and Schmidt, 2002). The diversity of Stx-encoding bacteriophages is, evolutionarily speaking, an important mechanism for the spread of toxin genes among isolates, because it may inhibit bacteriophage exclusion due to superinfection immunity, competition for integration sites and restriction-modification systems.

A study (Perna et al., 2001) comparing the genomic sequences of a pathogenic *E. coli* O157 strain and a laboratory-maintained *E. coli* K12 strain showed that 4.1 million base pairs of core sequences are conserved between the genomes, but that this backbone sequence is interspersed with sequences unique to one strain. The *E. coli* O157 strain contained 1.3 million base pairs of strain-specific DNA, most of which encompassed bacteriophage and pathogenicity island DNA (Fig. 8.1). A comparative genomic analysis (Ohnishi et al., 2001) of *E. coli* O157 strain Sakai and EDL933 also revealed that both have very similar bacterial DNA content but

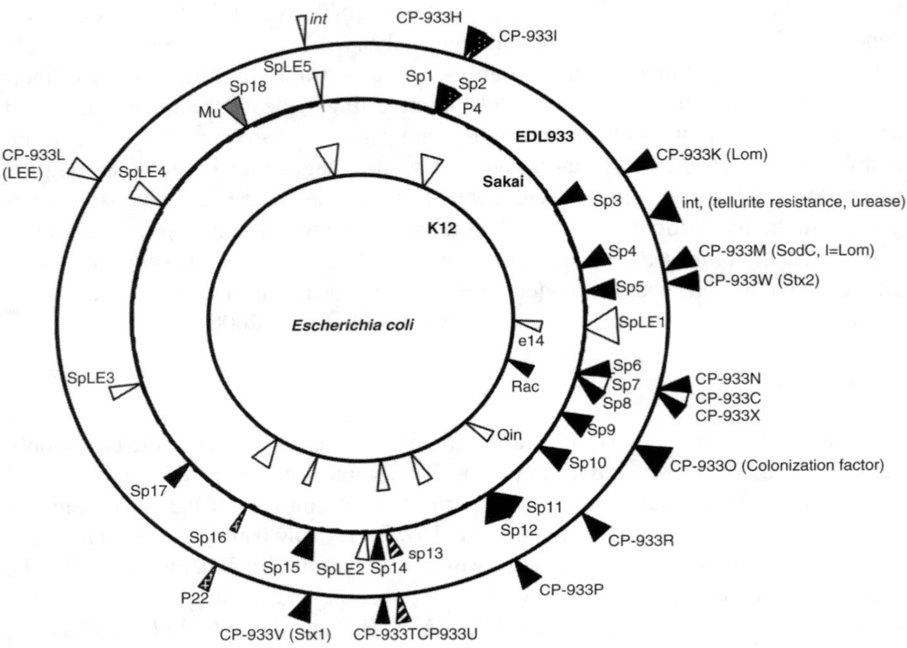

FIGURE 8.1 Comparative genomic maps of the prophage and prophage-like elements of *E. coli* strains EDL933, Sakai, and K12. The outer circle shows the prophages and prophage remnants (triangles) in strain EDL933 of *E. coli* serotype O157 (Perna et al., 2001), the middle circle represents the Sakai strain of *E. coli* serotype O157 (Yokoyama et al., 2000), and the inner circle is the genomic map of the nonpathogenic laboratory strain *E. coli* K12 (Blattner et al., 1997). Lambdoid-like prophages are indicated as black triangles, PAIs encoding integrases are indicated as white triangles, P2 prophages are indicated as striped triangles, P4 prophages are indicated as dotted triangles, P22 prophages are indicated as checkered triangles, and a Mu prophage is indicated in gray.

differ in their prophage content. In strain EDL933, 12 prophage sequences were identified; whereas, in strain Sakai, 18 prophages were identified, 13 of which were lambda-like (Fig. 8.1). A dotplot matrix of the different lambda-like prophages revealed large regions of high sequence identity (Boyd and Brüssow, 2002). Also, *in silico* genomic analysis of the two *E. coli* O157 serotypes strains identified numerous virulence factors encoded by their prophages (Ohnishi et al., 2001). They included (i) Stx-1, a cytotoxin, and the Bor and Lom proteins (conferring serum resistance and cell adhesion, respectively) encoded by bacteriophages Sp3-Sp5, Sp8-12, Sp14, and Sp15, (ii) an intestinal colonization factor encoded by bacteriophages 933 and O, and (iii) tellurite resistance gene products and superoxide dismutase (SodC) encoded by bacteriophages Sp 4 and Sp10 (Ohnishi et al., 2001). It is tempting to speculate that the pathogenic potential of the two isolates correlates with their prophage content, and that acquisition of the bacteriophage-encoded virulence factors may have played a decisive role in the emergence of O157 as a foodborne pathogen. The laboratory *E. coli* K12 strain contains the inducible lambda phage as well as numerous lambda prophage remnants (DLP-12, e14, Rac and Qin) (Fig. 8.1). Phage lambda contains both *bor* and *lom* genes, which have been associated with virulence.

8.2.1.3. *V. cholerae*

A recent addition to the list of bacteriophage-encoded toxins is *V. cholerae*'s CT, which is the only known toxin that is encoded by a filamentous bacteriophage (CTXϕ) (Waldor and Mekalanos, 1996). CT, similar to Shiga toxin, is an A-B toxin consisting of one active subunit and five binding subunits. However, CT is an enterotoxin that affects membrane permeability by disrupting the normal flow of ions in the small intestines. The A-B subunits, encoded by the *ctxAB* genes, share extensive sequence homology with heat-labile enterotoxins produced by enteropathogenic *E. coli* isolates. Also, the CTXϕ, which carries *ctxAB*, is similar in size, structure, and synteny to other filamentous coliphages, such as M13 and f1 (Fig. 8.2). CTXϕ has the typical gene organization of about ten DNA replication, coat, and morphogenesis genes with sequence homology. However, CTXϕ differs from M13 and other filamentous coliphages because it lacks a putative bacteriophage export gene (gene IV) (Fig. 8.2). Instead, CTXϕ and CT exploit the same pathway for export out of the cell, a type II extracellular protein secretion system (Davis et al., 1999). In addition, CTXϕ unlike other coliphages, is maintained as a prophage within the *V. cholerae* chromosomes (Waldor and Mekalanos, 1996). The CTXϕ genome can be divided into two functional domains: the 4.7 kb core region and the 2.4-kb repeat sequence (RS2) region (Fig. 8.3). The CTXϕ core region genes (*cep*, *orfU*, *ace* and *zot*) correspond to gVIII, gIII, gVI, and gI of M13; whereas *ctxAB* do not share sequence similarity with known genes and are not essential for phage production (Fig. 8.2). The genes in the RS2 region (*rstR, rstA* and *rstB*) encode regulation, replication, and integration functions (Waldor and Mekalanos, 1996; Waldor et al., 1997). The Ace and Zot proteins are also enterotoxins (Trucksis et al., 1993; Fasano et al., 1991). However, since Ace and Zot are primarily structural and assembly proteins of CTXϕ enterotoxicity is likely a secondary trait, a side consequence of bacterial invasion of the human host. An analysis (Davis et al., 1999) of the CTXϕ repressor gene (*rstR*) involved in superinfection immunity showed considerable sequence variation in closely related strains, which suggests a possible mechanism for polylysogeny in *V. cholerae*. Interestingly, unlike other integrated prophages, CTXϕ does not excise from the chromosome to form extrachromosomal CTXϕ particles. Instead, CTXϕ is generated by a replicative process that requires tandem elements, either CTX-CTX prophages or CTX-RS1 prophages, within the chromosome (Fig. 8.3) (Davis et al., 2000).

RS1ϕ is a satellite filamentous phage that (i) contains *rstR, rstA* and *rstB* genes homologous to the RS2 region genes of CTXϕ (ii) encodes an additional gene (*rstC*), and (iii) depends on the CTXϕ for packaging and transmission (Faruque et al., 2002; Davis et al., 2002). Examination of the diversity of CTXϕs derived from a variety of *V. cholerae* natural isolates revealed the presence of two distinct lineages of CTXϕs within the classical and El Tor biotypes of *V. cholerae* O1 serogroup isolates, the

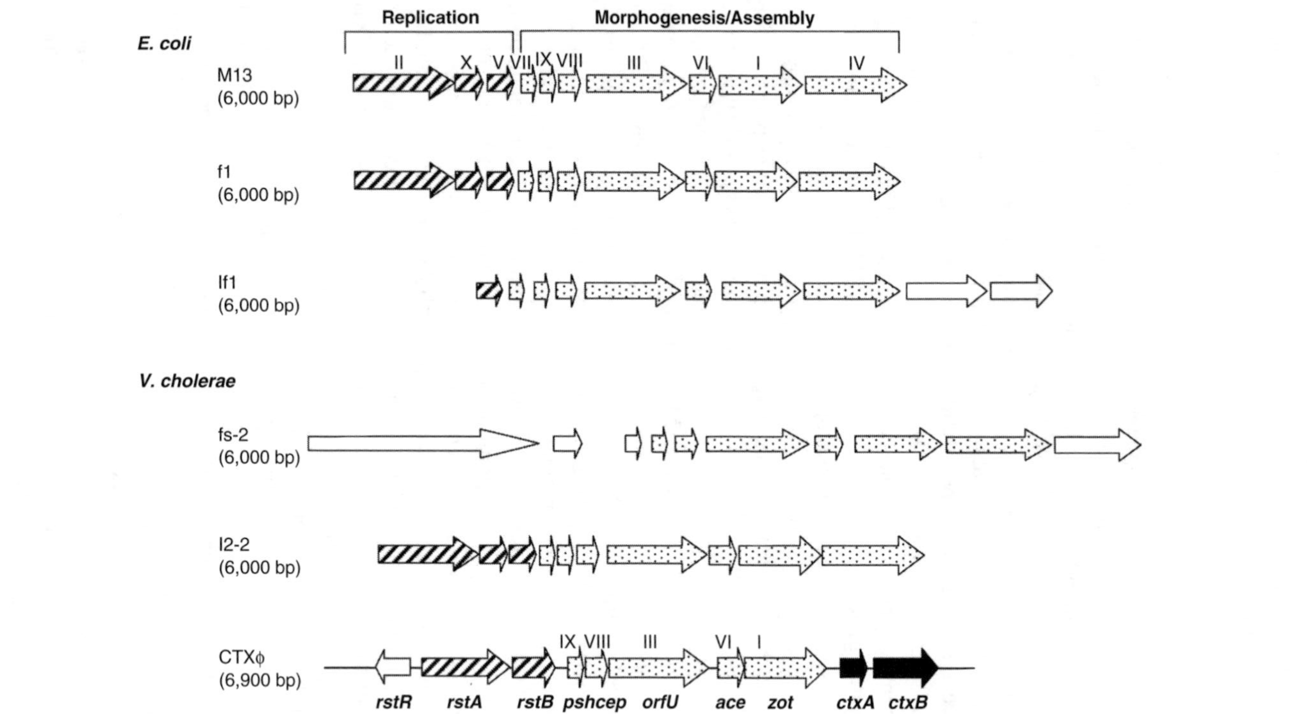

FIGURE 8.2 Schematic representation of ssDNA filamentous bacteriophages (f1 and M13) for *E. coli* and *V. cholerae*. Arrows indicate open reading frames (ORFs), and the direction of the arrows indicates the direction of transcription. The Roman numerals above the ORFs indicate gene designations. Genes encoding similar proteins have identical pattern arrows among the phages. DNA replication genes are represented as striped arrows, phage morphogenesis and assembly genes are represented as dotted arrows, and toxin genes are represented as black arrows. The Ff class of the filamentous phages are 98% homologous at the nucleotide level; therefore, the amino acid sequences of their proteins are virtually identical.

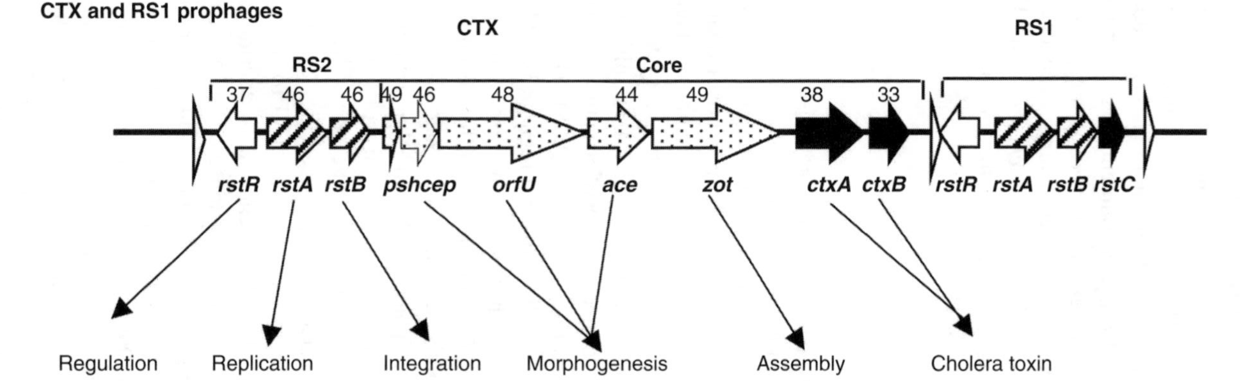

FIGURE 8.3 Gene organization and gene designation of the El Tor-derived CTXϕ and RS1ϕ genomes (Waldor and Mekalanos, 1996; Waldor et al., 1997). Open arrows represent CTXϕ ORFs and the direction of transcription of each gene. White triangles indicate end repeats. The numbers underneath the arrows indicate each gene's percent GC content. The *rstR*, *rstA* and *rstB* genes of both RS1ϕ and RS2 regions are almost identical. The *rstC* and *ctxAB* genes indicated in black encode an antirepressor and CT, respectively.

predominant cause of epidemic cholera (Boyd et al., 2000a). These data indicate independent acquisition of CTXϕ among isolates (Boyd et al., 2000a; Waldor and Mekalanos, 1996; Waldor et al., 1997), which is not surprising because CTXϕ transduction is readily detectable *in vitro* and *in vivo* (Waldor and Mekalanos, 1996; Waldor et al., 1997). Nonetheless, the distribution of CTXϕ among *V. cholerae* natural isolates is sporadic; CTXϕ is present in most O1 and O139 serogroup isolates, but it is only occasionally isolated from non-O1/non-O139 serogroup isolates (Faruque et al., 1998a; 1998b). CTXϕ has been isolated from environmentally occurring isolates of *V. mimicus* (Boyd et al., 2000b), a close relative of *V. cholerae*, which indicates an additional reservoir for this bacteriophage. Also, a comparative sequence analysis (Boyd et al., 2000b) of several genes of CTXϕ from *V. cholerae* and *V. mimicus* showed near sequence identity, indicating recent transfer, evolutionary speaking, of this bacteriophage between the species.

8.2.2. Proteins Involved in Bacterial Attachment/Colonization

Proteins involved in bacterial attachment to host cells, a critical step in the establishment of a bacterial infection, are novel additions to the list of bacteriophage-encoded virulence factors (Table 8.1). In *V. cholerae*, the toxin co-regulated pilus (TCP) plays an essential role in intestinal colonization of the human host (Taylor et al., 1987). Therefore, the recent proposal that TCP genes are encoded on a phage (Karaolis et al., 1999) is highly significant and has been followed with great interest (Lee, 1999; Davis and Waldor, 2003; Miller, 2003; Faruque et al., 2003). Initially, the *tcp* gene cluster, which encodes TCP, was defined as part of a pathogenicity island (PAI), the *Vibrio* pathogenicity island (VPI) (Karaolis et al., 1998). The VPI conforms to the definition of a PAI; i.e., it is a large chromosomal region (39.5 kb) that encodes several virulence gene clusters and a phage-like integrase, it inserts adjacent to a tRNA-like gene (*ssrA*), and it has a GC content that differs from the host chromosome (Fig. 8.4). Subsequently, Karaolis and colleagues (Karaolis et al., 1999) found that TCP is encoded by a filamentous phage, VPIϕ, and that the major pilus protein (TcpA), monomers of which make up the TCP pilus structure, is also the VPIϕ coat protein. The filamentous phage itself appeared to be very different from other filamentous phages based on the nucleotide sequences (Fig. 8.2 and Fig. 8.4). In addition, the authors have been unable to show transfer of VPIϕ among *V. cholerae* O1 serogroup isolates, the main cause of epidemic cholera. Furthermore, Faruque et al. (2003a) recently were unable to confirm the existence of VPIϕ when they examined a collection of TCP-positive, VPI-containing, clinical and environmental isolates of *V. cholerae* under the same conditions (which included mitomycin- and UV-induction) reported by Karaolis et al. (1999). Although minor differences between strains and growth conditions may account for the discrepancy between the two reports, the study by Faruque et al. (2003a), together with the lack of confirmatory evidence to the existence of VPIϕ, suggests that additional studies are required to confirm or refute the existence of VPIϕ and to better understand its role in the pathogenesis of *V. cholerae*.

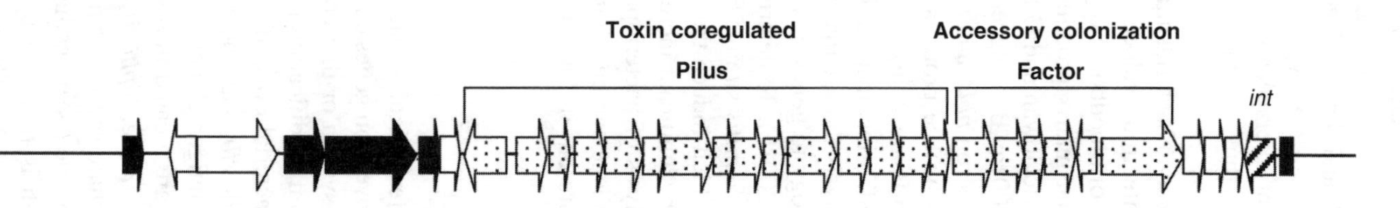

FIGURE 8.4 Schematic representation of the 39.5-kb VPI-1 (VC0819-VC0847) regions of *V. cholerae* strain N16961 (Heidelberg et al., 2000). ORFs are indicated by arrows that point in the direction of transcription. Virulence genes are colored black, *int* genes are indicated as striped arrows, proteins of known function are indicated as dotted arrows, hypothetical proteins are indicated as black arrows, and genes without a significant match are indicated as white arrows.

The distribution of the VPI region among *V. cholerae* is similar to the distribution of CTXϕ; i.e., VPI is present in most O1 and O139 serogroup isolates, and it is absent from most *V. cholerae* non-O1 and non-O139 serogroup isolates. Interestingly, several *V. mimicus* isolates carrying VPI have been identified, and their VPI genes show remarkable sequence identity to those of the VPI in *V. cholerae* biotype El Tor (Boyd et al., 2000b). These data suggest that there was recent interspecies lateral gene transfer of this region between *V. cholerae* and *V. mimicus*; thus, indicating that the VPI region was mobile in the recent past.

Another example of bacteriophage-encoded proteins involved in bacterial attachment and colonization is found in *S. mitis*, a leading cause of infectious endocarditis. Infection of the endocardium commences with the attachment of bloodborne organisms to deposits composed of platelets, fibrin, and extracellular matrix proteins on the damaged heart valve surface (Durack and Beeson, 1975; McGowan and Gillett, 1980). Recently, Bensing and colleagues (Bensing et al., 2001a; Bensing et al., 2001b) identified two distinct loci, *pblT* and a polycistronic operon *pblA* and *pblB*, in *S. mitis* strain SF100. The latter two genes encode large surface proteins that promote binding to human platelets. However, PblA and PblB are unusual since neither protein shows a strong similarity to known bacterial adhesins; instead, both resemble structural components of bacteriophages. Further analysis showed that *pblA* and *pblB* are clustered with genes showing significant homology to genes in streptococcal bacteriophages r1t, 01205 and Pp-1 (Bensing et al., 2001b). To determine whether *pblA* and *pblB* reside within a prophage, cultures of *S. mitis* strain SF100 were treated with either UV light or mitomycin C, both of which induce the lytic cycle in temperate bacteriophages. Both treatments resulted in significant increases in the transcription of *pblA*, and Southern hybridization analysis of bacteriophage DNA revealed that *pblA* and *pblB* were contained in the SM1 phage genome. Further experimentation indicated that the genes encoded proteins present in SM1 phage particles (Bensing et al., 2001b).

8.2.3. Proteins Required for Host Immune Avoidance

Once a pathogen has entered a human host, the first line of defense against bacterial infection is the innate immune response. Several lysogenic conversion genes encode proteins that alter bacterial recognition by the host immune system or that confer serum-resistance (Table 8.1). In this category are the O-antigen modification genes whose proteins alter bacterial antigenicity, thus facilitating bacterial evasion of the host immune system. The O-antigen modification genes include *rfb, oac*, and *gtr* of *E. coli* and *Shigella* spp., which express lipopolysaccharide O-antigen glucosylating, acetylating, and transferase proteins, respectively. The genes are located in several morphologically diverse bacteriophages of the families *Podoviridae*, *Siphoviridae*, *Myoviridae*, and *Inoviridae.* Genetically, the bacteriophages share several common features, such as the location of the O-antigen modification genes immediately downstream of the bacteriophage attachment site *attP*, which is preceded by the integrase *(int)* and excisionase *(xis)* genes in bacteriophages SfV, SfX, SfII, and Sf6 from *E. coli*, and in bacteriophage P22 from *S. enterica* (Allison and Verma, 2000; Vander Byl and Kropinski, 2000)) (Fig. 8.5). The integrase proteins of SfV, SfX,

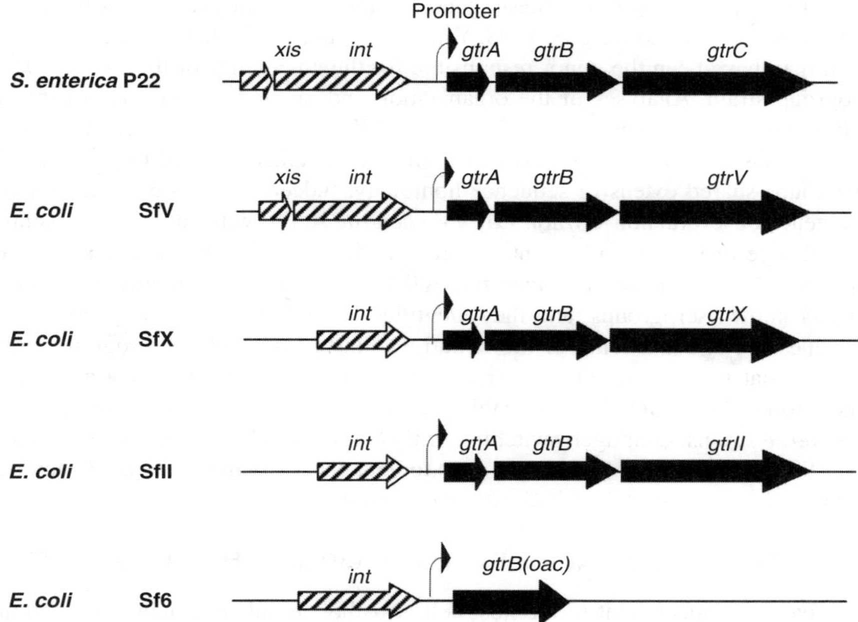

FIGURE 8.5 The O-antigen modifying, lysogenic conversion genes located between the phage lysin gene and the right attachment site in *Siphoviridae* phages for *E. coli* and *Salmonella*. The *int* and *xis* genes are are represented as striped arrows and the lysogenic conversion genes are black arrows. The promoter start point is indicated by a vertical arrow.

SfII, and SfI are very similar to the corresponding protein in phage P22, and the *attP* site of SfV, SfX, and SfII is identical to that of P22. The SfV phage's DNA packaging and head genes have the genetic organization of a lambda-like Siphovirus; however, its tail genes resemble those of a Mu-like Myovirus (Allison et al., 2002). Morphologically, SfV is a Myovirus; thus, chimeras between different classes of defined bacteriophages can also be carriers of lysogenic conversion genes.

The O-antigen plays a significant role in pathogenesis and is a major protective antigen, and the ability to switch O-antigen types has been acquired by many bacterial pathogens, such as *S. enterica, E. coli, V. cholerae, S. pneumoniae*, and *Neisseria meningitidis*. This capability can deleteriously affect vaccine development and efficacy, since vaccines may be rendered ineffective due to O-antigen switching. In *V. cholerae*, horizontal transfer of the O-antigen genes via a bacteriophage has been implicated in the emergence of the novel serogroup O139 that replaced the O1 serogroup as the major cause of epidemic cholera in Asia in 1992. Cholera caused by *V. cholerae* serogroup O139 infects more adults than the O1 serogroup, due to a lack of immunity in the adult population (Faruque et al., 2003c). Several studies have shown that the O139 serogroup strain was derived from an El Tor biotype *V. cholerae* O1 serogroup strain by the acquisition of O139 antigen genes (Bik et al., 1995; Waldor and Mekalanos, 1994; Mooi and Bik, 1997; Stroeher and Manning, 1997). The origin of the O139 serogroup biosynthesis genes is unknown; however, it has been suggested (Dumontier and Berche, 1998; Yamasaki et al., 1999) that an O22 serogroup strain may have been the donor responsible for the emergence of the Bengal O139 serogroup strain. Analyses of the organization and nucleotide sequence of the O-antigen biosynthesis genes of *V. cholerae* O1, O139, and O22 (Dumontier and Berche, 1998; Yamasaki et al., 1999) revealed a similar organization, and O139 and O22 serogroups shared extensive sequence homology. Indeed, there is evidence for the emergence of several non-O1/non-O139 *V. cholerae* strains with pathogenic potential by exchange of O-antigen biosynthesis regions in O1 El Tor and classical isolates (Li et al., 2002). The authors analyzed 300 *V. cholerae* strains representing all of the 194 known serogroups, and they identified several strains carrying *ctxAB* and *tcp* genes. DNA sequencing of the O-antigen cluster in one O37 serogroup strain revealed that most of the O1 serogroup *wbe* region was replaced by a novel *wbe* gene cluster, thus indicating a possible mechanism for the emergence of epidemic *V. cholerae* strains. O-antigen switching in *V. cholerae* has been proposed (Mooi and Bik, 1997; Stroeher and Manning, 1997) to be most likely mediated by a bacteriophage, as has been found with *E. coli* and *S. enterica.*

8.2.4. Proteins Essential for Bacterial Invasion of Host Cells

After bacterial attachment to the host cell, a pathogen either remains extracellular or invades the host cell to become intracellular. One example of an intracellular pathogen is *S. enterica*, a facultative pathogen that causes numerous infections, including typhoid fever, gastroenteritis, and septicemia. Many *S. enterica* serotypes are host-adapted; for example, serotype Typhi is restricted to humans, serotype Pullorum is strongly associated with chickens, and serotype Choleraesuis is strongly associated with pigs. The majority of human *S. enterica* isolates belong to subspecies I, whereas the other subspecies are mainly found in cold-blooded animals. *S. enterica* employs two specialized type III secretion systems that translocate effector proteins directly into the cytosol of its eukaryotic host cells, where they play a key role in the bacterial invasion process (Hansen-Wester et al., 2002). The type III secretion systems are encoded by two PAIs, *Salmonella* PAIs 1 and 2 (SPI-1 and SPI-2), which are found in all seven *S. enterica* subspecies (Hansen-Wester et al., 2002). Various combinations of effector proteins have been identified among *Salmonella* isolates, and many of those proteins, which are translocated either *via* SPI-1 (effector proteins SopB, SopD, SopE, SopE2, SspH1, and SlrP) or SPI-2 (effector proteins SspH1, SspH2, SlrP, GtgB, SseJ, SifA, and SifB), are encoded either adjacent to phage-like sequences (SopE2 and SspH2) or are found on lysogenic bacteriophages (SopE, GogB, SseI, and SspH1) (Fig. 8.6). The SopEϕ, a P2-like phage, encodes the effector protein SopE, which promotes invasion of tissue culture cells by *Salmonella* (Mirold et al., 1999). Among the lambda-like phages, Gifsy-1 encodes GogB, Gifsy-2 encodes GtgB (also called SseI or SfrH), and Gifsy-3 encodes SspH1 (Figueroa-Bossi et al., 2001). Also, the *sopE, gogB, gtgB*, and *sspH1* genes are located in the tail fiber-encoding regions of their respective bacteriophage genomes (Fig. 8.7). Lysogenic conversion by these and similar bacteriophages is probably responsible for the diversity of effector proteins observed among *S. enterica*, and the acquisition

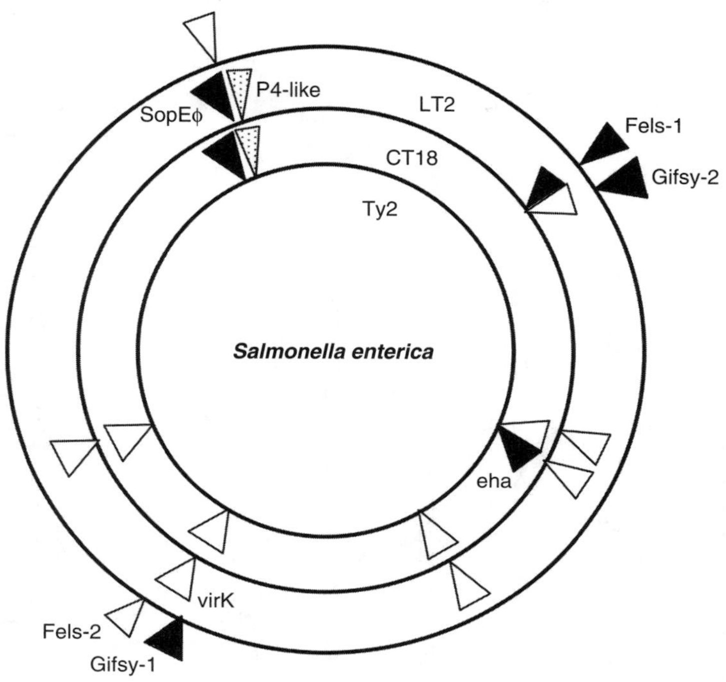

FIGURE 8.6 Distribution of prophages possessing virulence genes in *S. enterica* genomes. The outer circle represents the genome of serotype Typhimurium strain LT2, and the inner circles represents the genomes of serotype Typhi strain CT18 and Ty2, respectively. The prophages and suspected prophages are indicated as colored triangles (black triangles represent lambda-like phages and dotted triangles represent P4-like phages).

of various repertoires of lysogenic bacteriophage genomes could be an important step in the emergence of new epidemic clones and, perhaps, in the adaptation to new hosts.

In view of the many bacteriophages that encode effector proteins, it is not surprising to find a reassortment of bacteriophages among strains and a reassortment of effector proteins among bacteriophages (Hansen-Wester et al., 2002; Mirold et al., 2001). A case in point is the P2-like SopEϕ, which—over large parts of its genome—is closely related to Myovirus P2 from *E. coli*. At the *sopE* position, P2 encodes the bacteriophage resistance gene for protecting the lysogen against superinfection with coliphage T5 (Boyd and Brüssow, 2002). However, in other *Salmonella* strains, the *sopE* gene is found associated with lambda-like *Siphoviridae* (Mirold et al., 2001). The *sopE* integration site is occupied in other lambdoid *Salmonella* bacteriophages by still other lysogenic conversion genes (e.g., the type III effector *sseI*, the phagocytosis-activated *pagJ*, and the neuraminidase-encoding *nanH*) (Figueroa-Bossi et al., 2001). Sequence comparisons defined a *sopE* transfer cassette and *pagJ* flanked by a transposase (Mirold et al., 2001). Although there are a number of PAI-phage interactions among *S. enterica* serovar Typhimurium

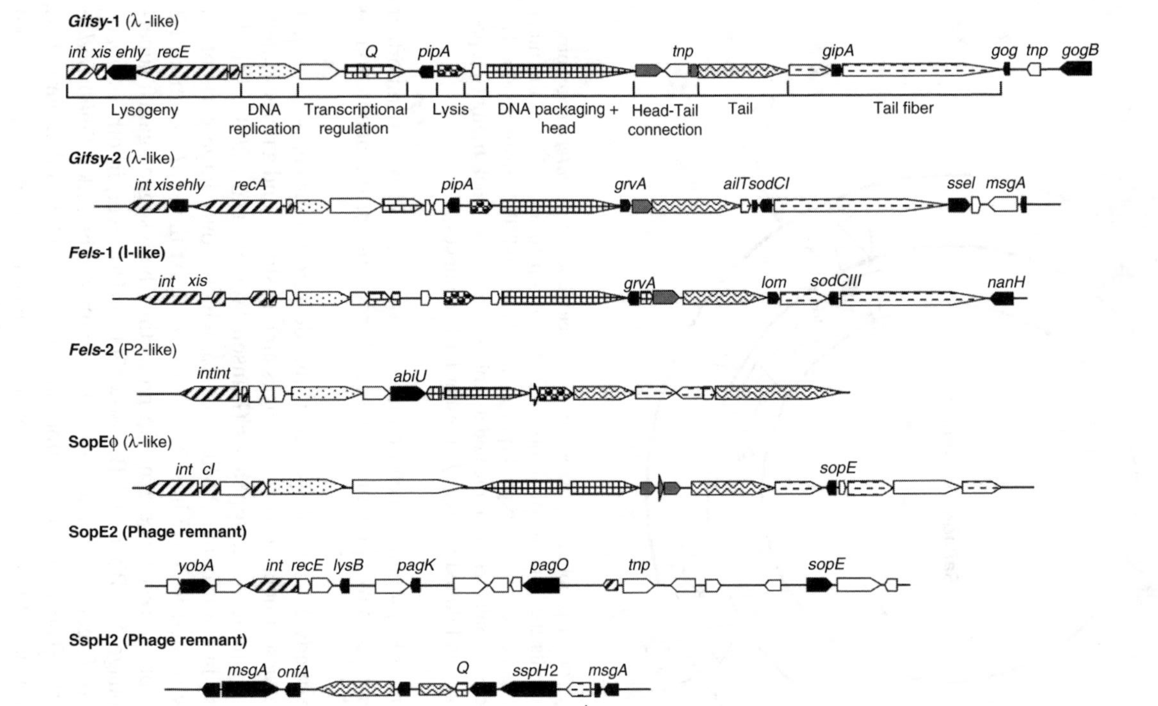

FIGURE 8.7 Genomic maps of five *S. enterica* prophages and two prophage remnants possessing genes that encode virulence factors. Prophages Fels-1 (STM0894-STM0928), Gifsy-2 (STM1005-STM1056), Gifsy-1 (STM2636-STM2584), and Fels-2 (STM2740-STM2694), and prophage remnants SspH2 (STM2230-STM2245) and SopE2 (STM1853-STM1870) are from the genome of *S. enterica* serotype Typhimurium (McClelland et al., 2001). The SopE (STY4645-4600) prophage is from the genome of serotype Typhi (Parkhill et al., 2001). To aid the comparison of the prophage maps, the encoding genes that may be attributed to modules are patterned: striped, for lysogenic conversion; dotted, for DNA replication; bricked, for transcription regulation; checkered, for DNA packaging and head proteins; gray, for head-to-tail proteins; wavy lined, for tail proteins; dashed lined, for tail fibers; bubbled, for lysis modules. Unattributed genes are not colored. Proven or suspected virulence genes are colored black and are annotated.

subspecies I isolates, in the other serovars and subspecies of *S. enterica* most of the Typhimurium-specific bacteriophages are absent or highly divergent (Porwollik et al., 2002; Boyd et al., 2003). This observation suggests that other serovars and subspecies of *S. enterica* may contain unique repertoires of phages that interact with SPI-1 and SPI-2. Analogous to the reassortment of *stx* genes among diverse bacteriophages in *E. coli*, reassortment of virulence genes among *Salmonella* bacteriophages may be an important mechanism for avoiding restrictions for bacteriophage-mediated gene transfer by superinfection immunity, competition for insertion sites, and restriction-modification systems.

8.2.5. Proteins Involved in the Survival of Intracellular Bacteria

Bacteriophages Gifsy-2 and Fels-1 encode an additional potential virulence factor, a periplasmic copper- and zinc-cofactored superoxide dismutase (Cu,Zn-SodC) (Fig. 8.7). The enzyme catalyzes the conversion of superoxide to hydrogen peroxide and molecular oxygen, thus protecting the bacterium from oxidation stress (Figueroa-Bossi et al., 2001). In *S. enterica*, three Cu,Zn-SodC-encoding genes, *sodCI, sodCII*, and *sodCIII*, have been identified (Figueroa-Bossi et al., 2001), and the amino acid sequence of SodCI shows 57% and 60% homology to that of SodCII and SodCIII, respectively. The chromosomally-encoded *sodCII* gene is present in all *Salmonella* isolates, and *sodCI* is carried on phage Gifsy-2 and is confined to strains belonging to the most pathogenic serotypes, which are exclusively subspecies I isolates (Fang et al., 1999; Figueroa-Bossi et al., 2001). The *sodCIII* gene was recently identified within the genome of phage Fels-1, in a location similar to that of *sodCI* in Gifsy-2 (Fig. 8.7) (Figueroa-Bossi et al., 2001). Carriage of SodC-encoding genes by *S. enterica* bacteriophages is not unique; e.g., a *sodC* gene has been identified (Ohnishi et al., 2001) in two lambda-like phages located at identical positions in the bacteriophage genome in *E. coli* O157 strain Sakai.

8.2.6. Putative Virulence Proteins

Neuraminidase (encoded by *nanH*) is another example of a possible phage-encoded protein that may play a role in bacterial virulence, by enhancing cellular survival of the bacterium. Phage Fels-1 from *S. enterica* serotype Typhimurium carries the *nanH* gene, which is located adjacent to the phage attachment site (Fig. 8.7) (Figueroa-Bossi et al., 2001). At the present time, the role of neuraminidase in the virulence of that species has not been rigorously evaluated. In the *V. cholerae* genome, the *nanH* gene is encoded on a 59-kb region designated VPI-2, which is unique to pathogenic isolates (Jermyn and Boyd, 2002). Interestingly, the VPI-2 region has a G+C content of 42%, which differs from the rest of the *V. cholerae* genome (47%), thus suggesting that the region was acquired by horizontal transfer. In addition, *nanH* is located adjacent to several genes that show significant homology to phage Mu (Fig. 8.8) (Jermyn and Boyd, 2002). The role of *nanH* in the pathogenesis of cholera has not yet been clearly defined either; however, the neuraminidase has been proposed (Galen et al., 1992) to act synergistically with CT (by converting higher-order

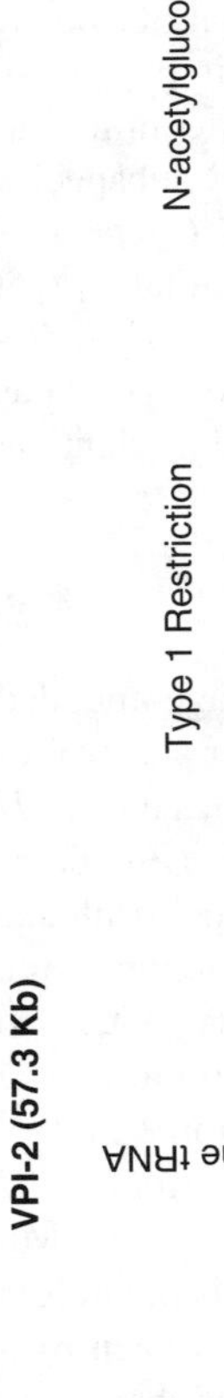

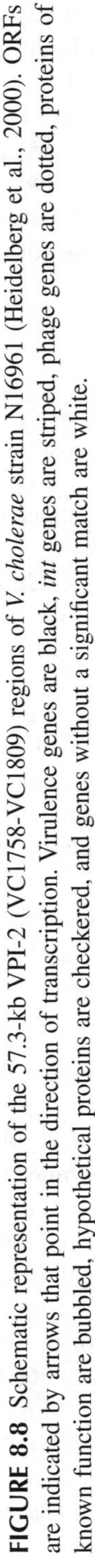

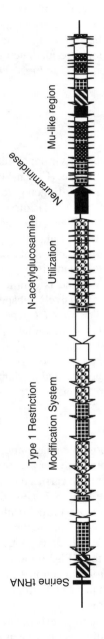

FIGURE 8.8 Schematic representation of the 57.3-kb VPI-2 (VC1758-VC1809) regions of *V. cholerae* strain N16961 (Heidelberg et al., 2000). ORFs are indicated by arrows that point in the direction of transcription. Virulence genes are black, *int* genes are striped, phage genes are dotted, proteins of known function are bubbled, hypothetical proteins are checkered, and genes without a significant match are white.

gangliosides to ganglioside GM1 [CT's receptor site]), thus increasing the binding of the toxin to host enterocytes. The *nanH* gene also is found in other bacterial genomes, where it also appears to have been acquired by horizontal transfer. For example, in *Clostridium perfringens, nanH* is integrated near an *attP* site, and it has a G+C content of 32%, compared to 27% for the *C. perfringens* genome.

8.3. MECHANISMS OF VIRULENCE GENE ACQUISITION BY BACTERIOPHAGES

In this section, three possible scenarios for the acquisition of virulence genes by bacteriophages are discussed (Table 8.2). As noted earlier in this chapter, the genes encoding the virulence factors may be (i) products of imprecise prophage excision from an ancestral host bacterium, (ii) transferable modules of extra bacterial DNA with their own promoters and terminators, or (iii) integral components of the bacteriophage genome.

8.3.1. Imprecise Prophage Excision

Most bacteriophage genes encoding bacterial virulence factors are not essential for bacteriophage replication, morphogenesis, or assembly. Additionally, since many virulence genes show an independent evolutionary history compared to the rest of the bacteriophage genome and are located near the bacteriophage attachment site, it is likely that bacteriophages acquired them by imprecise excision of the prophage from the bacterial host's genome. Imprecise prophage excision of toxin genes has been demonstrated for *S. pyogenes*; e.g., the genes encoding exotoxins SpeA and SpeC, encoded by phages T12 and CS112, respectively (Zabriskie, 1964; Goshorn and Schlievert, 1989), are located directly adjacent to the *attP* site, and they have their own promoter regions separate from the rest of the bacteriophages' regulatory control. Indeed, all sequenced *S. pyogenes* prophages possess lysogenic conversion genes at a specific position between the phage lysin and the right attachment site. A similar situation is found in *S. aureus* isolates, where the lysogenic conversion genes are located between the phage lysin-encoding site and the right attachment site (Desiere et al., 2001; Ferretti et al., 2001; Smoot et al., 2002). Further evidence in support of this model of independent acquisition of virulence genes via imprecise prophage excision comes from studies (Nakayama et al., 1999) of ϕCTX, a cytotoxin-converting bacteriophage of *P. aeruginosa.* The toxin gene *ctx* is located at the right bacteriophage attachment site and has a GC content that differs from the rest of the bacteriophage's genome. A potential variation of imprecise excision as a mechanism for virulence gene acquisition is imprecise prophage replication. In the CTXϕ genome, the *ctxAB* genes, which encode CT, are located adjacent to the bacteriophage's *attRS* site, and they have a distinct GC-content compared to the rest of the bacteriophage's genome (Fig. 8.3). However, since CTX virion production does not involve excision of the CTX prophage from the chromosome (Davis and Waldor, 2000), the pre-CTX genome most likely integrated adjacent to the *ctxAB* gene—and a CTXϕ possessing the *ctxAB* genes may have been generated *via* imprecise prophage replication. A precursor CTXϕ (pre-CTXϕ) has been identified in nontoxigenic isolates of *V. cholerae*, and it did not contain the *ctxAB* genes or the upstream control region containing the *ctxAB* promoter normally found 5′ of *ctxA* (Boyd et al., 2000a). Each pre-CTX prophage gave rise to a replicative plasmid form of the CTXϕ whose genomic organization was identical to that of the CTXϕ replicative form, except that it lacked the latter's *ctxAB* (Boyd et al., 2000a). The existence of both pre-CTXϕ and CTXϕ in *V. cholerae* populations raises the question of whether gene products other than CT encoded on CTXϕ impart a selective advantage to cells that carry the phage; i.e., perhaps the role of Ace and Zot in enterotoxicity is greater than previously appreciated.

8.3.2. Transferable Virulence Gene Cassettes

In contrast to the toxin-encoding genes described above, the putative virulence genes in *E. coli* and *Salmonella* prophages are not exclusively concentrated near the *att* sites; i.e, they are found inside the bacteriophage genome (Fig. 8.7). However, their position is not random, and most are found downstream of (i) anti-termination bacteriophage lambda N or Q homologs (Fig. 8.9), (ii) the lysis cassette (located at the center of the prophage map of lambda-like phages), (iii) the tail fiber-encoding genes involved in host recognition (which are frequently the target of recombination reactions), or (iv) the genes encoding the phage capsid protease. The virulence genes frequently represent separate transcriptional units flanked by their own promoter and terminator, which prevents transcription of parts of the prophage genome that have to be transcriptionally silent in order to maintain the prophage state. Since those virulence genes represent more genetic material than is found in comparable bacteriophages, the term *moron* has been proposed for this extra DNA (Hendrix et al., 1999).

DNA sequence analysis of the H-19B phage genome (Neely and Friedman, 1998b) revealed that the *stxAB* genes are downstream and in the same transcriptional orientation as a λ Q homology, which functions as a transcriptional activator at the late promoter P_R' of late phage genes. The *stx* genes either represent tightly controlled transcription units regulated by nutritional signals (low iron concentration, contact with eukaryotic host cells, etc.) or are coordinated with prophage induction. For example, Shiga toxin does not have transport signals but depends on bacterial lysis, after prophage induction in a small number of lysogenic cells, for export out of the cell (Wagner and Waldor, 2002). Thus, the location of *stx* between the Q antiterminator and the lysis cassette makes sense genetically (Fig. 8.9).

8.3.3. Integral Components of the Bacteriophage Genome

As stated previously, many bacteriophage genes that encode bacterial virulence factors appear to be remnants of an ancestral bacterial host, and they are not required for phage regulation and morphogenesis. An interesting exception is the PblAB phage's structural proteins that are also involved in attachment of *S. mitis* to human platelets, which may represent an example of proteins evolving dual functions (Bensing et al., 2001a; Bensing et al., 2001b). Similarly, in *V. cholerae* the CTXϕ's *ace* and *zot* gene products, which are phage morphogenesis and assembly proteins

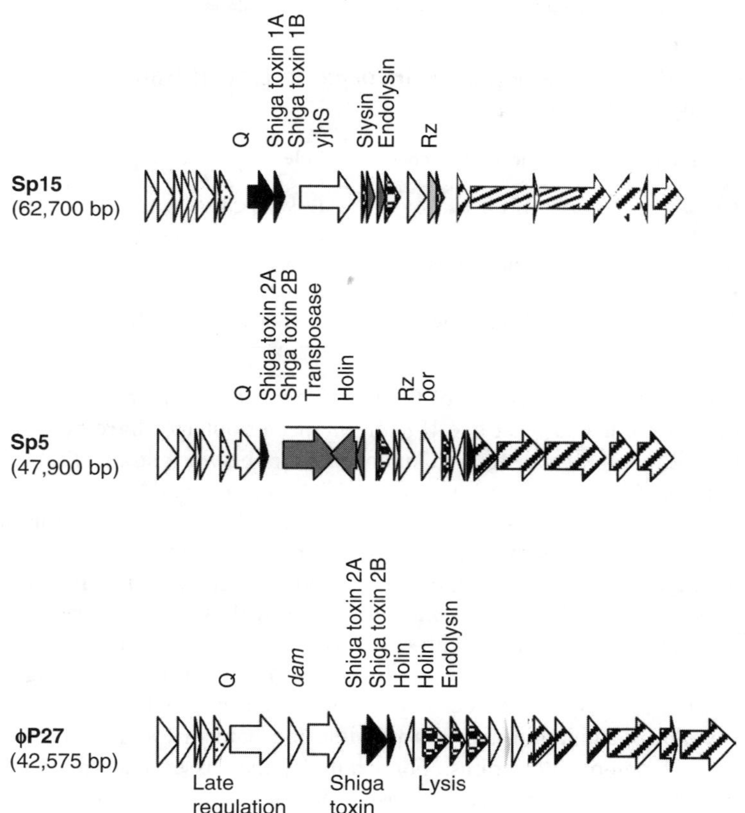

FIGURE 8.9 Genome location of the Shiga toxin genes (*stx1, stx2*) in three prophages of two *E. coli* O157 isolates (Recktenwald and Schmidt, 2002; Yokoyama et al., 2000). The antiterminator gene Q is indicated as a dotted arrow, the shiga toxin 1- and shiga toxin 2-encoding genes are black arrows, and the lysis genes are bubbled arrows.

(Waldor and Mekalanos, 1996), are also potent enterotoxins (Fasano et al., 1991; Trucksis et al., 1993).

8.4. ROLE OF BACTERIOPHAGES IN THE PATHOGENESIS OF BACTERIAL INFECTIONS

8.4.1. Indirect Involvement of Bacteriophages

Bacteriophages that encode virulence genes are generally considered passive vectors for the dissemination of those genes among bacterial populations, and they have been implicated in the dispersal of virulence factors and the emergence of new pathogenic strains in both Gram-positive and Gram-negative bacteria. For example, the CTXϕ is responsible for the horizontal transfer of the *ctxAB* genes between

TABLE 8.4
***V. cholerae* Bacteriophages Involved in Lateral Transfer of Virulence Factors**

Bacteriophage	Virulence Genes	Role
CTXϕ	*ctxAB, ace, zot*	Encodes CT, ACE, and ZOT toxins
RS1ϕ	*rstC*	Encodes antirepressor
KST-1ϕ	Helper phage	Transmits RS1ϕ
K139	*glo*	Encodes G-protein-like
CP-T1	Helper phage	Transmits RS1ϕ, CTXϕ, and VPI

V. cholerae isolates, and it is one of the key factors in the emergence of toxigenic strains. Indeed, to date at least five *V. cholerae* bacteriophages have been identified that contribute to lateral transfer of virulence factors between strains (Table 8.4). In addition, the *stxAB* genes in most *E. coli* STEC isolates reside in the genome of lambdoid prophages, and this probably accounts for the wide distribution of *stxAB* in more than 60 serotypes of *E. coli.* It is also possible that some of those strains arose via antigen switching, in which case a common pathogenic strain acquired a novel serotype. Also, *Shigella dysenteriae* strains with a clonal lineage of *E. coli* (i.e., they phylogenetically group with *E. coli* strains) have shiga toxin genes encoded on a lambdoid phage. Furthermore, the presence of numerous superantigen toxin-encoding genes in many streptococcal and staphylococcal prophages likely accounts for their wide dissemination and is a possible mechanism for the emergence and spread of new pathogenic strains. For example, the dissemination of the novel superantigen gene *speL*, in recent invasive and noninvasive *S. pyogenes* M3/T3 isolates in Japan, is likely to have been associated with bacteriophage carriage (Ikebe et al., 2002).

8.4.2. Direct Involvement of Bacteriophages

The bacteriophage life cycle can also be important in the pathogenesis of bacterial infections (reviewed in Wagner and Waldor, 2002). For example, phage-controlled regulation of the production and export of Stx toxins directly affects the pathogenesis of disease caused by STEC. In the *E. coli* O157 phage H 19B, the *stx* genes are located directly downstream of the PR′ promoter and upstream of the phage lysis genes (Neely and Friedman, 1998a; 1998b). Also, prophage induction by exogenous (e.g., mitomycin C and antibiotics) and endogenous (hydrogen peroxide released from human neutrophils) agents increases Stx production by their host strains (Wagner and Waldor, 2002). Functional studies (Neely and Friedman, 1998a) indicated that the Q protein of phage H-19B, acting in trans, directs high-level expression of Stx from repressed H-19B and 933W prophages. Transcription from the late phage promoter P_R', resulting from prophage induction, also is important for Stx production by the O157:H7 *E. coli* clinical isolate 1:361 (Wagner and Waldor, 2002).

CTX prophage induction also results in high toxin production by *V. cholerae.* CTXϕ either integrates into the bacterium's genome or replicates as an episome, and its replicative form (RF) yields high virion titers. Skorupski and Taylor (1997) have demonstrated that expression of *ctxAB* from CTX prophages is dependent on the transcriptional activators ToxR and ToxT. However, it has also been shown that *in vitro* and *in vivo* expression of CT by the CTXϕ RF does not require ToxR and ToxT (Wagner and Waldor, 2002). This observation suggests that CTX prophage induction may up-regulate toxin production during intestinal infection, by increasing the *ctxAB* copy number and by eliminating the requirement for ToxR and ToxT. This observation may also indicate the presence of as-yet-unidentified, alternative induction-related phage regulators of *ctx* transcription (Wagner and Waldor, 2002).

8.5. ROLES OF PHAGE-PHAGE INTERACTIONS IN BACTERIAL VIRULENCE

The previous section discussed the roles bacteriophages play in the pathogenic process, either as passive vectors for the dissemination of genes or as regulators of virulence gene expression. This section presents information concerning the roles of phage-phage interactions in bacterial virulence (Table 8.5) (reviewed in Boyd et al., 2001). The phage-phage interactions examined are divided into three categories, based on the requirement of a helper phage for (i) mobilization of another unrelated

TABLE 8.5
Various Categories of Phage-Phage Interactions Based on the Role of the Helper Bacteriophage

Helper Phage	Associated Phage	Bacterial Host	Reference
Helper phage supplies morphogenesis genes			
P2	P4	*E. coli*	(Lindqvist et al., 1993)
CTXϕ	RSIϕ	*V. cholerae*	(Faruque et al., 2002)
KST-1ϕ	RSIϕ	*V. cholerae*	(Faruque et al., 2003b)
Generalized transduction by helper phage			
80α	*tst* element	*S. aureus*	(Ruzin et al., 2001)
CP-T1	CTXϕ	*V. cholerae*	(Boyd and Waldor, 1999)
CP-T1	VPIϕ	*V. cholerae*	(O'Shea and Boyd, 2002)
Phage's host receptor is another phage			
VPIϕ	CTXϕ	*V. cholerae*	(Karaolis et al., 1999)
Helper phage potentates expression of virulence genes			
VPIϕ	CTXϕ	*V. cholerae*	(Lee et al., 1999)
RS1ϕ	CTXϕ	*V. cholerae*	(Davis et al., 2002)
Gifsy-2	*Gifsy*-1	*S. enterica*	(Figueroa-Bossi et al., 2001)
λ-like	λ-like	*E. coli*	(Sandt et al., 2002)

bacteriophage, (ii) entry of another bacteriophage into the host cell, or (iii) expression of bacteriophage-encoded virulence genes.

8.5.1. Helper Phage Facilitation of Bacteriophage Mobilization

8.5.1.1. *V. cholerae* CTXϕ Supplies Morphogenesis Genes for RS1ϕ

The classical example of the mobilization of a defective phage by a helper phage is the interaction between the *E. coli* lambdoid phages P2 (helper) and P4 (defective). P4 is dependent on P2, or on a related lambdoid bacteriophage, to supply the capsid, tail, and lysis genes required to assemble its capsid, package its DNA, and lyse the bacterial host cell. These requirements are essential for P4 to complete a lytic cycle and produce virions (Christie and Calendar, 1990; Lindquist et al., 1993). Therefore, P4 is a defective phage, which, once released as a virion, can—in the absence of P2 .—replicate and integrate its DNA within a host bacterium.

The requirement of RS1ϕ for the morphogenesis proteins of *V. cholerae*'s CTXϕ is analogous to the P2/P4 interaction in *E. coli. V. cholerae* strains responsible for contemporary cholera epidemics (i.e., biotype El Tor O1 serogroup and O139 serogroup strains) contain fully functional prophages that can produce the infectious RF of CTXϕ (Davis et al., 1999). In those isolates, the CTX prophage is always flanked by another genetic element, RS1. RS1 is identical to the RS2 region but contains an additional open reading frame, *rstC*, that encodes an anti-repressor that causes the RstR repressor to aggregate (Fig. 8.3), thus affecting the regulation and transmission of the CTXϕ (Davis et al., 2002). The production of CTXϕ virions by a lysogen depends on the presence of either a tandem array of CTX prophages or a single prophage followed by an RS1 element (Davis et al., 2000a). A recent study of RS1 demonstrated that it propagates horizontally as a filamentous bacteriophage that exploits the morphogenesis genes of CTXϕ (Faruque et al., 2002). The authors also confirmed the presence of an excised copy of RS1, and they showed that a kanamycin (Kn)-marked copy of that RF of RS1, pRS1-Kn, transduced *V. cholerae* isolates to Kn^r (Faruque et al., 2002). The isolates carried a single-stranded form of pRS1-Kn; therefore, they resembled the genome of a filamentous bacteriophage (RS1-Knϕ). Similar to CTXϕ, the RS1ϕ genome uses the *attRS* sequence to integrate site-specifically into the *V. cholerae* genome. However, only transductants of RS1-Knϕ, which already harbor the CTXϕ genome, can produce detectable RS1-Knϕ (Faruque et al., 2002). A second filamentous phage, KSF-1ϕ, which also provides functions required for RS1ϕ production, has been identified (Faruque et al., 2003b). However, unlike the CTXϕ, it does not require TCP for entry into *V. cholerae.* Thus, a hypothetical scenario for the evolution of CTXϕ would appear to have involved three main steps (Fig. 8.10): (i) a precursor-CTXϕ—pre-CTX) emerged when the RS1ϕ integrated adjacent to the core region sequence (presumably with the loss of *rstC*), thus forming the RS2 and core domains, (ii) the pre-CTXϕ infected an unknown bacterial host that contained the *ctxAB* genes, and (iii) the toxigenic CTXϕ was formed by a mechanism similar to imprecise excision. CTXϕ propagation does not involve excision

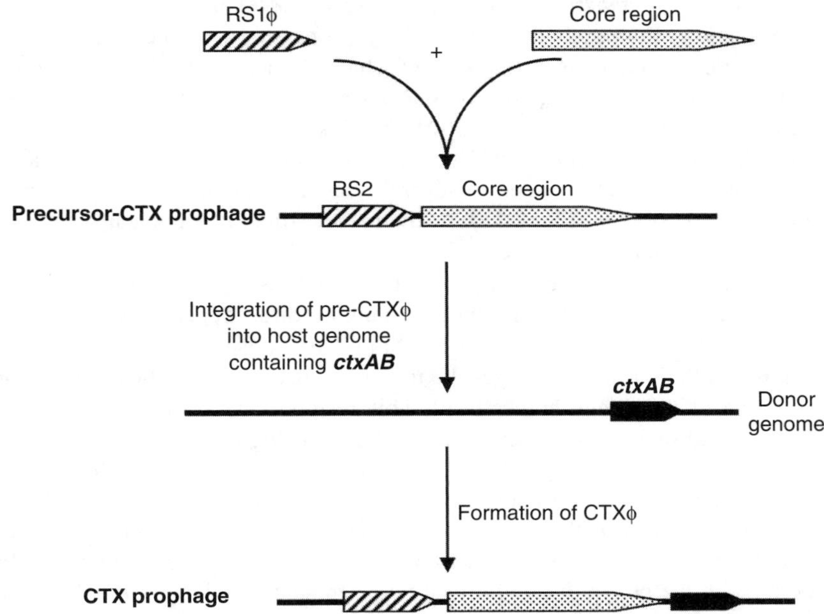

FIGURE 8.10 Model for the emergence of precursor-CTXϕ and CTXϕ. The 2.7-kb *rstR, rstA* and *rstB* genes of the RS1ϕ share near sequence identity with the 2.4-kb *rstR, rstA, and rstB* genes in the RS2 region of the CTX$\tilde{\phi}$. The RS1ϕ contains an additional gene (*rstC*) not present in the RS2 region. The precursor-CTXϕ probably arose when an RS sequence integrated next to the core region sequence in an unknown host. The pre-CTXϕ then infected and integrated next to the *ctxAB* genes in a new, unknown donor host. The RS region is indicated by an orange-filled arrow, the core region is indicated by a dotted arrow, and the CT-encoding genes are indicated by a black arrow.

of the CTX prophage from the chromosome but it is similar to propagation of replicative transposons or Mu phage (Davis and Waldor, 2000). As stated earlier, production of CXTϕ requires the presences of tandem elements, either CTX-CTX prophages or CTX-RS1 prophages. Hence, the CTXϕ is not present in isolates of the classical *V. cholerae* biotype, which do not contain tandem CTX prophages or the RS1 prophage (Davis et al., 2000b). However, the RS1 prophage is present in all *V. cholerae* El Tor isolates, which may explain the predominance of this biotype in current cholera epidemics. The interaction between CTX and RS1 prophages, which promotes the production of infectious CTXϕ, is important because it increases the dissemination of *ctxAB* among *V. cholerae* isolates.

8.5.1.2. *S. aureus* ϕ80 Generalized Transduction of the tst Element

The *tst* gene in *S. aureus* encodes toxic shock syndrome toxin-1 (TSST-1), a potent superantigen responsible for toxic shock syndrome associated with some *S. aureus* infections. In *S. aureus* strain RN4282, the 15-kb *tst* element is not capable of self-mobilization. Rather, it is mobilized by propagation of the staphylococcal generalized transducing phages 13 and 80α, which efficiently encapsidate and transduce the *tst* element (Lindsay et al., 1998; Ruzin et al., 2001). Novick and colleagues (Lindsay et al., 1998; Ruzin et al., 2001; Novick, 2003) have proposed that the 15-kb *tst* element in *S. aureus* RN4282 is a PAI, which they called SaPI-1, a term which has recently been extended to several related elements encoding superantigens in *S. aureus* isolates (Fig. 8.11). At the present time, it is unclear whether the SaPI-1 truly falls under the rubric of PAIs, because the *tst* elements it contains are characterized by certain phage-related features, such as encoding integrases, helicases and terminases, and the presence of flanking direct repeats (Novick, 2003). In strain RN4282, phage 80α induces the SaPI-1 to excise, replicate and encapsidated, at high efficiency, into phage-like infectious particles with heads about one-third the size of the phage helper head. This results in very high frequencies of transfer. In the absence of a helper phage the island is highly stable in the *S. aureus* genome. The high frequency, 80α-dependent mechanism outlined for SaPI-1's transfer in strain RN4282 probably is responsible for the horizontal spread of the *tst* gene among clinical isolates of *S. aureus*. The close association between SaPI-1 and its helper phage suggests that they are genetically related. Indeed, SaPI-1 may be a defective phage that requires a helper phage, similar to the P2/P4 interaction and the CTXϕ/RS1ϕ interaction (Boyd et al., 2001; Ruzin et al., 2001; Faruque et al., 2002). The SaPI-1 in *S. aureus* isolates is one of the first PAIs characterized in Gram-positive bacteria.

8.5.1.3. *V. cholerae* CP-T1 Generalized Transduction of VPI

Recently, O'Shea and Boyd (2002) reported that transfer of the VPI region between O1 serogroup strains could be mediated by generalized transduction with phage CP-T1. A Kn-labeled copy of the VPI was transferred, with high efficiency, from a *V. cholerae* El Tor biotype strain to four VPI-negative *V. cholerae* O1 serogroup strains via generalized transduction with phage CP-T (the entire 39.5-kb VPI region was transduced and integrated site-specifically at the same chromosomal position in the four recipient strains). These observations suggest that generalized transduction is the main mechanism for the transfer of VPI under natural conditions (O'Shea and Boyd, 2002).

8.5.1.4. *V. cholerae* CP-T1 Generalized Transduction of CTXϕ

Several studies (Faruque et al., 1998a; 1998b; Mekalanos et al., 1997; Waldor and Mekalanos, 1996) indicate that two sequential steps are essential for the evolution of pathogenic *V. cholerae* strains. The bacteria must first acquire the *tcp* operon, which encodes TCP (the receptor for CTXϕ, and they must then be lysogenized by CTXϕ (Fig. 8.12). A few *V. cholerae* O1 and non-O1 isolates that lack the *tcp* genes but contain the CTX prophage have been described (Ghosh et al., 1997; Said et al., 1995). Some authors (Faruque et al., 1998a; 1998b) have suggested that those isolates

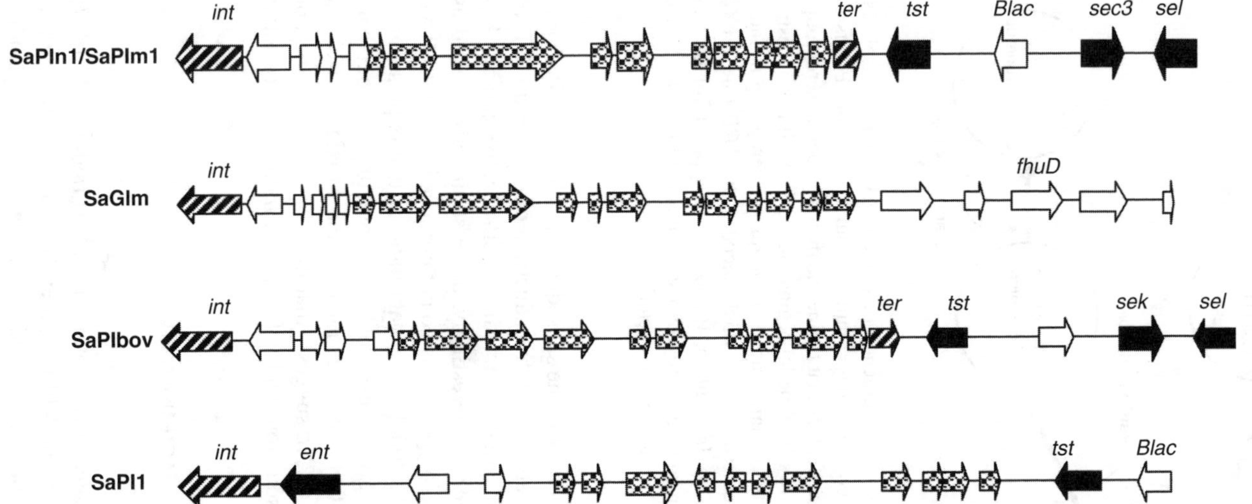

FIGURE 8.11 Genomic maps of three PAIs encoding the TSST1 toxic shock toxins of *S. aureus* (Kuroda et al., 2001). The *int* genes are represented by striped arrows, virulence genes as black arrows, and bacteriophage-like genes as dotted arrows. Genes of unknown function are indicated as white arrows. The SaGlm element does not encode any known virulence factor but it contains a central genomic region (indicated as bubbled arrows) that shares sequence identity with SaPI-1.

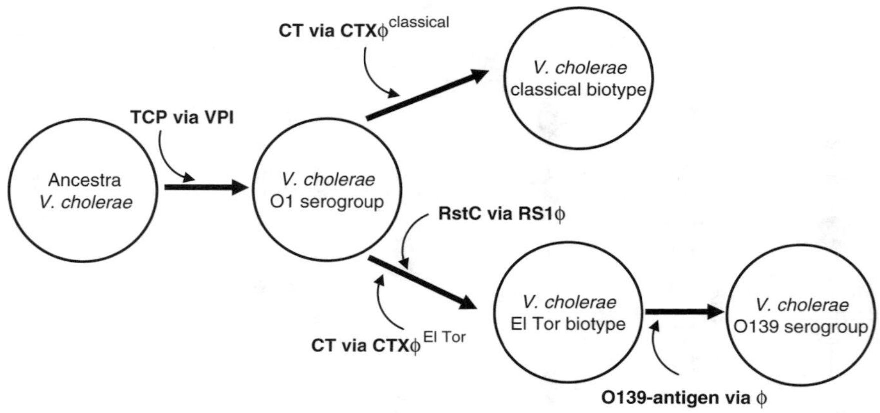

FIGURE 8.12 Evolutionary scenario for the emergence of epidemic *V. cholerae* isolates. From left to right, circles represent the core genome common to all *V. cholerae*. Points of acquisition of prophage-encoded virulence factors that define the major epidemic serogroups of *V. cholerae* are indicated by curved arrows. The model begins (on the left) with an ancestral *V. cholerae* progenitor, which gave rise to contemporary *V. cholerae* O1 serogroup isolates. The *V. cholerae* O1 progenitor first acquired VPI-1, which encodes the TCP (the CTX*ϕ* receptor), and were then infected by CTX*ϕ*. In addition, *V. cholerae* biotype El Tor isolates also acquired RS1*ϕ*. The *V. cholerae* Bengal O139 serogroup strain emerged from an El Tor strain by O-antigen switching probably mediated *via* a bacteriophage.

arose by TCP-mediated CTX*ϕ* infection, with the subsequent loss of the TCP region. However, CP-T1 phage-mediated generalized transduction of the entire CTX*ϕ* genome to both classical and El Tor isolates, including TCP-negative strains, has been reported (Boyd and Waldor, 1999). Interestingly, those strains also served as CTX*ϕ* donors, even though classical biotype strains do not produce infectious CTX*ϕ* (Boyd and Waldor, 1999; Davis et al., 2000b). The data indicate that expression of a specialized CTX*ϕ* receptor is not always essential for converting nontoxigenic strains to toxigenic strains. Also, the fact that the genomes of most toxigenic isolates of *V. cholerae* encode both TCP and CTX*ϕ* suggests that most of the strains did not arise as a result of generalized transduction, and it favors the model of sequential evolution of virulence.

8.5.2. Helper Phage Source of Bacteriophage Host Receptor

As stated previously, the TCP is *V. cholerae*'s receptor for CTX*ϕ* (Taylor et al., 1987; Waldor and Mekalanos, 1996). Thus, if the VPI region encodes a phage receptor, it appears that one bacteriophage's receptor is encoded within the genome of another bacteriophage. Thus far, the finding that the VPI region encodes the receptor (and perhaps is the receptor) for CTX*ϕ* is the only example of this type of phage-phage interaction, and this example may prove to be unique (reviewed in Lee, 1999).

8.5.3. Helper Phage Requirement for Virulence Gene Expression

8.5.3.1. RS1ϕ Encodes an Antirepressor Involved in CT Production Mediated by CTXϕ

As discussed above, the CTXϕ-related element RS1ϕ is a satellite phage whose transmission depends upon proteins produced by a helper CTX prophage. Unlike other satellite phages, RS1ϕ aids the CTX prophage, due to the RS1ϕ-encoded protein RstC. RstC is an antirepressor that counteracts the activity of the CTX repressor RstR. Therefore, RstC promotes transcription of CTX genes required for phage production and, hence, promotes transmission of both RS1ϕ and CTXϕ. In addition, RstC induces expression of *ctxAB* (which encode CT), thus contributing to *V. cholerae*'s virulence (Davis et al., 2002).

8.5.3.2. The VPI Encodes Regulatory Genes Required for CT Expression

The expression of *V. cholerae* genes in the CTX prophage and the VPI are regulated by a complex interaction of VPI genes (*toxT, tcpP,* and *tcpH*) and chromosomal genes (*toxS* and *toxR*) controlled by growth conditions and environmental signals. The three proteins ToxR, TcpP, and ToxT coordinately regulate transcription of the structural genes for CT and TCP. ToxR, a transcription activator, works synergistically with ToxS to activate expression of the VPI loci *tcpP, tcpH*, and *toxT*, and the CTXϕ loci *ctxAB* (reviewed in Davis and Waldor, 2003). ToxT amplifies its own expression and, together with TcpP and TcpH, directly activates production of CT and TCP (Davis and Waldor, 2003). Most of the experimental evidence for this virulence-regulating cascade is based on *in vitro* data. In that regard, *in vivo* studies (Lee et al., 1999) showed that the *in vivo* requirement for ToxR and TcpP, to regulate expression of CT and TCP, differs significantly from the *in vitro* requirement. An intimate interaction between VPI and CTXϕ gene expression has been demonstrated (Lee et al., 1999) using recombinase-based *in vivo* expression technology (RIVET) to monitor transcription of *ctxA* (encodes the catalytic subunit of CT) and *tcpA* (encodes the major TCP subunit) during infection in the infant mouse model of cholera. Production of TCP preceded CT production, and CT expression was induced only after increased levels of ToxR and ToxT appeared. The observed temporal progression suggests a mechanism that enables *V. cholerae* to delay CT release until the bacterium is close to host cells. The above-cited study (Lee et al., 1999) highlights the co-evolution of CTXϕ and VPI, and the importance of phage-phage interactions in the pathogenesis of cholera.

8.5.3.3. Interactions of *S. enterica* Gifsy-1 and Gifsy-2 Phages

As discussed in previous sections, the phages Gifsy-1 and Gifsy-2 encode several virulence genes essential for the pathogenesis of *Salmonella* infections (Figueroa-Bossi et al., 2001). Gifsy-2 encodes GtgB (a translocated effector protein) and SodC1, a superoxide dismutase. Curing *S. enterica* of the Gifsy-2 prophage has been reported (Figueroa-Bossi et al., 2001) to result in substantial attenuation of its mouse virulence, but under most circumstances, Gifsy-1-cured strains were fully virulent for mice. However, another study found that Gifsy-1 lysogens that contained *sodCI*, but lacked the remainder of Gifsy-2's genes, were more virulent than were isogenic strains lacking a Gifsy-1 prophage (Figueroa-Bossi et al., 2001). Also, the authors proposed that Gifsy-1 might encode virulence factors whose effects are dependent upon *sodCI*; that is, Gifsy-1 may contribute to virulence simply by enhancing expression of SodCI. Interestingly, four potential virulence genes are found in Gifsy-1: (i) *gogB*, a type III secretion system protein, (ii) *gogD*, a *pagJ* homolog that functions to increase *Salmonella* survival after phagocytosis, (iii) *gipA*, required for *in vivo* survival of *Salmonella* in Peyer's patches (Stanley et al., 2001), and (iv) *ehly*, an enterohemolysin (Fig. 8.7). The *gogD, pagJ,* and *gogD* genes are integrated adjacent to the right attachment site, and *pagJ* and *gogB* are on either side of a transposase, which indicates a possible mechanism of acquisition (Fig. 8.7). The e*hly* gene is integrated downstream of the *int* and *xis* genes, the same location where a similar gene is located in *E. coli* O157 prophage 933W (Plunkett et al., 1999). An antivirulence gene (*grvA*) is also carried on Gifsy-2, and both its deletion and overexpression increases virulence in a mouse model (Ho and Slauch, 2001). The complex interplay of partial redundancy (Gifsy-1 and Gifsy-2 prophages), and the possible interactions among the virulence factor-encoding genes (*grvA* and *sodC*), complicated a study (Ho and Slauch, 2001) to determine the contribution of each factor to animal virulence. Nevertheless, the results obtained by Ho and Slauch (2001) illustrate the potentially subtle ways in which interactions between phage-encoded proteins may contribute to bacterial virulence.

8.5.3.4. *E. coli* Prophage Enhancement of *eib* Expression

Four distinct *E. coli* immunoglobulin-binding (*eib*) genes are carried by separate prophages in *E. coli* strain ECOR-9 (Sandt et al., 2002). The Eib proteins confer resistance to human serum complement, and the expression of the *eib* genes is significantly enhanced by the overlapping genes *ibrA* and *ibrB*. The IbrA and IbrB proteins are very similar to proteins encoded within a prophage-like element in *E. coli* strain Sakai, and the genome segment containing *ibrA* and *ibrB* in strain ECOR-9 contains regions homologous to the Shiga-toxin-converting prophage (Sandt et al., 2002). These observations indicate an interdependence of gene products encoded by separate bacteriophages in *E. coli*. The *ibrAB* genes were found in all strains encoding Eib, and in most strains of the B2 phylogenetic lineage, which suggests that this phage-phage interaction is not a unique event.

8.6. BACTERIOPHAGES AND PATHOGENICITY ISLANDS (PAIS)

Many pathogenic bacteria encode PAIs (Hacker and Kaper, 2000), which possess a set of common unifying characteristics; for example, they are large chromosomal regions (35–200 Kb) encoding several virulence factors that, in general, are present in pathogenic strains and are absent from nonpathogenic strains. In many cases, they also encode a phage-like integrase (e.g., P4-like integrases (Table 8.6), integrate

TABLE 8.6
Bacteriophage-Related Integrases Associated with PAI and Genetic Islands

Phage-related Integrase	tRNA	Gene	PAI	Species	Reference
ϕR73, P4	*serV*	*vap*	PI	*D. nodosus*	(Cheetham et al., 1995)
SF6, P4, ϕR73	*selC, pheU*	*eaeB, esp*	PAI III	*E. coli*	(McDaniel et al., 1995)
ϕR73	*selC*	*hly*	PAI I	*E. coli*	(Blum et al., 1994)
P4	*leuX*	*hly, prf*	PAI II	*E. coli*	(Blum et al., 1994)
P4	*pheV*	*hly, pap*	PAI IV	*E. coli*	(Swenson, 1996)
P4	*pheR*	*hly, prs*	PAI V	*E. coli*	(Swenson, 1996)
P4	*pheR*	*afa*	PAIAL862	*E. coli*	(Lalioui et al., 2001)
P4	*asnT*	*fyuA-irp*	HPI-2	*E. coli*	(Schubert et al., 1998)
CP4	*ser*	Iha, urease	SpLE1	*E. coli* O157	(Ohnishi et al., 2001)
CP4	*ser*	LEE	SpLE4	*E. coli* O157	(Ohnishi et al., 2001)
P4	*pheU*	*aer*	SHI-3	*S. boydii*	(Purdy and Payne, 2001)
P4	*selC*	*aer*	SHI-2	*S. flexneri*	(Moss et al., 1999)
P4	*selC*	*aer*	SHI-2	*S. sonnei*	(Vokes et al., 1999)
CP4-57	*leu*	NA	HiGI1	*H. influenzae*	(Chang et al., 2000)
P4	*phe*	Symbiotic genes		*M. loti*	(Sullivan and Ronson, 1998)
Mu	NA	Phage genes	Region 3	*N. meningitidis*	(Klee et al., 2000)
L54a	-	*tst*	SaPI1	*S. aureus*	(Lindsay et al., 1998)
T12, T270	-	*tst, sec, sel*	SaPIbov	*S. aureus*	(Fitzgerald et al., 2001)
ϕR73	*selC*	Aerobactin	SHI-2	*S. flexneri*	(Moss et al., 1999)
P22	NA	O-antigen	NA	*S. flexneri*	(Adhikari and Berget, 1993)
Mu-like	*ser*	*nanH*	VPI-2	*V. cholerae*	(Jermyn and Boyd, 2002)
P4	*asnT*	*fyuA-irp*	HPI-1	*Y. enterocolitica*	(Carniel et al., 1996)
CP4	*asp*	*pgm,hms*	HPI-2	*Y. pestis*	(Cheetham et al., 1995)

adjacent to tRNA genes, and are flanked by repeat sequences). The percent G+C contents of PAIs usually differ from those of the host genome, which is indicative of foreign DNA acquired from another source (Hacker and Kaper, 2000). The mechanism(s) by which PAIs are transferred from one bacterial strain to another has not been elucidated, and PAIs do not encode gene products required for self-mobilization. Among *V. cholerae* isolates, the VPI region, which encodes TCP, has been proposed to correspond to the genome of filamentous phage named VPIϕ (Karaolis et al., 1999). However, as noted above, the existence of VPIϕ is a matter of controversy, and even the authors of the original report (Karaolis et al., 1999) did not observe VPIϕ transfer to *V. cholerae* O1 serogroups strains, the predominant cause of epidemic cholera. Nonetheless, the horizontal transfer of this region among strains is not in question, and recent sequence homology studies (Boyd et al., 2000) strongly suggest that the region was recently transferred between *V. cholerae* and *V. mimicus*. In addition, O'Shea and Boyd (2002) showed that a Kn-marked copy of the VPI from several *V. cholerae* isolates could be transferred to *V. cholerae* recipient strains by the generalized transducing phage CP-T1. Another example of phage-mediated transfer of a PAI involves SaPI-1, which encodes the toxic shock toxin in *S. aureus*. Staphylococcal phage 80a is required for excision, replication, and encapsidation of SaPI-1 (Lindsay et al., 1998). The observation that generalized transduction is the preferred mode of transfer for the SaPI-1 of *S. aureus* is consistent with the idea that generalized transduction of PAIs is the most plausible mechanism of transfer in nature (Lindsay et al., 1998; O'Shea and Boyd, 2002).

Incorrectly designating PAIs as phages (and vice-versa) is an impediment to our understanding of the origins and evolution of PAIs. The VPI clearly falls into the PAI category but the SaPI-1 element does not clearly conform to PAI criteria. Since many of the characteristics of PAIs can also be attributed to bacteriophages (for example, encoding an integrase, integration adjacent to a tRNA gene, flanking by repeat sequences, aberrant GC contents), it is possible that some PAIs may have a bacteriophage origin; however, most do not.

Some PAIs that are mosaic structures containing PAI and phage characterisitics have been identified; e.g., the VPI-2 region of *V. cholerae* (Fig. 8.8) (Jermyn and Boyd, 2002). The VPI-2 region encodes a restriction modification gene cluster, a gene cluster required for amino sugar utilization, and numerous genes encoding hypothetical proteins in the 5′ region of the island. Also, a 20-kb region in the 3′ end of the island contains several genes that show homology to Mu phage. Interestingly, only the amino sugar utilization genes are present in *V. mimicus* and *V. vulnificus*, and the only VPI-2 region present in most *V. cholerae* O139 isolates is the 20-kb region containing Mu-remnants (Jermyn W.S., personal communication). An additional factor complicating the separation of bacteriophages and PAIs is the interdependence between the two elements in many bacterial species (Boyd et al., 2001). For example, the effector protein GtgB in *S. enterica* is encoded by the Gifsy-2 phage, and it is secreted by the type III secretion system encoded by the SPI-2 PAI (Figueroa-Bossi et al., 2001).

A confusing aspect of the debate regarding the relationship between bacteriophages and PAIs is the tendency of some researchers to label cryptic prophages and prophage remnants as islands. Regions that predominantly encode phage structural, morphogenesis, and regulatory gene products should not necessarily be considered to be PAIs. The common features of PAIs and bacteriophages (e.g., they insert at tRNA sites, and they both contain a phage-like integrase) may reflect a general mode of transfer and a common integration mechanism rather than a common origin. For example, the fact that many bacteriophages integrate adjacent to tRNA genes may reflect the abundance and sequence conservation of those regions rather than some common ancestry with PAIs. Indeed, the presence of P4-like integrases within PAIs has led to the suggestion (Boyd et al., 2001) that some P2/P4 type interaction may have contributed to the integrases' acquisition and transfer. Finally, the ability of generalized transducing phages to transfer large regions of DNA in somePAIs may explain the close association between phages and PAIs.

8.7. CONCLUSIONS

This chapter focuses on the significant contributions of various bacteriophages to bacterial virulence and bacterial evolution. In recent years, our increased interest in the important roles that bacteriophages play in bacterial virulence has been driven by two very different fields of research. First, molecular microbial pathogenesis studies have discovered novel virulence factors and have elucidated their regulatory factors and mechanisms of action. Second, bacterial whole genome and comparative analyses have led to the discovery of numerous prophages within bacterial genomes (reviewed in Canchaya et al., 2003), many of which encode bacterial virulence factors. Our appreciation for the range of phage-encoded bacterial virulence factors, and the distribution and diversity of the bacteriophages involved, will increase as more bacteriophage genomes are characterized.

Two of the major themes emerging from comparative genomic analyses of both Gram-negative and Gram-positive bacteria are that prophages are quantitatively an important component of bacterial genomes, and that they are the major cause of strain differences in pathogenic isolates of various bacteria, including *E. coli*, *S. enterica* serovar Typhimurium and Typhi, *S. pyogenes*, and *S. aureus*. In addition, bacteriophages acting as vectors in the lateral transmission of bacterial virulence factors are potent factors in the emergence of new strains with pathogenic potential (Table 8.3 and Table 8.4). Although the phages' presence may potentially increase bacterial fitness, especially in certain niches such as the human intestines, the advantages or benefits of phage-encoded virulence factors for phages or bacteria are often unclear. Therefore, it is important to recognize that what we label as phage-encoded virulence factors may, in fact, have evolved for alternative and unrelated functions, and their roles in virulence may just be accidental consequences of their presence.

The advancement of knowledge concerning the roles of bacteriophages in the pathogenesis of bacterial infections also may be of significant practical importance. In that regard, the emergence of many antibiotic-resistant mutants of pathogenic bacteria recently has rekindled interest in using phages prophylactically and therapeutically in a variety of agricultural and clinical settings (for more details on this subject, refer to Chapters 13 and 14). Thus, improving our understanding of the roles of various phages and phage-encoded genetic loci in bacterial pathogenesis can be invaluable for developing safe and effective phage preparations free of "undesirable genes" (i.e., genes encoding bacterial virulence factors; see Table 8.1). Phage preparations developed for use in agricultural and clinical settings must be free of undesirable genes, in order to prevent or limit phage-mediated emergence of new pathogenic bacterial strains.

ACKNOWLEDGMENTS

I thank my colleagues who have provided ideas and suggestions that have made this review possible. Also, I thank members of my laboratory for their hard work and inspiration. I am also grateful to the Department of Microbiology, UCC, National University of Ireland, Cork for continued support. Due to space constraints, literature citations have been limited, in some cases to recent relevant reviews. Therefore, I apologize to those authors whose important work has not been included or cited. Research in my laboratory is supported by Enterprise Ireland basic research grants and a Higher Education Authority PRTLI-3 grant.

REFERENCES

Adhikari, P. and P.B. Berget (1993) Sequence of a DNA injection gene from *Salmonella typhimurium* phage P22. *Nucleic Acids Res*, 21, 1499.

Allison, G.E., Angeles D., Tran-Dinh N. and Verma N.K. (2002) Complete genomic sequence of SfV, a serotype-converting temperate bacteriophage of *Shigella flexneri*. *J Bacteriol*, 184, 1974–1987.

Allison, G.E. and N.K. Verma (2000) Serotype-converting bacteriophages and O-antigen modification in *Shigella flexneri*. *Trends Microbiol*, 8, 17–23.

Banks, D.J., S.B. Beres and J.M. Musser (2002) The fundamental contribution of phages to GAS evolution, genome diversification and strain emergence. *Trends Microbiol*, 10, 515–521.

Barksdale, L. and S.B. Arden (1974) Persisting bacteriophage infections, lysogeny, and phage conversions. *Annu Rev Microbiol,* 28, 265–299.

Barondess, J.J. and J. Beckwith (1990) A bacterial virulence determinant encoded by lysogenic coliphage lambda. *Nature*, 346, 871–874.

Bensing, B.A., C.E. Rubens and P.M. Sullam (2001a) Genetic loci of *Streptococcus mitis* that mediate binding to human platelets. *Infect Immun*, 69, 1373–1380.

Bensing, B.A., I.R. Siboo and P.M. Sullam (2001b) Proteins PblA and PblB of *Streptococcus mitis*, which promote binding to human platelets, are encoded within a lysogenic bacteriophage. *Infect Immun,* 60, 6186–6192.

Betley, M.J. and J.J. Mekalanos (1985) Staphylococcal enterotoxin A is encoded by a phage. *Science,* 229, 185–187.

Beutin, L., U.H. Stroeher and P.A. Manning (1993) Isolation of enterohemolysin (Ehly2)-associated sequences encoded on temperate phages of *Escherichia coli*. *Gene,* 132, 95–9.

Bik, E.M., A.E. Bunschoten, R.D. Gouw and F.R. Mooi (1995). Genesis of the novel epidemic *Vibrio cholerae* O139 strain: Evidence for horizontal transfer of genes involved in polysaccharide synthesis. *EMBO J,* 14, 209–16.

Blattner, F.R., G. Plunkett, 3rd, C.A. Bloch, N.T. Perna, V. Burland, M. Riley, J. Collado-Vides, et al. (1997) The complete genome sequence of *Escherichia coli* K-12. *Science,* 277, 1453–1474.

Blum, G., M. Ott, A. Lischewski, A. Ritter, H. Imrich, H. Tschape, and J. Hacker (1994) Excision of large DNA regions termed pathogenicity islands from tRNA-specific loci in the chromosome of an *Escherichia coli* wild-type pathogen. *J Bacteriol,* 62, 606–614.

Boyd, E.F., and H. Brüssow (2002) Common themes among phage-encoded virulence genes and diversity among the phages involved. *Trends Microbiol,* 10: 521–529.

Boyd, E.F., B.M. Davis and B. Hochhut (2001) Bacteriophage-bacteriophage interactions in the evolution of pathogenic bacteria. *Trends Microbiol,* 9: 137–144.

Boyd, E.F., K.L. Moyer, L. Shi and M.K. Waldor (2000b) Infectious CTXϕ and the Vibrio pathogenicity island prophage in *Vibrio mimicus*: evidence for recent horizontal transfer between *V. mimicus* and *V. cholerae. Infect Immun,* 68: 1507–13.

Boyd, E.F., S. Porwollik, F. Blackmer and M. McClelland (2003) Differences in gene content among *Salmonella enterica* serovar Typhi isolates. *J Clin Microbiol,* 41: 3823–3828.

Boyd, E.F., and M.K. Waldor (1999) Alternative mechanism of cholera toxin acquisition by *Vibrio cholerae*: Generalized transduction of CTXϕ by Bacteriophage CP-T1. *Infect Immun,* 67: 5898–5905.

Brown, R.C. and R.K. Taylor (1995) Organization of *tcp*, *acf*, and *toxT* genes within a ToxT-dependent operon. *Mol Microbiol,* 16: 425–439.

Buchrieser, C., R. Brosch, S. Bach, A. Guiyoule and E. Carniel (1998) The high-pathogenicity island of *Yersinia pseudotuberculosis* can be. *Mol Microbiol,* 30: 965–978.

Carniel, E., I. Guilvout and M. Prentice (1996) Characterization of a large chromosomal "high-pathogenicity island" in biotype 1B *Yersinia enterocolitica. J Bacteriol,* 178: 6743–6751.

Champion, G.A., M.N. Neely, M.A. Brennan and V. J. DiRita (1997) A branch in the *toxR* regulatory cascade of *Vibrio cholerae* revealed by characterization of ToxT mutant strains. *Mol Microbiol,* 23: 323–331.

Chang, C.C., J.R. Gilsdorf, V.J. DiRita and C.F. Marrs (2000) Identification and genetic characterization of *Haemophilus influenzae* genetic island 1. *Infect Immun,* 68: 2630–2637.

Cheetham, B.F., D.B. Tattersall, G.A. Bloomfield, J.I. Rood, and M.E. Katz (1995) A role for bacteriophages in the evolution and transfer of bacterial virulence determinants. *Gene,* 162: 53–58.

Christie, G.E. and R. Calendar (1990) Interactions between satellite bacteriophage P4 and its helpers. *Annu Rev Genet,* 24: 465–490.

Clark, C.A., J. Beltrame and P.A. Manning (1991) The *oac* gene encoding a lipopolysaccharide O-antigen acetylase maps adjacent to the integrase-encoding gene on the genome of *Shigella flexneri* bacteriophage Sf6. *Gene,* 107: 43–52.

Coleman, D.C., D.J. Sullivan, R.J. Russell, J.P. Arbuthnott, B.F. Carey and H.M. Pomeroy (1989) *Staphylococcus aureus* bacteriophages mediating the simultaneous lysogenic conversion of beta-lysin, staphylokinase and enterotoxin A: Molecular mechanism of triple conversion. *J Gen Microbiol,* 135: 1679–1697.

Davis, B.M., K.E. Moyer, E.F. Boyd and M.K. Waldor (2000) CTX prophages in classical biotype *Vibrio cholerae*: Functional phage genes but dysfunctional phage genomes. *J Bacteriol,* 182: 6992–6998.

Davis, B.M., H. Kimsey, A.V. Kane, and M.K. Waldor (2002) A satellite phage (RS1)-encoded antirepressor induces repressor aggregation and cholera toxin gene transfer. *EMBO J* 21: 4240–4249.

Davis, B.M., H.H. Kimsey, W. Chang and M.K. Waldor (1999) The *Vibrio cholerae* O139 calcutta CTXϕ is infectious and encodes a novel repressor. J Bacteriol, 181: 6779–6787.

Davis, B.M., and M.K. Waldor (2000) CTXϕ contains a hybrid genome derived from tandemly integrated elements. *Proc Natl Acad Sci U S A,* 97: 8572–8577.

Davis, B.M., and M.K. Waldor. "Mobile genetic elements and bacterial pathogenesis," In N.L. Craig, C.R., M. Gellert and L.A.M. (Eds.), *Mobile DNA II*. ASM Press, 2002.

Desiere, F., W.M. McShan, D. van Sinderen, J.J. Ferretti and H. Brüssow (2001) Comparative genomics reveals close genetic relationships between phages from dairy bacteria and pathogenic Streptococci: Evolutionary implications for prophage-host interactions. *Virology,* 288: 325–341.

DiRita, V.J., C. Parsot, G. Jander, and J.J. Mekalanos (1991) Regulatory cascade controls virulence in *Vibrio cholerae. Proc Natl Acad Sci U S A,* 88: 5403–5407.

Dumontier, S., and P. Berche (1998) *Vibrio cholerae* O22 might be a putative source of exogenous DNA resulting in the emergence of the new strain of *Vibrio cholerae* O139. *FEMS Microbiol. Lett.,* 164: 9198.

Durack, D.T. and P.B. Beeson (1975) Experimental bacterial endocarditis. I Colonization of a sterile vegetation. *Br J Exp Pathol,* 53: 667–678.

Fang, F.C., M.A. DeGroote, J.W. Foster, A.J. Baumler, U. Ochsner, T. Testerman, S. Bearson, et al. (1999) Virulent *Salmonella typhimurium* has two periplasmic Cu, Zn-superoxide dismutases. *Proc Natl Acad Sci U S A,* 96: 7502–7507.

Faruque, S.M., M.J. Albert, and J.J. Mekalanos (1998a) Epidemiology, genetics, and ecology of toxigenic *Vibrio cholerae*. Microbiol *Mol Biol Rev,* 62: 1301–1314.

Faruque, S.M., Asadulghani, M. Kamruzzaman, R.K. Nandi, A.N. Ghosh, G.B. Nair, J.J. Mekalanos, et al. (2002) RS1 element of *Vibrio cholerae* can propagate horizontally as a filamentous phage exploiting the morphogenesis genes of CTXϕ. *Infect Immun,* 70: 163–170.

Faruque, S.M., Asadulghani, M.N. Saha, A.R. Alim, M.J. Albert, K.M. Islam, and J.J. Mekalanos (1998b) Analysis of clinical and environmental strains of nontoxigenic *Vibrio cholerae* for susceptibility to CTXϕ: molecular basis for origination of new strains with epidemic potential. *Infect Immun,* 66: 5819–5825.

Faruque, S.M., J. Zhu , Asadulghani, M. Kamruzzaman and J.J. Mekalanos (2003) Examination of diverse toxin-coregulated pilus-positive *Vibrio cholerae* strains fails to demonstrate evidence for Vibrio pathogenicity island phage. *Infect Immun,* 71: 2993–2999.

Fasano, A., B. Baudry, D.W. Pumplin, S.S. Wasserman, B.D. Tall, J.M. Ketley and J.B. Kaper (1991) *Vibrio cholerae* produces a second enterotoxin, which affects intestinal tight junctions. *Proc Natl Acad Sci U S A,* 88: 5242–5246.

Ferretti, J.J., McShan W.M., Ajdic D., Savic D.J., Savic G., Lyon K., Primeaux C. et al. (2001) Complete genome sequence of an M1 strain of *Streptococcus pyogenes. Proc Natl Acad Sci U S A,* 98: 4658–4663.

Figueroa-Bossi, N., S. Uzzau, D. Maloriol and L. Bossi (2001) Variable assortment of prophages provides a transferable repertoire of pathogenic determinants in Salmonella. *Mol Microbiol,* 39: 260–271.

Fitzgerald, J.R., Monday S.R., Foster T.J., Bohach G.A., Hartigan P.J., Meaney W.J. and C.J. Smyth (2001) Characterization of a putative pathogenicity island from bovine *Staphylococcus aureus* encoding multiple superantigens. *J Bacteriol,* 183: 63–70.

Freeman, V.J. (1951) Studies on the virulence of bacteriophage-infected strains of *Corynebacterium diptheriae*. *J Bacteriol,* 61: 675–688.

Galen, J.E., Ketley J.M., Fasano A., Richardson S.H., Wasserman S.S. and J.B. Kaper (1992) Role of *Vibrio cholerae* neuraminidase in the function of cholera toxin. *Infect Immun,* 60: 406–415.

Ghosh, C., R.K. Nandy, S.K. Dasgupta, G.B. Nair, R.H. Hall, and A.C. Ghose (1997) A search for cholera toxin (CT), toxin coregulated pilus (TCP), the regulatory element ToxR and other virulence factors in non-01/non-0139 *Vibrio cholerae*. *Microb Pathog,* 22: 199–208.

Goshorn, S.C. and P.M. Schlievert (1989) Bacteriophage association of streptococcal pyrogenic exotoxin type C. *J Bacteriol,* 171: 3068–3073.

Hacker, J. and J. Kaper (2000) *Pathogenicity islands in enteric bacteria.* American Society of Microbiology, Washington D.C., 2000.

Hansen-Wester, I., B. Stecher and M. Hensel (2002) Analyses of the evolutionary distribution of Salmonella translocated effectors. *J Bacteriol,* 70: 1619–1622.

Hayashi, T., M. Makino, K. Ohnishi, K. Kurokawa, K. Ishii, K. Yokoyamma, C.G. Han, E., et al. (2001) Complete genome sequence of enterohemorrhagic *Escherichia coli* O157 and genomic comparison with a laboratory strain K12. *DNA Res,* 8: 11–22.

Heidelberg, J.F., J.A. Eisen, W.C. Nelson, R.A. Clayton, M.L. Gwinn, R.J. Dodson, D.H. Haft, et al. (2000) DNA sequence of both chromosomes of the cholera pathogen *Vibrio cholerae*. *Nature,* 406: 477–483.

Hendrix, R.W., M.C. Smith, R.N. Burns, M.E. Ford and G.F. Hatfull (1999) Evolutionary relationships among diverse bacteriophages and prophages: All the world's a phage. *Proc Natl Acad Sci U S A,* 96: 2192–2197.

Ho, T.D. and J.M. Slauch, (2001) Characterization of grvA, an antivirulence gene on the gifsy-2 phage in *Salmonella enterica* serovar typhimurium. *J Bacteriol,* 183: 611–620.

Hueck, C.J., (1998) Type III protein secretion systems in bacterial pathogens of animals and plants. *Microbiol Mol Biol Rev,* 62: 379–433.

Hynes, W.L., and J.J. Ferretti (1989) Sequence analysis and expression in Escherichia coli of the hyaluronidase gene of *Streptococcus pyogenes* bacteriophage H4489A. *Infect Immun,* 57: 533–539.

Ikebe, T., A. Wada, Y. Inagaki, K. Sugama, R. Suzuki, D. Tanaka, A. Tamaru, et al. (2002) Dissemination of the phage-associated novel superantigen gene *speL* in recent invasive and noninvasive *Streptococcus pyogenes* M3/T3 isolates in Japan. *Infect Immun,* 70: 3227–3233.

Inagaki, Y., F. Myouga, H. Kawabata, S. Yamai, and H. Watanabe (2000) Genomic differences in *Streptococcus pyogenes* serotype M3 between recent isolates associated with toxic shock-like syndrome and pasr clinical isolates. *J Infect Dis,* 181: 975–983.

Jermyn, W.S., and E.F. Boyd (2002) Characterization of a novel Vibrio pathogenicity island (VPI-2) encoding neuraminidase (nanH) among toxigenic *Vibrio cholerae* isolates. *Microbiol,* 148: 3681–3693.

Johansen, B., Y. Wasteson, P. Granum and S. Brynestad (2001) Mosaic structure of Shiga-toxin-2-encoding phages isolated from *Escherichia coli* O157: H7 indicates frequent gene exchange between lambdoid phage genomes. *Microbiol,* 147: 1929–1936.

Kaneko, J., Kimura T., Narita S., Tomita T., and K.Y. (1998) Complete nucleotide sequence and molecular characterization of the temperate staphylococcal bacteriophage ϕPVL carrying Panton-Valentine leukocidin genes. *Gene,* 215: 57–67.

Karaolis, D.K., J.A. Johnson, C.C. Bailey, E.C. Boedeker, J.B. Kaper, and P.R. Reeves (1998) A *Vibrio cholerae* pathogenicity island associated with epidemic and pandemic strains. *Proc Natl Acad Sci U S A,* 95: 3134–3139.

Karaolis, D.K., S. Somara, D.R. Maneval, Jr., J.A. Johnson and J.B. Kaper (1999) A bacteriophage encoding a pathogenicity island, a type-IV pilus and a phage receptor in cholera bacteria. *Nature,* 399: 375–379.

Klee, S.R., X. Nassif, B. Kusecek, P. Merker, J.L. Beretti, M. Achtman and C.R. Tinsley (2000) Molecular and biological analysis of eight genetic islands that distinguish *Neisseria meningitidis* from the closely related pathogen *Neisseria gonorrhoeae*. *Infect Immun.,* 68: 2082–2095.

Kuroda, M., T. Ohta, I. Uchiyama, T. Baba, H. Yuzawa, I. Kobayashi, L. Cui, et al. (2001) Whole genome sequencing of methicillin-resistant *Staphylococcus aureus*. *Lancet,* 357: 1225–1240.

Lalioui, L. and C. Le Bouguenec (2001) afa-8 Gene cluster is carried by a pathogenicity island inserted into the tRNA(Phe) of human and bovine pathogenic *Escherichia coli* isolates. *Infect Immun,* 69: 937–948.

Lee, C.A. (1999) *Vibrio cholerae* TCP: A trifunctional virulence factor? *Trends Microbiol,* 7: 391–392.

Lee, S.H., D.L. Hava, M.K. Waldor, and A. Camilli (1999) Regulation and temporal expression patterns of *Vibrio cholerae* virulence genes during infection. *Cell,* 99: 624–634.

Li, M., T. Shimada, J.G.J. Morris, A. Sulakvelidze and S. Sozhamannan (2002) Evidence for the emergence of non-O1 and non-O139 *Vibrio cholerae* strains with pathogenic potential by exchange of O-antigen biosynthesis regions. *Infect Immun,* 70: 2441–2453.

Lindqvist, B.H., G. Deho, and R. Calendar (1993) Mechanisms of genome propagation and helper exploitation by satellite phage P4. *Microbiol Rev,* 57: 1683–1702.

Lindsay, J.A., A. Ruzin, H.F. Ross, N. Kurepina, and R.P. Novick (1998) The gene for toxic shock toxin is carried by a family of mobile pathogenicity islands in *Staphylococcus aureus*. *Mol Microbiol,* 29: 527–543.

Masignani, V., M.M. Giuliani, H. Tettelin, M. Comanducci, R. Rappuoli, and V. Scarlato (2001) Mu-like prophage in serogroup B *Neisseria meningitidis* coding for surface-exposed antigens. *Infect Immun,* 69: 2580–2588.

McClelland, M. (2001) Complete genome sequence of *Salmonella enterica* serovar Typhimurium LT2. *Nature,* 413: 852–856.

McDaniel, T.K., K.G. Jarvis, M.S. Donnenberg and J.B. Kaper (1995) A genetic locus of enterocyte effacement conserved among diverse enterobacterial pathogens. *Proc Natl Acad Sci U S A,* 92: 1664–1668.

McGowan, D.A., and R. Gillett (1980) Scanning electron microscopic observations of the surface of the initial lesion in experimental streptococcal endocarditis in the rabbit. *Br J Exp Pathol,* 61: 164–171.

Mekalanos, J.J., E.J. Rubin, and M.K. Waldor (1997) Cholera: Molecular basis for emergence and pathogenesis. *FEMS Immunol Med Microbiol,* 18: 241–248.

Miller, J.F. (2003) Bacteriophage and the evolution of epidemic cholera. *Infect Immun,* 71: 2981–2982.

Mirold, S., W. Rabsch, M. Rohde, S. Stender, H. Tschape, H. Russmann, E. Igwe, et al. (1999) Isolation of a temperate bacteriophage encoding the type III effector protein SopE from an epidemic *Salmonella typhimurium* strain. *Proc Natl Acad Sci U S A,* 96: 9845–9850.

Mirold, S., W. Rabsch, H. Tschape, and W. D. Hardt (2001) Transfer of the Salmonella type III effector *sopE* between unrelated phage families. *J Mol Biol,* 312: 7–16.

Mooi, F.R. and E.M. Bik (1997) The evolution of epidemic *Vibrio cholerae* strains. *Trends Microbiol,* 4: 161–165.

Moss, J.E., T.J. Cardozo, A. Zychlinsky and E.A. Groisman (1999) The selC-associated SHI-2 pathogenicity island of *Shigella flexneri*. *Mol Microbiol,* 33: 74–83.

Nakayama, K., S. Kanaya, M. Ohnishi, Y. Terawaki, and T. Hayashi (1999) The complete nucleotide sequence of ϕCTX, a cytotoxin-converting phage of *Pseudomonas aeruginosa*: Implications for phage evolution and horizontal gene transfer via bacteriophages. *Mol Microbiol,* 31: 399–419.

Neely, M.N., and D.I. Friedman (1998a) Arrangement and functional identification of genes in the regulatory region of lambdoid phage H-19B, a carrier of a Shiga-like toxin. *Gene,* 223: 105–113.

Neely, M.N. and D.I. Friedman (1998b) Functional and genetic analysis of regulatory regions of coliphage H-19B: Location of shiga-like toxin and lysis genes suggest a role for phage functions in toxin release. Mol Microbiol, 28: 1255–1267.

Novick, R.P. (2003) Mobile genetic elements and bacterial toxinoses: The superantigen-encoding pathogenicity islands of *Staphylococcus aureus*. *Plasmid,* 49: 93–105.

O'Brien, A.D., J.W. Newland, S.F. Miller, R.K. Holmes, H.W. Smith, and S.B. Formal (1984) Shiga-like toxin-converting phages from *Escherichia coli* strains that cause hemorrhagic colitis or infantile diarrhea. *Science,* 226: 694–696.

O'Shea, Y.A., and E.F. Boyd (2002) Mobilization of the Vibrio Pathogenicity Island (VPI) between *Vibrio cholerae* O1 isolates mediated by CP-T1 generalized transduction. *FEMS Microbiol Lett,* 214: 153–159.

Ohnishi, M., K. Kurokawa and T. Hayashi (2001) Diversification of *Escherichia coli* genomes: Are bacteriophages the major contributors? *Trends Microbiol,* 9: 481–485.

Pacheco, S.V., O.G. Gonzalez, and G.L.P. Contreras (1997) The lom gene of bacteriophage lis involved in *Escherichia coli* K12 adhesion to human buccal epithelial cells. *FEMS Microbiol Lett,* 156: 129–132.

Parkhill, J. (2001) Complete genome sequence of a multiple drug resistance *Salmonella enterica* serovar Typhi CT18. *Nature,* 413: 848–852.

Perna, N.T., Plunkett G. 3rd, Burland V., Mau B., Glasner J.D., Rose D.J., Mayhew G.F., et al. (2001) Genome sequence of enterohaemorrhagic *Escherichia coli* O157:H7. Nature, 409: 529–533.

Plunkett, G.r., D.J. Rose, T.J. Durfee and F.R. Blattner (1999) Sequence of Shiga toxin 2 phage 933W from *Escherichia coli* O157:H7: Shiga toxin as a phage late-gene product. *J Bacteriol,* 181: 1767–1778.

Porwollik, S., R.M. Wong and M. McClelland (2002) Evolutionary genomics of Salmonella: Gene acquisitions revealed by microarray analysis. *Proc Natl Acad Sci,* 99: 8956–8961.

Purdy, G.E. and S.M. Payne (2001) The SHI-3 iron transport island of *Shigella boydii* 0-1392 carries the genes for aerobactin synthesis and transport. *J Bacteriol,* 183: 4176–4182.

Recktenwald, J. and H. Schmidt (2002) The nucleotide sequence of Shiga toxin (Stx) 2e-encoding phage fP27 is not related to other phage genomes, but the modular genetic structure is conserved. *Infect Immun,* 70: 1896–1908.

Reidl, J. and J.J. Mekalanos (1995) Characterization of *Vibrio cholerae* bacteriophage K139 and use of a novel mini-transposon to identify a phage-encoded virulence factor. *Mol Microbiol,* 18: 685–701.

Ruzin, A., J. Lindsay, and R.P. Novick (2001) Molecular genetics of SaPI1—a mobile pathogenicity island in *Staphylococcus aureus*. *Mol Microbiol,* 41: 365–77.

Said, B., H.R. Smith, S.M. Scotland, and B. Rowe (1995) Detection and differentiation of the gene for toxin co-regulated pili (tcpA) in *Vibrio cholerae* non-O1 using the polymerase chain reaction. *FEMS Microbiol Lett,* 125: 205–209.

Sandt, C.H., J.E. Hopper and C.W. Hill (2002) Activation of prophage eib genes for immunoglobin-binding poteins from the IbrAB genetic island of *Escherichia coli* ECOR-9. *J Bacteriol,* 184: 3640–3648.

Schubert, S., A. Rakin, H. Karch, E. Carniel and J. Heesemann (1998) Prevalence of the "high-pathogenicity island" of Yersinia species among *Escherichia coli* strains that are pathogenic to humans. *Infect Immun,* 66: 480–485.

Skorupski, K. and R.K. Taylor (1997) Control of the ToxR virulence regulon in *Vibrio cholerae* by environmental stimuli. *Mol Microbiol,* 25: 1003–1009.

Skorupski, K., and R.K. Taylor (1999) A new level in the *Vibrio cholerae* ToxR virulence cascade: AphA is required for transcriptional activation of the tcpPH operon. *Mol Microbiol,* 31: 763–771.

Smoot, J.C., Barbian K.D., Van Gompel J.J., Smoot L.M., Chaussee M.S., Sylva G.L., Sturdevant D.E., et al. (2002) Genome sequence and comparative microarray analysis of serotype M18 group A Streptococcus strains associated with acute rheumatic fever outbreaks. *Proc Natl Acad Sci U S A,* 99: 4668–4673.

Stanley, T.L., C.D. Ellermeier and J.M. Slauch (2000) Tissue-specific gene expression identifies a gene in the lysogenic phage Gifsy-1 that affects *Salmonella enterica* serovar typhimurium survival in Peyer's patches. *J Bacteriol,* 182: 4406–4413.

Sullivan, J.T. and C.W. Ronson (1998) Evolution of rhizobia by acquisition of a 500-kb symbiosis island that integrates into a phe-tRNA gene. *Proc Natl Acad Sci U S A,* 95: 5145–5149.

Swenson, D.L. (1996) Two pathogenicity islands in uropathogenic *Escherichia coli* J96: Cosmid cloning and sample sequencing. *Infect Immun,* 64: 3736–3743.

Taylor, R.K., V.L. Miller, D.B. Furlong and J.J. Mekalanos (1987) Use of *phoA* gene fusions to identify a pilus colonization factor coordinately regulated with cholera toxin. P*roc Natl Acad Sci U S A,* 84: 2833–2837.

Trucksis, M., J.E. Galen, J. Michalski, A. Fasano and J.B. Kaper (1993) Accessory cholera enterotoxin (Ace), the third toxin of a *Vibrio cholerae* virulence cassette. *Proc Natl Acad Sci U S A,* 90: 5267–5271.

Unkmeir, A. and H. Schmidt (2000) Structural analysis of phage-borne *stx* genes and their flanking sequences in Shiga toxin-producing *Escherichia coli* and *Shigella dysenteriae* type 1 strains. *Infect Immun,* 68: 4856–4864.

Uzzau, S., N. Figueroa-Bossi, S. Rubino and L. Bossi (2001) Epitope tagging of chromosomal genes in Salmonella. *Proc Natl Acad Sci U S A,* 98: 15264–15269.

Vander Byl, C., and A.M. Kropinski (2000) Sequence of the genome of Salmonella bacteriophage P22. *J Bacteriol,* 182: 6472–6481.

Verma, N.K., J.M. Brandt, D.J. Verma and A.A. Lindberg (1991) Molecular characterization of the O-acetyl transferase gene of converting bacteriophage SF6 that adds group antigen 6 to *Shigella flexneri. Mol Microbiol,* 5: 71–75.

Voelker, L.L. and K. Dybvig (1999) Sequence analysis of the *Mycoplasma arthritidis* bacteriophage MAV1 genome identifies the putative virulence factor. *Gene,* 11: 101–107.

Vokes, S.A., S.A. Reeves, A.G. Torres and S.M. Payne (1999) The aerobactin iron transport system genes in *Shigella flexneri* are present within a pathogenicity island. *Mol Microbiol,* 33: 63–73.

Wagner, P.L., D.W.K. Acheson and M.K. Waldor (1999) Isogenic lysogens of diverse shiga toxin2-encoding bacteriophages produce markedly different amounts of shiga toxins. *Infect Immun,* 67: 6710–6714.

Wagner, P.L., M.N. Neely, X. Zhang, D.W. Acheson, M.K. Waldor and D.I. Friedman (2001) Role for a phage promoter in Shiga toxin 2 expression from a pathogenic *Escherichia coli* strain. J Bacteriol, 183: 2081–2085.

Wagner, P.L. and M.K. Waldor (2002) Bacteriophage control of bacterial virulence. *Infect Immun,* 70: 3985–3993.

Waldor, M.K. and J.J. Mekalanos (1994) *Vibrio cholerae* O139 specific gene sequences. *Lancet,* 343: 1366.

Waldor, M.K. and J.J. Mekalanos (1996) Lysogenic conversion by a filamentous phage encoding cholera toxin. *Science,* 272: 1910–1914.

Waldor, M.K., E.J. Rubin, G.D. Pearson, H. Kimsey, and J.J. Mekalanos (1997) Regulation, replication, and integration functions of the *Vibrio cholerae* CTXϕ are encoded by region RS2. *Mol Microbiol,* 24: 917–926.

Weeks, C.R. and J.J. Ferretti (1984) The gene for type A streptococcal exotoxin (erythrogenic toxin) is located in bacteriophage T12. *Infect Immun,* 46: 531–536.

Wright, A. (1971) Mechanism of conversion of the Salmonella O antigen by bacteriophage epsilon 34. *J Bacteriol,* 105: 927–936.

Yamaguchi, T., T. Hayashi, H. Takami, K. Nakasone, M. Ohnishi, K. Nakayama, S. Yamada, et al. (2000) Phage conversion of exfoliative toxin A production in *Staphylococcus aureus*. *Mol Microbiol,* 38: 694–705.

Yamasaki, S., T. Shimizu, K. Hoshino, S.T. Ho, T. Shimada, G.B. Nair and Y. Takeda (1999) The genes responsible for O-antigen synthesis of *Vibrio cholerae* O139 are closely related to those of *Vibrio cholerae* O22. *Gene,* 237: 321–332.

Yokoyama, K., Makino K., Kubota Y., Watanabe M., Kimura S., Yutsudo C.H., Kurokawa K., et al. (2000) Complete nucleotide sequence of the prophage VT1-Sakai carrying the Shiga toxin 1 genes of the enterohemorrhagic *Escherichia coli* O157:H7 strain derived from the Sakai outbreak. *Gene,* 258: 127–130.

Yu, R.R. and V.J. DiRita (1999) Analysis of an autoregulatory loop controlling ToxT, cholera toxin, and toxin-coregulated pilus production in *Vibrio cholerae*. *J Bacteriol,* 181: 2584–2592.

Zabriskie, J.B. (1964) The role of temperate bacteriophage in the production of erythrogenic toxin by group A streptococci. *J Exp Med,* 119: 761–779.

9 Phage for the Detection of Pathogenic Bacteria

Catherine E.D. Rees[1] *and Martin J. Loessner*[2]
[1]School of Biosciences, University of Nottingham, Sutton Bonington Campus, Loughborough, Leicestershire, UK
[2]Institute of Food Science and Nutrition, Swiss Federal Institute of Technology, ETH Center, Zürich, Switzerland.

CONTENTS

9.1. Introduction 267
9.2. Recombinant Phage 268
9.2.1. Luciferase Reporter Phage (LRP) 268
9.2.2. Other Reporter Phage 272
9.3. Phage Display 273
9.4. Dual Phage 275
9.5. Phage Amplification Assays 277
9.6. Detection via Phage-Mediated Cell Lysis 279
9.7. Use of Phage and Phage Products as Binding Reagents 280
Acknowledgements 281
References 281

9.1. INTRODUCTION

The specificity of the interaction of a virus with its host cell immediately lends itself to methods for the identification of bacteria, in particular the pathogens. While many other procedures based upon antibodies (ELISA) or nucleic acid amplification (PCR) have been artificially developed to allow differentiation of bacterial cell structures, here we have a naturally evolved system in which the bacteriophage specifically recognizes and binds only to its own host cells. This interaction has been exploited in a number of different methods for the specific detection and differentiation of the individual host bacteria. One of the first uses of phage was in typing schemes, where a panel of phages with different lytic spectra is used to discriminate between different isolates of a bacterial species or genus, according to their ability to infect the isolate and form plaques. Differences in infectivity reflect differences in a number of cellular characteristics, such as cell surface receptors, the presence of restriction modification systems and related prophage, plasmid carriage, etc. When using an identical set of phages, typing results are highly reproducible between different laboratories, although

0-8493-1336-X/05/$0.00+$1.50
© 2005 by CRC Press LLC

it is important that the phages used are propagated under specified conditions. The ease of use and inexpensiveness of phage typing also contribute to the fact that it is still among the most widely used methods for strain identification of many bacteria—in particular, *Salmonella, Listeria,* and *Staphylococcus*.

Other detection techniques have been developed to directly harness the specific binding ability of bacteriophage. These include the production of fluorescently labelled antibodies directed against specifically adsorbing phage particles (Watson and Eveland, 1965), or fluorescently labelled phage prepared by cross-linking dyes to the phage surface (Hennes et al., 1995; Goodridge et al., 1999). In both of these methods target cells are detected following adsorption of phage to the host cells and then identification of this interaction by detection of the bound fluorescent signal. Although this basic host-binding identification step in phage-based assays remains the same in the various methods discussed in this chapter, our increased understanding of both phage genetics and phage structure has allowed new genetically engineered reagents to be developed.

9.2. RECOMBINANT PHAGE

9.2.1. Luciferase Reporter Phage (LRP)

The idea that phage can transduce genes into their host cell is not new. However the idea that reporter genes could be introduced into a phage so that the phage infection event could be readily detected was first proposed as a detection method by Ulitzur and Kuhn (1987). In this case the reporter of choice was the bacterial bioluminescence *(lux)* genes. Expression of these genes from a modified *lambda* phage was rapidly and sensitively detected using a luminometer (see Fig. 9.1). Initially a phage-based cloning vector containing the complete *lux* operon was used to demonstrate that as few as 10 *E. coli* cells could be detected following phage infection (Ulitzur and Kuhn, 1989). This group went on to produce luciferase reporter phage by random Tn*10*::*luxAB* mutagenesis of wild-type phage genomes and recovered recombinant $luxAB^{+}$ phage from bioluminescent plaques. These simple constructs termed Luciferase Reporter Phage (LRP) have been used to successfully detect the presence of enteric bacteria in meat (Kodikara et al., 1991) and *Salmonella* in eggs (Chen and Griffiths, 1996).

A more rational approach to the construction of an LRP was used by Loessner et al. (1996). In this case, genetic analysis of the virulent *Listeria* phage A511 (Loessner et al., 1994) allowed identification of the late region of the phage genome which encoded the major capsid and tail sheath proteins (Loessner and Scherer, 1995). As this construct was to be expressed in a Gram-positive bacterium, the fused *luxAB* genes from *Vibrio harveyi* (Hill et al., 1991) were used and introduced by recombination into the operon encoding the phage structural genes without disrupting any of the existing gene structures. This was achieved by first introducing a plasmid containing the modified phage operon into *L. monocytogenes* host cells. After plating the phage on these plasmid-containing hosts, the desired LRP recombinant clone was recovered from bioluminescent plaques, detected by exposing the plates to aldehyde vapour. Although the centre of the plaque is non-bioluminescent due to cell lysis, the infected cells at the edge of the plaque still express the phage-encoded

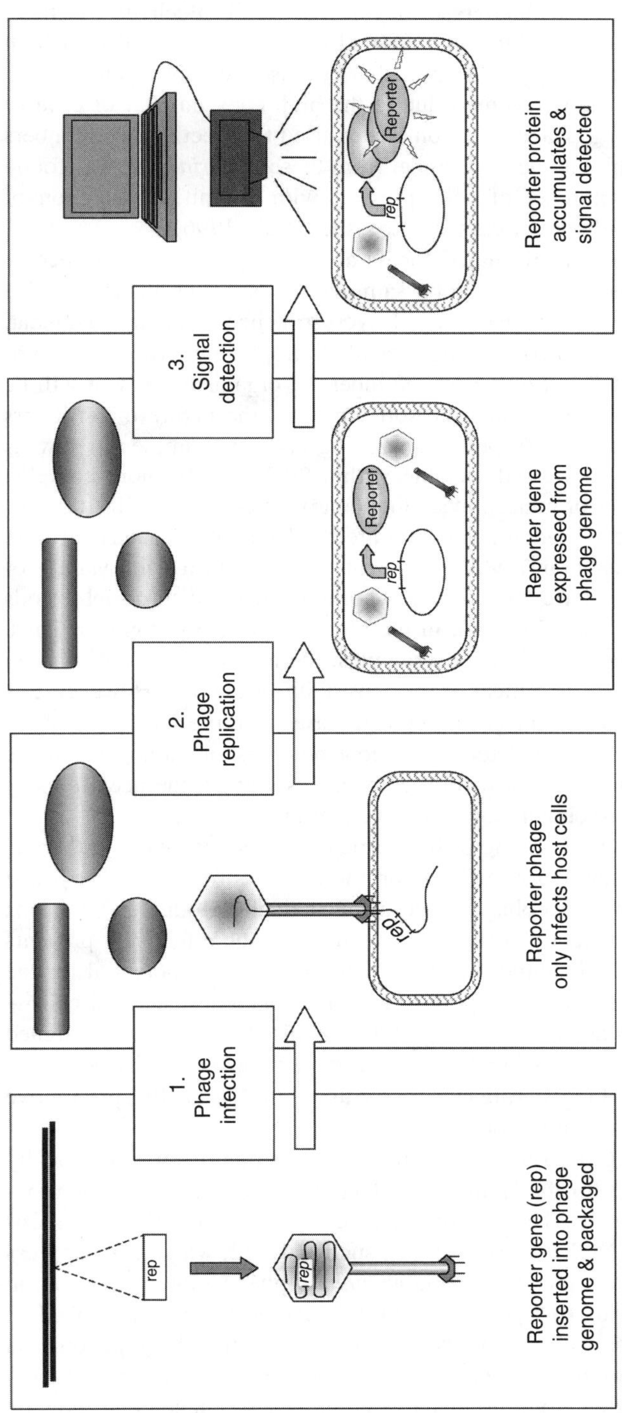

FIGURE 9.1 Reporter phage assay.
Reporter genes (e.g. *lux*, *luc*, *ina*, or *gfp*) are introduced into the phage genome and recombinant DNA packaged into phage particles (see text for details). Reporter phage are mixed with the sample but only infect susceptible hosts, removing the need for purification of the organism before detection. Following infection, reporter phage genomes are replicated and the reporter gene expressed. Once sufficient reporter protein has accumulated the expression of the reporter gene can be detected.

lux genes and the substrate for the luciferase enzyme—a volatile aldehyde—diffuses freely into the cells (Blisset and Stewart, 1989). LRP were recovered at a high frequency (5×10^{-4}; Loessner et al., 1996) probably because recombination between plasmid and phage sequences is favoured due to the high copy number of each of these DNA molecules during phage infection. When used to infect known numbers of *Listeria* host cells, the bioluminescent signal from these *Listeria* LRP was found to be proportional to the number of cells present, with a limit of detection of approximately 500 cells without enrichment (Loessner et al., 1996).

The *Listeria lux* phage was evaluated for its ability to detect the presence of *L. monocytogenes* in different types of food samples (Loessner et al., 1997). The three foods chosen for this study were cabbage leaves, soft cheese, and minced meat. The cabbage leaves have a flora comprising typical phylloplane microbes, predominantly pseudomonads and *Bacillus* species. Numbers of organisms present will be variable, depending on weather conditions, soil type, and the nature of fertilizers applied, and will be 10^5 to 10^7 cfu per gram. Using cabbage samples as few as 0.1–1 cell per gram of food were detectable within 20 hours. In more complex samples, such as soft cheeses or minced meat, the limit of detection was 10–100 cells per gram. In the cheeses the predominant flora are the lactic acid bacteria at high numbers (above 10^6 cfu per gram) while the meat will contain a wide variety of bacterial types, including enteric bacteria and Gram-negative psychrotrophic rods (again pseudomonads and related genera, in the range of 10^6 cfu per gram). Thus, as for many rapid methods applied to food samples, the nature of the food matrix and the competing microflora is critical to the sensitivity of the test. However, this study demonstrated the main advantages of a phage-based test; the specificity of the host-phage interaction allows the detection of low numbers of pathogens in the presence of much higher numbers of natural competitors without the need to carry out successive rounds of enrichment and selective plating.

Even though this provides a saving of both time and consumable costs, there is considerable resistance in many parts of the world to the use of recombinant phage in food analysis because these *lux* phage constitute Genetically Modified Organisms (GMOs). This is despite the fact that the use of virulent phage effectively prevents survival of infected cells, and should therefore also minimize the potential for the establishment of any phage-encoded reporter genes in replicating bacterial cells—one of the major concerns about the use of GMOs. While these tests are designed to be carried out only within the laboratory so that containment is easily achieved, in many countries the cost of registering labs for the use of GMOs is often prohibitive within the wider context of routine food testing.

Recently, Kuhn et al. (2002) have described the construction of a genetically "locked" *lux* phage derived for the *Salmonella* phage Felix 01 that cannot be propagated on wild-type host cells. The modified phage contain two *amber* mutations and will only grow on *supF*$^+$ *Salmonella* strains, such as K772, which can suppress such mutations. Since these phage are double *amber* phage mutants, reversion rates were very low (10^{-8} to 10^{-9} per generation). To allow these phage to infect wild type *Salmonella* cells, a *supF* gene was introduced into the phage genome on a DNA fragment in association with the *luxAB* genes and during the insertion process an essential gene was deleted from the phage genome. These phage could now be grown in a *supF*$^-$ host modified to express the deleted Felix-01 gene from a plasmid construct to provide the essential phage functions *in trans*. As long as these phage maintain the *luxAB-supF* segment, they will infect wild type *Salmonella* and produce bioluminescence following infection; however, no viable phage particles are produced at the end of the infection cycle due to the deletion in the essential gene. Interestingly, when tested against a range of different *Salmonella* isolates, it was found that a bioluminescent signal was produced from cells that will not propagate the wild type phage (i.e. the infection host range is wider than the replication host range; Kuhn et al., 2002), and this increases the versatility of the assay as the number of strains that can be detected is greater than predicted.

Although this strategy requires a more detailed understanding of the genetic organization of relevant regions of individual phage, the fact that the reporter phage produced are nonviable (and therefore cannot replicate to be released into the environment at the end of the assay) should address any concerns about the use of such GMOs in *in vitro* assays.

Although both the *Listeria lux* phage and the *Salmonella lux* phage have been successfully developed, they have not been extensively evaluated. In contrast, the recombinant phage developed for detection of *Mycobacterium tuberculosis* has received much attention. The reason for this is that the use of the reporter phage circumvents the need to culture this slow-growing organism before detection and therefore brings a huge benefit, reducing detection times from weeks to days. There also seems to be less concern about the use of GMOs in the context of medical diagnosis, but no commercialized test using LRP has yet been introduced.

The LRP developed for the detection of *Mycobacteria* contain the firefly luciferase (*luc* or *Fflux*) as the reporter gene (Jacobs et al., 1993). Initially, the lytic phage TM4 was used but the rapidity of the lytic cycle meant that only limited amounts of luciferase protein were produced and the limit of detection was 10^4 mycobacterial cells (Jacobs et al., 1993). When the same workers constructed a reporter phage based on the temperate phage L5, prolonged expression and accumulation of the luciferase protein was seen because the *luc* genes became associated with a constitutive chromosomal promoter when the phage integrated into the genome. This amplification of the signal allowed the limit of detection to be reduced to a minimum of 10^2 cells after a 40 h incubation period, or 10^3 cells after 20 hours incubation (Sarkis et al., 1995). However the limited host range of the L5 phage meant that it was not widely applicable for the detection of *M. tuberculosis* in a practical test. Further work has been carried out to improve the TM4-based LRP, including isolating various spontaneous mutants of the parent phage and changing both the site of insertion of the *luc* gene in the phage genome and the promoter used to expressed the gene. The most recently described TM4-based phage, phAE142, uses the P_{Left} promoter from phage L5 to express the *luc* genes to high levels, but is genetically unstable; propagation in wild type *M. smegmatis* leads to rapid loss of the *luc* genes. Hence a special propagating strain that constitutively expresses the L5 protein Gp71, the repressor of the P_{Left} promoter, is required for phAE142 (Bardarov et al., 2003). This promoter had previously been shown to express the *luc* gene to toxic levels in *Mycobacteria* (Brown et al., 1997), and the high level of *luc* expression in this phage construct has fortuitously created phage that again are unlikely to cause concern as a viable GMO. These phage have been used in combination with further refinements of the assay conditions, in particular processing steps for the clinical sputum samples to remove inhibitors of phage infection, and the assay has been shown to successfully detect mycobacteria in smear-positive sputum samples within 24–48 hours (Riska et al., 1997). It also has been shown to perform favourably in detection of positive mycobacterial specimens when compared with other standard microbiological testing methods such as MGIT (Bardarov et al., 2003).

The majority of work has been carried out with bioluminescence reporter genes, which, depending on their origin, have certain features that must be considered carefully. Bacterial LuxA and LuxB subunits from marine bacteria and fusions of the *luxAB* genes that create a single functional LuxAB polypeptide were found to be rather sensitive to temperatures above 30°C (Hill et al., 1991, Loessner et al., 1996), although the temperature stability of the enzymes varies depending on the host strain in which they are expressed (Mackey et al., 1994). This fact can limit their application to bacteria capable of growing and synthesizing the enzyme in a narrow temperature range. The problem can be addressed by using other luciferases from bacteria such as *Xenorhabdus luminescens* (Li et al., 1993).

The need for the delivery of specific substrates may also be a limiting factor, especially when using the firefly luciferases, which require rather large luciferin molecules that do not readily diffuse across bacterial membranes. In contrast, the small and lipophilic aldehyde used by the bacterial enzyme does freely diffuse across bacterial cell membranes but is toxic in high concentration to bacterial cells. Hence there is a need to deliver these substrates, which can also be short lived, at an optimal time point following phage infection to maximize the detection of the reporter gene being expressed from the phage.

9.2.2. Other Reporter Phage

The incorporation of other reporter genes into phage has been described. These reporters were chosen so the signal they generate can be detected without background noise from the sample, and hence the signal can be amplified to increase sensitivity. For instance the bacterial ice nucleation protein (*ina*) has been used to construct reporter phage for the detection of *Salmonella* (Wolber and Green, 1990). Very low levels of gene expression are sufficient to generate a detectable signal, and it has been reported that only one molecule of ice nucleation protein needs to be synthesized to allow a positive test result (Wolber, 1993). The Ina protein is detected using a phase-sensitive fluorescent dye that changes colour as the buffer freezes. When samples are cooled appropriately, only those containing infected *Salmonella* cells presenting Ina on their surface freeze, causing the dye to change colour, and this is readily detected using a spectrophotometer. This assay allowed the detection of samples containing only 2 cells/ml of *S. enteritidis* in a buffer system within 3 hours and the sensitivity was further increased by using salmonellae-specific immunomagnetic bead separation in combination with phage detection (Irwin et al., 2000).

A *lambda*-based reporter phage carrying the gene for green fluorescent protein (Gfp) has also been described and used to detect the presence of *E. coli* (Funatsu et al., 2002). As for the *ina* gene, no exogenous substrate is required, which simplifies the assay, and the fluorescent signal can be sensitively detected following excitation of the Gfp protein. Detection requires 4–6 hours of incubation since time for correct folding and oxidation of the Gfp polypeptide into the functional fluorophore is a rate-limiting step. However a benefit of this reporter is that, once formed, the Gfp protein is stable and the signal will be relatively long-lasting; the protein released from cells lysed during the phage infection will actually yield a better signal.

9.3. PHAGE DISPLAY

The development of phage display (Smith, 1985) has revolutionized the isolation of peptides with binding affinity for particular substrates. In this system, libraries of peptide variants (often produced using mutational PCR methods) are cloned into filamentous phage vectors as gene fusions to coat proteins. The cloned sequences are expressed along with the phage coat proteins and incorporated into the mature phage particle and are thus "displayed" on the surface of the mature virion. The size of the peptides that can be expressed are limited as phage structure cannot be disrupted so that the recombinant phage remain viable. The key to their success is that libraries of clones are easily screened because the peptides encoded by the cloned gene fragment are available for analysis (see Smith and Petrenko, 1997). For instance, clones expressing peptides with a particular affinity can be recovered from libraries of phage particles by biopanning experiments, where the target ligand is bound to a solid surface and then incubated with phage particles. Phage that do not bind tightly to the ligand are washed away, and the remaining bound fraction of the library, which is enriched for clones expressing peptides that bind to the target ligand, is recovered. Repeated rounds of biopanning lead to rapid identification of specific clones expressing peptides with the highest binding affinity (see Hayhurst and Georgiou, 2001, and Fig. 9.2).

This approach has led to the development of Phage Antibody technology where the peptide displayed is the antigen-binding domain of an antibody molecule. A library of such phage includes billions of clones expressing antibody-derived peptides with different binding affinities that can then be used in an affinity selection process (see Rader and Barbas, 1997). The most significant advance provided by this method for selecting antibody fragments is that there is no requirement for the target ligand to be immuno-reactive in a given host. Similarly, toxic substances, large amounts of which cannot be delivered into a living tissue to raise an antibody, can also be used as the target ligand.

Generally this type of technology has been used for medical or pharmaceutical applications, with the phage display system being used solely to identify a peptide with the desired binding characteristics. The cloned gene is then transferred to high-level expression systems to produce the antibody fragments in usable quantities. For use in detection assays, the molecule must also be labelled in some way so that the binding event can be quantified. This approach has been successfully used to develop a range of detection assays, including the identification of a variety of viruses (see Petrenko and Vodyanoy, 2003), the differentiation of *Candida* species in clinical specimens (Bliss et al., 2003) and to develop an assay for the species-specific detection of *Bacillus* spores (Zhou et al., 2002; Turnbough, 2003). A peptide-mediated magnetic separation technique has also been reported by Stratmann et al. (2002)

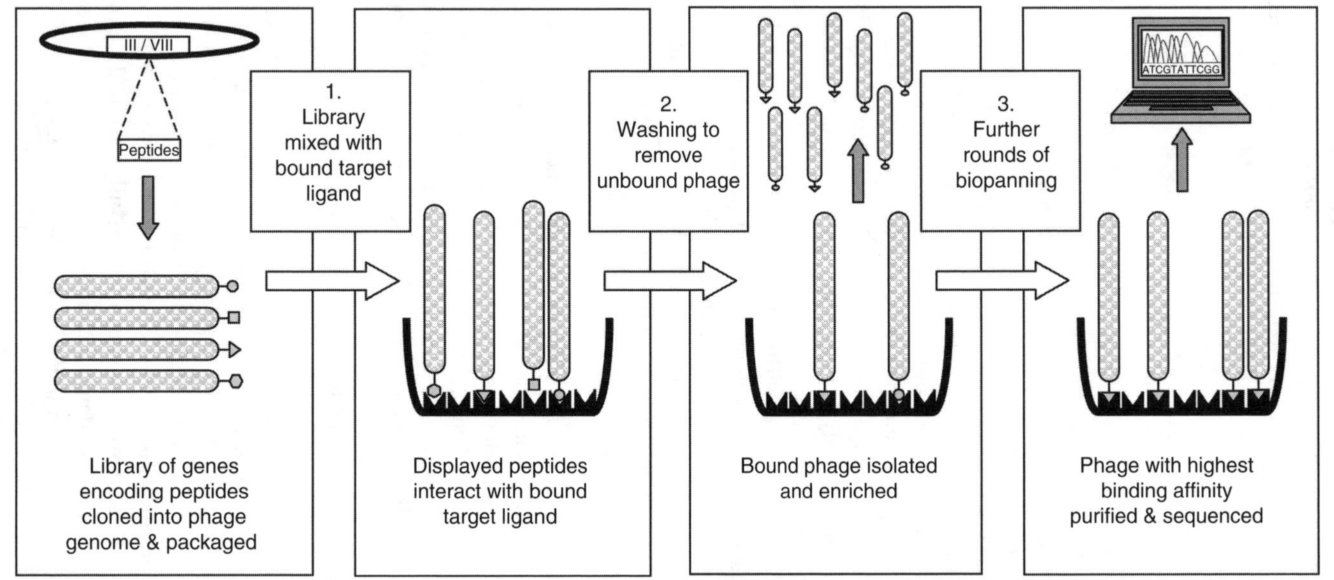

FIGURE 9.2 Phage display/biopanning.
Fragments of peptide libraries are cloned as protein fusions with either the Fd phage coat protein gene III (minor coat protein) or VIII (major coat protein). If peptides are fused with gene III protein, fewer copies of the peptide are displayed on the phage but functional phage are produced without the need for complementation. If peptides are fused with gene VIII protein, wildtype gpVIII must be provided *in trans* in the host cell to allow synthesis of viable phage particles which contain a mixture of native gpVIII and gpVIII fused to peptide sequences. Phage libraries are exposed to the target ligand bound to a solid support and unbound phage removed by washing. Bound phage are eluted and grown to enrich the number of phage present with the desired binding affinity. The biopanning process is repeated, using increasingly stringent washing procedures to finally isolate phage with the highest binding affinity. These are plaque purified and the cloned insert sequenced.

who used phage display to identify a peptide that specifically binds to *Mycobacterium avium* subspecies *paratuberculosis*. The peptide was coated onto paramagnetic beads and then used for magnetic capture of *M. paratuberculosis* from milk samples before carrying out PCR-based identification of the organism.

However, it has also been recognized that instead of purifying the peptide and labelling with a detectable marker, such as alkaline phosphatase, a fluorescent protein or dye, the phage particle itself can be used as a scaffold protein to stabilize the peptide in a detection assay. In this case the phage surface can be conjugated with dyes such as Cy5 or Alexa, either directly or via a fluorescently labelled monoclonal antibody raised against phage surface proteins. Ideally, the peptide is expressed as a fusion to the phage fd major coat protein, pVIII, so that up to 4000 copies of the target ligand binding peptide are present on the surface of the phage particle. This direct usage of phage display particles has been shown to be able to detect staphylococcal enterotoxin B with an efficiency comparative to that achieved for a conventional ELISA assay (Goldman et al., 2000). In addition, Turnborough (2003) showed that both the purified, labelled peptide and a labelled phage-display particle could be used to specifically identify *Bacillus* spores using fluorescent microscopy. However it was noted that the complex surface structure of the phage filament could interfere with the binding of the labelled phage-display particle to some types of spore samples, and this may represent a limitation of the direct utilization of peptide-display phage in detection assays.

9.4. DUAL PHAGE

A new application of phage in detection assays is Dual Phage technology (Wilson, 1999; Fig. 9.3). Like phage display detection methods, this again does not use the specificity of the host-phage relationship to achieve the detection event, but uses phage to detect the successful binding of an antibody to a specific antigen. Antibodies specific for the target molecule are covalently linked to the surface of two different transducing phages, each one conferring resistance to a different antibiotic. Antibody-labelled phage are mixed with the sample and allowed to bind to the antigen. Given the polyvalent nature of antibody binding to many targets, if this occurs the two different transducing phage become physically linked together. Host cells for the phage are then added and plated out on media selective for both antibiotic resistance genes; a low multiplicity of infection of phage is used so that the probability of co-infection by the free phage particles of different types is very low. The appearance of many doubly resistant colonies on the plates indicates that the phage particles were located close together while bound to the same antigenic particle, and this dual infection signals the binding of the antibodies to their cognate antigen.

In this assay, the amplification of the signal comes through the growth of a single doubly transduced cell into a visible antibiotic-resistant colony containing more than 10^8 cells, so the test has good sensitivity as each detection event becomes a visible signal. The ability to use two different antibodies—a different antibody can be linked to each of the transducing phage—also provides a high degree of specificity. This Dual Phage assay has the advantage that no phage engineering is required and it can be applied to any type of antigen for which an antibody already exists. Trials have

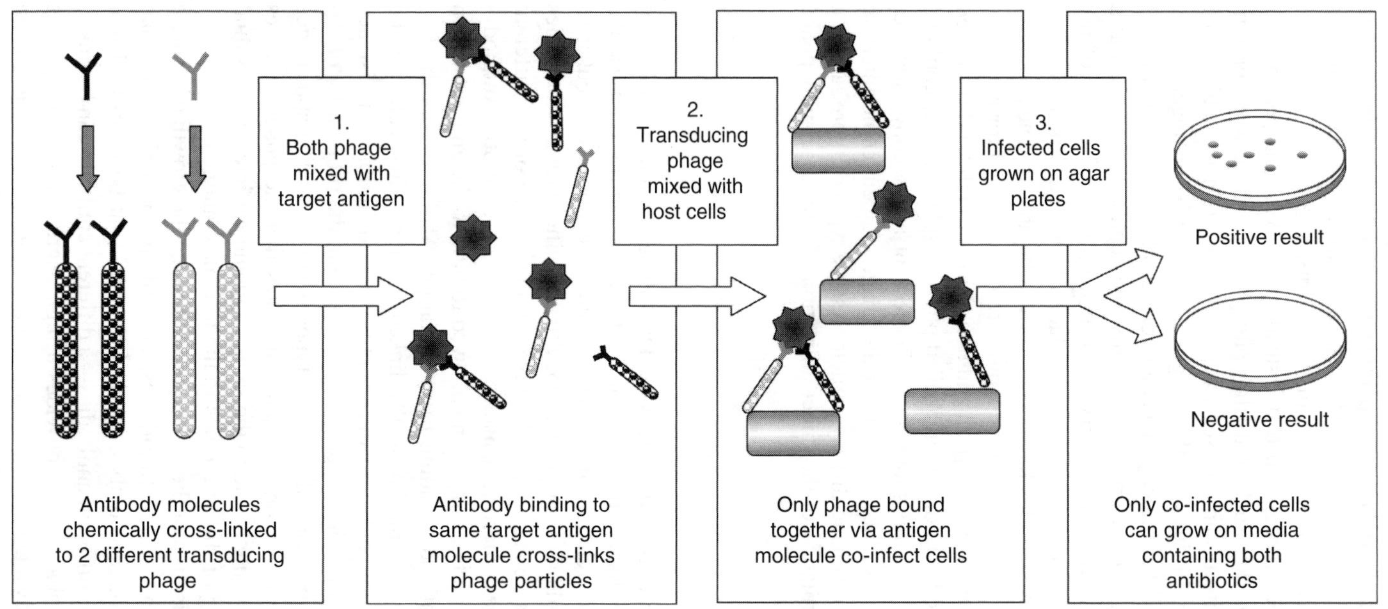

FIGURE 9.3 Dual phage detection assay.
In this assay two transducing phage are used which transduce resistance genes to two different antibiotics into a host cell. Antibodies to the target antigen are chemically bound to transducing phage; if antibodies to different epitopes of the target antigen are used this increases the specificity of the assay. The antibody-labelled phage are mixed with the target antigen and allowed to bind. The sample is mixed with host cells at a suitable dilution rate to ensure that random co-infection of host cells is a rare event. Infected host cells are plated on media containing both antibiotics to select for cells that have been doubly transduced. This only occurs if phage are located close together while bound to the same antigen particle (colonies on agar plate indicate a positive result).

been reported showing that the assay could sensitively detect HIV particles in blood samples, and it is also intended to develop sensitive dual phage assays for prions (Wilson et al., 2002), as prion-specific antibodies have been produced, presumably based on differences in peptide epitope conformations.

9.5. PHAGE AMPLIFICATION ASSAYS

Another method that has been established to detect pathogens using bacteriophage is the Phage Amplification technology. This assay does not use any modified phage, and the endpoint of detection is the formation of a plaque. Since no gene engineering is required, there are no GMO containment concerns, and new tests can be rapidly developed in response to the emergence of new problem bacteria. The technology also uses standard microbiological methods, avoiding the need for high level training in molecular methods and allowing the method to be easily established as part of more traditional microbiological testing regimes.

Such a test has been described by two separate groups and is termed either the Phage Amplification assay (Stewart et al.,1992; 1998) or Phage Amplified Biologically assay (PhaB; Wilson et al., 1997, McNerney et al., 1998). The word *amplification* refers to the fact that the initial infection of the target pathogen is allowed to develop into a plaque, thus providing amplification of the signalling event (see Fig. 9.4). The test begins by adding a phage specific for the target bacterium to a test sample. The samples are then incubated long enough for the phage to infect host cells and enter the eclipse phase. At this point a virucide is used to destroy all exogenous unadsorbed phage; a variety of virucidal compounds have been described, including acetic acid, ferrous ammonium sulphate, plant extracts, and essential oils. A short period of incubation is allowed to ensure complete killing of all the free phage particles, before the virucide is neutralized. The sample is mixed with plating bacteria of a propagating strain for the phage used (termed the *helper* bacteria or *sensor cells*) and the mixture is spread over the surface of an agar plate using a soft agar overlay. The replication cycle of the internalized phage can now be completed, with the formation of one plaque from each infected cell in the sample tested. Hence each plaque represents the presence of one target bacterium in the original sample, detected by using them to protect the phage from the action of the virucide. When testing samples that are likely to contain a high number of organisms overgrowth of the lawn can be a problem and some decontamination steps may be required to selectively enrich the target bacteria (see Mole and Maskell, 2001). However, because of the dilution of the initial sample throughout the assay, and the high number of helper cells added, problems are only encountered when the competitive microflora reaches a level of about 10^6 cfu per ml in the initial sample.

This technology has been shown to effectively detect bacterial pathogens such as *Listeria, Campylobacter, Pseudomonas, Salmonella,* and *Escherichia coli* (Stewart et al., 1998) but it has been most extensively exploited for the detection of *Mycobacterium tuberculosis* (FAST*Plaque*TB; for a review see Mole and Maskell, 2001). In this case, the phage amplification assay provides a great advantage over conventional testing methods since *M. tuberculosis* is a slow-growing organism taking up to 8 weeks to grow into visible colonies. In the FAST*Plaque*TB assay,

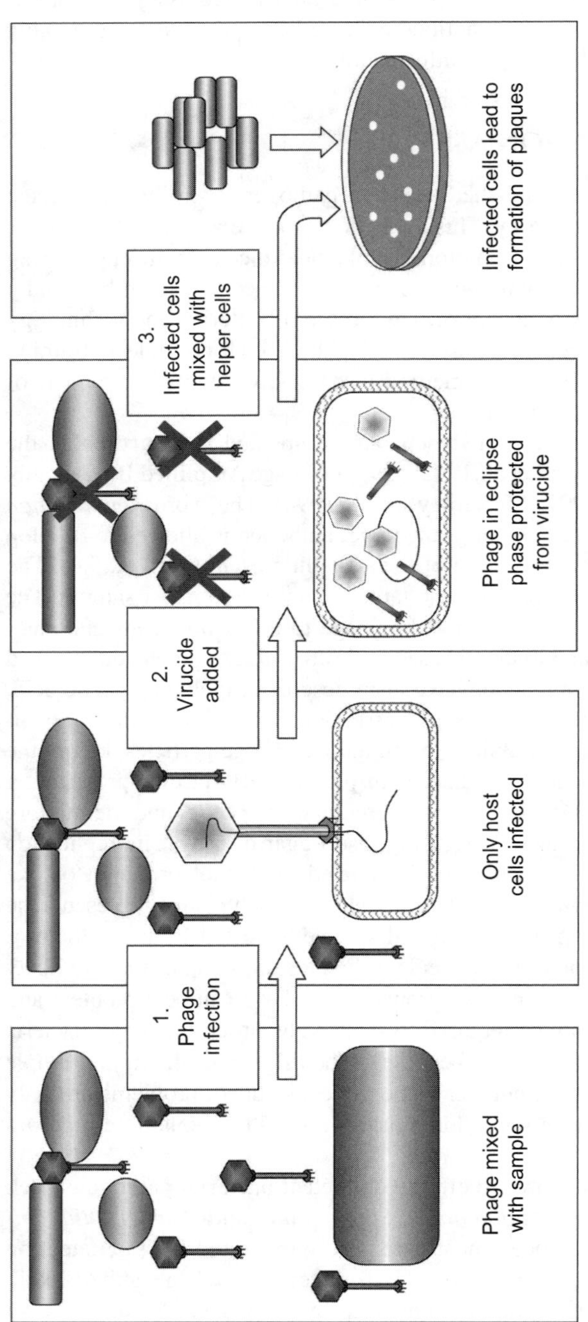

FIGURE 9.4 Phage amplification (or PhaB) assay.
No modification of phage are required for this assay. Phage for the target bacterium are mixed with the sample and allowed to infect susceptible cells. Once phage have entered the eclipse phase, a virucide is added which does not affect the target bacterium. External phage which have not infected cells are destroyed and the only viable phage which survive are those which successfully infect a host cell. The virucide is then neutralized and the sample containing the infected cells mixed with helper cells that will support phage replication (*these do not have to be the same cell type as the target bacterium*). The mixture is spread over agar plates and incubated to allow the lawn to develop. Phage within infected cells finish their replication cycle, lyse the host cell and progeny phage infect helper cells in the lawn, leading to formation of a plaque.

the helper organism used is the fast-growing *Mycobacterium smegmatis,* which is also sensitive to the phage but can form lawns within 12–48 hours. As for the detection of pathogens in food samples, the biggest challenge is dealing with the inhibitory effect of the sample matrix itself, in this case most commonly clinical sputum samples taken from patients. However, advances are being made to overcome this problem, and Park et al. (2003) report a new method that allows the detection of less than 100 viable *M. tuberculosis* cells in 1 ml of sputum sample.

9.6. DETECTION VIA PHAGE-MEDIATED CELL LYSIS

A commonly used rapid hygiene test for the detection of bacterial contaminants is the measurement of ATP released from bacterial cells by firefly luciferase (see Stanley, 1989). The linear relationship between the number of ATP molecules hydrolysed and the production of light by the enzyme allows the levels of ATP in a sample to be determined. Since the amount of ATP per bacterial cell in a given growth condition is quite constant (approximately 10^{-15} g per cell) the amount of light can be correlated to the number of bacterial cells present in a sample. These assays include a general lysing reagent to break open the bacterial cells and release the intracellular ATP; thus the results only give a measure of microbial load rather than the presence of specific pathogens within the microflora. By using either intact phage or recombinantly produced phage endolysins to replace the general lysing agent, the desired specificity can be added to this test. This was demonstrated using both intact phage (Sanders, 1995) and purified phage lysin (Stewart et al., 1996) for the detection of *Listeria monocytogenes*. More recently, Schuch et al. (2002) have used a cloned phage lysin for the ATP-based detection of *Bacillus anthracis* spores within 10 minutes of inducing germination; the assay could detect as few as 100 spores within 60 minutes of addition of the lysin.

Due to high backgrounds, the practical limit of detection for ATP assays of environmental or food samples is approximately 10^4 bacterial cells, and such high numbers of pathogens should not be present in food samples. However, using measurements of released adenylate kinase rather than ATP (see Corbitt et al., 2000) in combination with phage-mediated cell lysis to add specificity, the limit of detection has been reduced to less than 10^3 *E. coli* and *Salmonella* cells (Blasco et al., 1998; Wu et al., 2001). This compares favorably with other rapid methods of bacterial enumeration, such as immunoassays.

The use of phage to release a detectable intracellular marker has also been used to develop other rapid methods of detection. Chang et al. (2002) used a phage specific for *E. coli* O157:H7 to add specificity to a rapid method of detection based on conductance. Impedance is a rapid method that enables detection of bacterial growth by measuring the change in the electrical conductivity of the culture due to the transformation of uncharged or weakly charged molecules into highly charged species (e.g. conversion of polypeptides into individual amino acids). Detecting this change can be achieved either directly, by measuring the change in the conductivity of a liquid culture medium, or indirectly, by monitoring the change in the electrical conductivity of a reaction solution due to absorption of gases from the inoculated bacterial culture (for reviews see Silley and Forsythe, 1996; Wawerla et al., 1999).

The critical parameter when using such methods is the "time to detection" (i.e. when sufficient change in the medium has occurred that a difference in the impedance can be reliably measured). When phage were used in combination with this type of detection method, a positive detection event was a change (increase) in the time to detection for a growing culture when the phage was added to the sample. Simply, if the cells present in the test sample are lysed by the phage they do not grow and therefore do not change the conductance of the growth medium. By comparing a phage-treated sample with an untreated sample, it is therefore possible to identify the type of cells being detected by the conductance method. Again, the simplicity of this is that it uses standard microbiological methods and can quickly be modified to accommodate new bacterial groups of interest.

A similar approach has been taken by Neufeld et al. (2003) who used bacteriophage as a specific lysing agent to allow the detection of a cytoplasmic marker protein (in this case ß-galactosidase) by an amperometric method. This detection method is not an impedance technique per se, but uses a specialized biosensor to specifically detect the end product of the enzymic conversion of the ß-galactosidase substrate p-aminophenyl ß-D-galactopyranoside (PAPG) into p-aminophenol (PAP). Using this technique, they were able to detect *E. coli* cell numbers as low as 1 cfu/100 ml in less than 8 hours. Another variation of this idea has been reported by Takikawa et al. (2002). Here, no appropriate cellular marker protein was present, so cells of *Enterobacter cloacae* were transformed with a chitinase gene. Release of this enzyme from phage-sensitive cells was detected within 30 minutes by monitoring the hydrolysis of the chitinase substrate, 4-methylumbelliferon, using a simple spectrophotometric assay. This method of detection was proposed to track cells released into the environment for studies of bacterial ecology.

9.7. USE OF PHAGE AND PHAGE PRODUCTS AS BINDING REAGENTS

As knowledge of phage structure and function increases, more applications of phage and phage products can be used for the detection of pathogens. Phages have evolved to specifically recognize and bind to structures on the bacterial cell surface, and this interaction can be exploited to develop new reagents with specific binding characteristics. In its simplest form, the *Salmonella* phage Felix 01 (also referred to as phage *Sapphire*) has been used as a biosorbant to selectively separate cells from suspensions containing other related bacterial cell types (Bennett et al., 1997). Here, phage Felix 01 was attached onto polystyrene surfaces (both dipsticks and microtitre plates) by passive immobilization to produce biosorbent materials. These materials were then soaked in cultures of *Salmonella* and, after washing to remove nonspecifically bound cells, the presence of *Salmonella* cells bound to their surface was detected either directly (by fluorescent microscopy) or indirectly (by PCR). In both cases, *Salmonella* cells were found to be specifically bound to the biosorbent surfaces, and using the microtitre plate assay the *Salmonella* could be selectively enriched in the presence of equal numbers of competing Enterobacteriaceae. Strains of *Salmonella* that were not sensitive to this phage did not bind to the biosorbents. One of these strains (*Salmonella enterica* serovar Arizonae) is known to synthesize incomplete LPS molecules, supporting the conclusion that the specificity of the assay was due to phage-host interactions as the LPS forms the major part of the receptor for phage Felix 01. As more detail is discovered about the specific interactions that occur between phage and the host cell surface (see Ch. 9), phage proteins can be used to generate recombinant binding proteins, or protein complexes, with the desired binding properties—possibly even contributing to the development of novel families of affinity binding proteins. One such product, a tail fiber protein fragment of *E. coli* phage T4, is now commercially available (EndoTrap). The protein specifically recognizes and binds to a highly conserved component of the LPS of gram-negative bacteria. In a liquid chromatography format with the recombinant T4 tail spike proteins immobilized on a cross-linked agarose, it is used for endotoxin removal from medical and biological fluids.

The possibilities afforded by the recognition specificity have also been realized to some extent through the study of phage endolysins and an understanding of the detailed function of these proteins. The endolysins that destroy the bacterial cell wall at the end of bacteriophage multiplication are composed of at least two distinct domains, as discussed in Chapter 11. Besides the enzymatically active region of the protein, highly specific domains target the enzymes to sites within the bacterial cell wall. These cell wall binding domains (CBD) mediate tight (but non-covalent) and extremely specific binding of the enzymes to their substrate molecules (Loessner et al., 2002). This is an excellent example of how specific and fine-tuned the interaction of bacterial host and phage actually is. The extraordinary specificity and high affinity of these polypeptides renders them ideal tools for decoration of the target cells by fluorescently labelled CBD polypeptide domains (Loessner et al., 2002), and can also be employed for immobilization. Paramagnetic particles coated with anti-*Listeria* CBD molecules are very efficient tools for the separation and recovery of *Listeria* cells from various foods; model experiments indicated recovery rates of more than 90% even from dilute suspensions and mixed bacterial flora (Jan Kretzer and Martin Loessner, unpublished data).

ACKNOWLEDGEMENTS

Thanks are due to past and present members of our laboratories, who so enthusiastically turned to the "phage side"; we are especially grateful to Siegfried Scherer and the late Gordon Stewart for their support and encouragement and to Dr. Christine Dodd for helpful contributions to the manuscript.

REFERENCES

Bardarov S, Dou H, Eisenach K, Banaiee N, Ya S, Chan, J, Jacobs, WR, and Riska, PF (2003). Detection and drug-susceptibility testing of *M. tuberculosis* from sputum samples using luciferase reporter phage: Comparison with the Mycobacteria Growth Indicator Tube (MGIT) system. *Diag. Microbiol. Infect. Dis.*, 45: 53–61.

Bennett AR, Davids FGC, Vlahodimou S, Banks JG, and Betts RP (1997). The use of bacteriophage-based systems for the separation and concentration of Salmonella. *J. Appl. Microbiol.*, 83: 259–265.

Blasco, R, Murphy, MJ, Sanders, MF, and Squirrell DJ (1998). Specific assays for bacteria using phage-mediated release of adenylate kinase. *J. Appl. Microbiol.,* 84: 661–666.

Bliss JM, Sullivan MA, Malone J, and Haidaris CG (2003) Differentiation of Candida albicans and *Candida dubliniensis* by using recombinant human antibody single-chain variable fragments specific for hyphae. *J. Clin. Microbiol.,* 41: 1152–1160.

Blissett SJ, and Stewart GSAB (1989). In vivo bioluminescent determination of apparent Kms for aldehyde in recombinant bacteria expressing *luxAB*. *Letts. Appl. Microbiol.,* 9: 149–152.

Brown KL, Sarkis GJ, Wadsworth C, and Hatfull GF (1997). Transcriptional scilencing by the mycobacteriophage L5 repressor. *EMBO J.,* 16: 5914–5921.

Chang TC, Ding HC, and Chen SW (2002). A conductance method for the identification of Escherichia coli O157: H7 using bacteriophage AR1. *J. Food Protect.,* 65: 12–17.

Chen J, and Griffiths MW (1996). Salmonella detection in eggs using *lux*+ bacteriophages. *J. Food Protect.,* 59: 908–914.

Corbitt AJ, Bennion N, and Forsythe SJ (2000). Adenylate kinase amplification of ATP bioluminescence for hygiene monitoring in the food and beverage industry. *Letts. Appl. Microbiol.,* 6: 443–447.

Funatsu T, Taniyama T, Tajima T, Tadakuma H, and Namiki H (2002). Rapid and sensitive detection method of a bacterium by using a GFP reporter phage. *Microbiol. Immunol.* 46: 365–369.

Goldman ER, Pazirandeh MP, Mauro, JM, King, KD, Frey, JC, and Anderson, GP (2000). Phage-display peptides as biosensor reagents. *J. Molec. Recog.,* 13: 382–387.

Goodridge L, Chen J, and Griffiths M (1999). Development and characterization of a fluorescent-bacteriophage assay for detection of *Escherichia coli* O157:H7. *Appl Environ Microbiol.,* 65: 1397–1404.

Hayhurst A, and Georgiou G (2001). High-throughput antibody isolation. *Curr. Opin. Chem. Biol.,* 5: 683–689.

Hennes KP, Suttle CA, and Chan AM (1995). Fluorescently labeled virus probes show that natural virus poulations can control the structure of marine microbial communities. *Appl Environ Microbiol.,* 61: 3623–3627.

Hill PJ, Swift S, and Stewart GSAB (1991). PCR Based gene gggngineering of the *Vibrio harveyi lux* operon and the *Escherichia coli trp* operon provides for biochemically functional native and fused gene products. *Molec. Gen. Genet.,* 226: 41–48.

Irwin P, Gehring A, Tu SI, Brewster J, Fanelli J, and Ehrenfeld E (2000). Minimum detectable level of Salmonellae using a binomial-based bacterial ice nucleation detection assay (BIND). *J. AOAC Int.,* 83: 087–1095.

Jacobs WR, Barletta RG, Udani R, Chan J, Kalkut G, Sosne G, Kieser T, et al. (1993). Rapid assessment of drug susceptibilities of *Mycobacterium tuberculosis* by means of Luciferase Reporter Phages. *Science,* 260: 819–822.

Kodikara CP, Crew HH, and Stewart GSAB (1991). Near on-line detection of enteric bacteria using *lux* recombinant bacteriophage. *FEMS Microbiol. Letts.,* 83: 261–266.

Kuhn J, Suissa M, Wyse J, Cohen I, Weiser I, Reznick S, Lubinsky-Mink S, et al. (2002). Detection of bacteria using foreign DNA: The development of a bacteriophage reagent for Salmonella. *Int. J. Food Microbiol.,* 74: 229–238.

Li Z, Szittner R, and Meighen EA (1993). Subunit interactions and the role of the luxA polypeptide in controlling thermal stability and catalytic properties in recombinant luciferase hybrids. Biochim Biophys *Acta.,* 1158: 137–145.

Loessner MJ, Kramer K, Ebel F, and Scherer S (2002). C-terminal domains of Listeria bacteriophage peptidoglycan hydrolases determine specific recognition and high affinity binding to bacterial cell wall carbohydrates. *Molec. Microbiol.,* 44:335–349.

Loessner MJ, Krause IB, Henle T, and Scherer S (1994). Structural proteins and DNA characteristics of 14 Listeria typing bacteriophages. *J. Gen. Virol.,* 75: 701–710.

Loessner MJ, and Scherer S (1995). Organization and transcriptional analysis of the Listeria phage A511 late gene region comprising the major capsid and tail sheath protein genes *cps* and *tsh*. *J. Bacteriol.,* 177: 6601–6609.

Loessner MJ, Rees CED, Stewart GSAB, and Scherer S (1996). Construction of luciferase reporter bacteriophage A511::luxAB for rapid and sensitive detection of viable Listeria cells. *Appl. Environ. Microbiol.,* 62: 1133–1140.

Loessner MJ, Rudolf M, and Scherer S (1997). Evaluation of luciferase reporter bacteriophage A511::luxAB for detection of *Listeria monocytogenes* in contaminated foods. *Appl. Environ. Microbiol.,* 63: 2961–2965.

Mackey BM, Cross D, and Park SF (1994). Thermostability of bacterial luciferase expressed in different microbes. *J. Appl. Bacteriol.,* 7: 149–154.

McNerney R, Wilson SM, Sidhu AM, Harley VS, Al Suwaidi Z, Nye PM, Parish T, and Stoker NG (1998). Inactivation of mycobacteriophage D29 using ferrous ammonium sulphate as a tool for the detection of viable *Mycobacterium smegmatis* and *M. tuberculosis. Res. Microbiol.,* 149: 487–495.

Mole RJ, and Maskell TWO'C (2001). Phage as a diagnostic—the use of phage in TB diagnosis. *J. Chem. Technol. Biotechnol.,* 76: 683–688.

Neufeld T, Schwartz-Mittelmann A, Biran D, Ron EZ, and Rishpon J (2003). Combined phage typing and amperometric detection of released enzymatic activity for the specific identification and quantification of bacteria. *Anal. Chem.,* 75: 580–585.

Park DJ, Drobniewski FA, Meyer A, and Wilson SM (2003). Use of a phage-based assay for phenotypic detection of mycobacteria directly from sputum. *J. Clin. Microbiol.,* 41: 680–688.

Petrenko VA, and Vodyanoy VJ (2003). Phage display for detection of biological threat agents. *J. Microbiol. Meth.,* 53: 253–262.

Rader C, and Barbas CF (1997). Phage display of combinatorial antibody libraries. *Curr. Opin. Biotech.,* 8: 503–508.

Riska PF, Jacobs WR, Bloom BR, McKitrick J, and Chan J (1997). Specific identification of Mycobacterium tuberculosis with the luciferase reporter mycobacteriophage: Use of p-nitro-alpha-acetylamino-beta-hydroxy propiophenone. *J. Clin. Microbiol.,* 35: 3225–3231.

Sanders MF (1995). A rapid bioluminescent technique for the detection and identification of *Listeria monocytogenes* in the presence of *Listeria innocua*. In: Campbell, A.K., Kricka, L.J., and Stanley, P.E. (Eds), *Bioluminescence and Chemiluminescence: Fundamental and Applied Aspects.* John Wiley & Sons, Chichester. pp. 454–457.

Sarkis GJ, Jacobs WR, and Hatfull GF (1995). L5 Luciferase reporter mycobacterio-phages: A sensitive tool for the detection and assay of live Mycobacteria. *Molec. Microbiol.,* 15: 1055–1067.

Schuch R, Nelson D, and Fischetti VA (2002). A bacteriolytic agent that detects and kills *Bacillus anthracis. Nature,* 418: 884–889.

Silley P, and Forsythe S (1996). Impedance microbiology - A rapid change for microbiologists. *J. Appl. Bacteriol.,* 80: 233–243.

Smith GP (1985). Filamentous fusion phage—novel expression vectors that display cloned antigens on the virion surface. *Science,* 228: 1315–1317.

Smith GP, and Petrenko VA (1997). Phage display. *Chem. Rev.,* 97: 391–410.

Stanley PE (1989). A review of bioluminescence ATP techniques in rapid microbiology. *J. Biolumin. Chemilumin.,* 4: 375–380.

Stewart GSAB, Jassim SAA, Denyer SP, Park S, Rostas-Mulligan K, and Rees CED (1992). Methods for rapid microbial detection. Patent WO 92/02633.

Stewart GSAB, Loessner MJ, and Scherer S (1996). The bacterial *lux* gene bioluminescent biosensor revisited. *ASM News,* 62: 297–301.

Stewart GSAB, Jassim SAA, Denyer SP, Newby P, Linley K, and Dhir VK (1998). The specific and sensitive detection of bacterial pathogens within 4 h using bacteriophage amplification. *J. Appl. Microbiol.,* 84: 777–783.

Stratmann J, Strommenger B, Stevenson K, and Gerlach GF (2002). Development of a peptide-mediated capture PCR for detection of *Mycobacterium avium* subsp paratuberculosis in milk. *J. Clin, Microbiol.,* 40: 4244–4250.

Takikawa Y, Mori H, Otsu Y, Matsuda Y, Nonomura T, Kakutani K, Tosa Y, et al. (2002). Rapid detection of phylloplane bacterium *Enterobacter cloacae* based on chitinase gene transformation and lytic infection by specific bacteriophages. *J. Appl. Microbiol.,* 93: 1042—1050.

Turnbough CL (2003). Discovery of phage display peptide ligands for species-specific detection of Bacillus spores. *J. Microbiol. Meth.,* 53: 263–271.

Ulitzur S, and Kuhn J (1987). Introduction of lux genes into bacteria; a new approach for specific determination of bacteria and their antibiotic susceptibility. In: Slomerich, R., Andreesen, R., Kapp, A., Ernst, M., and Woods, W.G. (Eds), *Bioluminescence and Chemiluminescence new perspectives.* John Wiley & Sons, New York, pp 463–472.

Ulitzur S, and Kuhn J (1989). Detection and/or identification of microorganisms in a test sample using bioluminescence or other exogenous genetically introduced marker. US Patent 4,861,709.

Watson BB, and Eveland WC (1965). The application of the phage-fluorescent antiphage staining system in the specific identification of *Listeria monocytogenes. J. Infect.,* 115: 363–369.

Wawerla M, Stolle A, Schalch B, and Eisgruber H (1999). Impedance microbiology: Applications in food hygiene. *J. Food Protect.* 62: 1488–1496.

Wilson SM (1999) Analytical method using multiple virus labelling. PCT Patent WO99/63348.

Wilson SM, Al Suwaidi Z, McNerney R, Porter J, and Drobniewski F (1997). Evaluation of a new rapid bacteriophage-based method for the drug susceptibility testing of *Mycobacterium tuberculosis. Nat. Medicine.,* 3: 465–468.

Wilson SM, Lane A, Oliver J, and Stanley C (2002). Investigation of the potential of "Dual Phage," a novel ultra-sensitive detection label. *Clin. Chem.,* 48: B89 Part 2 Suppl. S.

Wolber PK (1993). Bacterial ice nucleation. *Adv. Micro. Physiol.,* 34: 203–237.

Wolber PK, and Green RL (1990). Detection of bacteria by transduction of ice nucleation genes. *Trends Biotechnol.,* 8: 276–279.

Wu Y, Brovko L, and Griffiths MW (2001). Influence of phage population on the phage-mediated bioluminescent adenylate kinase (AK) assay for detection of bacteria. *Letts. Appl. Microbiol.,* 33: 311–315.

Zhou B, Wirsching P, and Janda KD (2002). Human antibodies against Bacillus: A model study for detection of and protection against anthrax and the bioterrorist threat. *Proc. Natl. Acad. Sci. USA.,* 99: 5241–5246.

10 Control of Bacteriophages in Industrial Fermentations

Sylvain Moineau[1,2] *and Céline Lévesque*[1]
[1]Département de Biochimie et de Microbiologie, Faculté des Sciences et de Génie, Groupe de recherche en écologie buccale (GREB), Faculté de Médecine Dentaire, Université Laval, Québec, Canada
[2]Félix d'Hérelle Reference Center for Bacterial Viruses, Université Laval, Québec, Canada

CONTENTS

10.1. Introduction 285
10.2. Phages and the Dairy Industry 287
  10.2.1. Overview 287
  10.2.2. Phage Detection 288
  10.2.3. Sources of Phage Contamination 289
10.3. Phage Control Strategies 290
  10.3.1. Controlling Phage Proliferation 290
  10.3.2. Strain Selection 291
  10.3.3. Development/Selection of Bacteriophage-Insensitive Bacterial Mutants 291
  10.3.4. Genetically Engineering Phage-Resistant Bacterial Strains 293
Acknowledgments 294
References 295

10.1. INTRODUCTION

Bacteria are used in a variety of industrial fermentation processes because of their ability to convert a wide variety of substrates into complex products or specific molecules. In principle, any industries and technological processes relying on bacterial fermentation are vulnerable to bacteriophage infection, and phages represent a

0-8493-1336-X/05/$0.00+$1.50
© 2005 by CRC Press LLC

constant threat of serious economic losses for those industries. The literature describing such problems goes back several decades (reviewed by Ackermann and DuBow, 1987; Ogata, 1980; Wünsche, 1989), and faulty fermentations due to phages also are well recognized in many modern biotechnological industries commercializing or using various bacterial fermentation products (Table 10.1). However, many other cases of

TABLE 10.1
Examples of Faulty Fermentations Due to Bacteriophages

Bacterial Species	Product
Acetobacter sp.	Vinegar
Bacillus colistinus	Colistin
Bacillus polymyxa	Polymycin
Bacillus subtilis	Amylase, protease
Bacillus subtilis var. *natto*	Fermented soy beans
Bacillus thuringiensis	Insecticide (BT)
Brevibacterium lactofermentum	L-glutamic acid
Clostridium sp.	Acetone, butanol
Corynebacterium sp.	L-glutamic acid
Escherichia coli	Various biotechnology products
Gluconobacter sp.	Gluconic acid
Lactobacillus acidophilus	Fermented milk
Lactobacillus brevis	Lactic acid
Lactobacillus casei	Fermented milk
Lactobacillus delbrueckii subsp. *bulgaricus*	Yogurt
Lactobacillus delbrueckii subsp. *lactis*	Cheese
Lactobacillus fermentum	Sourdough bread
Lactobacillus helveticus	Cheese
Lactobacillus plantarum	Silage, sauerkraut
Lactococcus lactis	Buttermilk, cheese, sour cream
Leuconostoc mesenteroides	Sauerkraut, buttermilk, sour cream
Leuconostoc fallax	Sauerkraut
Oenococcus oeni	Malolactic fermentation
Propionibacterium freudenreichii	Cheese
Pseudomonas aeruginosa	2-Ketogluconic acid
Streptococcus thermophilus	Cheese, yogurt
Streptomyces aureofaciens	Tetracycline
Streptomyces endus	Endomycin
Streptomyces griseus	Streptomycin
Streptomyces kanamyceticus	Kanamycin
Streptomyces venezuelae	Chloramphenicol
Tetragenococcus halophila	Soy sauce
Xanthomonas campestris	Xantham

Updated from (Ackermann and DuBow, 1987; Jones, et al., 2000; Ogata, 1980; Wünsche, 1989)

phage contamination are not reported in scientific publications, and microbiologists learn of them primarily via confidential reports and personal communications (Ackermann and Moineau, unpublished data).

One industry that has openly acknowledged phage infection of their bacterial strains is the dairy industry. In fact, dairy microbiologists have attempted for more than 70 years to eliminate, or at least to bring under better control, the bacteriophages that interfere with the manufacture of many fermented milk products (Moineau et al., 2002). As a result, faulty milk fermentations are, undoubtedly, the best-documented examples of virulent bacteriophage infections in the industrial settings. The purpose of this chapter is to present an overview of the current strategies used by the dairy industry to curtail bacteriophage attacks. The general approaches described here are likely to be effective in many other similar fermentation processes (Table 10.1). In addition, more specific information about the ecology of industrial phages, including phages relevant to the dairy industry and to various non-dairy food fermentation processes (e.g., the preparation of sauerkraut) is presented in Chapter 6. Also, comprehensive reviews about dairy phages and their control strategies have recently been published (Boucher and Moineau, 2001; Coffey and Ross, 2002).

10.2. PHAGES AND THE DAIRY INDUSTRY

10.2.1. Overview

Nonsporulating, Gram-positive microorganisms known as the lactic acid bacteria (LAB) are used in the manufacture of yogurt and various cheeses. The LAB include members of several bacterial genera and species; e.g., *Lactococcus lactis*, *Lactobacillus* sp., *Leuconostoc* sp., and *Streptococcus thermophilus* (Table 10.1). Large-scale milk fermentations typically begin after inoculation with carefully selected cultures containing a blend of a few LAB strains at a level of 10^7 cells per ml of milk (Moineau et al., 2002). These commercial fermentative processes are vulnerable to virulent bacteriophages ubiquitous in the dairy environment, which can lyse the LAB, thereby delaying lactic acid production and even stopping the fermentation process. Typically, milk fermentation is compromised when phage titers are higher than a critical threshold of 10^4 plaque-forming units (PFU)/ml. When the titer rises beyond 10^5–10^6 PFU/ml, a fermentation failure is the likely consequence. The milk inoculated with the starter will not coagulate and the product's desired properties are altered or entirely lost. The production line must then be carefully cleaned to eliminate the phages. It should be noted that no level of bacteriophage contamination is acceptable in other biotechnology industries because of the potential for process and product adulteration (G. Bogosian, unpublished data). Dairy virulent phages may spread very rapidly (they have a latent period and a burst size of approximately 30 min and 100, respectively) within a dairy plant. Their exponential multiplication is of even more concern in any process that allows the bacteria to multiply for several generations, such as cheese manufacture. Based on the authors experiences, between 0.1% and 10% of milk fermentations within any given dairy factory may be negatively affected by phages.

10.2.2. Phage Detection

Virulent phages can proliferate quickly; therefore, constant monitoring is critical. A dependable phage assay should be available for any fermentation process that relies on bacteria to convert a substrate into a commercial product. The presence of virulent phages in dairy samples is usually detected using conventional microbiological methods such as plaque assays (see the Appendix section of this book for a detailed methodology). Even if starter cultures are composed of a blend of strains, plaque assays should be performed with pure cultures whenever possible. The presence of plaques at various sample dilutions confirms the presence of phages. However, it should be remembered that plaques formed by some phages may be difficult to observe on the bacterial lawn. In such cases, increasing or decreasing the incubation temperature, varying the multiplicity of infection, adding co-factors (e.g., Ca^{++} or Mg^{++}), or adding a bacterial cell wall-weakening agent (e.g., 0.25% glycine) may increase plaque size (Lillehaug, 1997).

If a reliable plaque assay is not available, tests assessing bacterial activity may be used to monitor the effect of fermentation by-products on the metabolic activity of bacteria. In this context, possibly the most common assay is the milk activity test. Briefly, 10 ml of milk (in each of two test tubes) are inoculated with aliquots (ca. 1 ml) of a starter culture. One tube is also inoculated (2%, final concentration) with a filtered (0.45 μm) sample to be tested for phages. After incubation (the temperature and time depend on the starter and process), the pH of the milk in the tubes is measured. If the pH of the milk containing the filtrate is 0.2 unit higher than that of the milk in the other tube (indicating reduced production of lactic acid), phage contamination is suspected. A pH indicator; e.g., bromocresol purple (100 μg/ml) may also be added to the milk, to avoid direct pH measurements. In such an assay, when the milk's pH drops below 5.4, the indicator turns from purple to yellow. Thus, the absence of a color change during incubation indicates the presence of an agent (e.g., phages) that interferes with lactic acid production in the inoculated milk. If many samples need to be processed daily, the availability of a pH indicator is particularly useful for the development of a miniaturized assay enabling simultaneous testing of multiple samples in, for example, 96-well microtiter plates.

Another approach for detecting bacteriophages in dairy products is to use an immunochemical assay (i.e., an ELISA). The ELISA technique uses antibodies raised against specific phage structural proteins and is usually fairly specific; thus, although that type of assay can be useful for detecting specific phages, it may not detect bacteriophages having different structural proteins. Also, the sensitivity of an ELISA for phage detection in dairy samples is usually low (Chibani Azaiez et al., 1998; Moineau et al., 1993). The availability of complete phage genome sequences has increased the popularity of using molecular DNA techniques (e.g., PCR and DNA hybridizations) to detect bacteriophages rapidly (Labrie and Moineau, 2000; Moineau et al., 1992). However, under some circumstances, the high sensitivity of DNA-based methods may yield false-positive results (Labrie and Moineau, 2000; Moineau et al., 1992). Thus, the choice of the phage detection method will depend on several factors, including the type of laboratory equipment available and the required detection sensitivity. Although of limited interest to most industries, additional information about the detected phages may be provided by electron microscopic observations, and the determination of host ranges, protein profiles, and genomic restriction patterns. The dairy industry is far ahead of other industries in that regard, as hundreds (if not thousands) of LAB phages have been characterized around the world (Ackermann, 2001; Brüssow, 2001; Brüssow and Desiere, 2001).

10.2.3. Sources of Phage Contamination

After the presence of phages is confirmed, the next step consists of identifying the source of contamination. This natural biological threat is either endogenous (i.e., a prophage) or exogenous (it is present in the substrate or environment). The use of a lysogenic strain is seldom warranted in any industrial process, because prophage(s) may be induced by the manufacturing conditions. Moreover, serially passaging a temperate phage may result in its replacement by a virulent, derivative deletion mutant (Bruttin and Brüssow, 1996). Accordingly, in order to eliminate strains containing inducible prophages, bacterial strain selection programs should include a screening assay for lysogeny. Genome sequencing projects have revealed that prophages are abundant in many bacterial species (Canchaya et al., 2003). Although the presence of prophages may be, at least in theory, beneficial in protecting the bacteria against closely related phages, or in allowing their timely lysis (e.g., during cheese ripening), the benefit of this superinfection immunity is limited because of the natural diversity of phage populations.

In the dairy industry, the most probable sources of virulent phages are the substrate and the factory environment (McIntyre et al., 1991). Indeed, most dairy fermentations are susceptible to phage infections because raw milk naturally contains lytic phages, albeit at low titers (between 10^1 and 10^4 PFU/ml). Two basic types of ecological situations are distinguishable in the industrial milk fermentation environment: the yogurt factory and the cheese factory. Milk for yogurt production undergoes treatment at 90°C, which usually kills all phages (Quiberoni et al., 1999); however, raw or pasteurized milk is used in cheese fermentation, and many phages remain viable after pasteurization (Chopin, 1980). Also, yogurt production is a relatively aseptic process, and the fermented product is minimally exposed to the factory environment. In contrast, cheese factories daily experience aerosol contaminations during whey separation in the open vats (Budde-Niekiel et al., 1985). Although the presence of phages is not a regular observation during yogurt production, it is sufficiently frequent to make phages the primary source of fermentation problems during that process. The problem usually manifests itself as fermentation delays or product alterations, and, occasionally, as a complete loss of product. In contrast, the potential problems that may occur during the fermentation process in cheese factories result from the persistent coexistence of phages and bacterial starter cultures in milk.

The diversity of phages in the milk used to manufacture fermented dairy products probably results from its being a mixture of milk obtained from several farms, each with its own agricultural management practices. Moreover, the fluid nature of milk and fermentation by-products (e.g., whey) ensures good phage dispersion in the substrate. In contrast, phages are usually trapped and localized in cheese. Also, if the same starter cultures are continuously used, they may provide permanent hosts for phage proliferation (Boucher and Moineau, 2001). Thus, the source of phage contamination will dictate the appropriate phage control strategies.

10.3. PHAGE CONTROL STRATEGIES

10.3.1. Controlling Phage Proliferation

Phages are omnipresent in the dairy environment; therefore, during the last several decades, considerable efforts were aimed at controlling their proliferation rather than trying to eradicate them (Moineau, 1999). The control of phage proliferation relies on the development and implementation of a variety of practical approaches at both the factory and laboratory levels (Table 10.2). Controlling phage proliferation in production facilities primarily depends on the risks and sources of phage infection associated with the manufacturing process used, as illustrated earlier for yogurt and cheese production. These control strategies include adequate factory hygiene, an adapted manufactory design, regular cleaning and disinfecting of the equipment and clothes, and an appropriate airflow. More specifically, cross-contamination should be avoided between the bacterial medium, the fermentation by-products and the factory equipment. Sodium hypochlorite and peracetic acid should be used in the cleaning and disinfecting procedures because they both rapidly eliminate phage particles

TABLE 10.2
Various Measures to Control Phages within a Dairy Food Factory

- Use starter cultures that do not contain strains with similar phage sensitivity patterns
- Perform daily tests for phage detection
- Aseptically prepare the starter cultures
- Use an anti-phage starter medium
- Directly inoculate milk with concentrated cultures
- Rotate bacterial starter strains
- Air filtration (HEPA) and positive pressure from starter room to product packaging
- Use footbaths (cleaned daily) containing disinfecting agent at the entry of rooms
- Limit the production of aerosols
- Clean and chlorinate vats between fills
- Use separate rooms for starter production and cheese manufacture
- Proper planning of the production pipelines to avoid contact between substrate (milk) and wastes (water, whey)
- Avoid using the same equipment (e.g., centrifuges) for milk and whey (which may contain high levels of phages)
- Clean the floor drains in the starter and fermentation rooms
- Regularly inspect equipments' surfaces to remove biofilms and milkstone deposits
- Spray the environment with adequate disinfecting agents (e.g., chlorine-containing compounds)
- Rinse the vats with water and not with by-products such as whey
- Rotate the manufacturing processes

(Binetti and Reinheimer, 2000). Phages are also usually sensitive to acid (pH 4 and lower) and alkaline (pH 11 and above) conditions (Adams, 1959). The air should be filtered and the efficiency of the air filters should be checked regularly because phage infections may arise from damaged filters. In that regard, phages have been detected (10^5 PFU/m^3) in the air of a dairy plant (Neve et al., 1994). Thus, to prevent contamination by phages from the outside environment, bacterial starter cultures also should be prepared in a positively pressurized room. In addition, plant personnel should be aware of the importance of phage control procedures and trained in their application. Other phage control measures are available at the factory level, and they will vary according to the type and size of the production facilities (e.g., yogurt production facilities are usually smaller than their cheese counterparts). These and other approaches have been reviewed by Champagne and Moineau (1998).

10.3.2. Strain Selection

The use of effective bacterial starter cultures is critical to obtain quality products. Strain selection is a long and expensive operation based on several microbiological, biochemical, and technical criteria (Moineau, 1999). The nature of these criteria will depend primarily on the product. Whenever possible, a mixture of two or three well-characterized bacterial strains should be used to prevent the possible phage infection of the entire starter culture and subsequent fermentation or process failures. In this context, one successful anti-phage strategy that is widely applied by cheese-makers involves rotating bacterial strains. A basic operation is to determine the phage host range with a collection of potential starter bacteria. This knowledge then allows the rational design of a starter rotation system. The impact of phages on the fermentation process can be limited if a starter strain with a susceptibility pattern A is alternated with a starter possessing a non-overlapping susceptibility pattern B. This strategy usually will maintain phage titers at levels that do not interfere with the fermentation process. However, the rotation is based on the availability of a sufficient number of performing strains. Difficulties in identifying phage-insensitive strains, and the variable fermentations caused by strain differences, are among the problems associated with rotation schemes (Moineau, 1999). Also, from an evolutionary point of view, the use of different strains may also favor the emergence of a heterogeneous phage population, which may enhance the potential for genetic recombination within the phage gene pool. Thus, if possible, it may be worthwhile to rotate the type of fermented dairy products manufactured in the same facility. For example, the manufacture of a cheddar cheese (by means of mesophilic bacteria) could be followed by the production of a mozzarella cheese, which uses thermophilic bacteria. If the population of a specific mesophilic phage has amplified during the cheddar cheese making, it is likely to be diluted out after the mozzarella production process.

10.3.3. Development/Selection of Bacteriophage-Insensitive Bacterial Mutants

When an industrial bacterial strain is found sensitive to a phage, the first step consists of replacing it with a phage-resistant strain possessing the same functional characteristics during the fermentation process. If such a strain is not readily available, the entire strain selection procedure must be repeated. This strategy is time-consuming and costly; therefore, the preferred approach may be to develop/select for phage-resistant bacterial mutants. The development of bacteriophage-insensitive mutants (BIMs) is generally the first approach used to transform a phage-sensitive strain into a phage-resistant mutant. At the present time, BIMs are obtained in the laboratory by a protocol which is very similar to the classical Delbrück-Luria phage challenge experiment (1943). After long-term exposure to lytic phages, a few outgrowing cells that have lost the capacity to adsorb the challenge phage, most likely because of a spontaneous mutation in the chromosomal gene encoding the phage receptor, are selected for further characterization. Phage-resistant mutants are not selected for in the factory because the fermentation process is restarted each time with a frozen standard culture. BIMs may also result from mutations that alter the bacterium's carbohydrate composition, protein profile, or concentration of lipoteichoic acid, which lead to modifications in its cell surface characteristics (Forde and Fitzgerald, 1999). Although the process for obtaining BIMs is relatively simple, it may have substantial disadvantages; e.g., a high frequency of phenotype reversion, and changes in bacterial physiology (e.g., a reduced growth rate). Nonetheless, many BIMs have been isolated and are being used successfully under industrial conditions.

Another strategy to develop a phage-resistant strain relies on the so-called phage defense mechanisms. Bacteria possess natural defense mechanisms enabling them to resist bacteriophage attacks. For example, more than 50 naturally occurring *L. lactis* phage-resistance systems have been characterized, which are divided into four groups according to the time at which they act during the lytic cycle (for a recent review, see Coffey and Ross, 2002). They include (i) inhibition of phage adsorption, (ii) blocking of DNA injection, (iii) restriction/modification systems (R/M), and (iv) abortive infection mechanisms (Abi). The genes encoding these natural defense mechanisms may be plasmid- or chromosomally encoded. One useful feature is that several plasmids are transferable by conjugation, enabling the development of phage-resistant strains without affecting their technologically useful properties. Moreover, these natural anti-phage systems are usually effective against several phage groups (Moineau, 1999). Anti-phage barriers have important commercial value, and many of them have been patented worldwide.

The above-noted four mechanisms of phage resistance have been studied to various degrees. Interference with phage adsorption is most likely achieved via the masking of bacterial receptors. This phenomenon has been attributed to the production of various substances, including galactosyl-containing lipoteichoic acid, a galactose/rhamnose or galactose/glucuronic acid polymer, and cell wall proteins (Coffey and Ross, 2002). However, the genetic determinants responsible for this natural defense mechanism remain to be determined. The second defense mechanism (blocking of DNA injection) is still poorly understood. It prevents the translocation of phage DNA into the bacterial cell without affecting phage adsorption. The genetic loci responsible for this bacterial phenotype have yet to be determined. The R/M systems have been identified in various bacterial groups, including LAB. They are characterized by the presence of two types of enzymes, a site-specific restriction endonuclease and a methylase. R/Ms eliminate a wide variety of phages by cutting their genomes. Finally, the Abi systems include all of the resistance mechanisms that act after phage adsorption/nucleic acid injection and lead to phage-elicited bacterial death. They include defense mechanisms that interfere with phage genome replication, transcription, translation, packaging, or phage particle assembly. These systems are more widespread than originally thought; e.g., more than 20 distinct Abi systems have been already characterized in lactococci (Coffey and Ross, 2002). Each system generally requires one or two genes for expression, and few similarities have been observed between the genes. However, little is known about the exact mode of action of the Abi proteins.

The potential of an anti-phage protective mechanism needs to be thoroughly evaluated because not all systems may be able to withstand phage attacks in industrial settings. Ideally, a system's efficacy of protection (EOP = the phage titer obtained with the resistant host containing the anti-phage system divided by the titer obtained with the sensitive host) should be tested against the main phage groups. A strong phage resistance mechanism will give EOPs in the range of 10^{-7} to 10^{-9}, with 10^{-9} being the maximum, technically measurable strength. An EOP in the range of 10^{-4} to 10^{-6} is considered to be a medium potency system, whereas a weak anti-phage barrier has an EOP of 10^{-1} to 10^{-3}. Strong and medium systems are sought by the dairy industry because weak mechanisms may not provide sufficient protection during fermentation processes (Moineau, 1999). Another way of measuring the effectiveness of an anti-phage system is via a milk-based assay. Briefly, an improved strain will go through (in the presence of phages) a time-temperature curve that mimics a fermentation process. If its anti-phage system protects the strain against a high initial concentration of phages ($>10^6$ PFU/ml), it is likely to be effective in a factory environment. Interestingly, the efficacy of anti-phage mechanisms may be influenced by temperature, with some mechanisms being less efficient at higher temperatures (>37°C). Therefore, the type of dairy fermentation for which the modified strain is designed will also influence the overall evaluation of a resistance barrier.

Multiple, natural anti-phage defense mechanisms may be combined to strengthen the overall phage resistance of a particular strain. In addition, the rotation of an isogenic strain's derivatives containing various phage resistance mechanisms has been proposed (Sing and Klaenhammer, 1993). However, because of their remarkable evolutionary capacity (via point mutations or by exchanging modules with other phages or the host DNA), phages can overcome these host defense mechanisms (Bouchard and Moineau, 2000). Hence, novel phage defense strategies are needed to provide long-term phage resistance in agriculturally important bacterial strains.

10.3.4. Genetically Engineering Phage-Resistant Bacterial Strains

If the use of BIMs or of LAB possessing natural phage resistance mechanisms is not satisfactory, it is now possible to construct novel anti-phage systems. For example, during the last decade, some phage genetic elements have become a source of resistance traits (Moineau, 1999). One such type of system, called Per (phage-encoded resistance), involves the presence of the origins of replication of phages supplied on a high copy number plasmid in the bacterial cell (Hill et al., 1990;

O'Sullivan et al., 1993). These origins of replication interfere with phage multiplication by competing for replication factors, thus providing false replication signals. However, the Per systems are only effective against the phage group from which the origin is derived.

Gene silencing by antisense RNA technology is an alternative means of generating engineered phage-resistant strains (Walker and Klaenhammer, 2000). This technology has rapidly progressed because of the emergence of complete nucleotide sequences of phage genomes and the characterization of individual phage genes. A variety of phage genes have been targeted by antisense RNAs, including genes encoding a (i) putative subunit of DNA polymerase, (ii) single-stranded binding protein, (iii) putative transcription factor, (iv) major tail protein, (v) terminase, (vi) major capsid protein, and (vii) putative helicase. Since they are highly conserved among bacteriophages, the genes involved in DNA replication functions are, perhaps, the most obvious targets for antisense technology.

Triggered suicide is another bacteriophage defense strategy that has been designed as a genetically engineered form of an abortive infection (Djordjevic et al., 1997). This approach uses a bacteriophage-inducible promoter to trigger the expression of a suicide system that simultaneously kills the bacterial cell and the infecting phages. Lethal genes made of restriction gene cassettes are cloned on a high copy plasmid without their corresponding methylase gene downstream of an inducible promoter.

Unfortunately, these engineered resistance mechanisms are only effective against the phage group from which the targeted genetic element was derived. Thus, the systems are only likely to find an application with specific strains of high commercial value. It is also noteworthy that many industrially important strains are refractory to genetic manipulation. Therefore, there is a need for optimizing gene transfer methods and developing novel expression vectors that will be suitable for practical approaches.

In conclusion, several phage control measures are currently available, and their use largely depends on the type and size of the manufacturing facilities. These procedures include an adapted factory design, cleaning and disinfection, and appropriate air control. Although process failure is not always caused by phages, an efficient phage-detection method should be available. In addition to characterizing the contaminating phages, effective phage control measures often require that the source of contamination must be identified. The bacterial strain selection program also should include an assay for lysogeny as well as provide several phage-resistant strains. Fortunately, there are now many options for transforming a phage-sensitive strain possessing desirable fermentation properties into a phage-resistant derivative. Finally, the phage population within a factory is likely to change over time; therefore, control measures should be periodically revisited.

ACKNOWLEDGMENTS

We thank H.-W. Ackermann, H. Brüssow, S. Labrie, and D. Tremblay for critical reading of the manuscript. S.M. is grateful to the following organizations for funding his bacteriophage research program: Natural Sciences and Engineering Research

Council of Canada, Fonds de recherche sur la nature et les technologies du Québec, Agropur, Novalait, Inc., and Rhodia Food.

REFERENCES

Ackermann, H.W., Frequency of morphological phage descriptions in the year 2000. Brief review. *Arch Virol* 146: 843–857, 2001.

Ackermann, H.W. and DuBow, M.S., *Viruses of Prokaryotes. General Properties of Bacteriophages*. CRC Press, Boca Raton, FL, 1987.

Adams, M.H., "Effects of physical and chemical agents (except radiation) on bacteriophage particles," pp. 49–62 in *Bacteriophages*. Interscience Publishers, Bacteriophages, 1959.

Binetti, A.G. and Reinheimer, J.A., Thermal and chemical inactivation of indigenous *Streptococcus thermophilus* bacteriophages isolated from Argentinian dairy plants. *J Food Prot* 63: 509–515, 2000.

Bouchard, J.D. and Moineau, S., Homologous recombination between a lactococcal bacteriophage and the chromosome of its host strain. *Virology* 270: 65–75, 2000.

Boucher, I. and Moineau, S., Phages of *Lactococcus lactis:* an ecological and economical equilibrium. Recent Res Devel Virol 3: 243–256, 2001.

Brüssow, H., Phages of dairy bacteria. *Annu Rev Microbiol* 55: 283–303, 2001.

Brüssow, H. and Desiere, F., Comparative phage genomics and the evolution of *Siphoviridae*: insights from dairy phages. *Mol Microbiol* 39: 213–222, 2001.

Bruttin, A. and Brüssow, H., Site-specific spontaneous deletions in three genome regions of a temperate *Streptococcus thermophilus* phage. *Virology* 219: 96–104, 1996.

Budde-Niekiel, A., Moeller, V., Lembke, J. and Teuber, M., Ecology of bacteriophages in a fresh cheese factory. *Milchwissenschaft* 40: 477–481, 1985.

Canchaya, C., Proux, C., Fournous, G., Bruttin, A. and Brussow, H., Prophage genomics. *Microbiol Mol Biol Rev* 67: 238–276, 2003.

Champagne, C.P. and Moineau, S., "Baćtériophages," pp. 89–116 in *Production de ferments lactiques dans l'industrie laitière*, C.P. Champagne (Ed.). Agriculture et Agroalimentaire Canada. Bibliothèque Nationale du Canada, 1998.

Chibani Azaiez, S.R., Fliss, I., Simard, R.E. and Moineau, S., Monoclonal antibodies raised against native major capsid proteins of lactococcal c2-like bacteriophages. *Appl Environ Microbiol* 64: 4255–4259, 1998.

Chopin, M.C., Resistance of 17 mesophilic lactic *Streptococcus* bacteriophages to pasteurization and spray-drying. *J Dairy Res* 47: 131–139, 1980.

Coffey, A. and Ross, R.P., Bacteriophage-resistance systems in dairy starter strains: Molecular analysis to application. *Antonie Van Leeuwenhoek* 82: 303–321, 2002.

Djordjevic, G.M., O'Sullivan, D.J., Walker, S.A., Conkling, M.A. and Klaenhammer, T.R., A triggered-suicide system designed as a defense against bacteriophages. *J Bacteriol* 179: 6741–6748, 1997.

Forde, A. and Fitzgerald, G.F., Bacteriophage defence systems in lactic acid bacteria. *Antonie Van Leeuwenhoek* 76: 89–113, 1999.

Hill, C., Miller, L.A. and Klaenhammer, T.R., Cloning, expression, and sequence determination of a bacteriophage fragment encoding bacteriophage resistance in *Lactococcus lactis*. *J Bacteriol* 172: 6419–6426, 1990.

Jones, D.T., Shirley, M., Wu, X. and Keis, S., Bacteriophage infections in the industrial acetone butanol (AB) fermentation process. *J Mol Microbiol Biotechnol* 2: 21–26, 2000.

Labrie, S. and Moineau, S., Multiplex PCR for detection and identification of lactococcal bacteriophages. *Appl Environ Microbiol* 66: 987–994, 2000.

Lillehaug, D., An improved plaque assay for poor plaque-producing temperate lactococcal bacteriophages. *J Appl Microbiol* 83: 85–90, 1997.

Luria, S.E. and Delbruck, M., Mutations of bacteria from virus sensitivity to virus resistance. *Genetics* 28: 491–511, 1943.

McIntyre, K., Heap, H.A., Davey, G.P. and Limsowtin, G.K.Y., The distribution of lactococcal bacteriophage in the environment of a cheese manufacturing plant. *Int Dairy J* 1: 183–197, 1991.

Moineau, S., Applications of phage resistance in lactic acid bacteria. *Antonie Van Leeuwenhoek* 76: 377–382, 1999.

Moineau, S., Bernier, D., Jobin, M., Hébert, J., Klaenhammer, T.R. and Pandian, S., Production of monoclonal antibodies against the major capsid protein of the lactococcal phage ul36 and development of an ELISA for direct detection of phages in whey and milk. *Appl Environ Microbiol* 59: 2034–2040, 1993.

Moineau, S., Fortier, J. and Pandian, S., Direct detection of lactococcal bacteriophages in cheese whey using DNA probes. *FEMS Microbiol Lett* 92: 169–174, 1992.

Moineau, S., Tremblay, D. and Labrie, S., Phages of lactic acid bacteria: from genomics to industrial applications. *ASM News* 68: 388–393, 2002.

Neve, H., Kemper, U., Geis, A. and Heller, K., Monitoring and characterization of lactococcal bacteriophages in a dairy plant. *Kieler Milchwirtschaftliche Forschungsberichte* 46: 167–178, 1994.

Ogata, S., Bacteriophage contamination in industrial processes. *Biotechnol Bioengin* 22: 177–193, 1980.

O'Sullivan, D.J., Hill, C. and Klaenhammer, T.R., Effect of increasing the copy number of bacteriophage origins of replication, in trans, on incoming-phage proliferation. *Appl Environ Microbiol* 59: 2449–2456, 1993.

Quiberoni, A., Suarez, V.B. and Reinheimer, J.A., Inactivation of *Lactobacillus helveticus* bacteriophages by thermal and chemical treatments. *J Food Prot* 62: 894–898, 1999.

Sing, W.D. and Klaenhammer, T.R., A strategy for rotation of different bacteriophage defenses in a lactococcal single-strain starter culture system. *Appl Environ Microbiol* 59: 365–372, 1993.

Walker, S.A. and Klaenhammer, T.R., An explosive antisense RNA strategy for inhibition of a lactococcal bacteriophage. *Appl Environ Microbiol* 66: 310–319, 2000.

Wünsche, L., Importance of bacteriophages in fermentation processes. *Acta Biotechnol* 5: 395–419, 1989.

11 Phage as Vectors and Targeted Delivery Vehicles

Caroline Westwater[1] *and David A. Schofield*[2]
[1]Department of Stomatology
[2]Department of Microbiology and Immunology
Medical University of South Carolina, South Carolina

CONTENTS

11.1. Introduction....297
11.2. Filamentous Phage-Based Vectors....298
 11.2.1. The Ff Class of Bacteriophages....298
 11.2.2. Phagemids....299
 11.2.3. Helper Phages....299
11.3. Bacteriophage Lambda as a Vector....301
 11.3.1. Lambda Cloning Vectors....301
 11.3.2. Selection of Recombinant Genomes....303
 11.3.3. Lambda Vector System Applications....304
11.4. Bacteriophage P1 Cloning System....305
 11.4.1. P1 Cloning System and Methods....305
 11.4.2. P1-Based Artificial Chromosomes....307
 11.4.3. P1 Phagemid System for Delivery of DNA to Gram-Negative Bacteria....308
11.5. Phages as Lethal Agent Delivery Vehicles....309
11.6. Phage Vectors for Targeted Gene Delivery to Mammalian Cells....311
Acknowledgments....313
References....313

11.1. INTRODUCTION

Phage research has played a dominant role in understanding the most fundamental secrets of life, and it has also established the foundation for molecular biology. Many of the tools and techniques used in modern research could not have been developed without the extensive knowledge obtained through studies of basic phage biology.

0-8493-1336-X/05/$0.00+$1.50
© 2005 by CRC Press LLC

A large number of researchers studying diverse biological problems now depend on phages or phage-derived products for the isolation, sequence analysis, mutagenesis, expression, and transduction of DNA in a variety of hosts. The astonishing success of DNA engineering has led to the development of phage-display technology (see Ch. 9) and to the possibility that phages may be useful as therapeutic gene delivery vehicles. The purpose of this chapter is to review briefly the three major phage-cloning systems (Ff, lambda, and P1), to discuss the many advantages and applications of these vectors, and to describe the potential use of phages as molecular therapy vectors for bacterial and mammalian cells.

11.2. FILAMENTOUS PHAGE-BASED VECTORS

11.2.1. The Ff Class of Bacteriophages

The members of the Ff class of bacteriophages (M13, fd, and f1) have been used extensively as vectors for recombinant DNA cloning. These filamentous, male-specific coliphages have two important features that aid in their use as cloning vectors; they do not lyse their hosts, and they allow the packaging of DNA that is longer than the phages' unit lengths (for a review of filamentous phage biology, infection and assembly, see Russel, 1991; Webster, 1996); however, perhaps the most significant advantage of this group of phages is that they replicate as double-stranded, supercoiled DNA but package only a circular single-stranded molecule. Consequently, single-stranded DNA (ssDNA) can be prepared quickly and in sufficient purity directly from phage particles. Therefore, these vectors are particularly well suited for rapid sequence analysis by the Sanger dideoxy chain termination method and have been used extensively in oligonucleotide-directed mutagenesis (Messing, 1983; Vieira and Messing, 1987; Zinder and Boeke, 1982).

Messing and coworkers were the first to demonstrate that the *Escherichia coli* alpha complementation fragment of the *lacZ* gene could be inserted into the intergenic region of the M13 genome without any loss of phage function (Messing et al., 1977). Although the hybrid phage, M13mp1, brought the promise of a new type of vector, cloning was cumbersome because of the lack of convenient restriction sites in a nonessential region of the phage genome. The vector system was, however, significantly improved by the use of synthetic cloning sites (Boeke, 1981; Boeke et al., 1979; Gronenborn and Messing, 1978; Rothstein and Wu, 1981), insertion of a polylinker (Messing et al., 1981; Messing and Vieira, 1982; Vieira and Messing, 1982), and the use of marker inactivation systems (ß-galactosidase α-peptide, histidine operon, and antibiotic resistance) for screening vector inserts (Barnes, 1979; Herrmann et al., 1980; Hines and Ray, 1980; Messing et al., 1977; Zacher et al., 1980). While these vectors are valuable for the production of ssDNA, using a filamentous phage as a vector has certain disadvantages in comparison to plasmid vectors. These limitations include: (i) the tendency of some large inserts to be deleted (Herrmann et al., 1980; Hines and Ray, 1980); (ii) not all fragments are cloned with the same frequency; (iii) some fragments are cloned in only one orientation, and (iv) the number of phage particles produced decreases with increasing insert size (Hines and Ray, 1980). In this context, the vectors allow insertion of up to 10 kb of foreign DNA into the f1 genome without destroying phage viability. Insertion of the 9 kb plasmid pSC101 into f1 yields a virion 2.5 times the length of f1. With the aid of a helper phage, defective phages much smaller than wild-type phage are efficiently packaged (e.g., an insert of about 40 kb has been cloned into M13mp2). Over a broad range of phage DNA length (from about 7 × larger than unit length to one-third of the unit length), DNA is not barred from packaging. Defective or mini phages, however, grow only in the presence of a helper phage.

11.2.2. Phagemids

The above limitations of filamentous phages have been overcome through the use of chimeric phage/plasmid vectors, also known as phagemids, which combine the advantages of both plasmid and ssDNA phage vectors (for a review, see Mead and Kemper, 1988). One of the most important factors that led to the construction of phagemids was the demonstration, in Norton Zinder's laboratory, that the f1 intergenic region contains all of the *cis*-required functions for phage DNA replication and packaging (Dotto et al., 1981; Cleary and Ray, 1980; Zinder and Horiuchi, 1985). After infection with a helper phage, plasmids carrying this intergenic region are induced to enter the f1 mode of replication, and the resulting single-stranded molecule is assembled efficiently and encapsulated into phage particles (Dotto et al., 1981; Zagursky and Berman, 1984); however, in the absence of any phage gene products, the phage origin of replication is inactive and the phagemid replicates normally from its own plasmid origin. Subsequently, several useful cloning vectors containing the IG region of filamentous phages (f1, fd, M13, and IKe) were developed; these include yeast shuttle vectors (Baldari and Cesareni, 1985) and derivatives of the pUC (Vieira and Messing, 1987), pBluescript (Alting-Mees and Short, 1989), pBR322 (Zagursky and Baumeister, 1987) and pEMBL (Dente et al., 1983) vector series. A disadvantage of both the Ff-based vectors and the phagemid vectors is that only one of the DNA strands (the strand containing the viral origin) is packaged during ssDNA production. The dual origin vector pKUN9, which contains the replication origins and morphogenetic signals of phages IKe and f1, circumvents this problem by allowing the separate production of both DNA strands in a single-stranded form (Peeters et al., 1986).

11.2.3. Helper Phages

Phagemids have met an important need in molecular biology; however, this cloning system is not without problems. One disadvantage encountered with the use of phagemids is the significant reduction in the amount of ssDNA produced during the helper-phage-assisted mode of growth, compared to that of phage vectors. Phagemids are thought to compete with the incoming helper phage for replication factors, which results in a reduction in phage copy number and, hence, a reduction in the phage gene products necessary for the production of ssDNA (Vieira and Messing, 1987). This phenomenon has been shown to occur with phagemids containing M13's (Cleary and Ray, 1980; Zagursky and Berman, 1984), f1's (Dotto et al., 1981) or IKe's

(Peeters et al., 1986) functional origin of replication. Several helper phages (rv1, IR1, R408, and M13KO7) with mutations that result in alterations in the level or activity of the gene II protein are resistant to this interference (Enea and Zinder, 1982; Levinson et al., 1984; Russel et al., 1986; Vieira and Messing, 1987). These helper phages, therefore, improve the yield of ssDNA and increase the ratio of phagemid to helper phage DNA.

The helper phage M13KO7 is an M13 derivative that contains a mutated version of gene II (derived from M13mp1) and two insertions; a kanamycin resistance gene from *Tn*903 and the origin of plasmid 15 in its intergenic region (Vieira and Messing 1987). Since M13K07 is able to replicate independently of the gene II protein through the p15A origin, it is able to maintain adequate copies of its genome when grown in the presence of a phagemid; however, the mutated gene II product encoded by M13K07 interacts less efficiently with the origin of replication carried on its own genome than with the wild-type origin cloned into the phagemid (Dotto et al., 1981). This results in the preferential production of (+) strands from the phagemid and ensures that the viral particles released from the host bacteria contain mostly phagemid single-stranded DNA. R408, another frequently used helper phage, carries an internal packaging signal deletion, an IR1 interference-resistant mutation, and a mutation in a phage morphogenetic protein required for phage assembly (Russel et al., 1986). Helper phage R408, therefore, packages any ssDNA containing a complete morphogenetic signal more efficiently than it packages its own viral DNA (Russel et al., 1986).

The yield of phagemid and helper phage ssDNA is usually comparable, irrespective of whether a wild-type phage or an interference-resistant mutant is employed as a helper phage; however, the ratio of phagemid to helper phage ssDNA and the amount of each DNA recovered can vary depending on the combination of helper phage used with a particular recombinant. Moreover, certain sequences, the particular orientation of the DNA fragment, and plasmid-derived transcription across the phage origin of replication can suppress the yield of ssDNA (Vieira and Messing, 1991; Zagursky and Baumeister, 1987). The yield of ssDNA is also influenced by the structure and type of phagemid carried in the cell and also by its size (Levinson et al., 1984); however, a phagemid of up to 9.5 kb is not thought to influence the efficiency of packaging (Mead and Kemper, 1988).

In summary, while Ff phage vectors are valuable for the production of large amounts of ssDNA, some DNA inserts are unstable, especially when they are large. Recombinant phage vectors that produce deletions have a growth advantage during phage replication and, therefore, tend to quickly outgrow the original recombinant DNA phage; however, large or unstable sequences can be stably maintained in phagemids because these double-stranded vectors are only exposed to the single-stranded mode of replication for short periods of time. Although the practical advantages of phagemids are considerable, these cloning vectors produce lower levels of ssDNA compared to filamentous phage vectors. Nevertheless, interference resistant helper phages can increase the yield of phagemid single-stranded DNA and can also improve the ratio of phagemid to phage ssDNA.

11.3. BACTERIOPHAGE LAMBDA AS A VECTOR

11.3.1. Lambda Cloning Vectors

Bacteriophage lambda has played a crucial role in the cloning and manipulation of DNA ever since its introduction in the early 1970s as a cloning vector (Chauthaiwale et al., 1992; Murray, 1983). The relatively large cloning capacity of lambda-derived vectors and the highly efficient DNA packaging and infection process make these vectors attractive vehicles for the construction of both genomic and complementary DNA (cDNA) expression libraries (Christensen, 2001). The high efficiency of lambda *in vitro* packaging systems (commercial extracts can exceed 10^9 PFU/μg of DNA) becomes essential when the source of DNA is limited or when rare clones need to be isolated. An added benefit of lambda vectors is that phage libraries can be rapidly screened for the presence of a certain gene by using nucleic acid hybridization, immunological assays, or genetic selections at a high density (as many as 10^3 plaques per plate). In addition, lambda libraries are easily amplified without loss or misrepresentation of sequences, and the entire collection can be stored for years without a significant decrease in titer.

Bacteriophage lambda has a linear double-stranded DNA (dsDNA) genome of 48,502 bp, nearly 40% of which is dispensable for lytic growth. Replacing the central one-third of the genome (the region between the *J* and *N* genes) with foreign DNA yields a biologically active phage (Thomas et al., 1974); however, only genomes in the 38–51 kb range are packaged with high efficiency and produce viable phage particles. Therefore, lambda's packaging requirement places both an upper and a lower limit on the size of DNA fragments that can be cloned in the phage's genome. Thus, the difference between the minimal amount of DNA needed to code for genes essential for lytic growth (about 30 kb) and the maximum amount of DNA accommodated by a lambda phage head determines the capacity remaining for foreign DNA.

The wealth of knowledge obtained from studies of the basic biology of lambda (Hendrix et al., 1983) has permitted its genome to be modified to suit the desired application of the lambda vector. During the last 30 years, a wide variety of lambda derivatives have been developed for the propagation of foreign DNA (Table 11.1). These vectors either have a unique restriction enzyme site at which DNA fragments are inserted (insertion vectors; e.g., λgt10 and λgt11), or they contain a pair of restriction sites flanking a dispensable "stuffer" fragment that is replaced with heterologous DNA (replacement vector; e.g., Charon and EMBL series). Insertion vectors are useful for cloning small DNA fragments (up to 10 kb) such as cDNA, while replacement vectors are convenient for cloning DNA fragments up to 24 kb in length. Therefore, they are particularly useful for the construction of genomic libraries. On the other hand, cosmids (small plasmids containing a lambda *cos* site) allow larger DNA fragments (40–50 kb) to be cloned (Collins and Hohn, 1978). Although this increased cloning capacity can make cosmids the preferred vehicles for analyzing eukaryotic genes or large contiguous regions of the genome, lambda cloning vectors have a greater efficiency in library construction and screening.

TABLE 11.1
Examples of Lambda Cloning Vectors and Their Relevant Features

Name	Type of Vector	Cloning Capacity[a]	Selection	Biological Features	Reference/ Source
λTriplEx2	Insertion	Up to 8.6 kb	Blue/White	Cloned cDNAs can be expressed in all three reading frames from one recombinant clone; *cre-lox* automatic subcloning	BD Biosciences-Clontech
λgt10	Insertion	Up to 7.6 kb	Turbid/Clear plaques; Hfl	Useful for preparing cDNA libraries; screen with nucleic acid probes	(Huynh et al., 1985)
λgt11	Insertion	Up to 7.2 kb	Blue/White	Expression vector; inserts cloned in-frame with *lacZ* will be expressed as a fusion protein; screen with antibody probes	(Young and Davis, 1983)
λZAP	Insertion	Up to 10.1 kb	Blue/White	Contains Bluescript SK^- phagemid; ß-galactosidase fusion polypeptides; helper phage automatic subcloning; T3 and T7 promoters	(Short et al., 1988)
λZAP Express	Insertion	Up to 12.0 kb	Blue/White	Helper phage automatic subcloning; fusion polypepides; allows both prokaryotic (*lac* promoter) and eukaryotic (CMV immediate early promoter) expression	Stratagene
λYES	Insertion	Up to 8.3 kb	Fill-in adaptor strategy	Contains yeast-*E. coli* shuttle vector; *cre-lox* automatic subcloning; *GAL1* and *lac* promoters for expression; *URA3* and *bla* gene for selection	(Elledge et al., 1991)
CHARON 40	Replacement	9.0 – 22.1 kb	Blue/White	Polystuffer (235 bp repeats containing *lac* operator); *gam* gene retained on right arm, allows propagation on *recA* hosts	(Dunn and Blattner, 1987)
EMBL 3/4	Replacement	8.6 – 21.7 bp	Spi phenotype	Derived from λ1059; pBR322 sequences replaced with a fragment carrying *E. coli trpE*	(Frischauf et al., 1983)
λ1059	Replacement	8.8 – 21.9 kb	Spi phenotype	*Red* and *gam* genes are located on central "stuffer" fragment; contains *chi*D sequence; carries pacI plasmid containing multiple *att* sites and ColE1 origin (phasmid)	(Karn et al., 1980)
λDASH II	Replacement	8.7 – 21.8 kb	Spi phenotype	T3 and T7 promoters flank cloning sites to facilitate preparation of probes; prevents multiple inserts by partial fill-in technique	Stratagene
λFIX II	Replacement	8.7 – 21.8 kb	Spi phenotype	T3 and T7 promoters allow production of end-specific transcripts; large number of cloning sites available	Stratagene

[a] The cloning capacity is given in kilobases (kb) and assumes that the packaging limits on the genome are between 78% and 105% of that of wild-type lambda (37.8–50.9 kb). Although larger genomes can be recovered, the viability of the phage decreases when its genome length is greater than 51 kb.

Extensive research has been directed toward the development of lambda cloning vectors and many technical advances have facilitated the construction and analysis of lambda libraries. *In vitro* packaging systems have enabled the easy recovery of recombinant molecules and the natural tropism of lambda phage to its host has made the transfer of clones to *E. coli* a highly efficient process (for further information on two-strain and single-strain packaging systems, see Hohn and Murray, 1977; Rosenberg et al., 1985). The introduction of amber mutations (*S*am100, *B*am100, and *W*am403) into the lambda genome and the use of specific amber suppressor hosts have ensured that cloning vectors and recombinants have poor survival rates in the natural environment, thus ensuring a level of biological containment (Blattner et al., 1977). In addition, many attractive features incorporated into lambda vectors have facilitated the cloning and manipulation of DNA sequences. These include the: (i) removal of restriction sites from essential regions of the genome; (ii) introduction of suitable polylinkers in the dispensable one-third of the lambda genome; (iii) introduction of strong promoters derived from bacteriophages T7 and T3, or from *Salmonella typhimurium* phage SP6, to allow *in vitro* transcription of cloned inserts; (iv) presence of regulated promoters for controlled expression; (v) ability to convert lambda clones into plasmid molecules through *automatic subcloning* (discussed later), and (vi) availability of techniques aimed at differentiating recombinants from non-recombinants.

11.3.2. Selection of Recombinant Genomes

The efficiency of the lambda cloning system relies heavily on the ability to distinguish or select recombinant clones from those phages that are merely reconstituted phage genomes. Removal of the central "stuffer" fragment from a replacement vector, prior to the addition of donor DNA, should result in a low background of non-recombinants since the resulting vector falls below the minimum packaging size required to produce a viable phage; however, because "size-selection" does not select or screen against reassembly of the parent genome, genetic selection methods have been developed to provide an additional level of enrichment for the recombinant phage population. For example, lambda vectors derived from λ1059 offer a strong positive selection for cloned inserts based on the Spi (**s**ensitivity to **P**2 **i**nterference) phenotype of lambda (Karn et al., 1980). This strategy makes use of the fact that phage replication is inhibited in P2 lysogens when the phage contains functional *red* and *gam* genes (Zissler et al., 1971). Recombinant phages harboring inserts in place of the stuffer fragment (which carries the *red* and *gam* genes) have a Spi$^-$ phenotype and are distinguished from the parental vector by plating on strains lysogenic for P2 (Karn et al., 1980). Wild-type phages are unable to form plaques on a P2 lysogen whereas recombinant phages are able to form plaques on strains lysogenic for phage P2. A disadvantage of this selection scheme is that growth of *red*$^-$ *gam*$^-$ phages depends on the host recombination function to generate a suitable substrate for packaging; therefore, the choice of bacterial strain is an important consideration during recovery of *red*$^-$ *gam*$^-$ recombinants (for a more detailed discussion, see Murray, 1983; 1991). Additional selection strategies based on a change in phenotype associated with substitution of the central fragment (e.g., blue/white selection) have also been developed (see review by Williams and Blattner, 1980); however, although they facilitate the recognition of recombinants, they do not select for recombinant phages or against parental phage vectors (Blattner et al., 1977).

The ability to discriminate between a recombinant phage and a parental vector becomes vital when cloning small fragments, since insertion vectors do not have a minimum size limit and the vector itself is therefore packageable. Cloning of foreign DNA into these vectors typically exploits the insertional inactivation of a phage gene (*c*I, *int*, and *redA*) or genetic marker (*lacZ*) whose function can be monitored. "Immunity insertion vectors" (Murray et al., 1977), such as λgt10, exploit a unique *Eco*RI site within the *c*I repressor gene of the immunity region of the lambdoid phage 434 (λimm^{434} Huynh et al., 1985). Successful cloning of DNA into the *c*I gene of λimm^{434} can be visually screened by a change from a turbid-plaque (cI^+) to a clear-plaque (cI^-) morphology. Moreover, recombinants can be directly selected on a host strain (such as NM514) carrying the *high frequency lysogeny* mutation (Hfl^-). In an Hfl^- strain, cI^+ parental phages repress the genes within their genomes to the extent that plaque formation is suppressed; however, recombinant phages (cI^-) cannot repress their genomes and therefore, they form plaques with normal efficiency (Hoyt et al., 1982). Thus, the use of high frequency lysogenic hosts, together with insertional inactivation of the *c*I gene, provides a means of efficiently eliminating non-recombinant phages from λgt10 cDNA libraries.

Although lambda libraries can be screened efficiently (with nucleic acid or antibody probes) when plated at high plaque density, the large size of lambda vectors complicates the restriction mapping and analysis of genes after cloning. Therefore, several lambda vectors (e.g., λEXLX and λZAP) have been modified to allow the easy conversion of lambda clones into plasmid molecules without the need for subcloning procedures. These vectors preserve the benefit of high-efficiency lambda library construction, while providing the convenience of plasmids for characterizing cloned DNA. This strategy, often referred to as *automatic subcloning,* allows a plasmid to be "popped out" of the lambda clone through the use of the bacteriophage P1 *cre/loxP* site-specific recombination system (Palazzolo et al., 1990), or by using an amber mutant of a filamentous helper phage (M13mp7-11, M13gt120 or Exassist) to excise an insert-containing phagemid from the lambda vector λZAP (Short et al. 1988). By incorporating heterologous elements that can function in different species, various lambda/plasmid hybrid shuttle vectors have been adapted for cDNA expression in *Arabidopsis* (Fuse et al., 1995), yeast (Brunelli and Pall, 1993; Elledge et al., 1991), filamentous fungi (Brunelli and Pall, 1994; Holt and May, 1993), and mammalian cells (λZAP Express, Stratagene).

11.3.3. Lambda Vector System Applications

Although it was originally designed for the efficient construction and screening of genomic or cDNA libraries, the lambda vector system is useful for a broad range of applications. Modified lambda vectors have been introduced into the chromosomes of transgenic mice in order to measure spontaneous and induced tissue-specific mutation rates (Kohler et al., 1991). In addition, the recombination-proficient phage vector λ2TK has been used as a targeting vector (TV) to introduce genomic mutations directly into embryonic stem cells, thus avoiding the requirement for plasmids to be 'rolled out' from primary phage clones (Tsuzuki and Rancourt, 1998). Recently, the construction of targeting vectors has been streamlined by using an approach in which modification cassettes and point mutations are shuttled to specific positions in phage TVs during a single (Aoyama et al., 2002; Unger et al., 1999) or double crossover recombination event with a plasmid (Tsuzuki and Rancourt, 1998). Moreover, using a retro-recombination screening-based strategy, phage TVs containing specific genomic clones can now be rapidly isolated from an embryonic stem cell genomic library in λ2TK by integrative and excisive recombination (Woltjen et al., 2000). Thus, the retro-recombination protocol allows the rapid isolation of phage-targeting vectors for specific gene knockout experiments.

11.4. BACTERIOPHAGE P1 CLONING SYSTEM

11.4.1. P1 Cloning System and Methods

Nat Sternberg first described the use of P1 as an alternative to cosmid and yeast artificial chromosome systems for the cloning of large chromosomal fragments (Sternberg, 1990). The P1 cloning system is relatively efficient (typically, 10^5 clones/μg of vector DNA) and accepts inserts in the 70–100 kb range (Pierce and Sternberg, 1992; Sternberg, 1992, 1994). Importantly, P1 libraries contain very few clones from noncontiguous segments of genomic DNA (the level of chimerism is estimated to be <5%), and library fidelity may be more faithful than that of other high molecular weight cloning systems (Sternberg, 1994). In addition, P1 recombinant clones are easily isolated and differentiated, using standard alkaline lysis methods, from the host's genetic background.

The original P1 cloning vector (pAd10) consists of two domains divided by flanking P1 *lox*P recombination sites (Sternberg, 1990; Sternberg et al., 1990). The first domain is characterized by the multicopy ColE1 replicon, the P1 *pac* site orientated so as to direct packaging counterclockwise, and a 10.6 kb adenovirus stuffer fragment inserted into the *bla* (Amp^R) gene of pBR322. The second domain contains the aminoglycoside 3'-phosphotransferase gene (Kan^R) from transposon *Tn*903, the tetracycline resistance gene (Tet^R) from pBR322, and two P1 replicons—(the P1 plasmid maintenance region that stably maintains plasmid DNA in *E. coli* at about one copy per cell and a LacI-regulated P1 lytic replicon). After activation by isopropyl ß-D-thiogalactoside (IPTG), the latter replicon amplifies the plasmid copy number 25-fold in six generations (Sternberg, 1990).

Detailed descriptions of the methods involved in the P1 cloning process, includes: (i) preparation of vector and insert DNA; (ii) packaging extract production; (iii) *in vitro* packaging of ligated DNA, and (iv) recovery of packaged DNA in bacterial hosts are contained in several technical reviews (Pierce and Sternberg, 1992; Sternberg, 1995). In brief, the P1 vector is first digested with restriction enzymes (*Bam*HI and *Sca*I) to generate two vector arms, and the arms are subsequently treated with phosphatase and ligated to size-selected DNA fragments obtained by partial digestion with

*Sau*3AI. The *pac* site on the ligated DNA is then cleaved (the stage 1 reaction) by an extract that contains the P1 pac cleavage proteins (PacA and PacB) and host integration factor IHF, but lacks phage heads and tails. The stage 1 extract is prepared by induction of a P1 lysogen carrying an amber mutation in gene 10 (*am10.1*). Gene 10 is required for the production of all P1 late proteins (including heads, tails, and phage lysozyme production), but it is not required for the production of proteins that recognize and cleave the *pac* site. In stage 2 of the packaging reaction, the *pac*-cleaved DNA is packaged unidirectionally into empty phage proheads and the filled head (about 110–115 kb) is converted to an infectious particle by the addition of phage tails. The head-tail extract is prepared by induction of a lysis-defective (Δ*lydAB*)-P1 lysogen carrying an amber mutation (*am131*) in the genes necessary for pacase function (Sternberg et al., 1994).

It is important to note that both the pacase and head-tail extracts are prepared from strains containing mutations in the P1 restriction-modification system (r^-m^-), the *E. coli* restriction systems (*mcrABC, hsdR, hsdM*), and the *E. coli* exonuclease V function. In addition, the temperature-sensitive C1 repressor mutation (*c1.100,* (Rosner, 1972) also allows rapid induction of the prophage's lytic phase, after shifting the temperature of an exponentially growing lysogenic culture from 32°C to 42°C. During the final step of the cloning process, the packaged DNA is transduced into a bacterial host expressing the Cre recombinase, where the DNA is re-circularized through recombination between the vector's two *loxP* sites. In order to minimize insert rearrangement, the DNA is delivered to a host strain containing a *recA* mutation (inactivates the major homologous recombination system), the *lacI^q* repressor mutation (inhibits activation of the P1 lytic replicon), and mutations in the methylation-restriction pathways. The utility of this cloning system was demonstrated by its ability to construct a 52,000 member human DNA library consisting of 26 pools of 2,000 clones, each with an average insert size of 75 to 100 kb (Sternberg et al., 1990); however, clones without insert DNA were found to represent as much as 80% of the total DNA isolated from the pools after amplification, despite the fact that they represented only a small number (10%–20%) of the initial clones. This major obstacle was overcome by constructing a positive selection vector, pAd10sacBII (Pierce et al., 1992a), based on the properties of the *sacB* gene from *Bacillus amyloliquefaciens*.

The *sacB* gene encodes levansucrase, an enzyme that converts sucrose to high molecular weight fructose polymers (levan). Cells expressing *sacB* die in media containing >2% sucrose because accumulation of levan in the periplasm of Gram-negative bacteria is toxic. In pAd10sacBII, *sacB* is expressed from a synthetic *E. coli* promoter that contains a P1 C1 repressor-binding site overlapping near-consensus promoter elements. Pierce et al. (1992a) have noted that it is important to maintain the plasmid in a strain containing the P1 C1 repressor because low levels of *sacB* expression can interfere with cell growth, even in the absence of sucrose, and can lead to deletions within the *sacB* gene or inactivation of its promoter (Pierce et al., 1992b). Foreign DNA cloned into the unique *Bam*HI site, which is located between the *E. coli* promoter and the *sacB* gene, results in recombinants with a higher sucrose resistance than non-recombinants; therefore insert-containing clones, but not insert-absent clones, can be grown in the presence of 5% sucrose.

The use of a counterselectable marker provides a ~70-fold discrimination between recombinant and non-recombinant clones (Pierce et al., 1992a). The pAd10sacBII vector also has several advantages over the original P1 cloning vector; e.g., it contains rare restriction sites (*Sfi*I, *Sal*I and *Not*I) and T7/Sp6 polymerase binding sites for the characterization of cloned DNA and the production of riboprobes. The improved P1 cloning process has been used to construct several libraries from mammalian genomes (human, mouse, and rat) and from other organisms (*Drosophila melanogaster, Drosophila virilis, Schizosaccharomyces pombe, Limulus polyphemus, Pneumocystis carinii,* and *Pinus radiata*) (Metcheva et al., 1996; Shepherd and Smoller, 1994).

11.4.2. P1-Based Artificial Chromosomes

Although the P1 cloning system has contributed significantly to the mapping and analysis of complex genomes (Shepherd and Smoller, 1994; Sternberg, 1994), it is restricted by the size of the capsid and requires an elaborate *in vitro* packaging system. The bacteriophage T4 *in vitro* packaging system is capable of packaging P1-based vectors with inserts up to 122 kb; however, this system does not circumvent the other limitations of the PI cloning system (Rao et al., 1992). Recently, P1-based artificial chromosomes (PACs), which combine the advantages of the BAC (bacterial artificial chromosomes) and P1 system, have significantly expanded the size of fragments that can be stably maintained as plasmid molecules in *E. coli* (Ioannou et al., 1994). The PAC cloning vector (pCYPAC1) retains the positive selection system and the two P1 replicons; however, the need for P1 headful packaging and site-specific recombination is eliminated because the circular ligation products are introduced into *E. coli* by electroporation. The new PAC system can provide cloning and stable maintenance of inserts in the 100–300 kb range, which is nearly twice the size of inserts recovered with the original P1 system. The large size and stability of the PAC clones has made them well suited for physical mapping and positional cloning of the mammalian genome. Therefore, several groups have subsequently retrofitted P1/PAC clones with eukaryotic reporter genes and selectable markers for functional analysis of cloned DNA in mammalian cells. Retrofitting procedures include: (i) subcloning inserts into a modified PAC vector (Mejia and Monaco, 1997); (ii) inserting a *loxP* expression cassette by Cre-mediated recombination (Kaname and Huxley, 2001); (iii) using a *loxP*-containing *Tn*10-based retrofitting procedure to generate nested deletions in P1 clones (Coren and Sternberg, 2001; Chatterjee and Coren, 1997; Chatterjee and Sternberg, 1996; Sternberg, 1994), and (iv) using Chi-stimulated homologous recombination to make deletion mutations (Nistala and Sigmund, 2002). Alternatively, P1/PAC clones can be converted into yeast artificial chromosomes (Poorkaj et al., 2000) or they can be captured by yeast-bacterial shuttle vectors (Bhargava et al., 1999; Criswell and Bradshaw, 1998) through homologous recombination in yeast. Several groups have subsequently constructed PAC shuttle vectors (pJCPAC-Mam1 and pPAC4) that allow stable maintenance in both bacterial and mammalian cells (Coren and Sternberg, 2001; Frengen et al., 2000). Libraries constructed with this new vector system will serve as important reagents for future functional genomic studies.

11.4.3. P1 Phagemid System for Delivery of DNA to Gram-Negative Bacteria

The ability to deliver genetic information to bacteria has become an important step in virtually all aspects of molecular genetics. Although the majority of bacteria cannot be transformed by natural competence (Lorenz and Wackernagel, 1994), several protocols have been established for inducing DNA uptake (Hanahan et al., 1991; Lorenz and Wackernagel, 1994). *E. coli* has become a universal host for initial cloning and analysis of recombinant DNA, partly because of the ease with which it can be transformed. DNA molecules can be introduced into *E. coli* K12, for example, by transformation of electrocompetent cells, phage transduction, and conjugational mating (Benedik, 1989; Dower et al., 1988; Hanahan et al., 1991). On the other hand, although a number of factors have been found to increase transformation efficiency, many bacterial species of clinical, environmental, and industrial importance remain recalcitrant to standard transformation techniques.

Recently, a new broad-host-range phage delivery system (Westwater et al., 2002) has been developed in an effort to facilitate the movement of DNA between various Gram-negative species. In contrast to transformation, this system should be an efficient means of introducing recombinant DNA into bacteria since phage infection normally occurs at a high frequency for host cells carrying the appropriate receptor. This approach utilizes a bacteriophage P1-based *in vivo* packaging system to deliver a broad-host-range vector to bacterial cells. The P1 delivery system consists of a plasmid (P1pBHR-T) that carries a P1 *pac* site to package the vector, a P1 lytic replicon to generate concatemeric DNA, a kanamycin resistance marker for transfer detection, and a broad-host-range origin of replication (Fig. 11.1). The plasmid is maintained in a P1 lysogen that provides all of the replication factors needed to activate

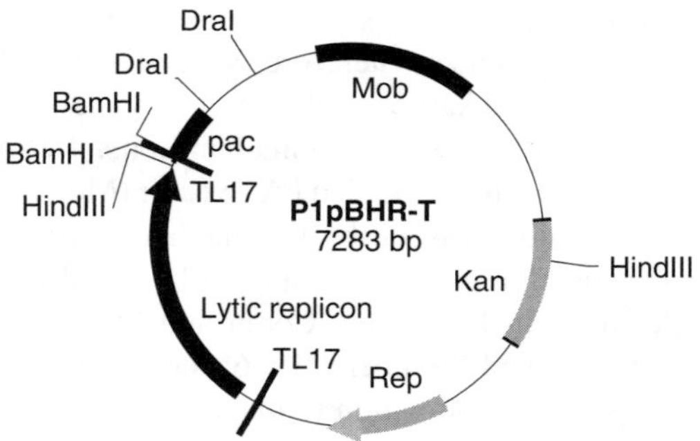

FIGURE 11.1 Map of the broad-host-range plasmid P1pBHR-T.
The genes encoding the mobilization protein (*mob*), replication protein (*rep*) and kanamycin resistance marker (*kan*) are derived from the broad-host-range cloning vector pBBR122. Sequences originating from the P1 bacteriophage include the packaging site (*pac*) and lytic replicon. The TL_{17} transcriptional terminator is a natural element derived from the *E. coli* alpha-operon. Only relevant restriction sites are indicated (from Westwater et al., 2002; used with permission of the Society for General Microbiology).

the P1 origin of replication on the plasmid and all of the structural components to form mature viral particles. The chloramphenicol-marked P1 prophage also carries a temperature-sensitive C1 repressor mutation (Rosner, 1972), which allows efficient prophage induction by a temperature shift. Induction of the lysogen by shifting the temperature from 32°C to 42°C results in activation of the lytic replicon in the plasmid, packaging of the plasmid concatemeric DNA into P1 phage particles and lysis of the lysogen. The system's versatility was demonstrated by its ability to deliver plasmid DNA to *E. coli, Shigella flexneri, Shigella dysenteriae, Klebsiella pneumoniae, Citrobacter freundii,* and *Pseudomonas aeruginosa* (Westwater et al., 2002). The ability to transduce additional bacteria should be possible since P1 can adsorb to and inject DNA into 23 Gram-negative species, including *Yersinia pestis, Myxococcus xanthus,* and *Agrobacterium tumefaciens* (Yarmolinski and Sternberg, 1988).

11.5. PHAGES AS LETHAL AGENT DELIVERY VEHICLES

During the last decade, the emergence of antibiotic resistance among bacterial and fungal pathogens has become a serious problem of global significance. Some infections are now recalcitrant to antibiotics of last resort and the apparent lack of new generation drugs in the developmental pipeline is even more worrisome. In response to this challenge, phage therapy (see Chapters 13 and 14) and lethal agent delivery systems (Westwater et al., 2003) are currently being evaluated as alternative therapies for the treatment of bacterial infections. Although both approaches use phage-targeting mechanisms to achieve delivery to the bacterial population, *lethal agent delivery systems* (LADS) do not kill their hosts via the production of phage-encoded proteins that degrade the cell wall peptidoglycan or block the peptidoglycan synthesis pathway. In contrast, LADS technology utilizes a bacteriophage-based *in vivo* packaging system to create delivery vehicles capable of injecting genetic information encoding antimicrobial agents into targeted pathogenic bacteria.

Westwater and coworkers recently tested the concept of using phages as molecular therapy vectors that provide an efficient and selective means of delivering lethal agents to bacteria *in vivo* (Westwater et al., 2003). In this proof-of-concept study, we exploited the ability of a non-lytic filamentous phage to package and deliver phagemids to strains of *E. coli* harboring the primary- and co-receptor for M13 (the F conjugative pilus and TolQRA complex, respectively). Although the phagemid can encode any protein that targets an essential cellular component, initial experiments were performed with the addiction toxins associated with the bacterial chromosome. The genes encoding the toxins Gef and ChpBK (Masuda et al., 1993; Poulsen et al., 1989) were placed under the transcriptional control of a LacI-regulated promoter and cloned into a plasmid containing the f1 intergenic region. Male *E. coli* cells carrying this phagemid were infected with the M13 helper phage R408, which resulted in preferential packaging of phagemid DNA over helper phage DNA (lysates contained 95% lethal-agent phagemid particles). The lysates were used to infect *E. coli in vitro* (multiplicity of infection = 3.6), and the cultures were grown under conditions that either turned the expression of the toxins on (with the addition of IPTG) or off (in the presence of the LacI repressor). Colony forming unit assays determined that phage delivery of the lethal agents in the presence of IPTG resulted in a 948- and

1579-fold reduction in viable cell counts for phagemid lysates pGef and pChpBK, respectively. This indicated that Gef and ChpBK exerted a bactericidal effect and the loss of viability was specifically due to the expression of the lethal agents. To investigate whether phages can function as an antimicrobial agent delivery system *in vivo,* a nonlethal bacteremic mouse model of infection was employed. Immunocompromised (cyclophosphamide-treated) mice were intraperitoneally injected sequentially within 5 minutes with a single dose of *E. coli,* LADS preparation and IPTG. Peripheral blood samples were taken at various time points, and viable bacterial counts were determined (Fig. 11.2). Mice treated with LADS showed a significant reduction in the number of circulating bacteria, compared to mice treated with phagemids lacking the gene encoding the lethal agent Gef or ChpBK. These results support the hypothesis that phages can be used efficiently to construct phagemids capable of expressing bactericidal proteins *in vivo,* and that phage delivery

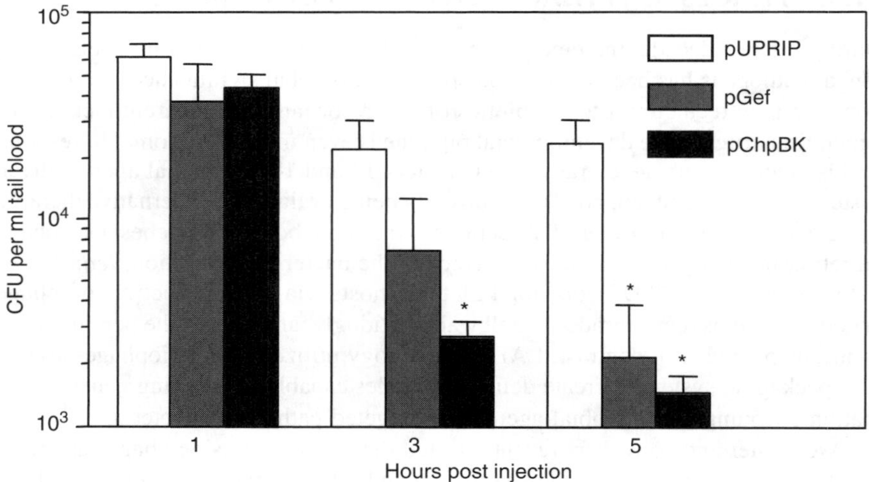

FIGURE 11.2 Phage-delivered lethal agents significantly reduce the level of viable bacteria in the blood of mice.
A single dose of *E. coli* ER2378 (200 μl, 5×10^8 CFU/ml), IPTG (100 μl, 250 mM), and a preparation of LADS (200 μl, 1.2×10^{10} phagemid-containing particles/ml) was administered by intraperitoneal injection to cyclophosphamide-treated mice. Mice were treated with a single dose of phage preparation containing phagemid pGef, pChpBK, or pUPRIP (multiplicity of infection of 3.6). The control vector pUPRIP is identical to the phagemids pGef and pChpBK except that it lacks a gene encoding a lethal agent. At the indicated time points, blood samples were taken and bacterial counts were determined by plating on Luria-Bertani plates containing tetracycline (20 μg/ml). Mice whose blood contained < 20 CFU/ml (lowest level of detection) at 1 hour were eliminated from the analysis. The viable bacterial counts in the blood are plotted as the mean ± standard deviation for each treatment group. Statistical analysis (unpaired *t* test) indicated significantly fewer viable bacteria in the blood at the times indicated for the experimental pGef and pChpBK groups, compared to the control pUPRIP group (*, $P \leq 0.05$, two-tailed, adjusted for multiple comparisons) (from Westwater et al., 2003; used with permission of the American Society for Microbiology).

systems may provide an alternative approach to conventional antibiotics for the treatment of bacterial infections.

11.6. PHAGE VECTORS FOR TARGETED GENE DELIVERY TO MAMMALIAN CELLS

A major limitation for the development of effective gene therapy is that vector systems must be capable of delivering therapeutic genes to specific cells while minimizing toxicity and expression in other tissues. Prokaryotic phage vectors have been proposed (Larocca and Baird, 2001; Larocca et al., 2002a; Larocca et al., 2002b; Monaci et al., 2001) as an alternative to existing animal viruses (adenoviruses, retroviruses, and lentiviruses) and nonviral DNA complexes. Phage-based vectors are attractive delivery vehicles because without targeting, phage particles have virtually no tropism for mammalian cells. However, the genetic flexibility of filamentous phages allows biologically relevant macromolecules, including antibodies, proteins, and peptides, to be displayed as fusions with the phages' capsid proteins (for details on phage display, see Chapter 9). Therefore, this relatively simple system allows phage display vectors to be targeted to specific cell surface receptors through the use of rational design and selection from combinatorial libraries. Moreover, modified phage particles can be conveniently purified from the supernatant fluids of bacterial cultures and are stable under a variety of harsh environments (e.g., extremes of pH).

Initial studies (Ishiura et al., 1982; Okayama and Berg, 1985; Yokoyama-Kobayashi and Kato, 1993; 1994) indicated that chemical transfection methods could be used to introduce both lambda and filamentous phage particles containing a reporter gene under the control of a eukaryotic promoter into cultured mammalian cells. These early experiments implied that phage particles were uncoated in eukaryotic cells and that phagemid single-stranded DNA could be converted to dsDNA, thus leading to reporter gene expression; however, phage-mediated transfection in conjunction with DEAE-dextran, cationic lipid, or calcium phosphate precipitation is severely limited by the lack of cell-type targeting elements. Hart and coworkers subsequently demonstrated that phage particles engineered to display multiple copies of an integrin-binding peptide (GGCRGDMFGC) as a pVIII fusion could be targeted and internalized through receptor-mediated endocytosis by mammalian cells (Hart et al., 1994). In addition, a lambda phage modified to display an RGD peptide as a C-terminal fusion to the major tail protein pV has recently been shown specifically to transfect COS cells with a *lacZ* reporter gene (Dunn, 1996).

Larocca et al. (1998) were the first to report that FGF2-targeted filamentous phages were capable of mediating transient and stable transduction of fibroblast growth factor (FGF) receptor-positive mammalian cells. In this system, phagemids carrying a cytomegalovirus driven reporter gene (green fluorescent protein [*GFP*] or *lacZ*) were packaged by the helper phage M13KO7 and targeted to cell surface-bound receptors using avidin-biotin-linked FGF2. This nongenetic method for targeted, phage-mediated transduction required FGF2 targeting and was phage concentration dependent; however, the transduction efficiency was low (~1%). Later studies demonstrated that phages genetically displaying growth factors (epidermal growth factor and FGF2) as fusions to their minor coat protein were capable of delivering a GFP expression cassette to cells bearing the appropriate receptors (Kassner et al., 1999; Larocca et al., 1999). Competition with free ligand or with a neutralizing anti-receptor antibody demonstrated that mammalian cell transduction was highly specific and that gene transfer by targeted phages was dependent on ligand, time, and dose. The ability of ligand-targeted phages to transduce mammalian cells was also confirmed with phages displaying an anti-ErbB2 (also known as HER2, human epidermal growth factor receptor 2), a single-chain antibody fragment (Poul and Marks, 1999), an HER2 ligand (Urbanelli et al., 2001), and the adenovirus penton base capsid protein (Di Giovine et al., 2001). Collectively, these studies demonstrate that filamentous phages modified to target cell surface receptors or cell surface antigens can bind and transduce cells *via* receptor-mediated endocytosis.

Although the results of the *in vitro* studies described earlier clearly supported the potential use of phages as gene therapy vehicles, the transduction efficiencies with targeted phages (1%–4%) were significantly lower than with other viral delivery systems; however, transduction levels were similar to antibody-targeted or ligand-targeted DNA-complex systems. Recent studies have shown that the number of targeting ligands displayed on the phage particles positively dictates the efficiency of cell-surface binding and internalization (Becerril et al., 1999; Ivanenkov et al., 1999a; Larocca et al., 2001; Poul and Marks, 1999). Multivalent display of targeting-molecules is thought to mediate either high-avidity binding or receptor dimerization, which provides a multi-point interaction between the phages and the cells, and in turn triggers internalization. Subsequently, transduction efficiencies as high as 8% were obtained in human prostate carcinoma cells (PC-3) with multivalent display (3 to 5 copies per phage) of EGF (Larocca et al., 2001). Since the majority of cells internalize targeted phages but do not necessarily express the reporter gene, several groups have suggested that the transduction frequency is limited by one or more post-uptake events (e.g., intracellular phage degradation, endosomal escape, nuclear translocalization, strand-conversion or transcription). Recently, genotoxic treatment, which increases the transduction efficiency of the single-stranded adeno-associated virus (presumably by enhancing the conversion of ssDNA to dsDNA), was also found to enhance phage-mediated transduction of mammalian cells (Burg et al., 2002). Indeed, inhibition of topoisomerase I increased EGF-mediated phage transduction in all cell lines tested and in prostate tumor xenografts (Burg et al., 2002). Although the transduction efficiency in response to genotoxic treatment varied (ranging from 0.2% to 40%), this study strongly suggests that modified phage-display vectors have potential as delivery vehicles.

The ability to identify ligands for specific cell types (Barry et al., 1996) and organ-homing peptides (Pasqualini and Ruoslahti, 1996; Rajotte et al., 1998), together with the knowledge that phages can be internalized by mammalian cells through an endocytotic pathway, has led to the selection of phages expressing targeting molecules that not only bind to cell surface receptors, but result in internalization of a phage delivery vehicle (Becerril et al., 1999; Ivanenkov et al., 1999a; Ivanenkov et al., 1999b) and, ultimately, gene expression (Ivanenkov et al., 1999b; Kassner et al., 1999; Poul and Marks, 1999). Bacteriophage vectors are relatively simple to genetically modify by rational design; however, the use of *ligand identification via expression* (LIVE) technology provides the ability to select directly improved phage delivery vehicles from highly diverse libraries through molecular evolution (Kassner et al., 1999). As more efficient phage delivery vehicles are selected for desirable characteristics, such as cell-specific targeting, stability *in vivo* and reduced immunogenicity, the specific mechanisms by which targeted phage particles successfully transduce cells will obviously require further investigation. Understanding the fate of phage-delivered DNA is likely to lead to the incorporation of receptor-binding molecules as well as sequences that enhance endosomal escape, nuclear localization, and transgene expression. Recently, lambda phage displaying a 32-mer nuclear localization signal of SV40 T antigen showed a higher affinity for the nucleus and induced marker gene expression much more efficiently than did wild-type phage particles, when delivered into the cytoplasm of human fibroblasts (Akuta et al., 2002). In conclusion, the genetic flexibility of filamentous phages will clearly allow the custom design of gene transfer vehicles for specific therapeutic applications, and it offers the prospect of targeted gene delivery with minimal toxicity.

ACKNOWLEDGMENTS

We thank Edward Balish, Josh Levinson, Jim Norris, and Michael Schmidt for their constructive comments during the preparation of this chapter.

REFERENCES

Akuta, T., Eguchi, A., Okuyama, H., Senda, T., Inokuchi, H., Suzuki, Y., Nagoshi, E., et al., Enhancement of phage-mediated gene transfer by nuclear localization signal. *Biochem Biophys Res Commun* 297: 779–786, 2002.

Alting-Mees, M.A. and Short, J.M., pBluescript II: Gene mapping vectors. *Nucleic Acids Res* 17: 9494, 1989.

Aoyama, C., Woltjen, K., Mansergh, F.C., Ishidate, K. and Rancourt, D.E., Bacteriophage gene targeting vectors generated by transplacement. *Biotechniques* 33: 806–810, 812, 2002.

Baldari, C. and Cesareni, G., Plasmids pEMBLY: New single-stranded shuttle vectors for the recovery and analysis of yeast DNA sequences. *Gene* 35: 27–32, 1985.

Barnes, W.M., Construction of an M13 histidine-transducing phage: a single-stranded cloning vehicle with one EcoRI site. *Gene* 5: 127–139, 1979.

Barry, M.A., Dower, W.J. and Johnston, S.A., Toward cell-targeting gene therapy vectors: selection of cell-binding peptides from random peptide-presenting phage libraries. *Nat Med* 2: 299–305, 1996.

Becerril, B., Poul, M.A. and Marks, J.D., Toward selection of internalizing antibodies from phage libraries. *Biochem Biophys Res Commun* 255: 386–393, 1999.

Benedik, M.J., High efficiency transduction of single strand plasmid DNA into enteric bacteria. *Mol Gen Genet* 218: 353–354, 1989.

Bhargava, J., Shashikant, C.S., Carr, J.L., Bentley, K.L., Amemiya, C.T. and Ruddle, F.H., pPAC-ResQ: A yeast-bacterial shuttle vector for capturing inserts from P1 and PAC clones by recombinogenic targeted cloning. *Genomics* 56: 337-339, 1999.

Blattner, F.R., Williams, B.G., Blechl, A.E., Denniston-Thompson, K., Faber, H.E., Furlong, L., Grunwald, D.J., et al., Charon phages: safer derivatives of bacteriophage lambda for DNA cloning. *Science* 196: 161–169, 1977.

Boeke, J.D., One and two codon insertion mutants of bacteriophage f1. *Mol Gen Genet* 181: 288–291, 1981.

Boeke, J.D., Vovis, G.F. and Zinder, N.D., Insertion mutant of bacteriophage f1 sensitive to EcoRI. *Proc Natl Acad Sci U S A* 76: 2699–2702, 1979.

Brunelli, J.P. and Pall, M.L., Lambda/plasmid vector construction by in vivo cre/lox-mediated recombination. *Biotechniques* 16: 1060–1064, 1994.

Brunelli, J.P. and Pall, M.L., A series of yeast/*Escherichia coli* lambda expression vectors designed for directional cloning of cDNAs and cre/lox-mediated plasmid excision. *Yeast* 9: 1309–1318, 1993.

Burg, M.A., Jensen-Pergakes, K., Gonzalez, A.M., Ravey, P., Baird, A. and Larocca, D., Enhanced phagemid particle gene transfer in camptothecin-treated carcinoma cells. *Cancer Res* 62: 977–981, 2002.

Chatterjee, P.K. and Coren, J.S., Isolating large nested deletions in bacterial and P1 artificial chromosomes by in vivo P1 packaging of products of Cre-catalysed recombination between the endogenous and a transposed loxP site. *Nucleic Acids Res* 25: 2205–2212, 1997.

Chatterjee, P.K. and Sternberg, N.L., Retrofitting high molecular weight DNA cloned in P1: Introduction of reporter genes, markers selectable in mammalian cells and generation of nested deletions. *Genet Anal* 13: 33–42, 1996.

Chauthaiwale, V.M., Therwath, A. and Deshpande, V.V., Bacteriophage lambda as a cloning vector. *Microbiol Rev* 56: 577–591, 1992.

Christensen, A.C., Bacteriophage lambda-based expression vectors. *Mol Biotechnol* 17: 219–224, 2001.

Cleary, J.M. and Ray, D.S., Replication of the plasmid pBR322 under the control of a cloned replication origin from the single-stranded DNA phage M13. *Proc Natl Acad Sci U S A* 77: 4638–4642, 1980.

Collins, J. and Hohn, B., Cosmids: A type of plasmid gene-cloning vector that is packageable in vitro in bacteriophage lambda heads. *Proc Natl Acad Sci U S A* 75: 4242–4246, 1978.

Coren, J.S. and Sternberg, N., Construction of a PAC vector system for the propagation of genomic DNA in bacterial and mammalian cells and subsequent generation of nested deletions in individual library members. *Gene* 264: 11–18, 2001.

Criswell, T.L. and Bradshaw, S., Transfer of P1 inserts into a yeast-bacteria shuttle vector by co-transformation mediated homologous recombination. *Nucleic Acids Res* 26: 3611–3613, 1998.

Dente, L., Cesareni, G. and Cortese, R., pEMBL: A new family of single stranded plasmids. *Nucleic Acids Res* 11: 1645–1655, 1983.

Di Giovine, M., Salone, B., Martina, Y., Amati, V., Zambruno, G., Cundari, E., Failla, C. M., et al., Binding properties, cell delivery, and gene transfer of adenoviral penton base displaying bacteriophage. *Virology* 282: 102–112, 2001.

Dotto, G.P., Enea, V. and Zinder, N. D., Functional analysis of bacteriophage f1 intergenic region. *Virology* 114: 463–473, 1981.

Dower, W.J., Miller, J.F. and Ragsdale, C.W., High efficiency transformation of *E. coli* by high voltage electroporation. *Nucleic Acids Res* 16: 6127–6145, 1988.

Dunn, I.S., Mammalian cell binding and transfection mediated by surface-modified bacteriophage lambda. *Biochimie* 78: 856–861, 1996.

Dunn, I.S. and Blattner, F.R., Charons 36 to 40: Multi enzyme, high capacity, recombination deficient replacement vectors with polylinkers and polystuffers. *Nucleic Acids Res* 15: 2677–2698, 1987.

Elledge, S.J., Mulligan, J.T., Ramer, S.W., Spottswood, M. and Davis, R.W., Lambda YES: A multifunctional cDNA expression vector for the isolation of genes by complementation of yeast and *Escherichia coli* mutations. *Proc Natl Acad Sci U S A* 88: 1731–1735, 1991.

Enea, V. and Zinder, N.D., Interference resistant mutants of phage f1. *Virology* 122: 222–226, 1982.

Frengen, E., Zhao, B., Howe, S., Weichenhan, D., Osoegawa, K., Gjernes, E., Jessee, J., et al., Modular bacterial artificial chromosome vectors for transfer of large inserts into mammalian cells. *Genomics* 68: 118–126, 2000.

Frischauf, A.M., Lehrach, H., Poustka, A. and Murray, N., Lambda replacement vectors carrying polylinker sequences. *J Mol Biol* 170: 827–842, 1983.

Fuse, T., Kodama, H., Hayashida, N., Shinozaki, K., Nishimura, M. and Iba, K., A novel Ti-plasmid-convertible lambda phage vector system suitable for gene isolation by genetic complementation of *Arabidopsis thaliana* mutants. *Plant J* 7: 849–856, 1995.

Gronenborn, B. and Messing, J., Methylation of single-stranded DNA in vitro introduces new restriction endonuclease cleavage sites. *Nature* 272: 375–377, 1978.

Hanahan, D., Jessee, J. and Bloom, F.R., Plasmid transformation of *Escherichia coli* and other bacteria. *Methods Enzymol* 204: 63–113, 1991.

Hart, S.L., Knight, A.M., Harbottle, R.P., Mistry, A., Hunger, H.D., Cutler, D.F., Williamson, R., et al., Cell binding and internalization by filamentous phage displaying a cyclic Arg-Gly-Asp-containing peptide. *J Biol Chem* 269: 12468–12474, 1994.

Hendrix, R.W., Roberts, J.W., Stahl, F.W. and Weisberg, R.A., *Lambda II.* Cold Spring Harbor Laboratory Press, New York, 1983.

Herrmann, R., Neugebauer, K., Pirkl, E., Zentgraf, H. and Schaller, H., Conversion of bacteriophage fd into an efficient single-stranded DNA vector system. *Mol Gen Genet* 177: 231–242, 1980.

Hines, J.C. and Ray, D.S., Construction and characterization of new coliphage M13 cloning vectors. *Gene* 11: 207–218, 1980.

Hohn, B. and Murray, K., Packaging recombinant DNA molecules into bacteriophage particles in vitro. *Proc Natl Acad Sci U S A* 74: 3259–3263, 1977.

Holt, C.L. and May, G.S., A novel phage lambda replacement Cre-lox vector that has automatic subcloning capabilities. *Gene* 133: 95–97, 1993.

Hoyt, M.A., Knight, D.M., Das, A., Miller, H.I. and Echols, H., Control of phage lambda development by stability and synthesis of *cII* protein: Role of the viral *cIII* and host hflA, himA and himD genes. *Cell* 31: 565–573, 1982.

Huynh, T.V., Young, R.A. and Davis, R.W., "Constructing and screening cDNA libraries in λgt10 and λgt11," pp. 49–78 in *DNA Cloning: A Practical Approach,* D.M. Glover (Ed.). IRL Press, Oxford, 1985.

Ioannou, P.A., Amemiya, C.T., Garnes, J., Kroisel, P.M., Shizuya, H., Chen, C., Batzer, M.A., et al., A new bacteriophage P1-derived vector for the propagation of large human DNA fragments. *Nat Genet* 6: 84–89, 1994.

Ishiura, M., Hirose, S., Uchida, T., Hamada, Y., Suzuki, Y. and Okada, Y., Phage particle-mediated gene transfer to cultured mammalian cells. *Mol Cell Biol* 2: 607–616, 1982.

Ivanenkov, V., Felici, F. and Menon, A.G., Uptake and intracellular fate of phage display vectors in mammalian cells. *Biochim Biophys Acta* 1448: 450–462, 1999a.

Ivanenkov, V., Felici, F. and Menon, A.G., Targeted delivery of multivalent phage display vectors into mammalian cells. *Biochim Biophys Acta* 1448: 463–472, 1999b.

Kaname, T. and Huxley, C., Simple and efficient vectors for retrofitting BACs and PACs with mammalian neoR and EGFP marker genes. *Gene* 266: 147–153, 2001.

Karn, J., Brenner, S., Barnett, L. and Cesareni, G., Novel bacteriophage lambda cloning vector. *Proc Natl Acad Sci U S A* 77: 5172–5176, 1980.

Kassner, P.D., Burg, M.A., Baird, A. and Larocca, D., Genetic selection of phage engineered for receptor-mediated gene transfer to mammalian cells. *Biochem Biophys Res Commun* 264: 921–928, 1999.

Kohler, S.W., Provost, G.S., Fieck, A., Kretz, P.L., Bullock, W.O., Sorge, J.A., Putman, D.L., et al., Spectra of spontaneous and mutagen-induced mutations in the lacI gene in transgenic mice. *Proc Natl Acad Sci U S A* 88: 7958–7962, 1991.

Larocca, D. and Baird, A., Receptor-mediated gene transfer by phage-display vectors: Applications in functional genomics and gene therapy. *Drug Discov Today* 6: 793–801, 2001.

Larocca, D., Burg, M.A., Jensen-Pergakes, K., Ravey, E.P., Gonzalez, A.M. and Baird, A., Evolving phage vectors for cell targeted gene delivery. *Curr Pharm Biotechnol* 3: 45–57, 2002a.

Larocca, D., Jensen-Pergakes, K., Burg, M.A. and Baird, A., Gene transfer using targeted filamentous bacteriophage. *Methods Mol Biol* 185: 393–401, 2002b.

Larocca, D., Jensen-Pergakes, K., Burg, M.A. and Baird, A., Receptor-targeted gene delivery using multivalent phagemid particles. *Mol Ther* 3: 476–484, 2001.

Larocca, D., Kassner, P.D., Witte, A., Ladner, R.C., Pierce, G.F. and Baird, A., Gene transfer to mammalian cells using genetically targeted filamentous bacteriophage. *Faseb J* 13: 727–734, 1999.

Larocca, D., Witte, A., Johnson, W., Pierce, G.F. and Baird, A., Targeting bacteriophage to mammalian cell surface receptors for gene delivery. *Hum Gene Ther* 9: 2393–2399, 1998.

Levinson, A., Silver, D. and Seed, B., Minimal size plasmids containing an M13 origin for production of single-strand transducing particles. *J Mol Appl Genet* 2: 507–517, 1984.

Lorenz, M.G. and Wackernagel, W., Bacterial gene transfer by natural genetic transformation in the environment. *Microbiol Rev* 58: 563–602, 1994.

Masuda, Y., Miyakawa, K., Nishimura, Y. and Ohtsubo, E., chpA and chpB, *Escherichia coli* chromosomal homologs of the pem locus responsible for stable maintenance of plasmid R100. *J Bacteriol* 175: 6850–6856, 1993.

Mead, D.A. and Kemper, B., Chimeric single-stranded DNA phage-plasmid cloning vectors. *Biotechnology* 10: 85–102, 1988.

Mejia, J.E. and Monaco, A.P., Retrofitting vectors for *Escherichia coli*-based artificial chromosomes (PACs and BACs) with markers for transfection studies. *Genome Res* 7: 179–186, 1997.

Messing, J., New M13 vectors for cloning. *Methods Enzymol* 101: 20–78, 1983.

Messing, J., Crea, R. and Seeburg, P.H., A system for shotgun DNA sequencing. *Nucleic Acids Res* 9: 309–321, 1981.

Messing, J., Gronenborn, B., Muller-Hill, B. and Hans Hopschneider, P., Filamentous coliphage M13 as a cloning vehicle: insertion of a HindII fragment of the lac regulatory region in M13 replicative form in vitro. *Proc Natl Acad Sci U S A* 74: 3642–3646, 1977.

Messing, J. and Vieira, J., A new pair of M13 vectors for selecting either DNA strand of double-digest restriction fragments. *Gene* 19: 269–276, 1982.

Metcheva, I.S., Stedman, T.T. and Buck, G.A., An arrayed bacteriophage P1 genomic library of *Pneumocystis carinii*. *J Eukaryot Microbiol* 43: 171–176, 1996.

Monaci, P., Urbanelli, L. and Fontana, L., Phage as gene delivery vectors. *Curr Opin Mol Ther* 3: 159–169, 2001.

Murray, N.E., "Phage lambda and molecular cloning," pp. 395–432 in *Lambda II,* R.W. Hendrix, J.W. Roberts, F.W. Stahl and R.A. Weisberg (Eds.). Cold Spring Harbor Laboratory Press, Cold Spring Harbor, 1983.

Murray, N.E., Special uses of lambda phage for molecular cloning. *Methods Enzymol* 204: 280–301, 1991.

Murray, N.E., Brammar, W.J. and Murray, K., Lambdoid phages that simplify the recovery of in vitro recombinants. *Mol Gen Genet* 150: 53–61, 1977.

Nistala, R. and Sigmund, C.D., A reliable and efficient method for deleting operational sequences in PACs and BACs. *Nucleic Acids Res* 30: e41, 2002.

Okayama, H. and Berg, P., Bacteriophage lambda vector for transducing a cDNA clone library into mammalian cells. *Mol Cell Biol* 5: 1136–1142, 1985.

Palazzolo, M.J., Hamilton, B.A., Ding, D.L., Martin, C.H., Mead, D.A., Mierendorf, R.C., Raghavan, K.V., et al., Phage lambda cDNA cloning vectors for subtractive hybridization, fusion-protein synthesis and Cre-loxP automatic plasmid subcloning. *Gene* 88: 25–36, 1990.

Pasqualini, R. and Ruoslahti, E., Organ targeting in vivo using phage display peptide libraries. *Nature* 380: 364–366, 1996.

Peeters, B.P., Schoenmakers, J.G. and Konings, R.N., Plasmid pKUN9, a versatile vector for the selective packaging of both DNA strands into single-stranded DNA-containing phage-like particles. *Gene* 41: 39–46, 1986.

Pierce, J.C., Sauer, B. and Sternberg, N., A positive selection vector for cloning high molecular weight DNA by the bacteriophage P1 system: Improved cloning efficacy. *Proc Natl Acad Sci U S A* 89: 2056–2060, 1992a.

Pierce, J.C., Sternberg, N. and Sauer, B., A mouse genomic library in the bacteriophage P1 cloning system: organization and characterization. *Mamm Genome* 3: 550–558, 1992b.

Pierce, J.C. and Sternberg, N.L., Using bacteriophage P1 system to clone high molecular weight genomic DNA. *Methods Enzymol* 216: 549–574, 1992.

Poorkaj, P., Peterson, K.R. and Schellenberg, G.D., Single-step conversion of P1 and P1 artificial chromosome clones into yeast artificial chromosomes. *Genomics* 68: 106–110, 2000.

Poul, M.A. and Marks, J.D., Targeted gene delivery to mammalian cells by filamentous bacteriophage. *J Mol Biol* 288: 203–211, 1999.

Poulsen, L.K., Larsen, N.W., Molin, S. and Andersson, P., A family of genes encoding a cell-killing function may be conserved in all gram-negative bacteria. *Mol Microbiol* 3: 1463–1472, 1989.

Rajotte, D., Arap, W., Hagedorn, M., Koivunen, E., Pasqualini, R. and Ruoslahti, E., Molecular heterogeneity of the vascular endothelium revealed by in vivo phage display. *J Clin Invest* 102: 430–437, 1998.

Rao, V.B., Thaker, V. and Black, L.W., A phage T4 in vitro packaging system for cloning long DNA molecules. *Gene* 113: 25–33, 1992.

Rosenberg, S.M., Stahl, M.M., Kobayashi, I. and Stahl, F.W., Improved in vitro packaging of coliphage lambda DNA: a one-strain system free from endogenous phage. *Gene* 38: 165–175, 1985.

Rosner, J.L., Formation, induction, and curing of bacteriophage P1 lysogens. *Virology* 48: 679–680, 1972.

Rothstein, R. and Wu, R., Modification of the bacteriophage vector M13mp2: Introduction of new restriction sites for cloning. *Gene* 15: 167–176, 1981.

Russel, M., Filamentous phage assembly. *Mol Microbiol* 5: 1607–1613, 1991.

Russel, M., Kidd, S. and Kelley, M.R., An improved filamentous helper phage for generating single-stranded plasmid DNA. *Gene* 45: 333–338, 1986.

Shepherd, N.S. and Smoller, D., The P1 vector system for the preparation and screening of genomic libraries. *Genet Eng* (N Y) 16: 213–228, 1994.

Short, J.M., Fernandez, J.M., Sorge, J.A. and Huse, W.D., Lambda ZAP: A bacteriophage lambda expression vector with in vivo excision properties. *Nucleic Acids Res* 16: 7583–7600, 1988.

Sternberg, N., Bacteriophage P1 cloning system for the isolation, amplification, and recovery of DNA fragments as large as 100 kilobase pairs. *Proc Natl Acad Sci U S A* 87: 103–107, 1990.

Sternberg, N., "Library construction in P1 phage vectors," pp. 81–101 in *DNA Cloning: A Practical Approach*. IRL Press, New York, 1995.

Sternberg, N., The P1 cloning system: Past and future. *Mamm Genome* 5: 397–404, 1994.

Sternberg, N., Ruether, J. and deRiel, K., Generation of a 50,000-member human DNA library with an average DNA insert size of 75–100 kbp in a bacteriophage P1 cloning vector. *New Biol* 2: 151–162, 1990.

Sternberg, N., Smoller, D. and Braden, T., Three new developments in P1 cloning. Increased cloning efficiency, improved clone recovery, and a new P1 mouse library. *Genet Anal Tech Appl* 11: 171–180, 1994.

Sternberg, N.L., Cloning high molecular weight DNA fragments by the bacteriophage P1 system. *Trends Genet* 8: 11–16, 1992.

Thomas, M., Cameron, J.R. and Davis, R.W., Viable molecular hybrids of bacteriophage lambda and eukaryotic DNA. *Proc Natl Acad Sci U S A* 71: 4579–4583, 1974.

Tsuzuki, T. and Rancourt, D.E., Embryonic stem cell gene targeting using bacteriophage lambda vectors generated by phage-plasmid recombination. *Nucleic Acids Res* 26: 988–993, 1998.

Unger, M.W., Liu, S.Y. and Rancourt, D.E., Transplacement mutagenesis: a novel in situ mutagenesis system using phage-plasmid recombination. *Nucleic Acids Res* 27: 1480–1484, 1999.

Urbanelli, L., Ronchini, C., Fontana, L., Menard, S., Orlandi, R. and Monaci, P., Targeted gene transduction of mammalian cells expressing the HER2/neu receptor by filamentous phage. *J Mol Biol* 313: 965–976, 2001.

Vieira, J. and Messing, J., New pUC-derived cloning vectors with different selectable markers and DNA replication origins. *Gene* 100: 189–194, 1991.

Vieira, J. and Messing, J., Production of single-stranded plasmid DNA. *Methods Enzymol* 153: 3–11, 1987.

Vieira, J. and Messing, J., The pUC plasmids, an M13mp7-derived system for insertion mutagenesis and sequencing with synthetic universal primers. *Gene* 19: 259–268, 1982.

Webster, R., "Biology of the filamentous bacteriophage," pp. 1–20 in *Phage Display of Peptides and Proteins,* B. Kay, J. Winter and J. McCafferty (Eds.). Academic Press, New York, 1996.

Westwater, C., Kasman, L.M., Schofield, D.A., Werner, P.A., Dolan, J.W., Schmidt, M.G. and Norris, J.S., Use of genetically engineered phage to deliver antimicrobial agents to bacteria: An alternative therapy for treatment of bacterial infections. *Antimicrob Agents Chemother* 47: 1301–1307, 2003.

Westwater, C., Schofield, D.A., Schmidt, M.G., Norris, J.S. and Dolan, J.W., Development of a P1 phagemid system for the delivery of DNA into Gram-negative bacteria. *Microbiology* 148: 943–950, 2002.

Williams, B.G. and Blattner, F.R., "Bacteriophage lambda vectors for DNA cloning," pp. 201–229 in *Genetic Engineering*, J.K. Setlow and A. Hollaender (Eds.). Plenum Press, New York, 1980.

Woltjen, K., Bain, G. and Rancourt, D. E., Retro-recombination screening of a mouse embryonic stem cell genomic library. *Nucleic Acids Res* 28: E41, 2000.

Yarmolinski, M.B. and Sternberg, N., "Bacteriophage P1," pp. 291–418 in *The Bacteriophages,* R. Calendar (Ed.). Plenum, New York, 1988.

Yokoyama-Kobayashi, M. and Kato, S., Recombinant f1 phage particles can transfect monkey COS-7 cells by DEAE dextran method. *Biochem Biophys Res Commun* 192: 935–939, 1993.

Yokoyama-Kobayashi, M. and Kato, S., Recombinant f1 phage-mediated transfection of mammalian cells using lipopolyamine technique. *Anal Biochem* 223: 130–134, 1994.

Young, R.A. and Davis, R.W., Efficient isolation of genes by using antibody probes. *Proc Natl Acad Sci U S A* 80: 1194–1198, 1983.

Zacher, A.N., 3rd, Stock, C.A., Golden, J.W., 2nd and Smith, G.P., A new filamentous phage cloning vector: fd-tet. *Gene* 9: 127–140, 1980.

Zagursky, R. and Baumeister, K., Construction and use of pBR322 plasmids that yield single-stranded DNA for sequencing. *Methods Enzymol* 155: 139–155, 1987.

Zagursky, R.J. and Berman, M.L., Cloning vectors that yield high levels of single-stranded DNA for rapid DNA sequencing. *Gene* 27: 183–191, 1984.

Zinder, N.D. and Boeke, J.D., The filamentous phage (Ff) as vectors for recombinant DNA—a review. *Gene* 19: 1–10, 1982.

Zinder, N.D. and Horiuchi, K., Multiregulatory element of filamentous bacteriophages. *Microbiol Rev* 49: 101–106, 1985.

Zissler, J., Signer, E.R. and Schaefer, F., "The role of recombination in growth of bacteriophage lambda," pp. 445 in *The bacteriophage lambda,* A.D. Hershey (Ed.). Cold Spring Harbor Laboratory Press, New York, 1971.

12 The Use of Phage Lytic Enzymes to Control Bacterial Infections

Vincent A. Fischetti, Ph.D.
Rockefeller University, New York, NY

CONTENTS

12.1. Introduction 321
12.2. Enzyme Structure and Mode of Action 322
12.2.1. Enzyme Structure 322
12.2.2. Mode of Action 324
12.2.3. Specificity 325
12.2.4. Synergistic Effects 325
12.3. Practical Applications 327
12.3.1. *In Vivo* Colonization Experiments 327
12.3.2. Control of Sepsis and Bacteremia 327
12.3.3. Killing Biowarfare Bacteria 328
12.3.4. Use of Phage Lysins in Diagnostics 330
12.4. Immune Response and Bacterial Resistance to Phage-Encoded Enzymes 330
12.4.1. Immune Response to the Enzymes 330
12.4.2. Bacterial Resistance to the Enzymes 331
12.5. Concluding Remarks 332
Acknowledgments 333
References 333

12.1. INTRODUCTION

Bacteriophage lytic enzymes or lysins are highly evolved molecules that have been specifically developed by phages over millions of years to quickly and efficiently allow their progeny to be released from the host bacterium. These enzymes damage the cell wall's integrity by hydrolyzing the four major bonds in its peptidoglycan component; they include (i) endo-β-N-acetylglucosaminidases or N-acetylmuramidases (lysozymes), which act on the sugar moieties, (ii) endopeptidases, which act

0-8493-1336-X/05/$0.00+$1.50
© 2005 by CRC Press LLC

on the peptide cross bridge, and (iii) N-acetylmuramyl-L-alanine amidases (or amidases), which hydrolyze the amide bond connecting the sugar and peptide moieties (Young, 1992). Most of the lysins that have been characterized thus far are amidases; the reason for this is currently unclear. The lysins usually do not have signal sequences to translocate them through the cytoplasmic membrane and cleave their substrate in the peptidoglycan. Instead, the lysins' translocation is controlled by a second phage gene product, the holin (Wang et al., 2000). During phage development in the infected bacterium, lysin accumulates in the cytoplasm in anticipation of phage maturation. At a genetically specified time, patches of holin molecules inserted in the cytoplasmic membrane are activated, resulting in the formation of pores allowing the preformed lysin in the cytoplasm to access the peptidoglycan, there-by causing cell lysis and the release of progeny phages (Wang et al., 2000). The ability of phages to lyse their targeted bacterial hosts has been known for almost a century, and phages have been used to prevent and treat human and animal diseases of bacterial origin since the late 1910s (see Chapters 13 and 14 for more details on this subject). However, phage-encoded enzymes have only recently begun to be used for various practical applications; e.g., (i) the preparation of bacterial "ghost" vaccines (Szostak et al., 1990; 1996), and (ii) reducing bacterial contamination in dairy products (Gaeng et al., 2000). Also, the emergence of antibiotic-resistance has prompted studies to determine the efficacy of phage-encoded enzymes for the prophylaxis and treatment of bacterial diseases of humans.

Many bacterial species colonize mucous membranes lining various human body sites (e.g., the upper and lower respiratory tract, the gastrointestinal and urogenital tracts, and the inner surface of the eyelids). In addition, mucosal reservoirs of many pathogenic bacteria (e.g., pneumococci, staphylococci, and streptococci) often are the focus of infection in human populations (Eiff et al., 2001; Coello et al., 1994; De Lencastre et al., 1999). Thus, reducing or eliminating this human reservoir in the community and in controlled environments (e.g., hospitals and nursing homes) should markedly reduce the incidence of human disease caused by those bacteria. However, with the exception of polysporin- and mupirocin-containing ointments, antibiotics are not commonly used to control the colonization of mucous membranes with pathogenic bacteria (i.e., a *carrier state*), because of the concern of developing antibiotic resistance. In this context, we believe that phage-encoded enzymes may be very effective in reducing or eliminating colonization of human mucosal membranes with various pathogenic bacteria, and we and others have demonstrated that this approach may have merit. For example, we have identified and purified several phage-encoded lytic enzymes and are the first to use them therapeutically; we have found that they rapidly kill various pathogenic Gram-positive bacteria *in vitro* and *in vivo* (Loeffler et al., 2001; Nelson et al., 2001; Schuch et al., 2002).

12.2. ENZYME STRUCTURE AND MODE OF ACTION

12.2.1. Enzyme Structure

One main feature of the phage lysins that have been identified so far is their two-domain structure (Fig. 12.1) (Diaz et al., 1990; Garcia et al., 1990). Their catalytic

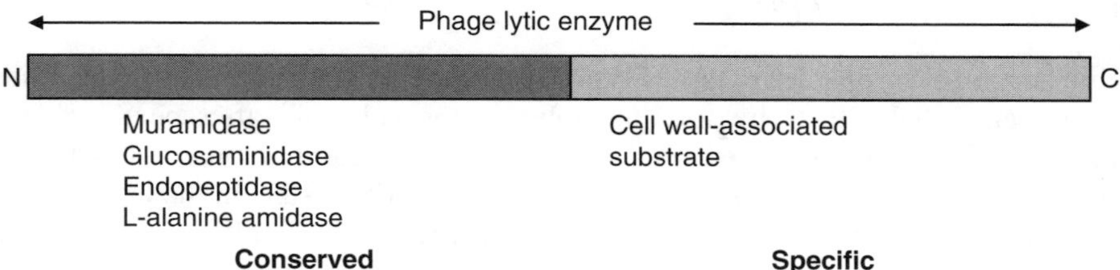

FIGURE 12.1 Basic structure of phage lytic enzymes.
The majority of the phage lytic enzymes that have been characterized so far contain two domains: an N-terminal catalytic domain and a C-terminal binding domain. The catalytic domain of each of the various enzymes cleaves one of the four major bonds in the bacterial peptidoglycan. Thus, each enzyme falls into one of four classes; it is a glucosaminidase, an N-acetylmuramidase, an endopeptidase, or an L-alanine amidase. At the present time, most of the identified enzymes are amidases. The C-terminal half of the molecule binds to a substrate in the targeted bacterium's cell wall. Sequence comparisons of enzymes in the same class indicate that the catalytic region is highly conserved, but that the C-terminal region is variable.

activity (i.e., muramidase, glucosaminidase, endopeptidase, or amidase activity) against the bacterial cell wall's peptidoglycan usually resides in their N-terminal domain, and their specificity is conferred by their C-terminal domain, which binds to a specific receptor in the host bacterium's cell wall (Lopez et al., 1997; Lopez et al., 1992; Garcia et al., 1988). The first crystal structure for a lytic enzyme's C-terminal/binding domain has recently been published for the choline-binding enzymes of *Streptococcus pneumoniae* (Fernandez-Tornero et al., 2001). Sequence comparisons between lytic enzymes of the same catalytic class show a great degree of identity at the N-terminal region but very little in the C-terminal region. The reason for the latter region's specificity was not apparent at first, since it seemed counterintuitive that the phage would design an enzyme that was specifically lethal for its host bacterium. However, as we learned more about how these enzymes function, a possible reason for this specificity has emerged (see section 12.4.2).

Because of their two-domain structure, it seemed plausible that the different domains in the enzymes could be changed, thus yielding an enzyme with a different bacterial specificity or a different substrate. Indeed, Garcia and colleagues (Garcia et al., 1990; Weiss et al., 1999) demonstrated that the catalytic domains of phage enzymes active against *S. pneumoniae* could be swapped, thus resulting in a new enzyme that recognized the same pneumococcal receptor/binding site but cleaved a different bond in the peptidoglycan. Their observation suggests the very important possibility of creating "designer" phage lysins possessing high specificity and equally high cleavage potential.

Lysin genes in double stranded DNA phage generally code for a single continuous polypeptide chain, however, some exceptions have been reported. For example, although introns have not been identified in bacterial host genomes, Foley et al. (2000) reported that 50% of the *S. thermophilus* strains they examined have lysogenic phage whose lysin genes are interrupted by a self-splicing (group) I intron.

12.2.2. Mode of Action

Thin-section electron microscopy of lysin-treated bacteria indicates that phage lysins exert their lethal effects by forming holes in the cell wall after peptidoglycan digestion. Compared to the external environment, a bacterium's interior is hypertonic; therefore, the cell wall's loss of structural integrity results in the extrusion of the cytoplasmic membrane and hypotonic lysis (Fig. 12.2A). Thin-section electron

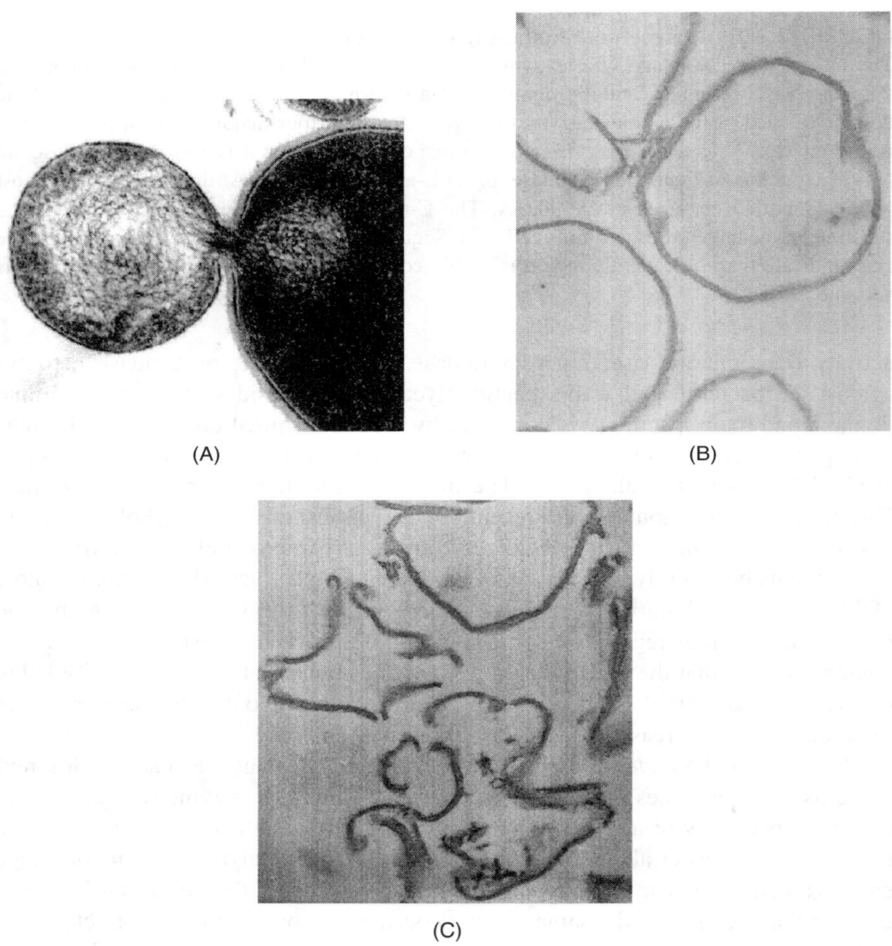

FIGURE 12.2 Effects of lysin on whole bacteria and cell walls.
(A) Thin section electron micrographs of whole group A streptococci treated with C1 phage lytic enzyme for 15 seconds. The cell walls of the bacteria have holes, which results in externalization of the cytoplasmic membrane and osmotic lysis of the cells. Isolated group A streptococcal cell walls pre-treatment (B) and post-treatment (C) with the phage enzyme. Enzyme-treated cells exhibit pieces of the cell wall, which indicates that structure's digestion by the enzyme.

micrographs of isolated streptococcal cell walls treated with lysin support the results obtained with lysin-treated bacteria; that is, they are indicative of cell wall destruction because of the large number of holes formed in the wall (compare Fig. 12.2B and 12.2C). Theoretically, a single hole should be able to kill a bacterium; however, it is currently uncertain whether this theoretical value is ever reached and, if so, how long it would take for death to occur after enzyme attachment.

12.2.3. Specificity

An interesting feature of bacteriolytic phage enzymes is their host specificity; e.g., lysins produced by streptococcal bacteriophages only kill streptococci, and enzymes produced by pneumococcal phages only kill pneumococci (Loeffler et al., 2001; Nelson et al., 2001). In that regard, the group A streptococcal lysin efficiently kills all group A streptococcal serotypes, and it kills group C and group G streptococci, but it does not affect streptococci normally found in the human oral cavity (Fig. 12.3).

Our studies examining the safety of using the *S. pyogenes*-specific lysin in two animal models, one involving mucosal tissue and the other utilizing skin tissue, did not reveal any deleterious changes when the enzyme was added to the models daily for 7 days, and the tissues were grossly and histologically examined (unpublished data). This observation was not surprising because the bonds cleaved by phage lysins are not present in eukaryotic cells. Thus, it is anticipated that these enzymes will be tolerated well by human mucous membranes. Another important observation is that the killing efficiency of a pneumococcal-specific lysin was the same against both penicillin-resistant and penicillin-sensitive pneumococci (Loeffler et al., 2001; Nelson et al., 2001). Thus, the significant advantage of using phage lysins to prevent and treat bacterial diseases include their efficacy against antibiotic-resistant bacteria and their specificity and low toxicity; they kill antibiotic-resistant, pathogenic bacteria without damaging the host's tissues and normal flora. On the other hand, numerous pathogenic bacterial mutants are now resistant to formerly effective antibiotics, and many antibiotics have toxic side effects (e.g., hepatotoxicity and nephrotoxicity) and kill normal floral bacteria in the mammalian body.

12.2.4. Synergistic Effects

Pneumococcal bacteriophages produce several bacteriolytic enzymes. We have isolated two enzymes: an amidase, which cleaves the bond between N-acetylmuramic acid and L-alanine, and a lysozyme, which cleaves the glycosidic bond between N-acetylmuramic acid and N-acetylglucosamine. Both enzymes have very different N-terminal catalytic domains; however, they have a very similar C-terminal domain, which binds to choline in the host bacterium's cell wall. Therefore, we examined whether the simultaneous use of both enzymes is competitive, or whether they, on the other hand, act synergistically to destroy the bacterial cell wall. We used three different protocols, all of which are standard methods (used in the pharmaceutical industry) to determine synergy: (i) time kill in liquid (ii) disk diffusion, and (iii) checkerboard analysis. The results we obtained with all three approaches revealed a clear synergistic effect on the efficiency of killing when both enzymes were simultaneously used

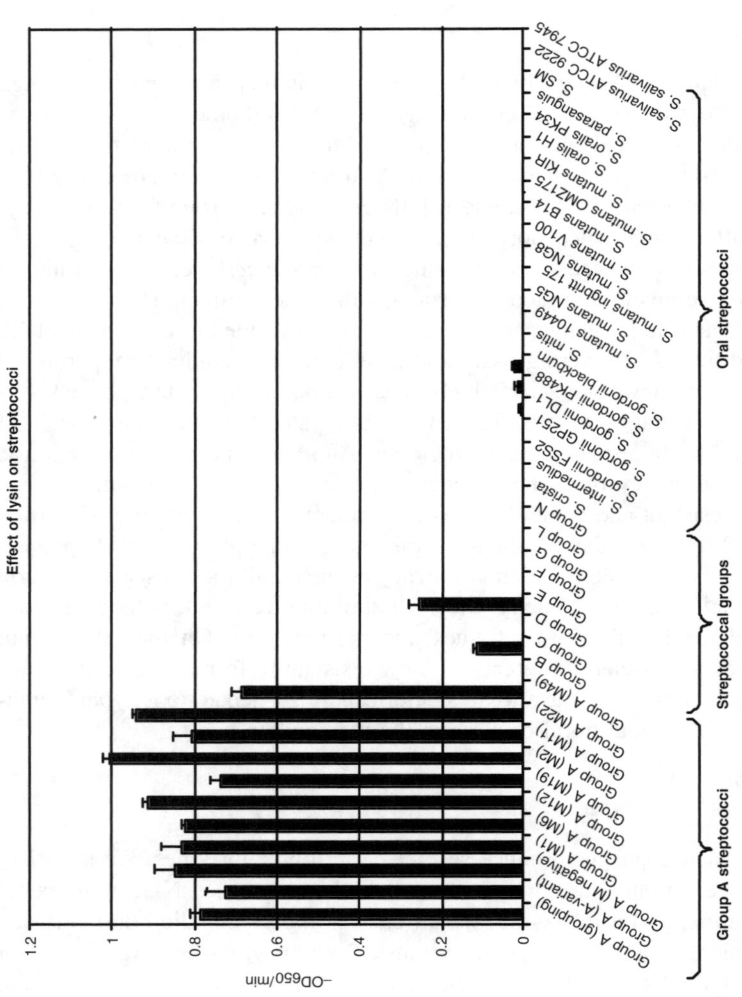

FIGURE 12.3 Lysin activity on various streptococci. Representative streptococcal strains were exposed to 250 units of purified lysin, and the OD_{650} of the bacterial suspension was monitored. The lysin's activity against each strain was reported as the initial velocity of lysis, in BOD_{650}/min, based on the time it took to decrease the starting OD by half (i.e., from an OD_{650} of 1.0 to an OD_{650} of 0.5). All assays were performed in triplicate and the data are expressed as the means of the standard deviations.

(Loeffler and Fischetti, 2003). Thus, in addition to increasing the killing efficiency, the simultaneous use of two phage lysins possessing different cleavage sites may significantly retard the emergence of bacterial resistance to the enzymes.

12.3. PRACTICAL APPLICATIONS

12.3.1. *In Vivo* Colonization Experiments

Two animal models have been used to determine the capacity of phage lysins to prevent or reduce the colonization of mucosal surfaces with pathogenic bacteria: an oral colonization model was developed for *S. pyogenes* (Nelson et al., 2001), and a nasal model was developed for pneumococci (Loeffler et al., 2001). All of the animals colonized with the bacteria and treated with a small amount of lysin specific for the colonizing organism were free of the bacteria 2 to 5 hours after lysin treatment (Figs. 12.4 and 12.5). During the *S. pyogenes* experiments, the animals were swabbed 24 and 48 hours after lysin treatment. At those time periods, most of the inoculated and lysin-treated animals remained negative for the colonizing streptococci, but one animal died and two others were again positive for streptococci (Fig. 12.4). A possible explanation for these results is that the positive animals became diseased during the first four days of colonization (during which time some organisms became intracellular) and, although the lysin was able to clear organisms found on the mucous membrane's surface, it was unable to kill intracellular bacteria. The bacteria that were present at 24 and 48 hours post-treatment with lysin were as sensitive to the lysin as were those in the original inoculum, ruling out the possibility that these bacteria were lysin-resistant mutants.

12.3.2. Control of Sepsis and Bacteremia

We have found that various lysins are active *in vitro* against their phages' host bacteria in human blood (unpublished data). This observation prompted us to determine whether the enzymes also were active *in vivo* in a sepsis or bacteremia model of infection. During our initial studies, we intravenously infected groups of Balb/c mice with a serotype 14 strain of *S. pneumoniae*, and we treated them 1 hour later with a single bolus of buffer (control group) or phage lysin Cpl-1 (2 mg) administered by the same route. All of the animals treated with Cpl-1 survived for 48 hours, whereas the median survival time of the buffer-treated mice was only ca. 25 hours, and only 20% of them survived for 48 hours ($p = 0.0003$). However, blood and organ cultures of the euthanized surviving mice showed that only one Cpl-1-treated animal was totally free of infection at 48 hours. Therefore, during a second experiment, we treated groups of mice with 2 doses of the Cpl-1 lysin (2×2 mg) or buffer at 5 and 10 hours postinfection, and we observed the animals for a longer period of time. The Cpl-1-treated mice and the buffer-treated mice had a median survival time of 60 hours and 30 hours, respectively ($p < 0.0001$). In addition, pharmacokinetic experiments (Loeffler et al., 2003) have revealed that the Cpl-1 enzyme has a half-life in the mouse of ca. 20 minutes. Thus, we anticipate that for intravenous applications, phage enzymes will need to be delivered by constant i.v. infusion in order to control bacteria under these conditions.

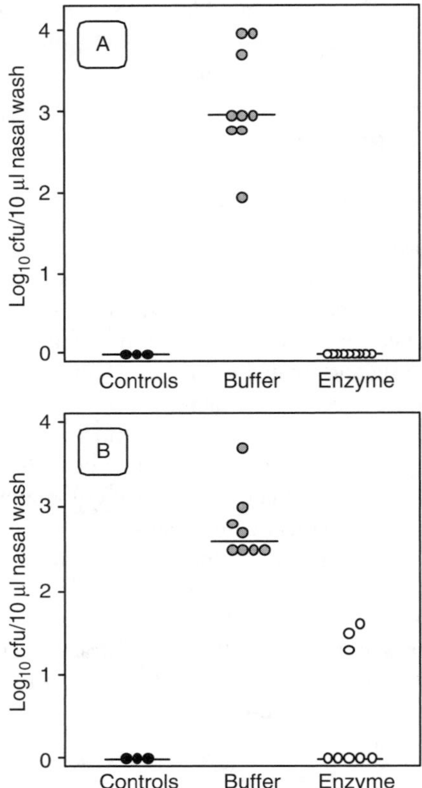

FIGURE 12.4 Rapid killing of streptococci.
Elimination of *S. pyogenes* in a mouse model of oral colonization. Mice were orally colonized with group A streptococci and followed for four days by oral swab. All animals were heavily colonized (>300 CFU/swab) for the four days. All animals were treated orally with a single dose of 500 Units of lysin and swabbed 2 hours later and found to be negative for colonizing streptococci. Animals swabbed 24 and 48 hours later were mostly still negative but two animals beginning to show colonizing streptococci, and one animal had died.

12.3.3. Killing Biowarfare Bacteria

The efficacy of phage lysins/bacteriolytic enzymes in killing pathogenic bacteria suggests that they may be a valuable tool in controlling biowarfare bacteria. To determine the feasibility of that approach, we (Schuch et al., 2002) identified a lytic enzyme produced by the gamma phage that is specific for *Bacillus anthracis* (Watanabe et al., 1975). In addition, we cloned the phage's gamma lysin gene and identified a ~700 bp ORF encoding a 26-kDa product very similar in size and features to various *Bacillus, Listeria*, and *Mycobacteria* phage lysins. The gamma lysin, referred to as PlyG, was purified to homogeneity by a two-step chromatography procedure, and its *in vitro* and *in vivo* activity against gamma phage-sensitive anthrax

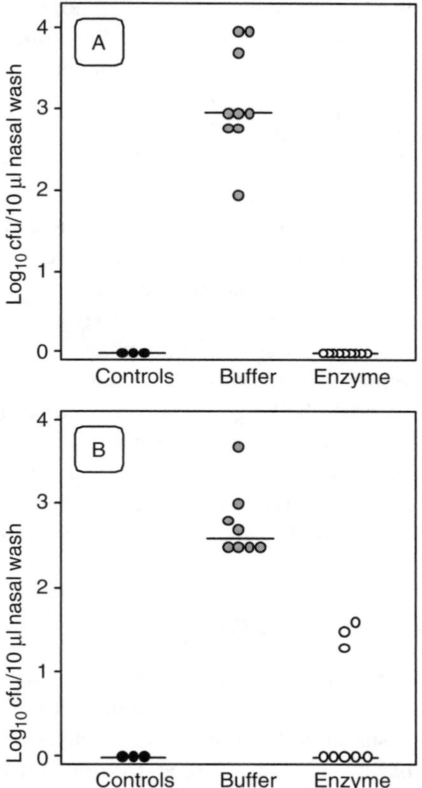

FIGURE 12.5 Killing pneumococci *in vivo*.
Elimination of *S. pneumoniae* serotype 14 in a mouse model of nasopharyngeal carriage. (A) After nasal and pharyngeal treatment with a total of 1400 units of Pal, pneumococci were not detectable in the nasal washings from the colonized mice; however, they were isolated from the nasal washings obtained from all of the colonized and buffer-treated mice ($p < 0.001$). Also, pneumococci were not isolated from non-colonized control mice. (B) After treatment with a total of 700 U Pal, pneumococci were completely eliminated from five of the eight colonized mice ($p < 0.001$), and their numbers were significantly reduced in the other three mice.

bacilli was characterized. During our initial *in vitro* studies, we observed that as little as 100 units of PlyG mediated (by three seconds after exposure) a 5000-fold decrease in the viable counts in *B. anthracis* suspensions containing ~10^7 CFU/ml. This lethal activity was observed in growth media, phosphate buffer, and to a lesser but significant extent in human blood. Subsequently, we found that the enzyme was active against ten *B. anthracis* strains from different clonal types isolated worldwide. In addition, the gamma-phage lysin was lethal for five mutant *B. anthracis* strains lacking either capsule or toxin plasmids, but was inactive against *B. anthracis* spores. Concerning the latter observation, it is well-established that the germination of *B. anthracis* endospores can be influenced by various amino acids (e.g., L-alanine, tyrosine, and adenosine), of which some (e.g., L-alanine) can trigger the germination of endospores (Hills, 1949; Ireland and Hanna, 2002; Titball and Manchee, 1987). Thus, it is possible that the combined *in vitro* application of a spore-germinating agent and PlyG may be used to kill new vegetative cells released by the elicited germination of *B. anthracis* endospores.

Research with virulent strains of *B. anthracis* is limited to a relatively small number of laboratories with the appropriate infrastructure (i.e., a BSL-3 level facility). Therefore, a genetically closely related species, *B. cereus*, is often used during preliminary investigations examining the value of various antimicrobial agents against *B. anthracis*. Thus, prior to determining PlyG's efficacy in treating *B. anthracis* infections *in vivo*, we examined its *in vivo* effect on a mouse-lethal *B. cereus* strain with a similar sensitivity to gamma phage as the wild-type *B. anthracis* isolate. Nine of ten mice injected i.p. with 10^7 CFU of the *B. cereus* strain died from septicemia within 4 hours post-challenge. However, 13 of 18 mice (72%) challenged with the bacteria and treated 15 minutes later with 100 μg of PlyG survived, and the death of three of the five animals that died was delayed >24 hours. Higher doses or multiple doses of the enzyme may have yielded even better results. Collaborative studies to determine the *in vivo* efficacy of PlyG against *B. anthracis* are currently in progress in a BSL-3 facility.

12.3.4. Use of Phage Lysins in Diagnostics

Bacteriophages have been used, since their discovery in the 1910s, to identify and classify bacterial species and strains. The basic approach utilizes a battery of phages to characterize the phage susceptibility profile of the bacterium being analyzed. More recently, this approach has been modernized, and various related methodologies have been proposed to identify specific bacteria rapidly in a complex bacterial mixture. One example of a more recently developed sensitive and rapid test to detect specific bacterial contaminants is to use firefly luciferase to quantitate ATP (Stanley, 1989) released from bacterial cells lysed by phages or phage lytic enzymes. The assay is based on the linear relationship between the number of ATP molecules hydrolyzed and the quantity of light photons subsequently produced. The bioluminescence may be rapidly measured in a bio-luminometer; thus, quickly identifying the bacterial contaminant in the sample. The assay is sensitive; it can detect and identify $<10^3$ organisms in a sample, which makes it very attractive for identifying low levels of pathogenic bacteria in a variety of food, clinical, and environmental samples. For more details about using phages and their lysins to detect pathogenic bacteria, refer to Chapter 9.

12.4. IMMUNE RESPONSE AND BACTERIAL RESISTANCE TO PHAGE-ENCODED ENZYMES

12.4.1. Immune Response to the Enzymes

A potential problem with the use of phage lytic enzymes for therapeutic purposes in humans and animals is the development of neutralizing antibodies. Unlike antibiotics, which are small molecules that are not immunogenic, enzymes are large proteins that are able to stimulate an immune response, especially when delivered systemically. However, small amounts (nanograms to a few micrograms) of mucosally delivered phage lysins are likely to be only weakly immunogenic, especially if they are delivered without an adjuvant. In this context, we recently characterized the effect of a rabbit hyperimmune serum (raised against pneumococcal-specific Cpl-1) against Cpl-1. Incubating (2 min at 37°C) 2 mg of Cpl-1 with a pneumococcal suspension (optical density [OD] = 0.5, at 600 nm in the presence of pre-immune/normal rabbit serum, decreased the suspension's OD by about 50%. However, when the enzyme was preincubated with the hyperimmune serum for 5 minutes and 1 hour at 37°C, it took 4.5 minutes and 10 minutes, respectively, to achieve a 50% reduction in the suspension's OD. These results indicate that the anti-Cpl-1 antibodies slowed but did not block Cpl-1's ability to kill pneumococci. Similar results were obtained during our *in vitro* experiments performed with antibodies directed against phage enzymes lytic for *B. anthracis* and *S. pyogenes*.

To test the *in vivo* relevance of our *in vitro* observations, (i) six mice that had received three i.v. administered doses of the Cpl-1 enzyme 4 weeks earlier (and that tested positive for IgG against Cpl-1; i.e., neutralizing titers of ca. 10, in five of six mice), and (ii) naïve control mice were challenged i.v. with pneumococci and were treated (by the same route) 10 hours postchallenge, with 200 μg of Cpl-1. The treatment reduced the bacteremic titer of Cpl-1-immunized mice and of naïve mice by a median $\log_{10}$ of 0.40 and 0.57, respectively. This difference is not statistically significantly; thus, our *in vivo* and *in vitro* data are in agreement.

Our results suggest that the C-terminal/binding domains of the examined lytic enzymes have higher affinities for their cell wall receptors than the antibodies have for the enzymes. This idea also is supported by the recent observation (Loessner et al., 2002) that the affinity of the cell wall binding domain of a *Listeria*-specific phage enzyme for its cell wall receptor is similar to that of the affinity of an IgG molecule for its specific antigen (i.e., both have nanomolar affinities). However, although this idea may explain the inability of antibodies to neutralize the lysins' binding domains, it does not explain why antibodies against the enzymes' catalytic domains do not neutralize those domains. In any event, our results are encouraging because they indicate that phage lysins may be used in chronic situations (i) controlling the colonization of pathogenic bacteria on the mucosal surfaces of susceptible populations (e.g., debilitated and/or immunocompromised people in hospitals, daycare centers, and nursing homes), and (ii) treating patients with septicemias and bacteremias caused by multiply antibiotic-resistant bacteria.

12.4.2. Bacterial Resistance to the Enzymes

Repeated exposure of nutrient agar-grown host bacteria (pneumococci and streptococci) to low concentrations of lysin did not lead to the emergence of lysin-resistant mutants. In addition, the bacteria isolated from colonies located at the periphery of a clear zone (created by a 10 μl drop of dilute lysin) were sensitive to the lysin even after the process was repeated 41 times. We were also unable to identify lysin-resistant mutants after several cycles of bacterial exposure to low concentrations of lysin in liquid cultures (Loeffler et al., 2001). Our results may be explained by the fact that the cell wall receptor for the pneumococcal lysin is choline (Fernandez-Tornero et al., 2001; Garcia et al., 1983), a molecule that is essential for pneumococcal viability. In that regard, polyrhamnose, a cell wall component of group A streptococci, is one of the components necessary for lysin binding to those bacteria, and polyrhamnose has also been shown (Yamashita et al., 1999) to be important for streptococcal growth. Therefore, it is possible that during phages' associations with bacteria over the millennia, to avoid becoming trapped inside the host, the binding domains of their lytic enzymes have evolved to target unique and essential molecules in the host bacteria's cell walls, thus making resistance to these enzymes a rare event. Since phages have performed the "high throughput assays" to identify the so-called "Achilles heel" of their host bacteria, we may be able to use that knowledge to identify the biosynthetic pathways for the lysins' cell wall receptors, and to identify inhibitors of the pathways. Theoretically, this approach might subsequently yield new antibacterial drugs which would be difficult to become resistant against.

12.5. CONCLUDING REMARKS

Although the ability of phage enzymes to lyse various bacteria *in vitro* has been known for some time (Sher and Mallette, 1953; Loessner et al., 1998; Gaeng et al., 2000; Fischetti et al., 1971), their use to control bacterial pathogens *in vivo* is a novel and intriguing approach (Loeffler et al., 2003; Nelson, et al., 2001). Similarly to traditional *phage therapy* approach (during which phages are used to lyse their targeted bacteria; see Chapters 13 and 14 for more details on this subject), using lytic enzymes allows specific killing of bacterial pathogens without affecting the surrounding normal (and often beneficial) microflora. As described in this chapter, purified phage lytic enzymes may be used to control pathogenic bacteria on mucous membranes and, therefore, they offer an intriguing opportunity to prevent (and, potentially, also to treat) various human infectious diseases that are preceded by the colonization of mucosal membranes. In addition, the enzymes may be of value in controlling various food-borne bacterial pathogens in the food processing industry; e.g., they may help reduce the use of antibiotics in livestock, or the use of harsh reagents to decontaminate food-processing facilities. Furthermore, because of the very serious problems caused by multiply antibiotic-resistant bacteria in hospitals, daycare centers and nursing homes, the enzymes may be of benefit in those environments.

Phage lytic enzymes probably have evolved over millions of years, during the co-evolution of bacteriophages and bacteria. Thus, advancing knowledge concerning the intricacies of this interrelationship will improve our understanding of the complex biological processes involved in phage-bacterial host interactions, and the evolutionary consequences of the interactions. The enzymes isolated thus far are relatively heat-stable (i.e., they tolerate temperatures up to 60°C), and fairly large quantities are relatively easy to obtain in a highly purified state. Thus, their real-life applications are technically feasible. In that regard, phage lytic enzymes may be invaluable in improving the prevention and treatment of various bacterial diseases, particularly those that are increasingly difficult to deal with using currently available antibiotics and other conventional antibacterial approaches.

ACKNOWLEDGMENTS

I wish to acknowledge the members of my laboratory who are responsible for much of the work described in this: Daniel Nelson, Jutta Loeffler and Raymond Schuch, and the excellent technical assistance of Shiwei Zhu, Mary Windels, and Ryann Russell. I also wish to thank Abraham Turetsky (Aberdeen Proving Grounds) and Leonard Mayer (CDC) for testing the gamma lysin against authentic *B. anthracis* strains. Our research was supported by grants from New Horizons Diagnostics and the Defense Advanced Project Agency (DARPA).

REFERENCES

Coello, R., Jimenez, J., Garcia, M., Arroyo, P., Minguez, D., Fernandez, C., Cruzet, F., et al., Prospective study of infection, colonization and carriage of methicillin-resistant *Staphylococcus aureus* in an outbreak affecting 990 patients. *Eur J Clin Microbiol Infect Dis* 13: 74–81, 1994.

De Lencastre, H., Kristinsson, K.G., Brito-Avo, A., Sanches, I.S., Sa-Leao, R., Saldanha, J., Sigvaldadottir, E., et al., Carriage of respiratory tract pathogens and molecular epidemiology of *Streptococcus pneumoniae* colonization in healthy children attending day care centers in Lisbon, Portugal. *Microb Drug Resist* 5: 19–29, 1999.

Diaz, E., Lopez, R. and Garcia, J.L., Chimeric phage-bacterial enzymes: a clue to the modular evolution of genes. *Proc Natl Acad Sci U S A* 87: 8125–8129, 1990.

Eiff, C.V., Becker, K., Machka, K., Stammer, H. and Peters, G., Nasal carriage as a source of *Staphylococcus aureus* bacteremia. *N Engl J Med* 344: 11–16, 2001.

Fernandez-Tornero, C., Lopez, R., Garcia, E., Gimenez-Gallego, G. and Romero, A., A novel solenoid fold in the cell wall anchoring domain of the pneumococcal virulence factor LytA. *Nat Struct Biol* 8: 1020–1024, 2001.

Fischetti, V.A., Gotschlich, E.C. and Bernheimer, A.W., Purification and physical properties of group C streptococcal phage-associated lysin. *J Exp Med* 133: 1105–1117, 1971.

Foley, S., Bruttin, A. and Brussow, H., Widespread distribution of a group I intron and its three deletion derivatives in the lysin gene of *Streptococcus thermophilus* bacteriophages. *J Virol* 74: 611–618, 2000.

Gaeng, S., Scherer, S., Neve, H. and Loessner, M.J., Gene cloning and expression and secretion of *Listeria monocytogenes* bacteriophage-lytic enzymes in *Lactococcus lactis*. *Appl Environ Microbiol* 66: 2951–2958, 2000.

Garcia, E., Garcia, J.L., Garcia, P., Arraras, A., Sanchez-Puelles, J.M. and Lopez, R., Molecular evolution of lytic enzymes of *Streptococcus pneumoniae* and its bacteriophages. *Proc Natl Acad Sci U S A* 85: 914–918, 1988.

Garcia, P., Garcia, E., Ronda, C., Tomasz, A. and Lopez, R., Inhibition of lysis by antibody against phage-associated lysin and requirement of choline residues in the cell wall for progeny phage release in *Streptococcus pneumoniae*. *Curr Microbiol* 8: 137–140, 1983.

Garcia, P., Garcia, J.L., Garcia, E., Sanchez-Puelles, J.M. and Lopez, R., Modular organization of the lytic enzymes of *Streptococcus pneumoniae* and its bacteriophages. *Gene* 86: 81–88, 1990.

Hills, G.M., Chemical factors in the germination of spore bearing aerobes. The effects of amino acids on the germination of *Bacillus anthracis*, with some observations on the relation of optical form to biological activity. *Biochem J* 45: 363–370, 1949.

Ireland, J.A.W. and Hanna, P.C., Amino acid- and purine ribonucleoside-induced germination of *Bacillus anthracis* delta sterne endospores: *gerS* mediates responses to aromatic ring structures. *J Bacteriol* 184: 1296–1303, 2002.

Loeffler, J.M., Djurkovic, S. and Fischetti, V.A., Phage lytic enzyme Cpl-1 as a novel antimicrobial for pneumococcal bacteremia. *Infect Immun* 71: 6199–6204, 2003.

Loeffler, J.M. and Fischetti, V.A., Synergistic lethal effect of a combination of phage lytic enzymes with different activities on penicillin-sensitive and- resistant *Streptococcus pneumoniae* strains. *Antimicrob Agents Chemother* 47: 375–377, 2003.

Loeffler, J.M., Nelson, D. and Fischetti, V.A., Rapid killing of *Streptococcus pneumoniae* with a bacteriophage cell wall hydrolase. *Science* 294: 2170–2172, 2001.

Loessner, M.J., Gaeng, S., Wendlinger, G., Maier, S.K. and Scherer, S., The two-component lysis system of *Staphylococcus aureus* bacteriophage Twort: a large TTG-start holin and an associated amidase endolysin. *FEMS Microbiol Lett* 162: 265–274, 1998.

Loessner, M.J., Kramer, K., Ebel, F. and Scherer, S., C-terminal domains of *Listeria monocytogenes* bacteriophage murein hydrolases determine specific recognition and high-affinity binding to bacterial cell wall carbohydrates. *Mol Microbiol* 44: 335–349, 2002.

Lopez, R., Garcia, E., Garcia, P. and Garcia, J.L., The pneumococcal cell wall degrading enzymes: a modular design to create new lysins? *Microb Drug Resist* 3: 199–211, 1997.

Lopez, R., Garcia, J.L., Garcia, E., Ronda, C. and Garcia, P., Structural analysis and biological significance of the cell wall lytic enzymes of *Streptococcus pneumoniae* and its bacteriophage. *FEMS Microbiol Lett* 79: 439–447, 1992.

Nelson, D., Loomis, L. and Fischetti, V.A., Prevention and elimination of upper respiratory colonization of mice by group A streptococci by using a bacteriophage lytic enzyme. *Proc Natl Acad Sci U S A* 98: 4107–4112, 2001.

Schuch, R., Nelson, D. and Fischetti, V.A., A bacteriolytic agent that detects and kills *Bacillus anthracis*. *Nature* 418: 884–889, 2002.

Sher, I.H. and Mallette, M.F., The use of bacteriophage in releasing two decarboxylases from *Escherichia coli* B. *J Biol Chem* 200: 257–262, 1953.

Stanley, P.E., A review of bioluminescent ATP techniques in rapid microbiology. *J Biolumin Chemilumin* 4: 375–380, 1989.

Szostak, M., Wanner, G. and Lubitz, W., Recombinant bacterial ghosts as vaccines. *Res Microbiol* 141: 1005–1007, 1990.

Szostak, M.P., Hensel, A., Eko, F.O., Klein, R., Auer, T., Mader, H., Haslberger, A., et al., Bacterial ghosts: non-living candidate vaccines. *J Biotechnol* 44: 161–170, 1996.

Titball, R.W. and Manchee, R.J., Factors affecting the germination of spores of *Bacillus anthracis*. *J Appl Bacteriol* 62: 269–273, 1987.

Wang, I.N., Smith, D.L. and Young, R., Holins: The protein clocks of bacteriophage infections. *Annu Rev Microbiol* 54: 799–825, 2000.

Watanabe, T., Morimoto, A. and Shiomi, T., The fine structure and the protein composition of gamma phage of *Bacillus anthracis*. *Can J Microbiol* 21: 1889–1892, 1975.

Weiss, K., Laverdiere, M., Lovgren, M., Delorme, J., Poirier, L. and Beliveau, C., Group A *Streptococcus* carriage among close contacts of patients with invasive infections. *Am J Epidemiol* 149: 863–868, 1999.

Yamashita, Y., Tomihisa, K., Nakano, Y., Shimazaki, Y., Oho, T. and Koga, T., Recombination between *gtfB* and *gtfC* is required for survival of a dTDP-rhamnose synthesis-deficient mutant of *Streptococcus mutans* in the presence of sucrose. *Infect Immun* 67: 3693–3697, 1999.

Young, R., Bacteriophage lysis: mechanism and regulation. *Microbiol Rev* 56: 430–481, 1992.

13 Phage Therapy in Animals and Agribusiness

Alexander Sulakvelidze[1,2] *and Paul Barrow*[3]
[1]Department of Epidemiology and Preventive Medicine, University of Maryland School of Medicine, Baltimore, MD
[2]Intralytix, Inc., Baltimore, MD
[3]Institute for Animal Health, Compton, Newbury, Berkshire, UK

CONTENTS

13.1. Introduction 336
13.2. Using Phages to Prevent and Treat Animal Infections 336
13.2.1. The First Known Use of Phages to Treat Bacterial Diseases of Animals 336
13.2.2. Early Studies Examining the Efficacy of Bacteriophages for Preventing and Treating Animal Infections 337
13.2.3. Recent Studies Examining the Efficacy of Bacteriophages For Preventing and Treating Animal Infections 338
13.2.3.1. *Salmonella* Infections 338
13.2.3.2. *Escherichia coli* Infections 339
13.2.3.2.1. Smith and Huggins' Studies 339
13.2.3.2.2. Experimental *E. coli* Infections in Chickens and Colostrum-Deprived Calves 342
13.2.3.2.3. *E. coli* Infections in Mice and "Long-Circulating" Bacteriophages 342
13.2.3.2.4. *E. coli* Respiratory Infections in Chickens 343
13.2.3.3. *Enterococcus faecium* Infections 343
13.2.3.4. *Vibrio cholerae* Infections 344
13.2.3.5. *Clostridium difficile* Infections 346
13.2.3.6. *Acinetobacter baumanii, Pseudomonas aeruginosa*, and *Staphylococcus aureus* Infections 347
13.2.4. Bacteriophages as a Potential Tool for Reducing Antibiotic Usage in Agribusiness 348

0-8493-1336-X/05/$0.00+$1.50
© 2005 by CRC Press LLC

13.3. Bacteriophages and Diseases of Aquaculture and Plants 350
13.3.1. Plant Infections 350
13.3.2. Infections of Aquacultured Fish 352
13.4. Phage-Elicited Bacterial Lysates as Vaccines 353
13.4.1. Using Phages to Produce Bacterial Vaccines 353
13.4.2. Using Phage-Generated Bacterial Lysates to Prevent and Treat Animal Infections 354
13.4.3. Phages and "Bacterial Ghost" Vaccines 355
13.4.5. Using Phage-Encoded Enzymes As Antibacterial Agents 358
13.5. Using Bacteriophages in the Food Production Chain.............................. 359
13.5.1. General Considerations 359
13.5.2. Using Phages to Reduce Contamination of Livestock with Foodborne Pathogens....................................... 360
13.5.3. Using Phages to Reduce Contamination of Raw Foods with Foodborne Pathogens... 361
13.5.4. Using Phages to Reduce Contamination of Ready-to-Eat Foods with Foodborne Pathogens... 363
13.5.5. Prevalence of Bacteriophages in Foods and Other Parts of the Environment 365
13.5.6. Food Spoilage and Bacteriophages 368
Acknowledgments... 371
References... 371

13.1. INTRODUCTION

Since the discovery of bacteriophages in 1915–1917, they have been used to prevent and treat various bacterial infections. Although *phage therapy* has been historically associated with the use of bacteriophages in human medicine, phages also have been extensively used in veterinary medicine and in various agricultural settings. The history and various aspects of phage therapy in humans are reviewed in Chapter 14 of this book. In this Chapter, we review the past and current use of phages to prevent and treat naturally occurring and experimentally induced infections of animals. In addition, we discuss the potential applications of phage therapy in various agricultural settings, including the potential value of bacteriophages for improving the safety of foods and preventing foodborne diseases of bacterial etiology, and their potential to reduce the use of antibiotics in livestock.

13.2. USING PHAGES TO PREVENT AND TREAT ANIMAL INFECTIONS

13.2.1. The First Known Use of Phages to Treat Bacterial Diseases of Animals

The first-known therapeutic use of phages in veterinary medicine is associated with Felix d'Herelle, the co-discoverer of bacteriophages. In the spring of 1919, large outbreaks of lethal fowl typhoid in chickens occurred in the Acris-sur-Aube region of France.

D'Herelle analyzed several dead animals from the outbreaks, and he isolated and identified *Salmonella gallinarum* as the etiologic agent of the disease. D'Herelle also isolated bacteriophages from the chickens, and he examined their efficacy in preventing and treating *S. gallinarum* infections in six experimentally infected chickens (d'Herelle, 1921, 1926). The results of the study were promising: Phage administration prevented the birds from succumbing to the bacterial infection, whereas the two control chickens not treated with phages died after a single dose of the challenge strain.

The promising results of the small pilot study prompted d'Herelle to almost immediately initiate larger trials, which he called *immunization experiments*. One hundred chickens were infected with *S. gallinarum* and 20 of them were treated ("immunized") with *S. gallinarum*-specific phage. The 20 phage-treated birds survived; whereas 60 (75%) of the phage-untreated birds died. Encouraged by these results (and the results of similar studies with rabbits and buffalo), d'Herelle subsequently used other phages in his collection to treat bacterial dysentery of humans (see Chapter 14). Also, the results of d'Herelle's studies examining phage prophylaxis and treatment of fowl typhoid prompted other investigators to begin examining the utility of bacteriophages in preventing or treating various naturally occurring and experimental bacterial infections in animals. As discussed below, the reported outcomes of those studies varied dramatically, depending on the infectious agents used, the animal models of infection, and the lytic potency of bacteriophages.

13.2.2. Early Studies Examining the Efficacy of Bacteriophages for Preventing and Treating Animal Infections

One of the earliest animal models used in several phage therapy studies was murine salmonellosis—a systemic disease produced in susceptible mice by several serotypes of *Salmonella*, including typhimurium, enteritidis, dublin, choleraesuis, and abortus. For example, Topley et al. (1925) used an *S.* Typhimurium strain to evaluate the efficacy of phage treatment in experimentally infected mice. After bacterial challenge, mice were administered phage *per os* and the mortality rates of, and bacterial shedding from, phage-untreated and phage-treated mice were analyzed. In contrast to d'Herelle's observations with *Salmonella*-infected chickens, the authors found that phage administration did not reduce mortality or bacterial shedding. Furthermore, a study by Topley, et al. (1925) revealed that intraperitoneal (i.p.) injection of their phages did not significantly reduce the spread of the *S.* Typhimurium infection among the mice. Several factors may have contributed to the failure of phage treatment in these two studies, including the use of phages that were not optimally effective against the challenge strain of *S.* Typhimurium; indeed, the phage apparently failed to completely lyse that strain *in vitro* (Topley and Wilson, 1925). In that context, Fisk (1938) reported that injecting antityphoid phages with good *in vitro* activity against the challenge strain into mice before challenge with typhoid bacilli conferred excellent protection. The phage-inoculated mice were protected for 24 h; phage therapy initiated 4 h after bacterial challenge also protected mice. Protection was not observed when the phage were heat-inactivated (70ºC, 50 min), thus indicating that viable phage were required for the preparation's efficacy.

In some early studies (Arnold and Weiss, 1926; Asheshov et al., 1937), authors simultaneously administered bacteria and phage to animals—an approach which may not be relevant to most real-life situations and may yield potentially misleading results. In this context, bacteria-phage interactions usually occur very rapidly; e.g., although bacterial lysis by phages may take 20–40 minutes, phage attachment to the bacterial membrane (followed by phage DNA injection) may occur within just a few seconds or minutes (see Chapters 3 and 7). Thus, mixing the bacteria with the phage before inoculating the animals with the mixture can yield results that primarily represent *in vitro* interaction of the bacteria in the inoculum, rather than results pertaining to the efficacy of phages in preventing or treating bacterial infections *in vivo*.

Early attempts at phage treatment of experimentally induced staphylococcal and streptococcal septicaemias in rabbits and mice were reported to be unsuccessful by several investigators (Clark and Clark, 1927; Krueger et al., 1932), including Giorgi Eliava (1930)—the co-founder of the Bacteriophage Institute in Tbilisi, Georgia (see Chapter 14). Also, many attempts to treat experimental plague in rabbits, guinea pigs, rats, and mice failed to influence the course of the disease (Naidu and Avari, 1932; Compton, 1930; 1928; Colvin, 1932). Moreover, in clear contrast to d'Herelle's earlier studies, Pyle (1926) reported phage therapy to be ineffective in treating fowl typhoid, even though he used phage with excellent *in vitro* activity against the infecting bacterium. On the other hand, injecting bacteriophages into the carotid artery has been claimed to significantly reduce the mortality of rabbits with experimental streptococcal meningitis (Kolmer and Rule, 1933). Also, *E. coli* cystitis in rabbits and guinea pigs has been reported (Marcuse, 1924; Larkum, 1926) to be cured or markedly alleviated by phage treatment. Furthermore, excellent results were reported by Dubos et al. (1943) who used intraperitoneal phage to treat cerebrally injected *S. dysenteriae* infections in mice. (These studies are discussed in some detail in Chapter 14.)

13.2.3. Recent Studies Examining the Efficacy of Bacteriophages for Preventing and Treating Animal Infections

13.2.3.1. *Salmonella* Infections

After the early work by d'Herelle (1921; 1926) and other investigators (1925; Topley and Wilson 1925; Fisk 1938), *Salmonella*-infected laboratory animals re-emerged as one of the most commonly used *in vivo* models to study the prophylactic and therapeutic value of bacteriophages. For example Berchieri, *et al.* (1991) isolated several bacteriophages lytic for *Salmonella* Typhimurium from chickens, chicken feed, and human sewage systems in England and evaluated their efficacy in treating experimental *S.* Typhimurium infections in chickens. The challenge strain produced a fatal infection in ca. 53% of the phage-untreated chickens, and administration of some of their phages produced a statistically significant reduction in chicken mortality. One of the most effective phages, $\phi 2.2$, also significantly reduced (i) the mortality caused by two other virulent strains of *S.* Typhimurium and (ii) viable numbers of the challenge strain in the alimentary tracts of chickens treated with the phage. For example, a 1 log reduction was observed in the number of challenge bacteria in the crop, small intestines and caeca at 12 h postchallenge and an 0.9 log reduction was seen in the liver at 24 and 48 h post-challenge. The authors hypothesized that, because of the challenge strain's relatively low LD_{50} dose of 10^3 CFU, that level of reduction might be clinically significant. A few phage-resistant colonies were isolated from chickens' caecal contents; however, they were of the rough (i.e., avirulent or less-virulent) phenotype. Also, no phage-neutralizing antibodies were detected in serum obtained from chickens sacrificed 32 days post challenge. The results also indicated that the inoculated phage, which was highly lytic *in vitro*, only persisted in the intestine as long as the *Salmonella* count remained high.

In all cases, administration of high-titer phage preparations was required for a positive therapeutic effect. The timing of the phage treatment was also important; i.e., initiating phage treatment shortly after the bacterial challenge was significantly more effective than was delaying the treatment. At the present time, the mechanism(s) for this time-dependence are not clear. However, since *Salmonella* is an intracellular pathogen, one possible explanation is that early administration of phages kills most of the salmonellae in the gut before they are internalized and, thus, protected from the lytic effect of phages. More research is likely to provide much needed information in that regard, and to generate critical data needed for the optimal design and implementation of phage-mediated prophylaxis and therapy of *Salmonella* infections in various agriculturally important animals.

13.2.3.2. *Escherichia coli* Infections

13.2.3.2.1. Smith and Huggins' Studies

E. coli can cause several types of noninvasive enteritis and septicemia in various animal species. Thus, initial studies examining the possible therapeutic value of phage in animals focused on *E. coli* infections, such as Larkum's 1926 work on phage therapy for *E. coli* cystitis in rabbits and guinea pigs (Larkum, 1926). After a long period of disinterest in phage therapy (see Chapter 14), *E. coli*-infected animal models once again emerged as a primary system to evaluate the possible efficacy of phage. Arguably the best-known studies of phage therapy in veterinary medicine were reported from the laboratory of Herbert Williams Smith and his colleagues at the Institute for Animal Disease Research in Houghton, Cambridgeshire, Great Britain. Their early work focused on *E. coli* septicemia experimentally induced in mice using a strain of *E. coli* O18:K1:H7 $ColV^+$ from a child with meningitis (Smith and Huggins, 1982). Some of the phages used during the study were specific for the capsular K1 antigen—a major virulence determinant in the challenge *E. coli* strain. A single intramuscular (i.m.) injection of one anti-K1 phage was more effective than were multiple i.m. injections of various antibiotics (tetracycline, ampicillin, chloramphenicol, or trimethoprim plus sulphafurazole) in protecting mice against a potentially lethal, i.m.- or intracerebrally (i.c.)-induced infection with the challenge *E. coli* strain. The phage treatment also was reported to be at least as effective as were multiple i.m. injections of streptomycin. Experiments enumerating the phages and bacterial pathogen in various animal tissues revealed that the phages persisted for

24 hours in the bloodstream, and that high phage titers (10^6 PFU/gram of tissue) persisted for several days in the spleen. A few phage-resistant mutants were identified during the study; however, they had the $K1^-$ phenotype, and thus were significantly less virulent than the wild-type $K1^+$ strain. Phage administration 3 to 5 days before the *E. coli* challenge also protected mice against the experimentally induced infection, although the protective effect varied among phages propagated on different bacterial host strains. The excellent results obtained during their initial study prompted the authors to expand their research to other animals.

During their second study, Smith and Huggins (1983) examined the therapeutic and prophylactic efficacy of *E. coli*-specific bacteriophage preparations in neonatal enteritis in calves, piglets, and lambs. During the experiments in calves, 71 newborn calves were infected orally with ca. 3×10^9 CFU of an O9:K30.99 enterotoxigenic strain of *E. coli.* Fifty-seven of the seventy-one calves were colostrum-fed; they were susceptible to diarrhea but resistant to septicemia. The calves were divided into 6 groups, 4 of which were treated with a mixture of two different *E. coli*-specific phages (Table 13.1). None of the nine calves treated with 10^{11} PFU of a phage mixture 8 h after bacterial challenge became ill. In addition, only 2 animals died in a group of 13 colostrum-deprived calves who received phages at the onset of diarrhea. In contrast, the mortality rate in the phage-untreated control groups was 93% among the colostrum-treated calves and 100% among the colostrum-free calves (Table 13.1). If administration of the phage treatment was delayed until the onset of diarrhea, the disease was not prevented; however, the severity of the illness was significantly reduced and so was the mortality. The phage did not totally eliminate the pathogenic *E. coli* strain from the gut, but appeared to reduce their numbers to a level which was below that required to produce disease. The phage persisted in the gut as long as their host strain's numbers remained high, and disappeared thereafter. Similar treatments with other specific phage preparations were effective in protecting piglets and lambs against challenge with enterotoxigenic strains of *E. coli* O20:K101.987P and *E. coli* O8:K85.99, respectively, and no phage-resistant mutants were identified in the phage-treated lambs. A few phage-resistant *E. coli* mutants were seen in the phage-treated calves and piglets; however, they had the $K30^-$ or $K101^-$ phenotype and, when tested in animals, appeared to have significantly lower virulence than the parental, wild-type strains.

In a subsequent series of studies, Smith et al. (1987b; 1987a) again used *E. coli*-infected calves to evaluate further the efficacy of phage treatment. During one of these studies (Smith et al., 1987b), they also evaluated the susceptibility of phages to various conditions likely to be encountered after their administration into animals. The authors noted that the low pH of the abomasum's contents affected the viability of orally administered phages, but that the deleterious effect was reduced if phages were administered shortly after milk feed. A similar effect was achieved by sodium bicarbonate administration in the feed just prior to or during oral phage treatment. In this context, it is noteworthy that neutralizing gastric acid by oral administration of bicarbonate mineral water shortly before phage administration is a common practice during human phage therapy studies in the former Soviet Union and Eastern Europe (see Chapter 14).

TABLE 13.1
Data from the 1983 Study by Smith and Huggins (1983). The Study Examined the Efficacy of Phage Therapy in Experimental *E. coli* Diarrhea in Calves, Piglets and Lambs

Animals/No of Animals	Colostrum Fed	Phage or Phage Mixture	Time Phage Given (after bacterial challenge)	Animals with Diarrhea	Dead Animals	Mortality
Calves		1×10^{11} PFU/dose				
6	No	B44/1 & B44/2	1 h	3	1	17%
8	No	No Phage	N/A	8	8	100%
9	Yes	B44/1 & B44/2	8 h	0	0	0%
21	Yes	B44/1 & B44/2	At onset of diarrhea	21	14	67%
13	Yes	B44/1 & B44/3	At onset of diarrhea	13	2	15%
14	Yes	No Phage	N/A	13	13	93%
Piglets		1×10^{10} PFU/dose				
7	N/A	P433/1 & P433/2	At onset of diarrhea		0	0%
7	N/A	No Phage	N/A		4	57%
Lambs[a]		1×10^{9} PFU/dose				
4	N/A	S13	8 h	2	0	0%
4	N/A	No Phage	N/A	4	1	25%
		1×10^{10} PFU/dose				
3	N/A	S13	8 h	2	0	0%
3	N/A	No Phage	N/A	3	1	25%

N/A—Not applicable

[a]The study was terminated at 24 h post bacterial challenge, at which time all remaining animals were sacrificed, and the levels of the challenge strain in their intestines was evaluated. The mortality for this group is based on the number of animal that died from infection under 24 h.

13.2.3.2.2. *Experimental* E. coli *Infections in Chickens and Colostrum-Deprived Calves*

The studies by Smith and Huggins, which have been reviewed by several authors (Barrow and Soothill, 1997; Sulakvelidze et al., 2001) and examined using mathematical models and statistical analyses (Levin and Bull, 1996), have stimulated the recent rekindling of interest in phage therapy in the West. For example, Barrow et al. (1998) used an *E. coli*-specific bacteriophage (previously isolated from sewage and found to attach to the K1 capsular antigen) to prevent septicemia and a meningitis-like infection in chickens caused by a $K1^+$ strain of *E. coli*. In untreated chickens, the experimental infection had a mortality rate of almost 100% after i.m inoculation with 10^6 CFU; however, a single i.m. injection of the phage preparation, prior to bacterial challenge, prevented morbidity and death. The phage treatment's efficacy was dose-dependent: the best protection was obtained when high doses (e.g., 10^6 PFU) of phages were injected. The injection of 10^4 PFU provided significant, albeit less, protection, and 10^2 PFU did not significantly protect the chickens. The protection also was obtained when phage administration was delayed until the signs of disease were evident.

Similarly encouraging results were obtained in the experiments with calves. For example, orally challenging (ca. 10^{10} CFU, by stomach tube) two colostrum-deprived calves with the $K1^+$ strain of *E. coli* elicited severe septicemic disease within 18–36 h post-challenge, at which time they were sacrificed for humane reasons. In contrast, three of four calves injected i.m. with the phage preparation (10^{10} PFU) 8 h post-challenge remained healthy, and the fourth calf appeared to be only slightly ill (Barrow et al., 1998). The number of calves examined during the study was too small for statistical analysis, but the phage treatment did appear to have had a positive therapeutic effect.

13.2.3.2.3. E. coli *Infections in Mice and "Long-Circulating" Bacteriophages*

An *E. coli*-infected mouse model recently was used to evaluate the *in vivo* efficacy of "long-circulating" bacteriophages as antimicrobial agents (Merril et al., 1996). The authors hypothesized that phage therapy may be deleteriously affected by the rapid elimination of phages by various mammalian host defense mechanisms, particularly by the reticuloendothelial system. Thus, in order to reduce phage elimination, they used a natural selection strategy (which they called the "serial passage" method) to obtain *E. coli* phages having an increased ability to remain in the bloodstream of mice. For example, to isolate λ phage mutants possessing that capability, mice were injected i.p. with 10^{11} PFU of λ phage W60, blood samples were obtained 7 h post-injection, the residual phages present in bloodstream were injected into mice, and the serial cycling of the phages (injection into animals, isolation, and regrowth) was repeated nine more times. Using this approach, the authors isolated *E. coli* phage λ W60 mutants and *S.* Typhimurium phage P22 mutants capable of staying in the circulation of mice for significantly longer than did the wild-type, parental phages. After their successful mutant isolation experiments, the authors compared the relative abilities of the wild-type λ phage and the "serially passaged" *E. coli* phage mutants to protect 1-week-old BALB/c mice against an experimental *E. coli* septicemia, and they found that a single i.p. injection of either of the two long circulating phages was more effective in rescuing bacteremic mice than was a single injection of the wild-type phage, which suggests that the persistence of high phage titers in the circulation has a positive impact on the efficacy of phage treatment (see Chapter 14 for more details about the study).

13.2.3.2.4. E. coli *Respiratory Infections in Chickens*

The ability of phage therapy to prevent fatal *E. coli* respiratory infections in broiler chickens was recently evaluated in a series of studies at the University of Arkansas. During their first study (Huff et al., 2002b), the authors performed three separate experiments. The first experiment involved challenging (air sac inoculation) groups of 3-day-old-chickens with mixtures containing (i) 10^3 CFU of *E. coli* and 10^3 or 10^6 PFU of the phages, and (ii) 10^4 CFU of *E. coli* and 10^4 or 10^8 PFU of the phages. In the second experiment, 1-week-old birds drank water containing 10^3 or 10^4 PFU of the phages/ml before being air sac-challenged with 10^3 CFU of *E. coli*, or they ingested water containing 10^4 or 10^6 PFU of phages/ml before challenge with 10^4 CFU of *E. coli*. During the third experiment, 1-week-old chickens were air sac-challenged with 10^4 CFU of *E. coli*, and they then drank water containing 10^5 or 10^6 PFU of phages/ml. In the first experiment, the chickens challenged with 10^3 CFU of *E. coli* exhibited a mortality rate of 80%; however, the mortality rate decreased to 25% and 5% when the birds were challenged with a mixture containing the bacteria and 10^3 or 10^6 PFU of phages, respectively. In addition, chickens challenged with 10^4 CFU of *E. coli* had a mortality rate of 85%, which decreased to 35% and 0% when they were challenged with a mixture containing the bacteria and 10^4 or 10^8 PFU of phages, respectively. On the other hand, phage administration did not protect the chickens in the second and third experiments. In their second study (Huff et al., 2002a), the authors reported that aerosol administration of bacteriophages was efficacious in preventing fatal *E. coli* respiratory infections in broiler chickens. The authors' third study (Huff et al., 2003b) compared the efficacy of aerosol administration and i.m. injection of bacteriophages in preventing fatal *E. coli* respiratory infections in broiler chickens. Their most recent study examined the efficacy of multiple vs. single i.m. injections of bacteriophage to treat a severe *E. coli* respiratory infection in broiler chickens (Huff et al., 2003a). As in their previous studies, the authors found that bacteriophages protected against a fatal respiratory challenge with *E. coli*, and that the outcome of the phage treatment was dependent on, among other factors, the route of phage administration (e.g., adding the bacteriophages to the drinking water did not protect the birds). Although the results of the above-described studies appear to be encouraging, they should be interpreted cautiously. For example, as described in section 13.2.2 of this Chapter, mixing phage with bacteria may result in phage infecting the bacteria prior to inoculation rather than *in vivo*. Thus, some or all of the reported positive results might be artifacts caused by the experimental protocols.

13.2.3.3. *Enterococcus faecium* Infections

Recently, the value of using bacteriophages for preventing or treating bacteremia caused by vancomycin-resistant enterococci (VRE) has been evaluated in an experimentally infected mouse model (Biswas et al., 2002). The authors developed an

E. faecium-infected mouse model to determine the efficacy of phage treatment in preventing a fatal septicemia elicited by i.p. injection of 10^9 CFU of a vancomycin-resistant *E. faecium* strain. A single i.p. injection (administered 45 min after the bacterial challenge) of 3×10^9 or 3×10^8 PFU of a phage preparation with potent *in vitro* lytic activity against the bacterium protected all of the infected mice; the efficacy of treatment was clearly dose-dependent (Fig. 13.1). When treatment was delayed until the animals were moribund (18 to 24 h postchallenge), ca. 50% of them still were rescued by a single injection of the phage preparation. In both instances, survival was associated with a significant decrease in the number of challenge bacteria in the animal's bloodstream.

They also explored whether their observed protective effect required viable phages that could grow in the bacterial host, or whether "phage rescue" might have been a function of non-specific immune response triggered by the injection of the phage preparation. A phage preparation (ca. 1×10^{10} PFU/ml) was heat-treated (80ºC, 20 min) and the destruction of viable phages confirmed by plaque assays. Four days after challenge with *E. faecium*, 80% of the mice treated with the viable, plaque-forming phage preparation survived; however, only 10% of the mice injected with the heat-inactivated phage preparation survived. The latter percentage was identical to that of the mice in the control group (i.e., mice treated with PBS). The results were statistically significant ($P < 0.0006$). The authors also addressed the possibility that the observed effect of heat-treatment may have resulted from inactivation or denaturation of other components (e.g., media components) of phage preparation that might have possessed therapeutic, immunostimulatory activity by comparing the *in vivo* protective abilities of two *E. faecium* phages with different *in vitro* lytic activities against the challenge strain of *E. faecium* (i.e., one phage formed plaques, and one strain did not form plaques, when grown on a lawn of the challenge strain). All of the infected mice treated with the lytic phage preparation survived, compared with survival rates of 20% and 50% for infected mice treated with the nonlytic phage preparation and with phosphate buffered saline (PBS) buffer, respectively ($P < 0.03$). This finding further supports the idea that the observed therapeutic effect in mice was due to the presence of viable phages with lytic activity against the challenge strain, rather than by specific or nonspecific immunostimulation. (The therapeutic mode of action of bacteriophages is discussed further in section 13.4 of this chapter and section 14.5 of Chapter 14.) The multiply injected mice did not exhibit anaphylactic reactions, fever, or any other side effects.

13.2.3.4. *Vibrio cholerae* Infections

Sarkar et al. (1996) examined the ability of *V. cholerae*-typing phages to reduce *in vivo* levels of *V. cholerae*, and to reduce fluid accumulation caused by cholera toxin, in the rabbit ileal loop (RIL) model. Ten phages (ATCC 51352 B1 to B10; $10^{10} - 10^{11}$ PFU/ml of each preparation) were injected, alone or together with 10^8 CFU of the challenge *V. cholerae* strain (MAK 757 or ATCC 51352), into ileal loop segments of outbred New Zealand rabbits. The PBS buffer was used as the negative control, and the toxigenic strain *V. cholerae* 569B Inaba was used as a positive control. The animals were sacrificed ca. 18 h post-treatment, and the fluid accumulation ratios

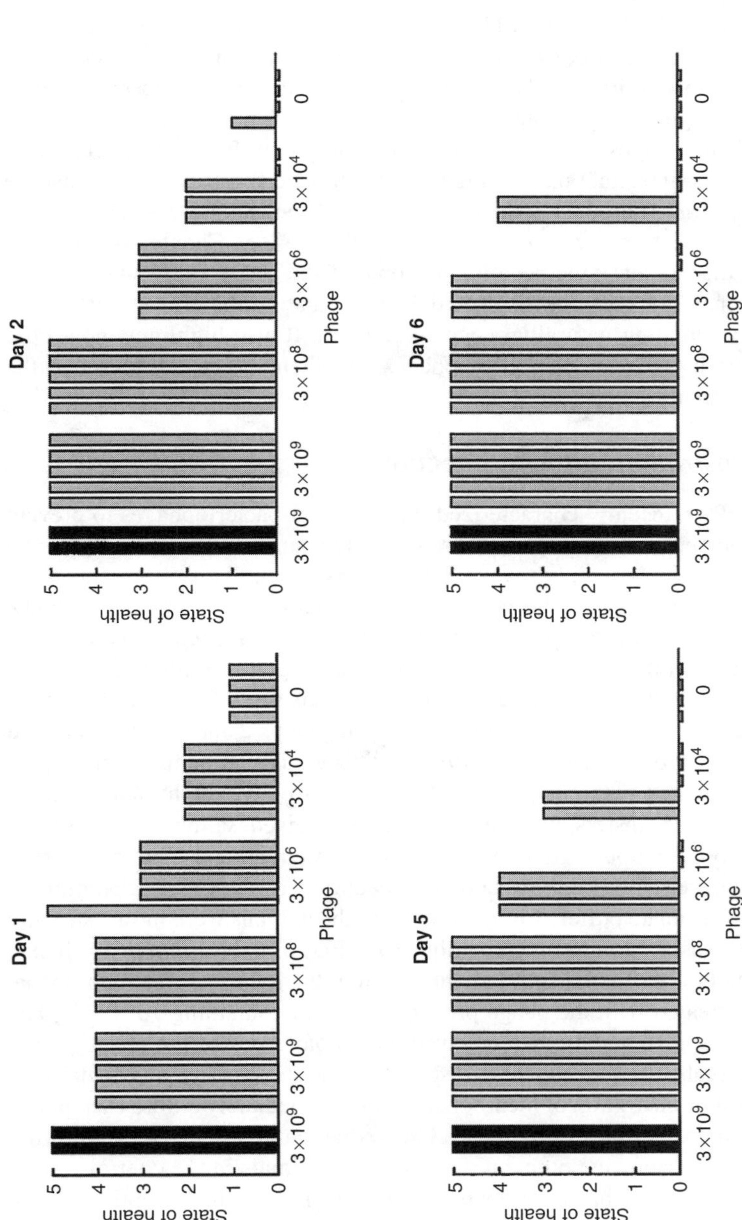

FIGURE 13.1 Efficacy and dose effect of an *E. faecium*-specific phage in protecting mice from a lethal VRE septicemia. The group on the far right (four mice) was an untreated control, which was injected i.p. with phage-free buffer instead of the phage preparation. The group on the far left (two animals) was a phage control, which was injected with the phage preparation and was not infected with *E. faecium* (from Biswas et al., 2002, reprinted with permission from the American Society for Microbiology, Washington, D.C., U.S.A.).

and the number of challenge bacteria in the RIL were determined. Phage administration was reported not to have a prophylactic effect, since it (i) did not reduce the number of challenge bacteria in the RIL, and (ii) did not reduce the fluid accumulation ratios in the RIL. The study did not specifically examine the *in vitro* ability of the phages to lyse the challenge bacteria; however, since a large collection of typing phages was used, it is likely that at least some of them would have had *in vitro* lytic activity against the challenge strain. The authors proposed that, although the anti-*V. cholerae* phages persisted in the RIL throughout the experiments, some component(s) of the intestinal milieu might have inhibited their activity against the challenge *V. cholerae* strain. Given the early history of the extensive and successful use of bacteriophages to prevent and treat human cholera (see Chapter 14 for more information on this subject), the negative outcome of the above study suggests that the RIL model may not be an optimal model for evaluating the efficacy of phages in preventing and treating naturally occurring cholera. It also highlights how little is known concerning the interaction of phages and their targeted bacteria in the mammalian intestinal tract.

13.2.3.5. *Clostridium difficile* Infections

Ramesh et al. (1999) recently characterized the ability of bacteriophages to prevent a fatal, *C. difficile* ileocecitis in experimentally infected hamsters. Their experimental protocols were designed to study many of the factors that influence the efficacy of phage therapy in animals; e.g., the persistence of phages in the mammalian body, the effect of gastric acidity on phage viability, optimal dosing regimens, etc. The bacteriophage used in the study was isolated from a lysogenic strain of *C. difficile*, and the experimental animals used during the study consisted of 26 adult hamsters. Groups 1a, 1b and 1c each contained 6 hamsters and groups 2 and 3 each contained 4 hamsters. In order to facilitate the *C. difficile* ileocecitis, the hamsters in groups 1 and 2 were pretreated with clindamycin (3 mg/100 g body weight, administered intragastrically); the hamsters in control group 3 received saline instead of the antibiotic. Twenty-four hours after the clindamycin-pretreatment step, the gastric acidity of the hamsters in all three groups was neutralized (via orogastric administration of 1 M bicarbonate buffer; 1 ml/animal), and the animals were intragastrically inoculated with a suspension (1 ml containing 10^3 CFU) of the *C. difficile* challenge strain. Immediately after the bacterial challenge step, the hamsters in subgroups la, 1b, and 1c were treated with the phage preparation (1 ml containing 10^8 PFU). The animals in subgroups 1b and 1c also received additional doses of the phage preparation, at 8 h intervals, for 48 h and 72 h, respectively (gastric acidity was neutralized with intragastric administration of bicarbonate buffer immediately before each phage treatment). The hamsters in groups 2 and 3 served as controls; i.e., they were not treated with the bacteriophage preparation. All of the animals in the control groups died within 96 hours after bacterial challenge; however, with the exception of one hamster, all of the phage-treated animals survived. The phages were not detectable in the animals' cecal contents shortly after they received the last dose of phage preparation. In addition, the phage therapy did not have a long-lasting protective effect in the hamsters; i.e., when the surviving hamsters were pre-treated with clindamycin and rechallenged with *C. difficile* (2 weeks after stopping phage therapy), they died within 96 hours postchallenge. The *C. difficile* phage used during the study has not been carefully characterized, and it may not have been an optimal choice for phage therapy because of its lysogenic potential and, thus, its possible less-than-optimal lytic activity. Nevertheless, the study's results indicate that *C. difficile* infections are amenable to phage therapy, and they highlight the importance of neutralizing gastric acidity prior to phage treatment via the oral route. As noted earlier, a gastric acidity neutralization protocol also has been suggested by other investigators (Smith et al., 1987b), and it was common practice during human phage therapy studies in the former Soviet Union and Eastern Europe (see Chapter 14).

13.2.3.6. *Acinetobacter baumanii, Pseudomonas aeruginosa,* and *Staphylococcus aureus* Infections

Soothill (1992) examined the efficacy of phages in treating experimental murine infections caused by *A. baumanii*, *P. aeruginosa,* and *S. aureus*. The *A. baumanii*- and *P. aeruginosa*-specific phages were isolated from sewage in Birmingham, United Kingdom, and the *S. aureus*-specific phage ϕ-131 was obtained from the Hirszfeld Institute of Immunology and Experimental Therapy in Wroclaw, Poland (see Chapter 14). The protective effects of phage were evaluated in groups of adult outbred mice injected i.p. with a predetermined lethal dose of bacteria inoculated simultaneously with the phage preparations. A decrease in body temperature (to 34°C) was found to be an accurate predictor of terminal illness; therefore, when the body temperature of the infected mice reached 34°C, they were sacrificed and their organs and tissues were analyzed for the presence of the challenge bacteria and bacteriophages. As few as 10^2 PFU of the *A. baumanii*-specific phage protected mice against 5 LD_{50} doses of a virulent strain of *A. baumanii*, and the phages multiplied in the mice during the course of the experiment. A positive outcome was also associated with negative *A. baumanii* cultures from the phage-treated mice. A much higher dose of the anti-*P. aeruginosa* phage preparation was required to protect mice against *P. aeruginosa* challenge, but the protective effect was significant against 5 and 10 LD_{50} doses of the bacterium (Table 13.2). In contrast, the *S. aureus*-specific phage (which was poorly lytic *in vitro*) did not protect mice from the lethal effects of the two challenge strains of *S. aureus*. Interestingly, the same *S. aureus* phage has been used extensively and successfully to treat human *S. aureus* infections in Poland (Slopek et al., 1987). The reason for the seeming discrepancy in results is not clear. Since one of Soothill's challenge strains was a clinical isolate obtained from Poland (where it also was used as a host strain for ϕ-131), host-strain differences do not explain the different results. The number of animals used during Soothill's studies was small, complicating rigorous statistical analysis of his data. Also, the phage preparations were injected simultaneously and by the same route as were the bacteria, and into a body cavity which gave a good opportunity for the phages to attach to the target bacteria; i.e., conditions which are not likely to be encountered in real-life settings involving phage therapy of established *A. baumanii* and *P. aeruginosa* infections. Thus, although Soothill's results are promising, they require confirmation in a more clinically relevant animal model. Interestingly, the *P. aeruginosa*-specific phage (BS24) used

TABLE 13.2
Data from the 1992 Study of Soothill (Soothill, 1992). The Study Examined the Value of Phage Therapy in Experimental Murine Infections Caused by *A. baumanii, P. aeruginosa* and *S. aureus*[a]

Phage Dose	Mortality	Phage Dose	Mortality	Phage Dose	Mortality
A. baumanii 1.5 × 10^8 CFU (Challenge with 8 × LD_{50}) Two mice per dose		*A. baumanii* 9.5 × 10^7 CFU (Challenge with 5 × LD_{50}) One mouse per dose		*A. baumanii* 5.6 × 10^7 CFU (Challenge with 3 × LD_{50}) Five mice per dose	
8.3 × 10^6 PFU	0%	1.0 × 10^2 PFU	20%	1.2 × 10^3 PFU	0%
1.7 × 10^6 PFU	0%	3.6 × 10^1 PFU	100%	1.2 × 10^2 PFU	0%
3.3 × 10^5 PFU	0%	1.2 × 10^1 PFU	100%	1.2 × 10^1 PFU	0%
6.6 × 10^4 PFU	0%	4 PFU	100%	1 PFU	100%
1.3 × 10^4 PFU	0%	No Phage	100%	No Phage	100%
No Phage	100%				
P. aeruginosa 1.5 × 10^8 CFU (Challenge with 10 × LD_{50}) Five mice per dose		*P. aeruginosa* 8.0 × 10^7 CFU (Challenge with 5 × LD_{50}) Five mice per dose			
2.9 × 10^8 PFU	0%	1.8 × 10^7 PFU	20%		
2.9 × 10^7 PFU	80%	6.0 × 10^6 PFU	100%		
5.8 × 10^6 PFU	100%	2.0 × 10^6 PFU	80%		
2.9 × 10^5 PFU	100%	6.7 × 10^5 PFU	100%		
No Phage	100%	No Phage	100%		

[a]The *S. aureus*-specific phage (ϕ-131) was poorly lytic *in vitro*, and it did not protect mice against *S. aureus*; thus, *S. aureus*-related data are not included in the table.

by Soothill has been later reported to prevent *P. aeruginosa*-mediated destruction of pigskin *in vitro* (Soothill et al., 1988) and to protect against *P. aeruginosa*-associated destruction of skin grafts *in vivo* (Soothill, 1994). The successful outcome of the latter study suggests that the local application of phage might be a useful tool for preventing and treating *P. aeruginosa* infections in burn grafts.

13.2.4. Bacteriophages As a Potential Tool for Reducing Antibiotic Usage in Agribusiness

An estimated 50% to 70% of the antibiotics used in the United States are given to farm animals (Gustafson, 1991), for three main purposes: (i) prophylactically, to prevent disease in flocks and herds, (ii) to treat sick livestock (the antibiotics used to treat the animals are usually the same as those used in human medicine, and they are administrated via drinking water or feed over a period of several days), and (iii) to improve digestion and utilization of feed, which often results in improved weight gain. Antibiotics used in the latter setting often are referred to as *growth-promoting antibiotics* or GPAs. Most GPAs are not commonly used in human medicine, and they are usually administered, in small amounts, to poultry and other livestock via feed. The use of antibiotics in livestock has become a major source of concern among public health officials, consumer groups, and livestock industry leaders because of the possibility that they contribute to the declining efficacy of antibiotics used to treat bacterial infections in humans (Smith et al., 2002). This concern caused the European Union to ban the use of four antibiotics (virginiamycin, bacitracin zinc, spiramycin, and tylosin phosphate) as additives in animal feeds, and a complete ban on all GPAs is likely to take effect in Europe in 2006 (Ferber 2003). Although no similar ban has yet been introduced in the United States, the US Food and Drug Administration has recently proposed regulations that could impose severe limitations on the future agricultural and farm-veterinary use of all antibiotics. Moreover, bills that would curb the use of animal antibiotics that are similar to human antibiotics have already been introduced in the U.S. Senate and House of Representatives (Ferber, 2003). Banning or markedly reducing the agricultural and farm-veterinary use of antibiotics may have a negative impact on the safety of foods and on the treatment of sick flocks or herds of domesticated livestock—unless effective, safe, and environmentally friendly alternatives can be developed. Bacteriophage-based antibacterial products may be one such alternative.

Phages are ubiquitous in the environment and their use in livestock is likely to provide one of the most environmentally friendly antibacterial approaches available today. In addition, phages' several important advantages over antibiotics (Pirisi, 2000) make their use in various livestock industries potentially very appealing. For example:

- Because of the specificity of phages, their use in agriculture is not likely to select for phage resistance in untargeted bacterial species; whereas, because of their broad spectrum of activity, antibiotics select for many resistant bacterial species, not just for resistant mutants of the targeted bacteria.
- Because the bacterial resistance mechanisms against phages and antibiotics differ, the possible emergence of resistance against phages will not affect the susceptibility of the bacteria to antibiotics used to treat humans—which is the key concern regarding the use of antibiotics in agribusiness.
- Unlike antibiotics (which have a long and expensive development cycle), phage preparations can readily be modified in response to changes in bacterial pathogen populations or susceptibility, and effective therapeutic phage preparations can be rapidly developed against emerging antibiotic- or phage-resistant bacterial mutants.

The range of bacterial pathogens that could be targeted with phages is quite wide, but initial work in this area is likely to start with bacterial pathogens known to be most problematic in poultry and other livestock industries. For example, since necrotic enteritis elicited by *Clostridium perfringens* is a significant cause of morbidity and mortality in chickens, initial studies to reduce the prophylactic and therapeutic use of antibiotics in the poultry industry may focus on determining the value of using phages against that pathogen. Phage preparations could also be developed against other bacterial pathogens that may be of concern for other domesticated livestock;

e.g., phage preparations for treating bovine mastitis should have a substantive practical applicability in the dairy industry. Such preparations (primarily targeting *S. aureus* strains) have been developed in the former Soviet Union, with preliminary results said to be encouraging (T. Gabisonia, personal communication). If initial studies in that direction generate similarly encouraging results in the United States and Western Europe, the approach may lead to the development of several phage-based therapeutic preparations for veterinary medicine. Further bacteriophage research related to agricultural applications could initially focus in two main directions: (i) developing phage preparations to be used as prophylactic or therapeutic antimicrobials (i.e., to directly lyse the targeted pathogens), and (ii) developing phage lysate-based preparations to be used as vaccines. The use of such preparations may help reduce the prophylactic and therapeutic use of antibiotics in farm animals. In addition, it may potentially have some growth-promoting effect in animals (e.g., by reducing stress caused by bacterial infections)—which may help reduce or eliminate the use of GPAs in various livestock industries.

Bacteriophages may also be of value in reducing the use of antibiotics in the aquaculture and farming industries. For example, fire blight, caused by the bacterium *Erwinia amylovora*, is a devastating disease of apple and pear trees in many countries around the world—and many commercial producers rely heavily on antibiotics (e.g., streptomycin) to prevent the disease by reducing the accumulation of epiphytic populations of *E. amylovora* on nutrient-rich stigmatic surfaces of blossoms (Schnabel and Jones, 2001). However, that practice has resulted in the emergence of streptomycin-resistant mutants of *E. amylovora* (Jones and Schnabel, 2000)—a good illustration of how the use of antibiotics in agribusiness may contribute to the emergence of antibiotic-resistant bacterial strains. Although they do not pose a direct health safety problem for humans, the spread of streptomycin-resistant *E. amylovora* strains may contribute to an increased spread of streptomycin-resistance genes (and, possibly, of other antibiotic-resistant genes) among various bacterial species, including those highly pathogenic for humans.

13.3. BACTERIOPHAGES AND DISEASES OF AQUACULTURE AND PLANTS

13.3.1. Plant Infections

Bacteriophages were first proposed as potential agents for controlling bacterial diseases of plants as early as 1926 (Moore, 1926; Okabe and Goto, 1963), and they were successfully used to control Stewart's Disease in corn by Thomas (1935). Phages were used by Civerolo et al., (1969) and by Civerolo (1973) to reduce *Xanthomonas oryzae* Bacterial Spot disease of peach seedlings by 86% to 100%. However, most of the published research concerning phages specific to bacterial plant pathogens has not been treatment-oriented; phages were often simply used as a tool for typing plant-infecting bacteria. Reviews about bacteriophages of various plant pathogens were published by Okabe and Goto (1963) and by Gill and Abedon (2003).

E. amylovora has been one of the most commonly studied plant bacterial pathogens, and several publications in the 1960s described the isolation, characterization, and use of various phages for typing *E. amylovora* strains. The first report in which the possible role of bacteriophages in the epidemiology of *E. amylovora*-associated fire blight was discussed came from Erskine (1973). Subsequently, *E. amylovora*-specific phage ϕEa1 was successfully used to treat/prevent fire blight in apple seedlings inoculated with *E. amylovora* (Ritchie and Klos, 1979; 1977). Treatment with the polysaccharide depolymerase encoded by ϕEa1 has also been reported to attenuate the symptoms of fire blight in pear fruit inoculated with *E. amylovora* (Hartung et al., 1988). Schnabel et al. (2001) recently reported isolating several *E. amylovora*-specific bacteriophages (some identical or closely related to ϕEa1) from various fruits and soil samples collected at sites displaying fire blight symptoms. More recently, Gill et al. (2003) reported isolating more than 40 *E. amylovora*-specific bacteriophages from sites in and around the Niagara region of southern Ontario and the Royal Botanical Gardens in Hamilton, Ontario. Molecular characterization of the phages with Polymerase Chain Reaction (PCR) and Restriction Fragment Length Polymorphism (RFLP) revealed that some of them were closely related to ϕEa1. A study of the host ranges of the phages revealed that certain types were unable to lyse some *E. amylovora* strains efficiently, and that the phages' lytic potential was not limited to that species (i.e., some phages also were capable of lysing the epiphytic bacterium *Pantoea agglomerans*). The authors indicated that investigating the potential of their phages as biocontrol agents would be the subject of their future research.

Such explorations have been conducted by other investigators for other bacterial diseases of plants and for other, perhaps somewhat unusual—but related—applications. In the latter context, colonization with *E. herbicola* and *Pseudomonas syringae* are thought to contribute to the susceptibility of some plant species to tissue damage caused by exposure to low temperatures, by acting as ice-forming nuclei at 1–3°C. Kozloff et al. (1983) received a US patent based on the idea that phages targeting those two species could be used to reduce their population on plant leaves and increase the frost-resistance of phage-treated plants. The results of their experiments suggested that plants infected with *E. herbicola* and treated with *E. herbicola*-specific phages sustained 20%–25% less damage than did plants experimentally challenged with *E. herbicola* but not treated with phages.

Jackson (1989) was granted a US patent for the use of phage preparations for (i) eliminating naturally occurring *P. syringae* from contaminated bean culls, and (ii) reducing the severity of disease symptoms in bean leaves experimentally infected with *P. syringae*. The same author subsequently developed phage preparations targeting *Ralstonia solanacearum* and *Xanthomonas campestris* pv. vesicatoria, the bacterial pathogens responsible for the two main diseases of tomato plants, known as Bacterial Wilt and Bacterial Spot, respectively. In the field trials, phage mixtures were typically poured over the soil at the base of 6- to 8-week-old plants; in the greenhouses, phage were applied in irrigation water (Fox, 2000). Two weeks after inoculation with the pathogen, 60% of the plants without the phage treatment had more than 12% defoliation, while defoliation in all phage-treated plants was less than 12%. The visual results were impressive (Fig. 13.2). The results of these studies are encouraging, and they suggest that bacteriophages may indeed be valuable tools for dealing with various plant diseases of bacterial origin. However, the studies have

FIGURE 13.2 Tomato plants treated with phages 3 and 5 days prior to inoculation with *Ralstonia solanacearum* (center and right plants, respectively), and an inoculated, phage-untreated control plant (left) (from Fox, 2000, reprinted with permission from the American Society for Microbiology, Washington, D.C., U.S.A.).

not yet been published in the peer-reviewed literature, and several critical details pertaining to the studies' design, the characteristics of the phages used, and so forth are not yet available for critical review.

13.3.2. Infections of Aquacultured Fish

Terrestrial animals and plants are not the only candidates for phage therapy, and the possible value of using phages to treat bacterial diseases of aquacultured fish has been gaining increased attention lately. One of the first reports in that area focused on using *Lactococcus garvieae*-specific phages for treating experimentally infected young yellowtail (*Seriola quinqueradiata*) (Nakai et al., 1999). Fish were experimentally infected by i.p. injection of the challenge *L. garvieae* strain, followed by i.p. or oral administration of phages with *in vitro* lytic activity against the challenge strain. All the infected fish that received the phage treatment survived, compared to only 10% of the phage-untreated fish. The strongest protective effect was observed with the fish that were treated with phages at the earliest time after infection with the bacterium. Protection was also observed in yellowtail that received phage-impregnated feed. The authors also analyzed the fishes' internal organs for the presence of phages, and they tested *L. garvieae* isolates recovered from dead fish for phage susceptibility. All of the isolates examined remained susceptible to the phages, and phage-neutralizing antibodies were not detected in serum samples obtained from the yellowtail (Nakai et al., 1999; Nakai and Park, 2002).

Similar studies were subsequently conducted for treating *Pseudomonas plecoglossicida*-caused bacterial hemorrhagic ascites disease in cultured ayu fish

(*Plecoglossus altivelis*) (Park et al., 2000; Park and Nakai, 2003). *P. plecoglossicida* is an opportunistic pathogen that can, under certain circumstances, cause disease with a high mortality rate in cultured ayu fish (Nishimori et al., 2000). Phages lytic for *P. plecoglossicida* were isolated from diseased ayu and the rearing pond water obtained from various fish farms in Tokushima Prefecture in Japan. In the first experiment, four groups of 20 ayu (10 g average weight) were fed commercial dry pellets impregnated with a live culture of *P. plecoglossicida* (10^7 CFU/g of pellet). After 15 min of feeding, two of the groups were immediately fed pellets impregnated with a phage suspension (10^7 PFU/g pellet). The control groups received regular feed without phage. The second experiment was designed in essentially the same manner, except that: (i) smaller fish were used (2.4 g average weight), (ii) the sample size was larger; i.e., each group contained 40–50 ayu, and (iii) one test group was treated with phages 1 h after feeding them bacteria-impregnated pellets, and another test group was treated with phages 24 h after feeding with the bacteria-impregnated pellets. The study duration in both cases was two weeks. In the first experiment, fish in the control (i.e., phage-untreated) group begin to die 7 days after the bacterial challenge, and the cumulative mortality rate in that group by the end of the experiment was ca. 65%. In contrast, the mortality rate in the phage-treated group was ca. 23% ($P < 0.001$). Phage treatment also reduced the mortality rate in the phage-treated group in the second experiment; noteworthy, phage administration 1 h after bacterial challenge was significantly more effective in preventing death than was phage administration 24 h after bacterial challenge (0% vs. 13% mortality, respectively). *P. plecoglossicida* was re-isolated from fish kidneys in the first experiment, and all of the cultures were found to be susceptible to the phages used in the study. In addition, phage-resistant mutants selected for *in vitro* appeared to be less virulent for ayu (Park et al., 2000).

The therapeutic efficacy of the anti-*P. plecoglossicida* phages used in these studies was recently further elucidated in a field trial, when phage-impregnated feed was administered to ayu in a pond where the disease occurred naturally. The daily mortality of the fish decreased at a constant level (ca. 5% per day), and by the end of the 2-week study period, it was approximately one-third of the mortality rate in the phage-untreated group. As with the previous studies, no phage-resistant organisms and phage-neutralizing antibodies were detected (Park and Nakai, 2003).

13.4. PHAGE-ELICITED BACTERIAL LYSATES AS VACCINES

13.4.1. Using Phages to Produce Bacterial Vaccines

An intriguing approach to using phages in agribusiness (and potentially for human therapy) involves using them to prepare bacterial lysates that can be used as vaccines. The approach is as old as conventional phage therapy (i.e., the approach of using phages to lyse etiologic agents *in vivo*), and it was developed in d'Herelle's Pasteur Institute laboratory shortly after the 1917 publication of his milestone paper (d'Herelle, 1917) on the discovery of bacteriophages. Beginning in 1914, d'Herelle was working to prepare bacterial vaccines for the Allied armies fighting in World War I (Summers, 1999). At that time, bacterial vaccine therapy was a very important research subject at the Pasteur Institute, and the Institute's director (Emile Roux, one of the developers of a very successful anti-diphtheria vaccine) strongly encouraged the Institute's staff to develop new vaccines and the methodologies needed for their preparation. As part of that effort, d'Herelle examined the ability of various *essences* to lyse pathogenic bacteria and to produce lysates effective as vaccines. Many exotic bacteriolytic substances were examined as the project progressed; e.g., during his work to develop a *Salmonella* vaccine, d'Herelle examined mustard, cinnamon, garlic, oregano, cloves, and thyme for their ability to lyse *Salmonella* (Summers, 1999). All of them lysed the bacterium, and d'Herelle (1916) reported that mustard-generated lysates protected mice against challenge with live *Salmonella* (interestingly, it was an extension of that study—trying to develop a vaccine against *Shigella*—that led to his co-discovery of bacteriophage in 1917; see Chapter 2). However, even though d'Herelle postulated that phages were immunostimulating agents (and actually referred to phage therapy as an "immunization;" see section 13.2.1), he never directly examined the efficacy of phage-elicited bacterial lysates as vaccines. The first investigator to study the possible active immunization property of phage-generated lysates was Tamezo Kabeshima, who was visiting d'Herelle's Pasteur Institute laboratory from Shiga's laboratory in Japan. Kabeshima (1919) used a phage-lysed *Shigella* preparation ("bactériolysat") to successfully immunize small laboratory animals in what was to become the first known use of phage lysates as protective vaccines. Several subsequent publications reported that preparations of phage-lysed bacteria were indeed very good immunizing agents whose protective effect was stronger than that of "regular" vaccines (e.g., vaccines prepared by heat- or chemical-inactivation of bacteria) (Arnold and Weiss, 1924; Larkum, 1929; Compton, 1928).

The underlying mechanism for the superior immunogenicity reported for phage-generated bacterial lysates is not clear, but it is possible that bacteriophage-mediated lysis is a more effective and gentler approach for exposing protective antigens of bacteria than are approaches used to prepare other "dead cell" vaccines. In this context, common methods used to inactivate bacterial pathogens for "dead cell" vaccines (e.g., heat-treatment, irradiation, and chemical treatment) may indeed deleteriously affect a vaccine's effectiveness by reducing the antigenicity of relevant immunological epitopes (Melamed et al., 1991; Holt et al., 1990; Lauvau et al., 2001). Interestingly, the idea that gentle lysis yields cell lysates possessing optimal immunogenicity was proposed by d'Herelle (1916). Using phage lysates as vaccines is likely to have continued unknowingly since the early experiments of Kabeshima, because most of the therapeutic phage preparations used as "direct antimicrobial agents" were contaminated with numerous bacterial antigens released from the lysed bacteria and, thus, in addition to their bacteriolytic effect, those phage preparations may also have inadvertently acted as vaccines.

13.4.2. Using Phage-Generated Bacterial Lysates to Prevent and Treat Animal Infections

Perhaps the best known phage-generated bacterial lysate currently available for sale in the United States is SPL, which is produced by Delmont Laboratories of Swarthmore, PA. SPL is prepared from broth cultures of at least two virulent strains of coagulase-positive *S. aureus* (types I and III) by lysing the bacteria with an excess of *S. aureus*-specific bacteriophages. The lysate contains cell wall and intracellular components released as a result of bacterial lysis, culture media ingredients, and viable bacteriophages. As described in Chapter 14 of this book, SPL was used to prevent or treat human infections in the United States during the period from the 1950s to the 1990s, but is currently solely used for veterinary applications. In one of the early publications describing the use of SPL, Esber et al. (1981) used a mouse model to evaluate the efficacy of SPL in preventing or treating S. *aureus* infections of animals. Treatment with SPL resulted in the survival of 80%–100% of the infected mice, compared to no survivors among the infected mice not treated with SPL. Survival was hypothesized to be due to enhancement of nonspecific immune resistance elicited by SPL-mediated activation of thymus-modulated lymphocytes and macrophages. In another study by Esber et al. (1985), SPL administration was noted to significantly increase the anti-staphylococcal IgG1, IgG2a, and IgG2b levels in the treated animals. Weekly injections of SPL have also been reported (Chambers and Severin, 1984) to be an effective treatment for chronic staphylococcal blepharitis in dogs; i.e., the preparation controlled the illness without any adverse side effects in the animals. Also, the simultaneous use of SPL and sodium oxacillin was reported (DeBoer et al., 1990) to be significantly ($P < 0.05$) more effective in treating idiopathic, recurrent, superficial pyoderma in dogs than was treatment with the antibiotic alone. However, not all reports have been similarly encouraging. For example, in at least one study (Giese et al., 1996), vaccination with SPL was not found to prevent the development of *S. aureus* blepharitis, phlyctenules, and catarrhal infiltrates in an experimentally infected rabbit model. Additional, rigorous studies are required to improve our understanding of what infections/clinical syndromes are most amenable to prophylaxis and treatment with SPL (and to prophylaxis and treatment with other phage-generated bacterial lysates) and the optimal dosing regimens, administration routes, and so forth.

13.4.3. Phages and "Bacterial Ghost" Vaccines

Using phage-encoded bacteriolytic enzymes, rather than viable bacteriophages, to prepare bacterial vaccines has recently been gaining increased attention, particularly in Western Europe (Szostak et al., 1990; Szostak et al., 1996; Szostak et al., 1997; Eko et al., 1994a; Mader et al., 1997; Panthel et al., 2003). Such preparations contain what are often called "bacterial ghosts," and they are obtained by using lysis gene *e* of bacteriophage *φ*X174 to lyse various Gram-negative bacteria (noteworthy, bacteriophage *φ*X174 has been injected into humans in the United States, during studies designed to determine the immune status of various immunocompetent and immunocompromised individuals, as discussed in Chapter 14). The genome of bacteriophage *φ*X174 encodes a single lysis protein E—an outer membrane protein that contains 91 amino acid residues (see Chapter 7), which can form a 40–200 nm diameter pore in the bacterial cell wall, though which the bacterium's intracellular constituents are expelled (Schon et al., 1995; Witte et al., 1990; Witte et al., 1992). The general methodology for obtaining "ghost" preparations can be briefly outlined on the example of a recent study in which *H. pylori* pHPC38 "ghost" has been

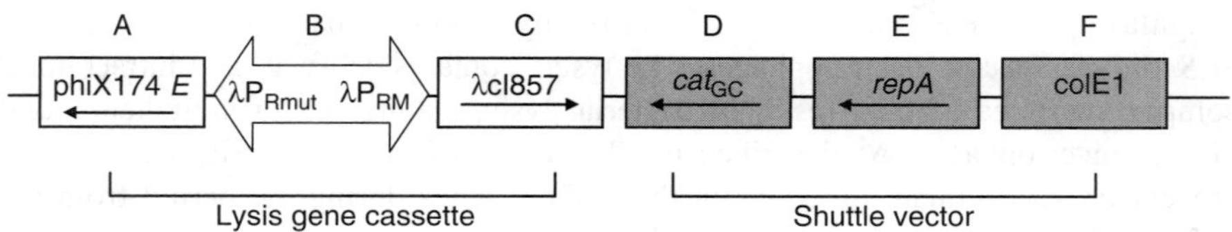

FIGURE 13.3 Schematic genetic map of *H. pylori* "ghost" pHPC38. (A) lysis gene *e* from bacteriophage *φ*X174; (B) promoter region of λ phage; (C) ts-repressor; (D) chloramphenicol acetyltransferase gene; (E) replication in *H. pylori*; (F) replication in *E. coli* (adapted from Panthel et al., 2003).

prepared and examined for immunogenicity in mice by Panthel et al. (2003). The authors used the *E. coli-H. pylori* shuttle plasmid pHel2 to construct *H. pylori* lysis plasmid pHPC38, which they introduced into *H. pylori* strain P79 by conjugation or by natural transformation. The *H. pylori* transformants or transconjugants carrying it were selected on a solid culture medium containing chloramphenicol; the shuttle vector contains the chloramphenicol acetyltransferase gene, cat_{GC}, so including chloramphenicol in the medium selects bacterial colonies that contain the shuttle vector encoding the resistance-conferring cat_{GC} gene (Fig. 13.3). The ghost preparation was subsequently used to vaccinate BALB/c mice orally (using an initial dose followed by two booster doses at 7-day intervals); control mice received PBS. Three weeks after receiving the last dose of the *H. pylori* ghost preparation, the vaccinated and control mice were orally challenged with a high dose (10^9 CFU) of the wild-type *H. pylori* strain P79. The mice were sacrificed 4 weeks later, and gastric colonization with the challenge *H. pylori* strain was quantitated. Vaccination with the ghost preparation reduced the number of *H. pylori* in the gastric samples by ca. 1000-fold compared to the control group. The study did not determine the length of the protective effect; however, the data strongly suggested that a ghost vaccine may be an effective tool for preventing at least initial colonization of the gastric mucosa with *H. pylori.*

In theory, it should be possible to use a similar strategy to prepare bacterial ghosts of all Gram-negative bacteria, provided that the E lysis cassette can be introduced into the recipient by an appropriate vector, thus allowing tight repression and induction control of the *e* gene (Jalava et al., 2002). Indeed, ghosts of several Gram-negative bacteria have been prepared using E protein-mediated lysis, including *E. coli* (Blasi et al., 1985), *Salmonella* serotypes Typhimurium and Enteritidis (Szostak et al., 1996), *V. cholerae* (Eko et al., 1994a; Eko et al., 2003; Eko, et al., 1994b), and *Pasteurella multocida* and *P. haemolytica* (Marchart et al., 2003). Since the ghosts retain their outer membranes with their immunostimulatory lipopolysaccharide (LPS) endotoxin structure essentially intact (with the exception of the pores formed by the E protein), they should provide protection similar to that obtained using whole-cell, attenuated vaccines—but without the associated safety concerns (Jalava et al., 2002; Witte et al., 1990). One example of using ghost vaccine to prevent disease in agriculturally important animals is the recent work with the *Actinobacillus pleuropneumoniae* ghost vaccine and infection of pigs. *A. pleuropneumoniae* is a major respiratory pathogen responsible for severe morbidity and mortality in pigs; although conventional *A. pleuropneumoniae* vaccines are currently available, they decrease mortality but are ineffective in reducing morbidity. Hensel et al. (2000) used an aerosol infection-pig model to study the protective potential of an *A. pleuropneumoniae* (serotype 9, reference strain CVI 13261) ghost vaccine. Pigs were intramuscularly vaccinated with the ghost vaccine or with a formalin-killed *A. pleuropneumoniae* preparation. Two weeks later, the vaccinated pigs and nonvaccinated placebo controls were challenged with *A. pleuropneumoniae* (10^9 CFU) by aerosol inhalation. The outcome was evaluated by clinical, bacteriological, serological, and post-mortem examinations. The pigs in the control (untreated) group developed fever and pleuropneumonia after challenge with *A. pleuropneumoniae*; in contrast, the pigs in both treated groups were fully protected against clinical disease. Significantly, the ghost vaccine (but not the formalin-killed bacterial lysate) also prevented colonization of the respiratory tract with *A. pleuropneumoniae*. Treatment with the two vaccines also significantly increased the amounts of IgM, IgA, IgG (Fc'), and IgG (H+L) antibodies reactive with *A. pleuropneumoniae.* Interestingly, higher titers of IgG (Fc') and IgG(H+L) were observed in pigs treated with the formalin-killed vaccine than in pigs vaccinated with the ghost preparation; on the other hand, prevention of the carrier state in ghost vaccine-treated pigs coincided with a significant increase in serum IgA when compared to formalin-killed vaccine. A subsequent publication of a similar study (Huter et al., 2000) confirmed that immunization with a *A. pleuropneumoniae* ghost vaccine (but not with a formalin-inactivated preparation) prevented colonization of pig lungs with *A. pleuropneumoniae.*

Ghost vaccines have also been reported to prevent infections caused by *K. pneumoniae* (causes severe infections in various agriculturally important animals, including dairy cows, poultry, ostriches, etc.), *E. coli* 078:K80 (causes colibacillosis in poultry, and septicemia in calves, piglets and lambs), *P. multocida* (causes fowl cholera, and pneumonia in pigs and cattle), and other bacterial pathogens. Comprehensive reviews about bacterial ghost vaccines have been published (Szostak et al., 1990; Szostak et al., 1996; Szostak et al., 1997), including a recent review on bacterial ghosts as vaccine candidates for veterinary applications (Jalava et al., 2002). The significant amount of bacterial LPS endotoxin in bacterial ghost preparations/vaccines could potentially limit the use of this type of vaccine. However, several studies found that effective doses of the ghost vaccines did not elicit appreciable side effects in any of the animals examined. Moreover, the LPS component has been proposed to play a critical role in immune stimulation associated with ghost vaccines. For example, Szostak et al. (1996) reported a significant correlation between the endotoxic activity of bacterial ghost preparations, as determined by the *Limulus* amoebocyte lysate (LAL) assay, and their capacity to stimulate the release of PGE2 and TNFα in mouse macrophage cultures. In addition, bacterial ghosts prepared from *E. coli* O26:B6 and *Salmonella typhimurium* have been shown (Mader et al., 1997) to elicit dose-dependent antibody responses against bacterial cells and their corresponding LPS when administered intravenously to rabbits via a standard immunization protocol. No side effects were observed in any of the rabbits that received the ghost vaccines, in doses of <250 ng kg-1. Szostak et al. (1996) found that the endotoxic activity of the bacterial preparations (analyzed by the LAL assay and the 2-keto-3-deoxyoctonate assay) correlated with the release of PGE2 and TNFα in mouse macrophage cultures and the endotoxic (i.e., fever) responses in rabbits—which suggests that the *in vitro* systems can be used to determine the potency of bacterial ghost vaccines (Mader et al., 1997).

At the present time, it is difficult to predict whether phage-generated bacterial lysates and/or ghost preparations will gain wide acceptance in veterinary medicine. Both approaches seem to be safe, effective, and relatively cost-efficient. However, phage lysates may have an advantage over ghost vaccines because they are simpler to prepare (e.g., they do not require the genetic engineering and optimal expression of *e* gene-containing constructs for bacterial lysis). The relative simplicity of preparing phage lysates compared to ghost preparations may be particularly important when complex vaccines need to be developed; e.g., vaccines against multiple serotypes of a given species when a vaccine based on one serotype does not provide adequate cross-protection across the species. The presence of viable phages may also serve as an additional efficacy-enhancing factor, increasing the effectiveness of a phage lysate via their antibacterial effect on the targeted bacterial pathogen.

13.4.5. Using Phage-Encoded Enzymes As Antibacterial Agents

In addition to the φX174-encoded E protein that has been used to prepare ghost vaccines, other phage-encoded enzymes have been used to lyse bacterial cells; e.g., the L protein of bacteriophage MS2 and the maturation protein A2 of phage Qβ (Bernhardt et al., 2001; Kastelein et al., 1982; Coleman et al., 1983), and the *B. amyloliquefaciens* phage endolysin. Also, some bacteriolysin-encoding genes and their applications have been patented; e.g., Auerbach and Rosenberg (1987) patented the use of cloned λ lysis genes as a method for bacterial cell disruption.

The applicability of the concept of using phage-encoded enzymes as antibacterial agents has been addressed in several recent studies. One example is the above-described use of the polysaccharide depolymerase encoded by ϕEa1 to attenuate the symptoms of fire blight in pear fruit inoculated with *E. amylovora* (Hartung et al., 1988). In another study, Loessner et al. (1998) cloned and sequenced the endolysin gene *plyTW* of *S. aureus* bacteriophage Twort (the gene encodes an ca. 53-kDa protein whose catalytic site is located in its amino-terminal domain). The cloned gene was over-expressed in *E. coli*, and the purified recombinant protein was shown to cleave staphylococcal peptidoglycan rapidly, thus suggesting that the endolysin may be an effective antibacterial agent against *S. aureus*. More recently, Gaeng et al., (2000) identified two endolysins (Ply118, an ca. 31-kDa L-alanoyl-D-glutamate peptidase, and Ply511, an ca. 37-kDa N-acetylmuramoyl-L-alanine amidase), encoded by the phage A118 (Loessner et al., 1995), which specifically hydrolyze the cross-linking peptide bridges in *Listeria* peptidoglycan, and they engineered a *Lactococcus lactis* strain that secreted both enzymes. This strain eliminated *L. monocytogenes* from dairy starter cultures used to produce cheese, which suggests that this approach may be of value in improving the safety of dairy products. More information about various lytic enzyme systems (including bacteriophage-encoded lytic enzymes) can be found in review articles by Dabora et al. (1990) and Young

(1992), and in Chapters 7 and 12. Bacteria-encoded enzymes capable of lysing bacteria (e.g., lysostaphin and lysozyme) have been studied from the 1960s through the 1990s as potential therapeutic agents in numerous animal models and in at least one human patient (Schuhardt and Schindler, 1964; Stark et al., 1974; Gunn and Hengesh, 1969; Oldham and Daley, 1991; Nuzov, 1984; Nakazawa et al., 1966). However, the idea that phage-encoded enzymes could, on their own, be used for therapeutic purposes was not extensively discussed until relatively recently (Loeffler et al., 2001; Nelson et al., 2001; Schuch et al., 2002; Morita et al., 2001).

13.5. USING BACTERIOPHAGES IN THE FOOD PRODUCTION CHAIN

13.5.1. General Considerations

Foodborne illnesses of microbial origin are serious food safety problems worldwide. The Center for Disease Control and Prevention (CDC) estimates that about 76 million cases of foodborne diseases (of which ca. 5000 are fatal) occur each year in the United States alone—and bacteria account for about 72% of all deaths associated with foodborne transmission (21% are due to parasitic infections and 7% are caused by viruses) (Mead et al., 1999). Among cases of foodborne illness, the leading causes of death are *Listeria*, *Toxoplasma*, and *Salmonella,* which together are responsible for more than 75% of foodborne deaths caused by known pathogens. In addition, Shiga toxin-producing *E. coli* (including strains of the 0157:H7 serovar) have recently emerged as a major food safety problem (in particular in ground beef) and they have caused several major outbreaks of disease with many fatalities. Also, several *Brucella, Yersinia, Shigella*, and other bacterial species are significant causes of foodborne disease, with *Campylobacter* species causing the largest number of cases in the United States (*C. jejuni* causes approximately 2.4 million illnesses/year in the United States) (Mead et al., 1999).

The epidemiology of foodborne diseases due to bacterial pathogens varies among the pathogens, as do the routes by which various bacteria contaminate food products. Some bacteria (e.g., *L. monocytogenes*) are environmental pathogens that usually contaminate foods in food processing/packaging plants. Other bacteria (e.g., *Salmonella* and *Campylobacter*) are part of the normal intestinal flora of many animals, and they contaminate foods during the slaughter or carcass processing cycle. Using bacteriophages to reduce contamination of foods with various bacterial pathogens will require an in-depth understanding of the epidemiology of the pathogen against which the phage preparation is to be used and the identification of critical intervention points in the processing cycle where phage application will be most beneficial (Stone, 2002). In this context, three possible areas of application for phage technology may be loosely identified: (i) Phages may be used to reduce intestinal colonization of live, agriculturally important animals that normally carry bacteria which present a foodborne disease risk; (ii) Phages may be applied directly onto raw foods, or onto environmental surfaces in raw food processing facilities, to reduce the levels of foodborne pathogens in raw foods; and (iii) Phages may be applied directly onto ready-to-eat foods (REF), or onto environmental surfaces in processing facilities for

REF, to reduce the levels of pathogenic bacteria in REF. Additional applications and combinational approaches (e.g., when phages are used in live animals and also during the processing) also can be utilized. At the present time, no such applications are utilized in industrial settings. However, as described below, the results of some published studies examining their efficacy suggest that the approaches have merit.

13.5.2. Using Phages to Reduce Contamination of Livestock with Foodborne Pathogens

Many bacterial pathogens capable of causing foodborne illness in humans are part of the normal flora of the gastrointestinal tract (GIT) of agriculturally important domesticated animals (cows, sheep, poultry, etc.). During the animals' slaughter and processing, there are multiple opportunities for the bacteria to contaminate the raw carcass, which increases the risk of the bacteria being subsequently ingested by humans and eliciting disease. Thus, approaches that eliminate or reduce the levels of foodborne pathogenic bacteria in foods have a potential to significantly improve the safety of food products. The use of bacteriophages may provide one such option.

As noted earlier in this chapter, several investigators have shown that oral administration of phages reduces the amount of targeted bacteria in, and reduces fecal shedding of the bacteria from, the intestines of the treated animals (Smith and Huggins, 1983; Smith et al., 1987a; Berchieri et al., 1991). The studies were not designed with food safety in mind—rather, they were conducted to determine whether oral administration of phages has a prophylactic effect on bacterial diseases in various animal species. Nevertheless, they provide indirect supporting evidence that oral administration of phages may indeed be effective in the context of food safety. The idea that phage-mediated reduction of intestinal colonization with pathogenic bacteria or of shedding of pathogenic bacteria may improve food safety has been advanced relatively recently (Brabban et al., 2003; Raya et al., 2003; Kudva et al., 1999). The preliminary data generated during these recent studies support the early observations already referenced and suggest that the approach has merit. For example, a single oral dose of a highly concentrated (4×10^{11} PFU) of an *E. coli* O157:H7-specific phage preparation recently was reported (Raya et al., 2003) to elicit, by two days post-administration, a marked reduction in intestinal levels of *E. coli* O157:H7 in sheep experimentally colonized with the bacteria (10^{10} CFU) three days prior to phage treatment. The authors proposed that "these results suggest that the protective effect of bacteriophages against *E coli* O157:H7 may contribute significantly to reducing the incidence of human infection if used in a preventive manner" (Raya et al., 2003). Although their initial findings were encouraging, the authors acknowledged the need to perform additional studies addressing several critically important issues not examined during their initial studies.

One such issue is the practical applicability of using phages to eliminate or significantly reduce the levels of foodborne pathogens in the GIT of agriculturally important animals. The composition of an animal's normal intestinal flora is very complex, and eliminating intestinal colonization by various bacteria (particularly if the bacteria are part of the GIT's normal flora) has proven to be very difficult. Thus, it is doubtful that phages can completely eradicate bacteria that commonly reside in the GIT of domesticated livestock and possess the ability to be foodborne pathogens of humans. Also, the value of reducing the levels of pathogenic foodborne bacteria in the GIT of domesticated livestock in terms of both environmental contamination and the impact on improving the safety of animal-based foods has yet to be determined. Furthermore, although phage resistance did not seem to be a major problem during short-term phage therapy studies in animals and fish (Smith and Huggins, 1983; Bull et al., 2002; Berchieri et al., 1991; Park and Nakai, 2003), long-term direct and continued exposure of phages and their targeted bacteria in the GIT may provide fertile grounds for selecting phage-resistant mutants (see Chapter 14). Thus, there may be advantages to using phages at an "epidemiological endpoint;" i.e., when cycling of the pathogen and phage in the environment and/or mammalian host does not occur or is minimized. An intriguing approach based on that idea was developed by Taylor et al. (1958), who injected phages into the interior of fertile eggs prior to incubation. Their hypothesis was that the phages would lyse the bacteria in the eggs, which in turn would improve the chick' hatch-rate—and some evidence of that effect was presented by the authors. For example, the hatch-rate of eggs experimentally infected with *Salmonella* Chittagong was 47% vs. 70% in phage-untreated and phage-injected eggs, respectively. The difference in rates was similar for other *Salmonella* serotypes (e.g., the hatch rate with eggs infected with *Salmonella* Pullorum was 44%, compared to 77% with eggs infected with the same bacterium and treated with a specific phage preparation). The authors did not discuss the potential food safety benefits of their observations, and the effect of treating the infected eggs with phage was solely evaluated by comparing the hatch rates; no rigorous microbiological examination of the eggs and hatched chicks was performed. However, it is tempting to speculate that, if the bacteriophages were effective in reducing the levels of pathogenic salmonellae present in the eggs, the approach may be of value in preventing human diseases caused by egg-borne bacterial pathogens. A benefit of the approach is that, since the phages injected into eggs will primarily be exposed only to bacteria in the eggs, a broad selective pressure against susceptibility to bacteriophages is unlikely to develop. Other possible food production uses for bacteriophages which are not likely to result in a selective pressure stimulating the emergence of phage-resistant bacteria, are discussed below.

13.5.3. Using Phages to Reduce Contamination of Raw Foods with Foodborne Pathogens

Another way that therapeutic phage may be used to improve food safety is to apply them directly onto raw food products or onto environmental surfaces in raw food processing facilities, in order to reduce the levels of foodborne pathogens in raw foods. As noted above, one of the most critical elements in that type of approach is to determine the optimal point(s) of intervention, or the time/step in the food processing cycle where exposure to phages is most beneficial and also minimizes the selective pressure for phage-resistance. A possible approach, suggested by the above-described studies of Taylor et al. (1958), would be to inject, rinse, or spray fertilized eggs with bacteriophage solutions before transferring them into incubators. That approach may be suitable for several egg-borne bacterial pathogens;

e.g., *Salmonella* and *C. jejuni*. Those bacteria often are present on the surface of fertilized eggs; because the temperature and humidity in incubators promote bacterial multiplication, the bacteria may increase in numbers during incubation, and chicks may become infected as they peck out of the eggs. Therefore, spraying phages onto the surface of eggs before transferring them into incubators may provide a gentle means of minimizing *Salmonella* contamination of eggs (and, perhaps, for increasing hatch-rates) which may subsequently lead to reduced levels/incidence of *Salmonella* in hatched chicks and to a reduction in *Salmonella* contamination of poultry products.

Alternatively, phages could be sprayed onto the chicken carcasses after post-chill processing (e.g., after the chlorine wash in chiller tanks in poultry processing plants in the United States, or after processing through air chillers in Europe). At that point, contamination with the usual foodborne bacterial pathogens should be minimal as a result of current *Salmonella* and *Campylobacter* reduction practices employed by all major poultry producers, and applying phages at that point in the processing cycle may provide a final means of product cleanup. In that context, Atterbury et al. (2003) recently reported that anti-*Campylobacter* bacteriophages are common commensals of retail poultry in the United Kingdom, and could survive commercial poultry processing procedures—which suggests that their application in real-life industrial settings may be technically feasible. Another important advantage of applying phages at this stage is that, since phages will not be carried to loci where they can readily be exposed to *Salmonella* or *Campylobacter* for a long period of time (e.g., to chicken houses), the risk of these bacteria developing resistance against the phages will be greatly reduced. Laboratory confirmation that applying phage onto chicken skin may be of value in reducing bacterial contamination has recently been published (Goode et al., 2003). The authors demonstrated that applying *C. jejuni*-specific phages onto chicken skin experimentally contaminated with the bacteria elicited a 10- to 100-fold reduction in the number of contaminating bacteria. For example, applying ca. 10^6 PFU of *C. jejuni* typing phage NCTC 12673/cm^2 of chicken skin reduced the levels of the test, contaminating *C. jejuni* strain on chicken skin by 95% ($t = 2.5$; $P = 0.04$). For *Salmonella* (which survived on untreated chicken skin better than did *C. jejuni*), the application of phages resulted in a significant ($P < 0.01$), ca. 99% reduction on phage-treated samples compared to the untreated controls. When the level of initial *Salmonella* contamination was low, phage treatment resulted in the samples being free of any recoverable *Salmonella* Enteritidis test strain organisms.

The results of these studies suggest that the approach of spraying specific phage preparations onto poultry carcasses after post-chill processing is efficacious in reducing carcass contamination with foodborne pathogens. The practical applicability of the approach may be complicated by narrow host range of phages—which may be a particular problem with *Salmonella*, a highly heterogeneous species containing more than 2400 serotypes (Popoff et al., 2000). However, it should be possible to target strains or serotypes known to be responsible for the majority of human illnesses, or which have increased virulence, such as *S*. Typhimurium definitive phage type 104 (DT104), which have rapidly emerged as major foodborne pathogens worldwide (Poppe et al., 1998). Assuming some flexibility from the regulatory agencies, it should also be possible to customize the phage preparations to provide coverage against additional *Salmonella* serotypes, or specific "problem" strains or serotypes in a flock or commercial production facility.

13.5.4. Using Phages to Reduce Contamination of Ready-to-Eat Foods with Foodborne Pathogens

An extension of the above-described approach is to apply specific phages onto various REF, in order to eliminate, or reduce the amount of, specific bacterial pathogens on those foods. Contamination of REF with foodborne pathogens is a potentially much more serious problem than is contamination of raw foods that are usually cooked before consumption, which can dramatically reduce the levels of pathogens in them. In contrast, and as the name implies, REF are often consumed without any additional processing; therefore, if they are contaminated with pathogens, the risk of human disease is high.

The value of using bacteriophages to eliminate or reduce the number of foodborne pathogens on various REF was first examined with fresh-cut fruits and vegetables by Leverentz et al. (2001). The authors examined the ability of phages to reduce experimental *Salmonella* contamination of fresh-cut melons and apples stored at various temperatures likely to be encountered in real-life settings. Directly applying the phage preparation (25 μl of 2×10^8 PFU/ml per fruit slice, applied by pipette) to the experimentally contaminated fruit reduced *Salmonella* populations by ca. 3.5 logs on honeydew melon slices stored at 5°C and 10°C, and by ca. 2.5 logs on slices stored at 20°C (Fig. 13.4), which was better than that achieved using commonly used chemical sanitizers. However, the phage preparation was less effective on fresh-cut Red Delicious apples than on the honeydew melon slices. Significantly, the titer

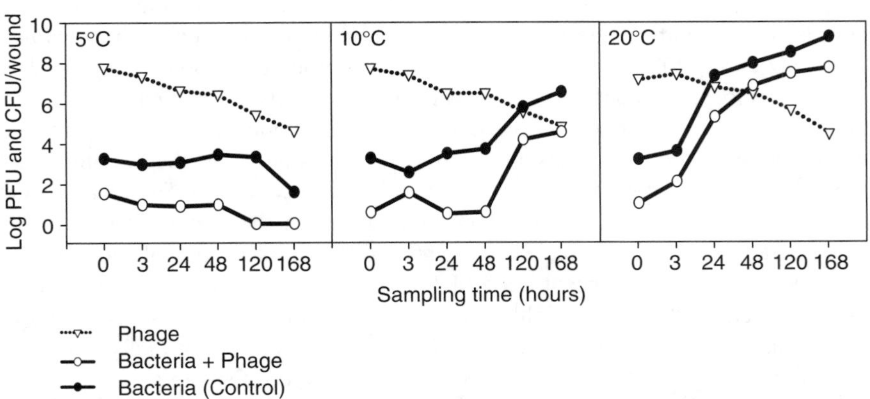

FIGURE 13.4 Populations of *Salmonella* Enteritidis on honeydew melon slices treated with a mixture of *Salmonella*-specific bacteriophages, and stored at 5°C, 10°C, and 20°C for 168 hours. Dotted line shows titer of the phage mixture re-isolated from slice of fruit stored at various temperatures (from Leverentz et al., 2001, reprinted with permission from *Journal of Food Protection*. Copyright held by the International Association for Food Protection, Des Moines, Iowa, U.S.A.).

of the phage preparation remained relatively stable on the melon slices, while on apple slices, it decreased to nondetectable levels in 48 h at all temperatures tested. The authors hypothesized that inactivation of the phages, possibly by the apple slices' more acidic pH (pH 4.2 versus pH 5.8 for melon slices), contributed to the inability of the phage preparation to reduce *Salmonella* contamination significantly in the apple slices. They also suggested that using higher phage concentrations and/or low-pH-tolerant phage mutants might be effective ways to increase the efficacy of phage treatment of fresh-cut produce with a low pH. The value of those suggested approaches has not yet been evaluated.

In another *Salmonella*-related study (Whichard et al., 2003), Felix O1 bacteriophage and its mutant possessing increased *in vitro* lytic activity against *Salmonella* Typhimurium DT104 have been used to treat chicken frankfurters experimentally contaminated with *Salmonella* Typhimurium DT104. The wild-type phage and the mutant phage elicited a 1.8 log and 2.1 log reduction, respectively, in *Salmonella* levels, compared to the levels in the untreated frankfurters ($P = 0.0001$). Although the mutant phage appeared to be more effective than the wild-type phage in lysing *Salmonella* lawns growing on a bacteriologic culture medium, the difference in the efficacies of the two phages in reducing the *Salmonella* content of the contaminated frankfurters was not statistically significant ($P = 0.5$). A similar approach—i.e., using specific phages to eliminate or significantly reduce the levels of contaminated bacteria on fresh-cut fruits and vegetables—also has been noted to be under investigation for *E. coli* 0157:H7 (Kudva et al., 1999).

Leverentz et al. (2003) recently reported the results of their second study examining the efficacy of phages, alone and in combination with the bacteriocin nisin, in reducing the levels of *L. monocytogenes* on experimentally contaminated melons and apples. The phage mixture, applied by pipeting or spraying, reduced *L. monocytogenes* populations by 2.0 to 4.6 log units on honeydew melons compared to controls treated with phage-free buffer. As previously observed with *Salmonella* (Leverentz et al., 2001), the reduction was less profound on apples (an ca. 0.4 log decrease compared to controls) than it was on melons. Using the phage preparation in combination with nisin decreased *L. monocytogenes* populations by up to 5.7 log units on honeydew melon slices and by up to 2.3 logs on apple slices, compared to the phage-untreated controls. In another parallel to their earlier *Salmonella* study, *L. monocytogenes*-specific phage titers were stable on the melon slices, but they declined rapidly on the apple slices. The effectiveness of the phage treatment depended on the initial concentration of *L. monocytogenes*; i.e., the greatest reductions were achieved with the highest phage/*L. monocytogenes* ratios. Since real-life contamination of fresh-cut produce with *L. monocytogenes* is not likely to occur at the artificially high bacterial levels used during the experiments (e.g., 2.5×10^5 and 2.5×10^6 CFU of *L. monocytogenes* per 0.785 cm^3 of fruit), the authors suggested that the phage preparation may be even more effective in real-life settings than during their experiments. Spray application of phages reduced the bacterial numbers at least as much as the earlier pipette application, which is of importance for commercial applications. Phage spray treatment has also been shown (S.L. Burnett, personal communication) to reduce the levels of *L. monocytogenes* on sliced cooked whole muscle cuts (red meat and poultry), sliced cooked cured whole muscle cuts (red meat and poultry), and whole muscle cuts (red meat and poultry) by at least 10-fold (and, on several foods, by more than 100-fold) when the products were stored refrigerated. In view of the recent USDA and FDA estimate (Anonymous, 2003) that a tenfold pre-retail reduction in contamination with *L. monocytogenes* would reduce the annual number of *L. monocytogenes*-caused deaths in the elderly population in the United States by nearly 50%, the aforementioned ability of specific phages to reduce significantly these levels on various REF foods may have very practical significance.

These studies suggest that using phages to reduce contamination of foods with various pathogenic bacteria is effective for at least some types of foods, and that the successful use of phages will be a matter of using the right phage in the right place and in the right concentration. However, being effective is only the starting point in the process of making this approach viable, and many additional important issues will have to be addressed before phages can be used to improve food safety in real-life settings. For example, phage preparations added to foods must meet stringent requirements for purity, which will necessitate the development of commercially viable protocols for the large-scale production of purified phage. Also, at the present time, it is not clear what the regulatory strategy will be for such phage-based products; e.g., it is unclear whether they will be treated as "direct food additives" (i.e., ingredients that have a long-lasting effect on the product; e.g., lactoferrin) or as "secondary direct food additives" (i.e., ingredients that do not have a long-lasting effect on the foods; e.g., ozone). Finally, consumers' acceptance of the idea that phages (i.e., kinds of viruses) may be added to their food is another unknown. Since the great majority of people are not aware of the naturally high prevalence of phages in the foods they consume daily (see section 13.5.5 below), they may be concerned by the idea of adding phages to some of their foods, even if its purpose is to eliminate harmful bacteria and make the foods safer for human consumption.

13.5.5. Prevalence of Bacteriophages in Foods and Other Parts of the Environment

As described in detail in Chapter 6, bacteriophages are the most ubiquitous and most diverse living entitites on Earth. The total number of phages on Earth is estimated to be 10^{30} to 10^{32}, and they are abundant in saltwater and freshwater, soil, plants, and animals; in humans, phages can be found on skin, in the GI tract, and in the mouth (Rohwer, 2003; Brüssow and Hendrix, 2002; Bergh et al., 1989; Gill et al., 2003; Bachrach et al., 2003; Yeung and Kozelsky, 1997). Bacteriophages also have been isolated from drinking water (Lucena et al., 1995; Armon et al., 1997; Armon and Kott, 1993; Grabow and Coubrough, 1986) and a wide range of food products, including ground beef, pork sausage, chicken, farmed freshwater fish, common carp and marine fish, oil, sardines, raw skim milk, and cheese (Hsu et al., 2002; Whitman and Marshall, 1969; 1971; Kennedy et al., 1986; Kennedy et al., 1984; Gautier et al., 1995). To give just a few examples, bacteriophages were recovered from 100% of examined fresh chicken and pork sausage samples and from 33% of delicatessen meat samples analyzed by Kennedy et al. (1984). The levels ranged from 3.3 to 4.4×10^{10} PFU/100 g of fresh chicken, up to 3.5×10^{10} PFU/100 g of fresh pork, and up to 2.7×10^{10} PFU/100 g of roast turkey breast samples. In another study

(Kennedy et al., 1986), samples of fresh chicken breasts, fresh ground beef, fresh pork sausage, canned corned beef, and frozen mixed vegetables were examined for the presence of coliphages. Although only three ATCC strains of *E. coli* were used as indicator host strains, coliphages were found in 48% to 100% of the various food samples examined. Several other studies have suggested that 100% of the ground beef and chicken meat sold at retail contain various levels of various bacteriophages (Hsu et al., 2002; Kennedy et al., 1986; Tierney et al., 1973).

Meats and vegetables are not the only foods in which the presence of phages has been well documented. For example, Gautier et al. (1995) recently reported that 50% of the Swiss cheese samples they analyzed contained phages lytic for *Propionibacterium freudenreichii* (dairy propionbacteria are used in the production of Swiss cheese because of their ability to produce the characteristic, desired flavor). The number of bacteriophages varied from 10 to 10^6 PFU/g of cheese, and the authors thought it was likely that their data was an underestimate, since only a few indicator strains were used during the screening. Bacteriophages have also often been found in various seafoods. For example, during studies examining the value of bacteriophages as surrogate markers for detecting pollution in shellfish, *B. fragilis* bacteriophages often were isolated from black mussels (Lucena et al., 1994). Bacteriophages also are often found in animal feed. In a recent study from Texas A&M University (Maciorowski et al., 2001), male-specific and somatic coliphages were detected in all animal feeds, feed ingredients, and poultry diets examined, even after the samples were stored at –20°C for 14 months.

The above data indicate that naturally occurring bacteriophages are commonly consumed by humans and animals; the daily ingestion of phages may be an important natural strategy for replenishing the phage population in the GIT and for regulating the colon's microbial balance. However, no accurate estimate of the amount of phage ingested daily by an average person is available at the present time. It is likely that only a small portion of the viable phage in foods—including phages intentionally added to foods, if any, to reduce the levels of foodborne pathogens—actually are or will be consumed by humans, because many commonly used food processing practices are detrimental for phage viability; e.g., microwaving for ≥1 min and/or boiling for ≥2 min reduces the levels of viable phage in various foods by >99.9% (A. Sulakvelidze, unpublished data). Thus, it is possible that most of the food-delivered phage found in the human GIT come from eating unprocessed foods. Furthermore, at least some of the ingested viable phage are likely to be killed by the acidic conditions in the stomach; probably only a small percent actually are capable of "colonizing" the intestines. The length of time that particular phage persists in the intestines also is unknown at the present time. However, data concerning the prevalence of phages in human feces and wastewater samples may provide some insight into their prevalence in the human GIT and in wastewater and sewage.

Phages capable of lysing *E. coli*, *B. fragilis*, and various *Salmonella* serotypes have been isolated from human fecal specimens in concentrations as high as 10^5 PFU/100 g of feces (Calci et al., 1998; Cornax et al., 1994; Furuse et al., 1983a; Kai et al., 1985). Importantly, because of the specificity of phages, employing a specific bacterial host for phage enumeration will only enable enumeration of the phages that can infect that particular host strain; thus, the value of such an approach for enumerating the entire phage population is limited. An interesting approach to counteract this limitation has recently been explored by Breitbart et al. (2003). The authors used partial shotgun sequencing to perform metagenomic analyses of an uncultured viral community in human feces. Bacteriophages were found to be the second most abundant category (after bacteria) in the uncultured fecal library, containing an estimated 1200 diverse genotypes, 80% of them siphoviridae. It is likely that further optimization of this approach, and further development of microarray-based technologies, will provide much information about the diversity of phage populations, and the peculiarities of phage-bacterial interactions, in the GIT.

A yet-unidentified proportion of the phages "colonizing" the human GIT is continuously being released into the environment via wastewater. Many studies have been performed to determine the amounts of various bacteriophages in wastewater samples (Lasobras et al., 1999; Hantula et al., 1991; Puig et al., 1999; Ketranakul and Ohgaki, 1989; Cornax et al., 1990; Havelaar and Hogeboom, 1984; Moce-Llivina et al., 2003; Leclerc et al., 2000; Duran et al., 2002; Furuse et al., 1981; 1979; Osawa et al., 1981b; Tartera and Jofre, 1987; Tartera et al., 1989), and some of them have been recently reviewed (Leclerc et al., 2000). To give just a few examples, one study (Lasobras et al., 1999) determined the levels of somatic coliphage, F-specific RNA phage, and *B. fragilis* phage in two water treatment plants in Spain. The protocols employed for phage isolation specifically focused on enumerating phage in sludge solids rather than on determining the total amounts of phage in wastewater samples. Also, it is likely that a proportion of the phage in the samples were not detected because of the methodological approaches employed; for example, sludge samples were centrifuged for a brief period of time before initiating phage isolation, supernatant fluids (possibly containing large amounts of phage) were discarded, and only the harvested sediments were processed for phage enumeration. The concentration of phage in the samples analyzed was still found to be fairly high (e.g., ca. 10^5 PFU of somatic coliphage/g-1 were recovered from primary sludge samples obtained from a biological treatment plant). Another study from Spain, by Puig et al. (1999), determined the prevalence of *B. fragilis* phage and *Salmonella* serotype Typhimurium phage in urban sewage samples and in wastewater samples from animals. Bacteriophage were isolated from all of the examined urban sewage samples and from 39%–100% of slaughterhouse wastewater samples (the percentage of positive samples varied depending on the host strain used for phage isolation). Bacteriophage concentrations varied depending on the samples, and they generally were higher in urban water samples than in samples from slaughterhouses. Similar observations were reported in a recent paper by Leclerc et al. (2000), in which the authors estimated daily per capita loadings of male-specific bacteriophage for various animal species. The authors also estimated the amounts of these phage released into the environment (PFU d-1) per animal host. The numbers ranged from a low of 9×10^2 PFU/day (canine) to a high of 2×10^7 PFU/day (horse). The sewage effluent loadings ranged from 1.1×10^{12} PFU/day to 2.0×10^{13} PFU/day for the same male-specific coliphage. Phages have also been commonly isolated from wastewater in Japan, Korea, and other countries (Furuse et al., 1979, 1981; Furuse et al., 1983a; Furuse et al., 1978; Furuse et al., 1983b; Havelaar et al., 1986; Havelaar et al., 1990; Kai et al., 1985; Osawa et al., 1981a; Osawa et al., 1981b).

As noted above, the amounts of phages released from human GIT into the environment have not been rigorously determined, but some estimates can be made. For example, Puig et al. (1999) reported that the average concentration of F-specific phages for *Salmonella* serotype Typhimurium WG49 in sewage from 13 locations in 7 countries in Europe was ca. 7.7×10^3 PFU/ml (Table 13.3). The average person in the United States generates approximately 135 gallons, or 511 liters, of sewage per day (Anonymous, 1972). Thus, assuming that the average number of F-specific phages in sewage in the United States is similar to that in Europe, the number of F-specific phages in raw sewage attributable to a single individual in the United States may be calculated as follows: 7.7×10^3 PFU/ml $\times$ 511 L. of sewage/person/day = 3.9×10^9 of phages specific for *Salmonella* serotype Typhimurium strain WG49 shed per person, per day. According to the U.S. Bureau of the Census, the resident population of the United States is approximately 292 million (as of October, 2003); thus, the amount of F-specific phages in U.S. sewage would be estimated at a mind-boggling 1.1×10^{18} PFU per day. Of course, this is a very rough estimate, based on many possibly incorrect assumptions. For example, it assumes that the F-specific phages do not multiply in sewage, which may be an appropriate assumption (based on previously published literature (Novotny and Lavin, 1971; Havelaar et al., 1990)) and that the phages' life-cycle in sewage is 24 h or less (which is not the case for the F-specific phages, shown (Lasobras et al., 1999) to remain viable in sewage samples for at least several days). Significantly, the above estimate is based on the number of only F-specific phage in sewage, and it does not account for any other phages that are also likely to be present there. Similar estimates based on phage prevalence data obtained during studies of two water treatment plants in Spain (Lasobras et al., 1999) suggest that the amount of somatic coliphages shed per person per day may be even higher than that of the F-specific phages. On the other hand, the number of *B. fragilis* bacteriophages shed per person per day is much lower than the amount of the F-specific phages and coliphages (Lasobras et al., 1999). The observed difference in prevalence between the *B. fragilis* phages and the F-specific phages and somatic coliphages may be due to a lower prevalence of *B. fragilis* phages in the human GIT. Another possible explanation for the observed difference might be that the methodological approach employed to enumerate the phages in sewage samples may have enabled better enumeration of the coliphages and F-specific phages than the phages specific for the more fastidious, anaerobic *B. fragilis*. However, more rapid inactivation of *B. fragilis* bacteriophages in sewage (compared to coliphages and F-specific phages) probably is not a contributing factor, since *B. fragilis* phages and somatic coliphages have been reported (Lucena et al., 1994) to have relatively low decay rates in sewage, the lowest among all phages examined.

13.5.6. Food Spoilage and Bacteriophages

Perhaps one of the least explored applications of bacteriophages in agribusiness is their possible efficacy in reducing food spoilage caused by various bacteria. Food spoilage involves a complex sequence of events involving the interaction of a combination of microbial and biochemical activities (reviewed in Borch et al., 1996; Huis in 't Veld, 1996). Microorganisms are the major cause of spoilage of most food

TABLE 13.3
Levels of *B. fragilis*- and F-specific Phages in Urban Sewage and Animal Wastewater Samples from Various Countries (adapted from (Puig et al., 1999))

Country	Bacteriophage Level (PFU/ml)					
	B. fragilis HSP40[a]		*B. fragilis* RY2056[a]		*S.* Typhimurium WG49[a]	
	Urban Sewage	Animal Wastewater	Urban Sewage	Animal Wastewater	Urban Sewage	Animal Wastewater
The Netherlands	1	0	4.6×10^2	3.4×10^3	5.1×10^3	3.0×10^4
The Netherlands	2.6	0	7.7×10^2	8.6×10^1	7.6×10^3	3.8×10^3
Ireland	1.4	0	3.0×10^2	0	4.6×10^3	4.8×10^2
Ireland[b]	1.6	0	4.4×10^2	0	6.9×10^3	1.0×10^1
Austria	8.5	ND	8.1×10^2	ND	1.6×10^3	ND
Austria	0.5	ND	6.1×10^2	ND	2.2×10^3	ND
Portugal	0.4	0	1.8×10^2	0.8	1.8×10^4	2.2×10^2
Portugal	0.1	ND	1.0×10^2	ND	3.8×10^4	ND
Portugal	0	0	1.0×10^2	ND	5.5×10^3	ND
Germany	2.2	0	7.8×10^2	1.3	4.8×10^3	5.7×10^3
Germany	1.3	ND	6.0×10^2	ND	2.2×10^3	ND
Sweden	0.9	ND	2.2×10^1	ND	1.9×10^3	ND
France	3.1×10^1	ND	2.3×10^2	ND	1.2×10^3	ND
South Africa	1.1×10^2	0	1.8×10^2	0	1.2×10^4	2.0×10^2
South Africa	4.5×10^2	0	ND	0	5.9×10^4	5.0×10^3
South Africa	1.2×10^2	ND	5.4×10^2	ND	1.7×10^4	ND
South Africa	2.3×10^2	ND	5.0×10^2	ND	2.4×10^4	ND

[a]Host strains used for phage enumeration
[b]The number of phages indicated for animal wastewater samples are for Denmark
ND—Not determined

products, and 25% of all post-harvest foods have been estimated to be lost due to microbial-elicited food spoilage (Gram and Dalgaard, 2002). However, only a few members of the microbial community, so called *specific spoilage organisms* (SSO), give rise to the offensive aromas and flavors associated with food spoilage. SSO are found in various types of foods, and various bacterial species are responsible for the different stages of food deterioration (Gram and Dalgaard, 2002; Samelis et al., 2000; Gamage et al., 1997; Jay et al., 2003).

Numerous approaches have been examined and used/are being used to reduce the number of SSO in foods, including hydrodynamic pressure processing, gamma irradiation, and pasteurization (Williams-Campbell and Solomon, 2002; Gamage et al., 1997; Roberts and Weese, 1998; Gould, 2000). Early attempts to use antibiotics to extend the shelf life of various foods were successful (Bernarde and Littleford, 1958); however, the use of antibiotics in agribusiness is not an optimal approach because it may promote the emergence and spread of antibiotic-resistance among various bacterial species (see section 13.2.4). An alternative to antibiotics may be disinfecting solutions and bacteriocins; e.g., nisin and lysozyme have been used, with some success, to control meat spoilage (Nattress et al., 2001; Thomas et al., 2002). Also, lactic acid bacteria have been used to inhibit the growth of pathogenic bacteria and SSO in various foods (Hernandez et al., 1993). However, lactic acid bacteria do not eliminate SSO, but, rather, inhibit their growth by competing for nutrients in the meat being treated. Moreover, lactic acid bacteria may contribute to the middle and late stages of food spoilage (Leroi et al., 1998). Therefore, biocontrol agents that do not contribute to food spoilage and can specifically target the SSO—without disturbing the normal, beneficial microflora of foods—may be an attractive modality for improving the shelf life of various food products. Bacteriophages seem to fit these requirements; however, the idea of using phages to reduce the levels of SSO in various foods has not been rigorously pursued and only a few publications are available on the subject.

One of the first publications examining the ability of phages to extend the shelf life of foods is the study by Ellis et al. (1973). The authors demonstrated that *P. fragi*-specific bacteriophages (originally isolated from ground beef (Whitman and Marshall, 1971) reduced the number of *P. fragi* in refrigerated raw milk and increased the milk's shelf life. More recently, the ability of phages to reduce beef spoilage was evaluated by Greer et al. (1990). Although the authors did not observe an appreciable extension of the shelf life of the examined beef products, they only targeted one bacterial species, which probably was not the major SSO responsible for beef spoilage. Indeed, when the same group of authors specifically targeted *B. thermosphacta* (the species known to be responsible for the development of unpleasant odors in spoiled pork tissues), the storage life of the meat increased from 4 days in the controls to 8 days in the phage-treated samples (Greer and Dilts, 2002). Similar results were reported for beefsteaks; i.e., the application of *Pseudomonas*-specific phages almost doubled the steaks' shelf-life (Greer, 1986). Thus far, the ability of bacteriophages to increase the shelf life of seafood products has not been rigorously addressed (Delisle and Levin, 1969).

The above-cited publications suggest that bacteriophages specifically targeting SSO can effectively reduce the number of SSO on various foods and extend their shelf life. However, the practical applicability of that approach is not clear. For example, given the narrow specificity of phages and the complexity of the bacterial flora involved in food spoilage, it may be challenging to develop effective anti-SSO phage preparations for industrial use. The problem may be further compounded by the fact that the precise identity of SSOs still has not been clearly determined for many foodstuffs. These, however, are not insurmountable tasks, and should become increasingly realistic as our understanding of the microbiology of food spoilage of various foodstuffs improves as the result of the ongoing scientific research in the field.

ACKNOWLEDGMENTS

We thank Gary Pasternack, Roy Voelker, and Eliot Harrison for their input pertaining to the prevalence of phages in foods and the environment, and Arnold Kreger for his editorial assistance.

REFERENCES

Anonymous,. Manual of Design Procedure and Criteria, pp. 5–6. Dept. of Public Works Bureau of Water and Waste Water, Baltimore, Maryland, 1972.

Anonymous,. Quantitative assessment of the relative risk to public health from foodborne *Listeria monocytogenes* among selected categories of ready-to-eat foods. Center for Food Safety and Applied Nutrition, Food and Drug Administration; Food Safety and Inspection Service, U.S. Department of Agriculture; and Centers for Disease Control and Prevention, 2003.

Armon, R., Araujo, R., Kott, Y., Lucena, F. and Jofre, J., Bacteriophages of enteric bacteria in drinking water, comparison of their distribution in two countries. *J Appl Microbiol* 83: 627–633, 1997.

Armon, R. and Kott, Y., A simple, rapid and sensitive presence/absence detection test for bacteriophage in drinking water. *J Appl Bacteriol* 74: 490—496, 1993.

Arnold, L. and Weiss, E., Antigenic properties of bacteriophage. *J Infect Dis* 34: 317–327, 1924.

Arnold, L. and Weiss, E., Prophylactic and therapeutic possibilities of the Twort-d'Herelle bacteriophage. *J Lab & Clin Med* 12: 20–31, 1926.

Asheshov, I.N., Wilson, J. and Topley, W.W., The effect of an anti-Vi bacteriophage on typhoid infection in mice. *Lancet* 1: 319–320, 1937.

Atterbury, R.J., Connerton, P.L., Dodd, C.E., Rees, C.E. and Connerton, I.F., Isolation and characterization of *Campylobacter bacteriophages* from retail poultry. *Appl Environ Microbiol* 69: 4511–4518, 2003.

Auerbach, J.I. and Rosenberg, M., Externalization of products of bacteria, U.S. patent 4,637,980. SmithKline Beckman Corporation, U.S.A, 1987.

Bachrach, G., Leizerovici-Zigmond, M., Zlotkin, A., Naor, R. and Steinberg, D., Bacteriophage isolation from human saliva. *Lett Appl Microbiol* 36: 50–53, 2003.

Barrow, P., Lovell, M. and Berchieri, A., Jr., Use of lytic bacteriophage for control of experimental *Escherichia coli* septicemia and meningitis in chickens and calves. *Clin Diagn Lab Immunol* 5: 294–298, 1998.

Barrow, P.A. and Soothill, J.S., Bacteriophage therapy and prophylaxis: rediscovery and renewed assessment of potential. *Trends Microbiol* 5: 268–271, 1997.

Berchieri, A., Jr., Lovell, M.A. and Barrow, P.A., The activity in the chicken alimentary tract of bacteriophages lytic for *Salmonella typhimurium*. *Res Microbiol* 142: 541–549, 1991.

Bergh, O., Borsheim, K.Y., Bratbak, G. and Heldal, M., High abundance of viruses found in aquatic environments. *Nature* 340: 467–468, 1989.

Bernarde, M.A. and Littleford, R.A., Antibiotic treatment of crab and oyster meats. *Appl Microbiol* 5: 368–372, 1958.

Bernhardt, T.G., Wang, I.N., Struck, D.K. and Young, R., A protein antibiotic in the phage Qbeta virion: Diversity in lysis targets. *Science* 292: 2326–2329, 2001.

Biswas, B., Adhya, S., Washart, P., Paul, B., Trostel, A.N., Powell, B., Carlton, R., et al., Bacteriophage therapy rescues mice bacteremic from a clinical isolate of vancomycin–resistant *Enterococcus faecium*. *Infect Immun* 70: 204–210, 2002.

Blasi, U., Henrich, B. and Lubitz, W., Lysis of *Escherichia coli* by cloned phi X174 gene E depends on its expression. *J Gen Microbiol* 131 (Pt 5): 1107–1114, 1985.

Borch, E., Kant-Muermans, M.L. and Blixt, Y., Bacterial spoilage of meat and cured meat products. *Int J Food Microbiol* 33: 103–120, 1996.

Brabban, A., Callaway, T., Dutta, G., Dyen, M., Edrington, T., Kutter, E., Raya, R., et al., Characterization of a new T-even bacteriophage with potential for reducing *E. coli* O157:H7 levels in livestock, p. 59 in *103rd General Meeting of the American Society for Microbiology*. American Society for Microbiology, Washington, D.C., 2003.

Breitbart, M., Hewson, I., Felts, B., Mahaffy, J.M., Nulton, J., Salamon, P. and Rohwer, F., Metagenomic analyses of an uncultured viral community from human feces. *J Bacteriol* 185: 6220–6223, 2003.

Brüssow, H. and Hendrix, R.W., Phage genomics: Small is beautiful. *Cell* 108: 13–16, 2002.

Bull, J.J., Levin, B.R., DeRouin, T., Walker, N. and Bloch, C.A., Dynamics of success and failure in phage and antibiotic therapy in experimental infections. *BMC Microbiol* 2: 35, 2002.

Calci, K.R., Burkhardt, W., 3rd, Watkins, W.D. and Rippey, S.R., Occurrence of male-specific bacteriophage in feral and domestic animal wastes, human feces, and human-associated wastewaters. *Appl Environ Microbiol* 64: 5027–5029, 1998.

Chambers, E.D. and Severin, G.A., Staphylococcal bacterin for treatment of chronic staphylococcal blepharitis in the dog. *J Am Vet Med Assoc* 185: 422–425, 1984.

Civerolo, E.L., Relationship of *Xanthomonaspruni* bacteriophages to bacterial spot disease in Prunus. *Phytopathology* 63: 1279–1284, 1973.

Civerolo, E.L. and Kiel, H.L., Inhibition of bacterial spot of peach foliage by *Xanthomonas pruni* bacteriophage. *Phytopathology* 59: 1966–1967, 1969.

Clark, P.F. and Clark, A.S., Bacteriophage active against virulent hemolytic streptococcus. *Proc Soc Exper Biol & Med* 24: 635–639, 1927.

Coleman, J., Inouye, M. and Atkins, J., Bacteriophage MS2 lysis protein does not require coat protein to mediate cell lysis. *J Bacteriol* 153: 1098–1100, 1983.

Colvin, M.G., Relationship of bacteriophage to natural and experimental diseases of laboratory animals, with special reference to lymphadenitis of guinea pigs. *J Infect Dis* 51: 17–29, 1932.

Compton, A., Immunization in experimental plague by subcutaneous inoculation with bacteriophage. Comparison of plain and formaldehyde-treated phage-lysed plague vaccine. *J Infect Dis* 46: 152–160, 1930.

Compton, A., Sensitization and immunization with bacteriophage in experimental plague. *J Infect Dis* 43: 448–457, 1928.

Cornax, R., Morinigo, M.A., Gonzalez-Jaen, F., Alonso, M.C. and Borrego, J.J., Bacteriophages presence in human faeces of healthy subjects and patients with gastrointestinal disturbances. *Zentralbl Bakteriol* 281: 214–224, 1994.

Cornax, R., Morinigo, M.A., Paez, I.G., Munoz, M.A. and Borrego, J.J., Application of direct plaque assay for detection and enumeration of bacteriophages of *Bacteroides fragilis* from contaminated-water samples. *Appl Environ Microbiol* 56: 3170–3173, 1990.

DeBoer, D.J., Moriello, K.A., Thomas, C.B. and Schultz, K.T., Evaluation of a commercial staphylococcal bacterin for management of idiopathic recurrent superficial pyoderma in dogs. *Am J Vet Res* 51: 636–639, 1990.

Delisle, A.L. and Levin, R.E., Bacteriophages of psychrophilic pseudomonads. I. Host range of phage pools active against fish spoilage and fish-pathogenic pseudomonads. *Antonie Van Leeuwenhoek* 35: 307–317, 1969.

d'Herelle, F., *The Bacteriophage and Its Behavior.* Williams and Wilkins, Baltimore, Maryland, 1926.

d'Herelle, F., Contribution à l'étude de l'immunité. *Comptes rendus Acad Sci* (Paris) 162: 570–573, 1916.

d'Herelle, F., *Le bactériophage: son rôle dans limmunité.* Masson et Cie, Paris, 1921.

d'Herelle, F., Sur un microbe invisible antagoniste des bacilles dysentériques. *Compt Rend Acad Sci* (Paris) 165: 373–375, 1917.

Dubos, R., Straus, J.H. and Pierce, C., The multiplication of bacteriophage in vivo and its protective effect against an experimental infection with *Shigella dysenteriae. J Exp Med* 20: 161–168, 1943.

Duran, A.E., Muniesa, M., Mendez, X., Valero, F., Lucena, F. and Jofre, J., Removal and inactivation of indicator bacteriophages in fresh waters. *J Appl Microbiol* 92: 338–347, 2002.

Eko, F.O., Hensel, A., Bunka, S. and Lubitz, W., Immunogenicity of *Vibrio cholerae* ghosts following intraperitoneal immunization of mice. *Vaccine* 12: 1330–1334, 1994a.

Eko, F.O., Schukovskaya, T., Lotzmanova, E.Y., Firstova, V.V., Emalyanova, N.V., Klueva, S.N., Kravtzov, A.L., et al., Evaluation of the protective efficacy of *Vibrio cholerae* ghost (VCG) candidate vaccines in rabbits. *Vaccine* 21: 3663–3674, 2003.

Eko, F.O., Szostak, M.P., Wanner, G. and Lubitz, W., Production of *Vibrio cholerae* ghosts (VCG) by expression of a cloned phage lysis gene: potential for vaccine development. *Vaccine* 12: 1231–1237, 1994b.

Eliava, G., Au sujet de ladsorption du bactériophage par les leucocytes. *Comp rend Soc de biol* 105: 829–831, 1930.

Ellis, D.E., Whitman, P.A. and Marshall, R.T., Effects of homologous bacteriophage on growth of *Pseudomonas fragi* WY in milk. *Appl Microbiol* 25: 24–25, 1973.

Erskine, J.M., Characteristics of *Erwinia amylovora* bacteriophage and its possible role in the epidemology of fire blight. *Can J Microbiol* 19: 837–845, 1973.

Esber, H.J., DeCourcy, S.J. and Bogden, A.E., Specific and nonspecific immune resistance enhancing activity of staphage lysate. *J Immunopharmacol* 3: 79–92, 1981.

Esber, H.J., Ganfield, D. and Rosenkrantz, H., Staphage lysate: An immunomodulator of the primary immune response in mice. *Immunopharmacology* 10: 77–82, 1985.

Ferber, D., Antibiotic resistance. WHO advises kicking the livestock antibiotic habit. *Science* 301: 1027, 2003.

Fisk, R.T., Protective action of typhoid phage on experimental typhoid infection in mice. *Proc Soc Exper Biol & Med* 38: 659–660, 1938.

Fox, J.L., Phage treatments yield healthier tomato, pepper plants. *ASM News* 66: 455–456, 2000.

Furuse, K., Ando, A., Osawa, S. and Watanabe, I., Continuous survey of the distribution of RNA coliphages in Japan. *Microbiol Immunol* 23: 867–875, 1979.

Furuse, K., Ando, A., Osawa, S. and Watanabe, I., Distribution of ribonucleic acid coliphages in raw sewage from treatment plants in Japan. *Appl Environ Microbiol* 41: 1139–1143, 1981.

Furuse, K., Osawa, S., Kawashiro, J., Tanaka, R., Ozawa, A., Sawamura, S., Yanagawa, Y., et al., Bacteriophage distribution in human faeces: continuous survey of healthy subjects and patients with internal and leukaemic diseases. *J Gen Virol* 64 (Pt 9): 2039–2043, 1983a.

Furuse, K., Sakurai, T., Hirashima, A., Katsuki, M., Ando, A. and Watanabe, I., Distribution of ribonucleic acid coliphages in south and east Asia. *Appl Environ Microbiol* 35: 995–1002, 1978.

Furuse, K., Sakurai, T., Inokuchi, Y., Inoko, H., Ando, A. and Watanabe, I., Distribution of RNA coliphages in Senegal, Ghana, and Madagascar. *Microbiol Immunol* 27: 347–358, 1983b.

Gaeng, S., Scherer, S., Neve, H. and Loessner, M.J., Gene cloning and expression and secretion of *Listeria monocytogenes* bacteriophage-lytic enzymes in *Lactococcus lactis*. *Appl Environ Microbiol* 66: 2951–2958, 2000.

Gamage, S.D., Faith, N.G., Luchansky, J.B., Buege, D.R. and Ingham, S.C., Inhibition of microbial growth in chub-packed ground beef by refrigeration (2 degrees C) and medium-dose (2.2 to 2.4 kGy) irradiation. *Int J Food Microbiol* 37: 175–182, 1997.

Gautier, M., Rouault, A., Sommer, P. and Briandet, R., Occurrence of *Propionibacterium freudenreichii* bacteriophages in swiss cheese. *Appl Environ Microbiol* 61: 2572–2576, 1995.

Giese, M.J., Adamu, S.A., Pitchekian-Halabi, H., Ravindranath, R.M. and Mondino, B.J., The effect of *Staphylococcus aureus* phage lysate vaccine on a rabbit model of staphylococcal blepharitis, phlyctenulosis, and catarrhal infiltrates. *Am J Ophthalmol* 122: 245–254, 1996.

Gill, J.J., Svircev, A.M., Smith, R. and Castle, A.J., Bacteriophages of *Erwinia amylovora*. *Appl Environ Microbiol* 69: 2133–2138, 2003.

Goode, D., Allen, V.M. and Barrow, P.A., Reduction of experimental *Salmonella* and *Campylobacter* contamination of chicken skin by application of lytic bacteriophages. *Appl Environ Microbiol* 69: 5032–5036, 2003.

Gould, G.W., Preservation: Past, present and future. *Br Med Bull* 56: 84–96, 2000.

Grabow, W.O. and Coubrough, P., Practical direct plaque assay for coliphages in 100-ml samples of drinking water. *Appl Environ Microbiol* 52: 430–433, 1986.

Gram, L. and Dalgaard, P., Fish spoilage bacteria—problems and solutions. *Curr Opin Biotechnol* 13: 262–266, 2002.

Greer, G., Homologous bacteriophage control of *Pseudomonas* growth and beef spoilage. *J Food Protection*: 104–109, 1986.

Greer, G.G. and Dilts, B.D., Control of *Brochothrix thermosphacta* spoilage of pork adipose tissue using bacteriophages. *J Food Prot* 65: 861–863, 2002.

Greer, G.G. and Dilts, B.D., Inability of a bacteriophage pool to control beef spoilage. *Int J Food Microbiol* 10: 331–342, 1990.

Gunn, L.C. and Hengesh, J., The use of lysostaphin in treatment of staphylococcal wound infections. *Rev Surg* 26: 214, 1969.

Gustafson, R.H., Use of antibiotics in livestock and human health concerns. *J Dairy Sci* 74: 1428–1432, 1991.

Hantula, J., Kurki, A., Vuoriranta, P. and Bamford, D.H., Ecology of bacteriophages infecting activated sludge bacteria. *Appl Environ Microbiol* 57: 2147–2151, 1991.

Hartung, J.S., Fulbright, D.W. and Klos, E.J., Cloning of a bacteriophage polysaccharide depolymerase gene and its expression in *Erwinia amylovora*. *Mol Plant-Microbe Interact* 1: 87–93, 1988.

Havelaar, A.H., Furuse, K. and Hogeboom, W.M., Bacteriophages and indicator bacteria in human and animal faeces. *J Appl Bacteriol* 60: 255–262, 1986.

Havelaar, A.H. and Hogeboom, W.M., A method for the enumeration of male-specific bacteriophages in sewage. *J Appl Bacteriol* 56: 439–447, 1984.

Havelaar, A.H., Pot-Hogeboom, W.M., Furuse, K., Pot, R. and Hormann, M.P., F–specific RNA bacteriophages and sensitive host strains in faeces and wastewater of human and animal origin. *J Appl Bacteriol* 69: 30–37, 1990.

Hensel, A., Huter, V., Katinger, A., Raza, P., Strnistschie, C., Roesler, U., Brand, E., et al., Intramuscular immunization with genetically inactivated (ghosts) *Actinobacillus pleuropneumoniae* serotype 9 protects pigs against homologous aerosol challenge and prevents carrier state. *Vaccine* 18: 2945–2955, 2000.

Hernandez, P.E., Rodriguez, J.M., Cintas, L.M., Moreira, W.L., Sobrino, O.J., Fernandez, M.F. and Sanz, B., Utilization of lactic bacteria in the control of pathogenic microorganisms in food. *Microbiologia* 9 Spec No: 37–48, 1993.

Holt, M.E., Enright, M.R. and Alexander, T.J., Immunisation of pigs with killed cultures of *Streptococcus suis* type 2. *Res Vet Sci* 48: 23–27, 1990.

Hsu, F.C., Shieh, Y.S. and Sobsey, M.D., Enteric bacteriophages as potential fecal indicators in ground beef and poultry meat. *J Food Prot* 65: 93–99, 2002.

Huff, J.P., Grant, B.J., Penning, C.A. and Sullivan, K.F., Optimization of Routine Transformation of *Escherichia coli* with Plasmid DNA. *Biotechniques* 9: 570, 1990.

Huff, W.E., Huff, G.R., Rath, N.C., Balog, J.M. and Donoghue, A.M., Bacteriophage treatment of a severe *Escherichia coli* respiratory infection in broiler chickens. Avian Dis 47: 1399–1405, 2003a.

Huff, W.E., Huff, G.R., Rath, N.C., Balog, J.M. and Donoghue, A.M., Evaluation of aerosol spray and intramuscular injection of bacteriophage to treat an *Escherichia coli* respiratory infection. *Poult Sci* 82: 1108–1112, 2003b.

Huff, W.E., Huff, G.R., Rath, N.C., Balog, J.M. and Donoghue, A.M., Prevention of *Escherichia coli* infection in broiler chickens with a bacteriophage aerosol spray. *Poult Sci* 81: 1486–1491, 2002a.

Huff, W.E., Huff, G.R., Rath, N.C., Balog, J.M., Xie, H., Moore, P.A., Jr., and Donoghue, A.M., Prevention of *Escherichia coli* respiratory infection in broiler chickens with bacteriophage (SPR02). *Poult Sci* 81: 437–441, 2002b.

Huis in 't Veld, J.H., Microbial and biochemical spoilage of foods: An overview. *Int J Food Microbiol* 33: 1–18, 1996.

Huter, V., Hensel, A., Brand, E. and Lubitz, W., Improved protection against lung colonization by *Actinobacillus pleuropneumoniae* ghosts: Characterization of a genetically inactivated vaccine. *J Biotechnol* 83: 161–172, 2000.

Jackson, L.E., Bacteriophage prevention and control of harmful plant bacteria, U.S. patent 4,828,999, 1989.

Jalava, K., Hensel, A., Szostak, M., Resch, S. and Lubitz, W., Bacterial ghosts as vaccine candidates for veterinary applications. *J Control Release* 85: 17–25, 2002.

Jay, J.M., Vilai, J.P. and Hughes, M.E., Profile and activity of the bacterial biota of ground beef held from freshness to spoilage at 5–7 degrees C. *Int J Food Microbiol* 81: 105–111, 2003.

Jones, A.L. and Schnabel, E.L., "The development of streptomycin resistant strains of *Erwinia amylovora*," pp. 235–251 in *Fire Blight: The Disease and Its Causative Agent Erwinia amylovora*, J.L. Vanneste (Ed.). CAB International, Wallingford, Oxon, United Kingdom, 2000.

Kabeshima, T., Recherches expérimentale sur la vaccination préventive contre le bacille dysentérique de Shiga. *Comptes rendus Acad Sci* (Paris) 169: 1061–1064, 1919.

Kai, M., Watanabe, S., Furuse, K. and Ozawa, A., Bacteroides bacteriophages isolated from human feces. *Microbiol Immunol* 29: 895–899, 1985.

Kastelein, R.A., Remaut, E., Fiers, W. and van Duin, J., Lysis gene expression of RNA phage MS2 depends on a frameshift during translation of the overlapping coat protein gene. *Nature* 295: 35–41, 1982.

Kennedy, J.E., Jr., Wei, C.I. and Oblinger, J.L., Methodology for enumeration of coliphages in foods. *Appl Environ Microbiol* 51: 956–962, 1986.

Kennedy, J.E.J., Oblinger, J.L. and Bitton, G., Recovery of coliphages from chicken, pork sasuage, and delcatessen meats. *J Food Protection* 47: 623–626, .

Ketranakul, A. and Ohgaki, S., Indigenous coliphages and RNA-F-specific phages associated to suspended solids in activated sludge process. *Water Sci Technol* 21: 73–78, 1989.

Kolmer, J.A. and Rule, A., A note on the treatment of experimental streptococcus meningitis of rabbits with bacteriophage. *J Lab & Clin Med* 18: 1001–1003, 1933.

Kozloff, L.M. and Schnell, R.C., Protection of plants against frost injury using ice nucleation-inhibiting species-specific bacteriophages, U.S. patent 4,375,734, 1983.

Krueger, A.P., Lich, R. and Schulze, K.R., Bacteriophage in experimental staphylococcal septicemia. *Proc Soc Exper Biol & Med* 30: 73–75, 1932.

Kudva, I.T., Jelacic, S., Tarr, P.I., Youderian, P. and Hovde, C.J., Biocontrol of *Escherichia coli* O157 with O157–specific bacteriophages. *Appl Environ Microbiol* 65: 3767–3773, 1999.

Larkum, N.W., Bacteriophage as a substitute for typhoid vaccine. *J Bacteriol* 17: 42, 1929.

Larkum, N.W., Bacteriophagy in urinary infection; bacteriophagy in bladder. *J Bacteriol* 12: 225–242, 1926.

Lasobras, J., Dellunde, J., Jofre, J. and Lucena, F., Occurrence and levels of phages proposed as surrogate indicators of enteric viruses in different types of sludges. *J Appl Microbiol* 86: 723–729, 1999.

Lauvau, G., Vijh, S., Kong, P., Horng, T., Kerksiek, K., Serbina, N., Tuma, R. A., et al., Priming of memory but not effector CD8 T cells by a killed bacterial vaccine. *Science* 294: 1735–1739, 2001.

Leclerc, H., Edberg, S., Pierzo, V. and Delattre, J.M., Bacteriophages as indicators of enteric viruses and public health risk in groundwaters. *J Appl Microbiol* 88: 5–21, 2000.

Leroi, F., Joffraud, J.J., Chevalier, F. and Cardinal, M., Study of the microbial ecology of cold-smoked salmon during storage at 8 degrees C. *Int J Food Microbiol* 39: 111–121, 1998.

Leverentz, B., Conway, W.S., Alavidze, Z., Janisiewicz, W.J., Fuchs, Y., Camp, M.J., Chighladze, E., et al., Examination of bacteriophage as a biocontrol method for *Salmonella* on fresh-cut fruit: A model study. *J Food Prot* 64: 1116–1121, 2001.

Leverentz, B., Conway, W.S., Camp, M.J., Janisiewicz, W.J., Abuladze, T., Yang, M., Saftner, R., et al., Biocontrol of *Listeria monocytogenes* on fresh-cut produce by treatment with lytic bacteriophages and a bacteriocin. *Appl Environ Microbiol* 69: 4519–4526, 2003.

Levin, B. and Bull, J.J., Phage therapy revisited: the population biology of a bacterial infection and its treatment with bacteriophage and antibiotics. *Am Naturalist* 147: 881–898, 1996.

Loeffler, J.M., Nelson, D. and Fischetti, V.A., Rapid killing of *Streptococcus pneumoniae* with a bacteriophage cell wall hydrolase. *Science* 294: 2170–2172, 2001.

Loessner, M.J., Gaeng, S., Wendlinger, G., Maier, S.K. and Scherer, S., The two-component lysis system of *Staphylococcus aureus* bacteriophage Twort: A large TTG–start holin and an associated amidase endolysin. *FEMS Microbiol Lett* 162: 265–274, 1998.

Loessner, M.J., Wendlinger, G. and Scherer, S., Heterogeneous endolysins in *Listeria monocytogenes* bacteriophages: a new class of enzymes and evidence for conserved holin genes within the siphoviral lysis cassettes. *Mol Microbiol* 16: 1231–1241, 1995.

Lucena, F., Lasobras, J., McIntosh, D., Forcadell, M. and Jofre, J., Effect of distance from the polluting focus on relative concentrations of *Bacteroides fragilis* phages and coliphages in mussels. *Appl Environ Microbiol* 60: 2272–2277, 1994.

Lucena, F., Muniesa, M., Puig, A., Araujo, R. and Jofre, J., Simple concentration method for bacteriophages of *Bacteroides fragilis* in drinking water. *J Virol Methods* 54: 121–130, 1995.

Maciorowski, K.G., Pillai, S.D. and Ricke, S.C., Presence of bacteriophages in animal feed as indicators of fecal contamination. *J Environ Sci Health B* 36: 699–708, 2001.

Mader, H.J., Szostak, M.P., Hensel, A., Lubitz, W. and Haslberger, A.G., Endotoxicity does not limit the use of bacterial ghosts as candidate vaccines. *Vaccine* 15: 195–202, 1997.

Marchart, J., Dropmann, G., Lechleitner, S., Schlapp, T., Wanner, G., Szostak, M.P. and Lubitz, W., *Pasteurella multocida*- and *Pasteurella haemolytica*-ghosts: New vaccine candidates. *Vaccine* 21: 3988–3997, 2003.

Marcuse, K., Grundlagen und aufgaben der lysintherapie (d'Herelle's bakteriophagen). *Deutsche Med Wchnschr* 50: 334–336, 1924.

Mead, P.S., Slutsker, L., Dietz, V., McCaig, L.F., Bresee, J.S., Shapiro, C., Griffin, P.M., et al., Food-related illness and death in the United States. *Emerg Infect Dis* 5: 607–625, 1999.

Melamed, D., Leitner, G. and Heller, E. D., A vaccine against avian colibacillosis based on ultrasonic inactivation of *Escherichia coli*. *Avian Dis* 35: 17–22, 1991.

Merril, C.R., Biswas, B., Carlton, R., Jensen, N.C., Creed, G.J., Zullo, S. and Adhya, S., Long-circulating bacteriophage as antibacterial agents. *Proc Natl Acad Sci U S A* 93: 3188–3192, 1996.

Moce-Llivina, L., Muniesa, M., Pimenta-Vale, H., Lucena, F. and Jofre, J., Survival of bacterial indicator species and bacteriophages after thermal treatment of sludge and sewage. *Appl Environ Microbiol* 69: 1452–1456, 2003.

Moore, E.S., D'Herelle's bacteriophage in relation to plant parasites. *South Afr J Sci* 23: 306, 1926.

Morita, M., Tanji, Y., Orito, Y., Mizoguchi, K., Soejima, A. and Unno, H., Functional analysis of antibacterial activity of *Bacillus amyloliquefaciens* phage endolysin against Gram-negative bacteria. *FEBS* Lett 500: 56–59, 2001.

Naidu, B.P.B. and Avari, C.R., Bacteriophage in the treatment of plague. *Ind Jour Med Res* 19: 737–748, 1932.

Nakai, T. and Park, S.C., Bacteriophage therapy of infectious diseases in aquaculture. *Res Microbiol* 153: 13–18, 2002.

Nakai, T., Sugimoto, R., Park, K.H., Matsuoka, S., Mori, K., Nishioka, T. and Maruyama, K., Protective effects of bacteriophage on experimental *Lactococcus garvieae* infection in yellowtail. *Dis Aquat Organ* 37: 33–41, 1999.

Nakazawa, S., Itagaki, M., Yokota, T., Otani, Y. and Miwa, M., Basic studies on the antibiotic action of lysozyme. *J Antibiot [B]* 19: 34–47, 1966.

Nattress, F.M., Yost, C.K. and Baker, L.P., Evaluation of the ability of lysozyme and nisin to control meat spoilage bacteria. *Int J Food Microbiol* 70: 111–119, 2001.

Nelson, D., Loomis, L. and Fischetti, V.A., Prevention and elimination of upper respiratory colonization of mice by group A streptococci by using a bacteriophage lytic enzyme. *Proc Natl Acad Sci U S A* 98: 4107–4112, 2001.

Nishimori, E., Kita-Tsukamoto, K. and Wakabayashi, H., *Pseudomonas plecoglossicida* sp. nov., the causative agent of bacterial haemorrhagic ascites of ayu, *Plecoglossus altivelis*. *Int J Syst Evol Microbiol* 50 Pt 1: 83–89, 2000.

Novotny, C.P. and Lavin, K., Some effects of temperature on the growth of F pili. *J Bacteriol* 107: 671–682, 1971.

Nuzov, B.G., Use of lysozyme and prodigiozan in the treatment of suppurative infections in diabetics. *Vestn Khir Im I I Grek* 132: 62–65, 1984.

Okabe, N. and Goto, M., Bacteriophages of plant pathogens. *Annu Rev Phytopathology* 1: 397–418, 1963.

Oldham, E.R. and Daley, M.J., Lysostaphin: use of a recombinant bactericidal enzyme as a mastitis therapeutic. *J Dairy Sci* 74: 4175–4182, .

Osawa, S., Furuse, K., Choi, M.S., Ando, A., Sakurai, T. and Watanabe, I., Distribution of ribonucleic acid coliphages in Korea. *Appl Environ Microbiol* 41: 909–911, 1981a.

Osawa, S., Furuse, K. and Watanabe, I., Distribution of ribonucleic acid coliphages in animals. *Appl Environ Microbiol* 41: 164–168, 1981b.

Panthel, K., Jechlinger, W., Matis, A., Rohde, M., Szostak, M., Lubitz, W. and Haas, R., Generation of *Helicobacter pylori* ghosts by PhiX protein E–mediated inactivation and their evaluation as vaccine candidates. *Infect Immun* 71: 109–116, 2003.

Park, S.C. and Nakai, T., Bacteriophage control of *Pseudomonas plecoglossicida* infection in ayu *Plecoglossus altivelis*. *Dis Aquat Organ* 53: 33–39, 2003.

Park, S.C., Shimamura, I., Fukunaga, M., Mori, K.I. and Nakai, T., Isolation of bacteriophages specific to a fish pathogen, *Pseudomonas plecoglossicida*, as a candidate for disease control. *Appl Environ Microbiol* 66: 1416–1422, 2000.

Pirisi, A., Phage therapy—advantages over antibiotics? *Lancet* 356: 1418, 2000.

Popoff, M.Y., Bockemühl, J. and Brenner, F.W., Supplement 1998 (no. 42) to the Kauffmann-White scheme. *Res Microbiol* 151: 63–65, 2000.

Poppe, C., Smart, N., Khakhria, R., Johnson, W., Spika, J. and Prescott, J., *Salmonella typhimurium* DT104: A virulent and drug-resistant pathogen. *Can Vet J* 39: 559–565, 1998.

Puig, A., Queralt, N., Jofre, J. and Araujo, R., Diversity of bacteroides fragilis strains in their capacity to recover phages from human and animal wastes and from fecally polluted wastewater. *Appl Environ Microbiol* 65: 1772–1776, 1999.

Pyle, N.J., Bacteriophage in relation to *Salmonella Pullora* infection in domestic fowl. *J Bacteriol* 12: 245–261, 1926.

Ramesh, V., Fralick, J.A. and Rolfe, R.D., Prevention of *Clostridium difficile* induced ileocecitis with bacteriophage. *Anaerobe* 5: 69–78, 1999.

Raya, R., Callaway, T., Edrington, T., Dyen, M., Droleskey, R., Kutter, E. and Brabban, A.,. *In vitro* and *in vivo* studies using phages isolated from sheep to reduce population levels of *Escherichia coli* O157:H7 in ruminants, p. 62 in *103rd General Meeting of the American Society for Microbiology*. American Society for Microbiology, Washington, D.C., 2003.

Ritchie, D.F. and Klos, E.J., Isolation of *Erwinia amylovora* bacteriophage from the aerial parts of apple trees. *Phytopathology* 67: 101–104, 1977.

Ritchie, D.F. and Klos, E.J., Some properties of *Erwinia amylovora* bacteriophages. *Phytopathology* 69: 1078–1083, 1979.

Roberts, W.T. and Weese, J.O., Shelf life of ground beef patties treated by gamma radiation. *J Food Prot* 61: 1387–1389, 1998.

Rohwer, F., Global phage diversity. *Cell* 113: 141, 2003.

Samelis, J., Kakouri, A. and Rementzis, J., The spoilage microflora of cured, cooked turkey breasts prepared commercially with or without smoking. *Int J Food Microbiol* 56: 133–143, 2000.

Sarkar, B.L., Chakrabarti, A.K., Koley, H., Chakrabarti, M.K. and De, S.P., Biological activity and interaction of *Vibrio cholerae* bacteriophages in rabbit ileal loop. *Indian J Med Res* 104: 139–141, 1996.

Schnabel, E.L. and Jones, A.L., Isolation and characterization of five *Erwinia amylovora* bacteriophages and assessment of phage resistance in strains of *Erwinia amylovora*. *Appl Environ Microbiol* 67: 59–64, 2001.

Schon, P., Schrot, G., Wanner, G., Lubitz, W. and Witte, A., Two-stage model for integration of the lysis protein E of phi X174 into the cell envelope of *Escherichia coli*. *FEMS Microbiol Rev* 17: 207–212, 1995.

Schuch, R., Nelson, D. and Fischetti, V.A., A bacteriolytic agent that detects and kills *Bacillus anthracis*. *Nature* 418: 884–889, 2002.

Schuhardt, V.T. and Schindler, C.A., Lysostaphin therapy in mice infected with *Staphylococcus aureus*. *J Bacteriol* 88: 815–816, 1964.

Slopek, S., Weber-Dabrowska, B., Dabrowski, M. and Kucharewicz-Krukowska, A., Results of bacteriophage treatment of suppurative bacterial infections in the years 1981–1986. *Arch Immunol Ther Exp* (Warsz) 35: 569–583, 1987.

Smith, D.L., Harris, A.D., Johnson, J.A., Silbergeld, E.K. and Morris, J.G., Jr., Animal antibiotic use has an early but important impact on the emergence of antibiotic resistance in human commensal bacteria. *Proc Natl Acad Sci U S A* 99: 6434–6439, 2002.

Smith, H.W. and Huggins, M.B., Effectiveness of phages in treating experimental *Escherichia coli* diarrhoea in calves, piglets and lambs. *J Gen Microbiol* 129 (Pt 8): 2659–2675, 1983.

Smith, H.W. and Huggins, M.B., Successful treatment of experimental *Escherichia coli* infections in mice using phage: its general superiority over antibiotics. *J Gen Microbiol* 128: 307–318, 1982.

Smith, H.W., Huggins, M.B. and Shaw, K.M., The control of experimental *Escherichia coli* diarrhoea in calves by means of bacteriophages. *J Gen Microbiol* 133 (Pt 5): 1111–1126, 1987a.

Smith, H.W., Huggins, M.B. and Shaw, K.M., Factors influencing the survival and multiplication of bacteriophages in calves and in their environment. *J Gen Microbiol* 133 (Pt 5): 1127–1135, 1987b.

Soothill, J.S., Bacteriophage prevents destruction of skin grafts by *Pseudomonas aeruginosa*. *Burns* 20: 209–211, 1994.

Soothill, J.S., Treatment of experimental infections of mice with bacteriophages. *J Med Microbiol* 37: 258–261, 1992.

Soothill, J.S., Lawrence, J.C. and Ayliffe, G.A.J., The efficacy of phages in the prevention of the destruction of pig skin in vitro by *Pseudomonas aeruginosa*. *Med Sci Res* 16: 1287–1288, 1988.

Stark, F.R., Thornsvard, C., Flannery, E.P. and Artenstein, M.S., Systemic lysostaphin in man—apparent antimicrobial activity in a neutropenic patient. *N Engl J Med* 291: 239–240, 1974.

Stone, R., Bacteriophage therapy. Food and agriculture: testing grounds for phage therapy. *Science* 298: 730, 2002.

Sulakvelidze, A., Alavidze, Z. and Morris, J.G., Jr., Bacteriophage therapy. *Antimicrob Agents Chemother* 45: 649–659, 2001.

Summers, W.C., "Bacteriophage discovered," pp. 47–59 in *Felix d'Herelle and the Origins of Molecular Biology.* Yale University Press, New Haven, Connecticut, 1999.

Szostak, M., Wanner, G. and Lubitz, W., Recombinant bacterial ghosts as vaccines. *Res Microbiol* 141: 1005–1007, 1990.

Szostak, M.P., Hensel, A., Eko, F.O., Klein, R., Auer, T., Mader, H., Haslberger, A., et al., Bacterial ghosts: non-living candidate vaccines. *J Biotechnol* 44: 161–170, 1996.

Szostak, M.P., Mader, H., Truppe, M., Kamal, M., Eko, F.O., Huter, V., Marchart, J., et al., Bacterial ghosts as multifunctional vaccine particles. *Behring Inst Mitt* 98: 191–196, 1997.

Tartera, C. and Jofre, J., Bacteriophages active against *Bacteroides fragilis* in sewage-polluted waters. *Appl Environ Microbiol* 53: 1632–1637, 1987.

Tartera, C., Lucena, F. and Jofre, J., Human origin of *Bacteroides fragilis* bacteriophages present in the environment. *Appl Environ Microbiol* 55: 2696–2701, 1989.

Taylor, W.I. and Silliker, J.H., "Hatching of eggs," U.S. patent 2,851,006, 1958.

Thomas, L.V., Ingram, R.E., Bevis, H.E., Davies, E.A., Milne, C.F. and Delves-Broughton, J., Effective use of nisin to control *Bacillus* and *Clostridium* spoilage of a pasteurized mashed potato product. *J Food Prot* 65: 1580–1585, 2002.

Thomas, R.C., A bacteriophage in relation to Stewart's disease of corn. *Phytopathology* 25: 371–372, 1935.

Tierney, J.T., Sullivan, R., Larkin, E.P. and Peeler, J.T., Comparison of methods for the recovery of virus inoculated into ground beef. *Appl Microbiol* 26: 497–501, 1973.

Topley, W.W.C. and Wilson, J., Further observations of the role of the Twort-d'Herelle phenomenon in the epidemic spread of murine typhoid. *J Hyg* 24: 295–300, 1925.

Topley, W.W.C., Wilson, J. and Lewis, E.R., Role of Twort-d'Herelle phenomenon in epidemics of mouse typhoid. *J Hyg* 24: 17–36, 1925.

Whichard, J.M., Sriranganathan, N. and Pierson, F.W., Suppression of *Salmonella* growth by wild-type and large-plaque variants of bacteriophage Felix O1 in liquid culture and on chicken frankfurters. *J Food Prot* 66: 220–225, 2003.

Whitman, P.A. and Marshall, R.T., Characterization of two psychrophilic *Pseudomonas* bacteriophages isolated from ground beef. *Appl Microbiol* 22: 463–468, 1971.

Whitman, P.A. and Marshall, R.T., Interaction between streptococcal bacteriophage and milk. *J Dairy Sci* 52: 1368–1371, 1969.

Williams-Campbell, A.M. and Solomon, M.B., Reduction of spoilage microorganisms in fresh beef using hydrodynamic pressure processing. *J Food Prot* 65: 571–574, 2002.

Witte, A., Blasi, U., Halfmann, G., Szostak, M., Wanner, G. and Lubitz, W., Phi X174 protein E–mediated lysis of *Escherichia coli. Biochimie* 72: 191–200, 1990.

Witte, A., Wanner, G., Sulzner, M. and Lubitz, W., Dynamics of PhiX174 protein E–mediated lysis of *Escherichia coli. Arch Microbiol* 157: 381–388, .

Yeung, M.K. and Kozelsky, C.S., Transfection of Actinomyces spp. by genomic DNA of bacteriophages from human dental plaque. *Plasmid* 37: 141–153, 1997.

Young, R., Bacteriophage lysis: Mechanism and regulation. *Microbiol Rev* 56: 430–481, 1992.

14 Bacteriophage Therapy in Humans

Alexander Sulakvelidze[1,2] *and Elizabeth Kutter*[3]
[1]Department of Epidemiology and Preventive Medicine, University of Maryland School of Medicine, Baltimore, MD
[2]Intralytix, Inc., Baltimore, MD
[3]Lab of Phage Biology, Evergreen State College, Olympia, WA

CONTENTS

14.1. Introduction 382
14.2. History of Phage Therapy in the West 383
14.2.1. First Therapeutic Applications 383
14.2.2. The "Bacteriophage Inquiry" 384
14.2.3. Phage Therapy in Europe, Canada, and the United States During the 1920s to 1940s 388
14.3. Factors Contributing to the Decline of Interest in Phage Therapy During Early Phage Therapy Research 393
14.3.1. Credibility Problems Caused by a Lack of Quality Control and of Properly Controlled Studies 393
14.3.2. Conflicting Reports about the Efficacy of Phage Therapy 395
14.3.3. Development of Antimicrobials: Broad-Spectrum Antibiotics vs. Narrow-Spectrum Phages 396
14.4. Phage Therapy in the Era of Antibiotics 397
14.4.1. Phage Therapy in Western Europe, Africa, and the United States 397
14.4.2. Phage Therapy in the Former Soviet Union 399
14.4.3. Phage Therapy in Georgia 406
14.4.4. Phage Therapy Centers in Russia 411
14.4.5. Phage Therapy in Poland 412
14.5. Key Considerations in Therapeutic Applications 413
14.5.1. Mode of Therapeutic Action of Phages 413
14.5.2. Phage Distribution, Pharmacokinetics, and Neutralizing Mechanisms in Mammals 414
14.5.2.1. Phage Movement between Biological Compartments 414
14.5.2.2. Nonspecific Mechanisms that Clear Phages from the Bloodstream 416

0-8493-1336-X/05/$0.00+$1.50
© 2005 by CRC Press LLC

14.5.2.3. The Development of Phage-Neutralizing Antibodies 417
14.5.2.4. Emergence of Bacterial Strains Resistant to Particular Phages 419
14.5.3. Safety Considerations 420
14.5.4. Comparison of Phages and Antibiotics: Advantages and Limitations of Phage-Based Therapies 423
Acknowledgments 426
References 426

14.1. INTRODUCTION

The discovery of bacteriophages is one of the most important milestones in the history of biomedical research—one that has led to many fundamental discoveries and breakthroughs in the life sciences. In the pre-antibiotic era of the early twentieth century, the potential of bacteriophages to be a powerful tool in dealing with infectious diseases of bacterial etiology also captured the imagination of the scientific and non-scientific communities, and it inspired many writers and journalists to write about bacteriophages and their possible therapeutic applications (Ho, 2001). Bacteriophage therapy also served as a leitmotif of *Arrowsmith*, a well-known novel by Sinclair Lewis (1925) that won the Pulitzer Prize. The novel, published during the early days of phage research, told the story of Martin Arrowsmith, a physician who used bacteriophages to deal with an outbreak of bubonic plague on a Caribbean island. However, for a variety of reasons (some discussed later in this chapter), the initially strong interest in phage therapy was short-lived and phage therapy was all but forgotten in the Western world shortly after antibiotics became widely available. The recent and rapidly increasing emergence of multiantibiotic-resistant mutants of pathogenic bacteria has rekindled the interest of the Western scientific community, industry, and general public in this almost century-old approach—as documented by the appearance of a number of phage therapy-related publications in the Western peer-reviewed literature (Alisky et al., 1998; Sulakvelidze et al., 2001; Biswas et al., 2002; Cerveny et al., 2002; Duckworth and Gulig, 2002) and media (Radetsky, 1996; Kuchment, 2001; Martin, 2003), and the formation of new biotechnology firms commercializing phage-based technology in the West (Sivitz, 2002; Thacker, 2003). Phages also recently resurfaced as the "saviors of humankind" in a best-selling novel—*Prey* by Michael Crichton (2002)—in which phages are used to destroy laboratory-escaped "bacterial nanoparticles" threatening life on earth.

Biomedical technology today is very different from what it was in the early days of phage therapy research, and our understanding of biological properties of phages and the basic mechanisms of phage-bacterial host interaction has improved dramatically since the days of early therapeutic uses of bacteriophages. These advances can have a profound impact on the development of safe therapeutic phage preparations having optimal efficacy against their specific bacterial hosts and on designing science-based strategies for integrating phage therapy into our arsenal of tools for preventing and treating bacterial infections. In this context, although bacteriophage therapy can probably be used to prevent or treat most bacterial infections, it is likely that antibiotic-resistant bacteria will become the first "clinical" targets of this reemerging technology. It is also likely that, despite phage therapy's long history in Eastern Europe and the former Soviet Union (and, in the pre-antibiotic era, in the United States, France, and other countries), therapeutic phage preparations will have to undergo a rigorous approval process by various regulatory agencies (e.g., the Food and Drug Administration in the United States) before they can be commercialized and used to treat humans in the Western world. On the other hand, therapeutic phages may find an easier acceptance in nonclinical settings, such as agribusiness and veterinary medicine. For example, they may be used to improve the safety of food products, or to help reduce the use of antibiotics in poultry and other livestock industries. This chapter primarily focuses on therapeutic applications of bacteriophages in humans, and the uses of phages in veterinary medicine and agricultural settings are described in more detail in Chapter 13.

14.2. HISTORY OF PHAGE THERAPY IN THE WEST

14.2.1. First Therapeutic Applications

From the time he first discovered bacteriophages (see Chapter 2), Felix d'Herelle was interested in their relationship to disease and their potential as therapeutic agents. Indeed, shortly after publishing his first paper about bacteriophages (1917), d'Herelle used phages to treat bacterial dysentery. The studies were conducted at the Hôpital des Enfants-Malades in Paris in the summer of 1919 (Summers, 1999), under the clinical supervision of Professor Victor-Henri Hutinel, a prominent French pediatrician and the hospital's chief of pediatrics. Although d'Herelle and members of his family previously had ingested large quantities of phage cultures, a dose 100-fold higher than the therapeutic dose was ingested by d'Herelle, Hutinel, and several hospital interns before initiating this anti-dysentery phage treatment—in what was probably the first formal "phase I human volunteer" phage safety trial ever conducted. One day after ingesting the phage preparation, none of the volunteers reported any side effects, and Hutinel signed off on the first known therapeutic use of bacteriophages in humans. The first patient treated—on August 2, 1919—was a 12-year-old boy hospitalized with severe dysentery (10 to 12 bloody stools per day). After taking pretreatment samples for microbiological analysis, d'Herelle orally administered 2 ml of his most potent anti-dysentery phage preparation. The patient's condition rapidly improved after phage ingestion: he passed three more bloody stools the same afternoon and one non-bloody stool during the night, and all symptoms disappeared by the next morning. Three additional patients with bacterial dysentery were successfully treated shortly afterwards: three brothers (3, 7, and 12 years old) were admitted in grave condition in September 1919, after their sister had died at home from the same illness. Each boy received one dose of the anti-dysentery phage preparation, and all three started to recover within 24 hours (Summers, 1999).

D'Herelle did not immediately publish the results of the Hôpital des Enfants-Malades studies. Instead, the first published report on the use of phages to treat a bacterial disease of humans was a short note by Richard Bruynoghe and Joseph Maisin (1921), who used bacteriophages to treat staphylococcal skin disease in six patients. The bacteriophage preparation (0.5 to 2 ml per injection; no titer was given, since no phage quantitation method was yet available) was injected into and around surgically opened lesions, and the authors reported regression of the infections within 24 to 48 hours. Additional, similarly successful studies to treat local staphylococcal infections with bacteriophages soon followed, among them a paper by André Gratia (1922), a former student of Jules Bordet—the director of the Pasteur Institute in Brussels, a Nobel Prize winner (for the theory of serological reactions), and one of d'Herelle's most outspoken critics who was convinced that the bacteriophage phenomenon was due to enzymes rather than to bacterial viruses, as suggested by d'Herelle. This controversy about the nature of bacteriophages raged until images from electron microscopes finally became available (Ruska, 1940), and it had a strong negative impact on the acceptance of d'Herelle's phage therapy ideas. In addition, there were many therapeutic failures as well as successes, and few topics in the biological sciences have created as much controversy as the debate concerning the nature of phages and their efficacy in treating bacterial infections. Some of the factors involved were explored in depth by Van Helvoort (1992) and others (Sulakvelidze et al., 2001; Summers, 2001); they are also discussed later in this chapter.

14.2.2. The "Bacteriophage Inquiry"

In the early years of his phage work, d'Herelle roamed the world exploring potential applications of phage therapy. One of his most publicized early successes was his 1925 treatment of four bubonic plague patients in Alexandria, Egypt. In 1920, while working at the Pasteur Institute branch in Saigon, d'Herelle had isolated anti-plague phage from rat feces in the village of Bac Lieu during a severe plague outbreak (Summers, 1999). He was eager to study these phages further, but his return to the Pasteur Institute in Paris was greeted by tightened physical conditions, personal conflict, and intellectual controversy. Therefore, he soon transferred to the Institute of Tropical Medicine in Leiden, and later accepted an offer to became a health officer for the League of Nations, to direct the bacteriological laboratory of the Egyptian Quarantine Service (the *Conseil Sanitaire, Maritime et Quarantenaire d'Egypte*), with special responsibility for controlling infectious diseases during the major Muslim pilgrimages, and on ships passing through the Suez Canal.

In July, 1925, three travelers on a ship passing through the Suez Canal were diagnosed with bubonic plague; an additional plague victim was later identified on another ship. Their clinical diagnosis of plague was confirmed by specific antiserum agglutination and guinea pig inoculation tests. All patients were quarantined in the station hospital and immediately treated by d'Herelle via the injection of 0.5 ml of his Bac Lieu "Pestis bacteriophage" preparation into each bubo. All four patients recovered dramatically, and the cases were presented in an article entitled "Essai de traîtement de la peste bubonique par le bactériophage" in the October 21, 1925 issue of *La Presse Médicale* (d'Herelle 1925). The report received much publicity, and d'Herelle was soon invited to conduct a large scale "Bacteriophage Inquiry" in India, in collaboration with Lt. Col. J. Morison, the Acting Director of the Haffkine Institute in Bombay. Although the study initially stipulated that the efficacy of phage treatment for both plague and cholera patients would be examined, it increasingly focused on cholera—to the extent that, by the time the results of the study were published, it was commonly referred to as the "Cholera Study."

The "Cholera Study"—or "Bacteriophage Inquiry" as it also was called—focused both on determining the efficacy of phage administration for the treatment of hospitalized patients and on the prophylactic use of anti-cholera phages in preventing or reducing the incidence of cholera. In the initial hospital studies, conducted in 1927 at the Campbell Hospital in Calcutta under the direction of d'Herelle, Major R. Malone, and Dr. M.N. Lahiri, they first tested 27 patients for cholera phages in carefully controlled fashion. During the study, orogastric administration of phages dramatically prevented cholera-associated fatalities in the hospital; the fatality rate dropped from 27%–30% to zero. Similarly good results were observed during the field trials: prophylactic and therapeutic administration of phages to 74 patients in villages in the Punjab region reduced the mortality rate to 8%, compared to 63% among 124 patients not treated with phages (d'Herelle et al., 1928). The results were very encouraging, and d'Herelle was asked to expand them to other parts of India. However, by that time, he had accepted a faculty position at Yale University, and he moved to the United States in 1928. The "Bacteriophage Inquiry" continued under the direction of Igor Asheshov, a Russian bacteriologist recommended by d'Herelle for the position.

Soon after Asheshov's appointment, anti-cholera phages were used in the first-of-its-kind, large-scale attempt to use phages prophylactically, by reducing or eliminating environmental contamination with *V. cholerae.* Phages were repeatedly poured into wells from which drinking water was obtained during the temporary lodging of pilgrims. The incidence of cholera among approximately 10,000 pilgrims who obtained their drinking water from phage-treated wells was ca. tenfold lower than the incidence of cholera among 70,000 pilgrims in neighboring regions where phages were not used. The outcome of the "environmental decontamination" approach was quite good; however, hospital evaluations conducted by Asheshov's team were not similarly unambiguous. The hospital trials were conducted in the Puri Cholera Hospital, where poor hospital staff cooperation and inaccurate diagnoses were noted by the researchers; in addition, the baseline mortality rate among cholera patients was relatively low (ca. 5%). That low mortality rate made it difficult to observe any statistically significant results with phage treatment, and evidence of a positive impact of phage administration was not obtained during the studies. On the other hand, during subsequent studies in the Patna Medical College Hospital, where the hospital staff was more cooperative, phage administration (together with other therapies) resulted in virtually no patients dying from cholera during the study period (Summers, 1999).

Asheshov gradually shifted the focus of his research from therapeutic phage applications to basic phage biology, and the clinical trials component of the "Bacteriophage Inquiry" continued under the direction of Dr. C.L. Pasricha at the Campbell Hospital. Bacteriophages were administered to patients in one ward, mainly by mouth, but also intravenously in some cases. The patients in the second ward were not treated with phages. Patients in both wards received the usual cholera treatment available at that time (most likely oral calomel and potassium permanganate, as well as intravenous administration of isotonic saline and bicarbonate) (Summers, 1999). Between 1933 and 1935, 1269 patients participated in the study, and approximately half of them were treated with 2-ml doses of phage every 4 hours. The difference in the overall mortality rate between the phage-treated and phage-untreated (control) groups was not significant (17% in the control group vs. 14% in the phage-treated group). However, for the 841 patients in whom the diagnosis of cholera was microbiologically confirmed, the difference was significant: 8% mortality among the phage-treated patients vs. 18% mortality in the control group. Also, the severity of disease in the phage-treated patients was significantly reduced compared to the patients in the control group; e.g., they were less toxic and less dehydrated, and their hospital stay was shorter (Pasricha et al., 1936).

Field trials also continued under the direction of Morison, who had recently been transferred from Bombay to direct the King Edward VII Pasteur Institute and the Medical Research Institute in Shillong in Asian Province (Summers, 1999). The trials primarily focused on Naogaon (Nowgong) and Habiganj villages—two widely separated, but comparable villages in Assam province. Both villages were endemic for cholera, and they had similar incidences of cholera outbreaks before the introduction of phages in 1928. In 1928, Morrison started his trial by distributing limited amounts of phage preparations to the inhabitants of Naogaon village. By the next year, phage availability increased greatly, and Morison started distributing them to many, if not all, village headmen in the vicinity of Naogaon. Cholera outbreaks did not occur in the Naogaon region during 1930–1935 when the "Cholera Study" was terminated. However, in the phage-untreated Habiganj village, there were spring and fall cholera outbreaks in both 1930 and 1931, and 473 cholera-attributed deaths occurred during the first half of 1932 (Fig. 14.1). The implications were staggering, and the Indian government promptly ordered that the anti-cholera phage preparations also should be distributed to the Habiganj region. After initiating phage treatment, the cholera mortality rate in that area dramatically declined to about 10 fatalities/year, and it remained at that level during the last 3 years of the study. However, in the

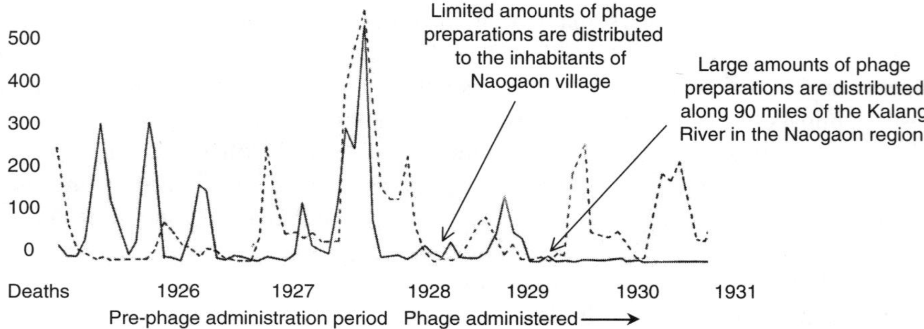

FIGURE 14.1 Mortality from cholera in two districts in Assam province (from data collected during the "Cholera study" in India). The *V. cholerae*-specific phage preparation was widely distributed in the Nowgong district, but it was not generally available in the Habiganj district (dotted line) (adapted from Summers, 1999).

nearby district of Sunanganj, where no phage preparations were administered, more than 1500 people died from cholera in 1933 (Summers, 1999).

The decision to start administering bacteriophages to the inhabitants of Habiganj village in 1932 very likely saved many lives in that region. However, the decision also became one of the main reasons for the controversial conclusion about the efficacy of phage treatment, because it effectively eliminated the control group from the Assam study. Indeed, in their evaluation of the "Cholera Study," the Cholera Advisory Committee concluded that, while the "balance of available evidence. . . indicates that the widespread use of bacteriophage for the control of cholera. . . has certain value," it was "most unfortunate that this trial was not carried out as a strictly controlled experiment and that the use of phage was not confined to the experimental areas only" (Summers, 1999). These lukewarm conclusions about the efficacy of phage treatment had a negative effect on the enthusiasm of funding agencies to continue their support of the Indian studies. Also, the departure of d'Herelle, Asheshov, and Morison—which coincided with rising Indian nationalism and the beginning of World War II—negatively impacted the situation. As a result, the "Bacteriophage Inquiry" was allowed to lapse before the controversy about the therapeutic value of anti-cholera phages was fully resolved. With regard to the "Cholera Clinical Inquiry" at Campbell Hospital, the 1936 Cholera Advisory Committee concluded that "it is recommended that this enquiry be now discontinued, a sufficiently extended series of cases of cholera having been treated with bacteriophage under controlled hospital conditions to provide information as to the value of the method," citing the 2.5-fold decrease in mortality.

In subsequent years (for over a decade), bacteriophage administration was one of the most popular prevention and treatment methods for cholera in India. In 1938, the Shillong Pasteur-Institut produced 400,000 doses of phage for the Bihar region alone, and the demand from other regions continued to rise (Hausler, 2003, p. 103). Duggal is quoted in a WHO cholera review (Duggal, 1949; Pollitzer et al., 1959) as saying that the method "caught the fancy of both the Doctors and laymen so much that the orthodox anti-cholera measures of disinfecting water supplies and anti-cholera inoculations were at the stage of being totally substituted when then the curve of the cholera incidence started rising again in 1937 and reached its culmination in big epidemics of 1944, resulting in 150,000 deaths. . . the medical practitioners do not (now) look upon it as a good anti-cholera measure." Thus, in 1944 the Cholera Advisory Committee of the Indian Research Fund reiterated its 1936 opinion that bacteriophage prophylaxis should not replace the orthodox methods of cholera control, such as vaccination, and it reemphasized that no single approach can be a "magic bullet" to deal with infectious diseases. Relatively more recent studies of anti-*V. cholerae* phages, which were performed in the 1960s and 1970s—in Pakistani hospitals, with Russian, US and WHO involvement—reported that treating cholera patients with the phages indeed shortened the duration of their illness. However, the amount of phages needed to elicit a beneficial effect suggested that they worked mainly by eliciting "lysis from without," rather than replicating in and lysing the etiologic agent in the small intestines. Also, some bacteria seemed to sit in pockets in the small intestines, where phages apparently did not penetrate well (Dutta, 1964; Marcuk et al., 1971; Monsur et al., 1970).

14.2.3. Phage Therapy in Europe, Canada, and the United States During the 1920s to 1940s

Therapeutic phage applications spread rapidly in response to the desperate need for treatments for bacterial disease, the lack of effective alternatives, and the early reports of successful outcomes. The pioneering studies (Bruynoghe and Maisin, 1921; Gratia, 1922) were soon followed by those of Spence et al. (1924), who used orally administered bacteriophage preparations to treat patients infected with Shiga (9 patients) and Flexner (11 patients) dysentery bacilli. Twelve patients treated with "conventional" methods served as the control group. The mortality was 30% lower, and the recovery time was about six days shorter, in the phage-treated group compared to the controls. Other studies also reported on the successful use of phages against dysentery (Burnet et al., 1930; Ridding, 1930; Taylor et al., 1930)

Oral, topical, and subcutaneous phage administration regimens were soon followed by more invasive routes of administration; e.g., phages were injected intravenously and intraspinally in patients with various bacterial infections. One of the first reports in which subcutaneous administration of phages was employed came from investigators in the United States, where the Michigan Department of Health conducted a study examining the efficacy of subcutaneous injection of bacteriophages in treating 208 patients with chronic furunculosis (Larkum, 1929b). Phage treatment apparently was successful: 162 (78%) of the patients did not exhibit recurring infections for at least 6 months after treatment, 19% of the patients showed some improvement, and improvement was not seen in 3% of the patients. Also, Schultz (1932) claimed "remarkable results in several cases" after intravenously injecting anti-staphylococcal phages into ten patients with staphylococcal septicemia. Similarly encouraging results also were reported by Schless (1932) after intraspinal injection of anti-staphylococcal phages. In the latter study, the phages were reported to reduce the severity of staphylococcal meningitis in about 24 hours, and to eliminate *S. aureus* in cerebrospinal fluid.

In the United States and Canada, phages were also used successfully—in the 1940s—to treat typhoid fever, which was refractory to treatment with the antibiotics available at that time. In one study, Walter Ward injected Vi-specific phages (i.e., phages that specifically recognize the Vi antigen, a major virulence factor of *Salmonella* Typhi) into mice challenged with *S.* Typhi, and he found that the mortality rate was 93% and 6% in the phage-untreated control group and the phage-treated group, respectively (Ward, 1942). These phages were subsequently used to treat human typhoid infections (Knouf, 1946). The treatment was effective, and phage administration reduced mortality from 20% in patients treated using more "common" therapeutic approaches, to approximately 5% in phage-treated patients. Also, the condition of the phage-treated survivors rapidly improved, from being largely comatose to full of vigor, with renewed appetite, in 24–48 hours. Shortly afterwards, Desranleau (1949) reported an even more effective use of phages for treating *S.* Typhi infections. The authors used various formulations of Vi-specific phages to treat a total of nearly 100 typhoid patients in the province of Quebec, Canada. One of their most potent formulations contained six phages, and its administration reduced the mortality from 20% to 2%. Numerous additional publications regarding

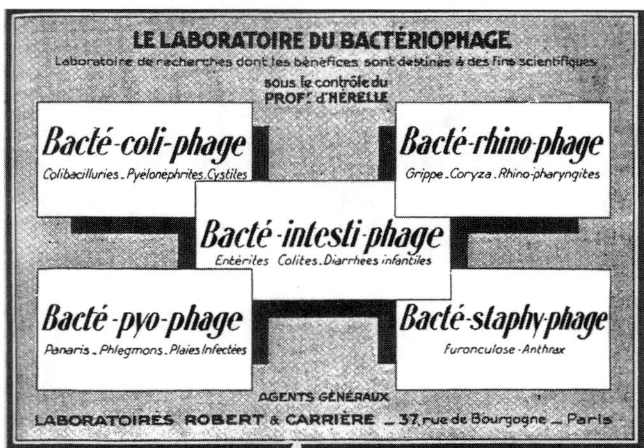

FIGURE 14.2 Advertisement for phage preparations from *Laboratoire du Bactériophage* in Paris, France, 1936. (Photograph from Dr. William Summers' archives)

the therapeutic use of phages have been published in the scientific literature, and some of the early phage therapy studies of that period are summarized in Table 14.1.

Many researchers isolated and grew up their own phages, using bacteria from their patients, but a number of companies also soon began to produce and market phage preparations for treating various diseases. D'Herelle, who had a strong entrepreneurial spirit and a thorough understanding of the special care and challenges involved, was among the pioneers. His commercial laboratory (*Laboratoire du Bactériophage*) in Paris produced several phage-based preparations targeted against various pathogenic bacteria. These preparations included *Bacté-rhino-phage* (for upper respiratory tract infections), *Bacté-intesti-phage* and *Bacté-coli-phage* (for diarrheal diseases), and *Bacté-pyo-phage* and *Bacté-staphy-phage* (for superficial infections) (Fig. 14.2). The laboratory also was apparently developing a preparation against intestinal gas-producing bacteria (*Bacté-gazzi-phage),* but it is not clear whether the preparation was ever fully developed or sold commercially (Summers, 2001).

Therapeutic phage preparations also were produced in the United States, in part by well-known pharmaceutical companies. In the 1930s, Eli Lilly produced at least seven phage products for human use, targeting staphylococci, streptococci, *E. coli,* and other bacterial pathogens. The preparations consisted of sterile-filtered phage-lysed broth cultures of the targeted bacteria (e.g., *Colo-lysate, Ento-lysate, Neiso-lysate,* and *Staphylo-lysate*), or the same preparations in a water-soluble jelly base (e.g., *Colo-jel, Ento-jel,* and *Staphylo-jel*). They were used to treat abscesses, suppurating wounds, vaginitis, mastoiditis, and acute and chronic respiratory tract infections. Other well-known companies involved in commercial therapeutic phage production included E.R. Squibb and Sons and Swan-Myers (a division of Abbott Laboratories) (Straub and Applebaum, 1933).

TABLE 14.1
Some of the Early Human Phage Therapy Studies Performed During 1920–1940

Study (References)	Infections	Comments
Boyce et al. (1933)	Infections of superficial and deep tissues	Bacteriophages were used to treat 200 patients, including patients with furunculosis and carbuncles.
Bruynoghe and Maisin (1921)	Skin infections	Bacteriophages were successfully used to treat *S. aureus* skin infections in six patients. This is the first published report of the therapeutic use of phages in humans.
Burnet et al. (1930)	Dysentery	Anti-dysentery phages were used for prophylaxis/treatment of dysentery.
Couvy (1932)	Plague	21 of 145 plague patients were treated with phages, and the remaining patients were treated with serum. The mortality rate was 33% and 65% in the phage-treated and untreated groups, respectively.
Dalsace et al. (1926)	Skin infections	154 patients with *S. aureus* infections were treated with local injections and wet dressings of bacteriophages. Permanent recovery was observed in ca. 76% of the patients, with chronic infections being the most difficult to treat.
D'Herelle (1925)	Plague	4 plague patients recovered dramatically after injecting bacteriophages into their bubos. This is the first published report of the use of phage therapy against plague. The success outcome was largely responsible for the subsequent initiation of large-scale studies of the efficacy of phages against plague and cholera in India (i.e., "the Bacteriophage Inquiry").
D'Herelle et al. (1928)	Enteric infections	Orogastric administration of phages dramatically reduced the cholera-associated fatality rate from 27-30% to zero, in the Campbell Hospital in Calcutta. During field trials, the prophylactic/therapeutic administration of phages to 74 patients in Punjabi villages reduced the mortality rate to 8%, compared to 63% among the 124 patients not treated with phages.
Kahn (1931)	Skin infections	Polyvalent bacteriophage preparations were used to treat 9 cases of acne and 5 cases of furunculosis. 50 to 90% of the patients improved, and 6 cases showed no significant improvement after phage administration.
Lampert et al. (1935)	Skin infections	Approximately 1,000 patients with wound infections (abscesses, cellulitis, furuncles and carbuncles) were treated with bacteriophages. The overall success rate was reported to be approximately 90%.

Larkum (1929a).	Skin infections	208 patients with *S. aureus*-infected furunculosis were treated by subcutaneous injection of bacteriophages. 162 (78%) of the phage-treated patients showed no recurrent infections for at least 6 months after treatment; appreciable improvement was not seen in 6 patients (3%), and the remaining patients showed some improvement.
Longacre et al. (1940).	Septicemia	Ninety patients septicemic with *S. aureus* participated in the study. Twenty-one of the patients were given small amounts of phage, 35 patients were given larger amounts of phage (what authors considered to be "adequate" amounts), and 54 patients were not given phages at all. The mortality rates in the phage-treated groups were 73%, and 29%, respectively. The combined mortality rate in the two groups treated with phages (low and high doses) was 47%, compared to 81% in phage-untreated, 3rd group.
MacNeal and Frisbee (1932)	Urinary infections	Subcutaneous injections of phage, together with irrigation of bladder and renal pelvis, were reported to be the best for treating urinary *S. aureus* infections. Other administration regimens included oral phage administration and intramuscular injections.
Melnik et al. (1935)	Bacterial dysentery	Large group of children were given phages prophylactically. Children in group # 1 were given 1 ml of phage (7 doses in 10 day intervals), children in group # 2 received the same regimen plus ox bile, and children in group # 3 received no treatment. The dysentery morbidity rates in the groups were 1.44, 0.15, and 6.38 per 100 children, respectively.
Mikeladze et al. (1936)	Bacterial dysentery	The mortality rate among phage-treated 21 dysentery patients was ca. 5%; whereas, the mortality rate in 63 patients receiving symptomatic treatment (but no phage) was ca. 15%.
Morrison and Gardner (1936)	Lung infection	A patient with *B. coli* lung infection was treated with installation of 30 ml of bacteriophage. The necrotic character of infection was gone in 24 h after phage administration, and the patient made uneventful recovery.
Rice (1930)	Suppurative infections	Bacteriophage filtrates were used to treat suppurative conditions in 300 patients. Best results where observed when phages were repeatedly injected into unopened abscesses with a fine needle. Many of the cases were previously treated with other therapeutic measures without much success.
Ridding (1930)	Bacterial dysentery	Anti-dysentery phages were used to treat acute bacillary dysentery in Khartoum province, Sudan.
Ruddell et al. (1933)	Surgical infections	Bacteriophages were used to treat 27 patients with ruptured appendix and having local or general peritonitis. Phage treatment was not successful in 3 of the patients all of whom died from peritonitis. The remaining 24 patients all recovered at various rates, depending on the severity of their illness.

(*Continued*)

TABLE 14.1
(Continued)

Study (References)	Infections	Comments
Schless (1932)	Bacterial meningitis	Intraspinal administration of bacteriophages eliminated *S. aureus* infection in spinal fluid cultures, and alleviated toxic syndromes in 24 hours.
Smith (1924)	Typhoid fever	Seven cases of typhoid fever were treated with bacteriophages. In five cases, improvement in symptoms was observed, and the remaining two cases showed no improvement. Clinical improvements were associated with negative blood cultures.
Spence and McKinley (1924)	Bacterial dysentery	Phages were administrated orally to 20 patients infected with Shiga (9 patients) and Flexner (11 patients) bacillus. Twelve patients, not treated with bacteriophage, but treated with the "conventional" methods, served as control group during the study. 30% less mortality was reported in phage-treated patients compared to phage-untreated patients.
Stout (1933)	Skin infections	Phage administration was reported to be successful in the treatment of 85% of simple boils, 75% of generalized furunculosis, and 60% of chronic furunculosis. Inconsistent results were observed during the treatment of carbuncles. Number of patients per disease was not delineated, but the whole study involved approximately 1,500 cases.

The successful outcomes of early therapeutic phage applications and a possibility of their commercialization led to dozens of entrepreneurs entering the phage therapy field, with much enthusiasm but little understanding of phage biology or bacterial pathogenesis. Not surprisingly, many of the early phage therapy trials were unsuccessful. As discussed below, in retrospect, there were a number of probable explanations for the failures.

14.3. FACTORS CONTRIBUTING TO THE DECLINE OF INTEREST IN PHAGE THERAPY DURING EARLY PHAGE THERAPY RESEARCH

14.3.1. Credibility Problems Caused by a Lack of Quality Control and of Properly Controlled Studies

One of the most important factors that contributed to the decline of interest in phage therapy in the Western world was a credibility problem. A paucity of appropriately conducted studies and the lack of well-established and standardized testing protocols, interfered with rigorously documenting the value of phage therapy. In addition, several companies started producing phages commercially and, in their eagerness to boost profits, some made exaggerated claims concerning the effectiveness of their products. For example, a preparation called *Enterophagos* was marketed as being effective against herpes infections, urticaria, and eczema (Barrow and Soothill, 1997)—conditions against which phages could not possibly be effective. Such factors discredited phage therapy in the eyes of several well-recognized and widely respected researchers of that time—which, in turn, had a far-reaching, negative impact on the further development and evaluation of phage therapy in the West.

Many production-related problems also complicated early phage therapy research. Before the availability of spray-dryers, freeze-dryers, refrigerators and freezers, the viability of phage preparations was a concern; phage titers could rapidly decline and render the preparations ineffective. Thus, several "stabilizers" and "preservatives" were used in an attempt to increase the viability of phages. However, in the absence of a good understanding of the biological nature of phages and their stability to various physical and chemical agents, many of the ingredients added to prolong phage shelf-life (e.g., phenol, which was included in some of the early therapeutic phage preparations) actually had a detrimental effect on phage viability and were toxic for humans. Indeed, when Straub et al. (1933) analyzed several lots of commercial phage preparations manufactured by Eli Lilly & Co, Swan-Myers, and E.R. Squibb & Sons in 1933, they found that several of the preparations had very low activity titers, presumably because they contained additives negatively affecting phage viability. Similar observations were noted by d'Herelle, who found that some commercial preparations did not contain any biologically active phages (Summers, 2001). The poor results obtained with nonviable preparations (or preparations containing viable phages in low concentrations) were likely to have contributed to skepticism concerning the possible value of phage therapy.

Many other phage production-related problems complicated early phage therapy research. For example, in an early effort to increase the target range of therapeutic phage preparations, mixtures of phages lytic against several bacterial pathogens were commercially prepared and manufactured. However, when Max Delbrück analyzed one of those polyvalent products, he found that it only contained one phage strain. Apparently, in an effort to streamline phage production, the laboratory personnel mixed together all of the phage strains and attempted to grow them as a cocktail. Not surprisingly, one of the phage strains outgrew the others, and the final preparation contained a predominant amount of that phage rather than the several phages it was supposed to contain. Such a preparation would not have been optimally effective against all of the multiple bacterial pathogens it was claimed to be active against, and its use probably would have yielded disappointing results and increased skepticism regarding the value of phage therapy. Interestingly, the "superphage" that Delbrück isolated from that "polyvalent" phage mixture has become quite famous in the phage community. It is currently known as T7, a member of the well-known T-phage family (Summers, 2001).

Another phage production-related problem was the purity of commercial phage preparations. Therapeutic phage preparations, including those from well-established companies in the United States and elsewhere, consisted of crude lysates of host bacteria, and thus contained bacterial components (including endotoxins) that may have had adverse effects on patients undergoing treatment, particularly if they were injected intravenously. Some examples of adverse reactions during phage therapy are found in the scientific literature of the 1920s and 1930s. For example, 43% of patients that received phage injections against staphylococcal infections were reported to have side effects ranging from mild to severe (Larkum, 1929b), and similar observations were described by Cipollaro (1932) in patients whose pyodermas were treated with phages. Interestingly, in many studies where phage therapy was found to cause some adverse effects, the efficacy of the treatments was still reported to be quite good. For example, phage therapy was reported by Cipollaro and Sheplar to have a positive effect in 79% of 67 patients suffering from furunculosis and carbuncles. Also, Larkum (1929b) claimed that phage therapy was effective in 78% of his 208 phage-treated patients.

A study by Naidu (1932) is sometimes cited as an example of possible ill effects from endotoxins. They reported that daily injections (i.v. and into the bubo, 2 times/day for 2 days) of a phage-produced *Y. pestis* lysate processed through a Chamberland filter candle *worsened* the clinical outcome; i.e., the percentage of plague patients surviving the disease after phage treatment was less than the percentage of patients recovering after treatment with serum. However, the study did not explain how patients were selected into the phage-treated vs. serum-treated groups, and the authors did not comparatively analyze specific side effects observed in the two treatment groups. Also, statistical analysis of the data was not performed, and it is not clear whether the observed outcome was statistically significant. Finally, the phage preparations apparently were not examined for the endotoxin contamination prior using them in humans—and they were likely to contain high levels of *Y. pestis* endotoxin, which may have contributed to the observed worsening of the clinical symptoms in phage-treated plague patients.

In retrospect, it is difficult to know what proportion, if any, of the adverse symptoms occasionally associated with the use of therapeutic phages was due to phage treatment (e.g., due to the phage-mediated bacterial lysis or impurities in phage preparations) or was a natural manifestation of the underlying clinical condition. At any rate, significant problems appear to have been very rare. Interestingly, residual bacterial antigens may potentially support the healing process by stimulating the host's immune system. In fact, the primary therapeutic effect of "staph phage lysate," a preparation currently marketed in the United States (for veterinary applications), has been proposed to be due to immune stimulation elicited by *S. aureus* antigens, rather than to the lytic activity of the *S. aureus*-specific bacteriophages it contains (see section 14.4.1, and also Chapter 13).

14.3.2. Conflicting Reports about the Efficacy of Phage Therapy

As noted above, few topics in the biological sciences have created as much controversy as the debate concerning the nature of phages and their efficacy in treating bacterial infections. To address the controversy surrounding the therapeutic value of phages, the Council on Pharmacy and Chemistry of the American Medical Association commissioned Stanhope Bayne-Jones, a respected physician and bacteriologist, and Monroe Eaton, a prominent physician specializing in infectious diseases, to prepare a comprehensive review of the available literature. They reviewed more than 100 papers on bacteriophage therapy, and in 1934 they published a detailed review (Eaton and Bayne-Jones, 1934). The report was one of the most detailed reviews of phage therapy ever published, and its conclusions were clearly not in favor of phage therapy. Interestingly, one of the key conclusions of the report was that the bacteriophage was not a virus capable of parasitizing bacteria: "d'Herelle's theory that the material is a living virus parasite of bacteria has not been proved. On the contrary, the facts appear to indicate that the material is inanimate, possibly an enzyme." This conclusion was extended to say that "since it has not been shown conclusively that bacteriophage is a living organism, it is unwarranted to attribute its effect on cultures of bacteria or its possible therapeutic action to a vital property of the substance." At the present time, it is clear that the above conclusions of the Eaton and Bayne-Jones report were incorrect, and, in retrospect, one might understand how the authors' disbelief in the animate nature of bacteriophages led to their negative conclusions concerning the efficacy of phage-based therapy.

The Eaton-Bayne-Jones report was generally well written, authored by two prominent researchers of their time, and sponsored by the AMA. Therefore, it had a dramatically negative impact on shaping the opinions about phage therapy of many physicians and researchers in the West. Indeed, 7 years after the Eaton-Bayne-Jones report, Krueger and Scribner (1941) published a sequel to the Eaton-Bayne-Jones report, and they closely followed the mindset and conclusions of their predecessors. They also rejected the idea that phages are bacterial viruses, and they boldly stated that the "nature of bacteriophage is no longer in question. It is a protein of high molecular weight and appears to be formed from a precursor originating within the bacterium." They further concluded that "it is equally evident that phage solutions possess no measurable degree of superiority over well known and accepted preparations." This apparent confirmation of the Eaton-Bayne-Jones report shortly before the start of the "antibiotic era" contributed significantly to further decline of interest in phage therapy in the United States. The decline of interest in phages as therapeutic agents was not affected by the continuing studies in the former Soviet Union and Eastern Europe, since few of those studies were available to the international scientific community. Also, the great popularity of phage therapy in the former Soviet Union may have further contributed to the decline of phage therapy research in the West, because phage therapy was by then associated with "Soviet medicine," with all its negative political connotations.

Some other potentially counterbalancing factors were also at work at the time, as discussed by Häusler (2003). In 1942, both the Lancet and the British Medical Journal published editorials about the apparently successful use of an anti-dysentery phage preparation by the Soviet military in the Middle East and Far East. Indeed, both the Soviet and German armies were actively using phage preparations to prevent and treat various infectious diseases among their troops. In the United States, the National Research Council's Committee on Medical Research (NRC/CMR) approached Morris Rakieten, a close associate of d'Herelle's during his appointment at Yale, to discuss the possibility of using phages to deal with bacterial infections occurring among U.S. troops. By November of 1942, the NRC was supporting anti-dysentery phage research by several top U.S. bacteriologists, including Rene Dubos at Harvard, Arthur Schade, and Leona Caroline at the Overly Biochemical Research Foundation in New York, and Morton and Perez Otero at the University of Pennsylvania. In striking similarity with many phage therapy-related studies conducted in the former Soviet Union, these investigators were also working under tight security conditions. However, several publications still ensued as a result of their studies—some of which are reviewed below and in Chapter 13.

14.3.3. Development of Antimicrobials: Broad-Spectrum Antibiotics vs. Narrow-Spectrum Phages

Several years before bacteriophages were discovered, arsenic compound 606 (trade named Salvarsan) was claimed, by Paul Erlich, to be the first "magic bullet" against infectious diseases (Pizzi, 2000). Salvarsan was effective in treating several bacterial infections, but its usefulness was limited because of its toxicity. Several other chemicals used to treat bacterial infections in those years were similarly toxic or had limited efficacy in killing bacteria. Therefore, the initial reports about the discovery of bacteriophages were met with great enthusiasm, and hopes were raised that infectious diseases were finally controllable. Indeed—and as described throughout this chapter—phages were used therapeutically almost immediately after their discovery, and numerous reports of their use in clinical settings were published in the 1920s. Then a paper entitled "On the antibacterial action of cultures of *Penicillium*, with special reference to their use in silation of *H. influenzae*" was published in the *British Journal of Experimental Pathology* by Alexander Fleming (1929), a Scottish physician/researcher and the discoverer of lysozyme. This publication started a chain of events that eventually led to the discovery of penicillin and to the extensive use of many different antibiotics (i.e., to the "antibiotic era")—which, in turn, had a strong negative impact on phage therapy research in the West. Interestingly and ironically, initial studies with penicillin were not actively pursued—in part because of bacteriophages. Indeed, the belief in the curative effect of bacteriophages was so strong at that time that George Dreyer, professor and chairman of the Sir William Dunn School of Pathology at Oxford University, thought that the bacteriolytic effect of penicillin was due to a bacteriophage—and Dreyer abandoned his research of penicillin after he established that it was not a bacteriophage (Friedman and Friedland, 1998). Also noteworthy, it was in the very same department that Howard Florey (who assumed the chairmanship of the department after Dreyer's death) reinitiated studies on penicillin several years later, which eventually led to the famous *Lancet* publication (Chain et al., 1940) that documented the therapeutic properties of penicillin and started the "antibiotic era."

Antibiotics could be used against various bacterial pathogens, whereas bacteriophages were only effective against limited (and specific) targets. Therefore, although phage therapy could be used to kill specific bacterial pathogens, it was also much more likely (than was treatment with broad-spectrum antibiotics) to fail if the etiologic agent had not been accurately determined or if the etiologic agent's *in vitro* susceptibility to the phage had not been properly established. The broad specificity of antibiotics often made them more effective in the absence of a microbiologically confirmed diagnosis, which favored their use in clinical settings. As a result, Western physicians soon lost interest in therapeutic phages, especially since many more effective antibiotics were discovered and marketed. Also, Western phage researchers turned their attention from medicine to studies of the fundamental nature of phages. In the process, they laid the foundations of molecular biology, as discussed in Chapter 2.

14.4. PHAGE THERAPY IN THE ERA OF ANTIBIOTICS

14.4.1. Phage Therapy in Western Europe, Africa, and the United States

Despite the apparent miracle of antibiotics, some phage-based prophylactic and therapeutic preparations continued to be produced commercially in Western Europe, Africa, and the United States during the 1950s and 1960s. For example, *P. aeruginosa*-specific phages were used, in Egypt, to treat burn wound sepsis and post-burn infections of humans during the 1980s and early 1990s (Abdul-Hassan et al., 1990). Also, Saphal (a small pharmaceutical firm located in Vevey, Switzerland) produced several phage-based preparations, called Coliphagine, Intestiphagine, Pyophagine, and Staphagine in drinkable and injectable forms, salves, and sprays. The preparations were licensed in Switzerland, and some medical insurance plans paid for their use (Häusler, 2003).

In France, the Pasteur Institute produced therapeutic phages into the 1970s. J.F. Vieu (1975) tabulated the 476 phages isolated and prepared by the Bacteriophage Service at the Pasteur Institute from 1969 through June 1974. Most of the preparations targeted *Pseudomonas, Staphylococcus, E. coli*, and *Serratia*, and they were used to treat various infections, including skin infections (e.g., furunculosis and carbuncles), septicemia, osteomyelitis, wound infections (e.g., after burns and surgery), pyelonephritis and urinary tract infections, and middle ear and sinus infections. In some cases, large quantities of phage mixtures targeting specific *E. coli* and *Salmonella* strains were used to decontaminate hospital units. Both stock phages (from the main collection) and phages specially adapted to a specific patient or hospital ward were employed. Also, Lang (1979) treated seven septicemic patients with bacteriophages during the 1970s, and they reported that the outcome was good in five cases, fair in one case, and unsuccessful in one case.

In the United States, therapeutic phage production was essentially nonexistent after antibiotics became widely available. However, phages were used in this country to prepare animal and human vaccines, at least some of which contained high titers of viable bacteriophages. As described in Chapter 13, one such preparation ("Staph phage lysate," or SPL) is still extensively marketed for veterinary applications by Delmont Laboratories; the preparation also was used therapeutically in humans from the 1950s to the 1990s. Safety trials of SPL were completed in 1959, and it was licensed for human therapeutic use shortly afterwards by the National Institutes of Health (Salmon and Symonds, 1963; Mudd, 1971). The preparation was administered to patients by a variety of routes, including (i) intranasal aerosol (0.2–0.5 ml, at 1 to 7 day intervals), (ii) topical applications and instillations (up to 1 ml/dose, with administration intervals ranging from a few hours to a few days), (iii) orally (0.1 ml for infants, and up to 1 ml for adults, once every 12 h), (iv) subcutaneously (0.05 ml to 0.5 ml, once every few hours or every few days), and (v) occasionally intravenously (not to exceed 0.1 ml, once every 0.5 h to 12 h) (Baker, 1963). The preparation was apparently very safe; i.e., only a few minor side-effects (e.g., minimal local erythema and swelling) were observed after administering more than 35,000 doses of SPL over a period of 12 years.

The safety and efficacy of SPL was documented (Salmon and Symonds, 1963) after using SPL to treat 607 cases of chronic staphylococcal infections in the United States during the late 1950s to early 1960s (the youngest a 1-week-old infant who had a breast abscess). Most of the patients treated with SPL had failed to respond to treatments with various antibiotics (tetracycline, penicillin, and chloromycetin), and a sulfonamide. The SPL doses and dosing regimens used were, in general, the same as those used by Baker (1963). The results were reported to be quite good: 80% of the patients were said to be "recovered," 18% were "improved," and 2% were "unchanged". Anaphylactic reactions were not observed in any of the patients, and, with the exception of a few patients who experienced "temporary discomfort," SPL was concluded not to have elicited side effects in the patients. Several additional studies in which SPL was used to treat patients with furunculosis, pustular acne, pyoderma, eczema, bronchial asthma, recurrent inflammation of Bartholin's gland, and upper respiratory infections have been published (Mudd, 1971; Kress et al., 1981; Dean et al., 1975; Buda et al., 1968; Vymola and Buda, 1977). Also, at least one double blind, randomized trial using SPL was conducted with 31 patients with chromic recurrent hidradenitis suppurativa (Angel et al., 1987). The outcome was evaluated based on the results of a complete blood count, determination of erythrocyte sedimentation rate, and biweekly bacteriological culturing of specimens from the lesions (the study's duration was 20 weeks). The treatment was reported to have produced "improvement" in 83% of the patients, compared to 13% of the patients in the placebo group treated with the bacteriologic broth in which the host bacteria used to prepare the SPL were grown ($P < 0.0001$).

The aforementioned reports have three common features: (i) they suggest that SPL is safe for humans and is efficacious in treating various infections caused by *S. aureus*, (ii) they were published in somewhat difficult-to-obtain journals and, thus, they are not widely known to the scientific community, and (iii) the authors acknowledged that the mechanisms by which SPL exercised its beneficial effect were not well delineated, but they were thought to involve nonspecific and specific immune stimulation (i.e., the preparation's therapeutic effect was attributed to its immunostimulating activity rather than to the direct antibacterial action of the *S. aureus*-specific phages in it). In this context, early studies by Dean et al. (1975) suggested that SPL stimulated lymphoproliferative responses in both T and B cell subpopulations. The results of more recent studies also support the idea that SPL stimulates antibody production; e.g., it has been reported to stimulate immunoglobulin production in human peripheral lymphocyte cultures (Lee et al., 1985b), to activate polyclonal B cells in mouse spleen cell cultures (Lee et al., 1982), and to elicit polyclonal antibody secretion by mouse spleen cells (Lee et al., 1985a).

None of the studies mention the presence of active phage in the lysate; the mindset was simply quite different and phage concentrations were not checked. However, the Kutter lab has tested SPL samples from several parts of the country and confirmed that they contained active staph phage at concentrations of 10^8 to 10^9 per ml, as might have been expected, and the probability must be considered that the active phage themselves played a major role in the observed therapeutic effects. Additional rigorously designed studies are required to determine the value of using the "phage lysate approach" to prevent and treat various bacterial infections, and to improve our understanding of the mechanisms by which phage-generated bacterial lysates may protect against bacterial infections. In the meantime, production of SPL for human therapeutic applications was suspended by Delmont Laboratories in the 1990s, and it is currently available only for veterinary applications.

The above examples show that, during the antibiotic era, phage therapy was occasionally utilized, on a small scale, in a handful of countries around the world. However, the situation was very different in the former Soviet Union and the countries of Eastern Europe, where therapeutic phage production reached its apogee in the 1960s–1980s. Although several institutions in those counties were involved in therapeutic phage research and production, the best known were centered at the Eliava Institute of Bacteriophage, Microbiology, and Virology (EIBMV) in Tbilisi, Republic of Georgia and, on a much smaller scale, at the Hirszfeld Institute of Immunology and Experimental Therapy in Wroclaw, Poland.

14.4.2. Phage Therapy in the Former Soviet Union

In the Soviet Union, phage therapy became deeply entrenched during and shortly after World War II; for political and economic reasons, phage therapy was extensively exploited for both diarrheal disease and the treatment of traumatic infections even after antibiotics became widely available. Therapeutic bacteriophages have been administered to humans by many routes, including (i) orally (in liquid formulations, or in special tablets designed to protect the phages from gastric juice (Meshalova et al., 1969; Orlova and Garnova, 1970) and rectally (via enemas), to treat gastrointestinal and other infections, (ii) topically (in creams, rinses, and tampons), to treat infections of the skin, eyes, ears, and nasal and vaginal mucosa, etc., (iii) via aerosols or intrapleural injections, to treat respiratory tract infections, and on rare occasions (iv) intravenously, to treat septicemias (Table 14.2). Therapeutic bacteriophages also were extensively used in newborns and small children, as reviewed by Samsygina (1984).

Development of phage preparations against dysentery was one of the top priorities of the Soviet therapeutic phage research (Krestovikova, 1947). A major focus was on the production of a stable, high-titer, anti-dysentery phage preparation in tablet form. The results of initial studies with phage-containing lysates obtained from bacterial broth cultures did not reproducibly yield phage preparations whose output was sufficiently high, and whose cost was sufficiently low, to make their production commercially feasible. However, an improved method was developed by Sergienko (1945), which involved (i) producing phage-containing lysates of bacteria growing on thin layers of a solid, 1%–2% agar-containing medium, (ii) killing the remaining viable bacteria by exposure to chloroform fumes (18 hours under 0.3 atmospheres), (iii) harvesting the chloroform-treated, phage-containing layer, and mixing it with starch and horse serum, and (iv) manufacturing tablets containing the mixture. The approach yielded crude phage preparations that were stable and had high titers. Animal experiments revealed no toxic side effects from the residual bacterial debris in the tablets, even after the tablets were dissolved and injected. Instead, enhanced immunization against the bacteria was observed.

The Alma-Ata branch of the Central Institute of Epidemiology and Microbiology used the aforementioned approach to produce more than 4 million tablets (equivalent to more than 40,000 liters of phage-containing bacterial lysates). These preparations were used in large-scale clinical trials shortly afterwards, when the Kazakh Institute of Epidemiology and Microbiology administered 1–3 tablets 3 times daily at 5-day intervals to 13,913 people, and compared the results with those obtained with a control group of 12,690 untreated people (Sergienko, 1945). Phage administration appeared to slightly reduce the incidence of dysentery in all study participants: Approximately 0.17% of persons in the treated group developed the disease, compared to 1.5% in the control group. The treatment worked best in children under the age of 9; in this group of persons, 0.3% vs. 5.2% developed the disease in phage-treated and phage-untreated groups, respectively (Table 14.3). Commercial production was quickly organized in Moscow, Sverdlovsk, Stalingrad, Tashkent, and other cities, and one million tablets were prepared in Stalingrad (now Volgograd) alone during the first quarter of 1943. This method of production was also cheaper than the traditional liquid approach, and it reportedly saved 3.5 million rubles in 1943 at that facility (Sergienko, 1945).

Another major target of phage therapy in the former Soviet Union was gas gangrene, and the wartime progress in that area was documented by Zhuravlev (1945). For example, highly potent phage preparations against *Clostridium perfringens* were isolated and found to be efficacious in preventing and treating experimentally induced gas gangrene in mice and guinea pigs. In addition, several surgeons claimed that they were able to prevent clostridial wound infections in humans if

TABLE 14.2
Some of the Major Human Phage Therapy Studies Performed in Poland and the Former Soviet Union

Study (References)	Etiologic Agents	Comments
Anpilov and Prokudin (1984)	*Shigella*	The study was conducted in a double-blinded manner. The incidence of dysentery in the phage-treated groups was reported to be ca. 10-fold less than that occurring in the control, phage-untreated group.
Babalova et al. (1968)	*Shigella*	17,044 children were treated with phage preparations vs. 13,725 children in control group. Based on clinical diagnosis, the incidence of dysentery was 3.8-fold higher in the placebo group than in the phage-treated group (6.7 and 1.76 per 1,000 children, respectively). Based on the culture-confirmed cases, the incidence of dysentery was 2.6-fold higher in the placebo group than in the phage-treated group (1.82 and 0.7, respectively).
Bogovazova et al. (1992)	*K. ozaenae, K. rhinoscleromatis,* and *K. pneumoniae*	Phages were tested for adverse side effects in laboratory animals. They were found to be non-toxic and non-allergic. The phages were subsequently used to successfully treat *Klebsiella* infections in all of the 109 patients examined during the study.
Cislo et al. (1987)	*Pseudomonas, Staphylococcus, Klebsiella, Proteus,* and *E. coli*	31 patients with chronically-infected skin ulcers were treated orally and locally with phages. The success rate was 74%. Minor side-effects (eczema, pain, or vomiting) were observed in 6 patients.
Ioseliani et al. (1980)	*Staphylococcus, Streptococcus, E. coli, and Proteus.*	Phages were successfully used, together with antibiotics, to treat lung and pleural infections in 45 patients.
Kochetkova et al. (1989)	*Staphylococcus* and *Pseudomonas*	131 cancer patients with postsurgical wound infections participated in the study. Of these, 65 patients received phages, alone or in combination with antibiotics, and the rest received antibiotics. Phage treatment was successful in 82% of the cases, and antibiotic treatment was successful in 61% of the cases.
Kucharewicz-Krukowska et al. (1987)	*Staphylococcus, Klebsiella, E. coli, Pseudomonas,* and *Proteus*	The immunogenicity of therapeutic phages was analyzed in 57 patients. The authors concluded that the phages' immunogenicity did not impede therapy.
Kwarcinski et al. (1994)	*E. coli*	Recurrent subphrenic abscess (after stomach resection) caused by an antibiotic-resistant strain of *E. coli* was successfully treated with phages.

(Continued)

TABLE 14.2
(Continued)

Study (References)	Etiologic Agents	Comments
Lazareva et al. (2001)	*Staphylococcus, Streptococcus,* and *Proteus*	Fifty-four patients (of the 94 patients examined) were treated with bacteriophages. Phage treatment was associated with fewer septic complications and better temperature normalization compared to no phage treatment. Microbiologically, phage treatment elicited a 2-fold reduction in the number of staphylococci and streptococci, and a 1.5-fold reduction in the number of *Proteus* recovered from burn wounds.
Litvinova et al. (1978)	*E. coli and Proteus*	Phages were successfully used, together with bifidobacteria, to treat antibiotic-associated dysbacteriosis in 500 low-birth-weight (premature) infants.
Markoishvili et al. (2002)	*Pseudomonas, Staphylococcus, Proteus,* and *E. coli*	The wounds/ulcers healed completely in 67 (70%) of 96 patients whose wounds were covered with a phage-containing biodegradable matrix. In 22 cases in which microbiologic data were available, healing was associated with the concomitant elimination of, or a reduction in, specific pathogenic bacteria in the ulcers.
Martynova et al. (1984)	*S. aureus* and *P. aeruginosa*	*S. aureus*- and *P. aeruginosa*-specific phages were used prophylactically (in forms of mouth rinses) 2 times/day for 3-5 days in 27 patients. Control group consisted of 10 healthy individuals. Phage treatment was associated with normalization of microflora in infected sites, and it also stimulated the production of secretory immunoglobulins (IgA in particular).
Meladze et al. (1982)	*Staphylococcus*	Phages were used to treat 223 patients with infections of the lung parenchyma and pleura, and the results were compared to those of 117 patients treated with antibiotics. Full recovery was observed in 82% of the patients in the phage-treated group, as opposed to 64% of the patients in the antibiotic-treated group.
Miliutina and Vorotyntseva (1993)	*Shigella* and *Salmonella*	The efficacy of treating salmonellosis using phages and a combination of phages and antibiotics was examined. The combination of phages and antibiotics was reported to be effective in treating cases where antibiotics alone were ineffective.
Perepanova et al. (1995)	*Staphylococcus, E. coli,* and *Proteus*	Adapted phages were used to treat acute and chronic urogenital inflammation in 46 patients. The efficacy of phage treatment was 92% (marked clinical improvements) and 84% (bacteriological clearance).

Sakandelidze et al. (1974)	*Staphylococcus, Streptococcus,* and *Proteus*	Phages administered subcutaneously or through surgical drains in 236 patients with antibiotic-resistant infections eliminated the infections in 92% of the patients.
Sakandelidze et al. (1991)	*Staphylococcus, Streptococcus, E. coli, Proteus,* enterococci, and *P. aeruginosa*	1,380 patients with infectious allergoses were treated with phages (360 patients), antibiotics (404 patients), or a combination of phages and antibiotics (576 patients). Clinical improvement was observed in 86, 48 and 83% of the cases, respectively.
Slopek et al. (1987)	*Staphylococcus, Pseudomonas, E. coli, Klebsiella,* and *Salmonella*	550 patients were treated with phages. The overall success rate of phage treatment was 92%.
Stroj et al. (1999)	*K. pneumoniae*	Orally administered phages were used successfully to treat meningitis in a newborn, after antibiotic therapy had failed.
Tolkacheva et al. (1981)	*E. coli* and *Proteus*	A combination of phages and bifidobacteria was used to treat dysentery in 59 immunosuppressed leukemia patients. The superiority of treatment with phage-bifidobacteria was reported.
Zhukov-Verezhnikov et al. (1978)	*Staphylococcus, Streptococcus, E. coli,* and *Proteus*	The superiority of treatment with adapted phages (phages selected against bacterial strains isolated from individual patients), compared to commercial phage preparations, was reported in 60 patients with suppurative infections.
Weber-Dabrowska et al. (1987)	*Staphylococcus* and various gram-negative bacteria	Study examined the persistence and distribution of orally-administered phages in 2 healthy volunteers and 56 patients with suppurative infections. On day 10 after phage administration, phages were recovered from 47 of the 56 blood samples, and 9 of the 26 urine samples examined.
Weber-Dabrowska et al. (2001)	*S. aureus, P. aeruginosa, Klebsiella,* and *E. coli.*	20 cancer patients refractory to treatment with commonly available antibiotics were treated by oral administration of specific phages (3 doses/day, for 2 to 9 weeks). Some patients were also treated by local administration of phages. Complete healing of the local lesions and termination of the suppurative process was reported for all phage-treated patients. Side effects were not observed in any of the patients subjected to phage treatment.

TABLE 14.3
The Incidence of Dysentery Among Persons Treated with Dry Bacteriophage and Among Untreated Controls (From Sergienko, 1945)

	Phage-Treated Group		Phage-Untreated Control Group	
Age (Years)	No. of Persons	No. of Persons who Developed the Disease	No. of Persons	No. of Persons who Developed the Disease
0.5–1	163	1	254	41
1–2	561	6	604	45
3–4	689	1	696	35
5–7	965	–	816	16
8–9	601	–	436	8
10–14	1,267	1	1,271	4
15–19	1,369	–	1,347	10
20–29	2,296	–	1,966	11
30–39	2,419	3	2,026	9
40–49	1,708	3	1,441	8
50 and older	1,875	8	1,833	2
Total	13,913	23	1,2690	189

they treated the wounds with the phage preparations within the first 24 hours post-wounding. Subsequently, teams of physicians went to the front lines in order to evaluate the prophylactic efficacy of the anti-gangrene phages under battlefield conditions. They used a cocktail containing lytic phages against *C. perfringens, C. oedematiens, C. histolyticum, C. putrifucus,* and "*Vibrio septique*" (now known as *C. septicum*), coupled with a second cocktail against aerobic bacteria (staphylococci and streptococci) capable of producing severe wound infections. Approximately 10,000 wounded soldiers on the Finnish front participated in the studies, and the number of deaths among the phage-treated soldiers decreased by approximately 66%. In another study, phage administration reduced mortality by approximately 19% (767 soldiers received the phage preparations) in comparison with 42% of those receiving only the conventional therapies (Krestovikova, 1947).

Shortly after World War II, a center for phage research was started in Gorky (formerly Nizhniy Novgorod)—at the Gorky Research Institute of Epidemiology and Microbiology. The center specialized in research with anti-dysentery phages, which was begun by the military during the war. The military also conducted some interesting comparative *in vitro* susceptibility studies with anti-dysentery phage preparations produced by the Gorky Institute and those produced by the EIBMV (Podolsky, 1960). The preparations produced in Gorky lysed ca. 96% of the major dysentery-causing species (including *S. boydii, S. flexneri,* and *S. zonnei*) found in the Gorky area, in contrast to the preparation produced by the EIBMV, which lysed ca. 62% of the same strains. The author suggested that the difference between the phage preparations may have been due to the fact that the bacterial strains used to screen the phage preparations were from the Gorky region—and that phages developed in Gorky against local strains would be likely to have better lytic activity against them, compared to the EIBMV preparations developed primarily against dysentery strains common in Georgia and the Caucasus region.

Several hundred peer-reviewed papers, abstracts, meeting proceedings, and doctoral dissertations reporting the high efficacy of phage therapy have been published in various former Soviet Republics; some of these studies have been reviewed in English (Sulakvelidze et al., 2001; Sulakvelidze and Morris, 2001; Alisky et al., 1998). To give just a few examples, phage therapy has been reported to be effective against staphylococcal lung infections (Ioseliani et al., 1980; Meladze et al., 1982), eye infections (Proskurov, 1970), neonatal sepsis (Pavlenishvili and Tsertsvadze, 1993), urinary tract infections (Perepanova et al., 1995), and wound infections (including surgical wound infections) (Pokrovskaya et al., 1942.; Peremitina et al., 1981) (Table 14.2). However, no appropriate controls were included in many of these trials, or controls were used but information needed for rigorous evaluation of the authors' conclusions was not provided. A good example of the latter situation was a large-scale study evaluating the efficacy of bacteriophages for prophylaxis and treatment of bacterial dysentery (Anpilov and Prokudin, 1984). The study was conducted in 1982–1983, apparently in a double-blinded manner, and included a large, unspecified number of Red Army soldiers stationed in four distinct but unspecified geographic regions of the former Soviet Union. The incidence of dysentery in the phage-treated groups was reported to be ca. tenfold less than that occurring in the control group which received calcium gluconate tablets instead of phages, and the *P* values were reported to be <0.0001. These results correlate well with those in other Soviet publications evaluating the efficacy of phage treatment against bacterial dysentery (Pavlova et al., 1973; Babalova et al., 1968; Solodovnikov Iu et al., 1970; Solodovnikov Iu et al., 1971). However, the authors did not specify where the study was conducted, and how many patients were enrolled in each arm of the study. In addition, the methods used to evaluate the results were only vaguely mentioned. The study was published in an army-affiliated journal during the era of secrecy and totalitarianism of the communist regime. Therefore, it is likely that some of the information was not disclosed because of censorship; nevertheless, the inability to independently analyze the authors' conclusions impeded rigorous evaluation of the efficacy of the phage treatment used in the study.

The styles of Soviet and Western scientific publications were also very different, and even studies published in Soviet journals not affiliated with the army were in a format very different from what most Western scientists required in scientific journals. This fact, together with the language barrier (very few, if any, of the Soviet papers were in English), created the situation were many phage therapy-related publications from the former Soviet Union went unnoticed by the Western scientific community, or they were noticed and dismissed as being of poor scientific quality—with the lack of critical controls being one of the most common criticisms. The failure to include critical controls was not a common practice in the Soviet scientific literature; however, many phage therapy-related publications did, indeed, lack phage-untreated control groups. This situation probably resulted from a strongly nationalistic and unquestioned belief in the efficacy of phage therapy. Therefore, when control groups were included during phage therapy trials, they often were not used to evaluate the effectiveness of the approach *per se*; rather, they were used to compare the efficacy of new or modified phage preparations to that of other antibacterials (including antibiotics) or commercially available phage preparations. One example of the latter approach is the study by Zhukov-Verezhnikov et al. (1978), in which the authors compared the effectiveness of "adapted" bacteriophages (i.e., phages specifically selected, by 2–3 serial passages *in vitro*, for potent lytic activity against specific bacterial strains isolated from individual patients) to that of commercially available, more "general" phage preparations. The authors used phage preparations to treat 60 patients with suppurative surgical infections. Thirty patients were treated with phages specifically adapted to strains isolated from each patient, and an equal number of patients was treated with commercially available phage preparations targeted against staphylococci, streptococci, enteropathogenic *E. coli*, and *Proteus*. The adapted bacteriophages were reported to be five- to sixfold more effective in curing suppurative surgical infections than were the commercially available preparations, presumably because of their improved specificity.

14.4.3. Phage Therapy in Georgia

What is now called the Eliava Institute of Bacteriophage, Microbiology and Virology (EIBMV) was one of the key centers of therapeutic phage research in the former Soviet Union and it remained the key center of phage research in the newly independent country of Georgia. The institute (initially called the Microbiology Institute) was founded in 1923 by Giorgi Eliava, a prominent Georgian bacteriologist and one of d'Herelle's closest collaborators after the discovery of bacteriophages. In 1917, Eliava observed "the bacteriophage phenomenon" in a manner similar to Ernest Hankin's (1896) observation made in India (see Chapter 2)—interestingly, also against *V. cholerae*—in water from the Mtkvari river which runs through Tbilisi, the capital of Georgia. Although he could not explain the nature of the phenomenon at that time, Eliava made a logical connection between his observations and those described by d'Herelle, when d'Herelle's paper "Sur un microbe invisible antagoniste des bacilles dysentériques" (d'Herelle, 1917) was published later the same year. Eliava and d'Herelle met each other in Paris in 1918, during Eliava's first visit to the Pasteur Institute, and his initial visit (during 1918–1921) was followed by additional visits during 1925–1927 and 1931–1932 (Shrayer, 1996), during which Eliava and d'Herelle became friends and formed close collaborative ties. Indeed, after leaving Yale University, D'Herelle traveled to Tbilisi on two occasions during 1933–1935, and he spent several months in Georgia collaborating with Eliava and other Georgian phage researchers (Fig. 14.3). D'Herelle apparently planned to move with his family to Tbilisi permanently, and the Soviet government readily built a cottage for his use on the Institute's grounds. Working jointly, d'Herelle and Eliava expanded the Microbiology Institute into the Institute of Bacteriophage Research—which later became the EIBMV and the center of therapeutic phage research for the former Soviet Union. However, in 1937, Eliava was arrested by Stalin's secret police, pronounced a "People's Enemy," and executed without trial. Frustrated and disillusioned, d'Herelle never returned to Georgia.

Nonetheless, the EIBMV survived and later became one of the largest facilities in the world engaged in the development and production of therapeutic phage preparations. Elena Makashvili (Fig. 14.3) played a major role in maintaining the

FIGURE 14.3 Elena Makashvili, Felix D'Herelle (seated), and Giorgi Eliava in Tbilisi, at what is now the EIBMV. The photograph was taken between 1933 and 1935, during one of d'Herelle's two trips to Tbilisi, Georgia (photograph from Dr. A. Sulakvelidze's archives).

Institute's intellectual caliber and the knowledge of phage and bacteria after the death of Eliava. She remained chief of the Institute's Bacteriophage Division into the mid-1950s, taking personal responsibility for training new phage investigators. Many people still at the Institute studied extensively under her firm and talented hand. Another student of Giorgi Eliava, Irakli Giorgadze (who helped care for the horses Eliava kept for serum work) eventually became the Institute's director through the challenging period from 1959 into the 1980s. The Institute became the center of therapeutic phage work for the former Soviet Union and, for a time, the largest facility in the world engaged in the development and production of therapeutic phage preparations. During its heyday, the EIBMV employed more than 1000 researchers and support personnel, and it produced phage preparations (often several tons a day, in tablet, liquid, and spray forms) against a number of bacterial pathogens, including staphylococci, streptococci, *Pseudomonas*, *Proteus*, and many enteric pathogens. Many of the Soviet papers describing the therapeutic or prophylactic use of phages utilized phage preparations developed by, and manufactured in, the EIBMV.

A significant achievement of EIBMV scientists was the development of an injectable form of staphylococcal phage preparation that was used to treat human systemic infections in the 1980s. The development efforts were headed by Teimuraz Chanishvili, the current director of the EIBMV. The phage-containing lysates were obtained from bacteria growing in a chemically defined medium, and the phages were purified by centrifugation and filtration, and toxicity-tested *in vivo*. The results of the efficacy studies demonstrated that the preparation was effective in treating experimental staphylococcal sepsis in rabbits. Also, subsequent human "safety studies" (conducted with 20 human volunteers) indicated that it did not cause any adverse side effects when injected intravenously (T. Chanishvili, personal communication). The staphylococcal phage preparation was subsequently used to treat 653 patients in various hospitals, using multiple daily injections (to reduce the risk of the possible development of phage-neutralizing antibodies and to speed-up the treatment process). The preparation was effective, and it was soon recommended for treating emergency cases of *S. aureus* infections throughout the former Soviet Union, including therapeutic uses in infants. For example, it was used to treat more than 300 infants during an outbreak of neonatal sepsis in Moscow, Russia (T. Chanishvili, personal communication). Here too, the preparation appeared to be effective: Only one of the infants treated with the staphylococcal phage preparation died, versus 8 in the group treated with commonly available antibiotics. The same phage preparation was also used successfully to treat female sterility caused by *S. aureus* infection of the fallopian tubes, and to treat osteomyelitis (by injecting the preparation into an artery serving the affected region) (Bogorishvili et al., 1985).

The Institute also had an ongoing basic research program directed toward delineating the mechanisms responsible for the increased therapeutic potential of phages in its collection. As described in Chapters 3 and 7, bacterial infection and lysis by phages is a complex process consisting of a cascade of events involving several structural and regulatory genes—and phage-encoded genes responsible for the increased therapeutic potential of various phages were identified in several bacteriophages by the EIBMV investigators in the 1980s and 1990s. The factors contributing to the increased lytic potential of phages varied. In some cases, phage gene products were reported to increase the lytic potential of the phage by increasing the ability of phage DNA to resist the restriction-modification defenses of the host bacterial strain (Andriashvili et al., 1986). In other phages, gene products were directly lethal for the host cell (Adamia et al., 1990). Similar studies have been recently published in the United States (Garcia et al., 1983; Nelson et al., 2001; Schuch et al., 2002); also, see Chapter 12). Further characterization of these and similar mechanisms is likely to yield new information about phage-bacterial cell interactions, and to be useful for identifying novel phage-encoded gene products of potential therapeutic value.

The break-up of the former Soviet Union and the resulting economic difficulties in many former Soviet republics (including Georgia) caused severe hardship for the EIBMV, and profoundly reduced the institute's ability to continue phage therapy research. The former extensive industrial division was perfunctorily privatized and largely diverted to other purposes. The vast markets in Russia and the other former Soviet republics became inaccessible. The animal facilities were no longer available for antiserum production or for the routine efficacy and toxicity testing. As a consequence of these events, the EIBMV's production of therapeutic phage preparations became largely limited to work at a handful of hospitals and private clinics in Georgia. However, even under the current conditions, the phage therapy concept has remained very much alive in Georgia, and investigators there have continued to develop new phage-based therapeutic preparations. One, arguably the most intriguing, example of such novel preparations is PhagoBioDerm.

A novel and intriguing approach to treat infected wounds with phages has been recently developed at the Center for Medical Polymers and Biomaterials (CMPB), in the Georgian Technical University in Tbilisi. A group of CMPB investigators headed by Ramaz Katsarava has developed a biodegradable, nontoxic polymer composite (Katsarava et al., 1999). In subsequent studies conducted in collaboration with EIBMV scientists, the polymer was impregnated with bacteriophages (together with ciprofloxacin and benzocaine). The resulting preparation, designated PhagoBioDerm, was licensed by the Georgian Ministry of Health in 2000. In its current form, PhagoBioDerm is a perforated, biodegradable, nontoxic polymer composite containing a mixture of lytic bacteriophages ("PyoPhage" or "Piobacteriophagum fluidum," ca. 1×10^6 PFU/cm^2), an antibiotic (ciprofloxacin, 0.6 mg/cm^2), an analgesic (benzocaine, 0.9 mg/cm^2), α-chymotrypsin (0.05 mg/cm^2), and sodium bicarbonate (3.75 mg/cm^2) (Markoishvili et al., 2002; Katsarava and Alavidze, 2004). The "PyoPhage" preparation (produced at the EIBMV) is available commercially in Georgia (Fig. 14.4), and a specially designed version is added to the polymer composite during the manufacturing process for PhagoBioDerm. The preparation contains lytic bacteriophages active against *P. aeruginosa, E. coli, S. aureus, Streptococcus,* and *Proteus.*

The first peer-reviewed report of the therapeutic efficacy of PhagoBioDerm was recently published (Markoishvili et al., 2002). Briefly, 107 patients with ulcers that had failed to respond to conventional therapy (i.e., therapy with commonly available, antibiotic-containing ointments and various phlebotonic and vascular-protecting agents) were treated with PhagoBioDerm alone or in combination with other interventions during 1999–2000. The wounds/ulcers healed completely in 67 (70%) of 96 patients for whom follow-up data were available. In 22 cases in which microbiologic data were available, healing was associated with the concomitant elimination of, or a marked reduction in the number of, specific pathogenic bacteria in the ulcers. More recently, PhagoBioDerm was used to treat two Georgian lumberjacks who developed infected radiation burns after they discovered Soviet-era thermal generators containing strontium-90 and, not knowing what they had found, used the generators to warm themselves during a cold December night in woods near their village. As described

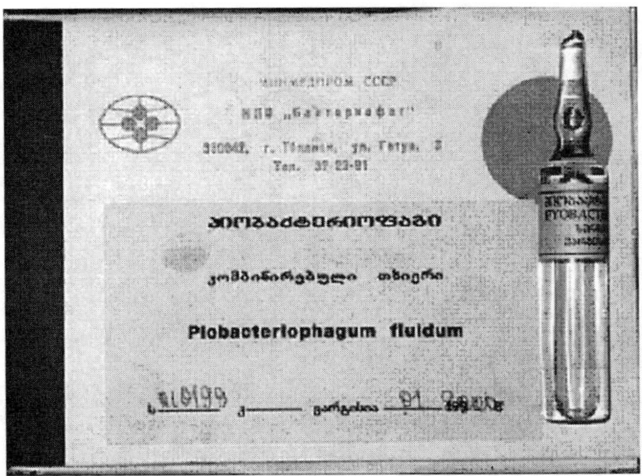

FIGURE 14.4 "Piobacteriophagum fluidum"—one of the polyvalent phage preparations produced by the EIBMV. The preparation targets a variety of bacterial pathogens, including *P. aeruginosa, E. coli, S. aureus, Streptococcus,* and *Proteus.*

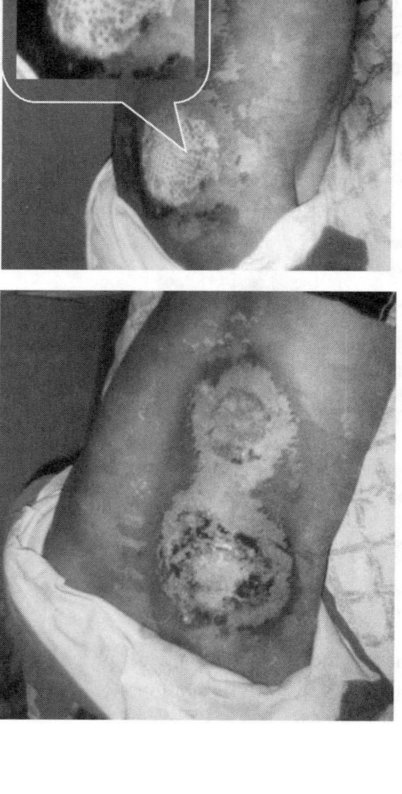

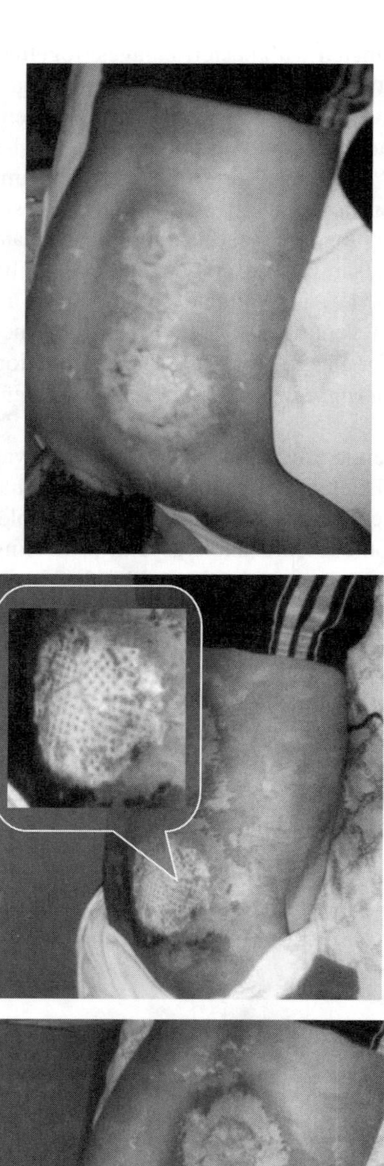

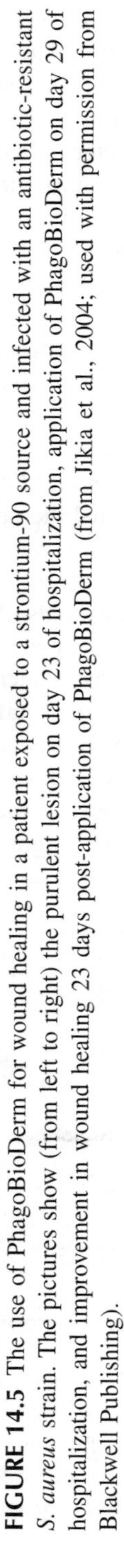

FIGURE 14.5 The use of PhagoBioDerm for wound healing in a patient exposed to a strontium-90 source and infected with an antibiotic-resistant *S. aureus* strain. The pictures show (from left to right) the purulent lesion on day 23 of hospitalization, application of PhagoBioDerm on day 29 of hospitalization, and improvement in wound healing 23 days post-application of PhagoBioDerm (from Jikia et al., 2004; used with permission from Blackwell Publishing).

by Stone (2002) and Jikia (2004), the lumberjacks developed severe local radiation burns that subsequently became infected with *S. aureus*. They were treated with various medications, including antibiotics and topical ointments; however, the treatments were only moderately successful, and their *S. aureus* infection could not be eliminated. Therefore, ca. one month after hospitalization, treatment with PhagoBioDerm was initiated. Purulent drainage stopped within 2 to 7 days, and the ulcers were almost completely healed after treatment for one month (Fig. 14.5). Clinical improvement was associated with rapid (within a 7-day time period) elimination of the etiologic agent, a strain of *S. aureus* resistant to many antibiotics (including ciprofloxacin included in PhagoBioDerm) but susceptible to the lytic bacteriophages contained in the PhagoBioDerm preparation. Further studies, including carefully designed, double-blinded clinical trials will be required to evaluate rigorously the efficacy of that novel wound dressing preparation, and to make it acceptable and available in countries outside Georgia. Meanwhile, other versions of PhagoBioDerm are apparently being developed in Tbilisi; e.g., a special version called "PhageDent" has been formulated for periodontal applications (R. Katsarava, personal communication).

14.4.4. Phage Therapy Centers in Russia

After the fall of the former Soviet Union, several research centers and private companies continued to develop and produce therapeutic phage preparations in Russia. In the 1990s, the phage production facilities at the Gorky Institute were acquired by the firm ImBio. According to the ImBio Web site (http://www.sinn.ru/~imbio/Bakteriofag.htm), the company is currently producing phage-based preparations in liquid, tablet, and cream formulations for treating bacterial dysentery ("Bacteriophagum dysentericum polyvalentum in tabulettis"), the early stages of salmonellosis ("Bacteriophagum salmonellae gr.ABCDE liquidum et siccum cum indumento acidoresistentis"), general gastrointestinal disorders ("Bacteriophagum coliproteicum liquidum"), *S. aureus* infections ("Bacteriophagum staphylococcus"), and *P. aeruginosa* infections ("Bacteriophagum *Pseudomonas aeruginosa* liquidum").

Immunopreparat is another large pharmaceutical company in Russia, which is located in the city of Ufa (approximately 60 miles west of the Ural Mountains). The company's subsidiary, Biophag, currently manufactures at least two complex phage preparations ("Bacteriophagum" and "Piobacteriaphagum") targeting various bacterial pathogens. Lazareva et al. (2001) recently described the use of "Piobacteriaphagum," a mixture of phages targeting staphylococci, streptococci, *Proteus,* and *E. coli*, to treat infected burn wounds. Of 54 phage-treated patients, 9 were treated only with phages, and 45 patients were treated with both phages and antibiotics; an additional 40 patients were treated only with antibiotics. In most of the phage-treated patients, phages were orally administered in tablets, 1–1.5 h prior to meals (2 tablets/dose, 3 doses/day, for 7 days); however, four patients ingested a liquid formulation containing the phages. Phage-related side effects were not observed in any of the phage-treated patients, and they also were reported to have fewer septic complications and more rapid temperature normalization, compared to the patients not treated with the phage preparations. Microbiologically, phage treatment elicited a twofold reduction in the number of staphylococci and streptococci, and a one-and-one-half-fold reduction in the number of *Proteus*, recovered from burn wounds.

Another large Russian company currently producing phages for various therapeutic applications is Biomed, in the city of Perm in the Urals. Biomed is a major Russian producer of pharmaceuticals and biological preparations, including vaccines, therapeutic sera, antitoxins, and various diagnostic reagents. The company first manufactured a bacteriophage preparation against dysentery between 1940 and 1950, and it began producing bacteriophage preparations targeting other bacterial pathogens in 1994—at the time when obtaining phage preparations from the EIBMV became difficult because of the disintegration of the Soviet Union.

14.4.5. Phage Therapy in Poland

Although therapeutic phages were used in several countries of Eastern Europe (Meitert et al., 1987; Pillich et al., 1978; Zilisteanu et al., 1973; Zilisteanu et al., 1971), the key research and development activities were centered at the Hirszfeld Institute of Immunology and Experimental Therapy (HIIET) in Poland. The Institute (named after the first director of the Institute, Ludwik Hirszfeld, a prominent Polish immunologist and microbiologist) was founded in Wroclaw, Poland, in 1952, and its investigators have been actively involved in phage therapy research since 1957, when therapeutic phages were first used to treat *Shigella* infections in Poland. The bacteriophage laboratory of the HIIET has been instrumental in developing and producing phage preparations for treating various bacterial diseases (e.g., septicemia, furunculosis, and pulmonary and urinary tract infections), and for preventing and treating post-operative and post-traumatic infections. The most detailed studies published in English in the "post-antibiotic" era (i.e., since the 1950s), on the use of phages in clinical settings, have come from Stefan Slopek and his colleagues at the HIIET. During 1983 to 1985, they published a series of six papers (Slopek et al., 1983b, a; Slopek et al., 1984; Slopek et al., 1985c; 1985b; 1985a) describing the use of phages against infections caused by several bacterial pathogens. In their seventh paper (Slopek et al., 1987), the authors summarized the results of treating 550 septicemic patients ranging in age from 1 week old to 86 years old, who were hospitalized in ten clinical departments and hospitals located in three different cities in Poland. Five hundred eighteen (94%) of the patients had been treated with commonly available antibiotics (they were not specified by the authors), and phage therapy was initiated after the antibiotic treatments were concluded to have been only moderately successful or unsuccessful.

Phage treatment was initiated after isolating the etiologic agent(s) from the patient, and selecting highly lytic bacteriophages for those particular bacteria from a collection of more than 250 bacteriophages. The phages were administered using various combinations of the following routes (i) orally, several times a day before eating and after neutralizing gastric acid by oral administration of baking soda or bicarbonated mineral water a few minutes prior to phage administration, (ii) locally, by applying moist, phage-containing dressings directly on the wounds or in pleural and peritoneal cavities, and (iii) by applying a few drops of phage suspension to the eye, middle ear, or nasal mucosa. The etiologic agents were staphylococci,

Pseudomonas, Escherichia, Klebsiella, and *Salmonella*, and during the course of phage treatment they were continuously monitored for phage susceptibility. If the expected reduction in bacterial counts of the targeted pathogen were not observed after phage administration, the phages were replaced with new bacteriophages lytic for the same bacterial strain. The duration of treatment varied from 7 days to about 3 months; in some cases phages were applied for up to 2 weeks after negative cultures were obtained. The outcome of therapy was evaluated based on the patient's physical recovery in conjunction with negative cultures, and the reported success rates ranged from 75% to 100% (92% overall). Control groups without phage treatment were not included in the study, but given the fact that most of the patients have been refractory to conventional treatment prior to initiating phage therapy suggests that phage therapy was indeed responsible for the observed therapeutic improvements in those patients.

In other publications from Poland (Table 14.2), phages were reported to be effective in treating cerebrospinal meningitis in a newborn (Stroj et al., 1999), skin infections caused by *Pseudomonas, Staphylococcus, Klebsiella, Proteus,* and *E. coli* (Cislo et al., 1987), recurrent subphrenic and subhepatic abscesses (Kwarcinski et al., 1994), shigellosis (Mulczyk and Slopek, 1974), and various chronic bacterial diseases (Kaczkowski et al., 1990). In addition to having a therapeutic effect by lysing disease-causing bacteria, administration of bacteriophages has been suggested (Weber-Dabrowska et al., 2000b) to stimulate the human immune system; e.g., it normalizes TNF-alpha serum levels and the production of TNF-alpha and IL-6 by blood cell cultures. An update/summary of the institute's phage therapy-related research has recently been published by Beata Weber-Dabrowska (who continued to lead the phage therapy work at the HIIET after the death of Stefen Slopek) and her colleagues (Weber-Dabrowska et al., 2000a).

14.5. KEY CONSIDERATIONS IN THERAPEUTIC APPLICATIONS

14.5.1. Mode of Therapeutic Action of Phages

Paradoxically, at the same time that phage therapy's efficacy was being severely questioned, there was also a controversy about the mechanisms by which phages might exert their putative therapeutic effects. D'Herelle had always contended that bacteriophages are viruses that parasitize susceptible bacteria, and that the efficacy of phage therapy was due to the bacteriolytic action of phages (d'Herelle, 1926). However, that view was challenged by researchers who believed that phages were enzymes or other inanimate substances that triggered autolysis by deleteriously affecting bacterial metabolism (Bordet, 1925; Arkwright, 1924). Also, during the early years of phage therapy, successful outcomes of phage treatment were sometimes attributed to antibacterial ingredients or contaminants found in therapeutic phage preparations, rather than to the lytic effect of phages (Eaton and Bayne-Jones, 1934). Indeed, early phage preparations were crude phage lysates of host bacteria, and they contained undefined amounts of media components, bacterial derivatives, secretory factors, preservatives, and other components that may have possessed some antibacterial activity and had a therapeutic effect. Some of these derivatives (e.g., bacterial LPS) could also potentially trigger specific immune response in the mammalian organism and provide protection via that mechanism. In this context, there have been several reports suggesting that phage lysates make particularly good immunizing agents and that their protective effect was even more profound than that of vaccines prepared by heat- or chemical-inactivation of bacteria (Arnold and Weiss, 1924; Larkum, 1929a). Also, as noted earlier in this chapter, the *S. aureus* phage-containing vaccine preparation ("Staph phage lysate" or SPL) has been used successfully to prevent and treat various infections of humans and animals in the United States during the 1950s–1990s.

Recent data support the idea that phage therapy is associated with the normalization of cytokine production (Weber-Dabrowska et al., 2000b; Weber-Dabrowska et al., 2002); thus, immune stimulation is indeed likely to play an important role during phage therapy. However, the relative contribution, to the observed prophylactic/ therapeutic effect elicited by phage preparations, of specific bacterial antigens present in the phage preparations vs. nonspecific, phage-mediated immune stimulation is not clear at the present time. Furthermore, it is not clear whether an effective "vaccination effect" will occur when highly purified (i.e., free of, or minimally contaminated with, bacterial antigens) phage preparations are used during future phage therapy studies, especially if the preparations are used after the onset of infection rather than administered repeatedly beforehand as a preventive measure. In this context, the presence of viable phages in the preparation seems to be the critical requirement for its efficacy during the treatment of already established bacterial infections (Dubos et al., 1943; Biswas et al., 2002). Many studies describing the efficacy of phages in treating experimentally infected animals have been published, and various mathematical models have been employed to analyze the complex kinetics of these "self-replicating pharmaceuticals" (Bull et al., 2002; Levin and Bull, 1996; Payne and Jansen, 2001; 2002, 2003; Payne et al., 2000). Some of these *in vivo* studies are discussed further in Chapter 13. A review of the phage's lytic cycle—which is ultimately responsible for phage-mediated lysis of bacteria—is given in Chapter 3, with many of the molecular details given in Chapter 7.

14.5.2. Phage Distribution, Pharmacokinetics, and Neutralizing Mechanisms in Mammals

14.5.2.1. Phage Movement between Biological Compartments

Interestingly, although many human trials probably were preceded by at least some preliminary testing in laboratory animals, and extensive preclinical animal testing was required for approving new phage formulations in the former Soviet Union, such an approach has been described in only a very few published studies. One example of a sequential preclinical and clinical approach was reported by Bogovazova (1991). They first used a mouse model to evaluate the safety and efficacy of phages produced by the Russian company Immunopreparat (see section 14.4.4) for treating infections by various *Klebsiella* species, and they determined the optimal phage concentration and administration route. Pharmacokinetic and toxicological studies were performed with mice and guinea pigs, using intramuscular, intraperitoneal, and intravenous administration of phages. Signs of acute toxicity or gross or histological changes were not observed, even when the injected dose/gm body weight was about 3500-fold higher than the projected human dose. They subsequently evaluated the safety and efficacy of the phages in treating 109 patients infected with *Klebsiella* (Bogovazova et al., 1992). The phage preparation was reported to be nontoxic for humans and to be effective in treating *Klebsiella* infections, as manifested by marked clinical improvement and bacterial clearance in the phage-treated patients. As shown in these studies—and in many other phage therapy studies—a number of factors may influence the outcome of phage therapy, many of which can be highly variable from system to system. Some of these factors are described in some detail below.

The results of several studies indicate that phages rapidly enter the mammalian circulatory system and are distributed throughout the body, whether administered intravenously, intramuscularly, intraperitoneally, or even orally, but they soon disappear from the blood and most organs unless they find bacteria in which they can multiply. Appelmans (1921) first reported that phages injected into the bloodstream of uninfected rabbits disappeared rapidly from the blood and internal organs, and persisted longest in the spleen; this observation was confirmed in a number of later studies, such as those by Evans (1933) and Geier (1973). Using ^{51}Cr-labelled phages, Inchley (1969) found that the liver phagocytized more than 99% of T4 phage within 30 minutes after i.v. injection. Noteworthy, many early experiments examining phage pharmacokinetics in animal models were done without the presence of specific host bacteria for the phages, which could have influenced the outcome. Indeed, more recent reports have shown that in the presence of host bacteria, therapeutic phages can be found in the mammalian circulatory system irrespective of the administration route; e.g., therapeutic phages administered orally to infected patients were recoverable from their bloodstream for several days (Babalova et al., 1968; Weber-Dabrowska et al., 1987). Also, studies with experimentally infected animals (Bogovazova et al., 1991; 1992) found that phages enter the bloodstream within 2 to 4 hours, and that they are still recoverable from the internal organs (liver, spleen, kidney, etc.) after ca. 10 hours postinjection.

One of the best pharmacokinetic studies comparing phage treatment of *infected* and *uninfected* animals was published by Dubos et al. (1943). The authors injected white mice intracerebrally with a dose of a smooth *S. dysenteriae* strain sufficient to kill more than 95% of the mice in 2–5 days, and they treated them by i.p. injection of a mixture of phages isolated from New York City sewage, grown in the same bacteria and purified only by sterile filtration. Pharmacokinetic studies showed that when the phages were given to *uninfected* mice, they appeared in the bloodstream almost immediately, but the levels started to drop within hours and very few were ever seen in the brain. In contrast, in the *infected* animals, brain levels soon greatly exceeded blood levels (10^7–10^9 phage/gram of brain tissue were often seen by 8 hours), and the concentration of phages in the brain started to decrease between 75 and 138 hours, as the infection was eliminated. After the first 18 hours, the concentration of phages in the blood were lower than in the brain, but they were still present at 10^4–10^5 per ml in those cases where the brain phage levels were still above 10^9 per gram. This observation suggests that phages are capable of crossing the blood-brain barrier and multiplying within intracerebral bacteria. The mechanism(s) by which phages cross the blood-brain barrier are unknown at the present time, but their ability to do so may have important practical implications for treating bacterial meningitis.

In the aforementioned studies, the presence of phages in the internal organs correlated with their therapeutic efficacy. Without treatment, or when treated with heat-killed phages or filtrates of bacterial cultures, only 3/84 (ca. 4%) of mice survived; whereas, 46/64 (ca. 72%) of those given 10^7–10^9 phages survived. The study of Dubos et al. (1943) was well designed and rigorously performed, and it addressed several important questions concerning the efficacy and pharmacokinetics of therapeutic phage preparations. For example, the authors clearly established that, with the treatment regimen employed, the lytic effect of phages, rather than the immune stimulant possibly present in the phage lysate, was responsible for the observed therapeutic effect (no efforts were made to immunize the mice with phage lysates repeatedly prior to the bacterial challenge and, thus, the study did not rule out the possibility that the phage lysate could cause active immunity against host bacteria). The study also showed that phages could (i) locate and multiply in foci of bacterial infections anywhere in the body, even in such privileged compartments as the brain, and (ii) be found in the circulation as long as they are replicating in a reservoir of infection somewhere in the body. Similar observations were reported shortly afterwards (Morton and Engely, 1945; Morton and Perez-Otero, 1944).

14.5.2.2. Nonspecific Mechanisms That Clear Phages from the Bloodstream

Besides being of academic interest, phage persistence in the mammalian organism has been proposed (Merril et al., 1996) to have an impact on the efficacy of phage treatment because rapid elimination of phages from the mammalian organism might reduce the number of phages to a level which is not sufficient to combat the infecting bacteria. Multiple mechanisms may potentially contribute to the disappearance of phages from the mammalian bloodstream and lymphatic tissues. Antibody responses would be one such mechanism (see section 14.5.2.3. for more information on this subject), but they may take several days to occur, even if there had been a previous exposure to the phages. The innate immune system (earlier called the reticuloendothelial system or RES) is another possible mechanism that is likely to be primarily responsible for removal of phages from the mammalian organism shortly after their administration. To address this issue, Merril (1996) used a natural selection strategy (which they called the "serial passage" method) for selecting phages having an increased ability to remain in the circulation of mice. The authors used rapidly disappearing virulent derivatives of phage λ (λ_{vir}) and P22 to select for hardier mutants of the phages by serially passing them through mice 10 times, initially injecting 10^{11} phage and regrowing the survivors after each step. After 10 serial passages, they identified two lambda phage mutants that persisted in mice longer than did the wild-type phage. A single i.p. injection of either of the two "long circulating" phages was more effective in rescuing bacteremic mice than was a single injection of the wild-type phage, which suggests that the persistence of high phage titers in the circulation has a positive impact on the efficacy of phage treatment.

Interestingly, the same change in amino acid composition was observed in both phages: from glutamic acid to lysine at position 158 of the main head protein; apparently the charge at this position has a strong effect on the susceptibility of phages to entrapment by the RES. The "long-circulating" phages and the mechanisms responsible for their ability to remain in the circulation for prolonged periods of time are of clear academic interest. However, the practical applicability of such phages is not clear. For example, the lambdoid phage used in the study is a member of a phage family that has been very closely associated with bacterial pathogenicity determinants (see Chapter 8), and it is highly unlikely to be suitable for use in therapeutic phage applications. Also, from a practical standpoint, using a multiple phage administration approach is likely to be technically more feasible than is "serially passaging" every phage through animals before using them therapeutically.

14.5.2.3. The Development of Phage-Neutralizing Antibodies

As noted above, the production of phage-neutralizing antibodies is another potential problem that could hamper phage therapy. Indeed, the development of neutralizing antibodies after parenteral administration of phages has been well documented, as discussed by Adams (1959a), and i.v. administration of phage ϕX174 has been used extensively to study details of the immune response in immunocompetent and immunocompromised patients (Fogelman et al., 2000; Lopez et al., 1975; Ochs, et al. 1993a; Ochs, et al. 1992; Ochs et al., 1971; Pyun et al., 1989). However, it is unclear whether the development of neutralizing antibodies actually can reduce the efficacy of phage therapy, especially when phages are administered orally or locally (the two administration routes where high titers of neutralizing antibodies are not likely to develop). In this context, in the studies performed in Poland (Kucharewicz-Krukowska and Slopek, 1987), antibodies against *S. aureus*-specific phages were detected, *before initiating phage therapy*, in 12 (ca. 21%) of the 57 patients examined—and also in 11% of healthy controls. After treatment, antibodies were seen in 54% of the patients. A detailed study on 30 patients looked at the anti-staph-phage antibody titer before and after phage administration and compared it with the efficacy of phage treatment, as shown in Fig. 14.6. On the one hand, the only patients who did not appear to benefit from phage therapy were two of the five patients with relatively high anti-phage antibodies before phage treatment, which is consistent with the idea that pre-existing antibodies could interfere with treatment. On the other hand, the clinical condition of the other three patients who had anti-phage antibodies prior to treatment markedly improved after phage treatment, and there was little correlation between the development of antibodies during treatment and its efficacy. Additional, more extensive studies are strongly indicated.

As discussed in Chapter 3, various parameters affect the development of phage-neutralizing antibodies. Levels of phage-neutralizing antibodies are low after a single injection or after a series of closely spaced injections, and most of the antibody that

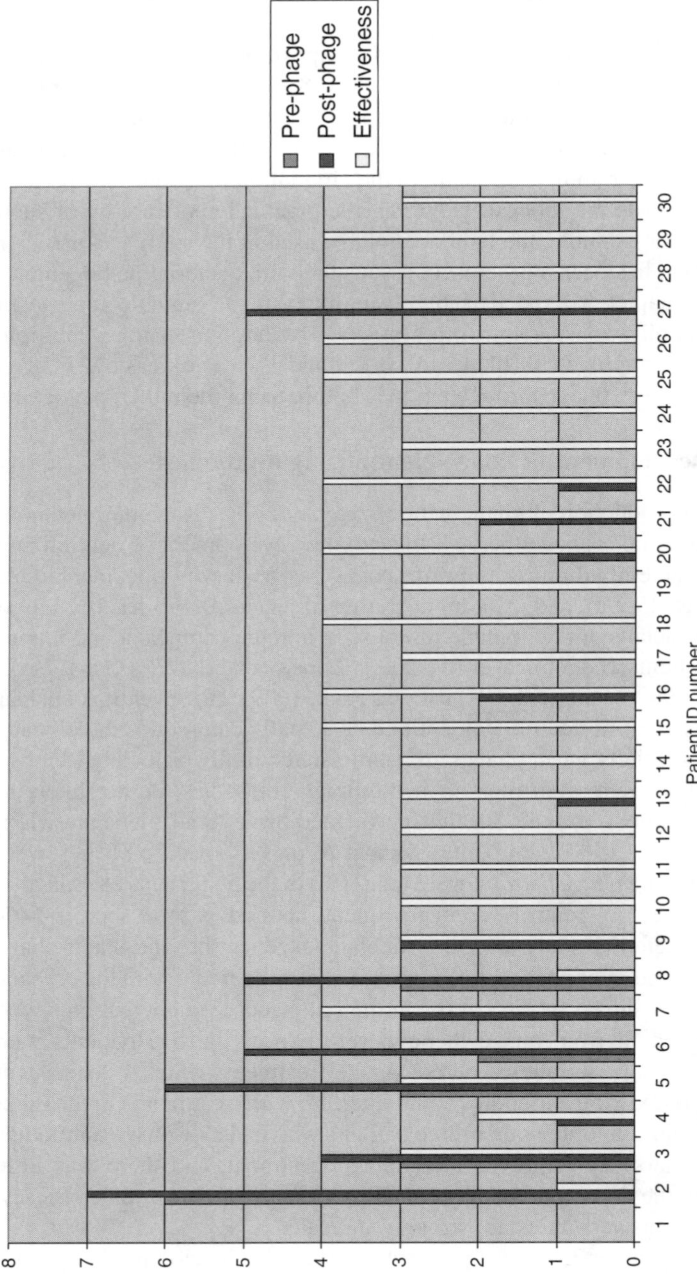

FIGURE 14.6 Individual titers of antibody against *S. aureus* phage before and after phage treatment, compared with the efficacy of the treatment (1 = no significant improvement, 2 = some improvement, 3 = no further sign of infection, 4 = complete healing). (This graph was created from the data presented by Kucharewicz-Krukowska and Slopek, 1987).

develops under these regimes is of the IgM type, whose neutralizing activity is low and largely reversible (Gachechiladze, Chapter 3 box and personal communication). High titers of long-lasting, higher-affinity IgG antibodies are seen only when multiple injections are spaced several weeks apart. Thus, the production of neutralizing antibodies (i. e., antibodies against the tail adhesins) should not be a significant obstacle during initial or relatively short-term therapeutic treatments, at least. Furthermore, the antigenic properties vary considerably between different phage families; for example, T4 phage is a good immunogen, but T1 and T5 phages are much less antigenic, and a *Bacillus subtilis* phage reportedly needs an adjuvant and repeated intraperitoneal injections to generate a detectable immune response (Adams, 1959a). Even under a "worst case scenario," when phages must be repeatedly administered intravenously or when high titers of neutralizing antibodies are present, it should be possible to use therapeutic phage preparations in an effective manner. For example, phages with decreased immunogenicity can be selected (Adams, 1959a) and phages lytic for the same target pathogen but having different antigenic profiles may be used. No cross-reactivity has been seen among *Podoviridae, Myoviridae*, and *Siphoviridae*, and even the closely related, immunogenic T4-like phages often show little or no neutralization cross-reactivity. A regimen involving closely timed repeated local phage administration may be of value, and phages may be efficaciously administered in local lesions or at high levels even when there are high titers of antibodies against the particular phage. Furthermore, neutralizing antibodies cannot affect already-attached phages (Adams, 1959a).

14.5.2.4. Emergence of Bacterial Strains Resistant to Particular Phages

Prior to the antibiotic era, bacterial mutants resistant to dyes and other lethal agents were noted, and Alexander Fleming observed the emergence of penicillin-resistant mutants soon after his discovery of that antibiotic (Livermore, 2000). The emergence of phage-resistant bacterial mutants was similarly observed soon after the discovery of bacteriophages, and the phenomenon was suggested (even by d'Herelle) to be a potential problem of phage therapy (Summers, 1999; d'Herelle, 1930). However, in real-life, the emergence of phage-resistant mutants was not noted to be a problem in studies reported from Poland and the former Soviet Union, and phage-resistant mutants have not been observed (or their percentage was very low) in the few *in vivo* and *in vitro* studies where phage treatment-surviving colonies were examined for phage susceptibility. For example, during the studies of Smith et al. (1983), who evaluated the efficacy of phages in treating experimental *E. coli* diarrhea in calves, piglets, and lambs, phage-resistant *E. coli* mutants were not recovered from the phage-treated lambs; a few phage-resistant mutants were identified in samples from the piglets and calves, but they seemed to be less virulent than the original strain (see Chapter 13). Bull (2002) recently repeated the Smith and Huggins study, and they estimated the number of phage-resistant mutants—by the end of treatment—to be less than 20% of the overall host strain population. Also, Leverentz et al. (2001) were not able to detect phage-resistant *Salmonella* mutants during their studies examining the value of using bacteriophages as a biocontrol method for *Salmonella* on fresh-cut fruits.

The frequency of the spontaneous mutations that may confer phage-resistance (e.g., by changing receptor configuration, so that phage tail fibers cannot attach to the bacterial cell membrane) have not been rigorously determined for various phages; however, in general, they are thought to be comparable to those seen for antibiotics (Drake et al., 1998; Carlton, 1999). Also, the phage resistance rates could be further reduced by using a mixture of phages. In this context, in many of the studies referenced earlier, mixtures of two or more phages (i.e., "phage cocktails") were used in order to increase the target range of the phage preparations, and to reduce or prevent selecting for phage-resistant mutants. The use of phage cocktails was also very common in the former Soviet Union and Eastern Europe, where complex cocktails were used (and are still being used) to treat various bacterial infections. The phage cocktails usually contain several lytic phages active against the same bacterial strain, and their combined use is similar to the rationale of using two or more antibiotics simultaneously, which is known to help reduce the emergence of antibiotic-resistant bacterial mutants.

14.5.3. Safety Considerations

From a clinical standpoint, phages appear to be very safe. This is not surprising, given that humans are exposed to phages from birth (and, possibly, even *in utero*). Indeed, as discussed in Chapters 6 and 13, bacteriophages are arguably the most ubiquitous organisms on earth; e.g., one milliliter of nonpolluted water has been reported (Bergh et al., 1989) to contain ca. 2×10^8 PFU of phages/ml, and the total number of phages on Earth has been estimated to be in the range of 10^{30}–10^{32} (Brüssow and Hendrix, 2002). Phages are abundant in saltwater, freshwater, soil, plants, and animals, and they have been shown (Merril et al., 1972; Moody et al., 1975; Milch and Fornosi, 1975; Geier et al., 1975) to be unintentional contaminants of some vaccines and sera commercially available in the United States. Also, phages are normal commensals of the human body, and they have been commonly found in the human gastrointestinal tract, skin, urine, and mouth, where they are harbored in saliva and dental plaque (Caldwell, 1928; Yeung and Kozelsky, 1997; Bachrach et al., 2003).

The abundance of phages in the environment—and the continuous exposure of humans to them—explains the extremely good tolerance of the human organism to phages. However, that situation does not imply that all phage preparations can be safely used therapeutically, and side-effects can develop due to the therapeutic action of phages itself, as well as a result of using insufficiently purified phage preparations (especially injectable forms) therapeutically. For example, some side effects during phage therapy may be triggered by the *in vivo* lytic activity of phages against the etiologic agent; e.g., liver area pain in a patient undergoing phage therapy in Poland was suggested to be related to extensive liberation of endotoxins from the lysed disease-causing bacteria (Slopek et al., 1987). Similar complications also may be observed during antibiotic therapy (Prins et al., 1994), and various approaches (e.g., corticosteroid treatment) used to reduce this problem during antibiotic therapy may also be effective during phage therapy.

Insufficient purity of phage preparations may also be problematic. As noted earlier in this chapter, systemic side effects associated with phage therapy were reported in the early days of phage therapy (Larkum, 1929b; Naidu and Avari, 1932), and most—if not all—of them appeared to be due to bacterial byproducts; e.g., endotoxins, contaminating the phage preparations (King et al., 1934). Therefore, using appropriately purified phage preparations should eliminate endotoxin-mediated side effects or complications, particularly when using injectable phage preparations. In this context, various groups have reported that the administration—including the intravenous injection—of phage preparations did not elicit noticeable side effects in animals or humans if the preparations were well purified (Uhr et al., 1962; Merril et al., 1996). In Russia, Bogovazova et al. (1991; 1992) reported that *Klebsiella*-specific phages purified by molecular sieve chromatography (with Superose-12) did not elicit side effects when tested in animals and used during human clinical trials. During the 1970s–1990s in the United States, purified phage phi X174 was extensively used to monitor humoral immune function in humans, including patients with Down syndrome (Lopez et al., 1975), Wiskott-Aldrich syndrome (Ochs et al., 1982), and immunodeficiency conditions (Ochs et al., 1992; 1993b; 1993a). Also, purified phages have been injected intravenously into HIV-infected patients (Fogelman et al., 2000), patients with other immunodeficiency diseases (Ochs et al., 1971), and healthy volunteers (Ochs et al., 1993a)—without any apparent side effects.

These observations strongly suggest that phage therapy may provide one of the safest as well as most environmentally friendly methods currently available for prophylaxis and treatment of bacterial infections. However, in order not to compromise the safe use of therapeutic phage preparations, they should be produced, purified, characterized, and tested using current, state-of-the-art biotechnology. This approach will help ensure the consistency and safety of therapeutic phage preparations, and it is also likely to increase their therapeutic and prophylactic potential. As discussed below, rigorous characterization of each phage to be used therapeutically—in particular, screening for potentially harmful genes in their genomes—should also be given serious consideration.

One safety-related issue associated with phage therapy is the importance of confirming that therapeutic phage cocktails contain "lytic" (i.e., virulent) phages rather than "lysogenic" phages (a term sometimes used, mistakenly, for temperate phages), because some lysogenic phages may transfer harmful bacterial genes (e.g., genes encoding bacterial toxins) from one bacterial host to another. Temperate phages are extremely common in the environment, human gut, human oral cavity, foods sold at retail, etc., and they have been found in almost all bacterial genera, including *Staphylococcus, Vibrio, Borrelia, Pseudomonas, Burkholderia, Salmonella, Shigella, Bacillus, Corynebacterium, Listeria,* and *Streptococcus* (Adams, 1959b; Schicklmaier and Schmieger, 1995; Eggers et al., 2001; Langley et al., 2003). As discussed in more detail in Chapter 8, several temperate phages have been shown to carry various bacterial toxin genes (Freeman, 1951; O'Brien et al., 1984; Weeks and Ferretti, 1984; Betley and Mekalanos, 1985; Waldor and Mekalanos, 1996). Although the possibility of added gene transfer events is highly unlikely to bring significant danger to individual patients using therapeutic phages, the presence of temperate phages in therapeutic phage preparations manufactured on an industrial scale could increase the overall risk of potentially harmful genes being acquired by new bacterial strains, and the emergence of new pathogenic bacteria. Therefore, their use should be avoided during modern therapeutic phage applications, be it in agribusiness or human therapeutic settings. However, this may be easier said than done, since it is not clear whether a clear-cut division between lysogenic and lytic phages exists in nature, and well-established criteria for differentiating virulent phages from temperate phages are not available at the present time. Nevertheless, several possible approaches, which are discussed below, may be considered.

Adams (1959b) suggested that plaque morphology and activity spectrum effectively differentiate virulent and temperate phages; i.e., clear plaques and a broad spectrum of activity is characteristic of virulent phages, and turbid plaques and a narrow target range is indicative of temperate phages. However, the target range of a given phage and the clarity of the plaques it produces are strongly influenced by multiple, variable factors, including the nature of the host strain and its physiology at the time of testing, and the temperature and pH of the media in which testing is conducted. Also, some phages with a relatively broad host range lysogenize their targeted bacteria (Strauch et al., 2001), and some well-known transducing phages (such as P1) form clear plagues (Ikeda and Tomizawa, 1965). Therefore, the criteria of clear plaque morphology and broad activity spectrum, while suggestive of a "lytic" nature of the phage, may not always provide accurate information about its transducing ability. Sander and Schmieger (2001) have developed a PCR-based protocol for detecting generalized transducing bacteriophages in natural habitats, which is based on the rationale that lysates of generalized transducing phages contain the entire genome of the bacterial hosts. Thus, PCR primers specific for such broadly studied genes as those for 16S RNAs can be used to determine whether host-cell DNA has been packaged into the phage and, therefore, to determine the phage's ability to transduce its host. The above protocol has been shown to work well with *Salmonella* phages, but it will need to be carefully optimized for other bacterial species, which limits its general usefulness at the present time.

One alternative approach would be to fully sequence each and every phage to be used in agricultural or human therapeutic settings, and to avoid using phages that contain genes directly associated with bacterial virulence or antibiotic resistance. Indeed, the availability of whole phage genome sequences will allow candidate therapeutic phages to be screened for the presence of all potentially harmful genes whose nucleotide sequences are included in available databases (e.g., GenBank)—and phages containing such genes can be excluded from therapeutic preparations. Nucleotide sequence data are robust, and they are not affected by environmental parameters that may influence a phage's life cycle or behavior. Furthermore, sequence data are unambiguous and virtually free of human error, and they can easily be compared by various laboratories. Finally, using phage sequencing with comparative genome analysis will provide an unprecedented level of upgradeability; i.e., it is ideally suited to keep up with last minute developments in the fields of phage biology and bacterial genetics. For example, large numbers of bacterial gene sequences are submitted almost daily to GenBank, and the delineation of the roles of various genes in bacterial metabolism are continuously being improved as a result of ongoing scientific discoveries. In addition, the increased availability of robotic sequencers and the associated rapidly decreasing cost of sequencing make the sequencing of whole phage genomes a technically feasible approach, which can be invaluable for identifying "undesirable" genes in all phage genomes. The introduction and common availability of rapid and high-throughput, sequence-based screening methodologies (e.g., microarrays) also is likely to be useful in designing approaches for rapidly screening phages for the presence of "undesirable" genes, and for further modernization of phage therapy.

14.5.4. Comparison of Phages and Antibiotics: Advantages and Limitations of Phage-Based Therapies

Some of the key differences between therapeutic phages and antibiotics are summarized in Table 14.4. One of the main differences between phages and antibiotics, as far as their antimicrobial activity is concerned, is their target range. Phages are very specific, and they will normally only lyse strains of one species or a subset of strains within a species. On the other hand, antibiotics usually have a much broader target range, and a single antibiotic can often be used to treat diseases caused by various bacterial pathogens. The narrow specificity of phages, while advantageous in general (because it permits targeting of specific pathogens without affecting desirable bacterial flora), is also likely to create some technical challenges during the modern era of phage therapy, as it did in the 1940s and 1950s. Indeed, as noted earlier, their narrow specificity (compared to antibiotics) was a main reason for the decline of interest in therapeutic phage research in the West after antibiotics became available. In this context, the successful use of therapeutic phages in modern clinical settings will depend strongly on the ability of clinical microbiology laboratories to rapidly identify the etiologic agent and to ensure that the phage preparations are highly lytic *in vitro* against the bacterial strains causing the disease (Pirisi, 2000). This, in turn, may require the development of standardized testing methodologies to enable rapid identification of the appropriate therapeutic phage cocktail(s) before starting treatment, and the establishment of infrastructures to support phage therapy in hospitals and clinics.

Several studies in which the efficacy of phage preparations and antibiotics were evaluated side-by-side suggest that, when phage preparations are carefully selected, they can be as effective as—and sometimes more effective than—antibiotics in treating certain infections in humans (Meladze et al., 1982; Kochetkova et al., 1989; Sakandelidze, 1991) and experimentally infected animals (Smith and Huggins, 1982). For example, Meladze et al. (1982) used *S. aureus*-specific phages to treat 340 patients having purulent disease of the lung parenchyma and pleura. Over a period of 2 to 4 weeks, phages were administered by various routes, including intravenous and intrapleural injections, inhalations, and tracheal catheters. The results were evaluated based on the general condition of the patients, examination of x-rays, reduction in the amount of purulent sputum, and microbiological analysis of blood and sputum. Complete recovery was higher among the phage-treated patients (82% recovered) than among the antibiotic-treated patients (64% recovered) (Table 14.2). In another study (Kochetkova et al., 1989), phages produced by the EIBMV were used, alone and in combination with antibiotics, to treat various bacterial infections in 65 immuno-compromised cancer patients, and an additional

TABLE 14.4
Comparison of the Prophylactic and Therapeutic use of Phages and Antibiotics

Bacteriophages	Antibiotics
Phages are highly effective in killing their targeted bacteria (i.e., their action is bactericidal).	Some antibiotics (e.g., chloramphenicol) are bacteriostatic; i.e., they inhibit the growth of bacteria rather then killing them.
The high selectivity of bacteriophages permits the targeting of specific pathogens, without affecting desirable bacterial flora (i.e., phages are unlikely to affect adversely the "colonization pressure" of patients).	Antibiotics attack not only the disease-causing bacteria, but also all susceptible microorganisms, including the normal—and often beneficial - microflora of the host. Thus, their non-selective action affects the patient's microbial balance, which may lead to various side effects.
Humans are exposed to phages throughout life, and well tolerate them. Only a few minor side effects have been reported for therapeutic phages, and they may have been due to the liberation of endotoxins from bacteria lysed *in vivo* by the phages. Such effects also may be observed when antibiotics are used.	Multiple side effects, including intestinal disorders, allergies, and secondary infections (e.g., yeast infections) have been reported for antibiotics.
Because of phages' specificity, their use is not likely to select for phage resistance in other (i.e., not-targeted) bacterial species.	Because of their broad spectrum of activity, antibiotics may select for resistant mutants of many pathogenic bacterial species, not just for resistant mutants of the targeted bacteria.
Natural co-evolution of bacteria and phages should provide unprecedented flexibility in obtaining new phages lytic against phage-resistant bacteria emerging as a result of exposure to other phages or because of natural shifts in bacterial populations.	Developing a new antibiotic (e.g., against antibiotic-resistant bacteria) is a time-consuming process and may take several years to accomplish.
Because of phages' specificity, their successful use for preventing or treating bacterial infections requires identification of the etiologic agent and determining its *in vitro* susceptibility to the phage preparation prior to initiating phage treatment.	Compared to phages, antibiotics have a higher probability of being effective when administered before the identity of the etiologic agent is known.

66 cancer patients were treated with antibiotics alone. Overall, 82% of the phage-treated patients recovered from their infections, compared to ca. 61% of those treated only with antibiotics. At present, however, there is no consensus as to whether the combined use of phages and antibiotics is beneficial. In fact, in situations where the antibiotics interfere with the bacterial host's ability to support replication of the phages, the combined use of phages and antibiotics was hypothesized to be detrimental to the efficacy of phage treatment and to the overall outcome of therapy (Merril et al., 2003). On the other hand, several studies in which phages and antibiotics were used simultaneously (e.g., to treat chromic pulmonary suppurations, infectious allergoses, etc.) reported that the combined use of phages and antibiotics improved the outcome of therapy (Ioseliani et al., 1980; Sakandelidze, 1991; 1990).

The complementary use of phages and antibiotics may also help reduce the problem associated with the emergence of antibiotic- or phage-resistant bacterial mutants (because antibiotic resistance and phage resistance arise through different mechanisms). The major antibiotic-resistance mechanisms include enzymatic inactivation (e.g., β-lactamases cleave the β-lactam ring of penicillin), limiting access to the target ("efflux resistance"), and alteration of the bacterial target (e.g., resistance to rifampin is often due to mutations that prevent its binding to bacterial RNA) (Yao and Moellering, 1995). On the other hand, the two major phage-resistance mechanisms are thought to involve modifying the bacterial receptors, so the phage's tail fibers cannot attach to them, and modifying the bacterial anti-phage DNA-restriction enzymes. Thus, the simultaneous availability of the two classes of antibacterials may provide a much-needed "safety net" while antibiotic-resistant bacteria are emerging and spreading.

Very different timeframes and financial considerations are involved in the development of new therapeutic phage preparations and antibiotics, which may markedly affect how strongly the emergence of resistance affects their usefulness. When resistance against an antibiotic develops, it takes several years and several million dollars to develop a new antibiotic (Silver and Bostian, 1993)—and by the time the new drug is approved for human therapeutic use (or shortly after introducing it into medical practice), bacterial mutants resistant to it may already be identified. An example of this phenomenon is linezolid, a new antibiotic with activity against various bacterial pathogens, including vancomycin-resistant enterococci (VRE). Linezolid was in development for several years, and it was approved for human therapeutic use—in the United States—in April 2000. However, VRE mutants resistant to linezolid were identified in clinics in the United States less than a year later (Prystowsky et al., 2001), and at least one linezolid-treated patient was reported to have developed resistance during therapy (Gonzales et al., 2001). On the other hand, the natural coevolution of bacteriophages and their hosts, and the vast number of phages in nature, make it easy to identify new phages against any phage-resistant bacteria, a reflection of the "endless cycles of bacterial defenses and phage counter defenses" described by Lenski (1984). Indeed, phage therapy provides an unprecedented flexibility in rapidly responding to changes in bacterial pathogen populations or susceptibility, and to quickly update phage preparations if and when resistance or strain population peculiarities hinder their effectiveness. Moreover, custom-tailoring of therapeutic phage cocktails for a given patient or clinical center is also a possibility, as has been demonstrated in several studies from the former Soviet Union and Poland (Zhukov-Verezhnikov et al., 1978; Cislo et al., 1987; Slopek et al., 1987).

The aforementioned updating and custom-development of phage preparations will require extensive ongoing monitoring of the major pathogenic bacterial strains of concern and a degree of flexibility from the regulatory agencies, so that the benefits of the rapid developmental cycle for therapeutic phages are not negated by an excessively lengthy regulatory approval process. In this context, in the former Soviet Union the usual practice was to approve the initial, "principal" phage preparation and its manufacturing and quality control protocols. Thus, the phages in the preparation could be substituted/updated with new phages as required, as long as all of the approved product specifications and quality control criteria were met. In order for phage therapy to be successful in the developed world, it is likely that a similarly flexible approach will have to be adopted in the West. That type of approach is not unprecedented, at least in the United States. For example, the United States' influenza vaccine is continuously updated with a new vaccine strain (to keep up with the natural shift in the flu virus population), and a highly expedited regulatory approval process is in place to facilitate rapid changes in the strains that need to be used, which allows the new vaccines to be produced in a timely fashion.

ACKNOWLEDGMENTS

We are grateful to the many scientists in Tbilisi, and to Beata Weber-Dabrowska and Andrzej Gorki in Wroclaw, who have shared with us the expertise and experience they have gained over many years, and to Elena Jones and Svetlana Kameneva, who have helped us to understand the current "phage situation" in Russia. We also very much appreciate the critical input, during the preparation of this chapter, of Burton Guttman, Raul Raya, Donna Duckworth, Arnold Kreger, Mzia Kutateladze, and the many students in the phage laboratory at Evergreen. We thank Gautam Dutta for his extensive help with the references and figures.

REFERENCES

Abdul-Hassan, H., El-Tahan, E., Massoud, E. and Gomaa, R., Bacteriophage therapy of pseudomonas burn wound sepsis. *Ann Medit Burn Club* 3: 262–264, 1990.

Adamia, R.S., Matitashvili, E.A., Kvachadze, L.I., Korinteli, V.I., Matoyan, D.A., Kutateladze, M.I. and Chanishvili, T.G., The virulent bacteriophage IRA of *Salmonella typhimurium*: cloning of phage genes which are potentially lethal for the host cell. J Basic Microbiol 30: 707–716, 1990.

Adams, M.H. (1959a). "Antigenic properties," pp. 97–119 in *Bacteriophages*. Interscience Publishers, Inc., London.

Adams, M.H., (1959b). "Lysogeny," pp. 365–380 in *Bacteriophages*. Interscience Publishers, Inc., London.

Alisky, J., Iczkowski, K., Rapoport, A. and Troitsky, N., Bacteriophages show promise as antimicrobial agents. *J Infect* 36: 5–15, 1998.

Andriashvili, I.A., Kvachadze, L.I., Bashakidze, R.P., Adamiia, R. and Chanishvili, T.G., Molecular mechanism of phage DNA protection from the restriction endonucleases of *Staphylococcus aureus* cells. *Mol Gen Mikrobiol Virusol*: 43–45, 1986.

Angel, M.F., Ramasastry, S.S., Manders, E.K., Ganfield, D. and Futrell, J.W., Beneficial effect of Staphage lysate in the treatment of chronic recurring hidradenitis suppurativa. *Surgical forum* 38: 111, 1987.

Anpilov, L.I. and Prokudin, A.A., Preventive effectiveness of dried polyvalent *Shigella* bacteriophage in organized collective farms. *Voen Med Zh*: 39–40, 1984.

Appelmans, R., Le bacteriophage dans l'organisme. *Comp Rend Soc de Biol* 85: 722–724, 1921.

Arkwright, J.A., Source and characteristics of certain cultures sensitive to bacteriophage. *Brit J Exper Pathol* 5: 23–33, 1924.

Arnold, L. and Weiss, E., Antigenic properties of bacteriophage. *J Infect Dis* 34: 317–327, 1924.

Babalova, E. G., Katsitadze, K.T., Sakvarelidze, L.A., Imnaishvili, N., Sharashidze, T.G., Badashvili, V.A., Kiknadze, G.P., et al., Preventive value of dried dysentery bacteriophage. *Zh Mikrobiol Epidemiol Immunobiol* 45: 143–145, 1968.

Bachrach, G., Leizerovici-Zigmond, M., Zlotkin, A., Naor, R. and Steinberg, D., Bacteriophage isolation from human saliva. *Lett Appl Microbiol* 36: 50–53, 2003.

Baker, A.G., Staphylococcus bacteriophage lysate topical and parenteral use in allergic patients. *The Pennsylvania Medical Journal* 66: 25–28, 1963.

Barrow, P.A. and Soothill, J.S., Bacteriophage therapy and prophylaxis: rediscovery and renewed assessment of potential. *Trends Microbiol* 5: 268–271, 1997.

Bergh, O., Borsheim, K.Y., Bratbak, G. and Heldal, M., High abundance of viruses found in aquatic environments. *Nature* 340: 467–468, 1989.

Betley, M.J. and Mekalanos, J.J., Staphylococcal enterotoxin A is encoded by phage. *Science* 229: 185–187, 1985.

Biswas, B., Adhya, S., Washart, P., Paul, B., Trostel, A.N., Powell, B., Carlton, R., et al., Bacteriophage therapy rescues mice bacteremic from a clinical isolate of vancomycin-resistant *Enterococcus faecium*. *Infect Immun* 70: 204–210, 2002.

Bogorishvili, E.V., Sharashidze, T.G. and Nadiradze, M.N., "Use of Bacteriophages in Women with Pelvic Inflamatory Disease Suffering from Infertility Due To Fallopian-Tube Infections," pp. 77–78 in *Proceedings of the Second Congress of Midwives and Gynecologists*, Tbilisi, Republic of Georgia, 1985.

Bogovazova, G.G., Voroshilova, N.N. and Bondarenko, V.M., The efficacy of *Klebsiella pneumoniae* bacteriophage in the therapy of experimental *Klebsiella* infection. *Zh Mikrobiol Epidemiol Immunobiol*: 5–8, 1991.

Bogovazova, G.G., Voroshilova, N.N., Bondarenko, V.M., Gorbatkova, G.A., Afanas'eva, E.V., Kazakova, T.B., Smirnov, V.D., et al., Immunobiological properties and therapeutic effectiveness of preparations from *Klebsiella* bacteriophages. *Zh Mikrobiol Epidemiol Immunobiol*: 30–33, 1992.

Bordet, J., Problem of bacteriophage or of transmissible bacterial autolysis. *Ann Inst Pasteur* 39: 717–763, 1925.

Brüssow, H. and Hendrix, R.W., Phage genomics: Small is beautiful. *Cell* 108: 13–16, 2002.

Bruynoghe, R. and Maisin, J., Essais de thérapeutique au moyen du bactériophage du Staphylocoque. *J Compt Rend Soc Biol* 85: 1120–1121, 1921.

Buda, J., Suk, K., Vymola, F. and Pillich, J., Treatment of recurrent inflammation of Bartholin's gland with staphylococcal lysate. *Cesk Gynekol* 33: 669–672, 1968.

Bull, J.J., Levin, B.R., DeRouin, T., Walker, N. and Bloch, C.A., Dynamics of success and failure in phage and antibiotic therapy in experimental infections. *BMC Microbiol* 2: 35, 2002.

Burnet, F.M., McKie, M. and Wood, I.J., Investigations on bacillary dysentery in infants with special reference to bacteriophage phenomena. *M J Australia* 2: 71–78, 1930.

Caldwell, J.A., Bacteriologic and bacteriophagic study of infected urines. *J Infect Dis* 43: 353–362, 1928.

Carlton, R.M., Phage therapy: past history and future prospects. *Arch Immunol Ther Exp* (Warsz) 47: 267–274, 1999.

Cerveny, K.E., DePaola, A., Duckworth, D.H. and Gulig, P.A., Phage therapy of local and systemic disease caused by *Vibrio vulnificus* in iron-dextran-treated mice. *Infect Immun* 70: 6251–6262, 2002.

Chain, E., Florey, H.W., Gardner, A.D., Heatley, N.G., Jennings, M.A., Orr-Ewing, J. and Sanders, A.G., Penicillin as chemotherapeutic agent. *Lancet* 2: 226, 1940.

Cipollaro, A.C. and Sheplar, A.E., Therapeutic uses of bacteriophage in pyodermias. *Arch Dermat Syph* 25: 280–293, 1932.

Cislo, M., Dabrowski, M., Weber-Dabrowska, B. and Woyton, A., Bacteriophage treatment of suppurative skin infections. *Arch Immunol Ther Exp* (Warsz) 35: 175–183, 1987.

Crichton, M., *Prey.* HarperCollins, New York, 2002.

Dean, J.H., Silva, J.S., McCoy, J.L., Chan, S.P., Baker, J.J., Leonard, C. and Herberman, R.B., *In vitro* human reactivity to staphylococcal phage lysate. *J Immunol* 115: 1060–1064, 1975.

Desranleau, J.-M., "Progress in the Treatment of Typhoid Fever with Vi Phages," pp. 473–478 in *Candian Journal of Public Health*, Novia Scotia, Halifax, Cananda, 1949.

d'Herelle, F., *The Bacteriophage and Its Behavior.* Williams and Wilkins, Baltimore, Maryland, 1926.

d'Herelle, F., *The Bacteriophage and Its Clinical Applications.* Charles C Thomas, Springfield, Illinois, 1930.

d'Herelle, F., Essai de traîtement de la peste bubonique par le bactériophage. *La Presse Méd* 33: 1393–1394, 1925.

d'Herelle, F., Sur un microbe invisible antagoniste des bacilles dysentériques. *Compt Rend Acad Sci* (Paris) 165: 373–375, 1917.

d'Herelle, F., Malone, R.H. and Lahiri, M., "The treatment and prophylaxis of infectious diseases of the intestinal tract and of cholera in particular," in *Transaction of the 7th Congress held in British India.* Far Eastern Association of Tropical Medicine, Calcutta, Thacker, 1928.

Drake, J.W., Charlesworth, B., Charlesworth, D. and Crow, J.F., Rates of spontaneous mutation. *Genetics* 148: 1667–1686, 1998.

Dubos, R., Straus, J.H. and Pierce, C., The multiplication of bacteriophage in vivo and its protective effect against an experimental infection with *Shigella dysenteriae*. *J Exp Med* 20: 161–168, 1943.

Duckworth, D.H. and Gulig, P.A., Bacteriophages: Potential treatment for bacterial infections. *BioDrugs* 16: 57–62, 2002.

Dutta, N.K., Prophylaxis and treatment of cholera based on experimental evidences. *Indian J Med Res* 52: 933–941, 1964.

Eaton, M.D. and Bayne-Jones, S., Bacteriophage therapy. Review of the principles and results of the use of bacteriophage in the treatment of infections. *J Am Med Assoc* 23: 1769–1939, 1934.

Eggers, C.H., Kimmel, B.J., Bono, J.L., Elias, A.F., Rosa, P. and Samuels, D.S., Transduction by phiBB-1, a bacteriophage of *Borrelia burgdorferi*. *J Bacteriol* 183: 4771–4778, 2001.

Evans, A.C., Inactivation of antistreptococcus bacteriophage by animal fluids. *Public Health Reports* 48: 411–426, 1933.

Fleming, A., On the antibacterial action of cultures of a Penicillium, with special reference to their use in the isolation of *B. influenzae*. *Brit J Exp Path* 10: 226–236, 1929.

Fogelman, I., Davey, V., Ochs, H.D., Elashoff, M., Feinberg, M.B., Mican, J., Siegel, J.P., et al., Evaluation of CD4+ T cell function In vivo in HIV-infected patients as measured by bacteriophage phiX174 immunization. *J Infect Dis* 182: 435–441, 2000.

Freeman, V.J.S., Studies on the virulence of bacteriophage-infected strains of *Corynobacterium diphtheriae*. *J Bacteriol* 61: 675–688, 1951.

Friedman, M. and Friedland, G.W., "Alexander Fleming and antibiotics," pp. 168–191 in *Medicine's Ten Greatests Discoveries*. Yale University Press, New Haven, Connecticut, 1998.

Garcia, P., Garcia, E., Ronda, C., Lopez, R. and Tomasz, A., A phage-associated murein hydrolase in *Streptococcus pneumoniae* infected with bacteriophage Dp-1. *J Gen Microbiol* 129 (Pt 2): 489–497, 1983.

Geier, M.R., Attallah, A.F. and Merril, C.R., Characterization of *Escherichia coli* bacterial viruses in commercial sera. *In Vitro* 11: 55–58, 1975.

Geier, M.R., Trigg, M.E. and Merril, C.R., Fate of bacteriophage lambda in non-immune germ-free mice. *Nature* 246: 221–223, 1973.

Gonzales, R.D., Schreckenberger, P.C., Graham, M.B., Kelkar, S., DenBesten, K. and Quinn, J.P., Infections due to vancomycin-resistant *Enterococcus faecium* resistant to linezolid. *Lancet* 357: 1179, 2001.

Gratia, A., La lyse transmissible du staphylocoque: Sa production, ses applications thérapeutiques. Compt Rend. *Soc de biol* 86: 276–278, 1922.

Hankin, E.H., L'action bactericide des eaux de la Jumna et du Gange sur le vibrion du cholera. *Ann de l'Inst Pasteur* 10: 511, 1896.

Hausler, T., *Gesund Durch Viren*, Germany, 2003.

Ho, K., Bacteriophage therapy for bacterial infections. Rekindling a memory from the pre-antibiotics era. *Perspect Biol Med* 44: 1–16, 2001.

Ikeda, H. and Tomizawa, J.I., Transducing fragments in generalized transduction by phage P1. I. Molecular origin of the fragments. *J Mol Biol* 14: 85–109, 1965.

Inchley, C.J., The activity of mouse Kupffer cells following intravenous injection of T4 bacteriophage. *Clin Exp Immunol* 5: 173–187, 1969.

Ioseliani, G.D., Meladze, G.D., Chkhetiia, N., Mebuke, M.G. and Kiknadze, N., Use of bacteriophage and antibiotics for prevention of acute postoperative empyema in chronic suppurative lung diseases. *Grudn Khir*: 63–67, 1980.

Jikia, D., Chkhaidze, N., Imedashvili, E., Mgaloblishvili, I., Tsitlanadze, G., Katsarava, R., Morris, J.G., et al., The use of a novel biodegradable preparation capable of the sustained release of bacteriophages and ciprofloxacin, in the complex treatment of multidrug-resistant Staphylococcus aureus-infected local radiation injuries caused by the exposure to Sr90. *Clin Exp Dermatol,* 2004, in press.

Kaczkowski, H., Weber-Dabrowska, B., Dabrowski, M., Zdrojewicz, Z. and Cwioro, F., Use of bacteriophages in the treatment of chronic bacterial diseases. *Wiad Lek* 43: 136–141, 1990.

Katsarava, R. and Alavidze, Z., Polymer blends as biodegradable matrices for preparing biocomposites, U.S. patent 6,703,040, Intralytix, Inc, U.S.A, 2004.

Katsarava, R., Beridze, V., Arabuli, N., Kharadze, D., Chu, C.C. and Won, C. Y., Amino acid-based bioanalogous polymers. Synthesis, and study of regular poly(ester amide)s based on bis(alpha-amino acid) alpha-alkylene diesters, and aliphatic dicarboxylic acids. *J Polymer Sci* 37: 391–407, 1999.

King, W.E., Boyd, D.A. and Conlin, J.H., The cause of local reactions following the administration of *Staphylococcus* bacteriophage. *Am J Clin Pathol* 4: 336–345, 1934.

Knouf, E.G., Ward, W.E., Reichle, P.A., Bower, A.G. and Hamilton, P.M., Treatment of Typhoid Fever with Type Specific Bacteriophage. *Journal of the American Medical Association* 132: 134–138, 1946.

Kochetkova, V.A., Mamontov, A.S., Moskovtseva, R.L., Erastova, E.I., Trofimov, E.I., Popov, M.I. and Dzhubalieva, S.K., Phagotherapy of postoperative suppurative-inflammatory complications in patients with neoplasms. *Sov Med*: 23–26, 1989.

Kress, D.W., Graham, W.P., Davis, T.S. and Miller, S.H., A preliminary report on the use of Staphage lysate for treatment of hidradenitis suppurativa. *Ann Plast Surg* 6: 393–395, 1981.

Krestovikova, V.A., Phage therapy and phage prophylaxis and their development through the work of Soviet researchers. *Zh Mikrobiol Epidemiol Immunobiol* 3: 56–65, 1947.

Krueger, A.P. and Scribner, E.J., Bacteriophage Therapy II. The bacteriophage: Its nature and its therapeutic use. *J Am Med Assoc* 19: 2160–2277, 1941.

Kucharewicz-Krukowska, A. and Slopek, S., Immunogenic effect of bacteriophage in patients subjected to phage therapy. *Arch Immunol Ther Exp* (Warsz) 35: 553–561, 1987.

Kuchment, A., "Superbug killers," pp. 42–43 in *Newsweek International,* 2001.

Kwarcinski, W., Lazarkiewicz, B., Weber-Dabrowska, B., Rudnicki, J., Kaminski, K. and Sciebura, M., Bacteriophage therapy in the treatment of recurrent subphrenic and subhepatic abscess with jejunal fistula after stomach resection. *Pol Tyg Lek* 49: 535, 23–24, 1994.

Lang, G., Kehr, P., Mathevon, H., Clavert, J.M., Sejourne, P. and Pointu, J., Bacteriophage therapy of septic complications of orthopaedic surgery. *Rev Chir Orthop Reparatrice Appar Mot* 65: 33–37, 1979.

Langley, R., Kenna, D.T., Vandamme, P., Ure, R. and Govan, J.R., Lysogeny and bacteriophage host range within the *Burkholderia cepacia* complex. *J Med Microbiol* 52: 483–490, 2003.

Larkum, N.W. Bacteriophage as a substitute for typhoid vaccine. *J Bacteriol* 17: 42, 1929a.

Larkum, N.W. Bacteriophage from Public Health standpoint. *Am J Pub Health* 19: 31–36, 1929b.

Lazareva, E.B., Smirnov, S.V., Khvatov, V.B., Spiridonova, T.G., Bitkova, E.E., Darbeeva, O.S., Maiskaia, L.M., et al., Efficacy of bacteriophages in complex treatment of patients with burn wounds. *Antibiot Khimioter* 46: 10–14, 2001.

Lee, B., Murakami, M. and Mizukoshi, M. (1985a). Polyclonal antibody secretion in mouse spleen cells induced by staphylococcal phage lysate (SPL). *Yakugaku Zasshi* 105: 747–750.

Lee, B., Murakami, M., Mizukoshi, M., Shinomiya, N. and Yada, J. (1985b). Effects of staphylococcal phage lysate (SPL) on immunoglobulin production in human peripheral lymphocyte cultures. *Yakugaku Zasshi* 105: 574–579.

Lee, B., Yamaguchi, K., Murakami, M. and Mizukoshi, M., Effect of staphylococcal phage lysate (SPL) on polyclonal B cell activation in mouse spleen cells *in vitro*. *Nippon Saikingaku Zasshi* 37: 787–788, 1982.

Lenski, R.E., Coevolution of bacteria and phage: are there endless cycles of bacterial defenses and phage counterdefenses? *J Theor Biol* 108: 319–325, 1984.

Leverentz, B., Conway, W.S., Alavidze, Z., Janisiewicz, W.J., Fuchs, Y., Camp, M.J., Chighladze, E., et al., Examination of bacteriophage as a biocontrol method for *Salmonella* on fresh-cut fruit: a model study. *J Food Prot* 64: 1116–1121, 2001.

Levin, B. and Bull, J.J., Phage therapy revisited: The population biology of a bacterial infection and its treatment with bacteriophage and antibiotics. *Am Naturalist* 147: 881–898, 1996.

Lewis, S., *Arrowsmith*. Grosset & Dunlap, New York, 1925.

Livermore, D.M., Antibiotic resistance in staphylococci. Int J Antimicrob Agents 16 Suppl 1: S3–10, 2000.

Lopez, V., Ochs, H.D., Thuline, H.C., Davis, S.D. and Wedgwood, R.J., Defective antibody response to bacteriophage phichi 174 in Down syndrome. *J Pediatr* 86: 207–211, 1975.

Marcuk, L.M., Nikiforov, V.N., Scerbak, J.F., Levitov, T.A., Kotljarova, R.I., Naumsina, M.S., Davydov, S.U., et al., Clinical studies of the use of bacteriophage in the treatment of cholera. *Bull World Health Organ* 45: 77–83, 1971.

Markoishvili, K., Tsitlanadze, G., Katsarava, R., Morris, J.G., Jr. and Sulakvelidze, A., A novel sustained-release matrix based on biodegradable poly(ester amide)s and impregnated with bacteriophages and an antibiotic shows promise in management of infected venous stasis ulcers and other poorly healing wounds. *Int J Dermatol* 41: 453–458, 2002.

Martin, R., "How ravenous Soviet viruses will save the world," pp. 164–177 in *Wired,* October, 2003.

Meitert, E., Petrovici, M., Sima, F., Costache, G. and Savulian, C., Investigation on the therapeutical efficiency of some adapted bacteriophages in experimental infection with *Pseudomonas aeruginosa*. *Arch Roum Pathol Exp Microbiol* 46: 17–26, 1987.

Meladze, G.D., Mebuke, M.G., Chkhetiia, N., Kiknadze, N. and Koguashvili, G.G., Efficacy of staphylococcal bacteriophage in the treatment of purulent lung and pleural diseases. *Grudn Khir*: 53–56, 1982.

Merril, C.R., Biswas, B., Carlton, R., Jensen, N.C., Creed, G.J., Zullo, S. and Adhya, S., Long-circulating bacteriophage as antibacterial agents. *Proc Natl Acad Sci U S A* 93: 3188–3192, 1996.

Merril, C.R., Friedman, T.B., Attallah, A.F., Geier, M.R., Krell, K. and Yarkin, R., Isolation of bacteriophages from commercial sera. *In Vitro* 8: 91–93, 1972.

Merril, C.R., Scholl, D. and Adhya, S.L., The prospect for bacteriophage therapy in Western medicine. *Nat Rev Drug Discov* 2: 489–497, 2003.

Meshalova, A.N., Kilesso, V.A., Kalianev, A.V., Rukhadze, E.Z. and Turchinskaia, M.V., Protection of bacteriophage from the destructive effect of gastric juice by means of pectin. *Zh Mikrobiol Epidemiol Immunobiol* 46: 67–72, 1969.

Milch, H. and Fornosi, F., Bacteriophage contamination in live poliovirus vaccine. *J Biol Stand* 3: 307–310, 1975.

Monsur, K.A., Rahman, M.A., Huq, F., Islam, M.N., Northrup, R.S. and Hirschhorn, N., Effect of massive doses of bacteriophage on excretion of Vibrios, duration of diarrhoea and output of stools in acute cases of cholera. *Bull World Health Organ* 42: 723–732, 1970.

Moody, E.E., Trousdale, M.D., Jorgensen, J.H. and Shelokov, A., Bacteriophages and endotoxin in licensed live-virus vaccines. *J Infect Dis* 131: 588–591, 1975.

Morton, H.E. and Engely, F.B., Dysentery bacteriophage: Review of the literature on its prophylactic and therapeutic uses in man and in experimental infections in animals. *J Am Med Assoc* 127: 584–591, 1945.

Morton, H.E. and Perez-Otero, E.J., The generation of dysentery bacteriophage in vivo during experimental infections with *Shigella paradysenteriae*, flexner, in mice. *J Bacteriol* 47: 475–476, 1944.

Mudd, S., Resistance against *Staphylococcus aureus*. *JAMA* 218: 1671–1673, 1971.

Mulczyk, M. and Slopek, S., Use of a new phage preparation in prophylaxis and treatment of shigellosis. *Acta Microbiol Acad Sci Hung* 21: 115–119, 1974.

Naidu, B.P.B. and Avari, C.R., Bacteriophage in the treatment of plague. *Ind Jour Med Res* 19: 737–748, 1932.

Nelson, D., Loomis, L. and Fischetti, V.A., Prevention and elimination of upper respiratory colonization of mice by group A streptococci by using a bacteriophage lytic enzyme. *Proc Natl Acad Sci U S A* 98: 4107–4112, 2001.

O'Brien, A.D., Newland, J.W., Miller, S.F., Holmes, R.K., Smith, H.W. and Formal, S.B., Shiga-like toxin-converting phages from *Escherichia coli* strains that cause hemorrhagic colitis or infantile diarrhea. *Science* 226: 694–696, 1984.

Ochs, H.D., Buckley, R.H., Kobayashi, R.H., Kobayashi, A.L., Sorensen, R.U., Douglas, S.D., Hamilton, B.L., et al., Antibody responses to bacteriophage phi X174 in patients with adenosine deaminase deficiency. *Blood* 80: 1163–1171, 1992.

Ochs, H.D., Davis, S.D. and Wedgwood, R.J., Immunologic responses to bacteriophage phi-X 174 in immunodeficiency diseases. *J Clin Invest* 50: 2559–2568, 1971.

Ochs, H.D., Lum, L.G., Johnson, F.L., Schiffman, G., Wedgwood, R.J. and Storb, R., Bone marrow transplantation in the Wiskott-Aldrich syndrome. Complete hematological and immunological reconstitution. *Transplantation* 34: 284–288, 1982.

Ochs, H.D., Nonoyama, S., Farrington, M.L., Fischer, S.H. and Aruffo, A. The role of adhesion molecules in the regulation of antibody responses. *Semin Hematol* 30: 72–79, 1993a.

Ochs, H.D., Nonoyama, S., Zhu, Q., Farrington, M. and Wedgwood, R.J. Regulation of antibody responses: The role of complement and adhesion molecules. *Clin Immunol Immunopathol* 67: S33–40, 1993b.

Orlova, Z.N. and Garnova, N.A., Use of tablet-form polyvalent bacteriophage with acid resistant coating in the treatment of dysentery in children. *Vopr Okhr Materin Det* 15: 25–29, 1970.

Pasricha, C.L., De Monte, A.J.H. and O'Flynn, E.G., Bacteriophage in the treatment of cholera. *Indian Med Gazette* 71: 61–68, 1936.

Pavlenishvili, I. and Tsertsvadze, T., Bacteriophagotherapy and enterosorbtion in treatment of sepsis of newborns caused by gram-negative bacteria. *Pren Neon Infect* 11: 104, 1993.

Pavlova, L.I., Sumarokov, A.A., Solodovnikov Iu, P. and Nikitiuk, N.M., Use of dysentery bacteriophage as a means of preventing dysentery. *Zh Mikrobiol Epidemiol Immunobiol* 50: 27–32, 1973.

Payne, R.J. and Jansen, V.A., Evidence for a phage proliferation threshold? *J Virol* 76: 13123; author reply 13123–13124, 2002.

Payne, R.J. and Jansen, V.A., Pharmacokinetic principles of bacteriophage therapy. *Clin Pharmacokinet* 42: 315–325, 2003.

Payne, R.J. and Jansen, V.A., Understanding bacteriophage therapy as a density-dependent kinetic process. *J Theor Biol* 208: 37–48, 2001.

Payne, R.J., Phil, D. and Jansen, V.A., Phage therapy: the peculiar kinetics of self-replicating pharmaceuticals. *Clin Pharmacol Ther* 68: 225–230, 2000.

Peremitina, L.D., Berillo, E.A. and Khvoles, A.G., Experience in the therapeutic use of bacteriophage preparations in suppurative surgical infections. *Zh Mikrobiol Epidemiol Immunobiol*: 109–110, 1981.

Perepanova, T.S., Darbeeva, O.S., Kotliarova, G.A., Kondrat'eva, E.M., Maiskaia, L.M., Malysheva, V.F., Baiguzina, F.A., et al., The efficacy of bacteriophage preparations in treating inflammatory urologic diseases. Urol Nefrol (Mosk): 14–17, 1995.

Pillich, J., Tovarek, J. and Fait, M., Ein beitrag zur behandlung der akuten hamatogenen und chronischen selcundaren osteomyelifiden bei kindem. *Z Orthop lhre Grenzgeb* 116: 40–46, 1978.

Pirisi, A., Phage therapy—advantages over antibiotics? *Lancet* 356: 1418, 2000.

Pizzi, R. A.,. "Salving with Science. Ten decades of drug discovery," pp. 35–49 in *The Pharmaceutical Century,* 2000.

Podolsky, B.M., Regarding the effectiveness of dysentery bacteriophage. *Voenno Medits Zh*: 38–39, 1960.

Pokrovskaya, M.P., Kaganova, L.C., Morosenko, M.A., Bulgakova, A.G. and Skatsenko, E.E.,. *Treatment of wounds with bacteriophages*. State Publishing House *Medgiz*, Moscow, USSR, 1942.

Pollitzer, R., Swaroop, S. and Burrows, W., Cholera. *Monogr Ser World Health Organ* 58: 1–1019, 1959.

Prins, J.M., van Deventer, S.J., Kuijper, E.J. and Speelman, P., Clinical relevance of antibiotic-induced endotoxin release. *Antimicrob Agents Chemother* 38: 1211–1218, 1994.

Proskurov, V.A., Use of staphylococcal bacteriophage for therapeutic and preventive purposes. *Zh Mikrobiol Epidemiol Immunobiol* 47: 104–107, 1970.

Prystowsky, J., Siddiqui, F., Chosay, J., Shinabarger, D.L., Millichap, J., Peterson, L.R. and Noskin, G.A., Resistance to linezolid: Characterization of mutations in rRNA and comparison of their occurrences in vancomycin-resistant enterococci. *Antimicrob Agents Chemother* 45: 2154–2156, 2001.

Pyun, K.H., Ochs, H.D., Wedgwood, R.J., Yang, X.Q., Heller, S.R. and Reimer, C.B., Human antibody responses to bacteriophage phi X 174: Sequential induction of IgM and IgG subclass antibody. *Clin Immunol Immunopathol* 51: 252–263, 1989.

Radetsky, P., "The good virus," *Discover Magazine,* November 11, 1996.

Ridding, D., Acute bacillary dysentery in Khartoum province, Sudan, with special reference to bacteriophage treatment. Bacteriological Investigation. *J Hyg* 30: 387–401, 1930.

Ruska, H., Die sichtbarmachung der bakteriophagen lyse im ubermikroskop. *Naturwissenschaften* 28: 45, 1940.

Sakandelidze, V.M., Combined antibacterial treatment of inflammatory processes, Ph.D. thesis, Zhordania Scientific Research Institute of Human Reproduction, Tbilisi, Georgia, 1990.

Sakandelidze, V.M., The combined use of specific phages and antibiotics in different infectious allergoses. *Vrach Delo* 3: 60–63, 1991.

Salmon, G.G. and Symonds, M., Staphage lysate therapy in chronic staphylococcal infections. *The Journal of the Medical Society of New Jersey* 60: 188–193, 1963.

Samsygina, G.A. and Boni, E.G., Bacteriophages and phage therapy in pediatric practice. *Pediatriia*: 67–70, 1984.

Sander, M. and Schmieger, H., Method for host-independent detection of generalized transducing bacteriophages in natural habitats. *Appl Environ Microbiol* 67: 1490–1493, 2001.

Schicklmaier, P. and Schmieger, H., Frequency of generalized transducing phages in natural isolates of the *Salmonella typhimurium* complex. *Appl Environ Microbiol* 61: 1637–1640, 1995.

Schless, R.A., *Staphylococcus aureus* meningitis: Treatment with specific bacteriophage. *Am J Dis Children* 44: 813–822, 1932.

Schuch, R., Nelson, D. and Fischetti, V.A., A bacteriolytic agent that detects and kills *Bacillus anthracis*. *Nature* 418: 884–889, 2002.

Schultz, E.W., Bacteriophage: possible therapeutic aid in dental infections. *J Dental Res* 12: 295–310, 1932.

Sergienko, F.E., "Dry Bacteriophages, Their Preparation and Use," pp. 116–123 in *Microbiology and Epidemiology*, B. Babsky, I.G. Kochergin, and V.V. Parin (Eds.). Medical Publications Ltd, London, 1945.

Shrayer, D.P., Felix d'Herelle in Russia. *Bull Inst Pasteur* 94: 91–96, 1996.

Silver, L.L. and Bostian, K.A., Discovery and development of new antibiotics: the problem of antibiotic resistance. *Antimicrob Agents Chemother* 37: 377–383, 1993.

Sivitz, L.,. "Beating Superbugs," pp. 26–27 in *Washington Techway,* January 7, 2002.

Slopek, S., Durlakowa, I., Weber-Dabrowska, B., Dabrowski, M. and Kucharewicz-Krukowska, A., Results of bacteriophage treatment of suppurative bacterial infections. III. Detailed evaluation of the results obtained in further 150 cases. *Arch Immunol Ther Exp* (Warsz) 32: 317–335, 1984.

Slopek, S., Durlakowa, I., Weber-Dabrowska, B., Kucharewicz-Krukowska, A., Dabrowski, M. and Bisikiewicz, R. (1983a). Results of bacteriophage treatment of suppurative bacterial infections. I. General evaluation of the results. *Arch Immunol Ther Exp* (Warsz) 31: 267–291.

Slopek, S., Durlakowa, I., Weber-Dabrowska, B., Kucharewicz-Krukowska, A., Dabrowski, M. and Bisikiewicz, R. (1983b). Results of bacteriophage treatment of suppurative bacterial infections. II. Detailed evaluation of the results. *Arch Immunol Ther Exp* (Warsz) 31: 293–327.

Slopek, S., Kucharewicz-Krukowska, A., Weber-Dabrowska, B. and Dabrowski, M. (1985a). Results of bacteriophage treatment of suppurative bacterial infections. IV. Evaluation of the results obtained in 370 cases. *Arch Immunol Ther Exp* (Warsz) 33: 219–240.

Slopek, S., Kucharewicz-Krukowska, A., Weber-Dabrowska, B. and Dabrowski, M. (1985b). Results of bacteriophage treatment of suppurative bacterial infections. V. Evaluation of the results obtained in children. *Arch Immunol Ther Exp* (Warsz) 33: 241–259.

Slopek, S., Kucharewicz-Krukowska, A., Weber-Dabrowska, B. and Dabrowski, M. (1985c). Results of bacteriophage treatment of suppurative bacterial infections. VI. Analysis of treatment of suppurative staphylococcal infections. *Arch Immunol Ther Exp* (Warsz) 33: 261–273.

Slopek, S., Weber-Dabrowska, B., Dabrowski, M. and Kucharewicz-Krukowska, A., Results of bacteriophage treatment of suppurative bacterial infections in the years 1981–1986. *Arch Immunol Ther Exp* (Warsz) 35: 569–583, 1987.

Smith, H.W. and Huggins, M.B., Effectiveness of phages in treating experimental *Escherichia coli* diarrhoea in calves, piglets and lambs. *J Gen Microbiol* 129 (Pt 8): 2659–2675, 1983.

Smith, H.W. and Huggins, M.B., Successful treatment of experimental *Escherichia coli* infections in mice using phage: Its general superiority over antibiotics. *J Gen Microbiol* 128: 307–318, 1982.

Solodovnikov Iu, P., Pavlova, L.I., Emel'ianov, P.I., Garnova, N.A., Nogteva Iu, B., Sotemskii Iu, S., Bogdashich, O.M., et al., The prophylactic use of dry polyvalent dysentery bacteriopathe with pectin in preschool children's institutions. I. Results of a strictly controlled epidemiologic trial (Yaroslavl, 1968). *Zh Mikrobiol Epidemiol Immunobiol* 47: 131–137, 1970.

Solodovnikov Iu, P., Pavlova, L.I., Garnova, N.A., Nogteva Iu, B. and Sotemskii Iu, S., Preventive use of dry polyvalent dysentery bacteriophage in preschool institutions. II. Principles of present-day tactics and application schedule of bacteriophage. *Zh Mikrobiol Epidemiol Immunobiol* 48: 123–127, 1971.

Spence, R.C. and McKinley, E.B.T., Therapeutic value of bacteriophages in treatment of bacillary dysentery. *South M J* 17: 563–568, 1924.

Stone, R., Bacteriophage therapy. Stalin's forgotten cure. *Science* 298: 728–731, 2002.

Straub, M.E. and Applebaum, M., Studies on commercial bacteriophage products. *JAMA* 100: 110–113, 1933.

Strauch, E., Lurz, R. and Beutin, L., Characterization of a Shiga toxin-encoding temperate bacteriophage of *Shigella sonnei*. *Infect Immun* 69: 7588–7595, 2001.

Stroj, L., Weber-Dabrowska, B., Partyka, K., Mulczyk, M. and Wojcik, M., Successful treatment with bacteriophage in purulent cerebrospinal meningitis in a newborn. *Neurol Neurochir Pol* 33: 693–698, 1999.

Sulakvelidze, A., Alavidze, Z. and Morris, J.G., Jr., Bacteriophage therapy. *Antimicrob Agents Chemother* 45: 649–659, 2001.

Sulakvelidze, A. and Morris, J.G., Jr., Bacteriophages as therapeutic agents. *Ann Med* 33: 507–509, 2001.

Summers, W.C., "Bacteriophage Discovered," pp. 47–59 in *Felix d'Herelle and the Origins of Molecular Biology*. Yale University Press, New Haven, Connecticut, 1999.

Summers, W.C., Bacteriophage therapy. *Annu Rev Microbiol* 55: 437–451, 2001.

Taylor, J., Greval, S.D.S. and Thant, U., Bacteriophage in bacillary dysentery and cholera. *Ind J Med Res* 18: 117–136, 1930.

Thacker, P.D., Set a microbe to kill a microbe: Drug resistance renews interest in phage therapy. *JAMA* 290: 3183–3185, 2003.

Uhr, J.W., Finkelstein, M.S. and Baumann, J.B., Antibody formation. III. The primary and secondary antibody response to bacteriophage phi X 174 in guinea pigs. *J Exp Med* 115: 655–670, 1962.

Van Helvoort, T., Bacteriological and physiological research styles in the early controversy on the nature of the bacteriophage phenomenon. *Med Hist* 36: 243–270, 1992.

Vieu, J.F., *Les bactériophages*. Flammarion, Paris, 1975.

Vymola, F. and Buda, J., Staphylococcal infections. *J Hyg Epidemiol Microbiol Immunol* 21: 468–472, 1977.

Waldor, M.K. and Mekalanos, J.J., Lysogenic conversion by a filamentous phage encoding cholera toxin. *Science* 272: 1910–1914, 1996.

Ward, W.E., Protective action of VI bacteriophage in *Eberthella typhi* infections in mice. *Journal of Infectious Disease*: 172–176, 1942.

Weber-Dabrowska, B., Dabrowski, M. and Slopek, S., Studies on bacteriophage penetration in patients subjected to phage therapy. *Arch Immunol Ther Exp* (Warsz) 35: 563–568, 1987.

Weber-Dabrowska, B., Mulczyk, M. and Gorski, A. Bacteriophage therapy of bacterial infections: An update of our institute's experience. *Arch Immunol Ther Exp* (Warsz) 48: 547–551, 2000a.

Weber-Dabrowska, B., Zimecki, M. and Mulczyk, M. Effective phage therapy is associated with normalization of cytokine production by blood cell cultures. *Arch Immunol Ther Exp* (Warsz) 48: 31–37, 2000b.

Weber-Dabrowska, B., Zimecki, M., Mulczyk, M. and Gorski, A., Effect of phage therapy on the turnover and function of peripheral neutrophils. *FEMS Immunol Med Microbiol* 34: 135–138, 2002.

Weeks, C.R. and Ferretti, J.J., The gene for type A streptococcal exotoxin (erythrogenic toxin) is located in bacteriophage T12. *Infect Immun* 46: 531–536, 1984.

Yao, J.D.C. and Moellering, R.C., "Antimicrobial Agents," pp. 1474–1504 in *Manual of Clinical Microbiology*, P.R. Murray, E.J. Baron, M.A. Pfaller, F.C. Tenover and R.H. Yolken (Eds.). American Society for Microbiology, Washington, D.C., 1995.

Yeung, M.K. and Kozelsky, C.S., Transfection of *Actinomyces* spp. by genomic DNA of bacteriophages from human dental plaque. *Plasmid* 37: 141–153, 1997.

Zhukov-Verezhnikov, N.N., Peremitina, L.D., Berillo, E.A., Komissarov, V.P. and Bardymov, V.M., Therapeutic effect of bacteriophage preparations in the complex treatment of suppurative surgical diseases. *Sov Med*: 64–66, 1978.

Zhuravlev, P.M. and Pokrovskaya, M.P., "The Phagoprphylaxis and Phagotherapy of Gas Gangrene," pp. 111–115 in *Microbiology and Epidemiology*, B. Babsky, I.G. Kochergin, and V.V. Parin (Eds.). Medical Publications Ltd, London, 1945.

Zilisteanu, C., Ionescu, H., lonescu-Dorohoi, T., Coman, O. and Badescu, M., L'utilisation du bacteriophage dans la prophylaxic et le traitement de la dysenterie bacillaire. *Arch Roum Pathol Exp Microbiol* 32: 193–198, 1973.

Zilisteanu, C., lonescu, H., lonescu-Dorohoi, T. and Mintzer, L., Considerations sur le traitement des infections urinaires par l'association bacteriophage-autovaccin-antibiotiques. *Arch Roum Pathol Exp Microbiol* 30: 195–207, 1971.

Appendix: Working with Bacteriophages: Common Techniques and Methodological Approaches

Karin Carlson
Department of Cell and Molecular Biology,
University of Uppsala, Uppsala, Sweden

CONTENTS

A.1. Introduction 439
A.2. Phage Stocks, Bacterial Cultures, and Common Microbiologic Media 439
A.2.1. Equipment 439
A.2.2. Working Procedures and Safety Precautions 440
A.2.3. Phage Stocks 440
A.2.4. Bacterial Cultures for Phage Work 441
A.2.4.1. Bacterial Stocks 441
A.2.4.2. Preparing Bacterial Cultures for Phage Infection 441
A.2.4.3. Determination of Bacterial Titers 442
A.2.4.4. Infection with Phage 443
A.2.5. Culture Media and Common Reagents 443
A.2.5.1. General Considerations 443
A.2.5.2. Solid Media 445
A.2.5.3. Media for Anaerobic Growth 447
A.2.5.4. Recipes for Commonly Used Bacteriologic Media and Reagents 448
A.3. Isolation of Phages from Natural Sources 449
A.3.1. General Considerations 449
A.3.2. Isolation of Phages from Environmental Sources by Direct Plate Selection 450
A.3.3. Concentration of Phages in Environmental Samples 450
A.3.3.1. Concentration by Ultrafiltration 450

0-8493-1336-X/05/$0.00+$1.50
© 2005 by CRC Press LLC

A.3.3.2. Concentration by Precipitation with Polyethylene Glycol 451
A.3.3.3. Concentration by Sequential Adsorption and Elution 451
A.3.4. Enrichment of Phages in the Source Material 451
A.3.4.1. Basic Enrichment Protocol 452
A.3.4.2. Multiple Enrichment 452
A.3.4.3. Obtaining Phages from Enriched Cultures 453
A.3.5. Obtaining Pure Stocks of Isolated Phage 453
A.3.6. Isolation of Phages from Bacteria That Do Not Form Lawns on Solid Media 453
A.3.7. Induction of Prophages 455
A.4. Phage Enumeration 455
A.4.1. Plaque Assays 455
A.4.1.1. Serial Dilutions 456
A.4.1.2. Agar Overlay Methods 457
A.4.1.3. Killer Titer 460
A.4.2. Most Probable Number (MPN) Assays 461
A.4.3. Enumerating Phage Particles by Microscopy 462
A.5. Preparing, Purifying and Storing Phage Stocks 462
A.5.1. General Considerations 462
A.5.2. Primary Lysates 463
A.5.2.1. Standard Liquid Culture 463
A.5.2.2. Plate Lysate 464
A.5.3. Secondary Lysates 466
A.5.4. Concentration and Purification of Isolated Phages 466
A.5.4.1. Concentration by Centrifugation 466
A.5.4.2. Concentration and Purification by Precipitation with Polyethylene Glycol (PEG) 467
A.5.4.3. Purification by Centrifugation in CsCl Gradients 467
A.5.4.4. Centrifugation in Sucrose Density Gradients 470
A.5.5. Storage of Phage 470
A.5.6. Sending Phage 471
A.6. Characterization of Phages 471
A.6.1. DNA Analysis 471
A.6.1.1 Isolation of Phage DNA 471
A.6.1.2. Digestion of Phage DNA with Restriction Endonucleases 472
A.6.1.3. Agarose and Polyacrylamide Gel Electrophoresis of Phage DNA 473
Materials 474
Procedure 474
A.6.1.4. Pulsed Field Gel Electrophoresis of Large Phage Genomes 475
A.6.1.5. Analysis of Phage DNA by Polymerase Chain Reaction and Sequencing 476

A.6.2. Biological Parameters of Phage Infection 476
A.6.2.1. Analysis of the Phage Infection Process 476
Materials 478
Procedure 479
Processing of Samples 480
A.6.2.2. Single Step Growth 481
A.6.3. Radiolabeling and Analysis of Phage Proteins by Polyacrylamide Gel Electrophoresis 482
A.6.3.1. Radiolabeling of Phage Proteins 483
Materials 484
Procedure 485
A.6.3.2. SDS-PAGE of Phage Proteins 487
A.6.4. Electron Microscopy of Phage Particles 488
Acknowledgments 490
References 490

A.1. INTRODUCTION

The information presented in this chapter is based largely on my experience and that of colleagues studying a limited number of phages infecting a limited number of hosts. Applying the methods to phages growing in bacteria with vastly different growth characteristics may require adjustments in incubation time, media, other growth conditions, etc. The chapter focuses primarily on the most common methodologies used for the isolation and initial characterization of phages. Additional methods for in-depth characterization of phages can be found in Adams (1959) and Carlson and Miller (1994), and many of the protocols for molecular analysis assembled by Sambrook and Russell (2001) and Ausubel et al. (2001) can be adapted for characterization of phages.

A.2. PHAGE STOCKS, BACTERIAL CULTURES, AND COMMON MICROBIOLOGIC MEDIA

A.2.1. EQUIPMENT

A laboratory in which basic phage studies are performed should contain the following equipment:

For aerobic cultivation: a thermostated water bath or a thermostated air incubator capable of gyratory agitation of up to about 400 rpm, with holders for 50- to 500-ml capacity Erlenmeyer flasks. Alternatively, aeration may be achieved in "bubbler tubes," culture tubes into which sterile-filtered air is blown through a capillary tube inserted into the culture medium.

For anaerobic cultivation: airtight jars purged with nitrogen after sealing allow easy liquid-culture work even with obligate anaerobes (Attebery and Finegold, 1969). Alternatively, a glove box with a suitable gas supply (e.g., 5% CO_2, 10% H_2, 85% N_2) can be used. Plates can be incubated in airtight jars with platinum catalysts and gas packs generating CO_2 and H_2; while cheaper than the first systems, these jars usually yield less reproducible conditions and may not work for strictly anaerobic bacteria.

Thermostated water baths, heating blocks or an incubator to hold flasks or tubes containing molten soft agar at the appropriate temperature for plating

Thermostated incubators (for plates)

Spectrophotometer or colorimeter

Centrifuges (large-capacity for phage purification; microcentrifuges for various analyses)

Light microscope equipped for phase contrast work. Access to an electron microscopic facility for determining phage morphology and taxonomical standing.

Miscellaneous supplies commonly found in all microbiology laboratories; e.g., test tubes and racks, microcentrifuge tubes, pipettes and pipetting devices, etc.

Depending on the research goals, additional equipment, supplies, and reagents may be required for special applications. Some of that equipment and supplies are described in the pertinent sections of this chapter. All items used for culturing and storage of phages must be sterilized.

A.2.2. Working Procedures and Safety Precautions

Standard microbiological aseptic techniques are adequate for most phage work, and sterile benches or hoods, while desirable, are not generally required. Since phages do not pose health hazards beyond those of their hosts, the laboratory safety level and decontamination procedures required for working with the host of a particular phage are sufficient for working with the phage itself. It is always a good habit to wear gloves when handling material isolated from natural sources and when handling hazardous chemicals such as radioisotopes, acrylamide and other mutagens, and when handling phage stocks and bacteria. U.S. guidelines for the safe handling of various bacterial species are available from the Centers for Disease Control and Prevention Web site (http://www.cdc.gov/od/ohs/biosfty/biosfty.htm).

A.2.3. Phage Stocks

Phages are fairly fragile and should be handled gently. Care must be taken not to expose them to osmotic shock, or to strong daylight or fluorescent light (which can cause DNA damage). Vigorous mixing of phage solutions with a pipette and aggressive vortexing should also be avoided. Depending upon the particular phage and the conditions used for preparation and storage of phage stocks (section A.5), the time during which phages maintain full viability will vary considerably. Therefore, the titers of phage stocks should be determined at the time they are prepared, and also shortly before their use. The procedures for determining phage titers are described in section A.4. A phage stock that has been stored cold for some time can be warmed to 37° for 5–10 min and gently mixed in order to disaggregate phage clumps. A phage stock whose titer has dropped considerably during storage should not be used for experiments, but phage can often be rescued from such stocks even 50 years later and regrown. Phage stocks (like stocks of microorganisms) should not be routinely propagated by serial passage, because variants that grow faster than the original strain will increase in frequency with every round of growth. Therefore, an archival or "master stock" of each phage strain should be maintained (in the most inert state possible for the particular phage, see section A.5.5), preferably in aliquots kept in different places as a precaution against laboratory accidents. New working stocks should be grown from this stock as required.

A.2.4. Bacterial Cultures for Phage Work

A.2.4.1. Bacterial Stocks

For most bacteria, stocks are most conveniently maintained at –70°C. To prepare bacterial stocks, a cryoprotectant such as glycerol (10%–30%, v/v) or dimethyl sulfoxide (DMSO, 5%–7%, v/v) is added to an exponential culture, and the samples then flash-frozen in ethanol-dry ice or liquid nitrogen for storage at –70°C. Samples mixed with any of the above cryoprotectants can also be simply placed into a –70°C freezer and allowed to freeze, with somewhat poorer recovery upon thawing. Alternatively, bacterial strains may be lyophilized (after suspension in an excipient such as 20% skim milk or 12% sucrose). Lyophilized samples can be stored (either refrigerated, frozen, or at room temperature) for several years without a noticeable decrease in titers. Bacterial stocks may also be stored as stab or slant cultures (refrigerated or at room temperature; conditions yielding the best survival need to be tested). However, bacteria stored this way need to be passaged regularly in order to maintain viability. This incurs a cost in labor and media, and may select for bacterial variants increasingly different from the original stock. Thus, this approach should only be used if no other alternatives are available for maintaining bacterial cultures. See Gherna (1981) for detailed procedures and recommendations for maintaining bacterial stocks.

A.2.4.2. Preparing Bacterial Cultures for Phage Infection

For phage work, the growth medium is a bacterial culture, generally in mid-exponential phase. Thus, it is imperative that bacterial cultures show properties as similar as possible in each experiment. The best way to achieve this is to grow fresh cultures in the same defined medium to the same growth phase for each experiment, using starter cultures that are diluted and regrown on the day of the experiment. A bacterial culture intended for infection should be monitored for a couple of generations before it is infected to ensure that it is growing exponentially, using a spectrophotometer. Although cell mass is then measured in units of optical density (OD) at 500–600 nm, it is not an optical density that is measured, but turbidity. Turbidity can also be measured using side arm flasks and a colorimeter. Alternatively, cells can be counted in special microscope counting chambers.

Starter cultures may be initiated with a single colony and grown overnight or longer under whatever conditions the bacteria favor. Such cultures normally enter stationary phase, and after dilution they will lag for an irreproducible period of time before resuming growth. Therefore, they need to be diluted at least 100-fold and go through several divisions to ensure that all cells are in the same growth phase at the time of infection. An alternative approach works well with bacteria that do not lose viability upon rapid chilling: a starter culture grown to mid-log-phase is immediately chilled in an ice bath and kept cold. Such cultures need only a 20-fold dilution for regrowth the next day, and the cells usually resume growth in a more reproducible manner than when stationary starter cultures are used. Under some growth conditions, bacteria may lag nonreproducibly irrespective of how the starter culture was grown or treated. After determining the growth characteristics of the host strain, one may circumvent the need for dilutions by starting several cultures at different densities the day before, so that one of them will be at a density appropriate for experiments the following morning. The starting material may be a previous liquid culture or colonies growing on a plate of solid medium. Our experience with enterobacteria in defined minimal media (generation times 40–80 min, aerobic or anaerobic growth) has shown that inoculating with 10^3–10^6 cells/ml from fresh colonies on plates of the same composition as the liquid growth medium can result in bacterial densities of about 1×10^8 cells/ml after about 18 hours incubation at 37°C.

A.2.4.3. Determination of Bacterial Titers

In many applications, the titer of the bacterial culture at the time of phage infection must be known, so that an appropriate number of phage particles can be added to infect all of the cells, yet not result in "lysis from without" (see Chapter 7), as may happen when too many phage attack simultaneously, resulting in immediate clearing of the culture. The turbidity at a given cell concentration varies with the bacterial strain, medium, temperature, and growth phase; therefore, turbidity measurements must be calibrated to cell counts, using the exact experimental conditions. Also, viable count estimations (determined by colony counts) may be misleading because, depending on how the bacterial culture has been grown and how the samples have been treated before plating, not all viable cells may give rise to colonies. In particular, cultures that have been grown under conditions vastly different from those on the plates, or that were chilled before plating, may yield low viable titers. Therefore, to avoid unpleasant surprises, it is prudent to use microscopy to determine the total number of cells before infection whenever possible. This also provides potentially relevant information about the state of the bacteria; the morphology and degree of aggregation of cells may vary significantly with growth conditions or other stresses.

To estimate the viable count, serially diluted (see section A.4.1.1) aliquots of bacterial cultures may be spread on the surface of pre-formed plates containing a suitable solid medium (section A.2.5.2), added together with the same medium in molten form to empty plates, or plated or spot-tested using the overlay technique as described for phage in section A.4.1.2. All procedures where phage or bacteria are enumerated using solid media cast in Petri dishes are commonly referred to as "plating". *Note*: Colonies growing within the agar will vary in size, and spot testing results in crowded conditions on the plate and therefore usually somewhat reduced titers. Several manufacturers currently offer automated systems for counting colony-forming units (CFU) growing on solid media in Petri dishes, including a digital camera and image analysis software. In addition to colony counting, many of these systems allow additional analyses, such as the determination of inhibition zone, colony size and shape analysis, phage plaque counts, etc.

A.2.4.4. Infection with Phage

When a liquid bacterial culture intended for phage studies reaches the required cell density, it should either be *chilled rapidly* (to be used as a plating indicator or starter culture) or, preferably, *infected immediately*. Also, sometimes it may be acceptable to chill exponential bacteria for some time, rewarm them for a few minutes, and then carry out infection studies. *Note*: Rapid chilling may have a detrimental effect on the viability of some bacteria. Thus, one needs to test the particular bacterial strain to determine whether phage infection proceeds in the same manner in such pre-chilled cultures as it does in cultures infected without the chilling step. Both phage and bacterial titers must be known reasonably well at the time of infection to achieve the desired ratio between infecting phage and infected bacteria (defined as the *multiplicity of infection* or MOI). Since the phages are randomly distributed over the bacteria, following the Poisson distribution, an MOI of ≥3 is necessary to ascertain that ≥95% of all bacteria are hit by a phages. If phages have low viability or adsorb poorly, or if the bacterial culture is so diluted that random diffusion will not bring all phage particles in contact with a host bacterium within the time allotted to the experiment, even higher MOI's are required for successful infection of most bacteria in the culture. Thus, pilot experiments are essential to determine the optimal MOI to be used for the particular experiment undertaken. Monitoring of the level of bacterial survivors will permit calculation of the *actual* MOI used (section A.4.1.3). See sections A.4 and A.2.4.3 for methods to determine titers of phage and bacterial suspensions, respectively, and sections A.3.6 and A.4.2 for additional discussions of the Poisson distribution and MOI.

A.2.5. Culture Media and Common Reagents

A.2.5.1. General Considerations

In general, phages can be grown in any medium under any conditions that support the growth of their host. However, synthetic (i.e., chemically defined) media have the most reproducible composition, are often cheaper than traditional rich micro-biologic media, and are likely to give the most reproducible results. Thus, whenever possible, the use of synthetic media is recommended for phage propagation and other phage-related work. Recipes for some commonly used media and diluents are given in Tables A.1–A.6; many pre-mixed powdered media are available from various vendors. It is prudent, and in some cases imperative, not to autoclave organic

TABLE A.1
Undefined Liquid Media

	Amount					
Component	**Luria Broth (LB)**[1]	**Tryptone Broth (TB)**[2]	**Tryptone Soy Broth**[3]	**Mannitol Salt Medium**[4]	**Brain Heart Infusion Broth**[5]	**Terrific Broth**
Tryptone	10 g	10 g				12 g
Yeast extract	5 g					24 g
Brain heart infusion					37 g	
Beef extract				1 g		
Pancreatic digest of casein			17 g			
Papaic digest of soybean meal			3 g			
Peptone or polypeptone				10 g		
Glycerol						4 ml
Glucose			2.5 g		18 g	
Mannitol				10 g		
NaCl	10 g	5 g	5 g	75 g	8.7 g	
KH_2PO_4					8.7 g	0.17M
K_2HPO_4						0.72M
Deionized or distilled water	To 1000 ml	To 1000 ml	To 1000 ml	To 1000 ml	To 1000 ml	To 1000 ml

[1] Also called Luria-Bertani broth. Glucose may be added to a final concentration of 0.1–0.5% (w/v). Adjust the pH to 7.0–7.5 (with HCl or NaOH, as required), and autoclave the broth (20–30 min, 2.1 bar (15 lb/sq inch), 121°C).

[2] Adjust the pH to 7.0–7.5 (with HCl or NaOH), and autoclave as described above.

[3] Commonly used for soil bacteria. Adjust the pH to 7.0–7.5 (with HCl or NaOH), and autoclave described as above.

[4] For staphylococci. Adjust the pH to 7.4 (with HCl or NaOH), and autoclave as described above.

[5] For streptococci. Adjust the pH to 7.4 (with HCl or NaOH), and autoclave as described above.

components of media. In particular, sugar solutions decompose easily during autoclaving. Such compounds should be filter-sterilized by passage through a 0.22 μm filter, or autoclaved separately at reduced pressure, and then added to the sterile medium. Phosphates of divalent cations are fairly insoluble at high temperatures and easily precipitate during autoclaving. Therefore, solutions of divalent metal salts and phosphates should be autoclaved separately and combined when cool. In addition, metal ions whose oxidation state is not maximal (e.g., Mn^{2+} or Fe^{2+}) may become

TABLE A.2
Undefined Agar Media

	Amount			
Component	**Luria Agar (LA)[1]**	**Tryptone Soy Bottom Agar (TSA)**	**Tryptone Bottom Agar (TBA)[2]**	**Tryptone Top Agar (TTA)[3]**
Trytptone	10 g		10 g	10 g
Yeast extract	5 g			
Agar	11 g	11 g	11 g	7 g
NaCl	10 g	5 g	8 g	10 g
Glucose	2 g	2.5 g	1 g	1 g
1 M $CaCl_2$	3 ml			
1 M NaOH			~1.2 ml	~1.2 ml
Pancreatic digest of casein		17 g		
Papaic digest of soybean meal		3 g		
Deionized or distilled water	To 1000 ml	To 1000 ml	To 1000 ml	To 1000 ml

[1] Adjust pH to 7.0–7.5 (with HCl or NaOH, as required), and autoclave as described for Table A.1.
[2] The pH after autoclaving (as described above) should be 7.0-7.2. If required, adjust the pH with NaOH or HCl.
[3] Also called "Soft Agar." Heat until all the solids are melted, mix thoroughly, adjust pH to 7.0 to 7.2, and dispense in 75–100 ml aliquots. Autoclave as described above.

oxidized during autoclaving. Thus, if maintaining their low oxidation state is important, solutions of their salts should be filter-sterilized. Manufacturers of bacteriologic media usually provide information on how to prepare the various media they sell, including instructions regarding any specific sterilization protocols, if autoclaving of certain ingredients is not recommended.

A.2.5.2. Solid Media

Different brands of agar and agarose have different gelling properties, so the actual brand on hand should be tested to determine the concentration of agar/agarose needed to produce plates with suitable consistency; plates should be firm but not hard. In principle, any liquid medium may be converted into a solid medium by adding an appropriate amount of agar/agarose to the liquid medium's recipe. However, common recipes for solid media often differ somewhat from those of their respective liquid partners. Preparation of a solid medium most commonly involves mixing the powdered ingredients with water, dissolving through stirring on a hot plate, and autoclaving. After the medium is removed from the autoclave, the flasks should be gently swirled to evenly distribute the melted agar or agarose throughout the solution. *Caution*: the fluid may be superheated and may boil over when swirled.

TABLE A.3
SL Medium for Lactobacilli

Component	Amount
Trypticase or tryptone peptone	10 g
Yeast extract	5 g
Agar	15 g
KH_2PO_4	6 g
$(NH_4)_2$-citrate	2 g
$MgSO_4 \cdot 7H_2O$	0.58 g
$MnSO_4 \cdot 4H_2O$	0.28 g
$FeSO_4$	0.03 g
Glucose	10 g
Arabinose	5 g
Sucrose	5 g
Tween 80	1 g
Na-acetate$\cdot 3H_2O$	2.5 g
Deionized or distilled water	1000 ml

The medium is made in four steps: (i) Mix the trypticase, yeast extract and agar in 500 ml of water, autoclave as described for Table A.1 and cool to 55°C; if required, keep the molten agar mixture in a water bath set at 55°C, to keep it from solidifying while the other ingredients are being prepared; (ii) Mix the phosphate, citrate, Mg- and Mn-salts, sugars and Tween 80 in 480 ml of warm sterile water, and filter-sterilize the solution; (iii) Dissolve the Na-acetate in 13–15 ml of sterile water, adjust to pH 5.4 with glacial acetic acid, adjust the volume to 20 ml with sterile water, and filter sterilize; and (iv) Mix the three solutions together, and adjust the pH to 5.4 (with HCl or NaOH), if required (Rogosa et al., 1951).

Thus, it is prudent to wait 5 to 10 min after opening the doors to the autoclave before swirling the flasks. Agar media used to pour plates ("bottom agar") should be allowed to cool to about 50°C before pouring, to avoid evaporation. If any thermolabile substances (e.g., antibiotics) are to be included in the solid medium, these should be added to the cooled medium, and the flasks should then be swirled again to distribute them evenly. About 25–30 ml of medium should be poured into each 90-mm diameter Petri dish. Plate-pouring machines, available from several vendors, produce plates with uniform properties containing exactly the same volume medium, and simplify the task of plate pouring considerably. If a large flask of medium is used to prepare a large number of Petri dishes, the flask should be gently swirled occasionally. It is important that the surface upon which the poured plates are resting is level because uneven agar thickness will affect both the plaque size and the efficiency of plating, reducing both on the thin end. When the medium has hardened completely, the plates should be inverted, incubated overnight at 37°C to dry and to let any bacterial contaminants show up as colonies, and stored at 4°C

TABLE A.4
M9, a Defined Minimal Medium

	Amount			
Component	**20 × M9 Salts**[1]	**M9 Glucose Liquid Medium**[2]	**M9 Glucose Agar (bottom agar)**[4]	**M9 Soft Agar**[5]
$Na_2HPO_4 \cdot 2H_2O$	120 g			
KH_2PO_4 (anhydrous)	60 g			
NaCl	10 g			
NH_4Cl	20 g			
20 M9 salts		5 ml	47.5 ml	47.5 ml
0.1 M $CaCl_2$		0.1 ml	0.95 ml	0.95 ml
0.1 M $MgSO_4$		1 ml	9.5 ml	9.5 ml
0.01 M $FeCl_3$		0.1 ml	0.95 ml	0.95 ml
1 M Na-citrate[6]			5.7 ml	5.7 ml
Thiamine HCl, 2 mg/ml, in 70% ethanol		0.05 ml	0.475 ml	0.475 ml
20% glucose		2.5 ml	9.5 ml	9.5 ml
Agar			9.5 g	6.5 g
Deionized or distilled water	to 1000 ml	to 100 ml	950 ml	950 ml

[1] Heat to dissolve. Dispense in 100 ml aliquots, and autoclave as described for Table A.1.

[2] Prepare from sterile solutions and sterile water, and do not autoclave the complete medium. Casamino acids may be added, to increase the growth rate, to a final concentration of 0.1%–1%—1% (M9CA medium).

[4] Divide the water equally among two flasks. Add the agar to one of them, and cool to 50°C after autoclaving. Add the sterile solutions of the remaining ingredients to the flask without the agar. Mix and pour the plates.

[5] Prepare as a bottom agar, and dispense in 75 ml aliquots.

[6] Optional

until needed. "Soft agar," required for plating with the overlay method (section A.4.1.2) contains 60%–65% of the agar concentration of the bottom agar. Soft agar is prepared the same way as bottom agar, but kept in aliquots, about 200 ml per flask, and used warm after reboiling to prepare over-layers on plates containing solidified bottom agar.

A.2.5.3. Media for Anaerobic Growth

Anaerobic growth of organotrophic bacteria may be fermentative or respiratory. In the former case, a fermentable carbon-and-energy source is needed, and, since many fermentations yield acid end products, the buffering capacity of the medium should

TABLE A.5A
G11[1]

Component	Amount
$NaNO_3$ (I)	1.5 g
$KH_2PO_4 \cdot 3H_2O$ (II)	0.04 g
$MgSO_4 \cdot 7H_2O$ (III)	0.075 g
$CaCl_2 \cdot 2H_2O$ (Iv et al., 1999)	0.036 g
Citric acid (Iv et al., 1999)	0.006 g
Ferric ammonium citrate (V)	0.006 g
EDTA (disodium magnesium salt) (V)	0.006 g
$NaCO_3$ (VI)	0.040 g
Trace metal mixture (VII)	1 ml
Distilled or deionized water	to 1000 ml

[1] Defined medium for cyanobacteria (Stanier et al., 1971).

TABLE A.5B
Trace Metal Mixture for BG11 (Component VII)

Component	Amount
H_2BO_3	2.86 g
$MnCl_2 \cdot 4H_2O$	1.81 g
$ZnSO_4 \cdot 7H_2O$	0.222 g
$CuSO_4 \cdot 5H_2O$	0.079 g
$Na_2MoO_4 \cdot 2H_2$	0.39 g
$Co(NO_3)_2 \cdot 6H_2O$	0.0494 g
Deionized or distilled water	to 1000 ml

Prepare stock solutions I–VI separately at 1000 strength, and autoclave as described for Table A.1. Use autoclaved and cooled water to prepare the medium; add 1 ml of solutions I–VII per 1 liter of medium. For nitrogen-fixing bacteria, the $NaNO_3$ may be omitted.

be sufficient to maintain the desired pH (usually a neutral pH) during bacterial growth. In our experience, Luria broth (LB) with 0.5% (w/v) glucose permits good fermentative yields of enterobacteria. Also, in M9 medium containing 0.5% glucose, the addition of Na_2CO_3 (50 mM, final concentration; starting pH of 7.5) permits fermentative growth to high cell densities without undue acidification, though the bacterial growth rate is somewhat reduced. Anaerobic respiratory growth requires a nonfermentable carbon source and a good electron acceptor, such as nitrate or fumarate, both typically at a final concentration of 0.5% (w/v). Since many bacteria also can ferment carbon sources commonly considered to be nonfermentable, growth in the same medium without the external electron acceptor should be monitored. For example, 0.5% (w/v) glycerol, glutamate, or pyruvate works well for many *E. coli* strains.

A.2.5.4. Recipes for Commonly Used Bacteriologic Media and Reagents

Some useful recipes are contained in Tables A.1–A.3 (undefined and semidefined media), A.4–A.5 (defined media) and A.6 (diluents). Additional recipes, including isolation methods and culture requirements for a large variety of bacteria, can be found in Stein (1973), Gerhardt et al. (1981), and Demain and Solomon (1986). Several of the media described here are commercially available.

TABLE A.6
Phage Diluents

Component	Concentration	
	TSG[1]	SM[2]
Tris-HCl (pH 7.4-7.5)	0.01M	0.05 M
NaCl	0.15M	0.1M
$MgSO_4$	—	10 mM
Gelatin, w/v	0.03%	1%

[1] For some applications, the $MgCl_2$ may be added to a final concentration of 10 mM (TSG-Mg). Autoclave as described for Table A.1, and store in 100 ml aliquots. If possible, the solutions should be autoclaved after aliquoting, to avoid the risk of possible contamination during aliquoting.
[2] Autoclave as described above, and store in 100 ml aliquots.

A.3. ISOLATION OF PHAGES FROM NATURAL SOURCES

A.3.1. General Considerations

As described in Chapter 6, phages are the most abundant life forms on earth, and they can be found in a large variety of environmental, plant, and animal sources. The best source of a given phage is material where its specific host is abundant. Thus, for phages infecting the human gut flora, municipal sewage is an excellent source; for cyanophages, seawater. Phages can be relatively easily isolated from various sources. The phage isolation procedures described here depend upon lytic growth of the phage, but may yield both virulent and temperate phages. They are suitable for isolating phages for many bacterial species, and similar procedures have been used to also isolate archaeal phages (Schleper et al., 1992; Nuttall and Dyall-Smith, 1993). Before starting isolation of a phage, its bacterial host needs to be isolated and grown in pure culture. During phage isolation, it is generally best to use host bacteria in the exponential growth phase, although some phages also appear able to propagate in bacteria in the stationary growth phase (Schrader et al., 1997; Wommack and Colwell, 2000). Unless specifically recommended to the contrary, the growth conditions (medium, oxygen level, temperature, incubation time, etc.) should be chosen according to the preference of the host bacterium.

Phages can be isolated from various aqueous (e.g., water) and nonaqueous (e.g., soil) sources. If phages are to be isolated from nonaqueous samples, the samples first should be suspended in a medium suitable for the growth of the targeted bacterial species. To remove solids, indigenous bacteria, and other organisms, the samples can be "clarified." This step can be accomplished by centrifugation ($1500 \times g$, 20 min), which will remove most of the bulk solids, after which the supernatant can be processed for phage isolation. Filtering the suspended samples through a 0.45 μm or 0.2 μm membrane filter (preferably a tangential flow filter, to avoid rapid clogging) is an alternative approach, but this step may still need to be preceded by a centrifugation step or by pre-filtering the suspensions through larger-pore filters. Sometimes samples can be processed directly for phage isolation without clarification. Once isolated, phages can be resuspended in phage diluent (TSG or SM), or growth medium. Chloroform is often used to break open infected cells, since only cells whose peptidoglycan layer has been weakened by phage-encoded lysozyme or endolysin will be lysed by chloroform. However, it should be omitted from all procedures where phage with lipid-containing envelopes are expected or sought. Recommendations for phage storage are described in section A.5.5.

A.3.2. Isolation of Phages from Environmental Sources by Direct Plate Selection

From liquid sources, serial dilutions of the sample are spotted or plated (see section A.4.1) on lawns of the desired host. Solid samples are mixed with or homogenized in growth medium or buffer (about 1 g solids per 10 ml medium) and allowed to sit at room temperature for a couple of hours before dilution and plating. If the suspensions contain much debris and bacteria, they may be clarified before plating. If the sample contains enough bacteriophage lytic for the targeted bacteria, the phages will form plaques. The theory behind plaque formation is discussed in section A.4.1, and methods to obtain pure phage stocks from single plaques in section A.3.5. Standard double-layer plating in 90-mm diameter Petri dishes accommodates about 0.2 ml of sample per plate. To directly plate a larger sample volume, an overlay composed of 1.5 ml of "2x soft agar" (Tables A.2, A.4), 1.5 ml of sample and 0.2 ml of a culture of the indicator bacteria may be used. Using thicker overlays, or thick monolayers, to accommodate even larger sample volumes is not recommended because those approaches may result in (i) interference with bacterial growth by some components in the sample, and (ii) poor oxygen diffusion, which will result in smaller plaques and reduced relative efficiency of plating when aerobic bacteria are the hosts. For dilute samples containing low numbers of phage particles, a concentration or enrichment step should be considered (sections A.3.3 and A.3.4).

A.3.3. Concentration of Phages in Environmental Samples

A.3.3.1. Concentration by Ultrafiltration

Phage in aquatic environments can be concentrated by ultrafiltration (Suttle et al., 1991; Wommack et al., 1996; Breitbart et al., 2002). Environmental samples usually are first clarified with 0.45 μm or 0.2 μm filters, although Suttle (1991) demonstrated that such a clarification step may adversely affect the viability of some phages. The clarified sample can be processed using any of the following (or similar) filter systems: (i) a spiral-wound ultrafiltration cartridge with a M_r cut-off of ca. 30,000-Da (Suttle et al., 1991), (ii) a tangential flow filter with a M_r cut-off of ca. 100,000-Da (Breitbart et al., 2002), and (iii) a disposable CX ultrafilter with a M_r cut-off of ca. 10,000-Da (Wommack et al., 1996). In general, filters should be selected based on their cut-off characteristics, the volume of the sample to be processed, and cost. For example, the first two filters mentioned above are suitable for volumes > 5 L and permit up to a 5,000-fold concentration (Breitbart et al., 2002), while the disposable filter is useful for volumes < 0.5 L and yields about a 10-fold concentration (Wommack et al., 1996).

A.3.3.2. Concentration by Precipitation with Polyethylene Glycol

Two-phase polymer separation techniques, such as polyethylene glycol (PEG) (Yamamoto et al., 1970; see A.5.4.2.) can be used to concentrate phage particles from natural sources. However, some environmental sources may contain components that interfere with the precipitation of phages by PEG. In such cases, 10-fold dilution of the sample before addition of PEG may improve precipitation. PEG-induced precipitation of phages from very diluted samples ($<10^5$ phage particles per ml) may be inefficient (Sealey and Primrose, 1982), and recovery in such cases may be improved by adding a "carrier phage" that is unable to infect the host bacteria, or by concentrating the samples by ultrafiltration. The methods that use PEG to concentrate phages in environmental samples are very similar to the methods that use PEG to concentrate phage stocks.

A.3.3.3. Concentration by Sequential Adsorption and Elution

Phages in aquatic environments may be adsorbed (after sample clarification, if necessary) by, and subsequently eluted from, various adsorbents: positively or negatively charged microporous filters, hydroxyapatite, protamine sulfate, aluminum hydroxide, iron oxide, and diatomaceous earth (Sealey and Primrose, 1982). Some phages are concentrated efficiently using this approach; Sobsey et al. (1990) and Mendez et al. (2002) describe applications using nitrate-acetate membrane filters in the presence of 0.05M $MgCl_2$. However, different phages have different adsorptive properties and the process is quite slow (Sealey and Primrose, 1982). Thus, unless there are specific reasons to concentrate samples by sequential adsorption-elution, ultrafiltration and PEG precipitation (in that order) are preferable approaches.

A.3.4. Enrichment of Phages in the Source Material

Enrichment of the phages in the source material often facilitates phage isolation. However, the outcome of an enrichment is unpredictable, depending on many factors beyond the control of the experimenter such as the abundance of the phage in the source material, the presence and concentration of other life forms, and constituents of the source material that may affect growth of either the host organism or the phages themselves. Therefore several different protocols should be examined to determine the best general procedure; the protocols described below should only serve as general guides. The ratio between source material and added (concentrated) growth medium should be varied to accommodate differences in abundance of phages and the nature of the source material, as should the addition of host bacteria. If one wants to enrich for particularly virulent, broad-spectrum phages that initially constitute a very small fraction of the phages in the sample, the protocol should utilize a larger volume of source material, several different host bacteria, and several cycles of selection. If the volume of the sample is large compared to the final culture volume, however, some of the sample's components may interfere with bacterial growth. That problem can often be circumvented if the volume of source material is $\leq 1\%–10\%$ of the culture's final volume. For simplicity, the enrichment procedures outlined below call for a 10% inoculum and rather small enrichment cultures. This should be scaled up to suit the circumstances at hand. If the sample constitutes more than 10% of the final volume, one should add enough 10 × culture medium to have the final enrichment culture contain 1 × culture medium. Alternatively, one may concentrate the phages in the sample (section A.3.3) before incubating the enrichment culture.

A.3.4.1. Basic Enrichment Protocol

Depending on the nature of the material and its expected phage content, various enrichment procedures may be used, as illustrated in Table A.7. All of the enrichment cultures should be incubated at a temperature and for a time period determined by the growth characteristics of the hosts (overnight for rapidly growing bacteria, longer for more slow-growing hosts) in loosely capped containers. The phages in enrichment culture A (Table A.7) will grow in and lyse the indigenous hosts, and the culture will yield a large variety of phages capable of infecting several different bacteria present in the environment from which the sample was taken. On the other hand, although some phages in enrichment culture B will proliferate in the indigenous bacteria, there will be a major enrichment of the phages that infect and lyse the added host(s). Finally, in enrichment cultures C, only those phages capable of infecting and lysing the added host(s) should proliferate and be enriched. When species-specific or strain-specific phages are sought, enrichment conditions C are preferable. After enrichment, cultures should be plated for plaques from which phages can be isolated, as described in sections A.3.2 and A.3.4.3. Some investigators prefer to terminate the enrichment by adding a few drops of chloroform to the enriched sample. However, this procedure is not suitable for phages with lipid envelopes.

A.3.4.2. Multiple Enrichment

To test large variations in the ratio of source material to host bacteria, or to enrich simultaneously for phages capable of infecting and lysing several different hosts,

TABLE A.7
Composition of Basic Enrichment Cultures

Component	Amount in Enrichment Culture		
	A	B	C
Phage source material	1 g or 1 ml	1 g or 1 ml	1 g or 1 ml, clarified[1]
Culture of the host bacterium	—	1 ml	1 ml
Bacteriologic culture medium[2]	10 ml	10 ml	10 ml

[1] Solid source material sho uld be homogenized, using a piston tissue homogenizer or similar device, in an appropriate culture medium (1 g per 10 ml of medium), and the suspension should be clarified as described in section 2.2.

[2] Use 9 ml for liquid source material

multiple mini-enrichment protocols may be examined in microtiter plates (allowing cultures of 50–150 μl) or honeycomb plates (allowing 400 μl cultures). To do this, the volumes mentioned in section A.3.4.1 are scaled-down so that each enrichment culture can be placed into a separate well. Different wells can be used to test different hosts, different dilutions of the sample, and different growth media; furthermore, different microtiter plates can be incubated under different conditions. The enrichments should be incubated overnight or for a couple of days (depending on the growth rate of the host bacteria), under conditions that prevent evaporation from the wells, and—for aerobic bacteria—under conditions that avoid oxygen depletion. Lysed cultures should be plated for plaques. If no visible lysis is observed, it still may be worthwhile to remove the contents of some randomly picked wells, treat them with chloroform, and assay for phages as described below.

A.3.4.3. Obtaining Phages from Enriched Cultures

If the bacteria are slow to lyse (even after adding chloroform), they can be chilled in an ice bath, followed by rapid warming to 37°C. If the resulting suspension is viscous it can be treated with pancreatic DNAse I (1 μg /ml) for 30 min at room temperature. If no visible lysis occurs, some of the cultures still should be processed, because the phages may be too few to lyse most of the bacteria, or phage-resistant bacteria may be present. (i) Sediment the bacterial debris by centrifugation (1500 × g, 20 min; if microtiter plates are used, a special rotor capable of centrifuging microtiter plates will allow semi-high-throughput processing), (ii) save the supernatant, and (iii) spot test (see section A.4.1.2) this supernatant to obtain plaques of phages present.

A.3.5. Obtaining Pure Stocks of Isolated Phage

If plaques are obtained from samples from natural sources, phages are picked from them with a sterile Pasteur pipette or toothpick for subculture. Often a number of different phages will grow. Thus, it is prudent to pick phages from several plaques (especially if the plaques' morphologies appear to be different) and to propagate them separately. Although each plaque obtained after direct plate selection or enrichment normally contains the progeny of a single bacteriophage particle, "spillover" by diffusion from adjacent portions of the lawn poses a considerable contamination risk, so that additional subculturing of the phage should be undertaken to ensure the uniformity of phage preparations. In subculturing, phages obtained from a plaque are resuspended in diluent or growth medium and replated under the same conditions used for the first plating. Stocks are grown (section A.5) after repeated subculturing, whereupon the phage can be characterized by methods such as those described in section A.6 to confirm its homogeneity and characterize its identity.

A.3.6. Isolation of Phages from Bacteria That Do Not Form Lawns on Solid Media

Some bacteria do not form lawns when grown on solid media, which complicates isolation of single phages. Should this occur, the following method may be used to circumvent the problem. First clarify, concentrate, and enrich the phage source material as described above. Then proceed as follows:

1. Prepare a serial dilution series of the phage preparation, so that the highest dilution is expected not to contain phage.
2. Grow the host bacteria to exponential phase in an appropriate culture medium. To permit phage growth before the culture reaches stationary phase, it is usually best to use diluted bacterial cultures (10^6 cells per ml). Distribute aliquots (10 μl = 10^4 cells) of the diluted culture into individual 100-μl wells of microtiter plates (if larger wells are used, volumes can be scaled up accordingly), at least 10 and preferably 20 wells per phage dilution to be assayed, and add 10 μl of the appropriate phage dilutions. The wells infected with phage from the same dilution are referred to as "a series."
3. After ~10 min for adsorption, add 60 μl of culture medium to each well. Incubate the cultures overnight—or for a couple of days—under conditions that prevent evaporation from the wells (yet avoid oxygen depletion for aerobic bacteria), with continuous monitoring for lysis. *Note*: If the number of phage particles produced per cell is low, a single phage infecting a culture containing about 10^4 bacteria may not produce enough new phages to cause visible clearing of the culture before it reaches a growth phase that no longer supports phage growth. Thus, if no visible lysis is seen, some of the cultures may still be processed in an attempt to isolate phages from the samples
4. To release lysis-inhibited phages, chloroform should be added to the cultures. If no visible lysis is induced by chloroform, or if lipid-containing phages are sought so that chloroform cannot be used, the culture series infected with the least diluted sample is collected, bacteria removed by centrifugation, and a new dilution series prepared for a new round of infection as described above.
5. Look for series where only one or two cultures exhibit lysis, where there is a good chance that the infection was initiated by a single phage, as explained below.

Phage particles are distributed in liquids according to the Poisson distribution, which has the form

$$P(k,m) = \frac{m^k \times e^{-m}}{k!}$$

where k is the actual frequency of the studied event (the frequency may be zero or a positive integral number), m is the average of k, and P is the frequency of samples for each value of k. If a phage dilution contains, on average, 0.1 phage particles per 10 μl ($m = 0.1$), each culture exposed to that volume of the dilution will be infected with, on average, 0.1 phage particles. Thus, a fraction of $e^{-0.1}$ (ca. 90%) of the 20 cultures incubated with 10 μl aliquots of the dilution (i.e., 18 cultures) will not be infected ($k = 0$) and, therefore, will not lyse. Also, a fraction of $0.1 \times e^{-0.1}$ (ca. 9%) of the 20 cultures (i.e., ca. 2 cultures) will be infected with one phage particle ($k = 1$), and the remaining 1% (i.e., <1 culture) will be infected with > 1 particle ($k \geq 1$). To obtain purer phage isolates, lysates from such series can be serially diluted and used to start a second round of growth as described above, again saving lysates from series where only one or two cultures lyse. Finally, phages are propagated from the lysed cultures as described in section A.5, using a small aliquot of the lysate rather than a plaque to start the culture.

A.3.7. Induction of Prophages

As discussed in Chapter 3, many bacteria are lysogenic; they carry one or more genomes of temperate phages (*prophages*) inserted in their chromosomes. Prophages can be induced to replicate by various techniques; however, successful isolation of the temperate phages released after prophage induction requires a permissive host strain in which the phages can proliferate in a lytic cycle. (The lysogenic bacteria remaining in the original culture are immune to the phage and cannot be used to propagate it). Spontaneous induction of a fraction of the lysogenic cells during growth may be sufficient to isolate the released phages. To try that approach, the lysogenic strain is spot-tested (section A.4.1.2.3) on a lawn of the permissive strain, and incubated under appropriate conditions. A zone of bacterial lysis developing around the lysogen's colonies suggests lytic propagation of induced prophages; however, the phenomenon may also be due to the release of bacteriocins. To determine if the observed lysis zones are due to phage infection—and to isolate the putative phages in the lysis zones—samples are picked using sterile Pasteur pipettes, suspended in a small volume of phage diluent (Table A.6), and various dilutions of the samples spot-tested on a lawn of the permissive host. If plaques form in the spots, lysis was caused by lytic propagation of induced prophages, and the released phages can now be propagated and characterized as described in sections A.5 and A.6, respectively. If this method does not yield phages, more aggressive prophage induction approaches such as UV radiation can be employed. The dose must be determined for the system on hand; for induction of prophage λ in *E. coli*, a dose of 0.1 lethal hits is appropriate (Arber et al., 1983). Alternatively, the culture can be treated with an antibiotic such as mitomycin C (0.15–5 μg/ml, depending on the host). Other inducing agents include ionizing radiation, organic peroxides, and nitrogen mustard. After treatment with an inducing agent, the lysogenic bacterial strain should be incubated for 1–2 generation times (to permit phage induction and phage growth), after which time phage development can be terminated by addition of a few drops of chloroform.

A.4. PHAGE ENUMERATION

A.4.1. Plaque Assays

The oldest but still most common and most useful method for enumerating phages is the plaque assay, which was first described by d'Herelle shortly after his discovery of bacteriophages (see Chapter 2). When a single phage particle encounters a permissive bacterium, it will infect it and later lyse it with the concomitant release of newly formed phage particles. When about 100–500 phages are mixed with about 10^8 bacterial cells that will support phage propagation and poured in a layer of soft agar on the surface of a solid medium supporting bacterial growth, the uninfected bacteria will resume growth and eventually reach stationary phase, forming a smooth opaque layer or *lawn* in the overlay. The phage particles will soon come into contact with bacteria, which they infect. The progeny phage from each infected bacterium will infect neighboring bacteria, and the lytic cycle will be repeated numerous times resulting in a growing zone of lysis, full of liberated phage, which eventually becomes visible to the naked eye as a "plaque" in the otherwise smooth lawn (see Fig. 4.2). Growth of the plaque is limited by slow diffusion of phage in the semi-solid soft agar, and (in most cases) by the fact that the host cells support phage growth only as long as they metabolize actively, so when cell growth stops, phage growth also ceases. Plaque size is primarily determined by the nature and size of the phage, thickness of bottom and top agar layers, concentration of agar in the top layer (and resultant diffusion rate), type of medium, plating cell density, distribution of adsorption times, and burst size. A production of about 10–15 phages per infected cell is generally sufficient for plaque formation. For a further discussion of parameters affecting plaque size, see Carlson and Miller (1994, p.428). Each phage particle that gives rise to a plaque is called a *plaque-forming unit* (PFU). The number of PFUs in a given volume of sample gives the viable phage concentration or *titer*. However, this is not necessarily an accurate estimate of the *absolute number* of phage particles present. The plaque-forming ability of phages may differ dramatically under various environmental conditions. A change in salt concentration can change the viable titer of coliphage T2 preparations ca. 1000-fold, and TBA plates (Table A.2) yield higher viable titers of coliphage T4 than do LA plates. Furthermore, assaying the same phage preparation against various host strains often will yield different titers with each host. In some cases this reflects the host on which the phage was initially grown and is the result of *restriction-modification* systems; in others, it reflects the concentration and properties of the particular host receptor under those conditions or the status of the tail fibers. For example, T4B phage strains require tryptophan to deploy the tail fibers, and the apparent titer is orders of magnitude lower in the absence of tryptophan. Thus, when specifying the phage titer, it is important to determine optimum plating conditions and identify confounding factors and to indicate the host strain used for the plaque assay and other parameters (e.g., the buffer in which the phages were suspended, the incubation media, etc.) that may have an impact on the assay's results. Sections A.5.4 and A.6.1 give methods for estimating the absolute number of phage particles. Standardized methods for enumeration of various phages in water are described by Anonymous (Anonymous 1995; 2000; 2001)

A.4.1.1. Serial Dilutions

To enumerate the large number of phage particles (and bacteria) encountered during culturing, serial dilutions are necessary. Phage diluent or bacteriologic culture medium can be used to prepare the dilutions; however, it is important that the medium is isotonic, in order to avoid phage rupture due to osmotic shock. As schematically

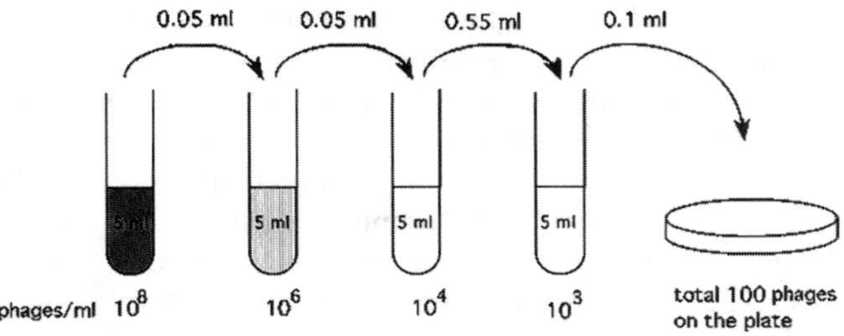

FIGURE A.1 Serial dilutions of phage solutions. For accurate results, it is important that the volumes are measured precisely and that a fresh pipet is used for each dilution step, as carryover from the previous dilution may compromise the accuracy of the results. To withdraw a volume for dilution, place the pipet tip just below the surface of the liquid, aspirate the desired volume, and wipe off excess liquid from the outside of the pipet tip by touching it against the inside of the tube as it is withdrawn. Blow the pipet's contents into the liquid in the next tube, and mix by gently swirling the tube.

outlined in Fig. A.1, a sequential series of $1/100 \times 1/100 \times 1/10$ dilutions of a phage stock originally containing 10^8 PFU/ml will give a final phage concentration of about 10^3 PFU/ml. Thus, 0.1 ml of the final dilution in this series will contain on the average 10^2 PFU, which is suitable for enumeration by a plaque-count on one plate. The precision of the assay increases with the number of plaques counted, since the phage are Poisson-distributed in the liquid (see section A.3.6). Thus, if the average number of PFUs in the aliquot plated is 100, about 70% of all plates will contain between 90 and 110 plaques (±10%). If, on the other hand, each plated aliquot contains on the average 10 PFU, less than 40% of all plates will contain between 9 and 11 plaques (±10%). Since the number of plaques (or bacterial colonies) required for a small sampling error cannot be counted on one plate, several plates (duplicates or triplicates) are required to obtain reasonably accurate titers; this also helps catch occasional other plating problems.

A.4.1.2. Agar Overlay Methods

Agar overlay methods are the most commonly used approach for determining phage titers. As seen in Tables A.2 and A.4, the soft-agar overlay contains relatively few nutrients, and the bottom layer's nutrients are the ones that primarily support growth of the indicator bacteria. Thus, unless very precise nutritional conditions are required for a specific host bacterium, the same soft-agar can be used for plates containing various bottom layers. Lower soft-agar concentrations give larger plaques. Before use, the soft-agar medium must be brought to a full boil to disrupt microcrystals; otherwise it will contain lumps that may interfere with plaque observation. After this, the medium should be equilibrated—in a thermostated incubator or water bath or in individual "plating tubes" in a heating block—at a temperature that the subsequently added phages and host bacteria can tolerate (usually about 45°C; at a lower temperature, low-melting-point agar or agarose may be needed to keep the soft-agar from gelling). It is important that the bench top used for pouring the soft-agar overlay (as well as for pouring the bottom layer) is *level*. An uneven overlayer will affect the plaque size and efficiency of plating. Also, since plaque formation depends upon diffusion in the agar layers, it is important that the plates are kept somewhat moist; i.e., not too wet and not too dry. Plates of media that are too wet (e.g., because they are freshly poured or stored in a very humid environment) are likely to have water droplets on the surface that may run and smear the plaques. Therefore, such plates should be dried before use. To avoid overdrying, place the plates with the bottom up on a shelf in the incubator (37°C or 42°C), lift off the bottom agar-containing part and place it so that it leans against the lid with the agar-side *down* for 15–30 min. Alternatively, leave the plates at 37°C overnight, bottom-side down and lids on. Plates of media that are too dry (i.e., old plates stored under conditions permitting evaporation) will yield small plaques that may be difficult to visualize. For the bottom layer, any solid medium permitting growth of the indicator bacteria can be used. However, the plaques formed in the presence of nutrient-poor media may be smaller and more difficult to enumerate than are the plaques formed on rich media, and different media may permit different numbers of phage to form plaques dependent upon the ratio of bacterial growth to phage growth.

Indicator bacteria: To visualize plaques, bacteria are needed to form a lawn on the plate. Such "indicator bacteria" need to be grown in liquid to a phase permitting phage adsorption and growth (normally mid-exponential phase, see section A.2.4). To produce a smooth lawn, 10^7 to 10^8 cells need to be added to each plate, in a volume of 0.1 to 0.2 ml for a 90-mm plate. Exponential cultures that have been chilled rapidly can often be used as plating indicators for some time if maintained at +4°C. However, the behavior of the host bacteria used needs to be checked in this respect, since some bacteria rapidly lose viability upon chilling.

Spot tests: The spot-test procedure only yields semiquantitative titers, but is useful for determining the host range of a particular phage, estimating approximate phage titers and following the course of complex experiments without using large numbers of plates. Prepare a plate containing a bottom layer of solid medium and an overlayer: 2.5–3 ml of soft-agar containing 0.2 ml of the indicator bacterial suspension for the typical 90-mm diameter Petri dish, or 4 ml of soft-agar and 0.3 ml of the indicator bacterial suspension for the disposable square plates that are particularly useful for such assays (see Fig. A.4). Place aliquots (5–10μl) of serial 100-fold dilutions of the phage suspension to be spot-tested in a row across the plate (to aid the spotting, one may draw circles on the plate's bottom ahead of time). With some practice, five rows (four spots per row) can be accommodated on a 90-mm diameter plate (Fig. A.2). Incubate the plate at a temperature that is appropriate for the indicator bacterium, until plaques are observed. After incubation, the presence of plaques in the spots where the phages were applied indicates that they are capable of infecting and lysing the bacterial strains examined. The number of plaques in the spot of an appropriate dilution will allow the calculation of an approximate phage titer, which can be verified by plating as described below. A phage that cannot form plaques under the assay conditions used may, nevertheless, kill the bacteria in spots containing very high phage concentrations, due to "lysis-from-without" (see section A.2.4.3 and Chapter 3). This phenomenon is easily distinguished from plaques caused

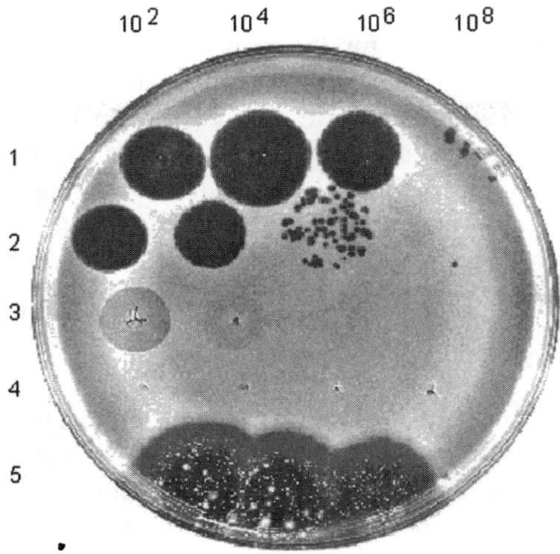

FIGURE A.2 Spot test of phage preparations on a 90-mm plate. Serial 100-fold dilutions of five different phage preparations were spotted in rows 1–5, as described in section A.4.1.2. The plate illustrates (a) good phage growth (rows 1,2,5), (b) killing without plaque formation (row 3), and (c) no growth due to host resistance (row 4; note the pipet tip's marks in the lawn, which shows that the absence of bacterial growth was not due to lack of sample application). The spot test also illustrates differences in plaque size and shape (rows 1,2,5).

by a phage's lytic growth cycle, since plaques are not formed in spots containing low phage concentrations, and <100 live phage can be recovered in a Pasteur-pipet-sized agar plug from the cleared spots; in a similar plug from a plaque produced by a lytic infection, >10^4 PFU are normally recovered even if the efficiency of plating is too low or plaque size too small to see individual plaques.

Plating in the soft-agar overlayer: Assay dilutions of the phage preparation (see section A.4.1.1) in duplicate or triplicate for increased accuracy, and expose the phages and bacteria to the warm soft agar for as short a period of time as possible. The phages and indicator bacteria may be added to a small test tube of warm soft-agar in a thermostated heating block, mixed gently and briefly, and poured onto the bottom layer in a Petri dish. Alternatively, to permit pre-adsorption of slowly adsorbing phages (which may help achieve uniform plaque sizes), incubate the mixture of phage and bacteria for 5–10 min before adding the 45°C soft agar. The mixture of soft-agar, indicator bacteria, and phage should be only lightly mixed before pouring because vigorous mixing may damage the phage particles and is likely to introduce air bubbles into the rather viscous agar, and air bubbles may look like plaques; flicking several times with the finger gives adequate mixing. Further mixing and spreading of the soft-agar overlay is achieved by sliding the plate in circles on the tabletop or by tilting it slightly in different directions immediately after pouring. The plate should be left on the bench until the layer solidifies; the plate can then be turned upside down and incubated (with the bottom agar side up—to prevent droplets of condensed water from falling onto the plaques and spreading the phage) for 24–48 hours under conditions suitable for the growth of the indicator bacteria. About 200–400 small plaques or 100 large plaques can be enumerated on a standard plate. If more are used, phages may compete with each other or plaques fuse together, thus yielding variable and inaccurate counts. Plaques can be counted using the naked eye or a magnifying glass, with a background and lighting suitable to facilitate plaque detection; simple counters are commonly used that record a count each time a plaque or colony (or the soft agar) is touched. Also, several manufacturers currently offer automated systems for counting plaques in Petri dishes.

A.4.1.3. Killer Titer

For a phage to form a visible plaque in a lawn of permissive bacteria, at least 10–15 progeny phages must be produced in each infected cell. Sometimes the host bacteria are "killed", i.e. no longer able to grow, divide, and form colonies, upon infection even though not enough new phages are produced to form a plaque. In such cases, the titer of killing particles (KP) can be determined by assaying the phages' ability to kill the host. An estimate of killing titer is also a useful test for the "quality" of an unknown phage suspension, since damage during storage is likely to reduce the viable titer (a measure of the ability to carry out productive infection) much more rapidly than the killing titer (largely a measure only of the ability to initiate infection); for example, T4 phage "ghosts", where the heads have been burst by osmotic shock and released the DNA, are still capable of killing bacteria. In addition, the procedure outlined below can be used to determine the "actual MOI" ("MOI_{actual}"), the average number of phage particles that have adsorbed to and killed the host bacteria. Under conditions where adsorption is poor this will give a better appreciation of actual infection conditions than the MOI as defined simply by the ratio between phage particles added and host cells present (see Kasman et al., 2002). If all phage particles in a preparation infect bacteria productively, the ratio between viable titer and killing titer, and between MOI_{added} and MOI_{actual}, both should be close to unity

To determine killer titers, it is important to know the exact bacterial concentration at the time of phage infection (see section A.2.4.3 for a discussion of procedures and pitfalls in such determinations) so that a precise MOI can be used. The best killing titer estimates are obtained using an MOI between 1 and 4, which necessitates assaying phage dilutions that differ by only a factor of 2 or 3. If the phage titer is unknown, a pilot experiment with serial tenfold dilutions will indicate the range of dilutions needed for a second more precise experiment. The steps involved in determining a phage's bacterial killing titer are:

1. Ahead of time, determine how long a time phage and bacteria need to be incubated together to achieve maximal phage adsorption to, and killing of, bacteria (the "adsorption time", A, see section A.6.3.1).
2. Prepare 3 serial twofold dilutions of the phage to get expected MOIs of about 1–4.
3. Grow an exponential culture of the bacterial host strain. Put 3 empty flasks into the same thermostated water bath.

4. Mix 0.05 ml of the exponential culture ("0-sample") with 5 ml of room-temperature diluent; quickly dilute to a final concentration of 1000–3000 expected CFU/ml and plate 0.1 ml in triplicate.
5. Immediately distribute 1 ml of the bacterial culture into empty flasks in the water bath.
6. At 1-min intervals, add 0.1 ml phage dilution to one 1-ml aliquot of the bacterial culture and continue incubation under the bacterial growth conditions.
7. At time A, take an aliquot (50 μl) from each infected 1-ml culture into 5 ml diluent, make a series of tenfold dilutions and plate 0.1 ml of each dilution in duplicate immediately.

Evaluation of results: Count the bacterial colonies on all plates, and calculate the fraction of surviving host bacteria (P_0) for each phage dilution used to infect the bacteria. Phages and bacteria are distributed in liquid according to the Poisson distribution (see section A.3.6). As applied to phages that infect bacteria, k is the number of phage particles adsorbing to and killing a bacterium, m (the "actual" multiplicity of infection, MOI_{actual}) is the average of k, and P is the fraction of bacteria infected by k phage particles. Therefore, the fraction of surviving host bacteria is:

$$P_0 = P(0,m) = e^{-m}$$

from which one obtains:

$$m = - lnP_0.$$

The killer titer of the phage is then calculated from m and the dilution of the phage used to infect the bacteria. For example, if 5% of the infected cells survived the infection, $P_0 = 0.05$; $m = -\ln 0.05 = 3$. If one obtains this fraction of survivors after infecting 1 ml of a bacterial culture containing 2×10^8 CFU/ml, the 0.1 ml phage dilution used for infection contains $10 \times 3 \times 2 \times 10^8$ killing particles (KP) per ml, that is, 6×10^9 KP/ml.

A.4.2. Most Probable Number (MPN) Assays

If the host bacteria do not form lawns on solid culture media, plaque assays cannot be used to estimate viable phage titers. In such cases, most probable number (MPN) assays can be used instead. However, they are less accurate than plating assays (see Koch, 1981) and require a large number of cultures for a reasonably accurate estimate. MPN assays are based on the Poisson distribution of phage in liquids (sections A.3.6 and A.4.1.3), with the Poisson parameters k and m referring to actual number of phage in a sample and the average number of phage per culture, respectively. To perform an MPN assay, grow host bacteria, prepare serial tenfold dilutions of the phage, and proceed as described in steps 1–5 in section A.3.6. Standardized methods (Anonymous, 1998) call for 16–64 wells per dilution tested. *Note*: If the number of phages produced per host bacterium is low, a single phage infecting a culture containing about 10^4 bacteria (as used in this example) may not yield enough phages to cause complete clearing of the culture before it reaches a growth phase that does not support bacterial growth. Thus, inconclusive results will be obtained under those conditions. In such cases, one should try starting with fewer bacteria per well. After inspecting the various series of infected bacterial cultures to detect series that contain both lysed and intact cultures, the fraction of intact cultures is calculated for each such series. This fraction corresponds to the 0 term (i.e., $k = 0$) in the Poisson distribution, and its estimation enables calculation of the average number of infecting phage particles per culture (m) in that series [$P_0 = P(0, m) = e^{-m}; m = -lnP$]. For example, if 3 out of 10 cultures have lysed, $P_0 = 0.7$ and $m = 0.35$. Thus, each 10 μl aliquot of the phage dilution used for that infection series contained, on average, 0.35 infectious particles, and that value can be used to calculate the approximate phage titer of the original suspension. Alternatively, many bacteriology handbooks contain MPN tables that can be consulted for titer estimates.

A.4.3. Enumerating Phage Particles by Microscopy

Phage particles may be enumerated by transmission electron microscopy (TEM) after negative staining or by epifluorescence microscopy after staining with DNA fluorochromes such as 4', 6'-diamidino-2-phenylindole (DAPI). The possible presence of bacterial debris and media components in the sample may occlude the phage particles and lead to an underestimate by both methods. Also, background fluorescence and possible binding of the fluorescent dyes to non-phage material may cause artifacts when using epifluorescence methods and some of the particles may be undetected. Enumeration by TEM eliminates the artifacts associated with epifluorescence microscopy, since one actually sees the particles. However, the equipment needed is bulky and expensive and there are also artifacts associated with TEM methods (Suttle, 1993). Fluorescence microscopy methods have been well described (Hennes and Suttle, 1995; Suttle, 1993; 1997). For a description of the methodology for electron microscopy, see section A.6.1.

A.5. PREPARING, PURIFYING AND STORING PHAGE STOCKS

A.5.1. General Considerations

Phage stocks can be prepared in any medium that permits growth of the bacterial host strain. For many bacteria, rich synthetic media—such as M9CA (Table A.4)—may give as good phage titers as broth media, and they may permit direct use of the phage lysate for radiolabeling experiments, without prior purification of the phage. The actual media and conditions required for optimal phage production vary, and they need to be determined for each phage. For example, some phages produce higher-titer stocks if grown for fewer lytic cycles than suggested below; i.e., by infecting at higher cell densities using a higher multiplicity of infection. On the other hand, in some cases (e.g., when working with *B. subtilis* phages), slow phage production necessitates infection at low cell densities, to prevent the bacteria from reaching the stationary growth phase before all of them are infected. Thus, the conditions described below only serve as guidelines that need to be modified for different phages and hosts.

Most phages studied to date propagate only in actively growing bacteria, although some can propagate in stationary cells (Schrader et al., 1997; Wommack and Colwell, 2000). During propagation, mutations will arise. The rates of appearance of spontaneous mutations per genome per generation are almost the same in bacteriophage and bacteria, which means that the rates of mutations per base pair per replication are about 100-fold higher in large dsDNA coliphages than in *E. coli* (Drake, 1991). Therefore, it is prudent to start new phage stocks from single plaques. On the other hand, there is a risk that the selected plaque was formed by a nondesired variant phage. Thus, an appropriate balance must be determined on a case-by-case basis between using single plaques to start cultures and picking 2–3 plaques to avoid the risk of accidentally selecting a variant. Ideally, an archival stock or "master stock" of each phage strain should be maintained, from which new working stocks are grown. In order to avoid the unintended selection of phage mutants or contaminants, it is advisable that the properties of new phage stocks be compared with the original stock's known properties.

Lytic phage progeny are harvested after they are released from their host bacteria by cell lysis. However, some phages (e.g., many T4-like phages) exhibit lysis-inhibition (see Chapter 7), the duration of which depends on the particular phage, host strain, and growth conditions. Cultures infected with such phages may be incubated for several additional hours (thus enhancing the yield of phages) before the phages' progeny are harvested. If the phages do not contain lipids, chloroform can be added to the infected cells to release the progeny phages and reduce their adsorption to bacteria and bacterial debris. A lysed culture is commonly centrifuged or filtered to remove bacterial debris, and the resulting suspension is commonly referred to as a "cleared lysate."

A.5.2. Primary Lysates

A.5.2.1. Standard Liquid Culture

1. Obtain plaques of the phage to be grown by plating an archival stock, new isolate or enrichment culture. For each phage, grow a small (ca. 10 ml) host strain culture to early exponential phase (about 7×10^7 CFU/ml).
2. Use a sterile Pasteur pipet to pick one or a few plaques. Carefully insert the pipet to the bottom of the dish—all the way around the plaque(s)—and rub the pipet gently against the bottom (in order to dislodge the agar plug). Withdraw the pipet and blow the agar plug into the bacterial culture; alternatively, elute the phage from the plug into 1 ml medium and use that to infect.
3. Incubate the phage-infected culture until lysis is observed or—if the infecting phage is lysis-inhibited—for 4–8 hr (for bacterial generation times around 20 min; longer for more slowly growing bacteria). Most lysis-inhibited phages will not cause complete lysis under these conditions. Longer incubation times often result in a reduced phage yield, probably due to secondary adsorption of the liberated phages.
4. Add a few drops of chloroform to the culture (unless the phage contains lipid), and mix gently (to induce or complete lysis). Generally, the highest viable phage titers are obtained if the lysate is left overnight at room temperature at this point.

5. *Optional*: Incubate the lysate with 10 μl of pancreatic DNase I (1 mg/ml, in phage diluent; Table A.6) for 30 min at room temperature. The purpose of this step is to complete degradation of residual bacterial DNA and unpackaged phage DNA, in order to avoid losses in the following step. (Endogenous nucleases often provide sufficient digestion of the free nucleic acids in the lysate, obviating the need for exogenous enzyme.)
6. Pour the mixture into a chloroform-resistant centrifuge tube (polypropylene or glass). Be sure not to get any chloroform into the tube. It would form a layer under the pellet, resulting in pellet dispersal when the supernatant is removed. Centrifuge 4000 × *g*, 10 min.
7. Decant or pipet the supernatant into another centrifuge tube (for a second centrifugation), or into a screw-cap glass tube (for storage). Carefully avoid dispersing the bacterial debris at the bottom of the tube.
8. *Optional*: Filter the ice-cold supernatant through a nitrocellulose membrane (0.6 or 0.45-μm pore-size), using a water aspirator or vacuum line. This produces a very clear, stable stock. *Note*: For this application, it is important that the pH is at or above 7.0; at a lower pH, the phage may easily stick to and clog the filters. Thus, if this option is chosen, the medium in which the bacteria are grown and infected with phage should be buffered to about 7.3–7.4. Also, it is important that the phage are kept cold prior to filtering, so that they are not lost due to adsorption to bacteria. Transfer the filtered phage suspension to a screw-capped glass tube.
9. Add a few drops of chloroform to the final phage suspension. Omit this step for lipid-containing phages.

The clarified phage lysate usually has a reasonably high titer (ca. 10^8–10^{10} PFU/ml) and can be used directly for many experimental applications, especially if it was prepared in a defined medium. For many phages, the clarified lysate is quite stable when stored cold and dark, and can be used for several months or longer without an appreciable reduction in the titer. For other phages, the lysates are considerably less stable (see section A.5.5).

A.5.2.2. Plate Lysate

Some phages that grow poorly in liquid culture grow better on solid media. Most commonly, plate stocks are grown using soft-agar overlays, as follows:

1. For 9-cm diameter Petri dishes, add 0.2 ml indicator bacteria culture (section A.4.1.2) to small dilution tubes. Larger size plates are convenient when making larger volumes of plate lysates.
2. Add about 10^3 to 10^5 PFU of phage to each tube. The idea is to add enough phage to produce confluent or almost confluent lysis on the plate. For lysis-inhibited T4-like phages, which produce fairly small plaques, one plaque usually will yield the appropriate number of PFU, and it can be transferred directly (with a Pasteur pipet) from the plate on which it grew. For phages that produce large plaques, a couple of dilutions need to be tested.

3. Add 3 ml of molten soft-agar to each tube, and quickly pour the mixture onto a bottom layer of solid culture medium. Incubate the plate at a suitable temperature for 6–8 hours, for bacteria that have generation times of about 20 min (longer incubation times may be required for slowly growing bacteria). For bacteria that have generation times of ca. 20 min, longer incubation times often result in a reduced phage yield, probably due to secondary adsorption of the liberated phages.
4. Use a spreading rod or spatula to scrape the soft-agar into a beaker, and rinse the bottom-agar's surface with a total of 5 ml of phage diluent for a set of 10 plates. Add a few drops of chloroform to the mixture, mix gently, and incubate at room temperature for about 30 min. *Note*: Some investigators prefer the simpler approach of flooding the plate's surface with diluent (a couple of ml per plate) together with 1–3 drops of chloroform, and, after a couple of hours (to allow the phages to diffuse into the liquid), decanting and saving the phage-containing liquid. The preparation obtained by that method contains less solubilized agar, but it usually is more dilute and is likely to yield fewer phages than the first method.
5. Transfer the liquid (and agar) to a chloroform-resistant centrifuge tube, taking care not to include the chloroform (because it will form a layer under the pellet, which may lead to pellet dispersal while the supernatant is removed).
6. Centrifuge (4000 × *g*, 10 min) the mixture.
7. Pipet the supernatant into a fresh centrifuge tube, and repeat the centrifugation step.
8. Filtering (see subsection A.3.3.2) is a good option at this point, because some agar—which can interfere with subsequent phage studies—may still be present in the preparation.
9. Transfer the final phage suspension into a screw-cap tube and add a few drops of chloroform (omit chloroform for lipid-containing phage). If possible, store the mixture for a few days before titering, to disengage phage particles that may have clumped together during the centrifugation step(s).

Such phage preparations are not as stable as those from liquid media, possibly because some of the phages are adsorbed and inactivated by solubilized agar.

Phage and bacteria can also be spread directly on a single, bottom layer of solid medium (Zierdt, 1959). This technique avoids contamination of the resulting phage stocks with agar. The method also is useful for phage-host systems that are sensitive to the temperature needed to keep soft-agar liquid for pouring, usually about 45°C.

1. Prepare a mixture of indicator bacterial culture and phage (ca. 0.5 ml total volume per plate) that will produce confluent bacterial lysis on the plate. Repeat for as many plates as needed.
2. Spread the mixture evenly on the plate(s), using a sterile glass spreader.
3. Incubate until lysis is complete (3–6 hours for bacteria with generation times of ca. 20 min).

4. Add culture medium or phage diluent (0.5–2 ml) to each plate, and spread evenly with a sterile glass spreader.
5. Slant the plate, pipet the liquid into a screw-capped tube and add a few drops of chloroform (omit chloroform for lipid-containing phage).

This methodology is slightly faster and the phage preparations contain less solubilized agar than those obtained using soft-agar, often enhancing shelf life, but the phage yield is usually lower.

A.5.3. Secondary Lysates

When relatively large quantities of phage are needed, it may be impractical to prepare phage-containing lysates from single plaques. Instead, a liquid stock, either a fresh primary lysate or an archival stock that has been checked recently for good viability, is used to infect a large bacterial culture. Refer to section A.5.1 for a discussion of variations in growth conditions, cell densities, and MOI to yield good phage production. For each phage lysate to be prepared, grow 100 to 500 ml cultures of the host strain to a predetermined cell density (e.g. 7×10^7 cells/ml) and infect the culture with a liquid stock, at a predetermined MOI (usually 0.01 to 0.1). Incubate the infected cells and obtain the phage as described in section A.5.2.1.

A.5.4. Concentration and Purification of Isolated Phages

Large phages, such as those in the T4, SPO1, and φKZ families, may be concentrated and purified rather easily by high-speed centrifugation (section A.5.4.1). Centrifugation is impractical for very large-volume preparations (unless a continuous flow centrifuge is available), but is convenient for other important applications; e.g., eliminating media components, purifying phage labeled with a radioisotope, or concentrating small- to medium-volume phage preparations that have low titers. Precipitation with polyethylene glycol (PEG) (section A.5.4.2) is gentler for phages than is centrifugation, works well with phages too small to collect easily by centrifugation, and is suitable for large-volume preparations; however, it is more time-consuming than is centrifugation. Both methods may be used to concentrate phage prior to their purification by centrifugation in CsCl gradients (section A.5.4.3). If a suspension medium that absorbs poorly at a wave-length of 260 nm is used and the DNA content of a single phage particle is known, the number of phage particles in a purified preparation may be estimated from the A_{260} of the preparation, since an A_{260} of 1 corresponds to ca. 50 μg of double-stranded nucleic acids and to ca. 40 μg of single-stranded nucleic acids. The viable titer of the concentrated phage preparations should be determined as described in section A.4.1. A comparison between the total particle titer (obtained from absorbance measurements) and viable titer gives a rough estimate of the quality of the phage preparation.

A.5.4.1. Concentration by Centrifugation

Phages with double-stranded (ds) DNA genomes of ≥40 kbp can be pelletted from a cleared lysate (see section A.5.1) by centrifugation at ca. $35{,}000 \times g$ for 20 min;

large phages (like T4, dsDNA genome of 170 kbp) may be sedimented at 18,000 × *g* for 60 min (see also box in section A.6.4). After centrifugation, elute the phage from the pellet by soaking it overnight in a small volume of diluent (Table A.6), and gently pipet or pour off the phage-containing supernatant. (Using vigorous pipetting or vortexing to suspend phages that have been sedimented by centrifugation is likely to damage the phages.) The phage can be titered and used immediately or stored (see section A.5.5) until required.

A.5.4.2. Concentration and Purification by Precipitation with Polyethylene Glycol (PEG)

Concentrating and purifying a phage preparation by precipitation with PEG involves the following steps (Yamamoto et al., 1970):

1. Dissolve NaCl in the preparation (to 0.5 to 1 M, final concentration) by continuous mixing (e.g., using a magnetic stirrer) at 4°C for 1 hour. The NaCl dissociates phage from bacterial debris and media components and improves the precipitation of phage by PEG. *Optional*: Remove the bacterial debris by centrifugation (11,000 × *g*, 10 min, 4°C), and transfer the phage-containing supernatant to a clean flask.
2. Maintaining the sample at 4°C, add PEG 8000 gradually, with constant stirring, to a final concentration of 8%–10%, w/v. Store the mixture at 4°C for at least an hour, to allow the phage particles to form a precipitate. Longer storage (even overnight) may improve the phage yield.
3. Sediment the precipitated phage by centrifugation (11,000 × *g*, 10–20 min, 4°C).
4. Carefully remove the supernatant, using a regular pipet or a pipet tip attached to vacuum suction and making sure not to touch the pellet. *Optional*: Spot-test the supernatant; if the PEG precipitation was successful, the supernatant should contain very few phage.
5. Add an appropriate volume of phage diluent (Table A.6) to the pellet, which is then left overnight (in the cold) to soften gradually before gently mixing the suspension. *Note*: Vigorous pipetting or vortexing is likely to damage the phage particles. *Optional*: After suspending the phages, any remaining debris is removed by low-speed centrifugation (1500 × *g*, 15 min, 4°C). Also, if desired, PEG can be removed from the preparation by chloroform extraction.

The above procedure will yield a semi-purified phage preparation useful for many applications. If desired, the preparation can be further purified by centrifugation in a CsCl gradient.

A.5.4.3. Purification by Centrifugation in CsCl Gradients

Equilibrium centrifugation in CsCl gradients separates phage particles according to their buoyant density rather than their size. The vast majority of phages that have been examined by electron microscopy (>95%, Ackermann, 2003) are tailed. Known tailed phages, belonging to all three families and infecting many different Gram-negative and Gram-positive bacteria, are all composed of approximately equal amounts of protein and DNA, giving them a buoyant density in CsCl between 1.45 and 1.52 g/ml (average 1.49; lipid-containing phages average 1.3 g/ml; (Fraenkel-Conrat, 1985)). Thus, a fairly standardized protocol can be employed to purify them. This has been exploited in recent global analyses of environmental phage sequences (Breitbart et al., 2002; 2003; 2004). Centrifugation in CsCl gradients can be used to process fairly large amounts of phage, and it yields preparations that are highly purified and are well suited for DNA extraction. In addition, it removes the bacterial lipopolysaccharide (i.e., endotoxin) that may contaminate phage preparations obtained from Gram-negative bacterial hosts. However, after being centrifuged in CsCl gradients, some phages (such as coliphage T4) do not form plaques unless the gradients are prepared in broth with a high $MgCl_2$ concentration, as in the protocol below. Therefore, in order to avoid unpleasant surprises, phages' behavior after exposure to CsCl should always be investigated prior to using the technique. Also, the saturated CsCl solution used to make the gradients should be prepared with buffer, and its pH should be determined before use, since solutions made from many commercial sources of CsCl become acidic during storage.

Most phages appear to be more sensitive to a rapidly decreased osmotic pressure than to a rapidly increased osmotic pressure; so, in most cases, CsCl can be added directly to the phage suspension. However, a gradual increase in the CsCl concentration may be needed for some sensitive phages. Unless a phage is known to tolerate a rapid decrease in osmotic pressure, the purified phage solution should be gently diluted or dialyzed to reduce or remove the CsCl. Dialysis buffer should be chosen depending on the planned use for the phage suspension; the buffer used in the procedure described below is suitable for DNA extraction. *Note*: Pretreatment of dialysis membranes with bovine serum albumin (BSA) prior to using them to dialyze phage preparations, can reduce losses caused by phage adsorption to the membrane. The procedure involves the following steps: (i) fill the dialysis bag with, or expose the membrane's surface to, sterile 1% (w/v) BSA in physiological saline, (ii) dialyze the bag, for ca. 30 min, against physiological saline (under the same conditions that will be used for the subsequent phage dialysis step), and (iii) rinse the membrane with sterile water (to remove the BSA), and remove as much liquid as possible before adding the phage suspension to the dialysis bag.

Materials needed for CsCl purification of phage include:

Concentrated phage suspension
LBTM = LB containing 0.2 M $MgCl_2$ and buffered with 0.01 M Tris-HCl (pH 7.4)
LBTM saturated with CsCl
3 M NaCl, 0.1M Tris-HCl (pH 7.4)
0.3M NaCl, 0.1M Tris-HCl (pH 7.4)
0.1 M Tris-HCl (pH 7.4)
Vaseline
Refractometer or analytical balance

Ultracentrifuge with swingout rotor and fitting polycarbonate or other transparent tubes
Syringe, 18 gauge needle
Dialysis tubing

1. Measure the volume of the phage preparation and add one volume of saturated CsCl in 1 × LBTM. For osmotic pressure-sensitive phages, the CsCl solution should be added gradually.
2. Adjust the preparation's density with the saturated CsCl solution (or with LBTM, if the CsCl concentration is too high) so that it has a density of 1.50 g/ml. Ideally, the density should be determined by measuring the refractive index (1.380 ± 0.002 at 20°C) using a sodium lamp. If no refractometer is available, accurately weigh the solution after determining its volume.
3. Transfer the mixture to clear polypropylene or polycarbonate ultracentrifuge tubes and centrifuge (100,000 × *g*) for at least 18 h. Shorter centrifugation times may be sufficient; however, longer centrifugation times usually give better results.
4. If the phage band (which appears bluish-white and opalescent) can be seen, it may be recovered from the side with a syringe, yielding a very pure phage preparation. If the phage concentration is low, positioning the centrifuge tube against a black background and shining a light from above may help to detect the band. In order to collect the phage band: (i) clean the outside of the centrifuge tube, (ii) smear some Vaseline (to avoid leakage around the needle) around the middle of an 18-gauge hypodermic needle attached to a syringe (a narrower gauge needle increases the risk for shear damage to phage particles) (iii) pierce the tube about 5–10 mm below the phage band, (iv) place the tip of the needle into the band, and *slowly* aspirate it into the syringe (being careful not to aspirate other bands that are visible in the gradient, as they contain bacterial debris and other impurities), and (v) remove the needle and gently expel the liquid into a fresh tube for temporary storage (at 4°C, in a tightly capped tube) before further processing. *Note*: If the tube cannot be punctured, collect the band from above, using a micropipette or Pasteur pipette.
5. If the phage band cannot be seen, sequential fractions of the gradient are obtained by puncturing the bottom of the centrifuge tube and allowing the gradient to drip into several small tubes. The phage-containing fractions are then located by OD_{260} measurements or by plaque assays (after gentle removal, by stepwise dilution, of the CsCl). This harvesting procedure will yield a less purified and more dilute phage preparation than will the protocol described above. If the starting preparation is very crude, the phage collected after the first centrifugation step can be subjected to a second centrifugation step.
6. Remove the CsCl from the phage suspensions by dialysis at 4°C. It is important to keep the dialysis time to a minimum (as mentioned above, phage particles may be lost by adsorption to the dialysis membrane).

Dialyze for 30 min against ca. 1000 volumes of 3 M NaCl, 0.1M Tris-HCl (pH 7.4) and then for 2 × 30 min against 1000 volumes of 0.3 M NaCl, 0.1M Tris-HCl (pH 7.4). It is a good idea to mix the dialysis buffer gently during the dialysis steps (e.g., by using a magnetic stirrer). After dialysis, remove the phage preparation from the dialysis bag and store in a sterile, tightly capped tube at 4°C until further analysis. For general storage recommendations, refer to subsection A.5.5.

A.5.4.4. Centrifugation in Sucrose Density Gradients

Phage also may be purified by centrifugation in solutions of sucrose or other substances that produce stable density gradients (Price, 1974). Such gradients separate according to size *and* shape; thus, prior knowledge of those phage parameters is necessary in order to obtain optimal results. For example, coliphages T7 and T4 have been successfully resolved (Serwer et al., 1978) by centrifugation (53,000 × *g*, 20 min) in a 5%–25% (w/v) sucrose gradient in 0.01M Tris buffer (pH 7.4) containing 0.1 M NaCl, 0.001M $MgCl_2$, and an unspecified concentration of ethidium bromide.

A.5.5. Storage of Phage

Phage may be stored as clarified lysates (section A.5.2, A.5.3) or as purified stocks (section A.5.4). Many phages are quite stable in various phage diluents (Table 2.6), if stored (i) highly concentrated (10^{10} to 10^{13} PFU/ml), (ii) free of bacterial debris, and (iii) at 4°C and protected from exposure to UV light (Wommack et al., 1996). Aqueous stocks of phages containing $>10^{13}$ particles per ml should be diluted to 10^{11} to 10^{12} PFU/ml, to avoid losses caused by aggregation or crystallization. Phages that lack lipid components are usually stored in diluents containing a few drops of chloroform (to prevent bacterial growth). However, since the sensitivity of various non-lipid-containing phages to chloroform may vary significantly, it is a good idea to determine the chloroform resistance of a given phage before including chloroform in the phage's storage solution. Some phages survive very poorly in refrigerated lysates or stocks (Clark and Geary, 1973), thus necessitating other storage conditions. Freezing at liquid nitrogen temperatures often will maintain a phage stock's viability (Clark and Geary, 1973), provided that it is mixed with a cryoprotectant (e.g., 5%–8% [v/v] dimethyl sulfoxide [DMSO]) before freezing. However, bacteriophages vary in their sensitivity to freezing and freeze-drying; e.g., large, complex and osmotic-shock-sensitive phages generally are more vulnerable than are small, osmotic-shock-resistant phages (Clark and Geary, 1973). Thus, appropriate storage conditions need to be experimentally determined for each phage. Ackermann (2005) has extensively reviewed the topic of long-term bacteriophage preservation based on experience at the Reference Center.

Freezing at –20°C in the presence of 10% glycerol (Engel et al., 1974; Mendez et al., 2002), or freeze-drying in the presence of various concentrations of skim milk (20%–100%), combinations of peptones (10%–20%), sucrose (10%), gelatin (0.5%), or sodium glutamate (2%–5%) (Davies and Kelly, 1969; Clark and Geary, 1973; Carne and Greaves, 1974; Engel et al., 1974; Zierdt, 1959; 1988) have yielded reasonably good survival of various bacteriophages. However, the best results have been obtained with 100% skim milk (for staphylococcal phages; (Zierdt, 1959; 1988), 5% peptone, 5% Na-glutamate + 0.5% gelatin (for mycobacteriophages; (Engel et al., 1974)), or 20% peptone in 10% sucrose (for corynephage H1; (Davies and Kelly, 1969)). The addition of sodium glutamate resulted in somewhat lower initial recoveries, but it improved the heat-resistance of the lyophilized phage preparations (Davies and Kelly, 1969). Thus, it is advisable to experimentally determine which cryoprotectant works best with a given phage. Also, the temperature and cooling rate during freezing or freeze-drying may be crucial for optimal viability; e.g., fast cooling (up to 450°C per min) and maintaining a very low temperature during freeze-drying have been reported (Davies and Kelly, 1969) to improve survival of many phages. Once phages are lyophilized, they can be stored (at room temperature, refrigerated or frozen) without a noticeable decrease in titer, for several years—if not indefinitely.

A.5.6. Sending Phage

Unless phage are mailed together with their host bacteria, special "infectious substances" paperwork is not required to mail them to, from or within the United States. Phage in a dried state (i.e., either freeze-dried or spray-dried) can be mailed in Eppendorf tubes, using an envelope with a soft liner. Phages in various buffers (i.e., aqueous phage suspensions) can be mailed similarly, but a significant loss of phage titer may occur during transport, so it is recommended that they be shipped on wet-ice or with commercially available cold packs using express mail services. Another possibility for many phages is to saturate glass-fiber filters with the phage suspension and mail them. Briefly: (i) place two glass-fiber filters on top of each other on a piece of plastic wrap, (ii) let 50–100 μl of the concentrated phage suspension soak into the filters, (iii) wrap the plastic wrap tightly around the filters, (iv) tape the labeled packet to a piece of cardboard or heavyweight paper on which identification information can be written, and (v) mail it in an ordinary envelope. Upon arrival, the filter papers are removed and placed in 1-ml of phage diluent (Table 6) for about 30 min, after which the phage are recovered by plating with an appropriate host.

A.6. CHARACTERIZATION OF PHAGES

A.6.1. DNA Analysis

In recent years, exploring the genetic information in various ways has taken on a primary role in phage characterization. Pulse field gel electrophoresis and restriction-digest studies provide means of rapid early classification, while PCR using primers from highly conserved genes of well-studied families and shotgun cloning and sequence analysis provide key, detailed information in this era of rapidly growing sequence banks and genomic analysis, as discussed in Chapter 5.

A.6.1.1. Isolation of Phage DNA

Methods to isolate DNA from phages are described in standard molecular biology manuals (e.g. Sambrook and Russell, 2001; Ausubel *et al.*, 2001); as well as by

Kricker and Carlson (1994); many vendors also sell kits suitable for isolating phage DNA. Methods avoiding the use of organic solvents, such as the hexadecyltrimethylammonium bromide (CTAB) procedure (Sambrook and Russell, 2001; Kricker and Carlson, 1994), yield somewhat less DNA of somewhat lower purity than procedures using organic solvents, but CTAB-purified DNA is suitable for such purposes as restriction digestion, cloning, sequencing and the polymerase chain reaction (PCR) and such methods are safer and have less environmental impact. *Note:* Phages must be suspended in approximately isotonic media or buffers for DNA extraction; thus, CsCl-purified phages must be dialyzed prior to extraction.

A.6.1.2. Digestion of Phage DNA with Restriction Endonucleases

Vendors of restriction endonucleases usually also supply the concentrated buffers needed for the digestion procedure and suggest cleavage conditions. *Note*: Phage DNA may be resistant to restriction cleavage because of modifications. Therefore, several enzymes may need to be tested to identify those capable of digesting a particular phage's DNA. The amount of phage DNA needed depends on the method used to visualize the results (see below). The volume of the enzyme added should not be more than one-tenth of the final volume of the mixture, in order to ensure that the glycerol contained in many commercial restriction enzyme preparations does not inhibit its activity. Restriction digestion of phage DNA may require longer incubation times than does digestion of similar amounts of plasmid DNA. Incubation for 4 h or overnight is likely to give the best results. After digestion, the DNA fragments can be analyzed by agarose or polyacrylamide gel electrophoresis (see Table A.8). The resulting patterns may be used to characterize and differentiate various phages. If the phage genome is very large

TABLE A.8
Separation of DNA Fragments in Agarose and Polyacrylamide Gels

Gel Matrix	Gel Concentration (w/v)	Approximate Separation Range (bp)
Agarose	0.5%	700–25,000
	0.7%	600–20,000
	0.9%	500–12,000
	1.5%	200–4,000
	2%	100–2,000
Sieving agarose	1% agarose + 3% sieving agarose	100–1,000
Polyacrylamide[1]	3,5%	1,000–2,000
	8%	60–400
	15%	25–150

[1] ratio acrylamide:bisacrylamide 30:1

(e.g., >100 kb), pulsed field gel electrophoresis (PFGE) may be used instead of standard agarose electrophoresis.

A.6.1.3. Agarose and Polyacrylamide Gel Electrophoresis of Phage DNA

Methods for agarose and polyacrylamide gel electrophoresis (PAGE) of DNA fragments are described in many laboratory manuals; for variations of the protocols outlined below, and for more detailed descriptions of procedures, see e.g., Sambrook and Russell (2001). Ethidium bromide (EtBr), a fluorescent and intercalating dye that is commonly used to visualize DNA in agarose gels, is a potent mutagen. Therefore, gloves must be worn when working with EtBr-containing solutions, and the solutions must be decontaminated prior their release into the environment; several manufacturers offer resins that bind and remove EtBr from dilute aqueous solutions. Many other DNA stains are available, offering lower toxicity (e.g., SYBR Safe) or higher sensitivity (e.g., SYBR Green or SYBR Gold) than EtBr. EtBr may be included in agarose gels or used in the staining bath after concluding electrophoresis; it interferes with the polymerization of acrylamide, so it cannot be included in polyacrylamide gels. Polyacrylamide gels are typically run in 1 × TBE (Tris-Borate-EDTA; for recipe, see below) at high voltage (~350V). Agarose gels are commonly run in 0.5–2x TBE or 1–2x TAE (Tris-Acetate-EDTA; see the recipe that follows). Double-stranded DNA migrates a little faster through TAE buffer than through TBE buffer. In general, high ionic strength, low voltage, and long run times are used to separate large fragments (5–20 kb), and low ionic strength, high voltage and short run times are used for small fragments (≤10 kb). To make agarose gels of higher concentration than 1%–1.5%, sieving agarose can be used. Also, gels containing high concentrations (4%–10%) of various types of less fragile, low gelling temperature agaroses can be used to analyze very small fragments (10–500 bp) of phage DNA.

Suitable size markers should be electrophoresed together with the test samples, to permit calculation of the DNA fragments' sizes. Approximate ranges for separation of linear DNA in gels containing various concentrations of agarose or acrylamide are summarized in Table A.8. Both agarose and polyacrylamide runs can be monitored by following the tracking dye. In 0.5%–1.5% agarose, gels run in 0.5 × TBE, bromphenol blue (BPB) will migrate at a rate approximately equal to that of linear ds DNA fragments with a size around 300 bp, and xylene cyanol (XFF) will migrate at approximately the same rate as ds DNA fragments of about 4 kbp. In 12% SDS-polyacrylamide gels run in the same buffer, these dyes comigrate with dsDNA fragments of approximately 20 (BPB) and 70 (XFF) bp, respectively. As little as 1 ng of DNA can be detected in agarose gels using EtBr; however, a long staining/destaining protocol (40–60 min) may be required to visualize such small amounts of DNA. The background fluorescence caused by unbound EtBr may be removed by soaking the gel (15–30 min, room temperature) in a 1 mM $MgSO_4$ solution. Inclusion of EtBr in gels slightly alters DNA mobility because its intercalation results in partial unwinding of the DNA. Because polyacrylamide quenches the fluorescence of EtBr, samples should contain at least 10 ng of DNA to be detectable in polyacrylamide gels stained with EtBr after electrophoresis. Use of a detergent-containing gel-loading reduces the risk of formation of protein aggregates that could interfere with DNA migration.

Materials

Purified phage DNA cleaved with (a) restriction endonuclease(s)
Size standards for electrophoresis (*e.g.* restriction digests of phage λ DNA)
Gel buffers:
5 × TBE (0.45 M Tris-borate, 0.01 M EDTA; pH 8); used for agarose and acrylamide gels
5 × TAE (0.2M Tris-acetate, 0.005 M EDTA; pH 8); used for agarose gels
Agarose or polyacrylamide gel prepared in running strength gel buffer
Ethidium bromide: 10 mg/ml (store dark at 4°C–25°C)—Agarose gels may include EtBr to 0.5 μg/ml
Electrophoresis set up for slab gels, with combs making wells holding at least 25 μl
Gel loading buffer, for example, 6 × SDS-gel load composed of: 60% (w/v) glycerol, 60 mM Tris-HCl (pH 7.5), 60 mM NaCl, 60 mM EDTA, 6% (w/v) sodium dodecyl sulfate [SDS], and 0.06% (w/v) bromphenol blue; warm to 37°C before use to dissolve the SDS.

Procedure

1. Mix (at room temperature) 20 μl of each sample, containing sufficient DNA for analysis (see below) with 4 μl of 6 × gel-load. Prepare suitable size standards the same way.
2. Slowly load the samples into the slots of the submerged gel, using disposable micropipets. Since the distortion of migration through the gel is minimal in its middle, loading the size standards there will permit more precise size estimates.
3. Start the electrophoresis. Although conditions for optimal separation need to be tested, a voltage of 1–5 V/cm distance between the electrodes is a good start.
4. After electrophoresis, remove the gel from the apparatus. If no EtBr was included in the gel, stain with EtBr (0.5 μg/ml in running strength gel buffer) for 10–45 min and rinse with deionized water until the background is clear. Inspect the gel using transmitted or incident UV light.
5. Photograph the gel using transmitted or incident UV light, an orange filter and a good camera. DNA–ethidium bromide complexes are best detected at a wavelength of 254 nm. However, since significant DNA nicking occurs at that wavelength, a wavelength of 302 nm, 312 nm, or 366 nm (in that order) is recommended for detecting EtBr-stained phage DNA that is to be extracted and used. *Caution*: Ultraviolet radiation is dangerous, particularly to the eyes. Therefore, UV light-blocking goggles or a full safety mask should be worn to minimize exposure when viewing gels in a UV transilluminator. Several companies manufacture gel documentation

systems that include UV transilluminators in an enclosed system, significantly reducing the exposure to UV light.

To calculate approximate lengths of DNA fragments, a standard curve is prepared from migration data for the fragments in the size standard. No mathematical transformation will yield an absolutely linear relationship between DNA size and distance migrated. For DNA fragments of 1–10 kb, approximate linearity is obtained by plotting size against the inverse of the distance migrated from the application point; for smaller DNA fragments better linearity is obtained by plotting size instead against the logarithm of the distance migrated. After determining the distance migrated by the unknown fragments, their sizes can then be extrapolated from the standard curve. Commercial gel documentation systems often contain software to calculate automatically sizes of unknown DNA fragments by comparing them to those of known standards.

A.6.1.4. Pulsed Field Gel Electrophoresis of Large Phage Genomes

DNA molecules that are too large to enter conventional agarose or polyacrylamide gels can be analyzed by pulsed field gel electrophoresis (PFGE) (Wrestler et al., 1996). PFGE is used to analyze restriction patterns of phage genomes or undigested phage DNA in the 100 to 200 kb range; however, DNA fragments up to 10 mb can also be analyzed by PFGE. During PFGE, pulsed, alternating, orthogonal electric fields are applied to an agarose gel, and DNA migrates along the new axis of a continuously changing electric field. Special equipment is required, and agaroses exhibiting minimal electroendosmosis, available from several vendors, provide the best separation. In order to perform PFGE of undigested phage DNA, intact phage particles are embedded in low melting-temperature agarose and then lysed *in situ* by soaking the agarose plugs in a solution containing detergent or protease (e.g. 1% Sarcosyl or 2 mg/ml proteinase K), followed by washing the plugs with TE buffer (10 mM Tris-HCl, 1 mM Na_2EDTA, pH 7.6). The DNA in the plugs can be restricted *in situ* with restriction endonucleases (using standard protocols); however, a slightly higher concentration of the restriction enzyme and longer digestion time than for aqueous restriction digestion of phage DNA may be required for optimal digestion. The plugs can be put directly into the slots of the agarose gel or melted at 65°C and the molten agarose containing the digested DNA added (using disposable micropipets) to the gel's slots. The optimal electrophoresis conditions for PFGE should be determined empirically for each preparation to be analyzed, by electrophoresing several gels under various conditions. Alternatively, if the expected size of the phage DNA or DNA fragments is known, some commercially available systems include algorithms that automatically derive optimal separation conditions for the expected size range of the DNA fragments. As with agarose gel electrophoresis, appropriate molecular weight markers need to be included during all PFGE experiments. Several markers specifically designed for PFGE are commercially available. The gels are stained and photographed as described above for agarose gels, and the resulting patterns are compared visually, or by using appropriate pattern analysis software.

A.6.1.5. Analysis of Phage DNA by Polymerase Chain Reaction and Sequencing

The polymerase chain reaction (PCR) can be used for comparative analysis of various phages, or for preparing sequencing templates (Tetart et al., 2001). Many commercial vendors provide kits for performing PCR and for sequencing PCR amplicons. Phage DNA usually does not have to be purified prior to its amplification by PCR but can be amplified *in situ* using lysed phages. The protocol involves the following steps: (i) Pick a phage plaque and suspend it in 50 μl water, leaving it at room temperature for 30–60 min to allow the phages to diffuse into the water. (ii) For a 50-μl PCR amplification, remove 25 μl of the phage suspension to a clean tube and denature by heating in a boiling water bath for a few minutes. (iii) Mix gently, add reagents and proceed as prescribed by the vendor of the PCR reagents. If this protocol does not work for a particular phage, phage DNA should be extracted (see section A.6.1.1). Alternatively, if phage-containing plugs are available from previous PFGE experiments, phage DNA can be rapidly extracted by two cycles of freezing and thawing of the plugs, followed by sedimenting the agarose by centrifugation ($1500 \times g$, 5 min). The resulting DNA-containing supernatant can be used directly for amplification by PCR. The amplification products can be analyzed by gel electrophoresis (as described in section A.6.1.3). In addition, the amplicons may be sequenced directly after treatment with alkaline phosphatase (to degrade unused nucleotides) and exonuclease (to degrade nonamplified, single-stranded DNA).

As an alternative to measuring absorbance at 260 nm, real-time PCR can be used to quantify the amount of phage DNA in a suspension, and thereby the number of phage particles, provided the amount of DNA per particle is known (Edelman and Barletta, 2003). Such estimates are useful in comparison to PFU titers to judge the quality of a phage preparation—what fraction of the phage can form a plaque under optimum conditions?

A.6.2. Biological Parameters of Phage Infection

A.6.2.1. Analysis of the Phage Infection Process

A number of parameters can be explored to check the efficiency and time course of a phage's ability to infect and develop in a particular host (cf. Fig. A.3, A.4, Ch. 3, Fig. 3.3). The optimal cell density at which a bacterial culture can be infected productively with phage depends on the host bacterium's growth characteristics. Generally, initial explorations are carried out in host cells in exponential growth, using an MOI >3 that results in infection of virtually all cells; most parameters of the infection process are generally very similar whether an individual cell has been infected with 1 phage or 8. Values that can then be measured include (1) number of surviving bacteria as a function of time after infection; (2) kinetics of phage adsorption and new phage production; (3) effects on the turbidity of the host and (4) number of "infective centers"—free phage plus phage-pregnant cells. These are important parameters to understand in most kinds of phage experiments.

Many well-studied phages generally infect rapidly and efficiently under standard laboratory conditions; both the number of surviving bacteria and the number of

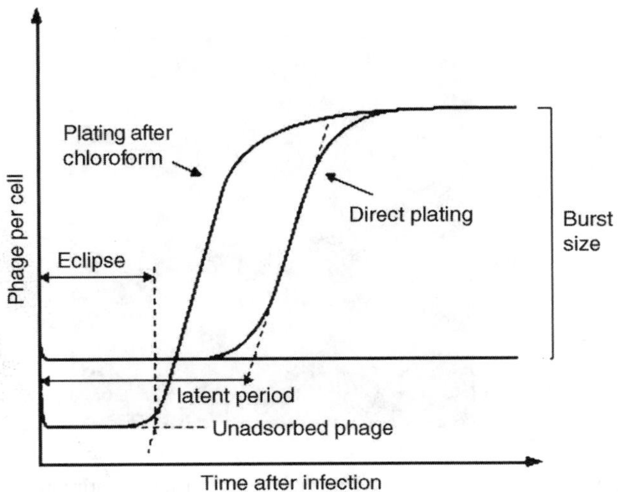

FIGURE A.3 Single step growth of bacteriophage (drawing by David Herthnek).

viable phage plummet two orders of magnitude within a few minutes for T4-like phages infecting *E. coli* B at an MOI of 5, for example. The cell mass, measured as turbidity, may continue to increase after infection, though usually more slowly than in a parallel uninfected culture (Fig. 3.3); the total amount of membrane material may also double (see Harper et al., 1994). The turbidity may remain high for a long time, due to lysis inhibition (Chapter 7)—as much as a day under some conditions, even though all the cells are infected and have made a number of phage. In such cases, adding *chloroform* will cause the cells to lyse and the turbidity to drop, and will free intracellular phage each of which can then make an individual plaque and be counted. Uninfected cells are killed but not lysed by chloroform, and cells that have not yet made much lysozyme will lyse slowly, but none of the cells actually infected by a virulent phage is likely to rid itself of the phage and be able to make a colony.

For many other phages, and for T4-like phages in some conditions, rates and even extents of attachment are often far slower, complicating detailed studies of the infection process. It is thus important to explore carefully the various parameters and look for the best hosts and conditions. The optimal cell density at which a bacterial culture can be infected productively with phage depends on the host bacterium's growth characteristics. Usually, infecting cells in the early exponential growth phase yields the best results. The results of phage infection experiments sometimes are a challenge to interpret. For example, in some cases, only a small fraction of the cells appear to be susceptible to the initial attack, even at an MOI of 10, yet it appears that the few progeny phage made in a first round of infection can then infect very efficiently, killing the cells and, somewhat later, causing them to lyse (Fig. A.3). (This and additional specific examples will be posted on our website, http://phage.evergreen.edu/methods.) In fact, newly made,

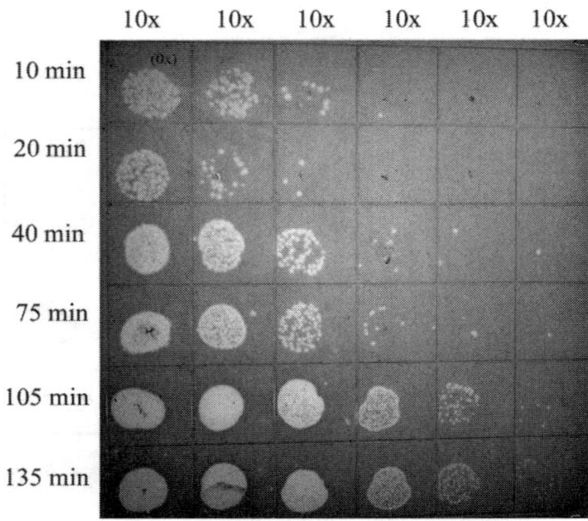

FIGURE A.4 Spot test analysis of phage concentrations during an entire infection on a single square plate.

so-called *nascent* phage may have increased infectivity and host range through as-yet poorly characterized mechanisms (Evans, 1940; Wollman, 1952; Brown, 1969; Simon, 1969; see Chapter 7).

In exploratory studies of phage infection, the number and timing of samples needed will depend on the particular phage-host system and will often take multiple attempts before a clear pattern emerges; spot testing (see Fig. A.2) makes such explorations manageable and is facilitated by using square plates (Fig. A.4). Initial experiments may include about 5–6 sample time points spread over two host doubling times; guided by results from such experiments more precise experiments can be set up using more time points as well as closer dilutions. When determining the infection characteristics of phages under aerobic conditions, the shaking of the water bath (ensuring aeration of the culture) should not be turned off (except for the minimal times necessary to remove samples), and flasks and tubes with growing or infected cells should not be taken out of the bath. Also, for optimal results, the exact titer of the bacterial culture at the time of infection must be known. (See section A.2.4.3 for a discussion of procedures and pitfalls in determining these cell titers.) In the following example bacteria are infected at 2×10^8 cells/ml using an MOI of 3, monitoring cell turbidity and survival as well as phage growth parameters. Media, diluents, methods for monitoring of bacterial growth and basic infection and plating techniques are described in sections A.2.4, A.2.5, and A.4.1.

Materials

Plates containing bottom agar, either 4 square plates labeled C, BS, S, and L or 7 round 90-mm plates labeled C, BS1, BS2, S1, S2, L1, and L2

Molten soft-agar, boiled and then kept at 45°C

Two 100-ml Erlenmeyer flasks (side-arm flasks if a colorimeter is available)—G (Growth) and C (Control)
Nine small dilution tubes, each containing 0.9 ml of diluent labeled "C1-C3" (uninfected Controls) and "S1-S6" (for Samples)
Six small dilution tubes, each containing 0.9 ml of diluent and 4 drops of chloroform—and labeled "L1-L6" (for Lysed cells); in an ice bucket (to reduce evaporation of the chloroform)
Thirty dilution tubes, each containing 5 ml of diluent
Pipets, test-tube racks and accurate timer
Host bacteria in liquid culture
Fresh indicator bacteria (see section A.4.1.2) kept on ice
Phage suspension containing $1.2 \times 10^{10} - 6 \times 10^{10}$ PFU/ml kept on ice

Determine what volume of the phage suspension contains 6×10^9 PFU, to give an MOI of ca.3. This volume should be between 0.1 and 0.5 ml, to permit accurate pipetting and to avoid excessive dilution of the bacterial culture upon infection. Calculate the sampling times (0.1–2 bacterial generation times). The samples in the S and C tubes should be processed as soon as possible—preferably as the experiment is going along.

Procedure

1. Grow the bacteria at an appropriate temperature, with good aeration for aerobic growth. Monitor growth until you expect about 2×10^8 cells/ml. If growth is monitored by microscopy, also determine the turbidity of the culture.
2. **About 5 min before** the culture reaches the desired concentration, place flask G and C into the shaking water bath and pipet 10 ml of the culture into each flask. Put the S and C tubes into a rack in the same water bath. (By sampling into warm diluent that is subsequently cooled, cold-shock of the bacteria is reduced). When the cells are ready, the experiment should be started immediately.
3. **When the culture reaches 2×10^8 cells/ml**, transfer 0.1 ml of the uninfected culture into tube C1 and place it on ice. Add the volume determined above of the phage suspension to flask G, mixing well as you put it in, and start the timer. Add the same volume of medium to flask C.
4. **After about 0.1 bacterial doubling time**, transfer 0.1 ml of the phage-infected culture from flask G to tube S1 (which is now put on ice), and transfer 0.1 ml from flask G to tube L1 (which is now put in a rack at room temperature, to let the chloroform lyse the infected cells).
5. **At about 0.33, 0.75, 1.2, and 1.8 doublings measure the turbidity in G and C**
6. **After about 0.25, 0.6, 1, 1.5 and 2 generation times,** transfer additional 0.1-ml samples from flask G to the remaining, appropriately labeled S and L tubes, and store the samples as in step 4.
7. **After 0.6 and 1.5 doublings, also transfer 0.1 ml samples from Control to C2 and C3**

Processing of Samples

The L tubes should be kept at room temperature for at least 30 min before being processed, to complete lysis of the cells; mix the contents of the tubes thoroughly, then let settle, and avoid the chloroform in the bottom when sampling. A sequential heat-cold-shock may facilitate lysis of recalcitrant cells, which can be a problem for some phage-host systems: put the L tubes in a 37°C water bath for a few minutes, then put them in an ice bath for a few minutes, and then return them to room temperature. The L tubes can usually be kept for several hours (overnight for some phages) without PFU losses.

Prepare a lawn of host bacteria (3 ml of soft-agar and 0.2 ml of an indicator culture, section A.4.1.2; 4 ml and 0.3 ml for square plates) on each S- plate; the C and BS plates are for determining bacterial titers and bacterial survivors, and do NOT get top agar or a lawn of bacteria.

Prepare six serial tenfold dilutions from each C and S tube, one sample at a time, moving each tube forward as you do the dilution and using a fresh pipet or tip for each. Since the first tube contains a tenold diluted culture, the samples to be spotted are tenfold to 10^7-fold dilutions of the cultures. Using a micropipettor, place 5 μl drops of each of the last five dilutions of samples C1–C3 in a row across plate C (see section A.4.1.2.3 and Figs. A.2 and A.4). Place 5 μl drops from the first dilutions of all S samples onto the BS plates, and of the last 5 dilutions of the same S samples on the S plates—starting at the top of the plate, so that the 3–5 rows of spots fit underneath each other—6 rows on square plates. By starting with the most dilute sample the same tip can be used for one whole series. Duplicates plates may be done for the phage for better accuracy and in case there are any problems.

Prepare a lawn in the same way as for S on the L plates, make similar serial dilutions from the L tubes (once the cells have had time to lyse—and being careful to avoid the chloroform), and spot the last 5–6 dilutions onto the L-plates. Store the plates right-side-up (on the benchtop), with their lids slightly ajar, until the drops have soaked into the agar, then put the lids back on and incubate the plates upside down at an appropriate temperature for the bacteria.

The plates are examined the next day, and the titers of bacteria (B plates), infective centers plus free phages (S plates), and intracellular plus free phages (L plates) are calculated from the number of observed colonies/plaques and the known dilution factors. The data can be used to calculate (i) the rate of phage adsorption to the host bacterium (from the drop in the free phage titer [noted on plate S] immediately upon infection), (ii) the bacterial viability and rate and efficiency of killing (from the titers on plate BS at different times after infection, relative to the titers on plate C1 and the cell density determined by microscope count or turbidity just before infection), (iii) the efficiency of formation of infective centers (from the early titers on the S plates with the early titers on the L plates subtracted, relative to the titer on plate C1 and the cell density at the time of infection) (iv) the eclipse period (from the rise in phage titers on plate L), (v) the latency period (from the rise in phage titers on plate S), and (vi) the burst size (the total phage production per infective center [from the L and S plate data] or per cell [microscope count before infection]). The extent of turbidity increase after infection and the comparison between the uninfected and infected cell turbidities will give you a better understanding of the phage's effect on general cell metabolism, including directly observing the time of lysis.

A.6.2.2. Single Step Growth

The eclipse and latency times of phages are important parameters that characterize phage-host interactions (Fig. A.3). The eclipse is the time from infection until the first new, infective phages particle have formed inside infected bacteria, while the latency is the time from infection until the first new, infective phage particles are released from host cells. Eclipse and latency times should be determined from singly infected cells, under conditions that prevent reinfection by the newly liberated progeny phages (Ellis and Delbrück, 1939; Doermann, 1952). This can be accomplished as follows:

1. Infect bacteria at a low multiplicity of infection (MOI, see section A.2.4.4). Since the phage particles and host bacteria are distributed in liquids according to the Poisson distribution (see section A.3.6), only about 10% of all bacterial cells will be infected if the MOI is 0.1 phage particles per bacterium. However, 90% of those infected cells will have been infected by a single phage particle.
2. Dilute the culture by at least 10^4-fold after infection. This reduces the probability that a phage released from a lysed bacterium will find a new cell to infect, and also reduces the likelihood of lysis inhibition (see Chapter 7). An additional advantage is that the culture may not need to be aerated after being diluted, because cells at this low density may obtain sufficient oxygen by diffusion.

The timing of the sampling is critical here, since you are plating one dilution of each sample rather than spot-testing several, and many time-consuming pipetting steps are required. Therefore, many approaches to facilitate the performance of this experiment are found in the literature; e.g., (i) performing the experiment at a lower temperature than that required for maximal growth of the host bacteria, (ii) chilling the host bacteria before infection, and (iii) performing the infection step in the presence of chloramphenicol or cyanide, which suppress bacterial growth (the chloramphenicol or cyanide are subsequently removed by dilution). The first two approaches usually work best. Dependent upon the phage and host used and the growth conditions employed, the timing of sampling appropriate to determine the infection parameters will vary and it is important to understand the general infection parameters well before starting.

In the following example, the bacteria in a 10-ml culture (containing ca. 2×10^8 cells/ml) are infected with a MOI of ca. 0.1 PFU per cell and the infected culture is sampled for phage at a number of intervals ranging from 0.1 to 2 bacterial generation times; no bacterial samples are taken. The actual timing, and the number of samples taken, will depend on the results from preliminary experiments. Therefore, the protocol presented below should serve only as a general guide. The supplies and reagents needed for the full-scale experiment are the same as those for the pilot experiment, except that no diluent is added to the sampling tubes (S and L). In addition, prepare

(i) two more growth flasks (G2 and G3) containing 10 ml and 20 ml of culture medium, respectively (to prepare additional dilutions of the infected culture), and (ii) more plates (with bottom agar), soft agar and and dilution tubes (to plate dilutions of all samples, instead of spot-testing them—see section A.4.1). Also, determine when the culture should be diluted (i.e., when most of the cells are infected [they plate as infective centers] or are "killed" [they no longer form colonies]). This time is referred to as "D" below. The eclipse period is referred to as "E," and the latency time as "La." Determine ahead of time what volume of the phage suspension contains 2×10^8 PFU, to yield an MOI of 0.1 (preferably 0.1–0.5 ml; dilute the suspension, if necessary). The experimental protocol involves the following steps:

1. Grow the bacteria under the same conditions and to the same density used for pilot experiments.
2. **About 5 min before** the culture reaches 2×10^8 cells/ml, place flask G into the same water bath as the bacterial culture, and pipet 10 ml of the culture into the flask. When the cells are ready, the experiment should be started immediately.
3. **When the culture reaches 2×10^8 cells/ml**, transfer 1 ml of the uninfected culture into tube S0, and place it on ice. Add the predetermined volume of the phage suspension, which contains ca. 2×10^8 PFU to flask G, start the timer, and place flasks G2 and G3 into the bath.
4. **At time D**, transfer an aliquot (0.1 ml) from G to G2 (yielding a 10^2-fold dilution), and transfer an aliquot (0.2 ml) from G2 to G3 (yielding a 10^4-fold dilution). All flasks should remain in the water bath. Transfer 1 ml from G3 to S1 (place on ice), and to L1 (store at room temperature).
5. **At time E – 5 min, E, E + 5 min, E + 15 min, etc.**, transfer 1 ml samples to appropriately labeled S and L tubes, and store as above. If the cells lysed during earlier experiments, samples also should be taken at times **La – 5 min, La, La + 5 min, La + 15 min, etc.**, and the experiment can be terminated at La + 30 min. If the cells didn't lyse during the pilot experiment, one may continue sampling for an extended time period.

The final dilutions to be plated (in order to calculate the bacterial titer at the time of infection, fraction surviving bacteria and infective centers and free phages in all samples) should be determined from the actual titers reached in the pilot experiment. Remember that the flask from which the S and L samples are taken already contains a 10^4-fold diluted culture, and that only about 10% of the bacteria will be infected in this full-scale experiment; therefore, bacterial survivors should not actually drop significantly until the cells lyse and a second round of infection occurs.

A.6.3. Radiolabeling and Analysis of Phage Proteins by Polyacrylamide Gel Electrophoresis

Analysis of phage structural proteins from the purified phage particles directly by the well-characterized techniques of one-dimensional sodium dodecyl sulfate-polyacrylamide gel electrophoresis (SDS-PAGE) (Laemmli, 1970; O'Farrell et al., 1973) is one traditional way to distinguish between different phages, get a sense of their complexity and help identify the structural-protein genes when the phage's genome is sequenced. The patterns of production of phage-encoded proteins may also be explored by one-dimensional as well as two-dimensional SDS-PAGE of samples pulse-labeled with radioactive amino acids after phage infection (Cowan et al., 1994; Dunn and Corbett, 1996), yielding detailed information concerning the amounts of various phage-encoded proteins, the course of an infection and the effects of various physiological conditions. Phage macromolecular synthesis begins soon after infection, and many phages efficiently turn off the synthesis of their hosts' proteins, so that phage products are formed almost exclusively.

A.6.3.1. Radiolabeling of Phage Proteins

Growth parameters such as appropriate host and titer for infection by the targeted phage, extent, and timing of host killing and the eclipse period need to be well characterized (see subsection A.6.2) before radiolabeling is attempted. To reduce labeling of host proteins, an MOI should be used that is as high as possible without danger of lysis from without (see Chapter 7)—usually an MOI of 8–12 is used. It is essential to monitor bacterial survival; host protein background may be too high due to poor phage adsorption or poor phage-induced inhibition of synthesis of host proteins. In the former case many bacteria survive as colony formers after phage infection, and in both cases labeled bacterial proteins may totally obscure most phage proteins. In such cases, pre-irradiation with UV light can be helpful. The bacteria must be in a medium that does not significantly absorb UV light, and they should be handled in the dark during and after irradiation (to avoid photoreactivation). Since kinetics of UV light inactivation of bacteria vary strongly dependent upon species and growth and irradiation conditions, host inactivation parameters must be determined under the conditions to be used. Also, phage production in cell populations irradiated to yield different levels of survival must be quantified beforehand, to determine the dose that prevents colony formation by most of the bacteria, but still permits virtually normal phage production. For instance, in our hands *E. coli* B cells irradiated so that about 15% of the cells survive as colony formers exhibit significantly reduced synthesis of bacterial proteins, but upon infection with phage T4 yield ca. 80% of the progeny phages produced by nonirradiated cells.

Since the protein production pattern shifts quite rapidly during the early stages of phage synthesis, it is crucial to time the start and end of each labeling period very precisely in order to obtain meaningful results and there should be no gaps between the first few infected-cell samples. A control host sample should be taken shortly before infection. Performing the radiolabeling at a temperature a few degrees below the host bacterium's optimal growth temperature will permit more leisurely sampling of the phage-infected culture but may not parallel the patterns at the higher temperature. Pilot experiments with different amounts of nonradioactive carrier amino acids may be needed to ensure that the radiolabel is incorporated throughout the pulse period. With *E. coli* B (a prototrophic bacterium) and coliphage T4 in a defined medium with little or no added amino acids (M9, Table A.4) at 37°C, the addition of an isotope preparation yielding $\leq 0.01\ \mu$M final concentration permits continuous incorporation throughout a 3-min pulse. The amount of radiolabel incorporated during each pulse is quantified upon transferring an aliquot (10–50 μl) to a filter paper disc that is subsequently treated with trichloroacetic acid (TCA, to precipitate proteins), washed with alcohol or acetone and dried before counting. An aliquot should also be transferred to a filter disc that is not treated with TCA to estimate the total amount of radioactivity added to each sample. TCA precipitation of the samples' proteins as they are removed from the culture prior to protein analysis (method I) reduces the possibility of problems caused by residual growth in cultures after sampling, or by cells lysing during centrifugation, but some investigators prefer to boil the samples in sample buffer (dissociating the ^{35}S -labeled amino acids from tRNA as well as denaturing the proteins), rather than precipitating them immediately (method II). In the example below, a limited number of samples are labeled in only one culture infected with one phage; often several infections are carried out at 1-min intervals to facilitate direct comparisons, and more samples are taken. To simplify the description, the sampling necessary to monitor bacterial survival is not included, but that *must* be done; please refer to section 6.2 for this procedure.

Materials

100-ml Erlenmeyer flask for infection (G).
Six 10-ml round-bottom tubes for radiolabeling (labeled 1–6)
500-ml beaker containing 100 ml 0.6% TCA, in separate ice bucket
Minimal medium (containing NH_4^+ as the nitrogen source; see however below for possible amino acid supplements)
^{35}S-labeled amino acid preparation ($\geq$ 1000 Ci/mmol)
Casamino acids solution (20% w/v in water)
0.6M and 0.3M TCA, on ice
Acetone or alcohol
8 paper filters discs that fit into scintillation vials; one per pulsed sample (labeled in pencil, not ink), including one for a blank (B) and one for input (i) spread out on a piece of aluminum foil to be ready for sampling.
Timer
Phage suspension containing $\geq 1.6 \times 10^{11}$ PFU/ml, kept on ice
Host bacteria growing in minimal medium
Indicator bacteria (see section A.4.1.2)

Some bacteria (e.g., B. subtilis and L. monocytogenes) will grow very slowly or not at all in minimal media. Should this be the case, the medium can be supplemented with a group of unlabeled, growth-enhancing amino acids (other than those ^{35}S-labeled) (see Table A.9); for E. coli B, adding 0.01% casamino acids speeds up growth to exponential phase substantially without interfering with protein labeling (Kutter, personal communication).

Decide on labeling times ("predetermined times" in the protocol below) and make a detailed flow chart. Determine what volume of the phage suspension contains ca. 1.6×10^{10} PFU (preferably 0.05–0.1 ml; dilute if necessary). Put the labeling tubes on ice, along with a tube containing about 10 ml of 0.6M TCA if you plan to TCA

TABLE A.9
Amino Acid Groups for Complementing Defined Media[1]

Amino Acid Group	Constituent Amino Acids[2]
1	Lysine, arginine, methionine, cystine
2	Leucine, isoleucine, and valine
3	Phenylalanine, tryptophan, tyrosine
4	Histidine, threonine, glutamic acid, proline, aspartic acid
5	Alanine, glycine, serine, hydroxyproline
6	(Optional Miscellaneous): nor-valine, nor-leucine, alpha-amino-butyric acid, cysteine.

[1] These supplements were originally described for *E. coli*, based on biochemical compatibility (what amino acid affects the biosynthesis of what other amino acid) (Lederberg 1950) and added as needed to 10 mg/L final concentration (except cystine (50 mg/L) and threonine (20 mg/L)). For other bacteria higher concentrations may give better growth promotion (in the range of 60 mg/mL for *Bacillus subtilis*; C. Stewart, pers. comm.), and cystine may be replaced by cysteine.

precipitate your samples before collecting them (Method I). Prepare a labeling mixture containing 70 μCi ^{35}S-labeled amino acids and cold amino acids as needed (to achieve continuous incorporation of the radioactive amino acid(s) throughout the pulse period) in 350 μl (dilute with growth medium), and dispense 50 μl aliquots of this mixture into each of the 6 labeling tubes, keeping them on ice. Transfer 2 μl of the labeling mixture onto the input filter ("i"). This filter should not be treated with TCA.

Procedure

1. Grow the bacteria under appropriate conditions, monitoring growth, preferably by microscopy, until the culture contains about 2×10^8 cells/ml.
2. **About 5 min before** the bacterial culture reaches that density, put the 100-ml Erlenmeyer flask ("G") and the labeling tube "1" into the shaking water bath.
3. Transfer a 1-ml aliquot of the bacterial culture to labeling tube C (control), and start the timer **(t [time] = 0 min)**. Pipet 10 ml of the bacterial culture into the infection flask G.
4. **At t = 3 min:** Add 50 μl casamino acids solution (final concentration 10 mg/ml, to chase all initiated peptide chains to completion) to labeling tube C and continue its incubation in the water bath. Transfer a 20 μl aliquot to a filter disc C. Place this disc in the TCA beaker, along with your blank disc.
5. **At t = 5 min:** Place labeling tube C in the ice bath; you may immediately add 1 ml of ice-cold, 0.6 M TCA to it (to stop growth and precipitate all proteins), swirl to mix and leave it on ice (method I).

6. Add 1.6×10^{10} PFU to the 10-ml bacterial culture in flask G (thus, yielding a MOI of ca. 8), **restart the timer (0 min)**, and continue incubating the culture in the shaking water bath.
7. **One min before the predetermined times:** Transfer the appropriately numbered labeling tube to the water bath.
8. **At the predetermined times:** Transfer a 1-ml aliquot of the infected culture to this labeling tube, and continue incubation with shaking.
9. **Three min later:** Add 50 μl casamino acids solution to the labeling tube in the water bath and continue its incubation there. To quantitate incorporation and be able to detect any later losses, transfer a 20 μl aliquot to the filter disc with the same number as the labeling tube. Place this disc in the TCA beaker on ice.
10. **Five min later** Remove the labeling tube from the water bath to the ice bath, where it will stay until centrifugation for Method II; for Method I, immediately add 1 ml of ice-cold, 0.6 M TCA to the labeling tube to precipate the labeled proteins and swirl to mix.

Processing of filters

Keep the TCA beaker on ice for 15 min after taking your last sample. Decant the liquid carefully into a beaker, and dispose the mildly radioactive acid appropriately. Add 100 ml cold 0.3 M TCA, and leave on ice for 15 min (occasionally swirling the beaker). Decant the TCA solution, and wash the filter disc again with cold 0.3 M TCA, leave on ice for 15 min. Finally wash the filter discs with cold acetone or alcohol, leave on ice for 30 min (occasionally swirl the beaker), and decant the acetone or alcohol. Allow the discs to dry on a piece of aluminum foil, crinkled to facilitate drying, and quantify the ^{35}S associated with them and with the non-acid-treated input filter.

Processing of samples for SDS-PAGE

For Method I: Keep the TCA-precipitated samples in the labeling tubes on ice until 30 min after the last sample was precipitated. Collect the precipitates by centrifugation ($3000 \times g$, 5 min), decanting the highly radiolabeled supernatants into a beaker and disposing appropriately. Wash the pellets with 1-ml of ice-cold acetone, vortex, and leave on ice for 30 min. Centrifuge as before. Dry the pellets in a vacuum chamber, or by blowing air or nitrogen over them. Dissolve the pellets in aliquots (200 μl) of the sample buffer used for SDS-PAGE (e.g., 41 mM Tris, 40 mM boric acid, 55 mM Na_2EDTA, 5% (w/v) β-mercaptoethanol, 5% (w/v) sucrose, 2% (w/v) SDS, 1% (w/v) bromphenol blue). Optional: boil the samples 5 min. Quick-freeze them in liquid nitrogen or in an acetone-dry ice bath. The solutions can be stored almost indefinitely at –70°C. For method II, centrifuge the samples at $5000 \times g$, 10 min as soon as possible after taking them. Resuspend in sample buffer as above, boil for 5 min and freeze and store as above. *Note:* Since heating at 100°C may aggregate large proteins, samples in which such proteins are likely to be present should be heated at a somewhat lower temperature (e.g., 95°C, 5 min).

A.6.3.2. SDS-PAGE of Phage Proteins

The proteins are resolved and characterized by SDS-PAGE, using any one of the Laemmli-type recipes and systems for discontinuous gel electrophoresis (Ausubel et al., 2001; Sambrook and Russell, 2001). The effective separation ranges of phage proteins in polyacrylamide gels depend on the concentration of polyacrylamide and the amount of cross-linking (see Table A.10). Most SDS-polyacrylamide gels are cast with a molar ratio of acrylamide:bisacrylamide of ca. 29:1, which enables one to resolve polypeptides whose M_r differ by ca. 3%. The stacking gel should be poured and allowed to polymerize the day the gels are used. *Caution*: Acrylamide and bisacrylamide are potent neurotoxins that can be absorbed through the skin. Therefore, gloves and a mask should be worn when weighing them, and pre-cast gels (available from many vendors, together with the electrophoresis equipment designed to accommodate them) should be used whenever possible. As with agarose electrophoresis of phage DNA fragments, appropriate molecular weight standards should be included in all polyacrylamide gels, to allow estimation of the M_r of the detected proteins, and to compare the protein band profiles of various phages analyzed on different gels.

Phage samples should be heated to about 95°C or boiled for 5 min before analyzing their proteins by SDS-PAGE. The gels are commonly run in a Tris-glycine electrophoresis buffer containing 25 mM Tris-HCl, 250 mM glycine and 0.1% SDS, pH 8.3. Proteins in polyacrylamide gels are most commonly stained with Coomassie Brilliant Blue and/or silver salts. In addition, fluorescent stains are available, such as the SYPRO, Ruthenium, and Deep Purple stains. The choice of staining procedure depends on the expected amount of protein in the phage samples, dye availability, and the personal preference of the investigator. In general, Coomassie Brilliant Blue staining is the least expensive approach, but it requires relatively long processing times and its staining intensity exhibits protein-to-protein variability. Silver staining is a more expensive and more labor-intensive procedure, and it has a similarly high—if not higher—degree of protein-to-protein variability. However, staining with silver salts (either ammoniacal silver solutions or silver nitrate solutions) is more sensitive

TABLE A.10
Separation of Proteins in SDS-Polyacrylamide Gels[1]

Polyacrylamide Concentration in Separating Gel (%, w/v)	Separation Range (kDa)
5%	55–210
7.5%	35–95
10%	20–80
15%	10–45

[1]A 1–2 cm stacking gel containing 2.5% polyacrylamide should be cast on top of the separating gel. Both stacking and separating gels are cast using a mixture with a ratio of acrylamide:bisacrylamide 29:1

than is staining with Coomassie Brilliant Blue, and it allows detection of 0.1–1 ng of poplypeptide/protein in a single band (staining with Coomassie Blue detects about 0.1 μg of protein in a single band). The fluorescent stains match silver staining in sensitivity, but require special equipment for visualization of the stained proteins. Detailed descriptions of the staining procedures are included in the manuals by Sambrook and Russell (2001) and Ausubel (2001), and similarly detailed instructions are usually included with the staining kits sold by various vendors. Polyacrylamide gels stained with Coomassie or silver stains may be photographed with the same equipment used for EtBr-stained DNA in agarose gels, with the exception that transmitted white light should be used instead of UV and no orange filter is required. For long-term storage, the stained gels may be dried on Whatman 3MM (or similar) filter paper, or air-dried between cellophane sheets. If the gels are stored in water for longer than 24 hours, adding 20% glycerol (final concentration) will decrease gel swelling and the associated band distortion. If radiolabeled phage proteins are electrophoresed, their location can be identified by autoradiography. Alternatively, an imaging screen can be used for digitized detection.

A.6.4. Electron Microscopy of Phage Particles

BOX 2: ELECTRON MICROSCOPY

Hans W. Ackermann

Dept. of Medical Biology, Laval University, Quebec, Canada

Electron microscopy is often the fastest and easiest way for phage identification, and the only way, other than relevant sequence analysis, to attribute an unknown phage to a family. In many cases, especially in phages of enterobacteria, bacilli, pseudomonads, rhizobia, and vibrios where phage taxonomy is relatively advanced, it also allows species and genus diagnosis. A great number of purification and staining techniques have been devised, but we describe here only the most basic and useful techniques. A more complete description of stains and staining techniques may be found in Ackerman and Dubow (1987).

Preparation

Lysates must be freed from excess proteins, sugars, and salts. High-titer lysates (about 10^8 particles per ml) are centrifuged and washed (preferably twice) in a buffer, for example in 0.1 M ammonium acetate (pH 7.0). Centrifugation is done in swinging-bucket or fixed-angle rotors. The latter allows for a considerable reduction of g forces and centrifugation times and can be done using relatively inexpensive high-speed centrifuges. We sediment tailed phages routinely in a fixed-angle rotor at $25{,}000 \times g$ for 60 min and increase this time to 90 min for small cubic and filamentous phages. Density gradient purification is not necessary; on the contrary, it will prove disastrous if phage concentrates have been insufficiently dialyzed. For simple identity controls, instant preparations may be made from lysed areas on agar.

Staining

Phages are stained by means of electron-dense salt solutions, generally sodium or potassium phosphotungstate (1%–2%, pH 7) or uranyl acetate (1%–2%, pH 4). Both stains are stable for at least a year. Ammonium molybdate, which is rarely used, produces the same type of staining as phosphotungstate. Fixation is unnecessary. A drop of phage suspension is deposited with a Pasteur pipette on an electron microcopic grid of 200–400 mesh provided with a carbon-coated Formvar film. A drop of stain is added and gently mixed with the phage. After 1 minute, the fluid is withdrawn with a filter paper and the grid is allowed to dry. Phages appear as "negative" translucent structures against a dark background. To make instant preparations from an agar dish, one deposits 1–2 drops of stain on a lysed area, agitates gently to make the phages float off, and touches the stain drop with an electron microscopic grid.

Effect of Stains

1. *Phosphotungstate* tends to cause disruption or rounding of phage heads, adheres rather poorly to grids, often gives poorly contrasted preparations, and coalesces into drops in humid weather. Preparations generally deteriorate after 1 month.
2. *Uranyl salts* have a strong affinity for dsDNA and tend to cause positive staining of phage heads. Positively stained capsids appear black and shrunken and should not be measured. Further artifacts are swelling of protein structures and, upon exposure to the electron beam, the appearance of a dark halo around phage heads that may be mistaken for an envelope. Uranyl salts give excellent contrast, tend to crystallize on the grid, and act as fixatives, so that preparations may be kept for up to 10 years. Phage heads are well preserved and generally angular.

Magnification Control

The magnification of electron microscopes is inconstant and varies with lens currents and other factors, sometimes within one and the same session. It is controlled by adjusting the photographic enlarger during the printing of electron micrographs. Latex spheres and diffraction gratings are unsuitable at high magnification. Instead, catalase crystals or T4 phage tails may be used at appropriate final magnifications (e.g., 300,000 times; Ackermann, 1987; Luftig, 1967).

Dos and Do nots

Measure enough particles (e.g. 20) per phage; measure on prints, not on the screen or negatives; measure at sufficiently high magnification (200–300,000 times). Provide complete dimensions. As a rule, they are complete if other people can build a model. Do not use uranyl acetate at a pH above 4 and do not make instant preparations from crude lysates.

For negative staining (Hall, 1955; Brenner and Horne, 1959) ammonium acetate is preferred, it is volatile and does not interfere with the stains. It is important to use highly purified water when preparing the reagents and grids for electron microscopy, and to work in a dust-free environment. The support film can be prepared from a number of plastic compounds (Speiss et al., 1987) followed by "glow-discharging" a thin layer of carbon under vacuum onto its surface. A detailed procedure for preparation of grids can be found in Carlson and Miller (1994); carbon-coated grids are also commercially available.

An alternative method for preparing phages for electron microscopy is to use centrifugation to attach them to a carbon-parlodion-coated electron microscope grid. For this, an insert is needed that fits snugly into an ultracentrifuge tube for a swing-out rotor. The adapter should have one rounded side fitting the rounded bottom of the tube, and a very flat surface that will be perpendicular to the long axis of the tube when the adapter is inserted into it. Attach the grids (carbon-parlodion-side up) to the flat surface of the adapter using double-sided tape. Then place the adapter + grids in a centrifuge tube which is filled with a sample of the phage-containing suspension. Centrifuge the phage particles onto the grid (the required centrifugal force and centrifugation time depends on the nature of the sample and the size of the phage particles; a good starting point is $180{,}000 \times g$ for 3 h). After centrifugation, aspirate the liquid from the tube, and recover the grids using forceps. Any remaining liquid is blotted off by gently touching the grid with tissue paper, whereupon the specimen is negatively stained as described above, and examined by transmission electron microscopy.

ACKNOWLEDGMENTS

Many of the protocols described in this chapter have circulated in the community of phage researchers for so long that proper attribution of credit for their development is impossible. Other protocols are modified versions, used in my laboratory, of common methods used by other investigators around the world. Thus, I would like to thank the entire phage community for generously sharing ideas, methods and recipes, which has made possible the compilation of this chapter. I am especially thankful to Alexander Sulakvelidze, Raul Raya, and Charles Stewart for their valuable input concerning phages of Gram-positive bacteria, and to Elizabeth Kutter, A. Sulakvelidze, Hans-Wolfgang Ackermann, Liana Gachechiladze, and Burton Guttman for their helpful contributions, discussions and support during preparation of this chapter.

REFERENCES

Anonymous, ISO 9308-3. Water quality. Detection and enumeration of *Escherichia coli* and coliform bacteria in surface and waste water—Part 3: Miniaturized method (Most Probable Number) by inoculation in liquid medium. International Organization for Standardization, Geneva, Switzerland, 2000.

Anonymous, ISO 10705-1: Water quality. Detection and enumeration of bacteriophages—Part 1: Enumeration of F-specific RNA bacteriophages. International Organization for Standardization, Geneva, Switzerland, 1995.

Anonymous,. ISO 10705-2: Water quality. Detection and enumeration of bacteriophages—Part 2: Enumeration of somatic coliphages. International Organization for Standardization, Geneva, Switzerland, 1998.

Anonymous, ISO 10705-4: Water quality. Detection and enumeration of bacteriophages—Part 4: Enumeration of bacteriophages infecting *Bacteroides fragilis*, International Organization for Standardization, Geneva, Switzerland, 2001.

Ackermann, H.W., Bacteriophage observations and evolution. *Res Microbiol* 154: 245–251, 2003.

Ackermann, H.W., "Classification of Bacteriophages," in *The Bacteriophages*, R. Calendar (Ed.). Oxford University Press, New York, 2005.

Ackermann, H.W., and Dubow, M., *Viruses of Prokaryotes*. CRC Press, Boca Raton, FL, 1987.

Adams, M.H., "Methods of study bacterial viruses," pp. 443–519 in *Bacteriophages*. Interscience Publishers, Inc., London, 1959.

Arber, W., Enquist, L., Hohn, B., Murray, N.E. and Murray, K., "Experimental methods for use with lambda," pp. 433–466 in *Lambda II*, R.W. Hendrix, J.W. Roberts, F.W. Stahl, and R.A. Weisberg (Eds.). Cold Spring Harbor Laboratory Press, Cold Spring Harbor, New York, 1983.

Attebery, H.R. and Finegold, S.M., Combined screw-cap and rubber-stopper closure for Hungate tubes (pre-reduced anaerobically sterilized roll tubes and liquid media). *Appl Microbiol* 18: 558–561, 1969.

Ausubel, F., Brent, R., Kingston, R.E., Moore, D.D., Seidman, J.G., Smith, J.A. and Struhl, K., *Current protocols in molecular biology*. John Wiley and Sons, New York, New York, 2001.

Breitbart, M., Hewson, I., Felts, B., Mahaffy, J.M., Nulton, J., Salamon, P. and Rohwer, F., Metagenomic analyses of an uncultured viral community from human feces. *J Bacteriol* 185: 6220–6223, 2003.

Breitbart, M., Salamon, P., Andresen, B., Mahaffy, J.M., Segall, A.M., Mead, D., Azam, F., et al., Genomic analysis of uncultured marine viral communities. *Proc Natl Acad Sci U S A* 99: 14250–14255, 2002.

Breitbart, M., Wegley, L., Leeds, S., Schoenfeld, T. and Rohwer, F., Phage community dynamics in hot springs. *Appl Environ Microbiol* 70: 1633–1640, 2004.

Brenner, S. and Horne, R.W., A negative staining method for high resolution electron microscopy of viruses. *Biochim Biophys Acta* 34: 103–110, 1959.

Brown, D.T., and Anderson, T.F., Effect of Host Cell Wall Material on the Adsorbability of Cofactor-requiring T4. *J Virol* 4: 94–108, 1969.

Carlson, K. and Miller, E.S., "Experimental protocols," pp. 421–483 in *Molecular Biology of Bacteriophage T4*, J.D. Karam (Ed.). American Society for Microbiology, Washington, D.C., 1994.

Carne, H.R. and Greaves, R.I.N., Preservation of corynebacteria by freeze-drying. *J Hyg Camb* 72: 467–470, 1974.

Clark, W.A. and Geary, D., Proceedings: Preservation of bacteriophages by freezing and freeze-drying. *Cryobiology* 10: 351–360, 1973.

Cowan, J., D'Acci, K., Guttman, B. and Kutter, E., "Gel analysis of T4 prereplicative proteins," pp. 520–527 in *Molecular Biology of Bacteriophage T4*, J.D. Karam (Ed.). American Society for Microbiology, Washington, D.C., 1994.

Davies, J.D. and Kelly, M.J., The preservation of bacteriophage H1 of *Corynebacterium ulcerans* U103 by freeze-drying. *J Hyg* (Lond) 67: 573–583, 1969.

Demain, A.L. and Solomon, N.A., *Manual of Industrial Microbiology and Biotechnology*. American Society for Microbiology, Washington, D.C., 1986.

Doermann, A.H., The intracellular growth of bacteriophages. I. Liberation of intracellular bacteriophage T4 by premature lysis with another phage or with cyanide. *J Gen Physiol* 35: 645–656, 1952.

Drake, J.W., A Constant Rate of Spontaneous Mutation in DNA-Based Microbes. *Proc Natl Acad Sci U S A* 88: 7160–7164, 1991.

Dunn, M.J. and Corbett, J.M., Two-dimensional polyacrylamide gel electrophoresis. *Methods Enzymol* 271: 177–203, 1996.

Edelman, D.C. and Barletta, J., Real-time PCR provides improved detection and titer determination of bacteriophage. *Biotechniques* 35: 368–375, 2003.

Ellis, E.L. and Delbrück, M., The growth of bacteriophage. *J Gen Physiol* 22: 365, 1939.

Engel, H.W., Smith, L. and Berwald, L.G., The preservation of mycobacteriophages by means of freeze drying. *Am Rev Respir Dis* 109: 561–566.

Evans, A.C., The Potency of Nascent Streptococcus Bacteriophage B. *J Bacteriol* 39: 597–604, 1940.

Fraenkel-Conrat, H., *The Viruses. Catalogue, Characterization and Classification.* Plenum Press, New York, 1985.

Gerhardt, P., Murray, R.G.E., Costilow, R.N., Nester, E.W., Wood, W.A., Krieg, N.R. and Phillips, G.B., *Manual of Methods for General Bacteriology.* American Society for Microbiology, Washington, D.C., 1981.

Gherna, R.L., "Preservation," pp. 208–217 in *Manual of Methods for General Bacteriology,* P. Gerhardt (Ed.). American Society for Microbiology, Washington, D.C., 1981.

Hall, C.E., Electron densitometry of stained virus particles. *J Biophys Biochem Cytol* 1: 1–12, 1955.

Harper, D., Eryomin, V., White, T. and Kutter, E., "Effects of T4 infection on membrane lipid synthesis," pp. 385–390 in *Molecular Biology of Bacteriophage T4*, J. Karam, J.W. Drake, K.N. Kreuzer, G. Mosig, D.H. Hall, F.A. Eiserling, L.W. Black et al. (Eds.) ASM Press, Washington, D.C., 1994.

Hennes, K.P. and Suttle, C.A., Direct counts of viruses in natural waters and laboratory cultures by epifluorescence microscopy. *Limnol Oceanogr* 40: 1050–1055, 1995.

Kasman, L.M., Kasman, A., Westwater, C., Dolan, J., Schmidt, M.G. and Norris, J.S., Overcoming the phage replication threshold: a mathematical model with implications for phage therapy. *J Virol* 76: 5557–5564, 2002.

Koch, A.L., "Growth measurements," pp. 188–191 in *Manual of Methods for General Bacteriology,* P. Gerhardt, R.G.E. Murray, R.N. Costilow, E.W. Nester, W.A. Wood, N.R. Krieg and G.B. Phillips (Eds.). American Society for Microbiology, Washington, D.C., 1981.

Kricker, M. and Carlson, K., "Isolation of T4 phage DNA," pp. 455–456 in *Molecular Biology of Bacteriophage T4*, J.D. Karam (Ed.). ASM Press, Washington, D.C., 1994.

Laemmli, U.K., Cleavage of structural proteins during the assembly of the head of bacteriophage T4. *Nature* 227: 680–685, 1970.

Luftig, R., An accurate measurement of the catalase crystal period and its use as an internal marker for electron microscopy. *J Ultrastruct Res* 20: 91–102, 1967.

M, G. m. and Antequera, F., Organization of DNA replication origins in the fission yeast genome. *Embo J* 18: 5683–5690, 1999.

Mendez, J., Jofre, J., Lucena, F., Contreras, N., Mooijman, K. and Araujo, R., Conservation of phage reference materials and water samples containing bacteriophages of enteric bacteria. *J Virol Methods* 106: 215–224, 2002.

Nuttall, S.D. and Dyall-Smith, M.L., HF1 and HF2: novel bacteriophages of halophilic archaea. *Virology* 197: 678–684, 1993.

O'Farrell, P.Z., Gold, L.M. and Huang, W.M., The identification of prereplicative bacteriophage T4 proteins. *J Biol Chem* 248: 5499–5501, 1973.

Price, C.A., Plant cell fractionation. *Methods Enzymol* 31: 501–519, 1974.

Rogosa, M., Mitchell, J.A. and Wiseman, R.F., A selective medium for the isolation and enumeration of oral and fecal lactobacilli. *J Bacteriol* 62: 132–133, 1951.

Sambrook, J. and Russell, D.W., *Molecular Cloning. A Laboratory Manual.* Cold Spring Harbor Laboratory Press, Cold Spring Harbor, New York, 2001.

Schleper, C., Kubo, K. and Zillig, W., The particle SSV1 from the extremely thermophilic archaeon *Sulfolobus* is a virus: demonstration of infectivity and of transfection with viral DNA. *Proc Natl Acad Sci U S A* 89: 7645–7649, 1992.

Schrader, H.S., Schrader, J.O., Walker, J.J., Wolf, T.A., Nickerson, K.W. and Kokjohn, T.A., Bacteriophage infection and multiplication occur in Pseudomonas aeruginosa starved for 5 years. Can J Microbiol 43: 1157–1163, 1997.

Sealey, N.D. and Primrose, S.B., The isolation of bacteriophages from the environment. *J Appl Bacteriol* 53: 1–17, 1982.

Serwer, P., Graef, P.R. and Garrison, P.N., Use of ethidium bromide fluorescence enhancement to detect duplex DNA and DNA bacteriophages during zone sedimentation in sucrose gradients: Molecular weight of DNA as a function of sedimentation rate. *Biochemistry* 17: 1166–1170, 1978.

Simon, L.D., The Infection of *Escherichia coli* by T2 and T4 Bacteriophages as Seen in the Electron Microscope. III. Membrane-Associated Intracellular Bacteriophages. *Virology* 38: 285–296, 1969.

Sobsey, M.D., Schwab, K.J. and Handzel, T.R., A simple membrane filter method to concentrate and enumerate male-specific RNA coliphages. *J Am Water Works Assoc* 82: 52–59, 1990.

Speiss, E., Zimmermann, H.P. and Lünsdorff, H., Chapter 6, pp. 147–166 in *Electrom Microscopy in Molecular Biology, a Practical Approach*, J. Sommerville and U. Scheer (Eds.). IRL Press, Oxford, 1987.

Stein, J.R., *Handbook of Phycological Methods, Cultural and Growth Measurements.* Cambridge University Press, Cambridge, United Kingdom, 1973.

Suttle, C.A., "Community structure: Viruses," pp. 272–277 in *Manual in Environmental Microbiology*, C.J. Hurst, G.R. Knudsen, M.J. McInerny, L.D. Stezenbach and M.V. Walter (Eds.). American Society for Microbiology, Washington, D.C., 1997.

Suttle, C.A., Enumeration and isolation of viruses, pp. 121–134 in *Current Methods in Aquatic Microbial Ecology*, P.F. Kemp, B.F. Sherr, E.F. Sherr and J.J. Cole (Eds.). Lewis Publ., Boca Raton, 1993.

Suttle, C.A., Chan, A.M. and Cottrell, M.T., Use of ultrafiltration to isolate viruses from seawater which are pathogens of marine phytoplankton. *Appl Environ Microbiol* 57: 721–726, 1991.

Tetart, F., Desplats, C., Kutateladze, M., Monod, C., Ackermann, H.W. and Krisch, H.M., Phylogeny of the major head and tail genes of the wide-ranging T4-type bacteriophages. *J Bacteriol* 183: 358–366, 2001.

Wollman, E.L., and Stent, G.S., Studies on Activation of T4 Bacteriophage by Cofactor. IV. Nascent Activity. *Biochemica Et Biophysica Acta* 9: 538–550, 1952.

Wommack, K.E. and Colwell, R.R., Virioplankton: Viruses in aquatic ecosystems. *Microbiol Mol Biol Rev* 64: 69–114, 2000.

Wommack, K.E., Hill, R.T., Muller, T.A. and Colwell, R.R., Effects of sunlight on bacteriophage viability and structure. *Appl Environ Microbiol* 62: 1336–1341, 1996.

Wrestler, J.C., Lipes, B.D., Birren, B.W. and Lai, E., Pulsed-field gel electrophoresis. *Methods Enzymol* 270: 255–272, 1996.

Yamamoto, K.R., Alberts, B.M., Benzinger, R., Lawhorne, L. and Treiber, G., Rapid bacteriophage sedimentation in the presence of polyethylene glycol and its application to large-scale virus purification. *Virology* 40: 734–744, 1970.

Zierdt, C.H., Preservation of staphylococcal bacteriophage by means of lyophilization. *Am J Clin Pathol* 31: 326–331, 1959.

Zierdt, C.H., Stabilities of lyophilized *Staphylococcus aureus* typing bacteriophages. *Appl Environ Microbiol* 54: 2590, 1988.

Index

A

A2 phage, 56
A118 phage, 101
Abedon, Steve, 3
Acholeplasma, 61
Acidianus, 63
Acinetobacter baumanii, 347
Acquaculture and phage therapy, 352–353
Acridine mutagenesis, 21
Actinobacillus pleuropneumoniae, 356–357
Actinomyces, 150
Adamia, Rezo, 55
Adams, Mark, 1, 33
Adenosine triphosphate (ATP), 330
Adhesin, 167–168, 174–176
Adsorption, 42–45, 72, 167–168, 451
Africa, phage therapy in, 397
African Swine Fever (ASFV), 85–86
Agar overlay methods for determining phage titers, 457–460
Ag43 protein and phage binding, 171
Agribusiness, reducing antibiotic usage in, 348–349
Alberts, Bruce, v–vii
American Medical Association (AMA), 11, 395
American Phage Group, 16, 17–18
American Society for Microbiology, 3
p-Aminophenol (PAP), 280
p-Aminophenylß-D-galactopyranoside (PAPG), 280
Amplification assays, 277–279
Amurins and lysins/lysis systems, 207, 208
Anaerobic growth of organotrophic bacteria, 447–448
Anaerobic respiration, 41
Anderson, Tom, 1
Animal-associated phages, 145–150
Animal infections, phages used to prevent/treat; *see also* Food production chain, using phages in
- *Acinetobacter baumanii,* 347
- agribusiness, reducing antibiotic usage in, 348–349
- *Clostridium difficile,* 346–347
- *Enterococcus faecium,* 343–345
- *Escherichia coli,* 338, 339–343
- first known use, 336–337
- *Pseudomonas aeruginosa,* 347–348
- *Pseudomonas plecoglossicida,* 352–353
- *Staphylococcus aureus,* 347
- studies, early, 337–338
- vaccines, phage-elicited bacterial lysates as, 354–355
- *Vibrio cholerae,* 344, 346

Antibiotics, 11, 322, 348–350, 396–397, 423–426
Antibodies, antiphage, 32–37
Antibodies, development of phage-neutralizing, 417–419
Antibody technique, phage, 273
Antigenic properties of phages, 32–37
Antiholins, 207, 210
Aptamers, 200–201
Archaea/archaephages, 31, 62–63, 70, 102
Arrowsmith (Lewis), v, 382
Asfarviridae, 85–86
Asheshov, Igor, 385
AS-1 phage, 58
Autoclaving, 444–445
Avian typhosis, 9

B

Bacillus
- detection of pathogenic bacteria, 270, 273, 275
- enzymes, phage lytic, 328
- replication and host-cell concentration, 133
- soil and plant-associated phages, 144–145

Bacillus amyloliquefaciens, 306
Bacillus anthracis, 279, 328–331
Bacillus cereus, 330
Bacillus gallinarum, 9
Bacillus megatherium, 46, 54
Bacillus subtilis
- classification of, 53
- cytoplasmic membrane, 38
- DNA transfer into the cell, 177
- evolution, phage, 105, 108, 111, 113–114, 121
- methodologies used when working with phages, 462
- morphogenesis, 201
- ø29 phage, 54
- ø105 phage, 55
- SPO1 phage, 55
- therapy, phage, 419

Bacteriophage Ecology Group, 3
"Bacteriophage Inquiry," 10, 384–387
Bacteriophage-insensitive mutants (BIMs), 292–293
Bacteriophages; *see also* Phage-phage interactions in bacterial virulence; Tailed phages; Tailless phages; *individual subject headings*
  focus on, reasons behind new, vi–vii
  ingestion of, 366
  numbers of, 92, 131–132, 365–369, 420
  overview, 1–3
  resistance, bacterial, 331–332, 419–420
Bacteriophages, The (Calendar), 2, 58, 166, 195
Bacteriophages (Adams), 1
Bacteroides, 147
Bacteroides fragilis, 366–368
Bayne-Jones, Stanhope, 395
Benzer, Seymour, 1, 21
Benzopyrene, 150
Binding reagents, phage/phage products as, 280–281
Biofilms, 137–139, 171
Bioinformatic analysis, 94–95
Biology, phage
  antigenic properties of phages, 32–37
  archaephages, 62–63
  bacterial structure/physiology relevant to infection process, 38–41
  chemical/physical agents, susceptibility of phages to, 33, 38
  host physiology and growth conditions, 49
  infection process, general overview of, 41–46
  lysogeny and its consequences, 46–48
  nature of bacteriophages, 30–32
  tailed phages, 49–58
  tailless phages, 58–62
Biowarfare bacteria and phage lytic enzymes, 328–330
Birnaviridae, 85
BK5-T phage, 100–102
Bloodstream, mechanisms that clear phages from the, 416–417
Bohr, Niels, 13
Bordet, Jules, 7–8, 384
Bovine hemorrhagic septicemia, 10
British Journal of Experimental Pathology, 396
British Medical Journal, 396
Brocothrix thermosphacta, 370
Bronfenbrenner, Jacques, 15
Brucella, 359
Bruynoghe, Richard, 383–384
Bubonic plague and early phage therapy applications, 10, 384
Bulloch, William, 6
Burnet, Frank M., 8
Burst size, 41, 132
Bxb1 phage, 105

C

C_1 phage, 54, 111
Campbell, Alan, 22
Campylobacter, 277, 359
Campylobacter jejuni, 362
Canada, phage therapy in, 388–393
Candida, 273
Capsids, 30, 70, 72–73, 85
Capsomers, 70
Caroline, Leona, 396
Caudovirales, 30, 69, 72, 85
Caulobacter, 59
Cell, 3
Cell wall binding domains (CBD), 281
Central Institute of Epidemiology and Microbiology, 400
Centrifugations and purifying/concentrating viruses, 12, 466–470
Chanishvilli, Teimuraz, 407
Chase, Martha, 18–19
Cheese production, 152–154, 289–290
Chelating agents and phage stability, 33
Chemical nature of phages, 12
Chemical/physical agents, susceptibility of phages to, 33, 38
Chesapeake Bay, 131
Chlorella, 85–86
Chloroform, 33, 38, 211, 470, 477
Cholera and early phage therapy applications, 10–11, 384–387; *see also* CTXø phage; *Vibrio cholerae*
Cholera vibrio, 5
Chp2 phage, 60
CI/II/III repressor proteins, 48
Citrobacter, 50
Classification, phage
  biological properties of prokaryote viruses, 77
  development of taxonomy, 68–70
  dimensions/physicochemical properties of prokaryote viruses, 76
  host range of prokaryote viruses, 78
  links between phages and other viruses, 85–86
  overview, 70, 71, 75
  problems, taxonomic, 81–84
  purpose of, 70–71
  tailed phages, 70–80
  tailless phages, 73–81
  viruses, phages and eukaryote, 80

Cloning systems, phage, *see* Vectors/delivery vehicles, phage as targeted
Clostridia putrifucus, 404
Clostridium botulinum, 227
Clostridium difficile, 346–347
Clostridium perfringens, 243, 349, 400, 404
Clostridium tetani, 227
Co-evolution of phages and their host bacteria, 32, 135–136
Cohen, Seymour, 19
Cold Spring Harbor Phage Meetings, 3, 15–16
Coliphagine, 397
Colonization experiments, phage lytic enzymes and in vivo, 327–329
Combinatorial biology with phages, 108
Concatemers, 72
Concentration of phages in environmental samples, 450–451, 466–470
Conditional lethal mutations, v
Conversion, phage, 224
Corticoviridae, 74–78, 142
Corynebacterium diphtheriae, 227
Council on Pharmacy and Chemistry, 11, 395
Counting colony-forming units (CFU), 443
CP-T1 phage, 250, 252
Credibility issues and decline of interest in phage therapy, 393–397
Cross-genus recombination, adhesin restructuring through, 174–176
Crystallographic studies, 192–193
CsCI gradients and purifying phages, 467–470
C-terminal binding domain and phage lytic enzymes, 323, 325, 331
CTXø phage
  CP-T1 phage mediated generalized transduction, 250, 252
  distribution of, 234, 236
  enterotoxins, 244–245
  horizontal transfer of genes between *Vibrio cholerae* isolates, 245–246
  imprecise prophage excision, 243–244
  independent acquisition, 234
  overview, 224, 231
  paradigm for lysogeny, 104
  RS1ø phage, morphogenesis genes supplied for, 248–249
  toxin production, high, 246–247
  VPIø genes, 253
Cyanomyoviridae, 58
Cyanophages, 57–58, 141, 142
Cystoviridae, 59, 74–78
Cystoviruses, 85
Cytokine production, 414
Cytoplasmic membrane, bacterial, 38

D

D2 phage, 203
D3 phage, 96, 102
D29 phage, 105
Dairy Council Meeting (2003), 3
Dairy phages/industry; *see also* Fermentations, controlling bacteriophages in industrial
  A2, 56
  contamination, sources of phage, 289–290
  detection, phage, 288–289
  ecology, phage, 130
  industrial phage ecology, 150–154
  J-1, 56
  LL-H, 56–57
  øadh, 57
  overview, 55–56, 96–97, 287
  Siphoviridae, Sfi11/21-like, 57, 100–101
  Streptococcus thermophilus and comparative genomics, 97–100
DDVI phage, 34–37
Deinococcus, 104
Delbrück, Max, 1, 2, 12–15, 17, 20, 46, 110, 394
Densities, phage reproduction and host, 133–134
Detection of pathogenic bacteria
  amplification assays, 277–279
  binding reagents, phage/phage products as, 280–281
  dairy industry, 288–289
  display, phage, 273–275
  dual phage, 275–277
  introduction, 267–268
  lysis, via phage-mediated, 279–280
  recombinant phage, 268–273
Detergents and phage stability, 33
d'Herelle, Felix
  commercial preparations, nonviable, 393
  debates over connection between phage and disease, 7–9
  discovery of phage, 1, 6–7
  ecology, phage, 130
  mode of action of phages, 413
  one-step growth experiment, 13
  taxonomy, development of phage, 68
  therapy, phage, 9–11, 336–337, 353, 383–384, 389
Diagnostics, use of phage lysins in, 330
Dilutions, enumerating phages using serial, 456–457
Display, phage, 273–275, 298
DNA, *see also* Classification, phage; Evolution, phage; Gene, nature and replication of the; Molecular mechanisms of phage infection; Vectors/delivery vehicles, phage as targeted
  animal-associated phages, 146, 147
  classification, phage, 81–84, 86

DNA (*continued*)
dairy phages, 56
ecology, phage, 131
fermentations, controlling bacteriophages in industrial, 294
gram-negative bacteria, 50–53
gram-positive bacteria, 53–57
lysogeny and its consequences, 47
marine phage ecology, 144
methodologies used when working with phages, 471–476
morphogenesis, 46, 203–204
penetration, 45
polyhedral phages, 73–78
sequence information, vi–vii
tailed phages, 53–57, 72
tailless phages, 59–61
transition from host to phage-directed metabolism, 45
virulence, bacterial, 229
Dreyer, George, 397
Drosophila, 13
Dual phage and detecting pathogenic bacteria, 275–277
Dubos, Rene, 11, 396
Duggal, Rajesh, 387
Dysentery and phage therapy, 383, 400, 404–405

E

E. R. Squibb and Sons, 389, 393
Eaton, Monroe, 395
Eclipse times characterizing phage-host interactions, 41–42, 481–482
Ecology, phage
animals, 145–150
biofilms, 137–139
host-phage relationships, 132–135
host physiology/nutritional status, 136–137
industrial, 150–155
introduction, 130–131
marine environments, 139–144
numbers in the natural world, 131–132
outlook, 155–156
overview, 7
soil and plants, 144–145
Education, need for a new type of science, vii
Efficacy of phage therapy, conflicting reports about, 395–396
Egypt, 397
Eib expression, 254
Electron microscopy, 9, 12, 68, 462, 488–490
Electron transport system (ETS), 41
Eliava, Giorgi, 406, 407
Eliava Institute of Bacteriophage, Microbiology, and Virology (EIBMV), 399, 404–405, 407–409
Eli Lilly, 389, 393
Ellis, Emory L., 13
Elution and adsorption, concentration of phages with, 451
Embryology, phage, 22–23
Encyclopedia of Virology (Granoff & Webster), 166
Endolysins, 206–211
Endopeptidases, 321–322
Endotoxins and phage-production problems, 394
Energetics, cellular, 40–41
Enrichment of phages in the source material, 451–453
Enteric phages, 146–149
Enterobacter agglomerans, 138
Enterobacter cloacae, 280
Enterococcus faecalis, 150
Enterococcus faecium, 343–345
Enterohemolysins (Hly), 229
Enterohemorrhagic E. *coli* (EHEC), 32
Enterotoxins, 244–245
Enumeration, phage
microscopy, 462
most probable number assays, 461–462
plaque assays, 455–461
Enzyme activities and metabolic changes in phage-infected cells, 19–20
Enzyme-linked immunosorbent assay (ELISA), 267, 288
Enzymes, particulate nature of phage vs. lytic, 7–9
Enzymes to control bacterial infections, phage lytic
antibacterial agents, 358–359
biowarfare bacteria, killing, 328–330
colonization experiments, in vivo, 327–329
conclusions/summary, 332
diagnostics, 330
evolution of, 117–120
immune response, 330–331
introduction, 321–322
mode of action, 324–325
resistance to the enzymes, bacterial, 331–332
sepsis and bacteremia, 327
specificity, 325, 326
structure, enzyme, 322–323
synergistic effects, 325, 327
Equipment/supplies used when working with phages, 439–440
Erlich, Paul, 396
Erwinia, 50
Erwinia amylovora, 350–351, 358
Erwinia herbicola, 351

Escherichia coli
animal-associated phages, 146, 147–149
animal infections, phages used to prevent/treat, 338, 339–343
classification problems, phage, 84
cytoplasmic membrane, 38
detection of pathogenic bacteria, 268, 277, 279, 281
DNA replication/repair/recombination, 185
ecology, phage, 135–136, 138
evolution, phage, 105, 106, 109
Ff phages as vectors of recombinant DNA cloning, 298
food production chain, using phages in, 359
lethal agent delivery systems, 309–311
lysins and lysis systems, 212, 213
methodologies used when working with phages, 463, 477
multiplication, phage, 13
Mu phage, 50
N4 phage, 50
pathogenic phenotype, turning into a, 32
P1 phage cloning system, 306
prevalence of bacteriophages, 366
recognition of host, 171
resistance to phages, bacterial, 419
therapy, phage, 397, 398, 406, 411, 413
T4 phage, 116
T7 phage, 43
virulence, bacterial, 229–231, 237, 246, 254
"Essai de traîtement de la peste bubonique par le bactériophage" (d'Herelle), 384
Eubacteria, 70
Euphotic zone and marine phage ecology, 140
Europe, phage therapy in, 389–392, 397–398
Evans, Earl, 18, 19
Evergreen International Phage Biology Meetings, 3
Evolution, phage
co-evolution of phages and their host bacteria, 135–136
enzymes, metabolic, 117–120
lytic phages
large virulent phages, 111–117
medium-sized virulent phages, 110–111
overview, 110
overview, 94–95
summary and outlook, 120–122
temperate phages
dairy phages, 96–101
hybrids, 109
Inoviridae, 104
lambda phage: dairy and lambdoid, 94–96, 101–103
large, 104–105
Mu phage, 104
overview, 95, 103
P2/P4 phage, 103
prophage genomics, 106–109
Exopolysaccharide (EPS), 138–139

F

FASTPlaqueTB assay, 277, 279
Feces and animal-associated phages, 146, 149
Félix d'Herelle and the Origins of Molecular Biology (Summers), 2
Fels-1 phage, 241
Fermentation and cellular energetics, 40–41
Fermentations, controlling bacteriophages in industrial
contamination sources of phage, 289–290
detection, phage, 288–289
ecology, phage, 150–155
faulty fermentations, examples of, 286
genetically engineering phage-resistant bacterial strains, 293–294
introduction, 285–287
mutants, development/selection of phage-insensitive bacterial, 291–293
overview, 287, 290–291
strain selection, 291
strategies, control, 290–291
Feulgen reaction, 12
Ff class of bacteriophages as vectors of recombinant DNA cloning, 298–300
FhuA receptor and *Siphoviridae,* 180–181
Filamentous phages, 74–79, 104, 298–300, 311–312
Fire blight and emergence of antibiotic-resistance bacterial strains, 350
Fish and phage therapy, 352–353
FITCH, 83
Flu, Paul-Christian, 8
Food industry and industrial phage ecology, 130, 150–155
Food production chain, using phages in
general considerations, 359–360
livestock contamination, 360–361
prevalence of bacteriophages in foods/other parts of the environment, 365–368
raw foods, contamination of, 361–362
ready-to-eat foods, 363–364
spoilage, food, 368, 370–371
Food web and marine phage ecology, 139–141
France, 397
Freezing as a phage storage method, 470–471
Fusselloviridae, 73, 75–79

G

G phage, 30, 54, 72
Gachechiladze, Ketevan, 34
Ganges River, 5
Gangrene and phage therapy, 400, 404
Gastrointestinal tract (GIT), reducing foodborne pathogens in the, 360–361
Gastrointestinal tract (GIT) into the environment, phages released from, 367–368
GC-content
  CTXø phage, 103
  lambda phage, 103
  mycobacteriophages, 105
  Myoviridae, 121
  SPO1 phage, 114
GenBank, 83
Gene, nature and replication of the; *see also* DNA
  embryology, phage, 22–23
  Hershey-Chase experiment, 18–19
  host-induced modification, 22
  intracellular steps in phage growth, 18
  Luria-Latarjet experiment, 16–17
  molecular genetics, Max Delbrück and development of, 12–15
  multiplicity reactivation, 17
  mutations, random occurrence of, 20–22
  parental-to-progeny label transfer, 18
  plasmids, 22
  regulation, gene, 16
  RNA and phage development, 22–23
  triplet nature of the genetic code, 21
General Aspects of Lysogeny (Campbell), 195
Gene Regulatory Circuitry of Phage λ (Little), 195
Genetically modified organisms (GMOs), 152, 270, 272, 277
Genetic reservoir, bacteriophages representing a huge untapped, vi
Genome sequencing, vi–vii, 81, 92–95, 142, 422–423; *see also* Evolution, phage
Georgia (independent country), phage therapy in, 406–411
German Physicians' Conference (2004), 3
Germany, 11
Gesund durch Viren: ein Ausweg aus der Antibiotika-krise (Häusler), 2
Gfp protein and detecting pathogenic bacteria, 273
Ghost vaccines, phages and bacterial, 355–358
Gifsy-1/2 phages, 241, 253–254
Giorgadze, Iraki, 407
Glassy transformation, 6, 8
Gorky Research Institute of Epidemiology and Microbiology, 404–405
Gram-negative bacteria; *see also individual subject headings*
  lambda phage, 49–50
  Mu phage, 50
  N4 phage, 50
  N15 phage, 50
  øKZ phage, 52
  P1 phage, 51
  P2/P4 phage, 51
  P22 phage, 51–52
  recognition, phage, 168–172
  T1 phage, 52
  T4 phage, 52
  T5 phage, 53
  T7 phage, 43
  virulence, bacterial, 227
Gram-positive bacteria; *see also individual subject headings*
  lactic acid bacteria, 55–57
  mycobacteria, 57
  overview, 53
  recognition of host, 172–173
  streptococci/staphylococci/*Listeria,* 54–55
  virulence, bacterial, 227
Gratia, André, 8, 334
Growth-promoting antibiotics (GPAs), 348–349
Guttaviridae, 75–79, 81

H

Haemophilus, 103, 104
Halobacterium, 185
Hankin, Ernest, 406
H-19B phage, 244
Head morphogenesis, 202–203
Helicobacter pylori, 355–356
Helper bacteriophages, 299–300; *see also* Phage-phage interactions in bacterial virulence
Hendrix, Roger, 202
Herpesviruses, 85
Hershey, Alfred, 1, 15, 18–19
HF1/2 archaephage, 62
High-affinity receptor proteins, 168
Hirszfield Institute of Immunology and Experimental Therapy (HIET), 399, 412
HK022 phage, 95, 96
HK097 phage, 95, 96, 203
Holins, 46, 206–211, 322
Holweck, Fernand, 15
Horizontal gene transfer, 93–94, 142–144, 224, 245–246
Host-induced modification, 22
Host-lethal proteins, 32

Host-phage relationships, dynamic, 132–135; *see also* Recognition of host and adsorption of tailed phages
Host physiology/nutritional status, 49, 136–137
Host specificity, 142–144, 173–176, 325, 326
Howe, Martha, 175
HP1 phage, 103
Human trials using phage therapy, first, 10–11; *see also* Therapy, phage
Hutinel, Victor-Henri, 383
Hybrid phages, 109
Hydrodynamic shear and gene separation, 19
5-Hydroxymethylcytosine, 20, 72

I

ImBio, 411
Immune avoidance, host, 236–238
Immune response and lytic phage enzymes, 330–331
Immunization experiments, d'Herelle's, 337
Immunoelectronmicroscopy, 33, 34
Immunopreparat, 414
India, early phage therapy in, 10–11, 384–387
Indicator bacteria and phage enumeration, 458
Industrial phage ecology, 150–155; *see also* Fermentations, controlling bacteriophages in industrial
Infection process, phage; *see also* Enzymes to control bacterial infections, phage lytic; Molecular mechanisms of phage infection
  adsorption, 42–45
  burst size, 41
  cytoplasmic membrane, 38
  energetics, cellular, 40–41
  latent/eclipse period, 41–42
  lysis, cell, 46
  methodologies used when working with phages, 443, 476–483
  morphogenesis, 46
  outer membrane/periplasmic space of gram-negative bacteria, 40
  overview, 44
  pathogenesis, 245–247
  penetration, 45
  peptidoglycan, 38–39
  single-step growth curve, 41–43
  transition from host to phage-directed metabolism, 45
Inoviridae
  classification of, 73, 74–78
  evolution, phage, 104
  nature of bacteriophages, 31
  small DNA phages, 60
  virulence, bacterial, 227, 236
Institute for Animal Disease Research, 339
International Committee for Nomenclature of Viruses (ICNV), 69
International Committee on Taxonomy of Viruses (ICTV), 69, 82, 84, 85
International Food Technology Association Meeting (2003), 3
International Virology Conference in Paris (2002), 3
Interscience Publishers, Inc., 1
Intestiphagine, 397
Intron ribozymes, 198–199
Invasion of host cells, proteins essential for, 238–241
IPTG (isopropyl-D-thiogalactopyranoside), 309–310

J

J-1 phage, 56
Jacob, Christian, 16
Jacob, Francois, 1, 16
Jermyn, W. S., 256
Join-cut-copy recombination-replication, 187
Jumna River, 5

K

K antigens, 168, 171, 176
K phage, 113
Kabeshima, Tamezo, 354
Kellenberger, Eduard, 1
Killers Within, The (Shnayerson), 2
Killer titer and phage enumeration, 460–461
Klebsiella, 413, 414
Klebsiella pneumoniae, 357
Kozloff, Lloyd, 18
Kutter, Elizabeth, ix
KVP40 phage, 115–117

L

L5 phage, 105
Lactic acid bacteria (LAB), 287; *see also* Dairy phages/industry
Lactobacilli, vaginal, 150
Lactobacillus, 96, 117, 145
Lactobacillus delbrueckii, 56
Lactobacillus plantarum, 112, 155
Lactobacillus salivarius, 154
Lactococcus, 96
Lactococcus casei, 56

Lactococcus garvieae, 352
Lactococcus gasseri, 57
Lactococcus lactis
enzymes, phage lytic, 358
evolution, phage, 97, 101, 102, 121
industrial phage ecology, 151, 154
LL-H phage, 57
MM1 phage, 54
overview, 56
recognition of host, 173
Lactococcus paracasei, 56
Lahiri, M. N., 385
Lambda phage (λ)
classification problems, phage, 84
detection of pathogenic bacteria, 272
DNA replication/repair/recombination, 186
enzymes, phage lytic, 358
evolution, phage, 94–96, 101–103
lysins and lysis systems, 211
lysogeny and its consequences, 47
morphogenesis, 204
nature of bacteriophages, 32
overview, 49–50
P22 phage and, controversy between, 82–83
prototypic lysogenic phage, 16
recognition of host, 168, 171, 176
transcription, 195
vectors/delivery vehicles, phage as targeted, 301–305
virulence, bacterial, 231
Lancet Journal, 396
La Presse Médicale, 10, 384
Latarjet, Raymond, 16–17
Latency times characterizing phage-host interactions, 41, 481–482
Le bacteriophage: Son rôle dans l' immunité (d'Herelle), 9
Lederberg, Esther, 16, 20, 47
Lederberg, Joshua, 20, 22
Lethal agent delivery systems (LADS), 309–311
Leuconostoc, 145, 155
Leviviridae, 31, 58, 74–78, 207
Leviviruses, 85
Lewis, Isaac M., 20
LHT system, 68
Ligand identification via expression (LIVE) technology, 312–313
Limulus amoebocyte lysate (LAL) assay, 357–358
Lindegren, Carl, 13
Linezolid, 425
Lipid-containing phages, 60–61
Lipopolysaccharide (LPS), 40, 167, 168, 281, 357
Lipothrixviridae, 73, 75–78
Listeria
detection of pathogenic bacteria, 277
enzymes, phage lytic, 328, 358
evolution, phage, 101
food production chain, using phages in, 359
Listeria lux, 271
Listeria monocytogenes
detection of pathogenic bacteria, 268, 270, 279
ecology, phage, 138
enzymes, phage lytic, 358
food production chain, using phages in, 364–365
MM1 phage, 54
Literature on phage, 1–2, 15–16
Livestock contamination, phages used to reduce, 360–361
LL-H phage, 56–57
LP65 phage, 112–114, 117
LPP-1 phage, 58
Luciferase reporter phage (LRP), 268–272
Luria, Salvador, 15, 16–17, 20
Lwoff, André, 16
Lysates and methodologies used when working with phages, 463–466; *see also* Vaccines, phage-elicited bacterial lysates as
Lysins and lysis systems; *see also* Enzymes to control bacterial infections, phage lytic; Lytic cycle; Virulent phages
holins/endolysins and lysis timing, 208–211
infection process, overview of, 46
overview, 206–207
single-protein systems of small phages, 207–208
T-even phages, 211–213
Lysogenic cycle; *see also* CTXø phage; Lambda phage; Prophage; Temperate phages
classification, phage, 72
consequences of lysogeny, 46–48
debate over connection between phage and disease, 8
ecology, phage, 133, 134–135
nature of bacteriophages, 32
repressor protein, 32
virulence, bacterial, 224
Lysogeny, 16
Lysozymes, 321
Lytic cycle; *see also* lytic phages *under* Evolution, phage
detection of pathogenic bacteria, 279–280
ecology, phage, 132–133
infection process, overview of, 46
marine phage ecology, 139
nature of bacteriophages, 32

M

M2 archaephage, 62
M13 phage, 104
M13K07 phage, 300
Maaløe, Ole, 18
Maisin, Joseph, 383–384
Makashvili, Elena, 406–407
Malone, R., 385
Maltose, 168
Mammalian cells, phage vectors for targeted gene delivery to, 311–313
Marine phage ecology
  cyanophages, 141
  food web, 139–141
  genomic analysis, 142
  host specificity and horizontal gene transfer, 142–144
  number of phages in natural world, 131
MD4 genes, 173
Meat fermentation and industrial phage ecology, 155
Media (culture) used in phage studies
  for anaerobic growth, 447–448
  general considerations, 443–445, 462
  recipes for, 443–449
  solid media, 445–447
Medicine/commercial possibilities for phage therapy, discovery of early, 9–11; *see also* Therapy, phage
Metabolic description of phage reproduction, 19–20
Methodologies used when working with phages
  analysis of the phage infection process, 476–483
  concentration/purification of isolated phages, 466–470
  DNA analysis, 471–476
  electron microscopy, 488–490
  enumeration, phage, 455–462
  equipment, 439–440
  infection with phage, 443
  introduction, 439
  isolation of phages from natural sources, 449–455
  lysates, primary/secondary, 463–466
  media and common reagents, 443–449, 462
  preparing bacterial cultures for phage infection, 441–442
  prophages, induction of, 455
  safety precautions, 440
  sending phage, 471
  sodium dodecyl sulfate-polyacrylamide gel electrophoresis, 482–488
  stocks, phage and bacterial, 440–441
  storage of phage, 470–471
  titers, determination of bacterial, 442–443
Microarray analysis, 108
Microbial Genetics Meetings, 3
Microscopy, *see* Electron microscopy
Microviridae, 31, 58–60, 73–78
Military support for phage therapy, 11, 404
Milk production, 153–154, 289–290
MM1 phage, 54
Model organism approach, v
Modular classification, 83–84
Modular theory of phage evolution, 93
Molecular Biology of Bacteriophage T4, The (Carlson & Miller), 3
Molecular Biology of Bacteriophage T4 (Karam et al.), 166
Molecular genetics, Max Delbrück and development of, 12–15
Molecular mechanisms of phage infection
  DNA replication/repair/recombination
    lambda phage, 186
    Mu phage, 189
    N15 phage, 190
    ø29 phage, 189–190
    repair/recombination and mutagenesis, 185–186
    replication process, 182–185
    T4 phage, 186–188
  DNA transfer into the cell
    FhuA receptor and *Siphoviridae,* 180–181
    overview, 177
    PRD1 and PM2 phages, 181–182
    T4 phage, 179–180
    T7 phage, 177–179
  introduction, 166
  lysins and lysis systems, 206–213
  morphogenesis and DNA packaging, 201–206
  recognition of host and adsorption of tailed phages, 167–176
  RNA, phage-related: structures/unexpected functions
    aptamers and SELEX, 200–201
    intron ribozymes, 198–199
    ribosome binding sites, 197–198
    translational bypassing, 199–200
  transcription, 190–196
Monod, Jacques, 16
Mononegavirales, 85
Morison, J., 384, 385
Morphogenesis
  DNA packaging, 203–204
  head formation, 202–203
  overview, 46, 201
  tail, 204–206
Most probable number (MPN) assays, 461–462

MS2 phage, 208
Mtkvari River, 406
Muller, Herman J., 13, 14
Multiplication, d'Herelle's concept of phage, 13
Multiplicity of infection (MOI), 443, 460, 477
Multiplicity reactivation (MR), 17
Mu phage
  DNA replication/repair/recombination, 189
  evolution, phage, 104
  nature of bacteriophages, 32
  overview, 50
  recognition of host, 176
Mutants, phage-resistance bacterial, 331–332, 419–420
Mutants and controlling bacteriophages in industrial fermentations, 291–293
Mutations, random occurrence of gene, 20–22
Mutations and model organism approach, v
MX-1 phage, 137
Mycobacteria, 271, 328
Mycobacteriophages, 57, 105
Mycobacterium paratuberculosis, 275
Mycobacterium smegmatis, 279
Mycobacterium tuberculosis, 271, 277, 279
Mycoplasma, 61
Mycoplasma genitalium, vi
Mycoplasmas, 61–62
Myoviridae
  classification, phage, 72–73, 75, 79
  evolution, phage, 103, 109, 114, 121
  industrial phage ecology, 155
  marine phage ecology, 141
  morphogenesis, 204
  nature of bacteriophages, 30–31
  therapy, phage, 419
  virulence, bacterial, 227, 236
Myxococcus xanthus, 137

N

N4 phage, 43, 45, 50, 171, 194
N15 phage, 50, 95, 96, 190
N-acetylglucosamine (NAG), 38, 168
N-acetyl muramic acid (NAM), 38, 210
N-acetylmuramidases, 321
N-acetylmuramyl-L-alanine amidases, 322
Nanozooplankton production, 139
National Academy of Sciences, vii
National Research Council's Committee on Medical Research (NRC/CMR), 396
National Science Education Standards, vii
Nature, 131
Neisseria, 104
Neisseria meningitidis, 237
Neuraminidase, 214, 243
Newcome, Howard B., 20
NfrA receptor and N4 phage, 171
Nidovirales, 85
N-terminal catalytic domain and phage lytic enzymes, 323, 325
Nucleic acid amplification, 267
Nucleocapsid, 30
Nucleoside triphosphates (NTPs), 40
Nutritional status of host, 136–137

O

ø1 phage, 34
ø6 phage, 59
ø11 phage, 54
ø29 phage, 54–55, 111, 189–190
ø80 phage, 168, 174
ø105 phage, 55
øadh phage, 57
øA1122 phage, 110
øH archaephage, 62
øKZ phage, 52, 111–112
øXI74 phage, 60, 85, 207
øYe03-12 phage, 110
O antigens, 168, 171, 236–238
Okazaki fragments, 183, 184
Omp receptors and phage recognition, 168, 171–172, 174
One-step growth experiment, 13
"On the antibacterial action of cultures of *Penicillium* with special reference to their use in silation of *Haemophilus influenzae"* (Fleming), 396
Oral cavity and animal-associated phages, 149–150
ORFs (open reading frames)
  C_1 phage, 54
  evolution, phage, 94, 122
  G phage, 54
  links between phages and other viruses, 86
  LP65 phage, 113
  mycobacteriophages, 57, 105
  P22 phage, 52
Origin-dependent replication, 187
Otero, Morton, 396
Otero, Perez, 396
Outer membrane of gram-negative bacteria, 40

P

P1 phage
  evolution, phage, 104–105
  nature of bacteriophages, 32

overview, 51
recognition of host, 171, 176
vectors/delivery vehicles, phage as targeted, 305–309
P2/P4 phage
evolution, phage, 103
lambda cloning vectors, 303–304
overview, 51
recognition of host, 175
virulence, bacterial, 239
P7 phage, 51, 104–105
P22 phage
lambda phage and, controversy between, 82–83
morphogenesis, 201, 202
overview, 51–52
recognition of host, 171
P60 phage, 110, 111, 142
Pakistan, 11, 387
Pantoea agglomerans, 351
PaP3 phage, 110–111
Parental-to-progeny label transfer, 18
Pasricha, C. L., 385
Pasteurella haemolytica, 356
Pasteurella multocida, 10, 356
Pasteur Institute, 354, 397
Pathogenicity islands (PAIs), 238, 254–257
Patna Medical College Hospital, 385
PblA/B phage, 236, 244
PBSX phage, 108
Penetration and infection process, 45
Penicillin, 396–397
Peptidoglycan, 38–39, 168, 210
Periplasmic space of gram-negative bacteria, 40
Per (phage-encoded resistance) system, 293–294
"Phage and the Origins of Molecular Biology" (Cairns et al.), 2
Phage and Virus Assembly Meetings, 3
Phage Ecology (Goyal), 131
"Phage Information Service," 15–16
Phagemids, 299, 308–309
Phage-phage interactions in bacterial virulence (helper phages)
gene expression, 253
mobilization, phage, 248–252
overview, 247–248
receptor, host, 252
PhagoBioDerm, 408–411
pH and phage stability, 33
Pharmacokinetic studies on phage therapy, 415–416
Photobacterium, 115
Phycodnaviridae, 85–86
Pili, 58–59
Plant diseases and phage therapy, 350–352
Plants and soil-associated phages, 144–145
Plaque assays
agar overlay methods, 457–460
dilutions, serial, 456–457
killer titer, 460–461
overview, 455–456
Plaque morphology mutants, 21
Plasmaviridae, 73, 75–79
Plasmids, 22
Pleomorphic phages, 75–79, 81
PM2 phage, 182
Podoviridae
classification, phage, 72–73, 75, 79
ecology, phage, 132
evolution, phage, 110
nature of bacteriophages, 30–31
P22 phage, 51
therapy, phage, 419
virulence, bacterial, 227, 236
Poisson distribution of phage in liquids, 461
Poland, phage therapy in, 412–413
Polyacrylamide gel electrophoresis (PAGE), 473–475, 482–488
Polyethylene glycol, concentration of phages with, 451, 467
Polyhedral DNA phages, 73–78
Polyhedral RNA phages, 74–78
Polylysogeny, 108
Polymerase chain reaction (PCR), 200, 267, 422, 476
Polysaccharide depolymerase, 138
Polysaccharide region, 168
Polythetic species concept, 69
Porins, 40
Portal protein initiator complex, 202
PRD1 phage, 60–61, 181–182
Prey (Crichton), 382
Prokaryote viruses, overview of, 75; *see also* Classification, phage
Prophage (quiescent state), 106–109, 230, 243–244, 455; *see also* CTXø phage; Lambda phage; Lysogenic cycle; Temperate phages
Propionibacterium freudenreichii, 366
PROTDIST, 83
Protein machine mechanisms, v
Proteins and bacterial virulence
attachment/colonization, 234–236
immune avoidance, host, 236–238
invasion of host cells, 238–241
putative virulence proteins, 241–243
survival of intracellular bacteria, 241
Proteins radiolabeled/analyzed by polyacrylamide gel electrophoresis, 482–488
Protein synthesis, inhibitors of, 20
Proteus, 406, 411, 412

Proton motive force (PMF), 41
Provisional Committee on Nomenclature of Viruses (PCNV), 68
Pseudoalteromonas, 142, 182
Pseudomonas
detection of pathogenic bacteria, 277
ecology, phage, 133, 134, 136, 145
evolution, phage, 96, 102, 109
food production chain, using phages in, 370
marine phage ecology, 144
morphogenesis, 203
Mu phage, 50
therapy, phage, 397, 413
Pseudomonas aeruginosa
animal infections, phages used to prevent/treat, 347–348
ecology, phage, 138
evolution, phage, 108, 110–112
therapy, phage, 397, 411
Pseudomonas plecoglossicida, 352–353
Pseudomonas putida, 95, 109
Pseudomonas pyocyanea, 16
Pseudomonas syringae, 59, 351
Pulsed field gel electrophoresis of large phage genomes, 475
Puri Cholera Hospital, 385
Putnam, Frank, 18
Pyophagine, 397

Q

Quantum mechanical model of the gene, 13

R

Radiobiology, 15, 16–17
Radiolabeling/analysis and polyacrylamide gel electrophoresis, 482–488
Rakieten, Morris, 396
Ralstonia solanacearum, 351
RB49 phage, 115
RB69 phage, 115
Recognition of host and adsorption of tailed phages
gram-negative bacteria, 168–172
gram-positive bacteria, 172–173
overview, 167–168
specificity, ranges of host, 173–176
summary, 176
Recombinant genomes and lambda cloning vectors, 303
Recombinant phage and detecting pathogenic bacteria, 268–273
Recombinase-based in vivo expression technology (RIVET), 253
Recombineering, 185
Regulation of λ Gene Expression by Transcription Termination and Antitermination (Friedman & Court), 195
Renaux, E., 8
Reoviridae, 85
Repair/replication, DNA, *see* Molecular mechanisms of phage infection
Replicative forms (RFs), 59, 60
Replicative transposition, 189
Reporter genes and detecting pathogenic bacteria, 268–273
Repressor protein, 32, 47, 48
Reproduction (phage) and phage ecology, 133–135
Research, bacteriophage: early history
chemical and physical nature of phages, 12
community, role of the phage, 15–16
debate regarding the nature of bacteriophages, 7–9
Delbrück (Max) and development of molecular genetics, 12–15
discovery of bacteriophages, 6–7
gene and of mutations, nature of the, 16–23
introduction, 5–6
practical applications, first: veterinary/medicine, 9–11
Resistance to phages, bacterial, 331–332, 419–420
Respiration and cellular energetics, 40–41
Restriction endonucleases, digestion of phage DNA with, 472–473
Rhodobacter capsulatus, 108
Ribosome binding sites, 197–198
"Riddle of Life, The" (Delbrück), 14
rII genes and lysins/lysis systems, 212, 213
Rita, Geo, 15
RNA; *see also* Gene, nature and replication of the; Molecular mechanisms of phage infection
animal-associated phages, 147, 149
classification, phage, 82–85
development, phage, 22–23
evolution, phage, 118
fermentations, controlling bacteriophages in industrial, 294
gram-negative bacteria, 53
gram-positive bacteria, 54–55, 57
polyhedral phages, 74–78
tailless phages, 58–59
transition from host to phage-directed metabolism, 45
Roseobacter, 142
Roux, Emile, 354

RS1ø phage, 231, 248–249, 253
Rudiviridae, 74–79, 85–86
Rumens of sheep/cattle and animal-associated phages, 146
Russia, phage therapy in, 411–412

S

Safety precautions, 420–423, 440
Salami production and industrial phage ecology, 155
Salmonella
  animal infections, phages used to prevent/treat, 337, 338–339
  detection of pathogenic bacteria, 268, 270–272, 277, 280
  enteric phages, 147
  food production chain, using phages in, 359, 361, 362, 364
  Mu phage, 50
  P1 phage, 105
  P22 phage, 52
  prevalence of bacteriophages, 366–368
  recognition of host, 171
  resistance to phages, bacterial, 419
  therapy, phage, 388–392, 398, 413
  vaccines, phage-elicited bacterial lysates as, 354, 356
  virulence, bacterial, 238, 239, 241, 254
Salmonella cholerasuis, 7
Salmonella enterica, 236, 238–241, 272
Salmonella gallinarum, 337
Salmonella typhimurium, 51, 303, 357
Salvarsan, 396
Saphal, 397
Sauerkraut/sausage production and industrial phage ecology, 155
Sb-1 phage, 55
Schade, Arthur, 396
Science education, need for a new type of, vii
SELEX (Systematic Evolution of Ligands by Exponential Enrichment), 200–201
Selfish gene concept, 107
Sequencing, genome, vi–vii, 81, 92–95, 422–423; *see also* Evolution, phage
Seriola quinqueradiata, 352
Serratia, 138, 145
SF153b phage, 138
Sfi11 phage, 57, 101, 102
Sfi21 phage, 57, 100–102
Shearing process and gene separation, 19
Shewanella, 104
Shiga-toxins (Stx1/Stx2), 229, 246
Shigella, 359, 412
Shigella dysenteriae, 229, 246, 338
Shillong Pasteur-Institute, 387
Single-step growth curve, 13, 41–43
Siphoviridae
  classification, phage, 72–73, 75, 79
  ecology, phage, 132
  evolution, phage, 100–103, 105, 121
  FhuA receptor, 180–181
  lambda-P22 controversy, 82
  marine phage ecology, 141
  nature of bacteriophages, 30–31
  therapy, phage, 419
  virulence, bacterial, 227–229, 236, 239
Siphoviruses, 57
Smith, Herbert W., 339
Smoking and vaginal lactobacilli
SM1 phage, 236
Sodium dodecyl sulfate-polyacrylamide gel electrophoresis (SDS-PAGE), 482–488
Soil and plant-associated phages, 144–145
SopEø phage, 239
Soviet Union, phage therapy in the former, 11, 399–406
Specificity, host, 142–144, 173–176, 325, 326
Specific spoilage organisms (SSO), 370–371
SPI-1/2 effector proteins, 238
Spiroplasma, 61
Spoilage, phages used to reduce food, 368, 370–371
SPO1 phage
  evolution, phage, 112, 113–114
  industrial phage ecology, 155
  overview, 55
  penetration, 45
  transcription, 195
Sporulating bacteria, 137
Spot tests and phage enumeration, 458–459
SPß phage, 55, 105
SpV4 phage, 60
Staphagine, 397
Staphylococcal phage lysate (SPL), 354–355, 398–399, 407–408, 414
Staphylococcus, 44, 133, 397, 413
Staphylococcus aureus
  animal infections, phages used to prevent/treat, 347
  ecology, phage, 138
  enzymes, phage lytic, 358
  evolution, phage, 109, 113
  ø11 phage, 54
  prophage, 108
  Sb-1 phage, 55
  staphylococcal phage lysate, 399
  therapy, phage, 408, 411
  virulence, bacterial, 228–229, 249–250, 256

Staphylococcus carnosus, 155
Stent, Gunther, 1
Storage of phage, 470–471
Strain collection and controlling bacteriophages in industrial fermentations, 291
Strand transfer complex (STC), 189
Streisinger, George, 1
Streptococcus, 96
Streptococcus agalactiae, 95
Streptococcus mitis, 236, 244
Streptococcus pneumoniae
  animal-associated phages, 149–150
  enzymes, phage lytic, 323, 327
  MM1 phage, 54
  virulence, bacterial, 237
Streptococcus pyogenes
  animal-associated phages, 150
  enzymes, phage lytic, 325, 327, 331
  evolution, phage, 109
  MM1 phage, 54
  virulence, bacterial, 228
Streptococcus thermophilus
  enzymes, phage lytic, 323
  evolution, phage, 97–101
  industrial phage ecology, 152–154
  MM1 phage, 54
  recognition of host, 173
  Sfi11/21 phages, 57
Streptomyces, 145
Stx toxins, 229, 246
Subcloning, automatic, 303, 304
Sucrose density gradients and concentrating/purifying phages, 470
Suicide as a bacteriophage defense strategy, triggered, 294
Sulakvelidze, Alexander, x
Sulfolobus, 63, 75–79, 81
Sunlight and marine phage ecology, 140–141
"Sur un microbe invisible antagoniste des bacilles dysentériques" (d'Herelle), 406
Swan-Myers, 389, 393
Switzerland, 397
Synechoccus, 141, 142
Synergistic effects, lytic enzymes and, 325, 327
Systematic Evolution of Ligands by Exponential Enrichment (SELEX), 200–201

T

T-phages (T1-7) as authorized phages, 15, 110
T1 phage, 33, 37, 52
T2 phage, 171, 456
T4 phage
  adsorption, 43
  animal-associated phages, 148
  antigenicity, 34, 37
  aptamers and SELEX, 200
  classification problems, 84
  DNA replication/repair/recombination, 186–188
  DNA transfer into the cell, 179–180
  ecology, phage, 133, 136
  enumeration, phage, 456, 460
  evolution, phage, 114–120
  infection cycle, overview of, 44
  intron ribozymes, 198–199
  lysins and lysis systems, 208, 211–213
  marine phage ecology, 142
  methodologies used when working with phages, 463, 477
  morphogenesis, 201, 205
  overview, 52
  penetration, 45
  P1 phage cloning system, 307
  recognition of host, 171, 174, 175
  ribosome binding sites, 197–198
  transcription, 193, 195–196
T5 phage, 37, 53
T7 phage
  classification problems, 84
  development of, 394
  DNA transfer into the cell, 177–179
  ecology, phage, 136
  evolution, phage, 110–111, 118
  nature of bacteriophages, 31
  overview, 43
  penetration, 45
  transcription, 193–194
Tailed phages; *see also* Recognition of host and adsorption of tailed phages; *individual subject headings*
  classification of, 70–80
  cyanophages, 57–58
  gram-negative bacteria, 49–53
  gram-positive bacteria, 53–58
  morphogenesis, 204–206
  overview, 49
Tailless phages; *see also individual subject headings*
  classification of, 73–81
  lipid, 60–61
  mycoplasmas, 61–62
  nature of bacteriophages, 31
  RNA, 58–59
  small DNA phages, 59–60
Targeting gene delivery, 305, 311–313
Target theory, 15

Tectiviridae, 31, 60, 74–78
Teichoic acid, 39
Temperate phages; *see also under* Evolution, phage; Lysogenic cycle; Prophage
  classification of, 72
  derivatives of, 32
  lysogeny and its consequences, 47–48
  nature of bacteriophages, 32
  therapy, phage, 421–422
  vertical evolution vs. horizontal gene transfer, 93–94
Temperature-sensitive phage mutations, v
Terminase ATPases, 203–204
Therapy, phage; *see also* Animal *listings;* Food production chain, using phages in; Methodologies used when working with phages
  acquaculture, 352–353
  antibiotics compared to, 423–426
  antibodies, development of phage-neutralizing, 417–419
  "Bacteriophage Inquiry," 384–387
  bloodstream, mechanisms that clear phages from the, 416–417
  decline of interest in, 393–397
  ecology, phage, 130–131
  Europe/Africa/United States, 397–399
  first therapeutic applications, 9–11, 383–384
  Georgia (independent country), 406–411
  introduction, 336, 382–383
  mammalian cells, targeted gene delivery to, 311–313
  mode of action, 413–414
  movement between biological compartments, phage, 414–416
  1920s to 1940s, Europe/Canada/U.S. in the, 388–393
  pharmacokinetics, 415–416
  plant diseases, 350–352
  Poland, 412–413
  resistance, bacterial, 419–420
  Russia, 411–412
  safety considerations, 420–423
  Soviet Union, the former, 399–406
  vaccines, lysates as, 353–359
Thermoproteus, 63
Thymidylate synthase (td or TS), 118–120
Tikhonenko, Tamara, 33
Timofeev-Ressovsky, Nicolai V., 13, 15
Titer of bacterial culture at time of infection, 442–443
TM4 phage, 271
Tobacco and vaginal lactobacilli, 150
Togaviruses, 85
Tol system, 40
Tombusviruses, 85
Toxic shock syndrome, 249–251
Toxin co-regulated pilus (TCP), 234, 252, 253
Toxins, bacterial virulence and extracellular
  Escherichia coli, 229–231
  overview, 227–228
  Streptococcus and *Staphylococcus,* 228
  Vibrio cholerae, 231–234
Toxoplasma, 359
Transcription
  crystallographic studies, 192–193
  lambda phage, 195
  N4 phage, 194
  overview, 190–193
  repressor protein, 32
  RNA polymerases, 190–192
  SPO1 phage, 195
  T4 phage, 193, 195–196
  T7 phage, 193–194
Transduction, 47, 144
Translational bypassing, 199–200
Transmission electron microscopy (TEM), 462
Tree-generating computer programs, 83
Triplet nature of the genetic code, 21
tRNa genes and bacterial virulence, 256, 257
Tryptophan, 44
Tuc2009 phage, 102
Tula phage, 171
Twort, Frederick, 1, 6–8
Typhosis, avian, 9

U

Ultracentrifuge, 12
Ultrafiltration, concentration of phages using, 450
Ultraviolet light, 33, 38, 47
United States phage therapy in the, 388–393, 398–399
Urovirales, 68

V

Vaccines, phage-elicited bacterial lysates as
  animal infections, preventing/treating, 354–355
  enzymes as antibacterial agents, 358–359
  ghost vaccines, 355–358
  producing vaccines, 353–354
Vagina and animal-associated phages, 150
Vancomycin-resistant enterococci (VRE), 425
Vectors/delivery vehicles, phage as targeted
  filamentous phage-based vectors, 298–300
  introduction, 297–298
  lambda phage, 301–305

Vectors/delivery vehicles (*continued*)
lethal agent delivery vehicles, phages as, 309–311
mammalian cells, 311–313
P1 cloning system, 305–309
Vertical gene transfer, 93–94, 224
Veterinary applications, early phage therapy in, 9–10, 130; *see also* Animal infections, phages used to prevent/treat
Vibrio, 115
Vibrio cholerae; see also CTXø phage
animal infections, preventing/treating, 344, 346
therapy, phage, 387
virulence, bacterial, 231–234, 237, 238, 246, 248–249
Vibrio harveyi, 268
Vibrio mimicus, 234, 236
Vibrio parahaemolyticus, 142
Vibrio septique, 404
Vieu, J. F., 397
Virion-encapsulated RNA polymerase (vRNAP), 194
Virions, 30, 70, 72, 85
Virulence, bacterial
acquisition of virulence genes by phages, 227, 243–245
attachment/colonization, 234–236
conclusions/summary, 257–258
immune avoidance, host, 236–238
introduction, 224–227
invasion of host cells, 238–241
pathogenesis of bacterial infections, 245–247
pathogenicity islands, 254–257
phage-phage interactions, 247–254
putative virulence proteins, 241–243
survival of intracellular bacteria, 241
toxins, extracellular, 227–234
Virulent phages; *see also* Enzymes to control bacterial infections, phage lytic; Lysins and lysis systems; Lytic cycle; lytic phages *under* Evolution, phage
classification, phage, 72
ecology, phage, 135
host-lethal proteins, 32
lysogeny and its consequences, 47
nature of bacteriophages, 32
therapy, phage, 421
toxins, extracellular, 227–234
vertical evolution vs. horizontal gene transfer, 93–94
VPIø phage, 234, 250, 252, 253, 256

W

Watson, Jim, 1, 17, 18
Wollman, Élizabeth, 16
Wollman, Eugène, 16
World Health Organization (WHO), 11
World War II, 11
Wounds treated using phage therapy, 400, 404, 408–411

X

Xanthomonas campestris, 351
Xanthomonas oryzae, 350
Xenorhabdus luminescens, 272
Xylella fastidiosa, 109

Y

Yersinia, 110, 359
Yogurt production and industrial phage ecology, 151–153

Z

Zimmer, K. G., 13, 15
Zinc-cofactored superoxide dismutase, 241

U.W.E.L. LEARNING RESOURCES